XIII International Conference on Applied Mathematics and Mechanics in the Aerospace Industry (AMMAI'2020)

IOP Conference Series: Materials Science and Engineering
Volume 927

Alushta, Russia
6 - 13 September 2020

Editor:

Anna Bogdan

ISBN: 978-1-7138-4765-6
ISSN: 1757-8981

Printed from e-media with permission by:

Curran Associates, Inc.
57 Morehouse Lane
Red Hook, NY 12571

Some format issues inherent in the e-media version may also appear in this print version.

Institute of Physics
Dirac House, Temple Back
Bristol BS1 6BE UK

Phone: 44 1 17 929 7481
Fax: 44 1 17 920 0979

techtracking@iop.org

Additional copies of this publication are available from:

Curran Associates, Inc.
57 Morehouse Lane
Red Hook, NY 12571 USA
Phone: 845-758-0400
Fax: 845-758-2633
Email: curran@proceedings.com
Web: www.proceedings.com

TABLE OF CONTENTS

IOP Conf. Series: Materials Science and Engineering **927** (2020) 011001 doi:10.1088/1757-899X/927/1/011001

Preface

XIII International Conference on Applied Mathematics and Mechanics in the Aerospace Industry (AMMAI'2020)

http://www.npnj.ru/index_e.htm

6-13 September 2020
Alushta, Russia

The Thirteenth International Conference on Applied Mathematics and Mechanics in the Aerospace Industry (AMMAI'2020) will be held on 6-13 September 2020 at Alushta Health and Educational Center, Russia. AMMAI'2020 is dedicated to the 90th anniversary of Moscow Aviation Institute and the 100th anniversary of the birth of academician I.F. Obraztsov.

The conference deals with the topical issues of the aerospace industry, including methods of applied mathematics and mechanics, as well as computer modeling. The conference focuses on solving the problems associated with modeling the flows of liquid, gas and plasma in combustion chambers, nozzles, jets and flow paths of power plants, as well as the interaction of gas and gas-dust flows with the aircraft surface. The participants of AMMAI'2020 will discuss the numerical methods, computational algorithms and software systems that allow solving problems of general mechanics, mechanics of continuous media and deformable solids, tribology in application to aviation, rocketry and astronautics. Information systems and technologies, in particular, high-performance computing technologies, will be also discussed at the conference.

This year the conference was planned to be held on 24-31 May 2020, but due to COVID-19 it was put forward to September 2020. The conference will be organized partially by personal attendance of participants as well as virtually using Internet communication technologies (Zoom, Skype). The conference attendance will be organized in compliance with all sanitary standards and safety requirements. The majority of participants will take part in the Conference virtually. It is worth noting that the option of virtual participation allows expanding significantly the geography and increasing the number of participants. More than 320 participants from different regions of Russian Federation as well as from Asia and Europe will present their reports during the Conference sessions.

The conference has already received 194 submissions that were reviewed. Finally, 85 papers were accepted for publication in IOP Conference Series: Materials Science and Engineering. The papers cover the conference topics; they present results of theoretical and experimental studies in such areas as hypersonic aerothermodynamics, combustion, detonation, multiphase flows and heat and mass transfer processes, nonequilibrium processes in nozzles and jets, plasma aerodynamics, rarefied gas flows, interaction of flows with the surface, solid mechanics, tribology. Among the accepted papers, there are some devoted to modern methods of numerical modeling, high performance computing, artificial intelligence algorithms, machine learning and big data processing.

The Organizing Committee of AMMAI'2020 acknowledges the financial support from Russian Foundation for Basic Research and the technical support from Moscow Aviation Institute.

We would like to thank reviewers for their time and effort in reviewing the papers, as well as authors, which took into account the reviewers recommendations and improved their papers to insure high quality of the AMMAI'2020 Proceedings.

Chair of the AMMAI'2020 Technical Programs Committee: Dr. Natalia Severina

Editor of AMAI'2020 Conference Proceedings: Anna Bogdan

IOP Conf. Series: Materials Science and Engineering **927** (2020) 011002 doi:10.1088/1757-899X/927/1/011002

Peer review declaration

All papers published in this volume of *IOP Conference Series: Materials Science and Engineering* have been peer reviewed through processes administered by the Editors. Reviews were conducted by expert referees to the professional and scientific standards expected of a proceedings journal published by IOP Publishing.

- **Type of peer review: Single-blind (**We had a two-level review system: first, the members of Program Committee reviewed papers by the relevance to the conference topics and paper format requirements; the second review process involved independent reviewers who evaluated papers according to the following criteria as the relevance of study, scientific quality, references adequacy, paper structure, etc. , then papers were also checked by English-speaking persons)

- **Conference submission management system:** no special management system was used, the papers were submitted via conference e-mail.

- **Number of submissions received:** 194 submissions received

- **Number of submissions sent for review:** 194 submissions were sent for review

- **Number of submissions accepted:** 85 papers accepted

- **Acceptance Rate (Number of Submissions Accepted / Number of Submissions Received X 100):** 43/57

- **Average number of reviews per paper:** 2

- **Total number of reviewers involved:** 14

- **Any additional info on review process: -**

- **Contact person for queries: Anna Bogdan, email:** coalise@tpu.ru

IOP Conf. Series: Materials Science and Engineering **927** (2020) 012001 doi:10.1088/1757-899X/927/1/012001

State-resolved and two-temperature rate coefficients for $CO+CO=CO_2+C$ reaction

E Kustova and A Savelev
Saint Petersburg State University, 7/9 Universitetskaya nab., St. Petersburg, 199034, Russia
E-mail: e.kustova@spbu.ru and aleksey.schumacher@gmail.com

Abstract. Several state-resolved and two-temperature models for the rate coefficients of reaction $CO+CO=CO_2+C$ are assessed, and a new simple state dependent model is proposed. Various models for the two-temperature non-equilibrium factor are considered: the theoretical alpha-model and the models based on averaging the state-resolved rate coefficients over the Boltzmann distribution. It is shown that for the Starik model, taking into account vibrational excitation of the partner CO molecule as well as the reaction product does not affect the two-temperature reaction rate coefficients. If the vibrational excitation of only one CO molecule is taken into account, the Starik model and that proposed in this work with the parameter $U=\infty$ yield identical results. However, our model is more general since it is capable to account for the preferential reaction mechanism from high vibrational states. Two-temperature rate coefficients provided by the α-model agree well with the averaged state-to-state ones, but the parameter α cannot be fixed in the entire temperature range and has to be chosen according to the specific ratio T/T_v. Thus, our computationally efficient state-specific model is recommended for state-to-state flow simulations; when being averaged with different non-equilibrium vibrational distributions, it can be used for modeling two-temperature flows.

1. Introduction

For modeling highly non-equilibrium flows, an accurate description of the physico-chemical processes rate coefficients is very important. Three levels of non-equilibrium flow description are traditionally distinguished: 1) a one-temperature description, when chemical reactions and vibrational relaxation are uncoupled, and the rate coefficients of chemical reactions are given by the Arrhenius law; 2) multi-temperature models for partially coupled relaxation and chemical reactions; in this case, the rate coefficients are either expressed in terms of the non-equilibrium factor depending on the vibrational temperature [1, 2, 3, 4, 5, 6] or obtained by averaging the state-resolved rate coefficients over different vibrational distributions [7, 8]; 3) state-to-state models for fully coupled vibrational relaxation and chemical reactions; in this approach, the vibrationally state-resolved reaction rate coefficients are required [7, 8, 9, 10, 11, 12]. The one-temperature model is the simplest one from the computational point of view, but, at the same time, may be rather inaccurate under the strong non-equilibrium conditions. The state-to-state approach is the most accurate method, but in this case, large computational efforts are required, since it is necessary to solve the sets of master equations for the populations of all vibrational levels of all molecular species in the flow. Multi-temperature models are capable to catch some non-equilibrium effects and, under high-temperature conditions, may provide a computationally efficient alternative to the state-to-state models.

In the present paper, we study the state-resolved and two-temperature rate coefficients and non-equilibrium factors for the reaction:

$$CO(i) + CO(k) \leftrightarrow CO_2(i_1, i_2, i_3) + C, \tag{1}$$

where i, k are the vibrational levels of CO molecules acting as the reagents; i_1, i_2, i_3 are vibrational quantum numbers of the CO_2 symmetric, bending and asymmetric modes, respectively. Although this reaction is of importance for highly non-equilibrium flows, the data for the state-resolved rate coefficients of this reaction are scarce. The commonly used models basically do not take into account the vibrational temperature; or, alternatively, some modifications of the alpha-model [3] introducing a kind of effective vibrational energy parameter [13] are used. Moreover, the effect of vibrational excitation of both reaction reagents and products on the rate coefficients has not been studied up to now.

The objectives of the present study are: 1) to assess various state-resolved and two-temperature models for the rate coefficients of reaction (1); 2) to evaluate the impact of vibrational excitation of the reaction products and reagents on the reaction rate; and 3) to estimate the range of applicability for the models and give recommendations for their choice under various conditions.

2. Model description

For the simplest one-temperature approach, the reaction rate coefficients are given by the Arrhenius law:

$$k^{\mathrm{eq}}(T) = AT^n \exp(-E_a/k_B T), \tag{2}$$

here T is the gas temperature, k_B is the Boltzmann constant, E_a is the activation energy, parameters A and n are determined from experiments. For reaction (1), we use the values for A, n and E_a proposed by Park [14].

The next level of detail and accuracy is provided by the multi-temperature approach. For this approach, the exchange reaction rate coefficients, similarly to dissociation, can be defined by the equilibrium rate coefficient $k^{\mathrm{eq}}(T)$ multiplied by the non-equilibrium factor $Z(T, T_v)$ (T_v is the vibrational temperature) [6, 15]:

$$k(T, T_v) = k^{\mathrm{eq}}(T) Z(T, T_v). \tag{3}$$

The non-equilibrium factor can be calculated either using the formulas obtained on the basis of experimental data and simple semi-empirical models, or by averaging state-resolved rate coefficients with different vibrational energy distributions. For carbon dioxide flows, the experimental data under non-equilibrium conditions are scarce, and the alpha-model [3] is often used. The choice of the parameter α depends on the reaction temperature regime: for quasi thermo-neutral reactions, the value from 0.3 to 0.6 are selected; for endothermic reactions, the value from 0.9 to 1 are suggested. A significant drawback of this model is the limitation on the vibrational temperature: this model can only be used for flows with $T_v > T$, which does not allow its using in shock heated flows and planetary entry problems.

Another appropriate method for the evaluation of the multi-temperature reaction rate coefficients is averaging the state-dependent rate coefficients with various vibrational distributions. This latter method is not widely used for carbon dioxide due to the absence of theoretical models and reliable data on the state-resolved reaction rate coefficients.

Let us now consider the state-to-state approach. For exchange reactions involving diatomic species, a simple but reliable theoretical model is proposed by Starik [8]; the advantage of the model is that it takes into account the vibrational excitation of both reaction reagents and product. In the present study, we generalize this model for reactions involving CO_2. A significant drawback of the above model is its extremely high computational cost.

In our previous studies [16], we showed that for exchange reactions in air, the best agreement with the quasi-classical trajectory (QCT) data is provided by the model based on the Aliat model [15], see [16, 12]. In this case, the rate coefficients for reaction (1) are calculated by the formulas:

$$k_i(T,U) = \begin{cases} C(T,U)\,k^{\rm eq}(T)\exp\left(-\frac{E_{\rm a}}{k_B U}\right)\exp\left[\frac{\varepsilon_i}{k_B}\left(\frac{1}{T}+\frac{1}{U}\right)\right], & \varepsilon_i < E_{\rm a}, \\ C(T,U)\,k^{\rm eq}(T)\exp\left(\frac{E_{\rm a}}{k_B T}\right), & E_{\rm a} < \varepsilon_i < D, \end{cases} \tag{4}$$

$$C(T,U) = Z^{CO}_{\rm vibr}(T)\left[\sum_{\varepsilon_i<E_a}\exp\left(-\frac{E_{\rm a}-\varepsilon_i}{k_B U}\right) + \sum_{\varepsilon_i>E_a}\exp\left(\frac{E_{\rm a}-\varepsilon_i}{k_B T}\right)\right]^{-1}, \tag{5}$$

D is the reagent dissociation energy, $Z^{CO}_{\rm vibr}$ is the equilibrium vibrational partition function of CO molecules. The model parameter U specifies an increase in the reaction probability with rising vibrational quantum numbers. For this parameter, one of the following values is commonly used in the literature: $U = 3T$, $U = D/6k$ and $U = \infty$. The choice of the parameter is specified by the considered flow conditions. At the present time, the model (4) allows vibrational excitation of only one reagent. It can be easily generalized to take into account vibrational excitation of both reagents and products when reliable QCT data become available in order to assess the model parameter U for various vibrational states.

The two-temperature reaction rate coefficients can be obtained by averaging the corresponding state-dependent coefficients over the vibrational distributions of the reacting molecules and summing over all vibrational states of the product.

3. Results and discussion

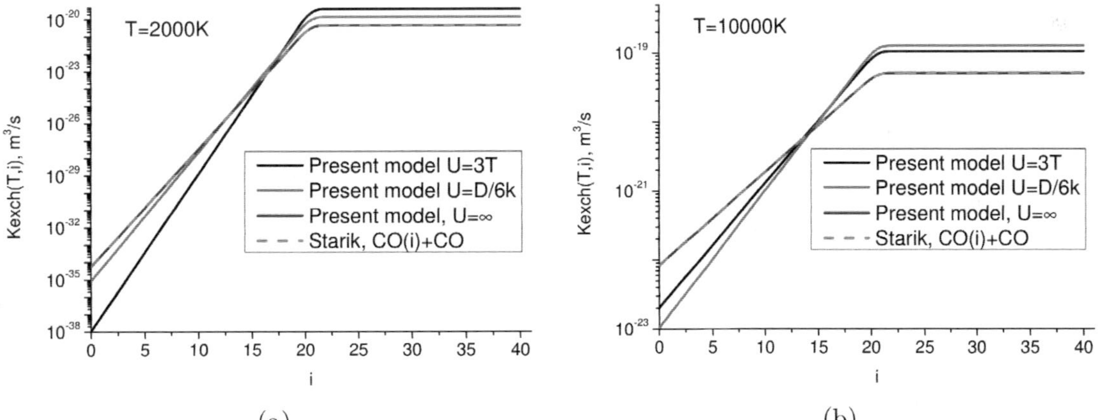

Figure 1. State-to-state reaction rate coefficients as functions of the CO vibrational level for $T = 2000$ K (a) and $T = 10000$ K (b).

First we compare the state-to-state reaction rate coefficients for a harmonic oscillator model calculated using our model (4) and that of Starik [8] taking into account the vibrational excitation of only one reagent. The results are presented in Figure 1. As we can see, for all temperatures, when taking into account the vibrational excitation of only one reagent, the results obtained using the Starik model [8] coincide with those obtained according to our model with the parameter $U = \infty$. At the same time, model (4) is considerably more efficient from the computational point of view and, moreover, allows one taking into account the preferential reactions from higher energy states by varying the parameter U. Thus, it can be easily fitted to the QCT data when they become available.

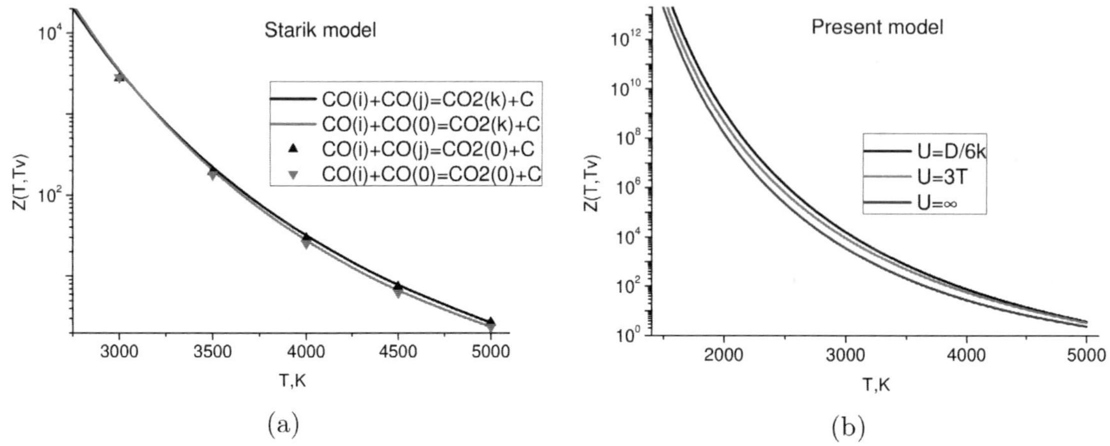

Figure 2. Two-temperature non-equilibrium factor as a function of T for the Starik model [8] with excited reaction reagents and products (a) and for our model with different U (b). $T_v = 5700$ K.

Now, let us evaluate the effect of vibrational excitation of the reaction products and reagents on the two-temperature non-equilibrium factors based on the state-resolved Starik model [8] and obtained by averaging $k_{i,k}^{i_1,i_2,i_3}$ with the Boltzmann distribution. Several combinations of the vibrational states of reagents and products are assessed; temperatures from 500 K to 5000 K with $T_v = 5700$ K are considered (vibrational temperature is selected based on the characteristic nozzle regimes, due to the limitations of the alpha-model considered further). The comparison of the results is presented in Figure 2(a). One can see that taking into account the vibrational state of the product does not have any effect on the non-equilibrium factors, and only accounting for the vibrational excitation of the second reagent has a small effect. For temperatures above 3000 K, taking into account the excitation of the second reagent yields slightly lower non-equilibrium factor, whereas for temperatures below 3000 K slightly higher ones.

In Figure 2(b), we show the non-equilibrium factors for the two-temperature reaction rate coefficients calculated using our model and different values of the U parameter. Using $U = \infty$ yields the smallest $Z(T, T_v)$ values compared to the values obtained for $U = 3T$ and $U = D/6k_B$ in the entire temperature range. The differences between the values of the non-equilibrium factor for different values of the model parameter are within the same order.

As is shown in Reference [13], for different initial excitation of CO molecules, reaction (1) can be considered as an endothermic or a quasi thermo-neutral one, which leads to a different choice of the alpha-model parameter. For endothermic reactions, the non-equilibrium factor varies within one order of magnitude under considered conditions and different α values. At the same time, for quasi thermo-neutral reactions, especially at low gas temperatures, the difference in the non-equilibrium factor can reach several orders of magnitude. Moreover, endothermic reactions are characterized by a much greater non-equilibrium factor at low temperatures.

In Figure 3, we present a comparison of non-equilibrium factors calculated for all models discussed above. The use of the α-model with α close to unit yields significantly higher non-equilibrium factors over the entire temperature range, however the values close to 0.9 can describe

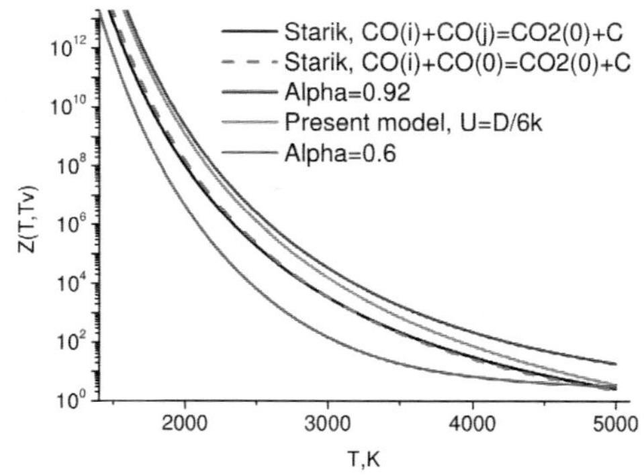

Figure 3. Two-temperature non-equilibrium factor as a function of T for all presented models and $T_v = 5700$ K.

the results obtained using our model quite well. At the same time, the values of α corresponding to quasi thermo-neutral reactions under-predict the non-equilibrium factor. For α values close to the upper boundary of the thermo-neutral reactions range, the difference in the rate coefficients is about two orders of magnitude, whereas for the lower limit of the thermo-neutral range, the discrepancy exceeds six orders of magnitude for temperatures below 2000 K.

Finally, we compare the non-equilibrium factors calculated according to the Starik model [8] and according to our model with different values of the parameter U at various vibrational temperatures. The results are presented in Figure 4. We can see that in the entire range of the gas temperature and the vibrational temperature, the use of the parameter $U = 3T$ leads to significant discrepancies compared to the other models, especially at low values of both temperatures. The use of the parameter $U = \infty$ of our model (as well as the use of the Starik model [8]) yields significantly greater non-equilibrium factor at $T > T_v$ and lower in the opposite case. With increasing the vibrational temperature, the effect of vibrational excitation of the second CO molecule on the non-equilibrium factor calculated according to the Starik model [8] decreases, but the results for all cases are within the same order of magnitude.

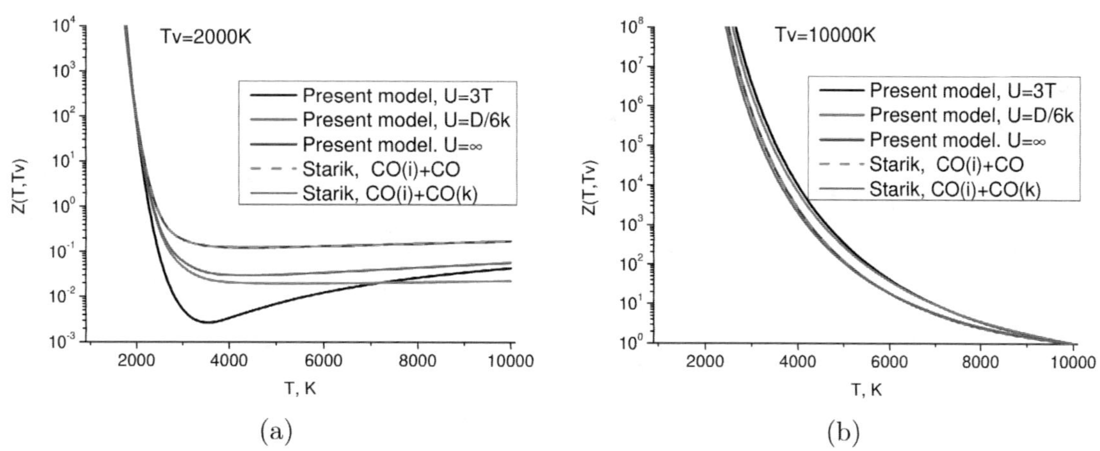

Figure 4. Two-temperature non-equilibrium factors as functions of T for the Starik model [8] and our model for different T_v.

Thus, using in the α-model the values of the parameter α characteristic for the quasi thermo-neutral reaction regime leads to a significant underestimation of the non-equilibrium factor under conditions when T and T_v are significantly different. On the other hand, the use of α recommended for the endothermic reaction regime leads to some over-prediction of the non-equilibrium factor for close values of T and T_v. Using the full Starik model accounting for the excitation of both reaction products and reagents does not make sense since the effect of such excitation is negligible; the reduced model that takes into account the vibrational excitation of only one reagent is much more efficient and provides a satisfactory agreement. The use of the parameter U different from $U = \infty$ in our model yields a more significant discrepancy in the rate coefficients, and makes it possible to take into account the preferential reaction mechanism for highly excited vibrational states. Moreover, our model is more flexible and can be easily fitted to the experimental data or the results of trajectory calculations.

4. Conclusions

Various models for the rate coefficients of reaction $CO+CO=CO_2+C$ are assessed, and a simple state dependent model is proposed. When calculating the two-temperature coefficients by averaging the state-resolved rate coefficients based on the Starik model, it is recommended to neglect the vibrational excitation of the reaction product (CO_2 molecule) and the partner CO molecule, since their excitation does not affect the rate coefficients. Neglecting the excitation of the partner and reaction product provides a significant reduction in the computational efforts. When excitation of only one CO molecule is taken into account, the two-temperature coefficients obtained on the basis of our model with $U = \infty$ coincide with those obtained on the basis of the Starik model. However, in this case both models yield under-predicted rate coefficients compared to the models taking into account the preferential reaction mechanism from high vibrational states (other values of U) or the α-model. The results provided by the α-model agree well with the results obtained by averaging the state-to-state rate coefficients, but the parameter α cannot be fixed in the entire temperature range and has to be chosen according to the specific ratio of T and T_v. Thus, for arbitrary conditions, we recommend using our computationally simple model averaging it with different non-equilibrium vibrational distributions.

Acknowledgments

This study is supported by the Russian Science Foundation, grant 19-11-00041.

References

[1] Hammerling P, Teare J and Kivel B 1959 *Phys. Fluids* **2** 422–426
[2] Marrone P and Treanor C 1963 *Phys. Fluids* **6** 1215–1221
[3] Rusanov V and Fridman A 1984 *Physics of Chemically Active Plasma* (Moscow: Nauka) (in Russian)
[4] Park C 1990 *Nonequilibrium Hypersonic Aerothermodynamics* (New York, Chichester, Brisbane, Toronto, Singapore: J.Wiley and Sons)
[5] Macheret S O, Fridman A A, Adamovich I V, Rich J W and Treanor C E 1994 *AIAA Paper* **94-1984** 6th AIAA/ASME Joint Thermophysics and Heat Transfer Conference, Colorado Springs, June 1994
[6] Knab O, Frühauf H and Messerschmid E 1995 *J. Thermophys. Heat Transfer* **9** 219–226
[7] Kustova E, Nagnibeda E, Oblapenko G, Savelev A and Sharafutdinov I 2016 *Chem. Phys.* **464** 1–13
[8] Arsentiev I V, Loukhovitski B I and Starik A M 2012 *Chem. Phys.* **398** 73–80
[9] Pogosbekian M and Sergievskaya A 2014 *Physical-Chemical Kinetics in Gas Dynamics* **15** p 227
[10] Kustova E V, Savelev A S and Sharafutdinov I Z 2016 *AIP Conference Proceedings* vol 1786 ed Ketsedever A and Struchtrup H (Melville, NY: American Institute of Physics) p 150007
[11] Kunova O, Kustova E and Savelev A 2016 *Chem. Phys. Lett.* **659** 80–87
[12] Kustova E, Savelev A and Armenise I 2019 *J. Phys. Chem. A* **123** 10529–10542
[13] Barreto P R P, de O Euclides H, Albernaz A F, Aquilanti V, Capitelli M, Grossi G, Lombardi A, Macheret S and Palazzetti F 2017 *The European Physical Journal D* **71** 259
[14] Park C, Howe J, Howe R, Jaffe R and Candler G 1994 *J. Thermophys. Heat Transfer* **8** 9–23
[15] Aliat A 2008 *Physica A* **387** 4163–4182

[16] Kustova E V, Savelev A S and Kunova O V 2018 *AIP Conference Proceedings* vol 1959 ed Kustova E and et al (Melville, NY: American Institute of Physics) p 060010

IOP Conf. Series: Materials Science and Engineering **927** (2020) 012002 doi:10.1088/1757-899X/927/1/012002

Effect of coaxially mounted disks on oscillations of a cylinder of large aspect ratio in air flow

D V Kaufman and A N Ryabinin

Department of Hydroaeromechanics, St. Petersburg State University, St. Petersburg, 198504, Russia

E-mail: a.ryabinin@spbu.ru

Abstract. The damped rotational oscillations of a cylinder whose length-to-diameter ratio is nine are considered. In the head part of the cylinder, a disc is mounted coaxially on the leg. The effect of the disk on damping of oscillations is described by aerodynamic rotational derivatives. Wind tunnel experiment is performed with the flow past a cylinder fixed on an elastic spring suspension. The diameter of the coaxial disk and the distance between the disk and the end of the cylinder vary. Deflected from the equilibrium position, the cylinder performs damping oscillations. The dependence of the amplitude of rotational vibrations on time is determined by the tensometric method. Coefficients of rotational pitch derivatives are found.

1. Introduction

In this paper, the dynamic characteristics of cylinders with coaxial disks are experimentally studied. Dynamic characteristics are described by dimensionless coefficients called aerodynamic derivatives. To determine the aerodynamic derivatives, the methods of free or forced oscillations are used [1, 2, 3, 4]. We use a simple method of free oscillations. It is known that the disk mounted coaxially at upstream end of the cylinder significantly reduces the drag [5]. However, the oscillations of cylinders with disks in the air flow are little studied. An exception is the study of rotational self-oscillations and damping oscillations of elastically fixed cylinders of small aspect ratio with disks [6]. The ratio of the length L to the diameter D of the cylinder was $\Lambda = L/D = 2$. In air flow, elastically mounted cylinder without disk performs rotational pitching oscillations with a steady amplitude. The same disk with coaxially affixed disk can oscillate with damping. In this paper, we consider a cylinder with aspect ratio $\Lambda = 9$.

2. Experimental method

The experiments were performed on the wind tunnel AT-12 of the Saint Petersburg state University, which has an open working part. The diameter of the output part of the round-section nozzle is 1.5 m. The test procedure is described in articles [7, 8].

Figure 1 shows the scheme of the experiment. The cylinder is suspended in the test section of the wind tunnel on a wire suspension, which allows it to rotate only around the axis perpendicular to the axis of the cylinder and the vector of the average velocity of the incoming flow. The axis of rotation pass through the center of the cylinder. The diameter of the cylinder $D = 0.1$ m. Wire diameter in wire suspension is equal to 0.6 mm. At 0.3 m from the center of the cylinder closer to the aft part of the cylinder, wire rods are attached to the top and bottom of the cylinder. The rods are connected to springs are located above and below the air jet. In the equilibrium position, the rods with springs hold the cylinder horizontally (angle of attack $\alpha = 0$).

The C-50 semiconductor strain gauge measured the tension of the lower spring. The voltage at the output of the strain gauge is applied to a Velleman-PCS500A PC oscilloscope, which converts an analog signal to a digital one. The oscilloscope is connected to a computer where the time dependence of the signal is recorded in a file. During experiments, the cylinder was deflected from the equilibrium position by an angle α_0, then released. Oscillations begin, which in all experiments are damped. The signal is recorded for 17 seconds. The spring tension values are read at a frequency of 100 Hz. Calibration of the device is carried out by suspending a known load to the cylinder at the point where the rods are attached. At the same time, the angle of attack α and the change in the strain gauge readouts are determined. As a result of the calibration experiment, the reduced spring rate also determined.

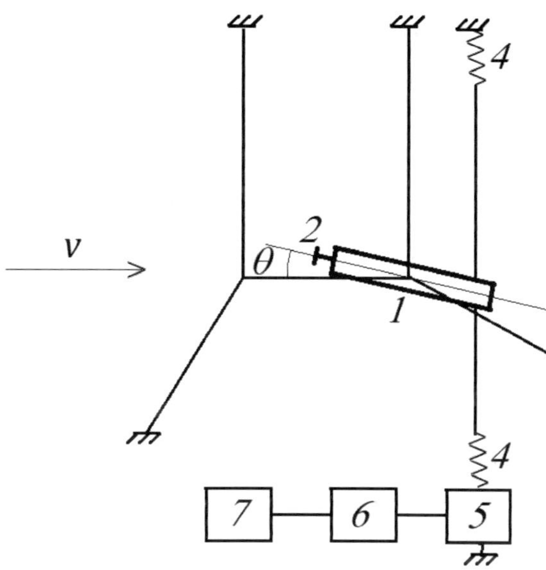

Figure 1. Scheme of the experimental installation. *1* — cylinder; *2* — disk; *3* — counterweight; *4* — springs; *5* — semiconductor strain gauge; *6* — PC-oscilloscope; *7* — computer.

3. Equation of motion

The equation of motion of the cylinder has the form:

$$I_z\ddot{\alpha} + r\dot{\alpha} + k\alpha = \left(m_z^{\omega} + m_z^{\dot{\alpha}}\right)\frac{qsL^2}{v}\dot{\alpha} + m_z^{\alpha}qsL\alpha, \tag{1}$$

where I_z is the moment of inertia, α is the angle of attack, r is the friction coefficient, $q = \rho v^2/2$ is velocity pressure, ρ is air density, k is the spring rate, $s = \pi D^2/4$ is the characteristic area, v is the air velocity, $m_z^{\omega}, m_z^{\dot{\alpha}}$ and m_z^{α} are the aerodynamic derivatives. In the right part of equation (1) there are terms that correspond aerodynamic moment. Let us introduce new symbols:

$$\omega^2 = \frac{k}{I_z}, \qquad \Omega = \frac{m_z^{\alpha}qsL}{I_z}, \qquad \mu = \left(m_z^{\omega} + m_z^{\dot{\alpha}}\right)\frac{qsL^2}{vI_z} - \frac{r}{I_z}.$$

Equation (1) is transformed:

$$\ddot{\alpha} + (\omega^2 - \Omega)\alpha = \mu\dot{\alpha}. \tag{2}$$

We seek the solution of this equation as a function $\alpha = \alpha_0 \exp(\lambda t)$, where $\lambda = n + ip$ is complex value. Substituting the expression for α in equation (2), and writing the imaginary and real parts separately, we get:

$$n = \mu/2, \qquad p^2 = \omega^2 - \Omega - \mu^2/4. \tag{3}$$

Substituting the expression for μ in the first equation of system (3), we get that n depends linearly on the flow rate: $n = a + bv$. The coefficients a and b are calculated using the least squares method. Sum of rotational derivatives $m_z^{\omega} + m_z^{\dot{\alpha}}$ can be written as:

$$m_z^{\omega} + m_z^{\dot{\alpha}} = \frac{4bk}{\rho s \omega^2}. \tag{4}$$

In the last expression (4), the oscillation frequency ω is determined in experiments with damped oscillations with the wind tunnel fan turned off.

4. Data processing

It was assumed that the amplitude A of the oscillations changes so slowly that during one period of oscillation, this change can be ignored in comparison with the amplitude. The result of measuring the tilt angle of attack α_i at time t_i is described by the formula:

$$\alpha_i = A\cos(pt_i - \varphi) + E + \xi_i = B\cos pt_i + C\sin pt_i + E + \xi_i, \qquad i = 1, 2, 3, \dots, N$$

where ξ_i is a random variable corresponding to the measurement error, B, C and E are constants. For a sample containing n elements, the parameters B, C and E are determined using the least squares method. Finding the minimum residual sum of squares $\sum_{i=1}^{N} \xi_i^2$ is reduced to solving a system of linear equations:

$$\begin{pmatrix} \sum_{i=1}^{N} \cos^2 pt_i & \sum_{i=1}^{N} \cos pt_i \sin pt_i & \sum_{i=1}^{N} \cos pt_i \\ \sum_{i=1}^{N} \cos pt_i \sin pt_i & \sum_{i=1}^{N} \sin^2 pt_i & \sum_{i=1}^{N} \sin pt_i \\ \sum_{i=1}^{N} \cos pt_i & \sum_{i=1}^{N} \sin pt_i & N \end{pmatrix} \begin{pmatrix} B \\ C \\ E \end{pmatrix} = \begin{pmatrix} \sum_{i=1}^{N} \alpha_i \cos pt_i \\ \sum_{i=1}^{N} \alpha_i \sin pt_i \\ \sum_{i=1}^{N} \alpha_i \end{pmatrix}.$$

In this paper, we selected a time interval equal to the oscillation period. If in this case the number n is large, then the system of equations is simplified by the fact that some sums are close to zero:

$$\sum_{i=1}^{N} \cos^2 pt_i \approx \sum_{i=1}^{N} \sin^2 pt_i \approx N/2,$$

$$\sum_{i=1}^{N} \cos pt_i \sin pt_i \approx \sum_{i=1}^{N} \sin pt_i \approx \sum_{i=1}^{N} \cos pt_i \approx 0.$$

We write the simplified approximate equations as follows:

$$\begin{pmatrix} N/2 & 0 & 0 \\ 0 & N/2 & 0 \\ 0 & 0 & N \end{pmatrix} \begin{pmatrix} B \\ C \\ E \end{pmatrix} = \begin{pmatrix} \sum_{i=1}^{N} \alpha_i \cos \omega t_i \\ \sum_{i=1}^{N} \alpha_i \sin \omega t_i \\ \sum_{i=1}^{N} \alpha_i \end{pmatrix}.$$

The oscillation amplitude A is determined by the formula

$$A = (B^2 + C^2)^{0.5}.$$

In our case, the number of measurement points in one period was less than a hundred, and we did not replace the system of equations with an approximate one. Figure 2 shows the typical dependence of the amplitude of damped oscillations on time on a logarithmic scale. This dependence is close to linear one. This indicates that the exponential damping of oscillations predicted by our equations takes place.

5. Results

In our study, the diameter of the coaxial disk *d* and the distance between the disk and the end of the cylinder g vary. Figure 3 shows the dependence of the damped coefficient *n* on the velocity of the incoming flow for a disk with a diameter of $d = 0.875D$ at four distances between the disk and the end of the cylinder. As the model predicts, the points on the chart fall with some variation on straight lines. Approximation lines are constructed using the least squares method are also plotted.

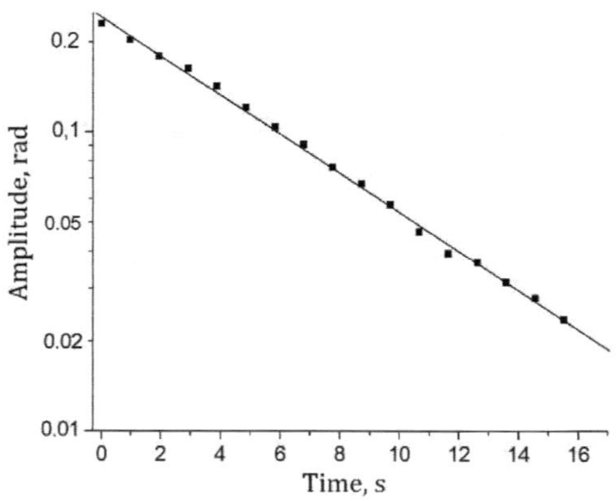

Figure 2. The dependence of the amplitude of damped vibrations on time.

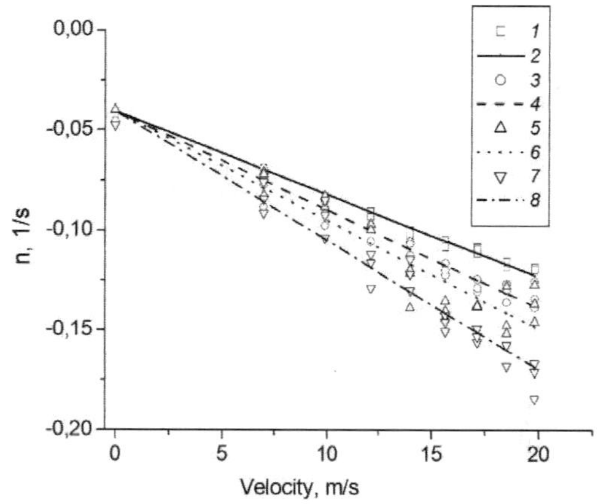

Figure 3. Dependence of the damping coefficient on the velocity of the incoming flow. Experiment and approximation. $d = 0.875D$. *1, 2* – $g/D = 0.5$; *3, 4* – $g/D = 0.6$; *5, 6* – $g/D = 0.7$; *7, 8* – $g/D = 0.8$.

Table 1 shows the values of aerodynamic derivatives calculated using the formula (4). The coefficient $m_z^{\omega} + m_z^{\dot{\alpha}}$ of a cylinder without disks is equal to -0.5 ± 0.06. Large-diameter coaxial disks do not affect the damping of cylinder oscillations. The greatest damping is provided by a disk with a diameter $d = 0.875D$, installed at a distance $g = 0.8D$.

Table 1. Rotational derivatives $m_z^{\omega} + m_z^{\dot{\alpha}}$ of the cylinder with the disk

d/D	$g/D = 0.5$	$g/D = 0.6$	$g/D = 0.7$	$g/D = 0.8$
1.25	-0.48 ± 0.02	-0.47 ± 0.02	-0.46 ± 0.02	-0.47 ± 0.02
1.0	-0.42 ± 0.02	-0.46 ± 0.02	-0.46 ± 0.02	-0.50 ± 0.02
0.875	-0.56 ± 0.03	-0.67 ± 0.03	-0.74 ± 0.03	-0.87 ± 0.03
0.75	-0.67 ± 0.03	-0.74 ± 0.03	-0.77 ± 0.03	-0.84 ± 0.03
0.625	-0.85 ± 0.04	-0.85 ± 0.04	-0.75 ± 0.04	-0.71 ± 0.04

6. Conclusions

The aerodynamic derivatives of the pitch coefficient $m_z^{\omega} + m_z^{\dot{\alpha}}$ of a cylinder with a head part in the form of a coaxial disk affixed on a leg at the upstream end of the cylinder are determined. Disks whose diameter exceeds the diameter of the cylinder do not have a strong effect on the damping of oscillations of an elastically fixed cylinder. Disks of smaller diameter promote rapid damping of the oscillations.

References

[1] Adamov N P, Puzyrev L N, Kharitonov A M, Chasovnikov E A, Dyadkin A A and Krylov A N 2013 Aerodynamic derivatives of the model of the jettisonable module of the emergency rescue system at hypersonic velocities *Thermophysics and Aeromechanics* 20 729-738
[2] Adamov N P, Gurin A M and Chasovnikov E A 2018 Methodical investigations of aerodynamic derivatives of a sharp cone on a setup with free oscillations and supersonic velocities 2018 *AIP Conference Proceedings* **2027** 030151
[3] Liu J, Song Y and Hu J 2019 Investigation on dynamic derivative test technique in hypersonic wind tunnel Proc. 2019 *Asia-Pacific Int. Symp. on Aerospace Technology. Lecture Notes in Electrical Engin.* **459** 883-891
[4] Zimper D and Huber K C 2020 Experimental and numerical investigations of unsteady aerodynamic derivatives for a generic lambda wing UCAV configuration *CEAS Aeronautical J.* 11 475-485
[5] Koenig K and Roshko A 1985 An experimental study of geometrical effects on the drag and flow field of two bodies separated by a gap *J. Fluid Mech.* **156** 167-204
[6] Kiselev N and Ryabinin A 2018 The study of shielding influence of the disks placed coaxially on rotational oscillations of the cylinder in the airflow *AIP Conference Proceedings* **1959** 050016
[7] Ryabinin A N and Kiselev N A 2017 Rotational oscillation of a cylinder in air flow *ARPN J. Engin. Applied Science* **12** 6803–6808
[8] Ryabinin A and Bogomolov R 2018 Rotational oscillation study of the cylinders with end plates in airflow *AIP Conference Proceedings* **1959** 050027

IOP Conf. Series: Materials Science and Engineering **927** (2020) 012003 doi:10.1088/1757-899X/927/1/012003

Development of power processing unit for a low-power ablative pulsed plasma thruster

A V Bogatyi, V I Bogatyi and S V Gordeev

Research Institute of Applied Mechanics and Electrodynamics of the Moscow Aviation Institute (RIAME MAI), 5 Leningradskoe shosse, 125080, POB 43 Moscow, Russia

E-mail: svyatoslavgordeev@mail.ru

Abstract. Research Institute of Applied Mechanics and Electrodynamics of the Moscow Aviation Institute (RIAME MAI) develops ablative pulsed plasma thrusters with low power consumption for small spacecraft. The paper considers the development of power processing unit (PPU) for such thrusters. PPU under discussion should meet the following requirements: design simplicity, small overall dimensions, ensuring the required power and frequency of thruster operation and high energy efficiency.

1. Introduction

One of the trends in the modern space industry is an increased interest in the development of space systems based on small spacecraft (SSC). Such spacecraft is characterized by relatively low cost of production and operation. Launch into orbit of the SSC constellations opens up new opportunities for the developers of space systems, such as global communications and the Internet, the operational solution of Earth remote sensing problems, etc. In addition, the use of SSC can reduce the time of space system development.

The effective operation of such spacecraft requires regular orbit correction, which makes it necessary to use propulsion systems (PS) on board. Such a propulsion system should have small dimensions and low power consumption, cost and weight. At the same time, it should have high efficiency to solve the tasks it faces in correcting the orbit of the SSC.

Taking into account above requirements, the use of propulsion systems based on the ablative [1] pulsed plasma thruster (APPT) as part of the SSC is promising. APPT is capable of providing accurate control for the thrust impulse, which makes it applicable for the spacecraft attitude control [2], [8]. Simplicity, reliability and relatively low cost can be attributed to the advantages of such propulsion systems (PS).

2. Ablative pulsed plasma thruster

RIAME MAI develops propulsion systems based on ablative pulsed plasma thrusters operating on the railgun principle with side propellant feeding. Teflon-4 is used as propellant. Figure 1 shows the schematic diagram of APPT.

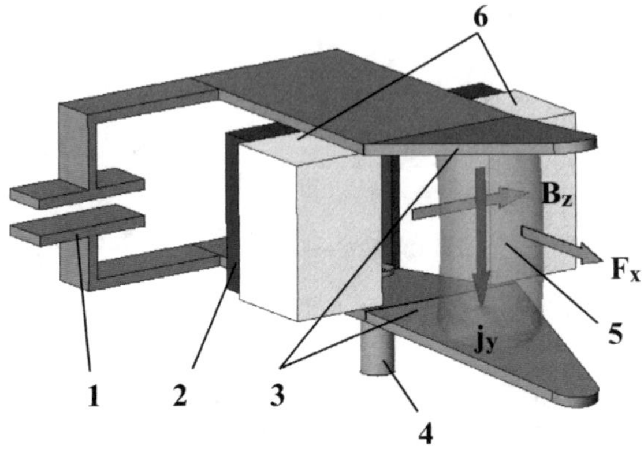

Figure 1. Schematic diagram of APPT of railgun geometry with side feed of propellant: 1 – energy storage capacitor unit (ESCU); 2 – end insulator; 3 – electrodes; 4 – igniter; 5 –plasma; 6 – propellant bar.

APPT operates as follows: the power processing unit charges the capacitors of energy storage unit (ESU). Then, the discharge initiation unit (DIU) starts the electric breakdown in the igniter, which is the ceramic insulator with metal electrodes inside. The electrons being the result of the breakdown in the igniter are accelerated in the direction of the positive electrode of the APPT discharge channel and close the circuit. As a result of the discharge of ESCU capacitors, an electrical breakdown occurs between the electrodes of the APPT discharge channel. The energy released during the breakdown leads to ablation, ionization and subsequent acceleration of a part of the propellant by the electromagnetic and gas-dynamic forces. The APPT operation is pulsed – the discharge duration is on the order of 10 ... 15 μs, the frequency of the thruster operation is defined by the power processing unit, so the average power consumption depends on the frequency of thruster operation only.

3. Power processing unit

The APPT power processing unit converts the onboard supply voltage of small spacecraft (12V – 27V) to a high voltage (1000 V and above) applied to the input of the ESCU and DIU.

The power processing unit of the APPT is actually a single high-voltage source operating from the onboard power supply. In addition to the thruster itself, which comprises the propellant storage and feed system, discharge channel and ESCU, the propulsion system based on the APPT includes PPU and DIU.

Figure 2 shows the schematic diagram of the APPT-based propulsion system for SSC.

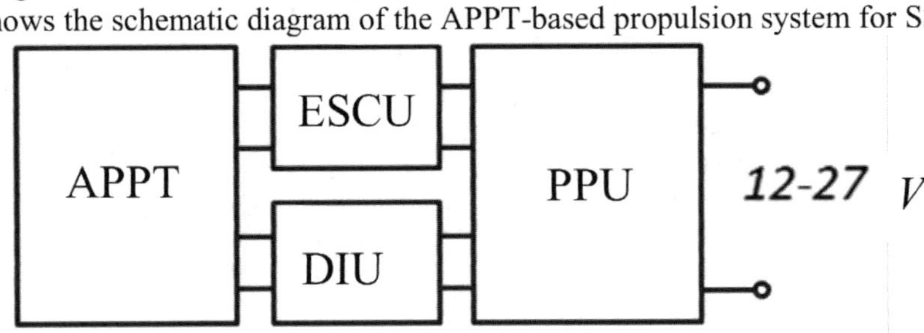

Figure 2. The schematic diagram of the propulsion system based on the APPT.

The following requirements are imposed on the power processing unit of the considered APPT:

- small size;
- low power consumption;
- high efficiency;
- high reliability.

In the work considered, a PPU was developed without the possibility of external control for power consumption. The nominal operating mode of the described PPU is achieved at the supply voltage of 12 V. However, the PPU ensures the thruster operation at supply voltages from 9 to 14 V. The thruster operation frequency and its power consumption change with the variation of the PPU input voltage.

The typical discharge frequency in the APPT is 1-2 Hz [7]. The developed PPU should provide the operation frequency of the same level, while it should be as simple as possible and ensure the thruster operation at supply voltage of 9-14 V and its shut down when the power is turned off.

The schematic diagram of the PPU is shown in figure 3. The PPU operates as follows: when the supply voltage is applied to the input, an alternating voltage generator with a frequency of about 40 kHz is turned on; it controls the operation of transistor T1. As a result, conditions are provided for the flow of alternating current through the primary winding of the step-up transformer Tp that provides high voltage of the required level after rectification. The duty cycle of the control alternating signal defines the power taken from the power source, as well as the frequency of the APPT operation.

Thus, when the on-board power is supplied to the PPU input, the charging of the high-voltage capacitors in the energy storage unit and in the discharge initiation unit starts.

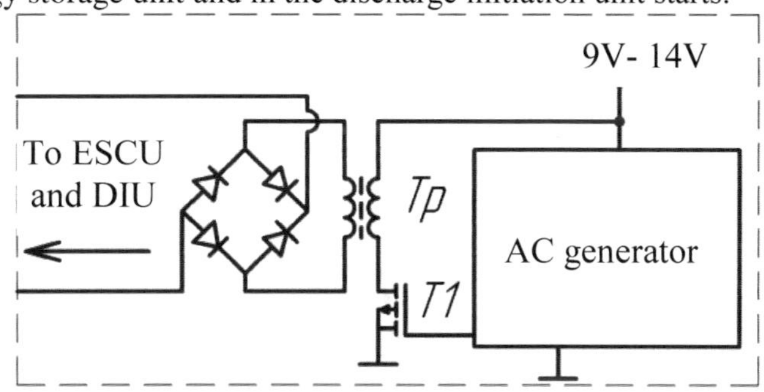

Figure 3. PPU schematic diagram.

When the voltage of about 1000 V is reached, the DIU is triggered; as a result, the APPT channel becomes conductive and the capacitor of energy storage is pulse-discharged, transmitting energy to the propellant (figure 4).

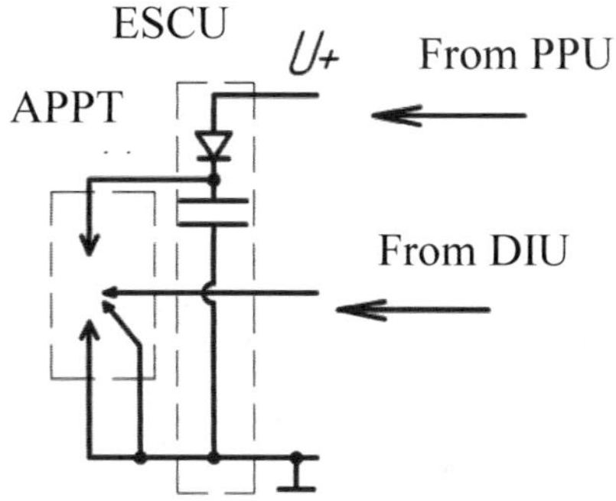

Figure 4. The schematic diagram of the APPT discharge channel.

4. Selection of the discharge initiation unit circuit

The voltage generated between the main discharge electrodes in the APPT discharge channel is not enough for the occurrence of spontaneous breakdown, therefore, in the APPT developed at RIAME MAI, a discharge initiating unit called as an "igniter" is used (figure 1) [7][7][7]. To initiate a discharge in the thruster channel, a voltage of 10 kV is applied to the electrodes of the igniter, which is generated by the discharge initiation unit.

The use of APPT on board the SSC requires the development of a lightweight, compact yet highly reliable DIU.

Two approaches were considered to create conditions for a breakdown in the igniter. In the first one, an alternating voltage with a frequency of about 10 kHz is supplied to the primary winding of the step-up transformer, and the secondary (high-voltage) winding is connected to the electrodes of the igniter. At each period of the electric discharge, a breakdown occurs between the igniter electrodes, and free electrons originate. Such circuit imposes restrictions on controlling the initial number of electrons, since it excludes the possibility for the accumulation of energy for the discharge in the igniter.

The second option implies the accumulation of energy in the capacitor with the following pulsed release of this energy in the igniter. Since such scheme allows one to vary the initial number of electrons by changing the energy stored in the capacitor, it was chosen for the developed DIU. Pulsed discharge of the capacitor requires switches capable of withstanding high voltage in the closed state and significant currents (tens of Amps) in the open state. Such a switch can be implemented with a thyristor or an arrester.

The DUI circuit with uncontrolled arrester, shown in figure 5, was chosen.

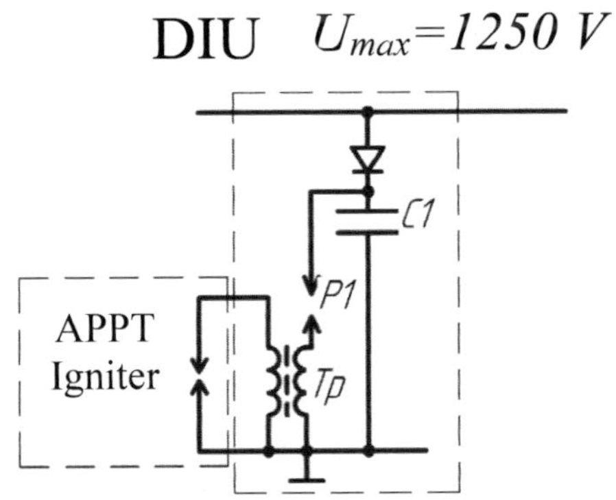

Figure 5. The DIU with uncontrolled arrester.

When the specified voltage is reached on the plates of the capacitor C1, a breakdown occurs in the uncontrolled arrester P1 and the capacitor discharges through the primary winding of the transformer Tr, creating high voltage between the electrodes of the APPT igniter and leading to an electric breakdown.

Using this circuit allowed us to increase the reliability of the DIU with a slight increase in its overall dimensions.

5. Results and discussion

The laboratory model of the power processing unit developed at the Moscow Aviation Institute provides the operation frequency of the APPT with discharge energy of 7.8 J in the range of 0.8 Hz to 3 Hz at the supply voltages from 9 V to 14 V. The appearance of the developed PPU is shown in figure 6. The power consumption was from 14 to 30 watts.

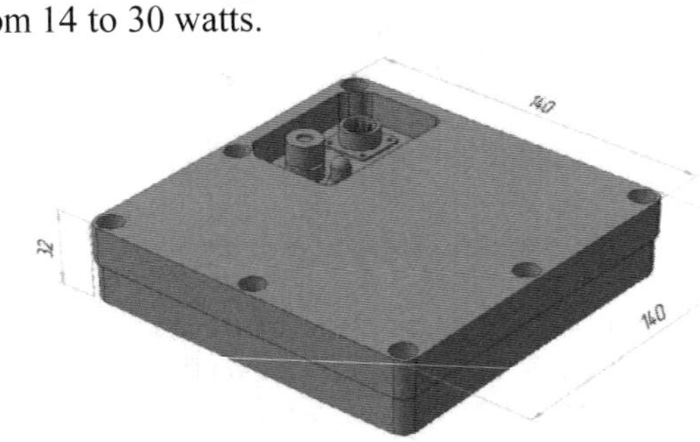

Figure 6. Appearance of the PPU for propulsion system based on low-power APPT.

When tested in air, the controlled arrester “RU-83” was used as a load simulating low-power APPT. The arrester is a sealed unit made of glass, filled with gas and having three metal electrodes. Two of them (anode and cathode) are connected to the positive and negative electrodes of ESCU. The third electrode is the control one. The output winding of the DIU transformer was connected to it and to the cathode. Thus, when the DIU is triggered, there is a high potential difference between the control electrode and the cathode and a breakdown occurs. The resulting electrons closed the electric circuit, leading to the discharge of ESCU capacitors.

The net power of developed PPU was determined by the relation:

$$P_{eff} = \frac{CU^2}{2} f; \qquad (1)$$

Here: C is the total capacitance of the ESCU and DIU capacitors (they are connected in parallel), U is the voltage, at which the breakdown occurs, and f is the frequency of breakdowns.

The total power consumption was estimated by (2):

$$P_{sum} = U_{sup} \frac{\int_0^T I_{sup}(t)dt}{T}; \qquad (2)$$

Here: U_{sup} is the PPU supply voltage, $I_{sup}(t)$ is the PPU supply current, $T = \frac{1}{f}$ is the APPT cycle.

The system efficiency was calculated by the following relation:

$$\eta = \frac{P_{eff}}{P_{sum}}; \qquad (3)$$

Figure 7 shows the oscillograms of the ESCU capacitors charging and discharging.

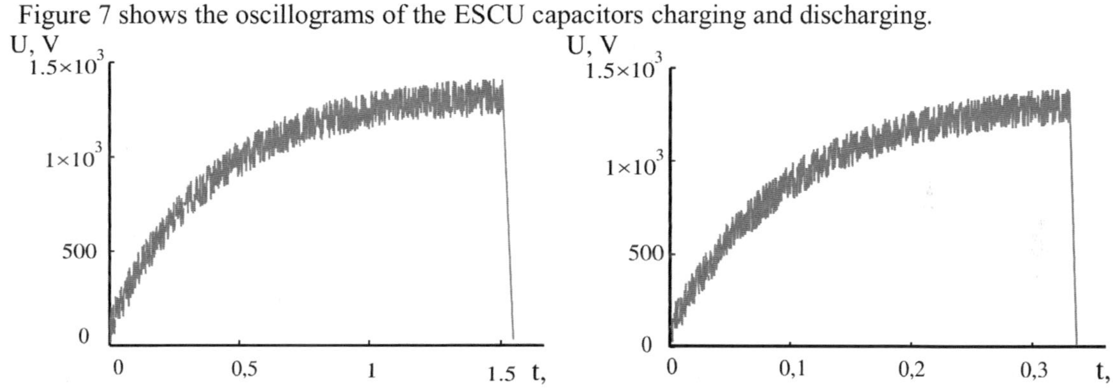

a) b)

Figure 7. Oscillograms of the ESCU capacitors charging and discharging: a) at the supply voltage of 9 V, b) at the supply voltage of 14 V. Discharge energy - 7.8 J.

As it can be seen from figure 7, the PPU under development for APPT provides the thruster operation frequency of 0.8 Hz (the period from the beginning of capacitors charging to the breakdown is 1.25 s) at the supply voltage of 9 V and of 3 Hz (the period - 0.33 s) at a supply voltage of 14 V.

The estimated efficiency of the power system at U_{sup}=9V is 0.81, which is a good value for systems with power consumption of tens of watts.

The developed circuit confirmed the operability and the ability to ensure the frequency of discharges in the APPT in the range of 0.8 - 3 Hz. At present, preparations are underway for vacuum tests using the APPT as a load.

6. Conclusion

A laboratory model of the power processing unit for the APPT was developed at the Moscow Aviation Institute. It provides the thruster operation frequency of 0.8 to 3 Hz at the supply voltage of 9-14 V,

IOP Conf. Series: Materials Science and Engineering **927** (2020) 012003 doi:10.1088/1757-899X/927/1/012003

respectively, and at a discharge energy of 7.8 J. The work done will reduce the development time for similar devises with other levels of power consumption.

The design of the discharge initiation unit for a low-power APPT is developed. When developing the discharge initiation unit circuit, two options were considered: the DIU with the thyristor, and the DIU with uncontrolled arrester. The first option showed low reliability at voltages above 1000 V and, therefore, the second one was chosen.

When tested in air, the controlled arrester RU-83 was used as the APPT model. The operability of the developed circuit was verified. Tests in a vacuum chamber with the APPT use as a load will allow us to evaluate the correctness of the selected circuitry and design solutions.

Acknowledgments

The work has been fulfilled within the frames of implementation of the Federal Target Program "Research and Development in the priority fields of the science and technology sector growth in Russia for 2014-2020" (Agreement No. 05.607.21.0308. The unique identifier of agreement: RFMEFI60719X0308)

References

[1] Minyushkin D N, Kryukov I A 2019 Calculation of aerodynamic heating and ablation of thermal protection system in axisymmetric formulation *AIP Conf. Proc.* **2181** 020025

[2] Mueller J 2000 Thruster Options for Microspacecraft: A Review and Evaluation of State-of-the-Art and Emerging Technologies *Micropropulsion for Small Spacecraft* Progress in Astronautics and Aeronautics **187**

[3] Spanjers G, Bromaghim D, Lake J, Dulligan M, White D, Schilling J and Bushman S 2001 AFRL MicroPPT Development for the TechSat 21 Flight *Proc. of 27th Int. Electric Propulsion Conf.* Pasadena CA IEPC-01-166

[4] Zakrzwski C, Benson S, Sanneman P and Hoskins A 2002 On-Orbit Testing of the EO-1 Pulsed Plasma Thruster AIAA 2002-3973

[5] Popov G A and Antropov N N 2006 Ablative PPT. New Quality, New Perspectives *Acta Astronautica* **59** 175-180

[6] Antropov N N, Bogatyy A V, Diakonov G A, Orlov M M, Popov G A, Tyutin V K and Yakovlev V N 2008 Development of Ablative Pulsed Plasma Thrusters at the Research Institute of Applied Mechanics and Electrodynamics *Kosmonavtika i raketostroyeniye* No. 3(52) M.: TsNIIMash. pp 28-34 (in Russian)

[7] Antropov N N, Bogatyy A V, Diakonov G A, Lyubinskaya N V, Popov G A, Semenikhin S A, Tyutin V K, Khrustalev M M and Yakovlev V N 2011 New Stage in the Development of Ablative Pulsed Plasma Thrusters at RIAME MAI *Bulletin of the S.A. Lavochkin Association* **5** 30-40 (in Russian)

[8] Diakonov G A, Lyubinskaya N V and Semenikhin S A 2009 Experimental research of models of APPT of low power *Aviacionnaya i kosmicheskaya tehnika i tehnologiya* **9/66** 136-8 (in Russian)

IOP Conf. Series: Materials Science and Engineering **927** (2020) 012004 doi:10.1088/1757-899X/927/1/012004

On stability of planar periodic motions of a satellite in vicinity of resonant rotation

T Churkina
Moscow Aviation Institute (National Research University), Volokolamskoe shosse 4, Moscow, 125993, Russia

E-mail: tatiana802@mail.ru

Abstract. The stability of planar resonant periodic rotations of a rigid body around its centre of mass in a central Newtonian field of forces is considered. The orbit of the centre of mass of the satellite is assumed to be an elliptical one, the body's inertia ellipsoid is general. Resonant rotations examined mean that the body makes 3 turns around its centre of mass during 2 revolutions of the centre of mass on its orbit (Mercury-type resonance 3:2). For the set of eccentricity value $e \in [0, 0.25]$ typical for planets' ones the regions of stability in the first approximation and regions of unstability are built in three-dimensional space of parameters (eccentricity and two inertial parameters).

1. Statement of the problem

The stability of planar periodic Mercury-type resonant rotations of a rigid body (satellite) around its centre of mass on elliptical orbit in a central Newtonian field of forces is considered. Mercury-type rotations mean that the body makes 3 turns around its centre of mass during 2 revolutions of the centre of mass on its orbit (resonance 3:2) [1, 2, 3].

Let $OXYZ$ be the orbital coordinate system such that the origin of coordinates O fixed to the centre of mass of the body and axes OX, OY, OZ directed along transversal, binormal to the orbit, and radius-vector relative to attracting centre respectively, and let $Oxyz$ be a coordinate system fixed in the body with axes directed along the principal central axes of inertia (moments of inertia A, B, C match axes Ox, Oy, Oz correspondingly). The orientation of the $Oxyz$ coordinate system with respect to the orbital one is defined by Euler angles ψ, θ, φ implemented in a standard way.

There exist planar rotations of a satellite around its centre of mass such that the satellite rotates with respect to the normal of the orbit such that $\psi = \pi$, $\theta = \pi/2$ and a change of angle φ meets a second order ordinary differential equation first derived by V.V. Beletsky [4]

$$(1 + e\cos\nu)\frac{d^2\varphi}{d\nu^2} - 2e\sin\nu\frac{d\varphi}{d\nu} + \gamma\sin\varphi\cos\varphi = 2e\sin\nu \qquad (1)$$

Here e – the eccentricity of the orbit, ν – the true anomaly, $n = 3(A-B)/C$ ($-3 \le n \le 3$) is parameter of inertia. While such motion axis Oz is normal to the orbit plane.

Mercury-type periodic rotations meet conditions

$$\varphi(4\pi) = \varphi(0) + 2\pi, \qquad \varphi'(4\pi) = \varphi'(0) \qquad (2)$$

Here prime indicates derivative with respect to ν.

Introduce a new variable $x = 2\varphi - \nu$ and equation (1) and conditions (2) take form [1,3,5]

$$(1+e\cos\nu)x'' - 2e\sin\nu x' + \gamma\sin(x+\nu) = 6e\sin\nu \tag{3}$$

$$x(4\pi) = x(0), \qquad x'(4\pi) = x'(0) \tag{4}$$

The boundary value problem (3), (4) was numerically investigated in [1,2,3,5,6]. In this paper, only symmetrical pure 2π –periodical solutions $x(\nu)$ are investigated (the case $x(0) = 0$). In [1] it was shown that conditions (4) are equivalent to condition

$$x(0) = x(\pi) = 0 \tag{5}$$

(for pure 2π –periodical symmetrical solutions).

In [1] areas of existence of the solution of boundary value problem (3), (5) are shown in plane of parameters (e, $x'(0)$). The number of existing considered periodic solution at fixed value of e depends on γ: there exist one ($\tilde{x}_1(\nu)$) or three ($\tilde{x}_i(\nu)$, i=1,2,3) periodic solutions (the case of two solutions gives a bifurcation). The numbering of solutions meets the condition $\tilde{x}_1(\nu) > \tilde{x}_2(\nu) > \tilde{x}_3(\nu)$ (in accordance with [3]). In this paper only solutions $\tilde{x}_1(\nu)$ denoted bellow by $\tilde{x}(\nu)$ are considered.

The couples of parameters ($x'(0)$, γ) giving 2π –periodical solutions $\tilde{x}(\nu)$ of the boundary value problem (3), (5) are represented in figure 1 (for a set of values of e taken from interval [0,0.25] and corresponding to most natural planet systems). Note that for greater values of e qualitative results obtained bellow are essentially different due to appearance of an extra regions of stability in the first approximation. The figure similar to figure 1 for only one value of $e = 0.2056$ corresponding to Mercury orbit is given in [3]. The coordinates of points from figure 1 are given in table 1.

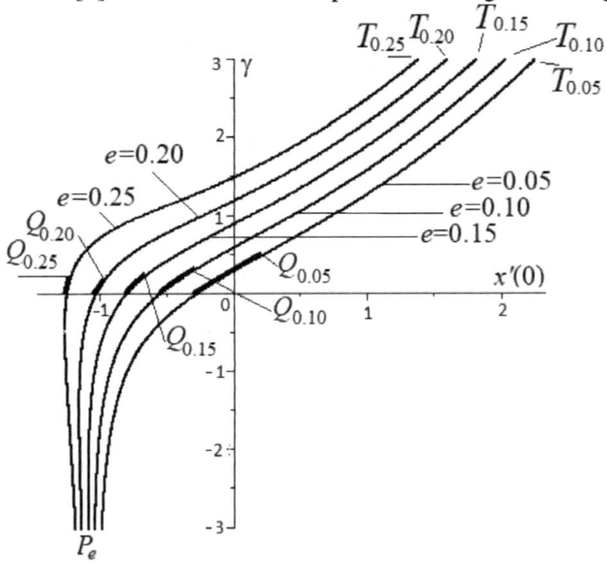

Figure 1. Initial conditions and values of parameter γ for 2π –periodical solutions.

In this paper, the problem of linear stability with regards to both plane and special perturbations for symmetrical pure 2π –periodic solutions $\tilde{x}_1(\nu)$ of boundary value problem (3), (5) is investigated.

Note that investigations dealt with linear and nonlinear analysis of similar problem (for special values of parameters) are in [7-11].

The Hamiltonian corresponding canonical equations describing the satellite motion on an elliptic orbit reads [3,12]

$$\begin{aligned}\Gamma=&\frac{\kappa\cos^2\varphi+\sin^2\varphi}{2\sin^2\theta(1+e\cos\nu)^2}p_\psi^2+\frac{\kappa\sin^2\varphi+\cos^2\varphi}{2(1+e\cos\nu)^2}p_\theta^2+\\&+\frac{1}{(1+e\cos\nu)^2}\left[(\kappa\cos^2\varphi+\sin^2\varphi)\mathrm{ctg}^2\theta+\frac{1}{\alpha}\right]p_\varphi^2-(\kappa-1)\frac{\sin\varphi\cos\varphi}{\sin\theta(1+e\cos\nu)^2}p_\psi p_\theta-\\&-(\kappa\cos^2\varphi+\sin^2\varphi)\frac{\mathrm{ctg}\,\theta}{\sin\theta(1+e\cos\nu)^2}p_\psi p_\varphi+(\kappa-1)\frac{\sin\varphi\cos\varphi\,\mathrm{ctg}\,\theta}{(1+e\cos\nu)^2}p_\theta p_\theta-\\&-\cos\psi\,\mathrm{ctg}\,\theta\,p_\psi-\sin\psi\,p_\theta+\frac{\cos\psi}{\sin\theta}p_\varphi+\frac{3}{2}\left[\left(\frac{1}{\kappa}\cos^2\varphi+\sin^2\varphi\right)\sin^2\theta-\alpha\sin^2\theta\right](1+e\cos\nu)\end{aligned}\qquad(6)$$

Here $\alpha=C/A$, $\kappa=3/(3-\gamma\alpha)$; p_ψ, p_θ, p_φ – dimensionless momentum associated with the coordinates ψ, θ, φ correspondingly, so ψ, θ, φ, p_ψ, p_θ, p_φ – canonical Hamiltonian variables. e, α and γ are parameter of the problem in use. Note that according to the physical sense of the problem $|\gamma|\le 3$, $\gamma\le -3+6/\alpha$.

Table 1. Coordinates of point from figure 1 (in plane ($x'(0)$, γ)).

E	Coordinates of points P_e	Coordinates of points Q_e	Coordinates of points T_e
0.05	$P_{0.05}(-0.996087, -3)$	$Q_{0.05}(0.217918, 0.504792)$	$T_{0.05}(2.238327, 3)$
0.10	$P_{0.10}(-1.049397, -3)$	$Q_{0.10}(-0.286118, 0.324409)$	$T_{0.10}(2.017976, 3)$
0.15	$P_{0.15}(-1.100214, -3)$	$Q_{0.15}(-0.657304, 0.244429)$	$T_{0.15}(1.801629, 3)$
0.20	$P_{0.20}(-1.148798, -3)$	$Q_{0.20}(-0.966455, 0.199969)$	$T_{0.20}(1.587167, 3)$
0.25	$P_{0.25}(-1.195368, -3)$	$Q_{0.25}(-1.236130, 0.173009)$	$T_{0.25}(1.372061, 3)$

Canonical equations with Hamiltonian (6) have the particular solution [1, 3]

$$\psi=\pi,\quad \theta=\frac{\pi}{2},\quad \varphi=\frac{1}{2}\nu+\frac{1}{2}\tilde{x}(\nu),\quad p_\psi=0,\quad p_\theta=0,\quad p_\varphi=\frac{\alpha}{3}(1+\cos\nu)^2\left(3+\frac{d\tilde{x}(\nu)}{d\nu}\right)\qquad(7)$$

2. Stability investigation

Introduce perturbations by formulas

$$\begin{aligned}&\psi=\pi+q_1,\quad \theta=\frac{\pi}{2}+q_2,\quad \varphi=\frac{1}{2}\nu+\frac{1}{2}\tilde{x}(\nu)+q_3,\\&p_\psi=p_1,\quad p_\theta=p_2,\quad p_\varphi=\frac{\alpha}{3}(1+\cos\nu)^2\left(3+\frac{d\tilde{x}(\nu)}{d\nu}\right)+p_3\end{aligned}\qquad(8)$$

Then after series expansion in a neighborhood of $q_i=p_i=0$ $(i=1,2,3)$ we obtain Hamiltonian of perturbed motion as $H=H_2+\ldots$ and $\tilde{H}=H_2$ meets linearized problem. Here H_2 is a quadratic part or Hamiltonian H.

Let $e=0$. In this case, the solution $\tilde{x}(\nu)$ can be obtained analytically through elliptic functions. The regions of stability in the first approximation and regions of unstability for such a case at spatial perturbations are obtained in [3] (see figures 2 and 3). At $\alpha<1$ there exist 4 bounded regions of stability in the first approximation (regions numbered from 1 to 4 in the figure 2). At $\alpha>1$ there exist 7 bounded regions of stability in the first approximation and six regions of unstability marked as lines in figure 3 as they a narrow. Regions of unstability originate from points (α_i ,0). The right curve in the figure 3 corresponds to the boundary of physically allowed parameters' area given by relation

$\gamma = -3 + 6/\alpha$ so the righter part is unallowed. α_i are the following: $\alpha_1 = 0.895178$, $\alpha_2 = 1.166761$, $\alpha_3 = 4/3$, $\alpha_4 = 1.386203$, $\alpha_5 = 1.600546$, $\alpha_6 = 1.680532$, $\alpha_7 = 1.976312$ [3].

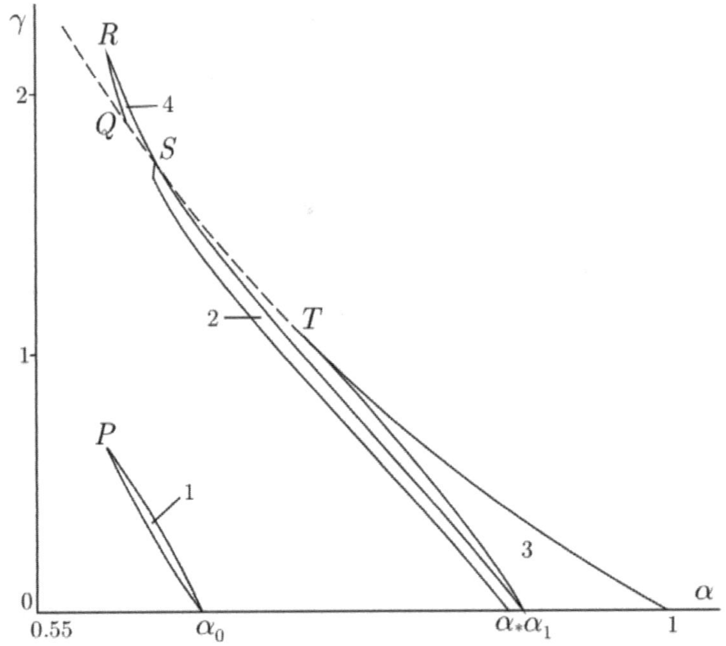

Figure 2. Regions of stability in the first approximation and regions of unstability ($\alpha < 1$, e=0) (P(0.598,0.665), Q(0.614,1.884), R(0.582,2.175), S(0.632,1.747), T(0.729,1.114), $\alpha_0 = 2/3$, $\alpha_* = 8/9$) [3].

Let $e \in (0, 0.25]$. Fundamental matrix $\mathbf{X}(\nu)$ of system with Hamiltonian $\tilde{H}$ satisfying the initial conditions $\mathbf{X}(\nu) = \mathbf{E}_3$ is constructed for fixed physically allowed values of e, α and γ, where $\mathbf{E}_3$ is the third order identity matrix. The characteristic equation reads

$$(\rho^2 - 2a\rho + 1)(\rho^4 - a_1\rho^3 + a_2\rho^2 - a_1\rho + 1) = 0 \tag{9}$$

If the inequalities

$$|a| < 1, \quad -2 < a_2 < 6, \quad 4(a_2 - 2) < a_1^2 < \frac{(a_2 - 2)^2}{4} \tag{10}$$

are satisfied then the linear system with Hamiltonian $\tilde{H}$ is stable [3]. In figure 1 the sets of parameter that give solutions satisfying condition $|a| < 1$ are marked with thick segments (stability in the first approximation with respect to plane perturbations).

The calculus show that the regions of instability go out into space $e > 0$ from instability regions given in figures 2, 3 and become wider with increase of e. Regions of stability in the first approximation at the same time go out into space $e > 0$ from stability regions 1, 2, 3 (figure 2) and stability regions presented in figure 3, major part of them become smaller with increase of e. The evolution of regions of stability in the first approximation with increase of parameter e is shown in figures 4-6. Stability region originating from region 1 (figure 2) fully vanishes with increase of e (at e=0.05 it still exists but is very small so it is not shown in figure 4). Instability regions in figures 4-6 are filled with solid black.

IOP Conf. Series: Materials Science and Engineering **927** (2020) 012004 doi:10.1088/1757-899X/927/1/012004

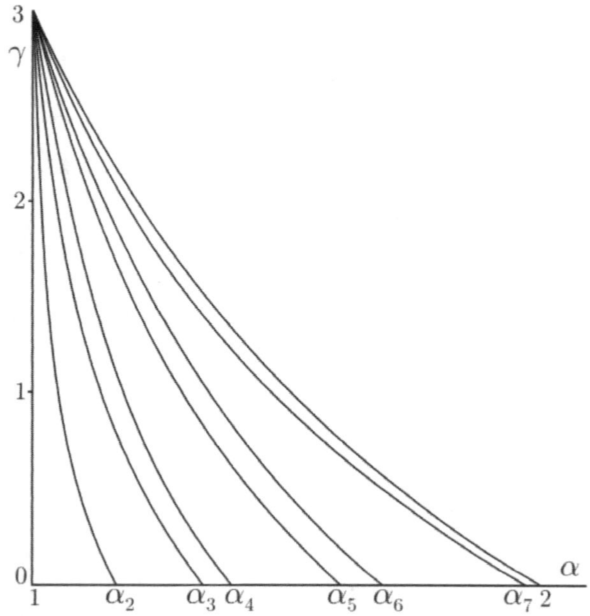

Figure 3. Regions of stability in the first approximation and regions of unstability ($\alpha > 1, e = 0$) [3]

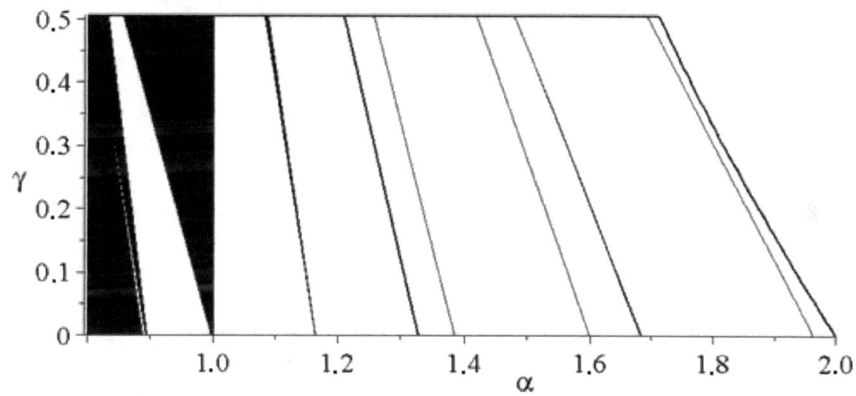

Figure 4. Regions of stability in the first approximation and regions of unstability ($e = 0.05$)

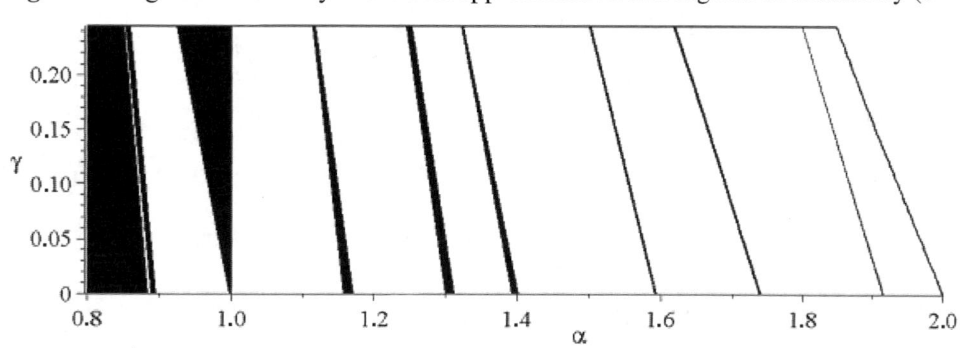

Figure 5. Regions of stability in the first approximation and regions of unstability (e = 0.15)

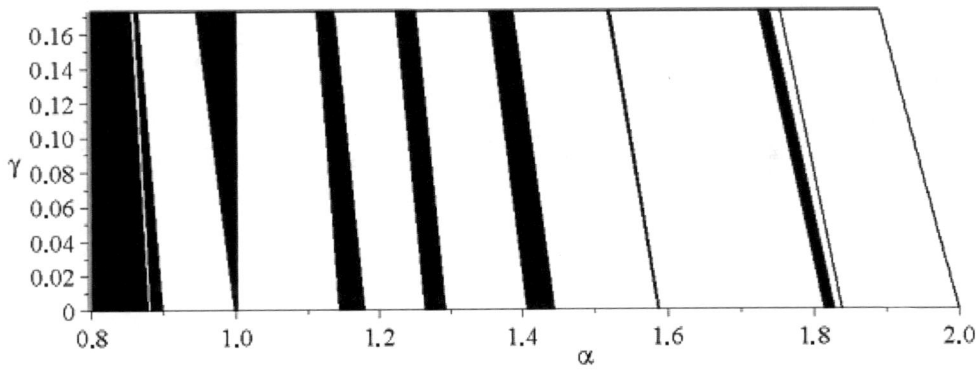

Figure 6. Regions of stability in the first approximation and regions of unstability (e = 0.25)

3. Conclusion

In the problem of linear stability with regards to both plane and special perturbations for symmetrical pure 2π–periodic Mercury-type solutions of Beletsky equation, there exist 10 numerically built regions of stability in the first approximation in three-dimensional space of parameters (eccentricity $e \in [0, 0.25]$ and two inertial parameters). One of these regions eliminates and others regions contract with the increase of parameter e.

Acknowledgments
This research was supported by a grant of the Russian Science Foundation (project No 19-11-00116).

References

[1] Beletskii V V and Lavrovskii E K 1975 On the theory of Mercury resonance rotation *Astronomy Journal* **52** 1299
[2] Bruno A D 2002 Families of periodic solutions to the Beletsky Equation *Cosmic Research* **40** 274
[3] Markeev A P 2009 *Linear Hamiltonian systems and particular problems on stability of satellite motion about its centre of mass* (Izhevsk: NITs "Regular and chaotic Dynamic")
[4] Beletskii V V 1959 On libration of the satellite *Artificial satellites of Earth* **3** 13
[5] Churkina T E, Stepanov S Y 2017 On the Stability of Periodic Mercury-type Rotations *Regular and Chaotic Dynamics* **22** 851
[6] Sarychev V A, Sazonov V V and Zlatoustov V A 1979 Periodic Rotations of a Satellite in the Elliptic Orbit Plane *Kosm. Issled.* **17** 190
[7] Bardin B S, Chekina E A and Chekin A M 2015 On the Stability of a Planar Resonant Rotation of a Satellite in an Elliptic Orbit *Regular and Chaotic Dynamics* **20** 63
[8] Bardin B S and Chekina E A 2016 On the stability of a resonant rotation of a satellite in an elliptic orbit *Nonlinear Dynamics* **12** 619
[9] Bardin B S and Chekina E A 2016 On stability of resonance rotation of a dynamically symmetric satellite in a plane of elliptical orbit *Trudy MAI* **89**
[10] Bardin B S and Chekina E A 2016 On the Stability of Resonant Rotation of a Symmetric Satellite in an Elliptical Orbit *Regular and Chaotic Dynamics* **21** 377
[11] Churkina T E 2014 Satellite rotation stability at a Mercurian type resonance *Mechanics of Solids* **49** 127
[12] Churkina T E 2007 Stability of a planar resonance satellite motion under spatial perturbations *Mechanics of Solids* **42** 507

IOP Conf. Series: Materials Science and Engineering **927** (2020) 012005 doi:10.1088/1757-899X/927/1/012005

The implicit fully coupled numerical method for flows in thermochemical nonequilibrium

A M Molchanov, M V Siluyanova and Yu M Kochetkov
Moscow Aviation Institute (National Research University), Volokolamskoe shosse, 4 125993 Moscow, Russia

E-mail: Alexmol_2000@mail.ru

Abstract. A new effective implicit fully coupled numerical method to compute a high-speed flowfield that is in thermochemical nonequilibrium has been developed. Such a flowfield is described by coupled partial differential equations for the conservation of mass, momentum, total energy, vibrational energies, rotational energy, mass conservation for each species, etc. The total number of equations can be very large, which greatly complicates the solution of general finite volume vector equation when using the standard fully implicit approach. The proposed method allows us to reduce operations with block tridiagonal matrices to inversion of 4x4 matrices and trivial matrix multiplication operations. Moreover, the approximation factorization error is minimized. The numerical procedures were applied to solve chemical and thermal nonequilibrium flow past a sphere body at Mach 27 and chemically nonequilibrium heterogeneous flow in the channel of a combined ramjet with a solid fuel gas generator. The results indicate that new method is 4-5 times more efficient than the standard fully implicit method.

1. Introduction

It is known that in chemically and thermally nonequilibrium flows, the characteristic times of the main processes can differ from each other by several orders of magnitude. Therefore, the system of equations is stiff, and its numerical solution requires large computer resources.

Implicit numerical methods are most often used to solve the stiffness problem. The use of these methods leads to the need of multiple inversion of matrices having a dimension corresponding to the number of equations at each time step in each grid node. In the case when the system contains a very large number of equations (tens or even hundreds), the direct use of fully implicit methods may become ineffective.

2. Numerical method

The basic system of equations describing the flow of a chemically and thermally nonequilibrium gas includes the equations of continuity, momentum, total energy, rotational energy, vibrational energies, electron energy, and mass conservation equations for each species. The fully implicit finite-volume representation of this system of equations is as follows [1,2]:

$$\mathbf{A}_{i,j}\delta\mathbf{U}_{i,j}^{n+1} + \mathbf{B}_{i,j}\delta\mathbf{U}_{i,j+1}^{n+1} + \mathbf{C}_{i,j}\delta\mathbf{U}_{i,j-1}^{n+1} + \mathbf{D}_{i,j}\delta\mathbf{U}_{i+1,j}^{n+1} + \mathbf{E}_{i,j}\delta\mathbf{U}_{i-1,j}^{n+1} = \Delta\mathbf{U}_{i,j}^{n+1} + \Delta t\mathbf{H}^{n}, \quad (1)$$

where $\delta\mathbf{U}_{i,j}^{n+1}$ is an implicit solution change of the main vector $\mathbf{U}$ at grid point i,j during the interval from time $n\Delta t$ to time $(n+1)\Delta t$; $\Delta\mathbf{U}_{i,j}^{n+1}$ is an explicit solution change of the main vector $\mathbf{U}$ due to viscous and inviscid fluxes; $\mathbf{H}$ is the source vector; $\mathbf{A}_{i,j},\mathbf{B}_{i,j},\mathbf{C}_{i,j},\mathbf{D}_{i,j},\mathbf{E}_{i,j}$ are block matrix elements whose size coincides with a number of equations in the main system. Here, to reduce the recording volume, a 2D case is presented; extension to a 3D problem does not introduce fundamental differences in the implementation of the method.

To obtain the system (1), Roe method [3] for splitting inviscid fluxes and Tysinger and Caughey approach [4] for implicit viscous fluxes were used.

The matrix $\mathbf{A}_{i,j}$ can be represented as $\mathbf{A}_{i,j}=\mathbf{I}+\alpha\Delta t\,\mathbf{a}_{i,j}-\alpha\Delta t\left(\partial\mathbf{H}/\partial\mathbf{U}\right)_{i,j}^{n}$, where $\mathbf{I}$ is the identity matrix; $\partial\mathbf{H}/\partial\mathbf{U}$ is the Jacobi matrix; the $\alpha\geq 0.5$ determines the order of accuracy of the time derivative. For $\alpha=0.5$ the time derivative is centered, Crank-Nicholson like, and second order accurate, otherwise it is of first order. At $\alpha=1$ the method is fully implicit and for $a>1$ it is over relaxed, but may have better convergence properties depending on the particular simulation.

The source $\mathbf{H}$ is associated with transitions between different energy modes, rates of formation of chemical components, etc. The time scales of these processes are very small and can cause the equation set to be extremely stiff. The cost of using fully implicit approach is expensive time steps and complicated coding. For example, if turbulent gas mixture contains 13 species, there are 9 vibrational modes and 1 rotational mode, we have the system of 28 equations: 1 continuity equation, 2 momentum equations, 1 total energy equation, 12 species equations, 9 vibrational energy equation, 1 rotational energy equation and 2 turbulence model equations. For this particular formulation, block tridiagonal solves are required, where each block has dimensions of 28x28. Another cost of the large equation set and fully implicit approach is a high memory requirement.

Therefore, we need to use a more economical approach, not associated with multiple inversions of 28x28 matrices at each grid point.

In [5], a convenient method for numerical implementation was proposed, based on sequential alternation at two half-steps in time of the explicit and implicit representation of viscous/inviscid members of the system and source. It is easy to show that this approach is equivalent to the following factorization of system (1):

$$\left[\mathbf{I}-0.5\Delta t\left(\frac{\partial\mathbf{H}}{\partial\mathbf{U}}\right)_{i,j}^{n}\right]\left[\left(\mathbf{I}+\alpha\Delta t\,\mathbf{a}_{i,j}\right)\delta\mathbf{U}_{i,j}^{n+1}+\mathbf{B}_{i,j}\delta\mathbf{U}_{i,j+1}^{n+1}+\mathbf{C}_{i,j}\delta\mathbf{U}_{i,j-1}^{n+1}+\mathbf{D}_{i,j}\delta\mathbf{U}_{i+1,j}^{n+1}+\mathbf{E}_{i,j}\delta\mathbf{U}_{i-1,j}^{n+1}\right] \quad (2)$$
$$=\Delta\mathbf{U}_{j}^{n+1}+\Delta t\mathbf{H}^{n}$$

The main advantage of this approach is that all the matrices that appear in the right bracket of the left side of equation (2), $\mathbf{a}_{i,j},\mathbf{B}_{i,j},\mathbf{C}_{i,j},\mathbf{D}_{i,j},\mathbf{E}_{i,j}$ and most importantly the matrices that appear in intermediate calculations, have a common block form:

$$\mathbf{\Phi}=\begin{pmatrix}\mathbf{\Phi}_{11} & \mathbf{\Phi}_{12}\\ \mathbf{\Phi}_{21} & \mathbf{\Phi}_{22}\end{pmatrix}=\begin{pmatrix}\mathbf{\Phi}_{11} & \mathbf{0}\\ \mathbf{\Phi}_{21} & \mathbf{D}\end{pmatrix}, \quad (3)$$

where the matrix blocks have the following dimensions: $\mathbf{\Phi}_{11}$ - 4x4; $\mathbf{\Phi}_{21}$ - N_ax4; $\mathbf{D}$ - diagonal matrix of size N_a. Here N_a is the number of additional equations, not including 4 basic equations (continuity, 2 momentum and total energy).

The main advantages of a matrix of the form (3) are as follows:

1) its inversion is reduced to a single inversion of the matrix $\mathbf{A}_{11}$ and trivial operations of matrix multiplication (Frobenius formula):

$$\mathbf{\Phi}^{-1}=\begin{pmatrix}\mathbf{\Phi}_{11}^{-1} & \mathbf{0}\\ -\mathbf{D}^{-1}\mathbf{\Phi}_{21}\mathbf{\Phi}_{11}^{-1} & \mathbf{D}^{-1}\end{pmatrix} \quad (4)$$

2) any necessary operations with such a matrix (inversion, multiplication) do not change its shape, i.e. the block $\mathbf{\Phi}_{12}$ remains zero, and the block $\mathbf{\Phi}_{22}$ remains a diagonal matrix.

Thus, even a significant increase in the number of additional equations (i.e., the number of chemical components) does not necessitate the use of large computer resources.
After multiplying the matrices, we obtain on the left-hand side of (2)

$$\left[\mathbf{I}+\alpha\Delta t\,\mathbf{a}_{i,j}-0.5\Delta t\left(\frac{\partial\mathbf{H}}{\partial\mathbf{U}}\right)^n_{i,j}-0.5\alpha(\Delta t)^2\left(\frac{\partial\mathbf{H}}{\partial\mathbf{U}}\right)^n_{i,j}\mathbf{a}_{i,j}\right]\delta\mathbf{U}^{n+1}_{i,j}+\left[\mathbf{I}-0.5\Delta t\left(\frac{\partial\mathbf{H}}{\partial\mathbf{U}}\right)^n_{i,j}\right]\mathbf{B}_{i,j}\delta\mathbf{U}^{n+1}_{i,j+1}$$
$$+\left[\mathbf{I}-0.5\Delta t\left(\frac{\partial\mathbf{H}}{\partial\mathbf{U}}\right)^n_{i,j}\right]\mathbf{C}_{i,j}\delta\mathbf{U}^{n+1}_{i,j-1}+\left[\mathbf{I}-0.5\Delta t\left(\frac{\partial\mathbf{H}}{\partial\mathbf{U}}\right)^n_{i,j}\right]\mathbf{D}_{i,j}\delta\mathbf{U}^{n+1}_{i+1,j}+\left[\mathbf{I}-0.5\Delta t\left(\frac{\partial\mathbf{H}}{\partial\mathbf{U}}\right)^n_{i,j}\right]\mathbf{E}_{i,j}\delta\mathbf{U}^{n+1}_{i-1,j} \quad (5)$$
$$=\Delta\mathbf{U}^{n+1}_{i,j}+\Delta t\mathbf{H}\left(\mathbf{U}^n_{i,j}\right)$$

A comparison of expressions (5) and (1) shows that the scheme (2) is a rather rough factorization of a fully implicit scheme (1): none of the original matrix elements are returned exactly. The coefficient at $\delta\mathbf{U}^{n+1}_{i,j}$ in (5) differs from the corresponding coefficient in (1) by a value of the second order of smallness. But the remaining members are now dependent on $\left[\mathbf{I}-0.5\Delta t\left(\partial\mathbf{H}/\partial\mathbf{U}\right)^n_{i,j}\right]$. This factorization error reduces accuracy and slows algorithm convergence.

The approximation factorization procedure can be significantly improved by reducing the decomposition error.

We can represent the matrix A $\partial\mathbf{H}/\partial\mathbf{U}$ as the sum of diagonal and off-diagonal components:

$$\frac{\partial\mathbf{H}}{\partial\mathbf{U}}=\mathbf{d}+\mathbf{T} \quad (6)$$

and consider the following scheme:

$$\left[\mathbf{I}-\alpha\Delta t\left(\frac{\partial\mathbf{H}}{\partial\mathbf{U}}\right)^n_{i,j}\right]\left(\mathbf{I}-\alpha\Delta t\mathbf{d}^n_{i,j}\right)^{-1}\times$$
$$\times\left[\left(\mathbf{I}+\alpha\Delta t\,\mathbf{a}_{i,j}-\alpha\Delta t\mathbf{d}^n_{i,j}\right)\delta\mathbf{U}^{n+1}_{i,j}+\mathbf{B}_{i,j}\delta\mathbf{U}^{n+1}_{i,j+1}+\mathbf{C}_{i,j}\delta\mathbf{U}^{n+1}_{i,j-1}+\mathbf{D}_{i,j}\delta\mathbf{U}^{n+1}_{i+1,j}+\mathbf{E}_{i,j}\delta\mathbf{U}^{n+1}_{i-1,j}\right]=\Delta\mathbf{U}^{n+1}_{i,j}+\Delta t\mathbf{H}^n \quad (7)$$

Here $\mathbf{d}$ is the diagonal part of the matrix $\partial\mathbf{H}/\partial\mathbf{U}$. In this case, the solution of system (7) is reduced to a single inversion of the matrix associated with the source and to the solution of the system in which all matrices have the form (3). In addition, the matrix $\partial\mathbf{H}/\partial\mathbf{U}$ is easily partitioned into blocks according to physical processes.

Let us analyze this scheme. After multiplying the matrices we get:

$$\left\{\mathbf{I}-\alpha\Delta t\left(\frac{\partial\mathbf{H}}{\partial\mathbf{U}}\right)^n_{i,j}+\alpha\Delta t\mathbf{a}_{i,j}-(\alpha\Delta t)^2\,\mathbf{T}^n_{i,j}\left(\mathbf{I}-\alpha\Delta t\,\mathbf{d}^n_{i,j}\right)^{-1}\mathbf{a}_{i,j}\right\}\delta\mathbf{U}^{n+1}_{i,j}+\left[\mathbf{I}-\alpha\Delta t\,\mathbf{T}^n_{i,j}\left(\mathbf{I}-\alpha\Delta t\,\mathbf{d}^n_{i,j}\right)^{-1}\right]\mathbf{B}_{i,j}\delta\mathbf{U}^{n+1}_{i,j+1}$$
$$+\left[\mathbf{I}-\alpha\Delta t\,\mathbf{T}^n_{i,j}\left(\mathbf{I}-\alpha\Delta t\,\mathbf{d}^n_{i,j}\right)^{-1}\right]\mathbf{C}_{i,j}\delta\mathbf{U}^{n+1}_{i,j-1}+\left[\mathbf{I}-\alpha\Delta t\,\mathbf{T}^n_{i,j}\left(\mathbf{I}-\alpha\Delta t\,\mathbf{d}^n_{i,j}\right)^{-1}\right]\mathbf{D}_{i,j}\delta\mathbf{U}^{n+1}_{i+1,j} \quad (8)$$
$$+\left[\mathbf{I}-\alpha\Delta t\,\mathbf{T}^n_{i,j}\left(\mathbf{I}-\alpha\Delta t\,\mathbf{d}^n_{i,j}\right)^{-1}\right]\mathbf{E}_{i,j}\delta\mathbf{U}^{n+1}_{i-1,j}=\Delta\mathbf{U}^{n+1}_{i,j}+\Delta t\mathbf{H}^n$$

And again, the coefficient at $\delta\mathbf{U}^{n+1}_{i,j}$ differs from the corresponding coefficient in (1) by a second order value relative to Δt. In addition, if diagonal elements play a decisive role in the matrix $\partial\mathbf{H}/\partial\mathbf{U}$, then

$$\left\|\alpha\Delta t\mathbf{T}^n_{i,j}\left(\mathbf{I}-\alpha\Delta t\mathbf{D}^n_{i,j}\right)^{-1}\right\|=\left\|\mathbf{T}^n_{i,j}\left(\mathbf{I}/(\alpha\Delta t)-\mathbf{D}^n_{i,j}\right)^{-1}\right\|=\left\|\mathbf{T}^n_{i,j}\left(\mathbf{I}/(\alpha\Delta t)-\mathbf{D}^n_{i,j}\right)^{-1}\right\|=\frac{\left\|\mathbf{T}^n_{i,j}\right\|}{\left\|\mathbf{I}/(\alpha\Delta t)-\mathbf{D}^n_{i,j}\right\|}\ll 1 \quad (9)$$

Thus, with sufficient accuracy, equation (8) is transformed into

$$\left\{\mathbf{I}-\alpha\Delta t\left(\frac{\partial\mathbf{H}}{\partial\mathbf{U}}\right)^n_{i,j}+\alpha\Delta t\mathbf{a}_{i,j}\right\}\delta\mathbf{U}^{n+1}_{i,j}+\mathbf{B}_{i,j}\delta\mathbf{U}^{n+1}_{i,j+1}+\mathbf{C}_{i,j}\delta\mathbf{U}^{n+1}_{i,j-1}+\mathbf{D}_{i,j}\delta\mathbf{U}^{n+1}_{i+1,j}+\mathbf{E}_{i,j}\delta\mathbf{U}^{n+1}_{i-1,j}=\Delta\mathbf{U}^{n+1}_{i,j}+\Delta t\mathbf{H}^n \quad (10)$$

This expression exactly matches the fully implicit scheme (1).

IOP Conf. Series: Materials Science and Engineering **927** (2020) 012005 doi:10.1088/1757-899X/927/1/012005

Thus, we obtain an implicit numerical scheme, in which, on the one hand, the problem of stiffness is completely solved (the scheme is unconditionally stable and fully coupled); on the other hand, the number of matrix inverses is reduced, and the matrices themselves are factorized into smaller ones, and their inversion requires significantly less computer resources than in a fully implicit scheme.

3. Numerical results

To test the numerical method just presented, we calculated a series of problems.

3.1. Test 1

Simulation of a nonequilibrium flow past the sphere head of a hypersonic aircraft was the first test case. Mach number M = 27, flight altitude = 75 km; 11 chemical components, 3 vibrational energy modes, 1 general rotational energy mode, zero catalytic activity.

The following boundary conditions were used: the radius of the spherical part R = 0.05m; freestream speed u_e = 7800 m/s; temperature and pressure in the surrounding space T_e = 208.4K, p_e = 2.388Pa; the standard atmospheric composition for a given altitude. Wall temperature T_w = 700K.

Figure 1 shows the heat flux density distribution on the surface of the head. Curve 1 - the calculation using a fully implicit scheme (equation (1)); curve 2 - the calculation using an implicit scheme with factorization (7).

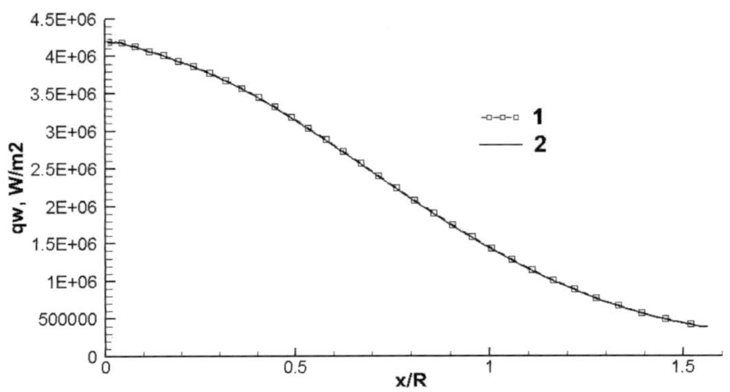

Figure 1. Heat flux density distribution on the surface of the head

Obviously, the results of both calculations completely coincide.

3.2. Test 2.

Chemically nonequilibrium heterogeneous flow in the channel of a combined ramjet with a solid fuel gas generator (Figure 1). Flight speed M = 7, altitude = 40 km; 14 species taken into account.

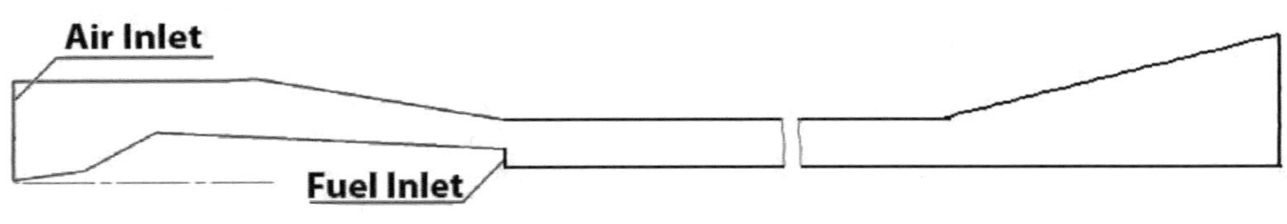

Figure 2. Schematic drawing of the combined ramjet

The following boundary conditions at the air inlet were used: flight speed u_e = 2222 m/s; ambient temperature and pressure T_e = 250.4K, p_e = 287.143Pa; the standard atmospheric composition for a given altitude. The parameters at the outlet of the gas generator (Fuel Inlet) are presented in table 1.

Table 1. Fuel Inlet parameters

Velocity, m/s	Temperature, K	Pressure, Pa	H	H2	H2O	CO	CO2	HCl	N2
1582	2026	8.E4	0.3438E-3	0.3775	0.1232E-4	0.3635	0.3157E-5	0.57227E-2	0.24203

The composition is set in mole fractions. The length of the combustion chamber = 21.5 m.
Figires 3,4 show the pressure distribution on the wall of the combustion chamber (Figure 3 – the calculation with considering the chemical reactions, Figure 4 - calculation without considering the chemical reactions)

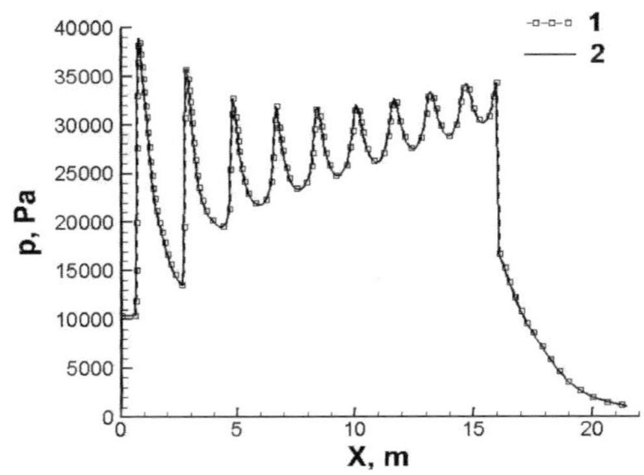

Figure 3. Pressure distribution on the wall of the combustion chamber (chemically reacting case)

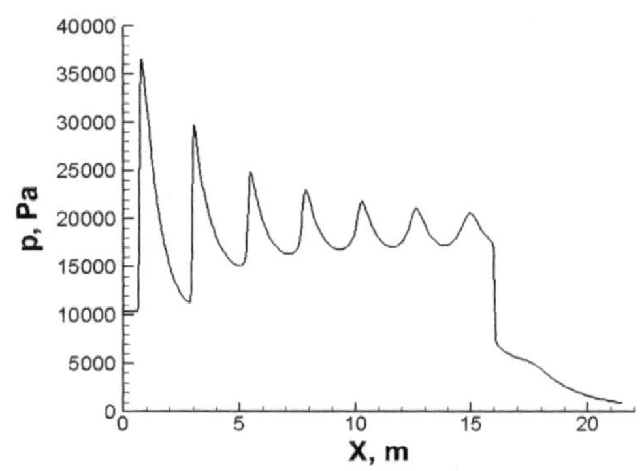

Figure 4. Pressure distribution on the wall of the combustion chamber (chemical reactions are not taken into account)

In Figure 3: curve 1 — the calculation using a implicit scheme Eq.(1); curve 2 - calculation using an implicit scheme with factorization (7). Both calculations are in good agreement.
Comparison of the effectiveness of various numerical schemes is presented in table 2:

Table 2. Comparison of the effectiveness of various numerical schemes

Scheme	Number of iterations	Computer time spent on one iteration, in relation to a standard fully implicit scheme
Fully implicit scheme (1)	600-800	1
Half-step alternating scheme (2)	1000-1500	0.2 - 0.25
Modified scheme (7)	650-900	0.2 - 0.25

Thus, the proposed scheme allows us to reduce the calculation time by 4-5 times in comparison with a fully implicit scheme. With a larger number of components and energy modes, this difference will be even more significant.

4. Conclusions

An implicit fully coupled numerical scheme has been developed for solving equations describing the flow of a thermochemically nonequilibrium gas. Using this numerical method, on the one hand, helped

IOP Conf. Series: Materials Science and Engineering **927** (2020) 012005 doi:10.1088/1757-899X/927/1/012005

to completely solve the stiffness problem (the scheme is unconditionally stable and fully coupled), and on the other hand, to reduce the number of matrix inversion. The matrices themselves are factorized into smaller ones, and their inversion requires significantly less computer resources than using a purely implicit scheme. The performed test calculations showed that the proposed numerical scheme is highly efficient.

References

[1] MacCormack R W 1997 AIAA Paper 97-2100
[2] Molchanov A M 2011 AIAA Paper 2011-3211
[3] Roe P 1986 *Annu. Rev. Fluid Mech.* **18** 337
[4] Tysinger T and Caughey D 1991 AIAA Paper 91-0242
[5] Molchanov A M 2018 *Fundamental research* **2** 28 DOI: 10.17513/fr.42069

IOP Conf. Series: Materials Science and Engineering **927** (2020) 012006 doi:10.1088/1757-899X/927/1/012006

Numerical modelling of the reflected waves from gas cavities and methane bombs in the heterogeneous media of the Northern regions

P Stognii, I Petrov and N Khokhlov
Moscow institute of physics and technology (National research university), Dolgoprudnyi, Institutskiy per.9, 141701, Russia

E-mail: stognii@phystech.edu

Abstract. The Arctic region contains lots of underground gas layers closed from the outside. In case of their opening, the gas is of great danger for the surface buildings and drilling equipment. In this work, we observe the main dangerous gas reservoirs – gas cavities and methane bombs. We build the mathematical models of these heterogeneities and presented the main mathematical equations allowing calculating the seismic waves spread through them correctly. The grid-characteristic method of the third order of accuracy is used in all the computations. We demonstrate the results of modelling the seismic waves spread through the heterogeneous medium with the gas cavities and through the homogeneous medium with the methane bomb. The analysis of the differences between the models is carried out. The problems of computing the models with gas cavities and with methane bombs need different approaches for the correct calculations, though the problem formulations are similar.

1. Introduction

The Arctic region contains a number of heterogeneities, which makes it difficult to explore the territories on the presence of hydrocarbon deposits [1]. Among these problematic inclusions there are fractures, ice formations, gas cavities and methane bombs.

Fractures may be of different forms, shapes, they are situated at different positions – all these parameters should be taken into account while carrying out the seismic works on the territory [2].

Then, ice formations significantly influence the seismic prospecting works as they bring in lots of seismic reflections into the seismograms, which are later analysed by the geologists [3].

Gas cavities [4] and methane bombs [5] are the heterogeneities, which are of great danger for the surface infrastructure. They differ only by the type of containing gas (consequently, different types of border conditions used in mathematical modelling). In case they are surrounded by the frozen soil and ice, they are of no danger. But when the drilling rig suddenly opens it or the ice, surrounding it, begins to melt, no longer keeping the gas layer closed, gas starts to spread all around and also towards the surface. The rushed gas can lead to the complete destruction of the drilling equipment, live buildings on the surface. Then, it is very important to detect gas layers in time and carry out the constant monitoring of them. This is not always possible due to the high cost of the seismic prospecting works and lack of the equipment, appropriate for the Arctic conditions. Then, the numerical modelling can be a great help in the solution of the problem of detection and monitoring of gas cavities and methane bombs.

Most research papers aim at examining the processes of developing such gas reservoirs, for example [6], as these gas layers contain large gas deposits. However, the gas cavities and methane layers are of great danger for surrounding objects [7]. We should study the characteristics of gas layers and the processes of its spread in space before starting the gas deposits development.

In this paper, we present the results of the numerical modelling of seismic waves spread through the heterogeneous media with the gas cavities and the methane bomb. The grid-characteristic method of the third order of accuracy is applied to compute all the models. We present the wave fields of the velocity distribution for the models with gas layers. The wave fields demonstrate the clearer results for the models with the methane bombs. However, the reflected waves from both heterogeneities – gas cavities and methane bombs are noticeable from the wave fields, obtained by the calculations using the grid-characteristic method.

2. The numerical method

For describing the seismic wave spread in the heterogeneous medium with gas layers, we used the equations for the linear-elastic medium [8]:

$$\rho\cdot(\partial\upsilon/\partial t) = (\nabla\cdot\sigma)^{T}, \tag{1}$$

$$\partial\sigma/\partial t = \lambda(\nabla\cdot\upsilon)I + \mu(\nabla\times\upsilon + (\nabla\times\upsilon)^{T}). \tag{2}$$

In (1), (2) υ is the speed of seismic waves, σ is the Cauchy stress tensor, ρ is the medium density, t is the time, λ and μ are the Lame parameters, I is the unit matrix.

We used the grid-characteristic method of the third order of accuracy to solve the system (1) numerically. Its description in detail can be found in [2, 9].

The descriptions of the gas layers were the following. For describing the gas cavities, we applied the system of equations (1), (2) as the sound velocities and the density of them are much bigger than zero. For computing the points on the boundaries of the gas cavity, we applied the transparent method, when the contact conditions between the outside medium and the gas cavity are not considered.

For describing the methane bombs, where the characteristic parameters are very small, we used the free-boundary condition [10], when the normal and tangential components of the stress tensor are equal to zero:

$$\sigma_{xx} = \sigma_{xy} = 0 \tag{3}$$

3. Model with the gas cavities

In this section we present the results of the numerical modelling of the seismic waves spread through the heterogeneous medium with the gas cavities. The gas cavity is the gas reservoir with high pressure inside [11]. It can be located only in fluid-saturated sand due to its low density. Therefore, we constructed the model containing the water layer on the surface – as it is usual for the Arctic conditions, several soil layers and the layers of fluid-saturated sand for the gas cavities accumulations.

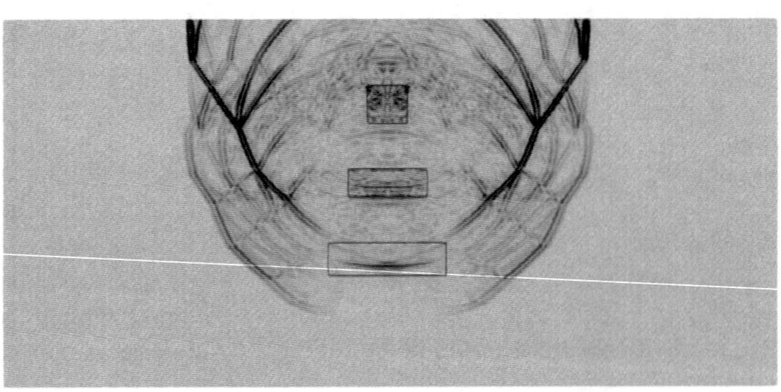

Figure 1. The wave field at time 0.7 sec.

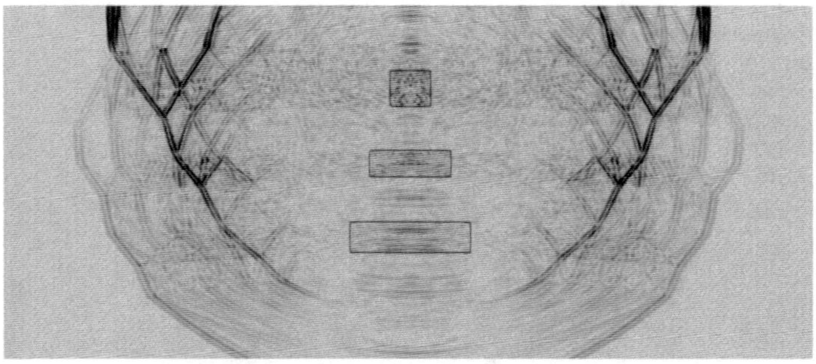

Figure 2. The wave field at time 1.0 sec.

The longitudinal sound velocity of the gas cavity was equal to 1000 m/s, the transverse sound velocity was 707 m/s, the density was 900 kg/m^3. In the calculations the step in time was equal to 10^{-4} s, the step in space was 1 m. We used the 30 Hz frequency source of impulse. We computed the model with three gas cavities, similar to the situation, when gas has already started to spread in space. The result wave fields of the normal velocity distribution inside the model are presented in figure1, 2. Figure 1 depicts the wave field at time 0.7 sec, figure 2 – at time 1.0 sec. The reflections from the gas cavities can be detected, though not very clearly.

4. Model with the methane bombs

In this section, we present the results of the numerical modelling of the seismic waves spread through the homogeneous medium with the methane bomb. The characteristics of the homogeneous medium were the following. The longitudinal sound velocity was equal to 5000 m/s in the medium, the transverse sound velocity was 3150 m/s, the density was 2600 kg/m^3. The time step was equal to 10^{-4} s, the space step was 1 m. The seismic impulse was described by the 30 Hz frequency. The result wave field of the normal velocity inside the model at different moments of time is presented in figure 3, 4, 5. Figure 3 depicts the wave field at 0.1 sec., figure 4 – at 0.2 sec., figure 4 – at 0.3 sec.

The seismic reflections from the methane bomb in figure 3 spread in space, including the surface, where they can be further detected by the seismic receivers. The seismic waves reflect several times from the heterogeneity, which should be taken into account when analysing the seismograms on the receivers. The reflections from the methane bomb are clearer and more easily to detect, if compare with the reflections from the gas cavity in figure 1, 2. It is connected with the deep location of the gas cavity (1,000 m from the surface) in comparison with the methane bomb location (200 m from the surface).

Figure 3. The wave field at time 0.1 sec.

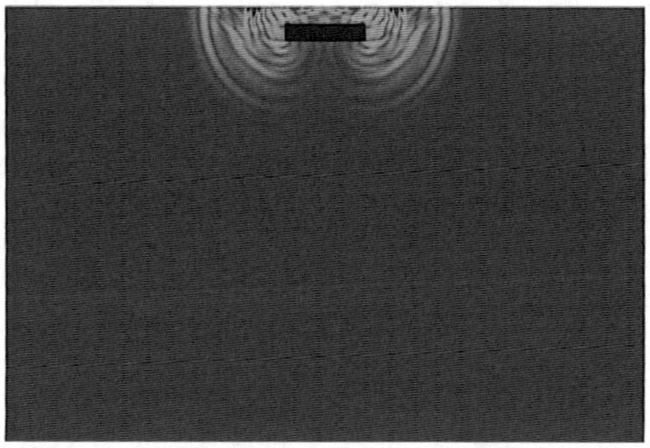

Figure 4. The wave field at time 0.2 sec.

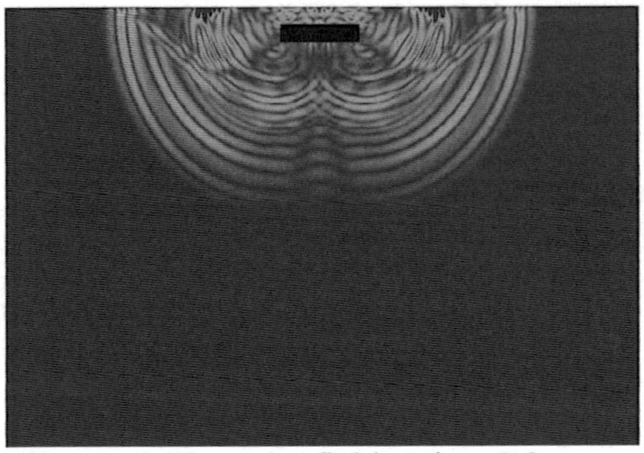

Figure 5. The wave field at time 0.3 sec.

5. Discussion

The previous works on the theme examined the gas-containing layers from the gas reservoirs formation view [12, 13]. The attention was mostly payed to the chemical structure of the gas layers. This work aims at solving the gas problem from the overall view, describing the gas reservoirs as the common layers.

The models with the gas cavities and the methane bombs are similar from the problem description view. However, they need different approaches for solving the problems. The gas cavity contains the gas of high density – 900 kg/m^3, whereas the methane bomb density is less than 1 kg/m^3. Therefore, the methane bomb cannot be described by the common linear-elastic layer, as it is done for the gas cavity. In order to solve this problem, we established the free boundary condition on all the borders of the methane layer. The result wave field from the gas cavity turned out to be less clear than the same wave field from the methane bomb. It is mainly connected with the deep location of the gas cavity – 1000 m from the surface, in comparison with 200 m from the surface of the methane layer location. Then, following from the first argument, the gas cavities are situated inside the medium, which is heterogeneous, while the methane layer, due to its location close to the surface, can be modelled inside the homogeneous medium. It makes the seismic reflections from the methane layer clearer, though the waves from the gas cavity are also observable due to the use of the grid-characteristic method of the third order of accuracy.

6. Conclusions

In this paper, we presented the results of the gas layers exploration in the Northern regions. We described the main equations used for characterizing the behaviour of the seismic waves in the media. The grid-characteristic method of the third order of accuracy was used in all the computations. The gas cavities were calculated as the linear-elastic media, the methane bombs were computed through establishing the free-boundary conditions on all the boundaries. The results of modelling the seismic waves spread through the heterogeneous medium with gas cavities and methane bombs demonstrated the possibility of detecting all these heterogeneities, though the reflections from the methane bombs were more noticeable.

Acknowledgements
The reported study was funded by RFBR according to the research project №19-07-00366.

References

[1] Gautier D L, Bird K J, Charpentier R R, Grantz A et al. 2009 Assessment of Undiscovered Oil and Gas in the Arctic *Science.* **324** 5931 1175-79
[2] Khokhlov N I and Stognii P V 2020 Novel approach to modeling the seismic waves in the areas with complex fractured geological structures *Minerals.* **10**

[3] Epifanov V P and Sazonov K E 2019 Effect of standing waves on the local strength of a modeled ice field *Doklady Physics.* **64** 12 14-18

[4] Stognii P V, Khokhlov N I and Petrov I B 2019 Numerical Modeling Of Wave Processes In Multilayered Media With Gas-Containing Layers: Comparison Of 2D And 3D Models *Doklady Mathematics.* **100** 3 586-588

[5] Glikson A 2018 The methane time bomb *Energy Procedia.* **146** 23-29

[6] Musakaev N G, Khasanov M K, Borodin S L 2018 The mathematical model of the gas hydrate deposit development in permafrost *International Journal of Heat and Mass Transfer.* **118** 455-461

[7] Blunier T 2000 Frozen methane escapes from the sea floor *Science.* **288** 68-69

[8] Savin G N, Rushchitskii Y Y and Novatskii V 1971 The theory of elasticity *Soviet Applied Mathematics.* **7** 808-811

[9] Muratov M V and Petrov I B 2019 Application of mathematical fracture models to simulation of exploration seismology problems by the grid-characteristic method *J. Computer Research and Modeling.* **11** 6 1077-82

[10] Favorskaya A, Golubev V and Khokhlov N 2018 Two approaches to the calculation of air subdomains: theoretical estimation and practical results *Procedia Computer Science.* **126** 1082-90

[11] Bogoyavlensky V I, Kerimov V Y, Olkhovskaya O O and Mustaev R N 2016 Improving the efficiency and safety prospecting, exploration and development of oil and gas in the sea of Okhotsk *Territorija "NEFTEGAS"[Oil and Gas Territory].* **10** 24-32

[12] Guang-Jin C and Tian-Min G 1998 A new approach to gas hydrate modelling *Chemical Engineering Journal.* **71** 2 145-151

[13] Matthew K D and Bruce A B 2001 A numerical model for the formation of gas hydrate below the sea floor *Journal of Geophysical Research: Solid Earth.* **106** B1 497-514

IOP Conf. Series: Materials Science and Engineering **927** (2020) 012007 doi:10.1088/1757-899X/927/1/012007

Contributions of VT and VV exchanges to the CO_2 relaxation in the full state-to-state modeling

V Gorikhovskii*, **E Nagnibeda****

Saint-Petersburg State University, 7/9 Universitetskaya nab., St. Petersburg, 199034 Russia

E-mail: *v.gorikhovskii@spbu.ru, **e_nagnibeda@mail.ru

Abstract. Numerical modeling of nonequilibrium state-to-state carbon dioxide kinetics is a computationally complex problem requiring the solution of a huge system of stiff differential equations. In the present study, we use parallel numerical scheme, based on the extended backward differential formula with adaptive timestep strategy. Using this scheme, we study the contribution of various types of energy exchanges to the solution of kinetics equations. The comparison of the results obtained with the use of the "cut" and full state-to-state approaches is also shown in the paper.

1. Introduction

Simulation of nonequilibrium kinetics of carbon dioxide relates the most computationally complex problems of the modern physical mechanics. Such modeling has many applications in various fields, such as plasma chemistry [1], laser technology [2], and simulation of spacecraft entering the atmospheres of Venus and Mars [3, 4]. In recent decades, the most rigorous approach based on the state-to-state description of vibrational and chemical relaxation was noticeably developed [5, 4, 6].

The state-to-state approach is based on the joint solution of the gas dynamics equations coupled to the equations for vibrational level populations of three types of vibrations of carbon dioxide molecules and other mixture molecules. Although this approach is the most accurate, it has huge arithmetic complexity. First of all, it is due to the solution of a stiff system consisting more than 6000 equations for vibrational level populations. In addition, intermediate calculations at an each step of the solver's computational scheme include the estimation of hundreds of thousands rate coefficients of vibrational energy transitions and chemical reactions. In some studies [1, 4], "cut" systems of vibrational levels and simplified models for vibrational energy exchanges were used for state-to-state modeling of carbon dioxide relaxation.

The numerical solution of stiff systems involves the use of methods with high accuracy and stability. However, the use implicit numerical methods occurs time and memory consuming. In the present paper, to solve the system of kinetic equations, we used the Explicit Backward Differential Formula (EBDF) [7]. This scheme is based on the predictions in superfuture points to go to an explicit scheme. Using the improved predictors can raise the accuracy of the solution without loss of stability [8]. This scheme allows us to solve rigid systems, whereas such calculations require a huge number of steps. The schemes based on EBDF make it possible to apply an adaptive timestep strategy using extrapolation of a local truncation error

IOP Conf. Series: Materials Science and Engineering **927** (2020) 012007 doi:10.1088/1757-899X/927/1/012007

[9]. This reduces the number of steps to thousands. This article explores the effectiveness of implementation of adaptive steps for the extended BDF approach.

Using such a scheme, numerical experiments were carried out to analyze the contribution of VT and VV exchanges to the solution of the system of kinetics equation. The exact solution of the full system is also compared with the results obtained using the "cut" models.

2. Governing equations

First of all, let us consider equations which describe vibrational relaxation in spatially uniform CO_2. i_1, i_2, i_3 are the vibrational quantum numbers corresponding to the symmetric, bending, and asymmetric CO_2 modes. The system of equations describing the change in population $n_{i_1,i_2,i_3}(t)$ of vibrational states (i_1, i_2, i_3) and the gas temperature $T(t)$ with the time has the form [5, 10]:

$$\frac{dn_{i_1,i_2,i_3}(t)}{dt} = R^{vibr}_{i_1,i_2,i_3}(t), \quad i_m = 0, \ldots, l_m, \quad m = 1, 2, 3, \tag{1}$$

$$\rho U = const, \tag{2}$$

where $R^{vibr}_{i_1,i_2,i_3}$ are the relaxation terms, ρ is the density and U is the total specific energy:

$$\rho U = \rho(E_{tr} + E_{rot} + E_{vibr}). \tag{3}$$

The total specific energy can be expressed through the translational (E_{tr}), rotational (E_{rot}) and vibrational (E_{vibr}) energies defined as follows:

$$\rho E_{tr} = \frac{3}{2}nkT, \quad \rho E_{rot} = nkT, \quad \rho E_{vibr} = \sum_{i_1,i_2,i_3} \varepsilon_{i_1,i_2,i_3} n_{i_1,i_2,i_3}, \tag{4}$$

where k is the Boltzmann constant, n is the number density, $\varepsilon_{i_1,i_2,i_3}$ is the vibrational energy of CO_2 molecule at the corresponding levels.

The following energy exchanges in the result of collisions of CO2 molecules are considered in the paper: VT exchanges of translational energy with vibrational energies of three CO_2 vibrational modes as well as VV exchanges of vibrational energy within an each CO_2 mode and between different modes of the carbon dioxide molecule colliding with an inert partner M. They are described by the relations:

$$VT_1 : CO_2(i_1, i_2, i_3) + M \rightleftharpoons CO_2(i_1 \pm 1, i_2, i_3) + M, \tag{5}$$

$$VT_2 : CO_2(i_1, i_2, i_3) + M \rightleftharpoons CO_2(i_1, i_2 \pm 1, i_3) + M, \tag{6}$$

$$VT_3 : CO_2(i_1, i_2, i_3) + M \rightleftharpoons CO_2(i_1, i_2, i_3 \pm 1) + M. \tag{7}$$

$$VV_{1-2} : CO_2(i_1, i_2, i_3) + M \rightleftharpoons CO_2(i_1 \pm 1, i_2 \mp 2, i_3) + M. \tag{8}$$

$$VV_{2-3} : CO_2(i_1, i_2, i_3) + M \rightleftharpoons CO_2(i_1, i_2 \pm 3, i_3 \mp 1) + M, \tag{9}$$

$$VV_{1-2-3} : CO_2(i_1, i_2, i_3) + M \rightleftharpoons CO_2(i_1 \pm 1, i_2 \pm 1, i_3 \mp 1) + M. \tag{10}$$

Vibrational spectra of three CO_2 modes are calculated based on the anharmonic oscillator model [11]. Chemical reactions are not taken into account in the present study.

Relaxation terms characterize the change of vibrational state populations in result of VT and VV-exchanges in CO_2 molecules:

$$R^{vibr}_{i_1,i_2^l,i_3} = R^{VT_1}_{i_1,i_2^l,i_3} + R^{VT_2}_{i_1,i_2^l,i_3} + R^{VT_3}_{i_1,i_2^l,i_3} + R^{VV_{1-2}}_{i_1,i_2^l,i_3} + R^{VV_{2-3}}_{i_1,i_2^l,i_3} + R^{VV_{1-2-3}}_{i_1,i_2^l,i_3}. \tag{11}$$

The relaxation terms for an each energy exchange can be expressed as:

$$R^{VT_1}_{i_1,i_2,i_3} = n_{i_1-1,i_2,i_3} k^M_{i_1-1\to i_1} + n_{i_1+1,i_2,i_3} k^M_{i_1+1\to i_1} - n_{i_1,i_2,i_3}(k^M_{i_1\to i_1-1} + k^M_{i_1\to i_1+1}), \quad (12)$$

$$R^{VT_2}_{i_1,i_2,i_3} = n_{i_1,i_2-1,i_3} k^M_{i_2-1\to i_2} + n_{i_1,i_2+1,i_3} k^M_{i_2+1\to i_2} - n_{i_1,i_2,i_3}(k^M_{i_2\to i_2-1} + k^M_{i_2\to i_2+1}), \quad (13)$$

$$R^{VT_3}_{i_1,i_2,i_3} = n_{i_1,i_2,i_3-1} k^M_{i_3-1\to i_3} + n_{i_1,i_2,i_3+1} k^M_{i_3+1\to i_3} - n_{i_1,i_2,i_3}(k^M_{i_3\to i_3-1} + k^M_{i_3\to i_3+1}), \quad (14)$$

$$R^{VV_{1-2}}_{i_1,i_2,i_3} = n_{i_1-1,i_2+2,i_3} k^M_{i_1-1,i_2+2\to i_1,i_2} + n_{i_1+1,i_2-2,i_3} k^M_{i_1+1,i_2-2\to i_1,i_2} - n_{i_1,i_2,i_3}(k^M_{i_1,i_2\to i_1-1,i_2+2} + k^M_{i_1,i_2\to i_1+1,i_2-2}), \quad (15)$$

$$R^{VV_{2-3}}_{i_1,i_2,i_3} = n_{i_1,i_2-3,i_3+1} k^M_{i_2-3,i_3+1\to i_2,i_3} + n_{i_1,i_2+3,i_3-1} k^M_{i_2+3,i_3-1\to i_2,i_3} - n_{i_1,i_2,i_3}(k^M_{i_2,i_3\to i_2-3,i_3+1} + k^M_{i_2,i_3\to i_2+3,i_3-1}), \quad (16)$$

$$R^{VV_{1-2-3}}_{i_1,i_2,i_3} = n_{i_1-1,i_2-1,i_3+1} k^M_{i_1-1,i_2-1,i_3+1\to i_1,i_2,i_3} + n_{i_1+1,i_2+1,i_3-1} k^M_{i_1+1,i_2+1,i_3-1\to i_1,i_2,i_3} - n_{i_1,i_2,i_3}(k^M_{i_1,i_2,i_3\to i_1-1,i_2-1,i_3+1} + k^M_{i_1,i_2,i_3\to i_1+1,i_2+1,i_3-1}), \quad (17)$$

where $k^M_{i_1,i_2,i_3\to i'_1,i'_2,i'_3}$ are the temperature-dependent rate coefficients of the energy transitions from the levels (i_1, i_2, i_3) to the levels (i'_1, i'_2, i'_3). To calculate the rate coefficients we use the formulas of the SSH theory generalized for CO_2 [12].

3. Numerical scheme

To solve the stiff system of equations (1)–(2) we use the explicit parallel methods based of the extended backward differential formula (EBDF) [7] with adaptive time-stepping strategy. Let us consider the system as an initial value problem:

$$\frac{dy}{dt} = R(y, t), \qquad y(t_0) = y_0, \quad (18)$$

where y is a vector of unknowns composed of populations and the gas temperature, y_0 is the initial condition.

The EBDF of the order k can be written as follows:

$$\sum_{j=0}^{k} \alpha_j y_{n+j} = \Delta t \beta_k R(y_{n+k}) + \Delta t \beta_{k+1} R(y_{n+k+1}), \quad (19)$$

where Δt is the timestep and the coefficients β are taken from [9].

Assuming that y_n,y_{n+1},...,y_{n+k-1} are available, one step of EBDF-scheme includes:

(i) Computation of the first predictor $\tilde{y}_{n+k}$ as a solution of the linear system:

$$\sum_{j=0}^{k} \alpha_j y_{n+j} = \Delta t R_{n+k}. \quad (20)$$

(ii) Computation of the second predictor $\tilde{y}_{n+k+1}$ as a solution of the linear system:

$$\sum_{j=0}^{k} \alpha_j \tilde{y}_{n+j+1} = \Delta t R_{n+k+1}. \quad (21)$$

(iii) Evaluation of the predictions of the relaxation terms($\tilde{R}_{n+k+1}$) by the solution of the equation(21):

$$\tilde{R}_{n+k+1} = R(t_{n+k+1}, \tilde{y}_{n+k+1}). \tag{22}$$

(iv) Solving (19) with the predictions of relaxation terms (22) to get the solution y_{n+k} of the EBDF-scheme.

It's the scheme of the order $(k+1)$ and the local truncation error (LTE) on the $n+k$ step is given by [8]:

$$LTE_{n+k} = \Delta t^{k+2}(\beta_{k+1}C_1(1 - \frac{\alpha_k - 1}{\alpha_k})\frac{\partial R}{\partial t}y^{(k+1)} + C_2 y^{(k+2)})(t_n) + o(\Delta t^{k+3}), \tag{23}$$

Adaptive time-stepping strategy is based on local truncation error estimations (23) [9]:

$$LTE_{n+k} = \Delta t^{k+2}\phi(y_k, t_k) + o(\Delta t^{k+3}). \tag{24}$$

Based on this estimation, a Richardson extrapolation can be constructed. This extrapolation allows to choose a largest timestep with error less or equal LTE_{n+k}. Suitable norm and a tolerance TOL is specified by the user. The result has to satisfy "error per step" criteria:

$$||LTE_{n+k}|| \leq TOL. \tag{25}$$

From this criteria, if Δt is reduced to $\gamma\Delta t$, then LTE_{n+k} is reduced to $\gamma^{k+2}LTE_k$. Thus, the largest γ for which the estimated local truncation error will pass the error test is approximately given by the expression:

$$\hat{\gamma} = \left(\frac{TOL}{||LTE_k||}\right)^{\frac{1}{k+2}}. \tag{26}$$

If the estimation of the norm LTE on the step $n+k$ is smaller than tolerance, timestep could be increased to $\Delta t_{n+k+1} = \hat{\gamma}\Delta t_{n+k}$. The global error of these scheme is less than $TOL \cdot M$, where M is the final number of steps.

For optimization of the numerical simulation of the kinetic equations it is necessary to use parallel data structures and distributed algorithms. The most expensive in time and memory in the $AT - EBDF$-scheme are the steps associated with: 1) calculation of relaxation terms and 2) computation of the predictors and unknowns y solving corresponding linear systems.

Evaluation of the relaxation terms is based on the effective calculation of the reaction rate coefficients. The efficient parallel data structure for storing and calculating the rate coefficients was proposed in our previous work [13]. This data structure enables efficient calculation of complete sets of rate coefficients using the SSH-theory formulas. To accelerate the algorithm of linear system solution, the submatrix method [14] was used.

4. Numerical experiment

Numerical experiments were carried out to take into account the contribution of VT and VV exchanges. All calculations were performed on an Intel® Core TM i7-4770k processor. During the experiments, a system with activated Hyper-Threading technology was used. The numerical schemes were implemented in C++ using POSIX streams.

The problem of modeling spatially homogeneous relaxation of carbon dioxide was considered. The equations of gas dynamics were solved jointly with the equations for relaxation terms. The system was considered taking into account all possible exchanges, and only exchanges that make the greatest contribution: VT_2, VV_{2-3}, VV_{1-2-3}. In the experiment, all calculations were performed using a fourth-order $EBDF$ scheme with an adaptive time step strategy and optimizations based on parallel calculations. Two cases of initial simulation data are used:

IOP Conf. Series: Materials Science and Engineering **927** (2020) 012007 doi:10.1088/1757-899X/927/1/012007

(i) the gas temperature $T_0 = 1000$ K; pressure $P_0 = 10000$ Pa; initial population is given by the Boltzmann distribution with the vibrational temperature $T_{V0} = 5000$ K;

(ii) the gas temperature $T_0 = 5000$ K; pressure $P_0 = 10000$ Pa; initial population is given by delta distribution with delta-peak in the level (0,0,6);

The contributions of an each type of exchanges W on one step of algorithm were calculated as

$$CON^W = \frac{\sum\limits_{(i_1,i_2,i_3)} e_{(i_1,i_2,i_3)} |R^W_{(i_1,i_2,i_3)}|}{\sum\limits_{(i_1,i_2,i_3)} e_{(i_1,i_2,i_3)} |R^{vibr}_{(i_1,i_2,i_3)}|}. \tag{27}$$

The contributions were determined at steps close to the beginning of the relaxation process ($t \cong 10^{-8}s$), steps with the main change in the gas temperature ($t \cong 10^{-5}s$), close to the equilibrium state of carbon dioxide ($t > 1s$).

The results for numerical experiment with the gas temperature $T_0 = 1000$ K, pressure $P_0 = 10000$, initial population is given by the Boltzmann distribution with the vibrational temperature $T_{V0} = 5000$ K are shown in the **Table 1**:

Table 1. The contributions of each type of exchanges for the gas temperature $T_0 = 1000$ K; pressure $P_0 = 10000$ Pa; initial population is given by the Boltzmann distribution with the vibrational temperature $T_{V0} = 5000$ K.

	$t \cong 10^{-8}s$	$t \cong 10^{-5}s$	$t > 1s$
CON^{VT_1}	0.011	0.009	0.010
CON^{VT_2}	0.374	0.302 ‘	0.221
CON^{VT_3}	0.005	0.004	0.004
$CON^{VV_{1-2}}$	0.012	0.025	0.035
$CON^{VV_{2-3}}$	0.227	0.312	0.353
$CON^{VV_{1-2-3}}$	0.371	0.348	0,377

The **Table 1** shows that the contributions of the VT_2, VV_{2-3} and VV_{1-2-3} exchanges are small. If relaxation terms R^{VT_2}, $R^{VV_{2-3}}$, $R^{VV_{1-2-3}}$ are excluded from the process of solving the system of kinetic equations, then changing the solution does not exceed the calculation error. This will reduce the computation time by about 40 percents. On the other hand, the use of a cropped model leads to a significant change in the solution.

On the other hand, the use of a cropped circuit leads to a significant change in the solution. In **Figure 1** the temperature change during relaxation is shown for the cropped and full model. For the "cut" model, relaxation is slower, which can lead to the accumulation of errors in modeling a spatially inhomogeneous gas. At the same time, a stripped-down model requires 8 times less computing resources.

Based on this, we can conclude that if necessary, accurate calculations should be done using the full level model, despite the computational complexity. To reduce the time spent on calculations, it is possible to use the model including only VT_2,VV_{2-3} and VV_{1-2-3} exchanges.

On the other hand, energy exchanges must be considered when studying relaxation with initial data containing delta distributions or mixture of distributions. The results for numerical experiment with the gas temperature $T_0 = 1000$ K, pressure $P_0 = 10000$, initial population is given by the delta distribution with delta-peak in the level (0,0,6) are shown in **Table 2**.

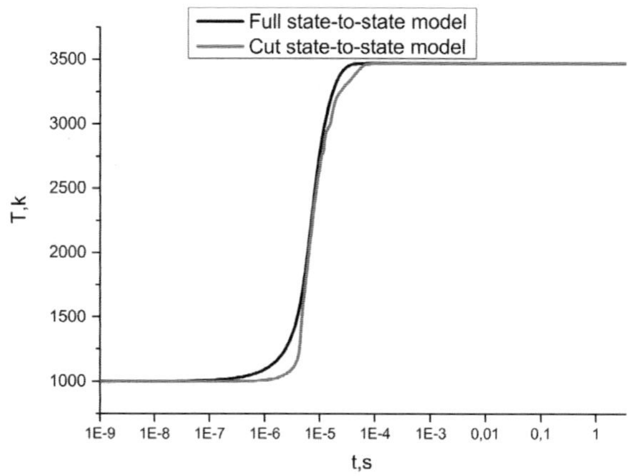

Figure 1. The gas temperature for full state-to-state and cut state-to-state models. In numerical experiment the gas temperature $T_0 = 1000$ K, pressure $P_0 = 10000$ Pa, initial population is given by the Boltzmann distribution with the vibrational temperature $T_{V0} = 5000$ K.

Table 2. The contributions of each type of exchanges for the gas temperature $T_0 = 1000$ K; pressure $P_0 = 10000$ Pa; initial population is given by the Delta distribution with the with delta-peak in the level (0,0,6).

	$t \cong 10^{-8}s$	$t \cong 10^{-5}s$	$t > 1s$
CON^{VT_1}	0.009	0.008	0.009
CON^{VT_2}	0.092	0.210	0.221
CON^{VT_3}	0.364	0.198	0.005
$CON^{VV_{1-2}}$	0.001	0.019	0.034
$CON^{VV_{2-3}}$	0.211	0.279	0.336
$CON^{VV_{1-2-3}}$	0.323	0.286	0,314

For such processes, the main energy transfer is highly dependent on the initial delta peak. In such a simulation, it is necessary to consider a complete energy exchange system.

5. Conclusions

Numerical modeling of the carbon dioxide nonequilibrium kinetics is a time-consuming computational task that involves solving a stiff system of more than 6000 differential equations. For solution of this problem optimized explicit methods are required. We propose several optimisation techniques for the EBDF scheme. The first optimisation is the transition from the fixed timestep to the adaptive timestep. The next improvement uses parallel computing to further reduce the computing time. Storing and evaluating rate coefficients in a parallel data structure, efficient calculation of relaxation terms, and the use of distributed computing in solving linear systems can further reduce the operating time in 4 times.

Numerical experiments on the modeling of spatially homogeneous carbon dioxide vibrational relaxation were performed for optimised computational schemes. On the basis of them, the influence of VT and VV exchanges on the solution of kinetics equations is investigated. In the case when the initial distribution of populations during the simulation is close to the Boltzmann distribution, we can consider a reduced model of energy exchanges: VT_2,VV_{2-3} and VV_{1-2-3}. However, if the initial distribution has peaks, it is necessary to consider the complete system of energy exchanges:VT_1, VT_2, VT_3, VV_{1-2}, VV_{2-3} and VV_{1-2-3}. The use of "cut" models of vibrational levels is not recommended, since they can lead to significant deviations in the solution.

Acknowledgments
The reported study was funded by RFBR, project numbers 19-31-90059 and 18-01-00493.

References

[1] Silva T, Grofulović M, Klarenaar B, Morillo-Candas A, Guaitella O, Engeln R, Pintassilgo C and Guerra V 2018 *Plasma Sources Sci. Technol* **27** 015019
[2] Nevdakh V and Ganjali M 2005 *Journal of Applied Spectroscopy* **72** 75–83
[3] Park C, Howe J, Howe R, Jaffe R and Candler G 1994 *J. Thermophys. Heat Transfer* **8** 9–23
[4] Kustova E V, Nagnibeda E A and Armenise I 2014 *The Open Plasma Physics Journal* **7** 76–87
[5] Kustova E and Nagnibeda E 2001 State-to-state theory of vibrational kinetics and dissociation in three-atomic gases *AIP Conf. Proc.* vol 585 ed Bartel T and Gallis M (Melville, NY: American Institute of Physics) pp 620–627
[6] Sahai A, Lopez B, Johnston C and Panesi M 2017 *AIAA SciTech Forum - 55th AIAA Aerospace Sciences Meeting* **AIAA 2017-0213**
[7] Cash J R 1980 *Numer. Math.* **34** 235–246 ISSN 0029-599X URL `https://doi.org/10.1007/BF01396701`
[8] Alberdi E and Anza J J 2011 *Journal of Computational Mathematics* **29** 199–214
[9] Shampine L F 2005 *J. Sci. Comput.* **25** 3–16 ISSN 0885-7474
[10] Kustova E and Nagnibeda E 2012 *Chem. Phys.* **398** 111–117
[11] Herzberg G 1951 *Infrared and Raman Spectra of Polyatomic Molecules* (New York: D.Van Nostrand Company, Inc.)
[12] Schwartz R, Slawsky Z and Herzfeld K 1952 *J. Chem. Phys.* **20** 1591
[13] Gorikhovskii V and Nagnibeda E 2019 *Vestnik of St Petersburg University. Mathematics. Mechanics. Astronomy* **6(64)** 659–671
[14] Lass M, Mohr S, Wiebeler H, Kühne T D and Plessl C 2018 A massively parallel algorithm for the approximate calculation of inverse p-th roots of large sparse matrices *Proceedings of the Platform for Advanced Scientific Computing Conference* PASC '18 (ACM) pp 7:1–7:11

IOP Conf. Series: Materials Science and Engineering **927** (2020) 012008 doi:10.1088/1757-899X/927/1/012008

Critical stress analysis for the basis of a denture prosthesis

S D Arutyunov[1], A D Nikitin[2], D I Grachev[1], G G Bagdasaryan[1]

[1]A I Evdokimov Moscow state university of medicine and dentistry, 20/1, Delegatskaya, Moscow, 127473, Russian Federation
[2]Institute for Computer Aided Design RAS, 19/18, 2nd Brestskaya st., Moscow, 123056, Russia

Email: nikitin_alex@bk.ru

Abstract. This paper is focused on the numerical simulations for the mechanical behavior of the prosthesis basis under typical chewing loads. The model is supplied by physically correct boundary conditions. The stress analysis in the prosthesis basis is performed for the cases of symmetric and not-symmetric loading conditions. The sets of numerical calculations were performed for all the stages of food chewing. It was found that not-symmetric loading has a higher influence on the structural integrity of the prosthesis. The significant stress amplitudes on the basis were found under the loading at the block I (incisors) and block II (fang). It is shown that the maximum stress due to typical chewing loading usually cannot produce the prosthesis fracture in one cycle. However, the stress level is high enough to produce the fatigue crack initiation.

1. Introduction

For a long time, the prosthesis shape and its implantation were designed based on the individual dentist experience and personal feelings of patients. Sometimes these criteria are not enough to guarantee the sufficient durability of the structure. It happens that in-service prosthesis fracture can appear much early compare to the average service life. Such premature failures are often related to the specific individual structure of the oral cavity and its morphology. The second reason is the natural structure of the dentition and the feeling of comfort for the patient associated with such structure. In the first case, the dentist should be guided by the general empirical knowledge and own experience. At the same time, there are no clear limits of permissible limits for varying the prosthesis geometry parameters. In the second case, the sole criterion of the appropriate shape for the prosthesis is a personal feeling of comfort for the patient. It happens that the artificial structure that is comfortable for the patient has an extremely low in-service life. The mathematically based limits for these parameter values are needed.

In recent years the general tendency in applied denture practice is the deep integration of mathematical modeling, experimental methods, and numerical simulations [1-3]. The numerical simulations are used for the problems of the structural integrity of the implants [4-6], prosthesis, design, treatments, and optimization [7], analysis of possible displacements of the basis under external loading. All these calculations are very difficult tasks because of the high variability of the connective tissue properties. Moreover, these properties can change in time and to be sensitive to the age, sex and other individual features of the patient. The majority of these simulations are aimed to build the individual model of the future prosthesis and show the 3D image of the final work. However, the knowledge about the basic aspects of the prosthesis mechanical behavior is an important task [8]. This

IOP Conf. Series: Materials Science and Engineering **927** (2020) 012008 doi:10.1088/1757-899X/927/1/012008

paper is focused on the study of basic mechanical behavior of the prosthesis-like structure under the loading at different locations. The aim of the research is to illustrate the general response of the structure to the chewing loads and find the least favorable for the structural integrity loading conditions. The analysis was performed in order to outline the difference between symmetric and not-symmetric loading conditions.

2. Model of the prosthesis and loading parameters

The 3D computer-aided design (CAD) model of the denture basis was reconstructed based on 2D cross-section drawings, figure 1. Some simplifying assumptions were used for the model such as mirror symmetry of the basis and uniformed thinness of the denture canvas.

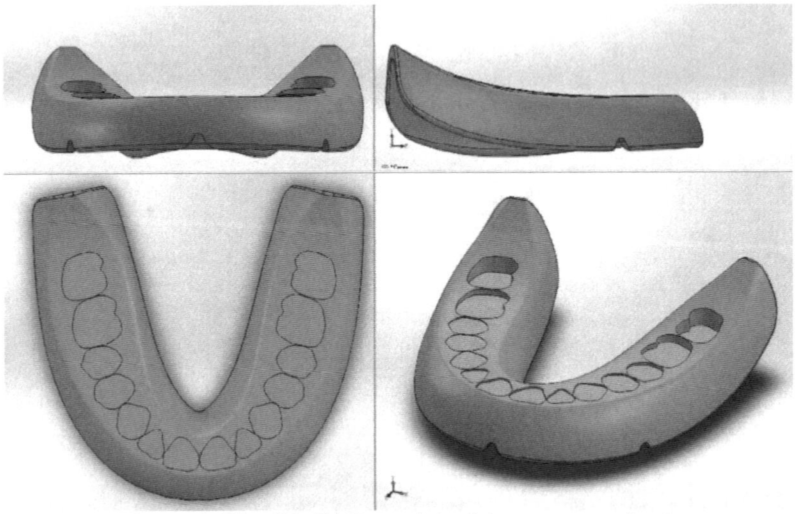

Figure1. The 3D CAD model of denture prosthesis

The geometry of the dental row is built based on typical shapes for artificial teeth [8]. All teeth have the bottom cylindrical part that is embedded into the prosthesis basis. The poly methyl methacrylate acryl (PMMA) is the most common material used to fabricate complete and partial dentures [9 - 11]. According to literature data [12] the ultimate tensile strength (UTS) for PMMA materials is vary from 47 to 79 MPa. The average value of UTS in the present study is taken equal to 60 MPa The mechanical properties of the materials for the prosthesis basis and teeth are presented in the table 1.

Table 1. Mechanical properties of the materials for teeth and prosthesis basis

#	Material location	Young's modulus, GPa	Density, kg. /m^3	Poisson's ratio
1	Prosthesis basis	1	1000	0.3
2	Teeth	2	1000	0.3

The contact of the teeth and prosthesis basis is made by using chemical bonds that allow a full contact of the polymer materials as one entail structure. Therefore, at the teeth-basis contact surface, the full clutch condition is applied. The prosthesis basis is installing on the soft tissues of the lower palate. Therefore, the special boundary conditions should be applied at the lower surface of the prosthesis basis. The soft tissues have a non-homogeneous distribution of the compliance. The compliance of the tissue is lower at the location of the alveolar ridge top and higher outside these locations. The corresponding zones at the prosthesis basis for simulation these boundary conditions are illustrated in figure 2.

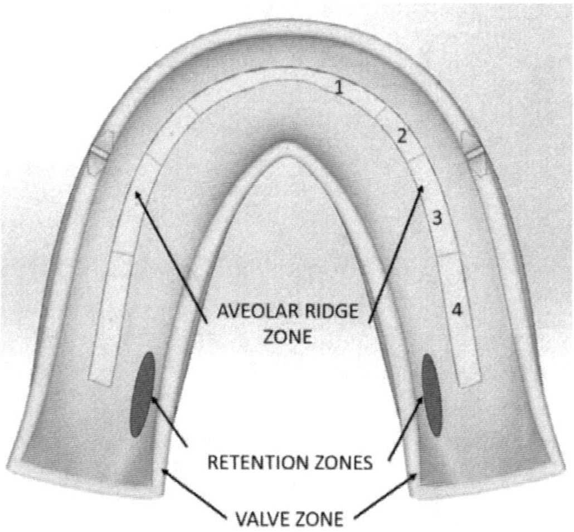

Figure 2. The schematic illustration of the applied boundary conditions

There are three distinguished zones: (1) the horseshoe-shaped of lower compliance follows the shape of the alveolar ridge; (2) two zones of retention forces; (3) valve zone that is located all around the perimeter of the denture prosthesis. The horseshoe-shaped area is divided into four sub-zones with different values of mechanical compliance. That is corresponding to clinically proven behavior. The distribution of the compliance within alveolar ridge provides the right position for the denture blade. In order to reach this objective, the numerical procedure for the compliance parameter optimization was applied. At the first stage of the optimization, the whole four zones of the horseshoe-shaped zone were assumed to have the same compliance. Then all teeth were loaded by homogeneous load i.e. the pressure at each dent was taken the same. At the result of this analysis some significant difference in vertical displacement was found for face and end parts of the denture bade. After this calculation, the compliance of the different subsection within the horseshoe-shaped area were adjusted. When the compliance of the model is balanced the all effects of stress-state will be related only to the loading conditions. The study is focused on the critical stress analysis induced by the chewing loads. It is assumed that maximum muscular force is limited by 100 N. The load schemes are presented in figure 3.

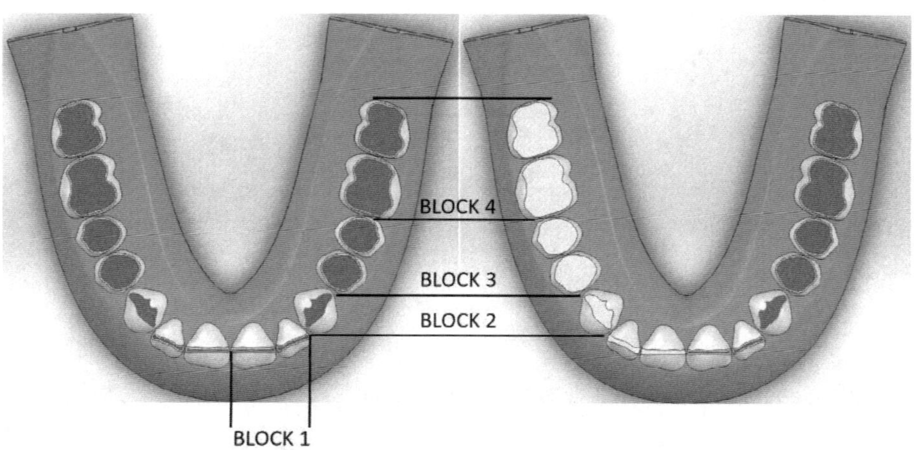

Figure 3. The scheme of loading and the results of calculation for the vertical displacements

The two different loading configurations are studied: symmetric or two-sides loading and nonsymmetric or one-side loading. The sequences of loading steps are following the natural chewing

IOP Conf. Series: Materials Science and Engineering **927** (2020) 012008 doi:10.1088/1757-899X/927/1/012008

process: (a) cutting load at block I location; (b) food breaking at the block II location; (c) and (d) chopping food at block III and block IV locations. The total force was assumed the same for both cases; however the pressure is different due to different contact surface at different blocks. The results of the calculation and analysis for the most critical loading conditions are presented in the next section.

3. Results and discussions

The loading was realized consistently on different blocks of teeth. There are four different blocks: block 1 is the pair of tooth incisors, block II is the canine tooth, block III is the pair of premolars, block IV is the pair of molars. The corresponding pressures on different blocks for symmetrical and not symmetrical loadings are given in the table 2.

Table 2. The loading parameters for symmetrical and non-symmetrical loadings

Loading type / location	Block I	Block II	Block III	Block IV
Symmetrical loading, pressure, MPa	5.2	4.2	0.9	0.5
Non-symmetrical loading, pressure, MPa	10.4	8.4	1.8	1

Among all the investigated cases (symmetric and non-symmetric loadings at blocks from I to IV) the most dangerous for the structural integrity of the denture prosthesis were the loads at face part of the prosthesis. The distribution of the stresses at the location of its extreme value due to symmetric loads on the first and second blocks is presented in figure 4. The stress bar is normalized to maximum stress and plotted with the isolines that are distant by 10 % difference. It can be clearly seen that maximum stresses are localized at the face part of the prosthesis. The stress distribution has an influence of the prosthesis shape mainly at the locations of technical notches. In the case of loading at the block I the stress distribution is less homogeneous in the face part of the prosthesis. There is one zone being at the prosthesis's symmetry line where the stress is found in the lower range. The downside views high stress values at the technical notches (for cords) and between incisors teeth. Therefore, the most likely crack initiation sides in this case are locations at the face part of prosthesis on the symmetry line or at the tips or technical notches. The maximum value of the stress is 18.8 MPa.

In the case of static loading on the block II the stress distribution is significantly changes. The face part of the prosthesis is less stressed. The maximum stresses are localized at the locations of the loaded teeth. Based on the downside view of the stress distribution it is clear that under such loading the contact area of the prosthesis with support (soft tissue) and tip of the technical notch are also having a high stress. Therefore, even under the loads at the block II the possible crack initiation can be localized at the technical notch tip.

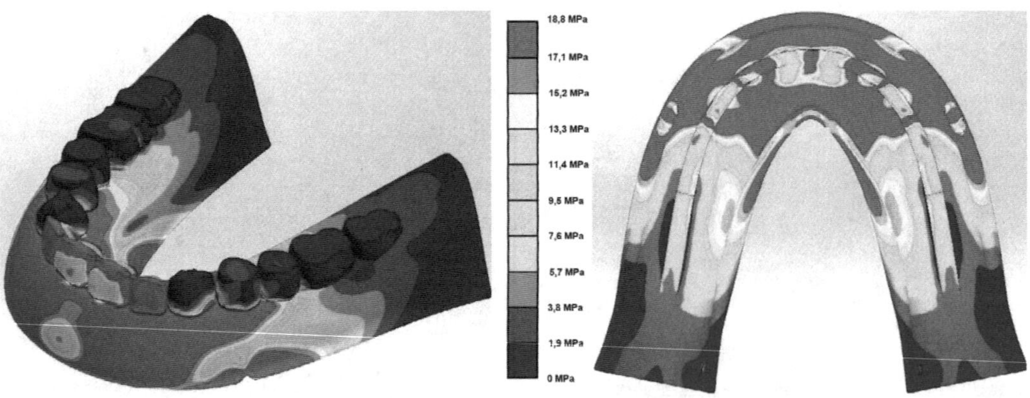

Stress distribution due to symmetric loading at the block

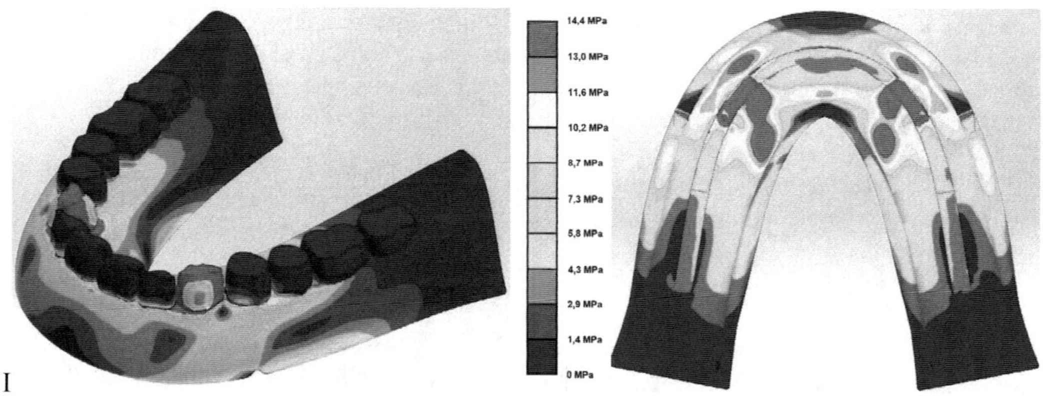

I

Stress distribution due to symmetric loading at the block II

Figure 4. The maximum stress distribution due to symmetric loads at different blocks: left – isometric view right – downside view

The analysis of the stress distribution in the case of the non-symmetric loading are presented in figure 5 for the block I and block II. The general difference between the results presented in figure 4 is higher stress values and the shift of the maximum stress zone to the prosthesis tail.

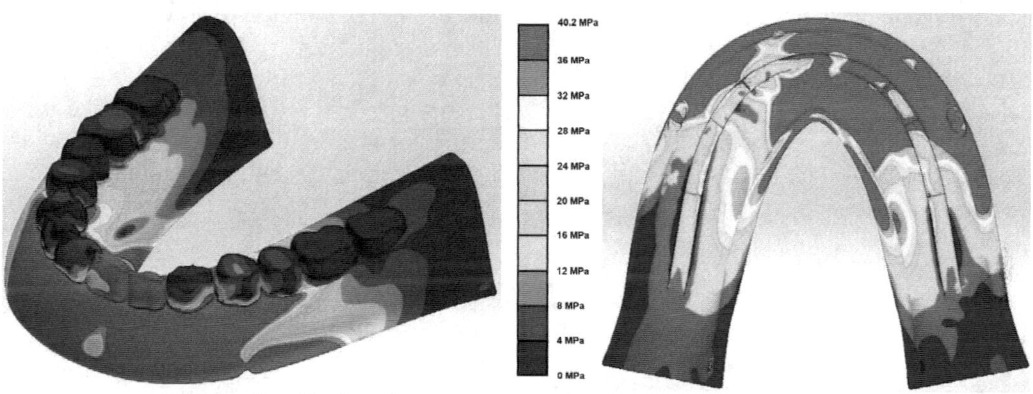

Stress distribution due to not symmetric loading at the block I

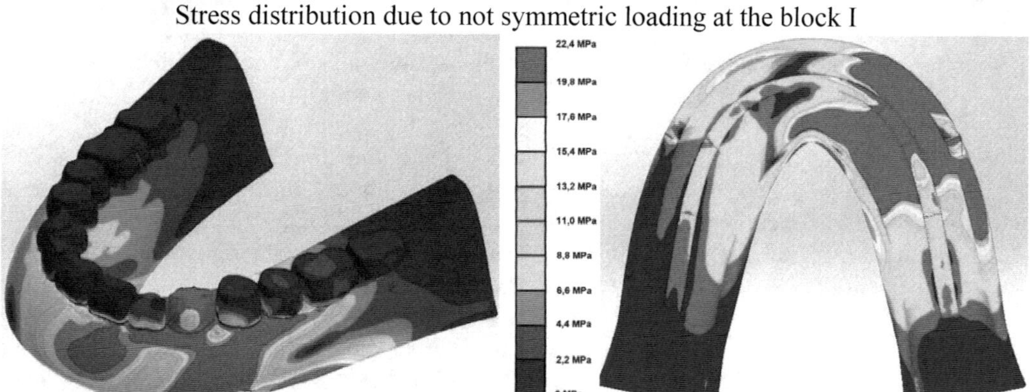

Stress distribution due to not symmetric loading at the block II

Figure 5. The maximum stress distribution due to not symmetric loads at different blocks: left – isometric view right – downside view

The area of a local minimum stress in the face part of the prosthesis is shifted to the loaded site. The results of maximum stress calculation due to the loading at different block are presented in the table 3. It is shown, that the critical for the structural integrity of the prosthesis loads appears due to first two stages of chewing: loading at block 1 and block 2.

Table 3. The results of maximum stress calculation due to loads at different blocks

Loading type / location	Block I	Block II	Block III	Block IV
Symmetrical loading, Max stress, MPa	18.8	14.4	2.93	1.1
Non-symmetrical loading, Max stress, MPa	40.2	22.14	5.67	2.34

The non-symmetric load on the second block produces the higher stress intensity compared to the case of the block I load. The maximum stress is localized at the teeth position and at the tip of technological notch. The notable stresses are found in the basis up to the molar position. The downside view shows a high stress values below the projection. The maximum stress amplitude for the non-symmetric loading is 40.2 MPa. The difference with the case of symmetric loading is about 2.15 times. Thus, the non-symmetric loading is most critical for the structural integrity of the prosthesis basis.

4. Conclusions

The balanced model of the prosthesis is introduced for the lower jaw. The boundary conditions are introduced according to the natural structure of the oral tissues. The model can be used for simulation of different applied situation and develop the recommendation on the chewing food rules for the patients with complete absence of teeth. The results of the calculations show that higher stress amplitudes can be found during the first two steps of chewing process: cutting food and its shredding. The one side chewing is more critical compared to the two sides case. However, in all the investigated cases the maximum stress is lower than ultimate tensile straight (UTS) of the material. The maximum stresses were in the range from 0.3 to 0.67 UTS. The result of maximum stress analysis show that the possible reasons for the cracking at prosthesis basis are related to (1) fatigue loading; (2) defects of material; (3) shape design errors; (4) local overloads due to specific food geometry.

Acknowledgement
The authors are grateful to Mr. Alexey Zhuravlev for the significant contribution in numerical simulations. The work was supported by State Contract of Institute for Computer Aided Design RAS.

References

[1] Iskanderov R M, Gvetadze R Sh, Butova V G, Andreeva S N and Timofeev D E 2019 The overall development strategy of dental laboratories equipped with CAD / CAM-systems *Stomotologia* **98(2)** 8 – 12
[2] Chumachenko E N, Arutyunov S D, Lebedenko I Yu and Ilinich A N 2002 Analysis of load distribution and the likelihood of irreversible changes in the jaw bone tissue during orthopedic treatment using dental intraosseous implants *Clinical Dentistry* **2** p 44
[3] Weisgeim L D, Scherbakov L N and Goncharov A A 2005 Influence of separate clinical aspects on the tensely deformed condition of the biomechanical system «combined prosthesis – supporting tissues» *Stomatologia* **4(16)** 18 – 20
[4] Al-Ali M A, Al-Ali M A, Takezawa A and Kitamura M 2017 Topology Optimization and Fatigue Analysis of Temporomandibular Joint Prosthesis *World Journal of Mechanics* **7(12)** 323 – 339
[5] Dubova L V, Tsarev V N, Zolkina Y S, Malik M V, Nikitin I S and Chuev V P 2019 Comparative assessment of milled materials for temporary unremovable dentures supported by the isoelastic implants according to the experimental study of their stress-strain states and microbial adhesion *Clinical dentistry* **3(87)** 74 – 78

[6] Lemeshevsky S V, Naumovich S A, Naumovich S S, Vabishchevich P N and Zakharov P E 2016 Numerical simulation of the stress–strain state of the dental system *AIP Conference Proceedings* **1773** 020004

[7] Shanidze Z L, Muslov S A, Arutyunov A S, Astashina N B and Arutyunov S D 2020 Biomechanical approach to dental orthopedic treatment of patients with postoperative maxillary defect *Russian Journal of Biomechanics* **24 (1)** 28 – 38

[8] Aruyunov S D, Grachev D I, Nikitin A D 2020 Mathematical modelling on the fracture of a laminar prosthesis basis under natural chewing loads *IOP Conference Series Materials Science and Engineering* **747** 012065

[9] Abdulrazzaq N S, Jafarzadeh K T, Behroozibakhsh M, Hajizamani H and Hajizamani S 2018 Recent advances and future perspectives for reinforcement of poly(methyl mathacrylate) denture base materials: a literature review *Journal of denture biomaterials* **5(1)** 490 - 502

[10] Oguzhan Y, Melik S and Kemal G U 2016 Biocompability of Dental *Polymers Polymer science: research advances, practical applications and educational aspects* 89 - 98

[11] Gautam R, Singh R D, Sharma V P, Siddhartha R, Chand P and Kumar R 2012 Biocompatibility of polymethylmethacrylate resins used in dentistry *J Biomed Mater Res B Appl Biomater* **100(5)** 1444-50

[12] Oleiwi J K and Hamad Q A 2018 Studying the Mechanical Properties of Denture Base Materials Fabricated from Polymer Composite Materials *Al-Khwarizmi Engineering Journal* **14(3)** 100- 111

On the inherent polycrystal stress concentration under external loading on example of single-phase α-Fe samples

V V Gulin, A D Nikitin

Institute for Computer Aided Design of RAS, 19/18, 2nd Brestskaya, 123056, Moscow, Russia

E-mail: kornet104@gmail.com

Abstract. It is known that most structural materials are polycrystals. Under any loading, the structural features of the polycrystal may lead to the appearance and accumulation of microplastic deformation due to stress concentrations. This statement is part of a multiscale approach to describing the fracture process. The present development of computer engineering allows modeling of complex non-homogeneous objects, such as polycrystal. In this paper, a numerical model is developed for analyzing of the polycrystal stress state under mechanical loading. It is shown that inherent stress concentration takes place. The localization and statistical estimation of inherent stress concentration is made on the example of α-Fe polycrystal.

1. Introduction

Knowledge of the physical nature of the structural destruction process can be considered as an integral part of the strength of materials. In last decades, research has been actively developing in the field of very high cycle fatigue (VHCF). It is shown that most structural materials with high durability (> 10^7 cycles) can fail at external load levels not exceeding the classical fatigue limit. According to general principles, the absence of plastic deformation exclude the formation of cracks. However, many researchers show [1-6], that under loading levels not exceeding the macroscopic elastic limit, local structural changes, called microplastic deformations, occur. Most researchers in the field of VHCF use this term without deciphering the very concept of «microplastic deformation». Empirical evidence indicates that the role of various defects, including grain boundaries, in the accumulation of microplastic deformation leading to fracture is different. However, a quantitative description of these differences in the literature is not given. Further research in the field of VHCF and material destruction as a whole leads to the fact that it is necessary to consider the behavior of the material in the framework of a multiscale approach with determination of the stress relaxation mechanisms. This direction is developing within the physical mesomechanics [7]. In [8] a three-level approach was proposed to describe the material as a self-organized hierarchical system. Theoretical foundation is to separate of macro-, meso- and microscopic scale levels. A similar approach also applied in [9]. In all these works, the most difficult is the formulation of the boundaries of applicability for phenomena describing at the chosen scale level, since there is no clear connection between the models at different scales.

One of the transition methods to the mesoscale representation is the consideration of polycrystal model. It is assumed that the material consists of grains, each of which is characterized by its own crystallographic symmetry – orientation. Particular attention in these models should be given to grain

IOP Conf. Series: Materials Science and Engineering **927** (2020) 012009 doi:10.1088/1757-899X/927/1/012009

boundaries, since they, within the continuum mechanics, are regions of an abrupt change in the physical properties of the material, and as a result, these regions lead to the formation of local stress concentrations. Henceforth, the paper proposes to consider the natural reaction of a polycrystalline structure to an external loading, and assume it as inherent quality.

2. Model of polycrystal structure

The modeling is based on the continuum mechanics. The problem is solved in the formulation of the linear theory of elasticity, which allows us to abstract from the choice of a specific system of units, so further we will consider only relative values. Within the volume of an individual grain, the medium is considered to be homogeneous, and anisotropic. Thus, its behavior can be described by means of the generalized Hooke's law (1), which determines the connection between stress tensor σ_{ij} and strain tensor ε_{kl} through the elastic modulus C_{ijkl}:

$$\sigma_{ij} = C_{ijkl}\varepsilon_{kl}\,, \quad i,j,k,l = 1,2,3 \tag{1}$$

2.1. Geometry creation

The polycrystal model is made in the form of a parallelepiped (cube in particular), divided into subregions – voxels (formed from words «**vo**lumetric pi**xel**»).

The size of grains in the volume is described by equivalent diameter of the sphere (ESD). For a finite space divided into grain subdomains (features), in the absence of discontinuities, the distribution of the equivalent sphere diameter is well described by the lognormal distribution (2) which is determined by the parameters μ and σ:

$$f_X(x) = \frac{1}{x\sigma\sqrt{2\pi}} e^{-(\ln x-\mu)^2/2\sigma^2} \tag{2}$$

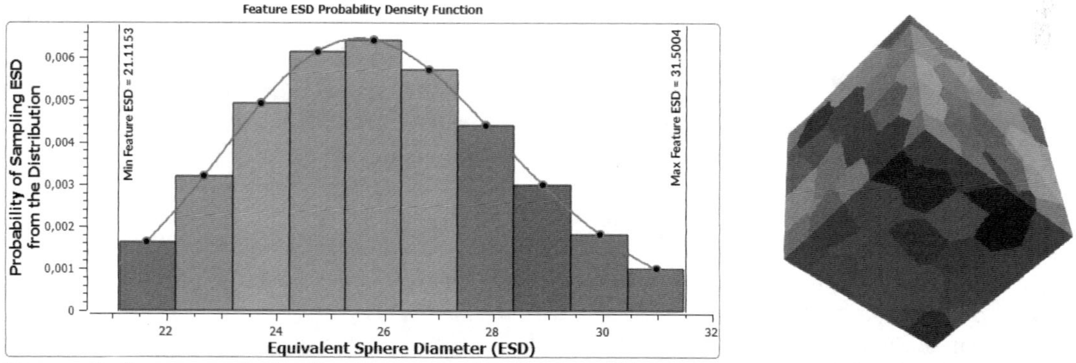

Figure 1. Lognormal distribution approximation used in current paper defined at σ=*0.1*, μ=*3.25* (left) and created polycrystal geometry, based on this particular distribution (right).

The following parameters have always been used in the current study: $\sigma = 0.1 = Const;\ 2.9 < \mu < 4$. The indicated range of parameters made it possible to place up to 300 grains in the 100×100×100 region. Continuous distribution function was approximated by the histogram in each particular case (Figure 1).

If we identify the grains with some equivalent ellipsoid, then in this case we can characterize them by the ratio between the semiaxes of this ellipsoid. In this article, only grains with an equiaxed equivalent ellipsoid were considered, that is, those whose ratios between the semiaxes are represented on average in the form 1:1:1. Separation of the modeled volume into regions was carried out using the DREAM.3D software.

Figure 2 presents examples of the created polycrystal geometry for the aforementioned distribution function.

a) b)

Figure 2. Polycrystal geometry with 304 – a) and 101 – b) numbers of grains. The color corresponds to the orientation through values of the inverse pole figure.

2.2. Finite element model creation

Numerical modeling and solution were performed in the ANSYS. The created geometric model consisting of voxels was converted so that each individual voxel became an elementary finite element (SOLID185).

A polycrystal from the alpha phase of iron α-Fe (ferrite) was considered in the article. This phase is a body-centered cubic lattice (BCC). The elastic constants of all grains are the same and correspond to the following values (Table 1) and are taken from [10]:

Table 1. Elastic constants of α-Fe at 300 K

C_{11}	C_{12}	C_{44}
230.37	134.07	115.87

These constants are defined for each grain in the individual Cartesian coordinate system, which were created by copying the main coordinate system with subsequent three rotations around the x, y and z axis by a random angle from 0° to 359°.

3. The choice of loads and boundary conditions

To study the mechanical reaction, which is caused only by the internal structure of the polycrystal itself, we used quasistatic uniaxial loading. Measures have also been developed to eliminate the effect of constraints on the stress-strain state.

To constrain the motion of the polycrystal as a solid, it is proposed to determine the position of six points of the polycrystal in space through springs of zero length. To minimize the influence to the stress-strain state, stiffness is taken that low, so: (a) the springs constrained movement of the polycrystal as an absolutely solid body and (b) influenced the values of stress-strain state characteristics in such a digit that would be less than the digit for which rounding is performed. In practice, it turns out to be sufficient to determine the stiffness of the springs by an order of magnitude 10^{-36} N/m.

4. Results of the analysis of regions of maximum stresses in a polycrystal structure

The results of numerical solution are presented in the form of (a) equivalent von Mises stress and (b) maximum tangential stress plots (Figure 3).

For further analysis, only the value of maximum stress concentration for particular configuration was used. In this case, it is possible to collect the results of many tests into a single diagram (Figure 4), which is the probability density to find the corresponding maximum value of the stress-strain state parameter.

For various polycrystal geometries, a statistical analysis of the maximum stress concentration were carried out. A total of 575 numerical experiments were carried out for various grain boundary geometry, relative orientation of the grains and also their number inside the volume 100×100×100. Numerical experiments were carried out with the 305, 192, 109, 66, 32 and 14 number of grains.

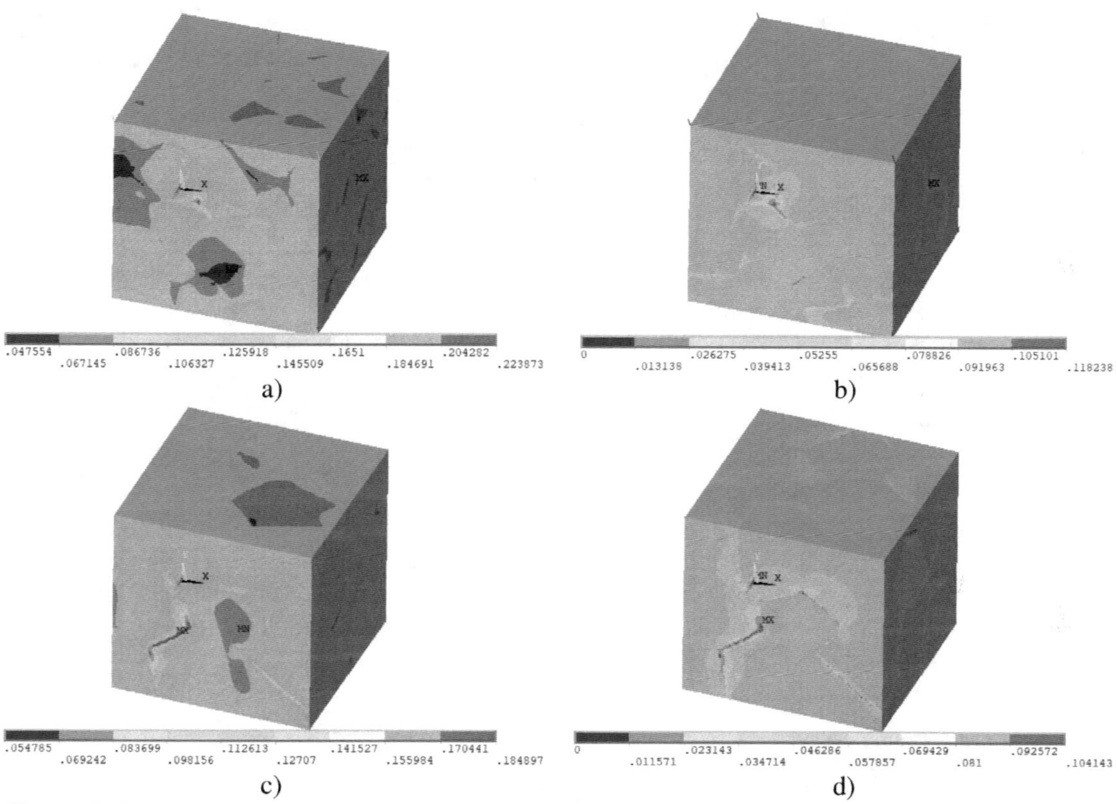

Figure 3. Examples of stress plot for different geometry and misorientation distribution. Von Mises (left column – a,c) and maximum tangential stress (right column – b,d). Applied tensile stress is equal to 0.1. Number of grains: 305 grains – a,b), 192 grains – c,d)

It was found that the values of the maximum concentration coefficients on average do not depend on number of grains (in the range of 14 to 305) in the volume. A theoretical minimum can be achieved if all grains have the same orientation and the case of a single crystal is obtained. In this case, the concentration coefficient is obviously equal to 1.

It was also found that the concentration of stresses lies in ranges (1.534, 2.595) and (1.489, 2.507), and the most probable (average over the data sampling) value of stress concentration is 1.867 and 2.086 for the von Mises stress and for maximum tangential stress, respectively. Thus, this model allows to estimate the value of the inherent stress concentration in a polycrystal at known values of the elastic constants of single crystals.

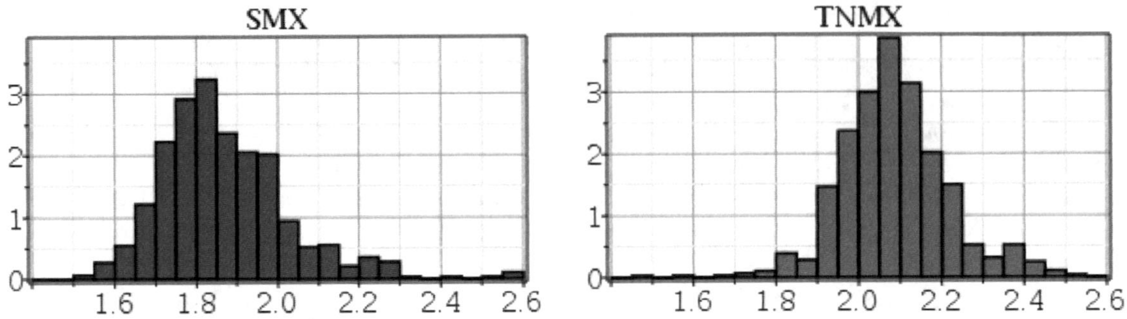

Figure 4. Histogram for maximum value of von Mises (SMX) and maximum tangent (TNMX) stress concentration over the entire polycrystal, area under each histogram is equal to 1.

These natural concentrations of inherent stresses, arising from the granular structure of the polycrystal, show that higher stress concentrations occur in real material than it is assumed in the isotropic consideration. The revealed concentration zones indicate that plastic deformation of such a structure is carried out not throughout the volume, but in narrow zones with a particularly high stress concentration. The development of these microdamages can lead to the occurrence of macrocracks by the very-high-cycle fatigue mechanism and subsequent destruction of the material. The results obtained make it possible to evaluate possible zones of microdamage nucleation and appropriate stress levels.

5. Conclusions

From the results of the study, it was found that in an absolutely elastic formulation, the natural reaction of the polycrystal is the inherent stress concentration, which for α-Fe on average exceeds twice the nominal. The perfectly elastic reaction in the study of the stress-strain state of a polycrystal, presented in this work, should be understood as a state preceding microplastic relaxation in areas of the highest stress concentration. The proposed model is of interest in the further development of multiscale approaches, since it can become the basis for the implementation of results obtained on different scales. It also gives a prediction of the inherent stress concentration in the polycrystal, at known values of the elastic constants of monocrystal.

Acknowledgements

This work was financially supported by RSF project № 19-19-00705.

Reference

[1] Bathias C, Drouillac L, Francois P Le 2001 How and why the S-N curve does not approach a horizontal asymptote, *International Journal of Fatigue* **23** 143-151

[2] Nikitin A, Palin-Luc T, Shanyavskiy A 2016 Crack initiation in VHCF regime on forged titanium alloy under tensile and torsion loading modes *International Journal of Fatigue* **93** 318-325

[3] Sakai T 2009 Review and Prospects for Current Studies on Very High Cycle Fatigue of Metallic Materials for Machine Structural Use *Journal of Solid Mechanics and Materials Engineering* **3(3)** 425-439

[4] Stanzl-Tschegg S E 1999 Fracture mechanisms and fracture mechanics at ultrasonic frequencies *Fatigue and Fracture of Engineering Materials and Structures* **22(7)** 567-579

[5] Burago N G, Nikitin I S, Nikitin A D, Stratula B A 2019 The assessment of fatigue durability and critical plane determination for multiaxial cyclic loading at an arbitrary shift of phases *PNRPU Mechanics Bulletin* **3** 27-36

[6] Wang C, Liu Y, Nikitin A, Wang Q, Zhou M 2019 A general scenario of fish-eye crack initiation on the life of high-strength steels in the very high-cycle fatigue regime *Fatigue and Fracture of Engineering Materials and Structures* **42(9)** 2183-94

[7] Panin V E, Egorushkin V E, Panin A V 2006 Physical mesomechanics of a deformed solid as a multilevel system. I. Physical fundamentals of the multilevel approach, *Physical Mesomechanics* **9(3-4)** 9-20

[8] Shanyavskiy A A, Banov M D, Zakharova T P 2010 Principles of physical mesomechanics of nanostructural fatigue of metals. Part II. Subsurface fracture of EP741 heat-resistant alloy *Physical Mesomechanics* **13(3-4)** 143-151

[9] McDowell D, Dunne F P E 2010 Microstructure-sensitive computational modeling of fatigue crack formation *International Journal of Fatigue* **32(9)** 1521-42

[10] Adams J J, Agosta D S, Leisure R G and Ledbetter H 2006 Elastic constants of monocrystal iron from 3 to 500 K *Journal of Applied Physics* **100(11)** 113530

IOP Conf. Series: Materials Science and Engineering **927** (2020) 012010 doi:10.1088/1757-899X/927/1/012010

Verification of numerical model of temperature calculations in radio-frequency ion thrusters

V Abgaryan, V Balashov, A Kupreeva and V Nigmatzyanov
Research Institute of Applied Mechanics and Electrodynamics of the Moscow Aviation Institute, Leningradskoye shosse 5, Moscow, 125080, POB 43, Russia

E-mail: vka.mai@mail.ru, kupreeva.mai@gmail.com

Abstract. A numerical model for temperature calculation in radio-frequency ion thrusters (RIT) is proposed. The equations used for heat fluxes determination differ from those used in the known computational models. A new way of distribution of a radiant power flux caused by the excited atoms and ions of the plasma discharge is proposed. In addition, a new radiant flux caused by recombination of ions and electrons on the surfaces bordering the discharge was taken into consideration. The numerical model was verified based on temperature measurements made with the thruster with an ion beam diameter of 8 cm operating at three different power modes. The temperature measurements were made by thermographic camera focused on blackened areas of the accelerating grid surface.

1. Introduction

Ion thrusters became widely used in space technology as low thrust engines for satellites stabilization and orbital altitude control, orbit raising and also as cruise propulsion in solar system exploration missions. Several designs of ion sources and, accordingly, ion thrusters based on them are sufficiently well developed and have been used in both space and ground-based applications. The most commonly used design is an ion thruster (IT) with a DC discharge [1]. Radio-frequency ion thrusters (RIT), in which the discharge plasma is formed by RF electromagnetic field was first proposed by H. Loeb [2]. Comparing to IT, RIT concept has some disadvantages as well as benefits. Some of the advantages of electromagnetic method for forming and maintaining of induction plasma discharge concern no requirement for electrode system inside the discharge chamber and the absence of magnetic system.

While ion thrusters operating, the construction elements are heated by the plasma components incident on surfaces bordering the discharge, that may influence the thruster operation. Therefore, for the preliminary definition of the thermal state on the design stage a numerical modeling of structural components heating is conducted. Given the thermal modeling is conducted in advance, an optimal choice of materials for design units, which is critical in terms of thermal loads, is available. Such design unit is an ion-extraction system (IES), which consists of two thin round plates with thickness of 0.3…2 mm perforated with a large number of apertures. The potential drop between two electrodes is several kV. The gap between these grids is about of 1 mm; therefore, the technical specifications of the thruster may suffer from a thermal deformation of the electrodes. The availability of an adequate thermal model makes it possible to simulate the grids deformation and to define the marginal operation modes of the thruster.

The research purpose was to develop a suitable numerical thermal model for its practical use at early stage of RIT design.

2. Heat fluxes appearing within the RIT

The numerical calculation models use heat fluxes from the plasma discharge components (ions, electrons and photons) that come onto RIT discharge chamber and IES electrodes surfaces as the Neumann boundary conditions. The expressions for the heat fluxes concerned were presented by Van Noord in [3] and used by him in the developed calculation thermal model for IT [4]. Different approaches to aircraft and spacecraft thermal modeling were considered also in [5-6]. The same expressions with some slight differences were used for thermal modeling for the ion sources with RF plasma heating in every known calculation model [7-8].

Figure 1 shows schematically the heat fluxes that appear in the RF discharge and in the ion beam for a RIT with a gas discharge chamber (GDC).

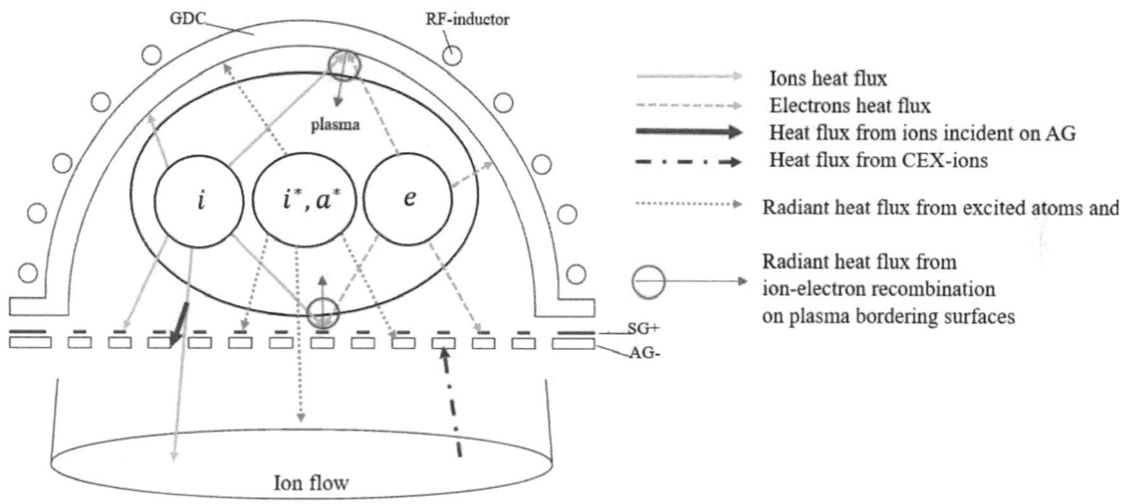

Figure 1. The scheme of heat fluxes in RIT: i – ions , i* – excited ions, a* – excited atoms, e – electrons, SG^+ – screen grid, AG^- – accelerating grid

The model uses analytical expressions for a number of heat fluxes components coming from the plasma discharge, which differ from those accepted in [3, 7-8]. The major differences of the presented calculation model relate to the radiant heat fluxes distributions. A short description for the major differences of the proposed model is given below. These distinctions are detailed in [9, 10].

2.1. The recombination radiant flux

Comparing to other calculation models, the radiant flux caused by recombination of ions and electrons on the surfaces bordering the discharge, which are GDC and screen grid (SG) surfaces, is taken into consideration (figure 2).

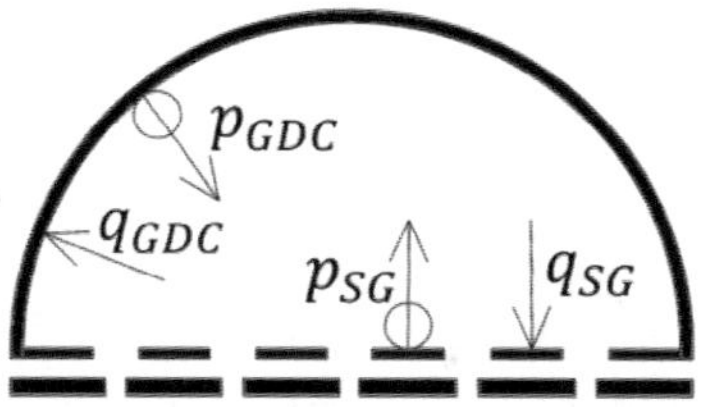

Figure 2. The recombination radiation heat fluxes in hemispherical GDC: p_{GDC} - power flux density coming from GDC, p_{SG} - power flux density coming from SG, q_{GDC} - power flux density absorbed by GDC, q_{SG} - power flux density absorbed by SG

The known models assume that the energy released as the result of the recombination is absorbed entirely by the point on the surface near which the recombination process took place. In fact, the spatial distribution of the photons produced by the recombination is more likely isotropic. That is why approximately half of the photons would move away from the surface near the points where photons are generated. Due to the surface roughness, the photons, which are coming out at acute angles to the surface, would be reflected or be absorbed on the cavities walls on the surface concerned, that leads to distortion of the isotropic indicatrix. Since there is no data about the form of the recombination radiation indicatrix in the publications, an assumption of radiation cosine indicatrix was made.

The models [3, 7] accept that the power density absorbed by GDC and SG surfaces in the result of the recombination is defined by the density of Bohm current j_B that is incident on the surfaces bordering plasma and equals the following:

$$A = j_B V_i \tag{1}$$

Here V_i is the ionization potential for the plasma-forming gas atoms. Here in after V_i =12.13 V as the propellant is accepted to be xenon.

However, in view of the mentioned above, the balance for this form of radiant power on the GDC and SG surfaces was reconsidered as the flux partially comes through the SG apertures and is incident on the acceleration grid (AG) or leaves the thruster.

In addition, in the model the partial reflection of this radiation from the GDC, SG and AG surfaces is taken into account. The recombination radiation consists of the quanta of ultraviolet light for more than 2/3. Therefore, the flux portions, absorbed by the thrusters surfaces, were defined according to the coefficients of reflection of UV radiation for materials of GDC and SG, R_{GDC} and R_{SG} respectively.

Given the assumption of the radiant cosine indicatrix and the hemispherical GDC, an accurate analytical solution for power fluxes densities p_{GDC}, p_{SG}, q_{GDC}, q_{SG} coming from surfaces of GDC and SC and absorbed by them (figure 2), is given in [9,10]. In the Table 1, the calculated values of relative flux densities for Al_2O_3 as the GDC material, Ti, and C as the SG material are shown. Values of coefficients R_{GDC} and R_{SG} are taken from [11]. Relative variables of incident and outgoing power fluxes are introduced for convenience:

$$\bar{p}_{GDC} = \frac{p_{GDC}}{A}, \bar{p}_{SG} = \frac{p_{SG}}{A}, \bar{q}_{GDC} = \frac{q_{GDC}}{A}, \bar{q}_{SG} = \frac{q_{SG}}{A} \tag{2}$$

Table 1. Relative densities of the recombination radiation power fluxes absorbed by and coming out of the GDC and SG surfaces

R_{GDC}	R_{SG}	$\bar{q}_{GDC}$	$\bar{q}_{SG}$	$\bar{p}_{GDC}$	$\bar{p}_{SG}$
0.12 (Al_2O_3)	0.11(C)	0.848	0.993	0.616	0.311
0.12 (Al_2O_3)	0.30 (Ti)	0.898	0.786	0.622	0.418

As it is seen in the column 3, GDC absorbs the heat power from this component, which is 10-15% less in compare with the values $\bar{q}_{GDC} = \bar{q}_{SG} = 1$, how it is accepted in [3, 7]. The screen grid also absorbs less heat power depending on the reflection coefficient R_{SG}. While using Ti, the flux $\bar{q}_{SG}$ is reduced by 22% that is caused by a significant magnitude of Ti reflection coefficient.

2.2 Radiant flux from excited-state atoms and ions

Another difference of the proposed model is related to the calculation methodology for the distribution on the GDC and both IES electrodes surfaces of the radiant flux that appears in the process of atoms and ions de-excitation. This flux is coming through the SG apertures onto AG. It is the principal determinant of the AG thermal state being the major contributor. In [3, 7] this power flux is assumed to be evenly distributed on the GDS and SG surfaces which, according to the authors, is the result of an approximation of 0-dimensional model of plasma. While developing the proposed model, the calculation of this power flux density within the same approximation of spatial plasma parameters uniformity was conducted. The integration over plasma volume showed that the flux density on the SG and AG surfaces decreases from the center to edges more than twice, and it decreases from apex to the bottom for about one and a half times on the GDC surface. The calculated distributions of this heat flux densities on GDC surface ($\bar{f}_{GDC}$) and both electrodes ($\bar{f}_{SG}$) are used in the numerical model and are shown in the figure 3, where H is a normalized height of the observation point on the GDC surface. The flux distribution on the AG is calculated with taking partial shadowing of its surface with SG into account.

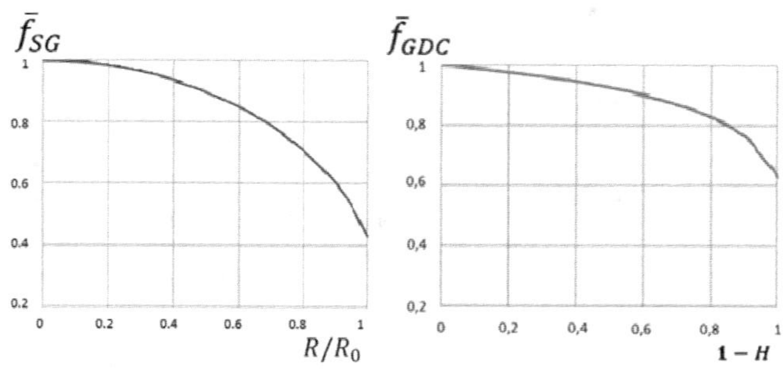

Figure 3. Reduced distribution functions for the radiant flux from atoms and ions deexcitation on GDC and electrodes surfaces

2.3. The radiant fluxes through IES

Another significant difference of the proposed model is a thorough consideration of the radiant fluxes coming from the discharge plasma through the apertures in IES. The components of the radiant heat fluxes in the elementary IES cell are calculated. The shares of radiant flux entering the cell, which are absorbed by IES screen and accelerating grids, as well as the shares, which are coming out of the thruster and back to GDC, can be determined. The shares of the fluxes are calculated taking in account the coefficients of reflection of UV radiation R_{SG}, $\mathrm{R_{AG}}$ and $\mathrm{R_{GDC}}$.

3. Results of the numerical thermal model verification

As the result of application of the proposed model the magnitude of the heat flux incident onto AG changed most significantly with an increase of 70% comparing to results of [3, 7]. Such increase is explained with more detailed calculations of the radiant flux shares, absorbed by AG, and also due to taking into consideration of the new radiant flux from the recombination of ions and electrons, which was not accounted for in any model before. This flux transports the heat power onto AG that is slightly less than the radiation from the atoms and ions deexcitation, but also markedly exceeds the heat power from what is incident onto AG with the charge-exchange ions.

The expressions for the heat fluxes carried by ions and electrons from the plasma onto the RIT surfaces are similar to the expressions explored in [3, 7].

The composed calculation model was verified by temperature measurements on the surface of the accelerating grid. The temperature measurements and the corresponding calculations of temperature distribution were conducted for three operating modes of the RIT-10 laboratory model with an ion beam diameter of 8 cm.

3.1. Thermal imaging of the temperatures

The temperature measurements were conducted with a thermal camera FLIR SC645. The camera was mounted on the vacuum test chamber to the viewport made of GeSi glass with high transmissivity in infrared spectrum. The angle of the sight beam of the device to the AG surface was around 40 degrees. The distance between the camera lens and the AG center amounted roughly 0.5 m, that allowed to obtain a thermogram with a sufficient angular resolution enough to temperature measurements, figure 4.

The readings of the temperature values were taken in the read points on acceleration grid, namely, in the centers of elementary triangles which apexes are neighboring apertures of the electrode, which is perforated in a hexagonal pattern. The measurements were performed in the grid area with preliminary application of carbon mark which emissivity is 0.95, figure 5.

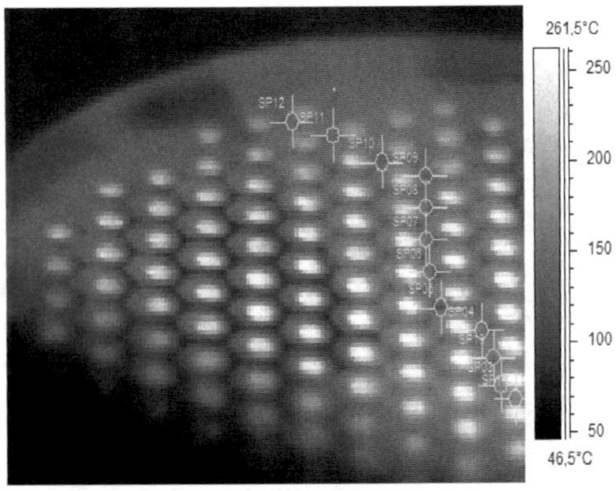

Figure 4. A thermogram from a thermal camera

Figure 5. Carbon marks on the accelerating grid.

The marks application allowed to avoid the main problem of thermal imaging that is related to the distortion of thermograms due to the reflected radiation fluxes.

3.2 Numerical calculations and measurement results for temperatures

The temperature calculations were performed for the simplified model of RIT-10. The implementation of the object model simplifications allows significantly reducing the amount of the calculation required which makes the proposed calculation model convenient for common use.

The power levels from the RF generator (RFG) on different operating modes are presented in table 2. The share of the original RF power released from the induction coil N_{ind} was determined with a Rogowski coil. It turned out to be 48% of the N_{RFG} power for all three modes that is shown in table 2. The rest portion of the power was lost in the RF path from the RFG to the induction coil that includes cables and a matching unit. The table 2 also shows the values of ion current I_0 coming out from RIT-10 and values of interception current on AG I_{int}.

Table 2. The values of N_{RFG}, N_{ind}, I_0 and I_{int} for different operating modes of RIT-10.

No.	RFG power N_{RFG}, W	RF-power on induction coil N_{ind}, W (48%N_{RFG})	Ion current I_0, mA	Interception current I_{int}, mA
1	60	28.8	27	0.8
2	82	39.4	49	1.8
3	98	47.0	72	11.4

Given the values of I_0, I_{int} and RF power absorbed by the discharge, the densities of all components of the heat fluxes coming from the plasma discharge onto the surfaces of the thruster units were determined. These densities of the heat fluxes were used as boundary conditions in a numerical calculation of a system of the stationary heat transfer equations for mesh elements of a simplified thruster design, which was conducted via Steady-State Thermal analysis in the ANSYS software environment [11].

The results of the temperature calculations and measurements in three operating modes of RIT-10 are shown in figure 6. The closest correspondence of the calculated and measured temperatures for all three modes was obtained with an assumption that the magnitude of RF power absorbed by the discharge is equal to 0.75 of RF power put into the induction coil. The red full line represents the calculated SG temperature profile along the radius of the grid.

IOP Conf. Series: Materials Science and Engineering **927** (2020) 012010 doi:10.1088/1757-899X/927/1/012010

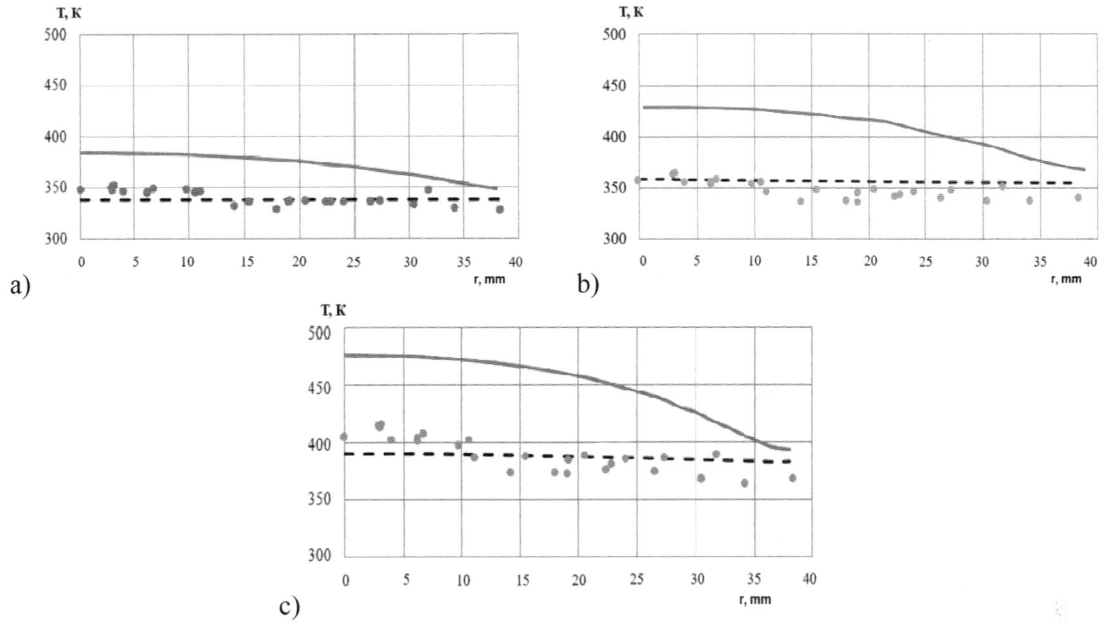

Figure 6. The calculated and experimental temperatures of ion extraction system electrodes for three operating modes of the thruster: a) 60 W, b) 82 W, c) 98 W.

According to our estimates, the accuracy of temperature measurements with a thermal camera was about 15°K. Henceforth it is planned to increase the accuracy of measurements. Despite it, satisfactory agreement between the calculated and measured temperature profiles of the AG surface has been obtained, as it is seen from figure 6. Therefore, the calculation thermal model is quite adequate.

Acknowledgments
The work was supported by the Ministry of Science and Higher Education of the Russian Federation, grant agreement No. 075-15-2019-1718, unique identifier of the agreement: RFMEFI60419X0211.

References

[1] Zhurin V V, Kaufman H R and Robinson R 1999 *Plasma Sources Science and Technology*, **8**(1) R1
[2] Löb H W 1962 *Acta Astronautica VIII*, **1** p 49
[3] Van Noord J L 1999 Thermal modelling of an ion thruster (*Diss. University of Michigan*)
[4] Van Noord J L 2007 *Proc. 43rd AIAA/ASME/SAE/ASEE Joint Propulsion Conference & Exhibit* (Cincinnati, OH) p 5218
[5] Formalev V F, Kolesnik S A and Garibyan 2019 *AIP Conference Proceedings* **2181**(1), p 020003
[6] Minyushkin D N and Kryukov I A 2019 *AIP Conference Proceedings* **2181**(1) 020025
[7] Dobkevicius M, Feili D 2016 *Eur. Phys. J. D.* **70**(11) p 227
[8] Gärtner W, Lotz B and Meyer B R 2013 *Proc. 33rd International Electric Propulsion Conference*, p 6
[9] Abgaryan V K, Nadiradze A B, Semenov A A and Troshin A E 2019 *Journal of Surface Investigation: X-ray, Synchrotron and Neutron Techniques*, **13**(6) 1054-60
[10] Abgaryan V K, Gidaspov V Y, Nadiradze A B and Semenov A A 2019 *Technical Physics Letters*, **45**(2) 123-125
[11] Werner W, Glantschnig K, Ambrosch-Draxl C 2009 *J. Phys Chem Ref. Data* **38** 1013-92

IOP Conf. Series: Materials Science and Engineering **927** (2020) 012011 doi:10.1088/1757-899X/927/1/012011

Smooth reconstruction method for surfaces defined by a continuous set of triangles

M Kh Abuziarov and D T Chekmarev

National Research Lobachevsky State University of Nizhni Novgorod, Nizhny Novgorod, Russia

E-mail: 4ekm@mm.unn.ru

Abstract. The paper considers the problem of triangular grids distortion in three-dimensional space, approximating the boundaries and interfaces of interacting solid, liquid and gaseous media. Unacceptable deformations of the mesh cells size and shape can occur during the numerical solution of continuous media dynamics problems. A method and algorithm for rebuilding grids using smooth piecewise polynomial interpolation by algebraic polynomials of the third degree is proposed.

1. Introduction.

In the numerical solution of the of elastoplastic, liquid and gaseous media dynamics problems using the Lagrangian-Eulerian methods [1-4], the calculations are carried out on a fixed uniform orthogonal Eulerian grid, through which the deformable body moves. The boundary of the body (computational domain) is defined as a polyhedral surface composed of continues triangular faces (STL file). In the process of solving the problem, the vertices of a polyhedral surface can experience significant displacements, which in many cases can lead to unacceptable deformation of triangular faces. In particular, edges that are too large may appear, too elongated or flattened, and the like. In this regard, the problem of local reconstruction of triangulated surfaces is relevant.

The paper proposes an approach to solving this problem. It is based on the method described in the theory of the finite element method for constructing the basic functions of high-order finite elements [5, 6] in combination with the spline approximation [7]. In general terms, this approach consists in the following sequence of actions.

1. Based on the available criteria, a local domain is determined in which a rebuild of the grid of triangles is necessary and on the existing faces points are marked - prototypes of future new vertex faces.
2. The tangent planes to the formed surface at the vertices of the faces are determined.
3. A smooth surface is constructed on each face in this local coordinate system, passing through the vertices of this face and tangent at these vertices to the planes defined above.
4. At the points marked on the initial faces, the coordinates of their images — the new points lying on the constructed surfaces — are calculated. These points are the vertices of the triangles of the new mesh.
5. If necessary, a new local mesh is docked with the original mesh and the resulting mesh is optimized using one of the known methods.

2. Construction of surfaces on triangular faces

We introduce the following notation:

$\overline{V}_i$ – vertices of a triangulated surface;

$\overline{E}_i$ – edges of faces;

G_i –triangular faces;

$\overline{n}_i$ – unit vectors normal to faces;

$\overline{N}_i$ –unit vectors normal at the vertices;

$\overline{x} = (x, y, z)$ –global Cartesian coordinate system;

$\overline{\alpha} = (\alpha, \beta, \gamma)$ –local Cartesian coordinate system;

$\overline{e}_i$ –basis vectors of the local coordinate system;

Let the local part of the triangulated surface, on which the grid is rebuilt, that contains many faces $\{G_1,...,G_k\}$ and many vertices $\{\overline{V}_1,...,\overline{V}_m\}$ of these faces, be selected. We define the unit normal vectors at the vertices using the relations

$$\overline{N}_i = \sum_{j \in S(i)} \frac{\overline{n}_j}{\rho(\overline{E}_i, \overline{x}_{cj})} \Bigg/ \left| \sum_{j \in S(i)} \frac{\overline{n}_j}{\rho(\overline{E}_i, \overline{x}_{cj})} \right| \tag{1}$$

Here $S(i)$ is the set of numbers of faces, incident vertices $\overline{E}_i$, $\overline{x}_{cj}$ are the coordinates of the center of the j-th face, $\rho(\overline{E}_i, \overline{x}_{cj})$ is the distance from the vertices $\overline{E}_i$ to the center of the j-th face $\overline{x}_{cj}$;

Having normal vectors $\overline{N}_i$, through each vertex we draw a plane orthogonal to a given vector, which is tangent to the smooth constructing surface. This plane in the global coordinate system is described by the following equation

$$\overline{N}_i \cdot (\overline{V}_i - \overline{x}) = 0 \tag{2}$$

After constructing the tangent planes, we can proceed to the construction of surfaces on the faces. We introduce the local coordinate system $\overline{\alpha} = (\alpha, \beta, \gamma)$ associated with the face. Let a face have vertices $\overline{V}_1, \overline{V}_2, \overline{V}_3$ and edges $\overline{E}_1 = \overline{V}_2 - \overline{V}_1, \overline{E}_2 = \overline{V}_3 - \overline{V}_1, \overline{E}_3 = \overline{V}_3 - \overline{V}_2$, $\overline{n}$ is the normal vector to the face. We place the origin at the vertex $\overline{V}_1$, basis vectors are defined by the relations:

$$\begin{aligned} \overline{e}_1 &= \overline{E}_1 / |\overline{E}_1| \\ \overline{e}_2 &= \overline{n} \times \overline{e}_1 \\ \overline{e}_3 &= \overline{n} \end{aligned} \tag{3}$$

In this coordinate system $\gamma = 0$, the face lies in the coordinate plane, and the formed surface can be represented by a full cubic polynomial in the local coordinate system

$$\gamma(\alpha, \beta) = C_0 + C_1\alpha + C_2\beta + C_3\alpha\beta + C_4\alpha^2 + C_5\beta^2 + C_6\alpha^2\beta + C_7\alpha\beta^2 + C_8\alpha^3 + C_9\beta^3 \tag{4}$$

Polynomial (4) can be constructed similarly to the Lagrange interpolation polynomial from its values at 10 points (see [5, 6]). These points are defined as follows (Figure 1):

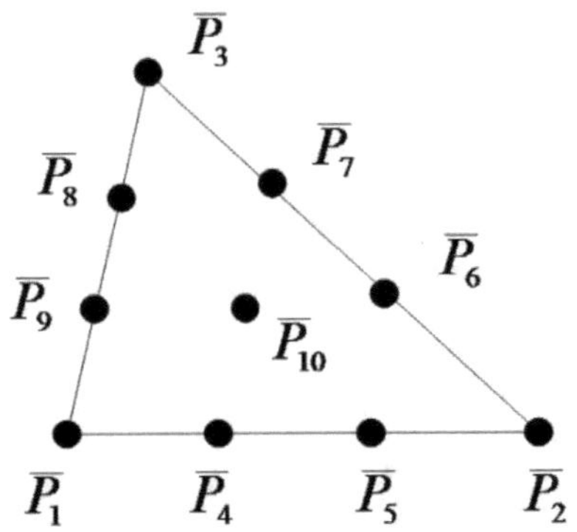

Figure 1. Interpolation nodes on the triangular face

$\overline{P}_1 = \overline{V}_1, \overline{P}_2 = \overline{V}_2, \overline{P}_3 = \overline{V}_3, \overline{P}_4 = (2\overline{V}_1 + \overline{V}_2)/3, \overline{P}_5 = (\overline{V}_1 + 2\overline{V}_2)/3, \overline{P}_6 = (2\overline{V}_2 + \overline{V}_3)/3,$
$\overline{P}_7 = (\overline{V}_2 + 2\overline{V}_3)/3, \overline{P}_8 = (\overline{V}_1 + 2\overline{V}_3)/3, \overline{P}_9 = (2\overline{V}_1 + \overline{V}_3)/3, \overline{P}_{10} = (\overline{V}_1 + \overline{V}_2 + \overline{V}_3)/3,$

Since the surface is to pass through the vertices of the triangle, the values of the function $U_i = \gamma(\overline{P}_i)$ at the points $\overline{P}_1, \overline{P}_2, \overline{P}_3$ are equal to zero: $U_1 = 0, U_2 = 0, U_3 = 0$. We will construct the values $U_4, U_5, U_6, U_7, U_8, U_9$ using the cubic spline of defect 2 (Hermitian interpolant) [7] on the edges $\overline{E}_1, \overline{E}_2, \overline{E}_3$. To do this, we use the condition that the tangent planes at the vertices $\overline{V}_1, \overline{V}_2, \overline{V}_3$ are tangent to the surface being formed. From here we obtain the values of the derivatives in the directions $\overline{E}_1, \overline{E}_2, \overline{E}_3$ at the corresponding vertices:

$$
\begin{aligned}
q_{11} &= \frac{\partial \gamma}{\partial \overline{E}_1}(\overline{V}_1) = -\frac{\overline{N}_1 \cdot \overline{e}_1}{\overline{N}_1 \cdot \overline{e}_3} & q_{12} &= \frac{\partial \gamma}{\partial \overline{E}_1}(\overline{V}_2) = -\frac{\overline{N}_2 \cdot \overline{e}_1}{\overline{N}_2 \cdot \overline{e}_3} \\
q_{21} &= \frac{\partial \gamma}{\partial \overline{E}_2}(\overline{V}_1) = -\frac{\overline{N}_1 \cdot \overline{E}_2}{|\overline{E}_2| \overline{N}_1 \cdot \overline{e}_3} & q_{23} &= \frac{\partial \gamma}{\partial \overline{E}_2}(\overline{V}_3) = -\frac{\overline{N}_3 \cdot \overline{E}_2}{|\overline{E}_2| \overline{N}_3 \cdot \overline{e}_3} \\
q_{32} &= \frac{\partial \gamma}{\partial \overline{E}_3}(\overline{V}_2) = -\frac{\overline{N}_2 \cdot \overline{E}_3}{|\overline{E}_3| \overline{N}_2 \cdot \overline{e}_3} & q_{33} &= \frac{\partial \gamma}{\partial \overline{E}_3}(\overline{V}_3) = -\frac{\overline{N}_3 \cdot \overline{E}_3}{|\overline{E}_3| \overline{N}_3 \cdot \overline{e}_3}
\end{aligned} \quad (5)
$$

Calculating the values of Hermite polynomials at points 4 through 9 we obtain the final formulas:

$$
\begin{aligned}
U_4 &= \frac{2|\overline{E}_1|}{27}(2q_{11} - q_{12}), U_5 = \frac{2|\overline{E}_1|}{27}(q_{11} - 2q_{12}), U_6 = \frac{2|\overline{E}_3|}{27}(2q_{32} - q_{33}), \\
U_7 &= \frac{2|\overline{E}_3|}{27}(q_{32} - 2q_{33}), U_8 = \frac{2|\overline{E}_2|}{27}(q_{21} - 2q_{23}), U_9 = \frac{2|\overline{E}_2|}{27}(2q_{21} - q_{23})
\end{aligned} \quad (6)
$$

The values in the center of the triangular face are determined using the condensation method described in [5, 6]:

$$
U_{10} = \frac{1}{4}\sum_{i=4}^{9} U_i - \frac{1}{6}\sum_{i=1}^{3} U_i \quad (7)
$$

Having the values of polynomial (4) at 10 points, we can compose a system of equations for its coefficients and solve it. It is more convenient to construct a polynomial by a formula similar to the formula of the Lagrange interpolation polynomial:

$$
\gamma(\alpha, \beta) = \sum_{i=1}^{10} U_i \varphi_i(\alpha, \beta) \quad (8)
$$

where functions of form are defined as follows. Let $L_{ij}(\alpha,\beta)=(\alpha-\alpha_i)(\beta_j-\beta_i)-(\beta-\beta_i)(\alpha_j-\alpha_i)$ is a linear function that vanishes to zero on the line connecting the points $\overline{P}_i=(\alpha_i,\beta_i,0)$ and $\overline{P}_j=(\alpha_j,\beta_j,0)$.

Then

$$\begin{aligned}
\varphi_1(\alpha,\beta)&=L_{23}(\alpha,\beta)\cdot L_{58}(\alpha,\beta)\cdot L_{49}(\alpha,\beta)/[L_{23}(\alpha_1,\beta_1)\cdot L_{58}(\alpha_1,\beta_1)\cdot L_{49}(\alpha_1,\beta_1)],\\
\varphi_2(\alpha,\beta)&=L_{13}(\alpha,\beta)\cdot L_{47}(\alpha,\beta)\cdot L_{56}(\alpha,\beta)/[L_{13}(\alpha_2,\beta_2)\cdot L_{47}(\alpha_2,\beta_2)\cdot L_{56}(\alpha_2,\beta_2)],\\
\varphi_3(\alpha,\beta)&=L_{12}(\alpha,\beta)\cdot L_{69}(\alpha,\beta)\cdot L_{78}(\alpha,\beta)/[L_{12}(\alpha_3,\beta_3)\cdot L_{69}(\alpha_3,\beta_3)\cdot L_{78}(\alpha_3,\beta_3)],\\
\varphi_4(\alpha,\beta)&=L_{13}(\alpha,\beta)\cdot L_{23}(\alpha,\beta)\cdot L_{58}(\alpha,\beta)/[L_{13}(\alpha_4,\beta_4)\cdot L_{23}(\alpha_4,\beta_4)\cdot L_{58}(\alpha_4,\beta_4)],\\
\varphi_5(\alpha,\beta)&=L_{13}(\alpha,\beta)\cdot L_{23}(\alpha,\beta)\cdot L_{47}(\alpha,\beta)/[L_{13}(\alpha_5,\beta_5)\cdot L_{23}(\alpha_5,\beta_5)\cdot L_{47}(\alpha_5,\beta_5)],\\
\varphi_6(\alpha,\beta)&=L_{12}(\alpha,\beta)\cdot L_{13}(\alpha,\beta)\cdot L_{47}(\alpha,\beta)/[L_{12}(\alpha_6,\beta_6)\cdot L_{13}(\alpha_6,\beta_6)\cdot L_{47}(\alpha_6,\beta_6)],\\
\varphi_7(\alpha,\beta)&=L_{12}(\alpha,\beta)\cdot L_{13}(\alpha,\beta)\cdot L_{69}(\alpha,\beta)/[L_{12}(\alpha_7,\beta_7)\cdot L_{13}(\alpha_7,\beta_7)\cdot L_{69}(\alpha_7,\beta_7)],\\
\varphi_8(\alpha,\beta)&=L_{12}(\alpha,\beta)\cdot L_{23}(\alpha,\beta)\cdot L_{69}(\alpha,\beta)/[L_{12}(\alpha_8,\beta_8)\cdot L_{23}(\alpha_8,\beta_8)\cdot L_{69}(\alpha_8,\beta_8)],\\
\varphi_9(\alpha,\beta)&=L_{12}(\alpha,\beta)\cdot L_{23}(\alpha,\beta)\cdot L_{58}(\alpha,\beta)/[L_{12}(\alpha_9,\beta_9)\cdot L_{23}(\alpha_9,\beta_9)\cdot L_{58}(\alpha_9,\beta_9)],\\
\varphi_{10}(\alpha,\beta)&=L_{12}(\alpha,\beta)\cdot L_{23}(\alpha,\beta)\cdot L_{13}(\alpha,\beta)/[L_{12}(\alpha_{10},\beta_{10})\cdot L_{23}(\alpha_{10},\beta_{10})\cdot L_{13}(\alpha_{10},\beta_{10})],
\end{aligned}\tag{9}$$

The final value of the function is obtained by the formula (8).

As a result, the construction of new vertices of the surface is carried out according to the following general algorithm:

1. A grid of triangles of the required size ("patch") is projected onto the initial triangulated surface
2. For each face containing at least one image of the node of the "patch", at the vertices of this face the vectors normal to the surface are constructed
3. Using the algorithm described above, the values of function are calculated at 10 points of the face
4. By the formula (8), the local coordinate of the new point of the surface is calculated
5. The global coordinates of the new vertex of the surface are calculated
6. The resulting vertices are combined into new faces
7. There is a docking of new and old faces of the corrected surface

3. Docking of surfaces built on faces

Surfaces constructed on faces using the algorithm described above (formula (8)) smoothly join at the vertices of the triangles. But they do not form a common continuous smooth surface, since if the two adjacent faces (having a common edge) have different normal vectors, then there will be a gap between them (see Figure 2).

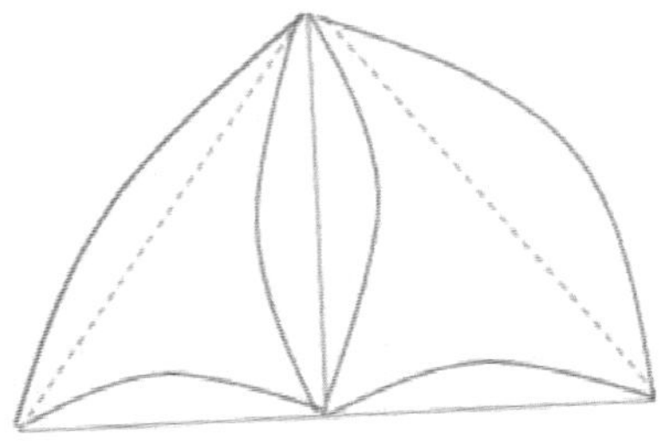

Figure 2. The gap between surfaces constructed on two neighbor faces

In accordance with the construction algorithm, none of the newly constructed points of the new surface will fall into any of the gaps, and the gaps will remain empty. In many cases, this problem can be ignored, since with a small difference in the normal vectors of adjacent faces, these discontinuities will be rather small and their existence will have little effect on the final result. However, we describe an algorithm for smooth conjugation of surfaces constructed on triangular faces. To do this, we select the zones on each triangular face (Figure 3).

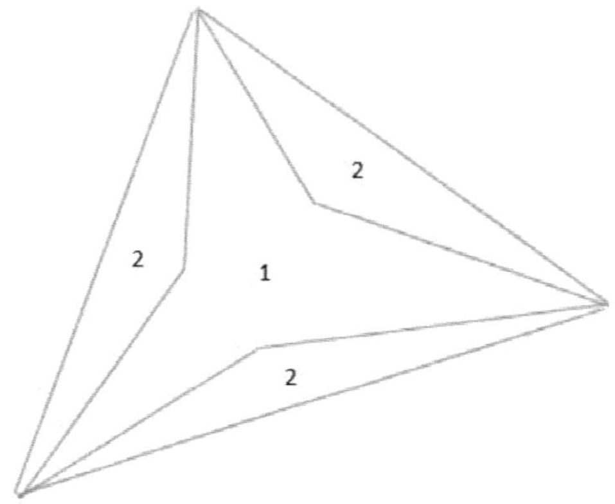

Figure 3. Division of a triangular face into zones

Zone 2 consists of three triangles adjacent to the edges of the faces. If the prototype of the new vertex falls into zone 1, then the new vertex is constructed according to the algorithm described above. If it falls into zone 2, then the following actions are performed.

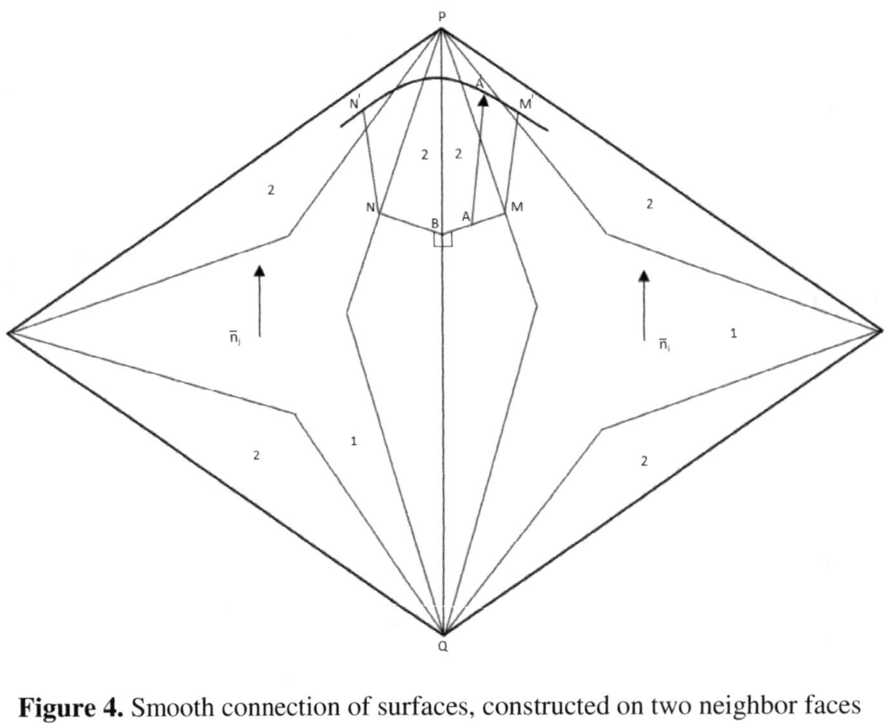

Figure 4. Smooth connection of surfaces, constructed on two neighbor faces

Let the image of the new vertex be the point $\overline{A}$ of the i-th face lying near the boundary of the j-th face (Figure 4). From this point, the perpendicular $\overline{A}\overline{B}$ is drawn to the common edge PQ with the j-th face. The same perpendicular extends to the border of zone 2 - the point $\overline{M}$. From the point $\overline{B}$ in the j-th face, the perpendicular $\overline{B}\overline{N}$ is restored to the edge $\overline{P}\overline{Q}$ until the boundary of zone 2. At points $\overline{M}$ and $\overline{N}$ using formula (8), points of a new surface $\overline{M}'$ and $\overline{N}'$ (each in its own local coordinate system of its face) and derivatives in directions $\overline{B}\overline{M}$ and $\overline{N}'$, respectively, are determined. Next, the values of the functions and derivatives are recalculated into a new local coordinate system with a basis $\overline{e}_1 = \dfrac{\overline{P}\overline{Q}}{|PQ|}, \overline{e}_2 = \overline{e}_3 \times \overline{e}_1, \overline{e}_3 = \dfrac{\overline{n}_i + \overline{n}_j}{|\overline{n}_i + \overline{n}_j|}$ and the Hermite interpolation polynomial is constructed from the values of the functions and derivatives at two points $\overline{M}$ and $\overline{N}$. The calculated value at the point $\overline{A}$ allows getting a new point on the surface $\overline{A}'$. Then it remains only to convert its coordinates into a global coordinate system.

4. Conclusion

The method and algorithm for reconstructing surface triangulation based on smooth interpolation are described. The method is based on spline interpolation using cubic polynomials. The surfaces built on the faces smoothly connected at the nodes and along the edges. The proposed method and algorithm can be useful in the numerical solution of three-dimensional dynamic problems of continuum mechanics with moving interfaces between moving solid, liquid and gaseous bodies.

Acknowledgments

The reported study was funded by RFBR and NSFC, projects number 19-58-53005, 19-0800320

References

[1] Puso M, Sanders J, Settgast R and Liu B 2012 *Computer Methods in Applied Mechanics and Engineering,* **245** 273-289

[2] Gilson L, Van Roey J, Gueders C, Gallant J and Rabet L 2012 A simple coupling of ALE domain with empirical blast load function in LS-DYNA *DYMAT 2012 conference*, Freiburg.

[3] Schwer L 2010 A Brief Introduction to coupling Load Blast Enhanced with Multi-Material ALE: The best of both worlds for air blast simulation *9th LS-DYNA Forum*, Bamberg.

[4] Abuziarov K M, Abuziarov M Kh, Glazova E G, Kochetkov A V, Krylov S V, Maslov E E and Romanov V I 2018 Numerical modeling of 3D problems of expansion of elastoplastic bodies by explosion. *The Problems of Strength and Plasticity*. Nizhny Novgorod. **80**(2) 255-266 (In Russian).

[5] Norrie D H and De Vries G 1978 *An introduction to finite element analysis* Academic Press, London, p 301

[6] Mitchell A R and Wait R 1977 *The Finite Element Method in Partial Differential Equations* London-New York-Sydney-Toronto. John Wiley & Sons.

[7] Kahaner D, Moler C and Nash S 1988 *Numerical methods and software* Englewood Cliffs, N J Prentice Hall, Prentice Hall series in computational mathematics. p 495

IOP Conf. Series: Materials Science and Engineering **927** (2020) 012012 doi:10.1088/1757-899X/927/1/012012

Stability Study of a Relative Equilibrium in the Planar Circular Restricted Four-Body Problem

B S Bardin[1, 2] and E V Volkov[1, 2]
[1]Moscow Aviation Institute (National Research University), Volokolamskoe shosse 4, 125993 Moscow, Russia
[2]Mechanical Engineering Research Institute of the Russian Academy of Sciences, 4, M. Kharitonyevskiy Pereulok, 101990 Moscow, Russia

E-mail: bsbardin@yandex.ru, evvolkov94@mail.ru

Abstract. We deal with the restricted planar four-body problem. That is, we consider motion of an infinitesimal small body (particle) under the Newtonian gravitational attraction of three bodies (primaries). It is supposed that primaries move in circular orbits forming Lagrange equilateral triangle; two of them have equal masses. By using the method of normal forms, we perform nonlinear stability study of a central configuration such that the particle is located in perpendicular bisector of the Lagrange equilateral triangle.

1. Introduction

The restricted four-body problem deals with the motion of a particle under the Newtonian gravitational attraction of three bodies, called primaries. The particle has negligibly-small mass, so that it does not affect the motion of the primaries, whose move on trajectories obtained as a solution of the three body-problem. In spite of the fact that restricted four-body problem has been considered from many points of view in numerous papers it continues to attract the attention of researchers. In particular, solutions describing the so-called central configurations of bodies are of particular interest. In papers [1-8], the existence, classification and stability of central configurations have been investigated. The central configuration in the restricted four-body problem is possible when primaries move in Keplerian orbits about their barycenter and form either equilateral triangle (Lagrange triangle solutions) or always stay on a straight line (Euler collinear solutions).

In this paper, we consider the planar circular restricted four-body problem, i.e. we suppose that three primaries rotate as a rigid equilateral triangle around their barycenter and the particle moves in the plane of primaries motion. We perform nonlinear study of stability of a central configuration, which is a relative equilibrium. In order words, we study an equilibrium position of the particle in the reference frame rotated with the equilateral triangle.

2. Problem statement and equations of motion

We denote by m_i $(i = 1, 2, 3)$ the masses of primaries P_i respectively. In what follows, we suppose that the masses of bodies P_1 and P_2 are equal: $m_2 = m_3$. We also suppose that Lagrange triangle configuration of the primaries P_i is stable. That is, the necessary Routh's stability condition [9] given by inequality, is satisfied.

$$m_1^2 - 50m_1m_2 - 23m_2^2 > 0 \tag{1}$$

To describe the motion of the particle P we introduce a rotating coordinate system $Oxyz$ with the origin in the barycenter O. The axis Oy passes through the body P_1, axis Oz is perpendicular to the plane of primaries motion and the axis Ox complements the coordinate system to right hand and orthogonal one (see Fig. 1).

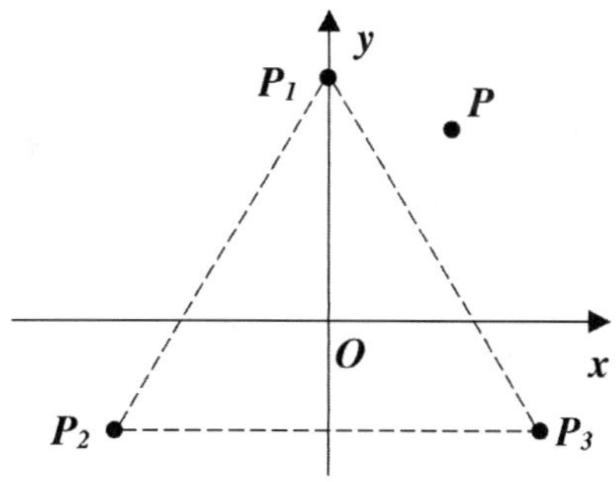

Figure 1. Coordinate system.

The equations of motions of the particle P can be written in the following canonical form [10].

$$\frac{d\xi}{d\tau} = \frac{\partial H}{\partial p_\xi}, \quad \frac{d\eta}{d\tau} = \frac{\partial H}{\partial p_\eta}, \quad \frac{dp_\xi}{d\tau} = -\frac{\partial H}{\partial \xi}, \quad \frac{dp_\eta}{d\tau} = -\frac{\partial H}{\partial \eta}. \tag{2}$$

The dimensionless coordinates ξ, η are introduced by means of formulae $x = r\xi$, $y = r\eta$, where r is the distance between the primaries. The independent variable ν can be regarded as the angle of the Lagrange triangular rotation. In the considered problem we have $\nu = \omega t$, where $\omega^2 = f(m_1 + 2m_2)r^{-3}$ and f is the gravitational constant . The Hamiltonian function reads

$$H = \frac{1}{2}(p_\xi^2 + p_\eta^2) + p_\xi \eta - p_\eta \xi - \frac{\mu}{\rho_1} - \frac{1-\mu}{2\rho_2} - \frac{1-\mu}{2\rho_3}, \tag{3}$$

where

$$\rho_1 = \left(\xi^2 + \left(\eta - \frac{\sqrt{3}}{2}(1-\mu)\right)^2\right)^{1/2}, \quad \rho_2 = \left(\left(\xi + \frac{1}{2}\right)^2 + \left(\eta + \frac{\sqrt{3}}{2}\mu\right)^2\right)^{1/2},$$
$$\rho_3 = \left(\left(\xi - \frac{1}{2}\right)^2 + \left(\eta + \frac{\sqrt{3}}{2}\mu\right)^2\right)^{1/2}, \quad \mu = \frac{m_1}{m_1 + 2m_2}. \tag{4}$$

The system (2) has the following steady state solution

$$\xi = 0, \quad \eta = \eta_*, \quad p_\xi = -\eta_*, \quad p_\eta = 0, \tag{5}$$

where $\eta_* = u(\mu) + \sqrt{3}(1-\mu)/2$ and $u(\mu)$ is a real root of the following equation

$$u + \frac{\sqrt{3}}{2}(1-\mu) - \mu\frac{\mathrm{sgn}(u)}{u^2} - \frac{(1-\mu)\left(u + \frac{\sqrt{3}}{2}\right)}{\left(u^2 + \sqrt{3}u + 1\right)^{3/2}} = 0. \tag{6}$$

The solution (5) describes equilibrium position located on the axis Oy of the coordinate system $Oxyz$. There exist three equilibrium positions of the above type. They are denoted in Fig. 2 by points A, B, C. The equilibrium position B exists for $\mu < 0.43242$, whereas the equilibrium positions A and C exists for any possible values of μ. In an equilibrium position the particle P form with the primaries P_i (i = 1, 2, 3) a rigid central configuration which rotates in the space with constant angular velocity ω.

It is clear that for stability such a central configuration the inequality (1) should be fulfilled. By using the parameter μ we can rewrite the inequality (1) in the form

$$\frac{1}{3} + \frac{4\sqrt{2}}{9} < \mu < 1. \tag{7}$$

IOP Conf. Series: Materials Science and Engineering **927** (2020) 012012 doi:10.1088/1757-899X/927/1/012012

For the central configuration $P_1P_2BP_3$ the inequality (1) is not fulfilled. Thus, it is always unstable. In linear approximation, the analysis of the configurations $P_1P_2AP_3$ and $P_1P_2CP_3$ has been performed in [1, 10]. It was established that the configuration $P_1P_2AP_3$ is unstable for any possible values of μ. The configuration $P_1P_2CP_3$ is unstable for $\mu < \mu_2 \approx 0.99458074$ and it is stable in linear approximation for $\mu_2 < \mu < 1$.

In this paper, we perform nonlinear stability analysis and obtain rigorous conclusions on stability of the central configuration $P_1P_2CP_3$ in the sense of Lyapunov. To this end we use the method of normal form and apply theorems on stability of autonomous Hamiltonian systems with two degrees of freedom.

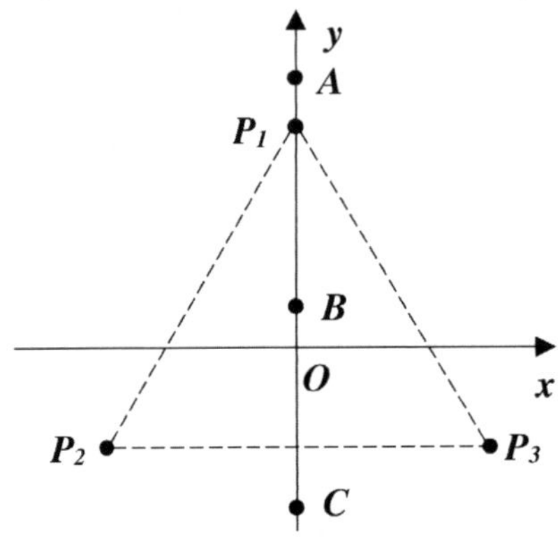

Figure 2. Equilibrium positions.

3. Linear normalization

Let us introduce local canonical variables q_1, q_2, p_1, p_2 in the neighborhood of the solution (5)

$$\xi = q_1, \quad \eta = \eta_* + q_2, \quad p_\xi = -\eta_* + p_1, \quad p_\eta = p_2 \tag{8}$$

and expand the Hamiltonian (1) in a power series with respect to the new canonical variables

$$H = H_2 + H_3 + H_4 + \cdots . \tag{9}$$

The quadratic part H_2 of the Hamiltonian reads

$$H_2 = \frac{1}{2}(p_1^2 + p_2^2) + \frac{1}{2}(aq_1^2 + bq_2^2) + p_1q_2 - p_2q_1, \tag{10}$$

where

$$a = \frac{4(1-\mu)\left(\left(2u+\sqrt{3}\right)^2 - 2\right)}{\left(1+\left(2u+\sqrt{3}\right)^2\right)^{5/2}} + \frac{\mu}{2|u|^3}, \qquad b = \frac{4(1-\mu)\left(1 - 2\left(2u+\sqrt{3}\right)^2\right)}{\left(1+\left(2u+\sqrt{3}\right)^2\right)^{5/2}} - \frac{\mu}{|u|^3}. \tag{11}$$

The expressions for cubic H_3 and quartic H_4 parts of the Hamiltonian are very awkward that is why we do not write down them here.

Conclusions on stability in the sense of Lyapunov can be obtained by using theorem on stability of autonomous Hamiltonian systems with two degrees of freedom [11-14]. In order to apply these theorems it is necessary to construct canonical change of variables, which brings the Hamiltonian into the simplest (normal) form. Usually it is enough to calculate the Hamiltonian normal form up to terms of the fourth degree with respect to canonical variables.

At the first stage of the normalization procedure, we normalize the quadratic part of the Hamiltonian. If $\mu_2 < \mu < 1$, then the characteristic equation of the linear system with the Hamiltonian H_2 has two pairs pure imaginary simple roots: $\pm i\omega_1, \pm i\omega_2$. In this case by means of the following linear change of variables

$$\begin{aligned} q_1 &= n_{11}Q_1 + n_{12}Q_2, \quad & q_2 &= m_{11}P_1 + m_{12}P_2, \\ p_1 &= m_{21}P_1 + m_{22}P_2, \quad & p_2 &= n_{21}Q_1 + n_{22}Q_2 \end{aligned} \tag{12}$$

we bring the Hamiltonian of the problem into the following form

$$H = \frac{1}{2}\omega_1(Q_1^2 + P_1^2) - \frac{1}{2}\omega_2(Q_2^2 + P_2^2) + H^{(3)}, \tag{13}$$

where

$$\omega_i = \left(\frac{1}{2}\left(a + b + 2 + (-1)^{i+1}\sqrt{(a-b)^2 + 8(a+b)}\right)\right)^{1/2}, \qquad (i = 1, 2)$$

$$n_{1i} = \frac{\chi_i(\omega_i^2 + b - 1)}{\omega_i}, \quad n_{2i} = -\frac{\chi_i(b\omega_i^2 - ab + a)}{\omega_i}, \quad m_{1i} = \delta_i\chi_i(\omega_i^2 - a - 1), \qquad m_{2i} = \delta_i\chi_i(a + b), \tag{14}$$

$$\chi_i^2 = \frac{\omega_i}{\delta_i\left(b\omega_i^4 - 2a(b-1)\omega_i^2 + (b-1)(a^2 + 2a + b)\right)}, \qquad (i = 1, 2).$$

By means of $H^{(3)}$ we denote terms of order higher than two.

If $\mu = \mu_2$, then the so-called second order resonance $\omega_1 = \omega_2 = \omega$ takes place, i.e. the characteristic equation of the linear system has two double roots: $\pm i\omega$, where $\omega \approx 0.69677051$. In this case the Hamiltonian can be brought into the following form

$$H = \frac{1}{2}(P_1^2 + P_2^2) + \omega(Q_1P_2 - Q_2P_1) + H^{(3)}. \tag{15}$$

The linear change of variables, which brings the Hamiltonian into form (15) reads

$$q_1 = n_{11}Q_2 + n_{12}P_1, \; q_2 = m_{11}Q_1 + m_{12}P_2, \; p_1 = m_{21}Q_1 + m_{22}P_2, \; p_2 = n_{21}Q_2 + n_{22}P_1, \tag{16}$$

where

$$n_{11} \approx -1.88075655, \; n_{12} \approx -0.96807745, \; n_{21} \approx -1.34905563, \; n_{22} \approx -2.00485196,$$
$$m_{11} \approx -0.76309333, \; m_{12} \approx -0.39278526, \; m_{21} \approx -0.54736236, \; m_{22} \approx -0.81344348. \tag{17}$$

If $\mu = 1$, then $\omega_2 = 0$, i.e. the first order resonance takes place. The following change of variables

$$q_1 = -\sqrt{3}Q_2 - 2P_1, \quad q_2 = Q_1 + \frac{2\sqrt{3}}{3}P_2, \quad p_1 = Q_1 + \frac{\sqrt{3}}{3}P_2, \quad p_2 = -\sqrt{3}Q_2 - P_1 \tag{18}$$

normalizes the Hamiltonian to the form

$$H = \frac{1}{2}(Q_1^2 + P_1^2) - \frac{1}{2}P_1^2 + H^{(3)}. \tag{19}$$

The normalization of the Hamiltonian part $H^{(3)}$ can be performed by means of a canonical near-identity transformation. The generating function for such a transformation has the form

$$S = Q_i v_i + S^{(3)}(Q_i, v_i), \qquad (i = 1, 2) \tag{20}$$

where $S^{(3)}(Q_i, v_i)$ is a convergent power series, which starts from terms of order three or higher.

In non-resonant case, the Hamiltonian normal can differ from ones in resonant cases. In particular, the normalization as well as the stability study in cases of resonances up to fourth order should be performed separately from the general case.

As we have already seen, the first ($\omega_2 = 0$) and second ($\omega_1 = \omega_2$) order resonances appear on the boundaries of the interval $\mu_2 < \mu < 1$. Inside of this interval the resonances of third and fourth orders take place: for $\mu = \mu_3 \approx 0.99648459$ we have the third order resonance, when $\omega_1 = 2\omega_2$ and for $\mu = \mu_4 \approx 0.99800139$ the fourth order resonance $\omega_1 = 3\omega_2$ takes place.

We have used the Deprit-Hori [15] method to calculate the coefficients of the generating function (20) and normal form of the Hamiltonian up terms of fourth order.

4. Nonlinear stability analysis

Let us start the nonlinear stability study from the general case when there are no resonances up to fourth order inclusively. That is, we suppose values of μ lie inside of the interval $\mu_2 < \mu < 1$ and $\neq \mu_3$, $\mu \neq \mu_4$. In this case, the Hamiltonian normalized up to fourth order term reads.

$$H = \omega_1 r_1 - \omega_2 r_2 + c_{20} r_1^2 + c_{11} r_1 r_2 + c_{02} r_2^2 + O\big((r_1 + r_2)^{5/2}\big), \tag{21}$$

where the canonical variables φ_i, r_i are introduced by formulae $u_i = \sqrt{2r_i}\sin\varphi_i$, $v_i = \sqrt{2r_i}\cos\varphi_i$.

If $\Delta(\mu) = c_{02}\omega_1^2 + c_{11}\omega_1\omega_2 + c_{20}\omega_2^2 \neq 0$, then it follows from the Arnold-Moser theorem [11-13] the stability of the equilibrium position of the system with Hamiltonian (21). The plot of the function $\Delta(\mu)$ obtained by numerical calculations is shown on Fig. 3.

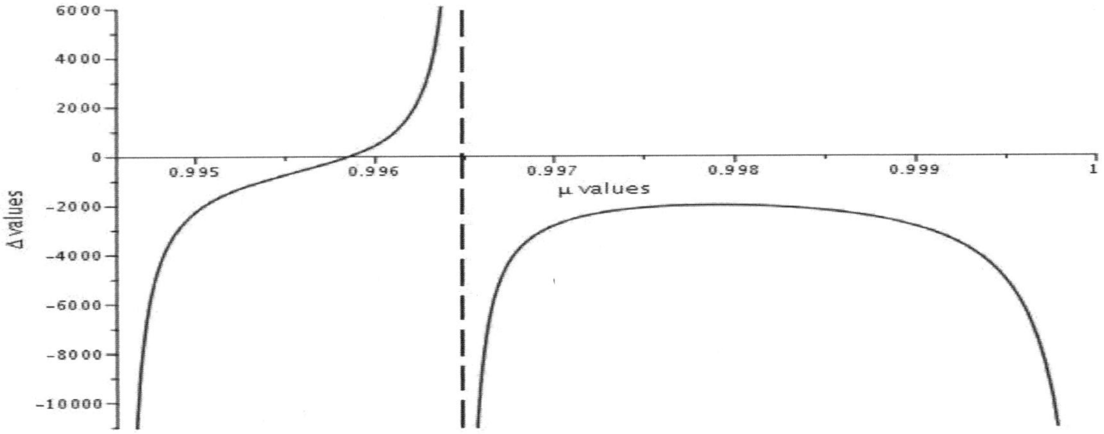

Figure 3. Dependence of Δ on μ in the interval 0.99458074<μ<1.

The function $\Delta(\mu)$ is equal to zero at $\mu = \mu_* \approx 0.99585655$ and has singular point at $\mu = \mu_3$. Thus, if $\mu \neq \mu_*$, then outside the resonant values of μ the configuration $P_1P_2CP_3$ is stable. For $\mu = \mu_*$ an additional study based on an analysis of terms of degree six (or higher) in the Hamiltonian expansion is necessary to solve the stability problem.

For $\mu = \mu_3$ the Hamiltonian normal form reads (third order resonance)

$$H = 2\omega_2 r_1 - \omega_2 r_2 - A r_2\sqrt{r_1}\sin(\varphi_1 + 2\varphi_2) + O((r_1 + r_2)^2). \tag{22}$$

The calculations have shown that $A \approx 17.63$. Since $A \neq 0$, in accordance with the Markeev theorem [14], the configuration $P_1P_2CP_3$ is unstable.

If $= \mu_4$, then the Hamiltonian normal form reads (fourth order resonance)

$$H = 3\omega_2 r_1 - \omega_2 r_2 + c_{20} r_1^2 + c_{11} r_1 r_2 + c_{02} r_2^2 + B r_2\sqrt{r_1 r_2}\sin(\varphi_1 + 3\varphi_2) + O\big((r_1 + r_2)^{5/2}\big). \tag{23}$$

In this case the stability conditions obtained in [14] reads: the equilibrium position is stable if $|3\sqrt{3}B| < |C|$; here $C = c_{20} + 3c_{11} + 9c_{02}$. Otherwise, if $|3\sqrt{3}B| > |C|$, then the equilibrium position is unstable. The calculations have shown that $B \approx 159.77$ and $C \approx -20560.00$. Thus, the configuration $P_1P_2CP_3$ is stable.

Now we pass to the boundary case $\mu = \mu_2$, when the second order resonance $\omega_1 = \omega_2$ takes place. In this case, the Hamiltonian normal form reads [14, 16]

$$H = \frac{1}{2}(P_1^2 + P_2^2) + \omega(Q_1P_2 - Q_2P_1) + (Q_1^2 + Q_2^2)(A(Q_1^2 + Q_2^2) + B(Q_1P_2 - Q_2P_1) + \\ + C(Q_1P_2 - Q_2P_1)) + H^{(6)}. \tag{24}$$

Our calculations have shown that $A \approx -47.00$. Since $A \neq 0$, in accordance with the Sokol'sky theorem [16, 17], the configuration $P_1P_2CP_3$ is stable.

At last we consider the limiting case $\mu = 1$, when the first order resonance $\omega_2 = 0$ takes place. In this case, Hamiltonian normal form reads [18-20].

$$H = \frac{1}{2}(Q_1^2 + P_1^2) - \frac{1}{2}P_1^2 + A_{21}(Q_1^2 + P_1^2)Q_2 + A_{03}Q_2^3 + A_{40}(Q_1^2 + P_1^2)^2 + \tag{25}$$

$$+A_{22}(Q_1^2+P_1^2)Q_2^2+A_{04}Q_2^4+H^{(5)}.$$

The calculation shows that $A_{21}=A_{03}=A_{40}=A_{22}=A_{04}=0$. It appears that in this case the equilibrium position is not isolated. Indeed, the equation of motion admit the following one parametric family of the equilibrium positions $p_\eta=\xi=\sin\alpha$, $p_\xi=-\eta=\cos\alpha$, where α is a parameter; $\alpha\in[0,2\pi)$. It means that in the case $\mu=1$ we meet the so-called transcendental situation [21, 22], when the stability problem cannot be solved by taking into account terms of any finite order in the Hamiltonian expansion. In [21] such a transcendental situation has been studied and it was shown that equilibrium position is unstable. Thus, $P_1P_2CP_3$ is unstable at $\mu=1$.

5. Conclusions

The results of our study can be summarized as is follows. For the values of parameter μ satisfying the inequalities $\mu_2\le\mu<1$, $\mu\ne\mu_*$, $\mu\ne\mu_3$, $\mu\ne\mu_4$, the central configurations denoted in Figure 2 by $P_1P_2CP_3$ is stable in the sense of Lyapunov. At $\mu=\mu_3$ this central configuration is unstable. We did not study the stability at $\mu=\mu_*$. In this case, an addition stability study is necessary to answer the stability question. Such a study can be performed based on normal form method by taking into account terms of order six (or higher) in the Hamiltonian expansion (9). In the limiting case $\mu=1$ the normal form method is not applicable. We have shown that in this case, the transcendental situation takes place and the central configuration $P_1P_2CP_3$ is unstable.

Acknowledgments
This work was supported by the grant of the Russian Scientific Foundation (project No. 19-11-00116) at the Moscow Aviation Institute (National Research University).

References

[1] Brumberg V A 1957 *Soviet Astronomy* **1** 57–79
[2] Simo C 1978 *Cel. Mech.* **18** 165–184
[3] Hampton M and Moeckel R 2006 *Invent. Math.* **163** 289–312
[4] Albouy A and Sun S 2012 *Proc. R. Soc. A.* **464** 1355–1365
[5] Albouy A 1996 *Contemporary Math.* **198** 131–135
[6] Budzko D A and Prokopenya A N 2010 *Programming and Computer Software* **36** 68–74
[7] Grebenikov E A, Gadomski L and Prokopenya A N 2007 *Nonlinear Oscillations* **10** 62–77
[8] Eduardo S G Leandro 2006 *J. Differential Equations* **226** 323–351
[9] Routh E J 1875 *Proc. London Math. Soc.* **6** 86–97
[10] Bardin B S and Esipov P A 2018 Investigation of Lyapunov stability of a central configuration in the restricted four-body problem *AIP Conference Proceedings* 1959, 040004 (2018)
[11] Arnold V I 1961 *Soviet Math.* **2** 247–249
[12] Arnold V I 1963 *Russ. Math. Surv.* **18** 85–191
[13] Siegel C L and Moser J K 1971 *Lectures on Celestial Mechanics* (New York: Springer)
[14] Markeev A P 1978 *Libration Points in Celestial Mechanics and Space Dynamics (in Russian)* (Moscow: Nauka)
[15] Giacaglia G E O 1972 *Perturbation Methoda in Non-Linear Systems* (New York: Springer)
[16] Sokol'sky A G 1975 On stability of autonomous Hamiltonian system with two degrees of freedom in the case of equal frequencies *J. Appl. Math. Mech.* vol 38 741–749
[17] Lerman L M and Markova A P 2009 On stability at the Hamiltonian Hopf bifurcation *Regular and Chaotic Dynamics* vol 14 **1** 148–162
[18] Sokol'sky A G 1977 On the stability of an autonomous Hamiltonian system with two degrees of freedom under first-order resonance *J. Appl. Math. Mech.* **41** 20–28
[19] Bardin B S and Chekina E A 2017 On the constructive algorithm for stability analysis of an equilibrium point of a periodic Hamiltonian system with two degrees of freedom in the second-order resonance case *Regular and Chaotic Dynamics* vol 22 **7** 808–824
[20] Bardin B S and Chekina E A 2018 On the constructive algorithm for stability investigation of

IOP Conf. Series: Materials Science and Engineering **927** (2020) 012012 doi:10.1088/1757-899X/927/1/012012

an equilibrium point of a periodic Hamiltonian system with two degrees of freedom in first-order resonance case *Mech. Solids* **53** 15–25

[21] Bardin B S and Maciejewski A J 2013 Transcendental case in stability problem of Hamiltonian system with two degrees of freedom in presence of first order resonance *Qualitative Theory of Dynamical Systems* vol 12 **1** 207-216

[22] Bardin B S 2018 On the stability of a periodic Hamiltonian system with one degree of freedom in a transcendental case *Dokl. Math.* vol 97 **2** 161–163

IOP Conf. Series: Materials Science and Engineering **927** (2020) 012013 doi:10.1088/1757-899X/927/1/012013

Study of methods of supplying a solid phase to a heterogeneous flow at high-pressures of a carrier gas

S A Borisov, Ju I Glukhovskaya, S V Dobrovolskiy, I M Platonov and I V Podporin
Moscow Aviation Institute (National Research University), Volokolamskoe highway 4, 125993, Moscow, Russia

E-mail: dobrovolskiy_s@mail.ru

Abstract. The study of the methods of supplying powder material to a supersonic heterogeneous high-pressure flow is carried out in order to determine the optimal type of dispenser that ensures the continuity of the supply of the solid phase to form a coating with anticorrosive properties on metal structures made of the following metal powders: stainless steel, aluminium and zinc or their mixtures using the low-temperature gasodynamic method. Various options for organizing the supply of powder to the main highway were also considered.

1. Introduction

The problem of reliability and durability of products, structures and mechanisms, economical use of materials, energy and labour resources has the connection with the effective protection of structures and mechanisms from erosion, corrosion and wear. Currently, about 30% of the annual metal smelting is spent on making up for losses from corrosion and wear.

The use of multifunctional protective coatings allows implementing a fundamentally new concept of using structural materials.

Currently, among the various technologies for applying protective coatings, high-tech gasodynamic methods are being developed in which the coating is formed from individual particles (powder) heated and accelerated using a high temperature gas jet (carrier gas) [1,2].

In the impact zone, there is a mutual shear deformation of crystal lattices in the materials of the particles and the substrate and the activation of the conjugated surfaces. In addition, dynamic effects arise in the contact patch as a result of a local pressure pulse and an increase in temperature at the moment of particle impact on the substrate. All these processes provide a high level of adhesive and cohesive properties of coatings.

A significant outflow rate of a heterogeneous stream carrying particles of the material from which the coating is formed, combined with heating and the possibility of mobile control of the component composition of the jet, contribute to the formation of a wide range of multifunctional coatings: from simple and refractory metals, oxides, carbides, borides, etc., to plastics. It allows applying coatings with a wide range of purposes: wear-resistant, corrosion-resistant, heat-protective, electrical insulating, etc.

The uniform thickness of the formed coating is one of the necessary conditions for the practical application of the gasodynamic coating method. It is ensured by the uniform supply of the solid phase into the heterogeneous flow.

At the same time, in addition to the uniformity of the powder supply, there is a number of other requirements for the dispensers used in research and industrial installations [3,4]:

- stability of powder supply to the stream;
- uniform regulation of the powder supply rate into the stream;
- low weight of the structure;
- simplicity, durability and compact design;
- possibility of remote automation in controlling the flow of powder;
- low cost;
- simple and inexpensive repairs [5].

2. Choosing the type of powder dispenser

In this study, several types of powder dispenser designs were considered:

- ejection: powder is transported due to the vacuum created by the moving gas stream;
- vibrational: powder is poured into the conveying highway under the action of its own mass and vibration and is supplied into the carrier gas stream;
- auger: powder is dosed by the movement of augers, which pour the powder into the conveying highway, after which the powder is supplied into the carrier gas stream;
- rotary type dispensers: powder is dosed by the movement of rotors, which pour the powder into the conveying highway, after which the powder is supplied into the carrier gas stream.

A study was conducted of the uniformity of powder supply by various types of dispensers (Figure 1 and Figure 2) [6].

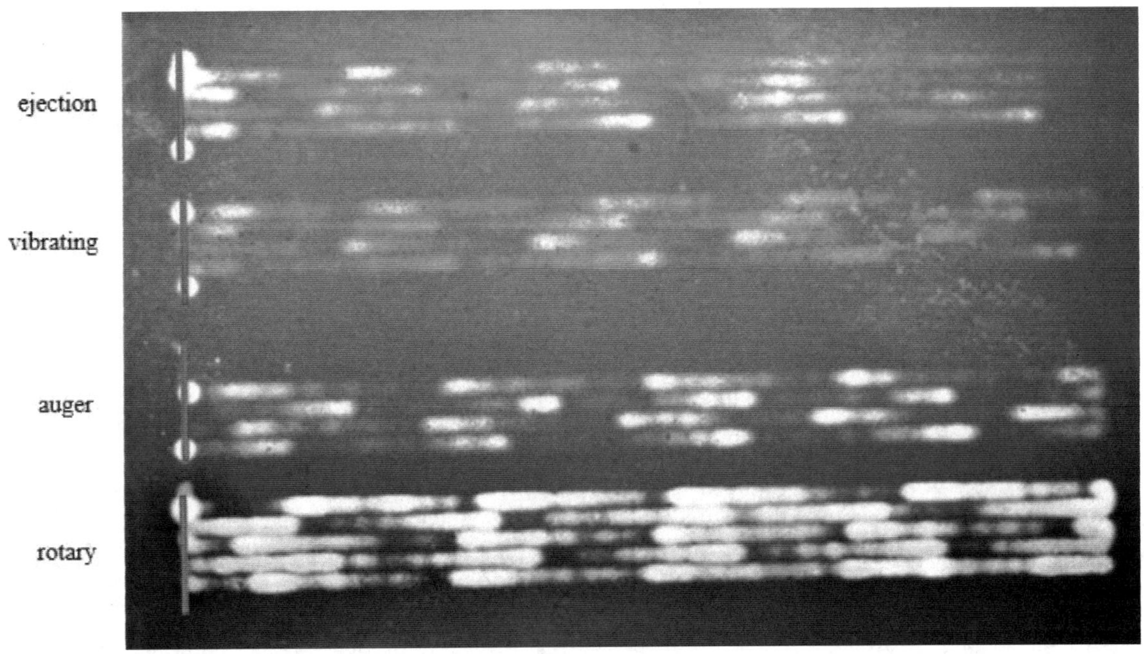

Figure 1. Determination of continuity of abrasive supply by type of dispenser.

Figure 1 shows that the most uniform and continuous supply of powder provides a rotary dispenser. The second place has the auger method of powder supply. Quite weak results were shown by ejection and vibration methods. However, under different operating conditions of the highway (maximum pressure and temperature of the working fluid), the latter methods may show greater efficiency, but this is the subject of another study.

When analysing the supply of powder by a rotary dispenser, it is possible to identify a small heterogeneity in the form of a sinusoidal wave, which is explained by the spiral shape of the screw. The percentage distribution of the powder supply continuity by the rotary dispenser is shown in the Figure 2.

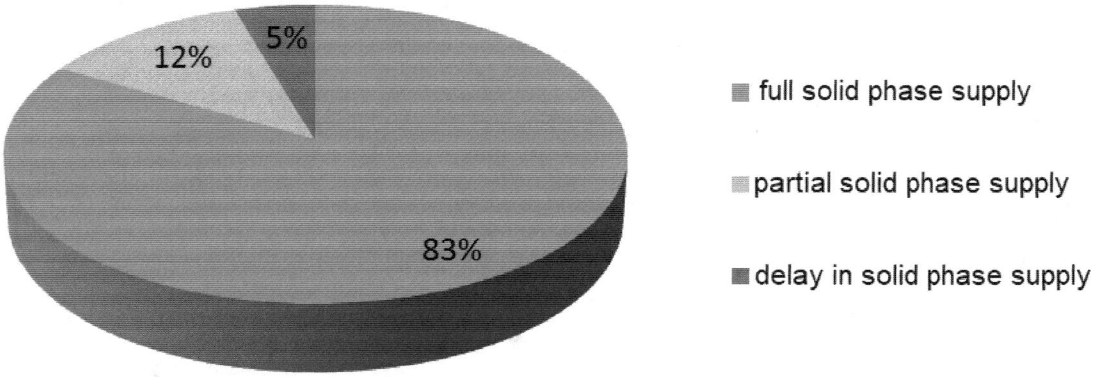

Figure 2. Percentage distribution of the continuity of solid phase supply when using a rotary dispenser.

Based on the results of the experiments, for the formation of coatings with anticorrosive properties from metal powders on metal structures using the gasodynamic method, a rotory-type dispenser was selected.

3. Organization of powder supply to the highway

After the most effective method of uniformly supplying powder to the highway has been selected, it is necessary to organize its introduction into the stream, uniform distribution of particles over the channel cross section and to ensure sufficient heating and the required speed for coating the surface.

The necessary level of energy is provided by communicating kinetic energy to the particles and increasing internal energy by heating them. The temperature of the particles is limited by the properties of the powder material due to the possibility of melting of the particles with too much heat. Particles can be heated at the stage of supplying into the dispenser, but this can lead to sintering of particles, which will not allow them to be applied to the surface, and in the worst case, damage to the substrate. Therefore, it seems most effective to heat the particles with a flow of carrier gas, accelerated to the desired speed.

Such a solution is a complex engineering task that requires a proven methodology for calculating the flow parameters, their interaction between themselves and powder particles. If we build this installation without preliminary calculations, there is a high probability of various problems with the gas flow and supply of particles of the solid phase, for example, blocking the flow or uneven distribution of particles along the channel. To prevent the occurrence of these problems, a computer simulation of the gas flow through the channel of the proposed installation was carried out [7].

An example of solving this problem is shown in the Figure 3. The geometry is selected taking into account the requirements for the parameters at the nozzle exit: pressure, speed and temperature. Carrier gas parameters are the following: working pressure is up to 50 bar; air consumption is up to 200 g/s; powder supply is up to 50 g/s; powder particle size is from 5 to 60 microns.

To obtain the necessary characteristics at the entrance to the channel, it is required to set sufficiently high values of pressure and temperature. However, the combination of high-pressure and temperature negatively affects the strength characteristics of used materials.

IOP Conf. Series: Materials Science and Engineering **927** (2020) 012013 doi:10.1088/1757-899X/927/1/012013

Figure 3. Geometry of the accelerator channel. Speed field.

Since the simultaneous heating and compression of gas is a really difficult task, a scheme for separating operations into two streams is proposed: high-pressure and low temperature stream and low pressure and high temperature stream. This approach allows to separately ensuring the reliability of the high-pressure pipeline and cooling / thermal insulation of the hot stream. When mixing two streams, it is possible to achieve the required parameters at the nozzle exit and ensure the reliability of the system. A typical view of this channel is shown in the Figure 4.

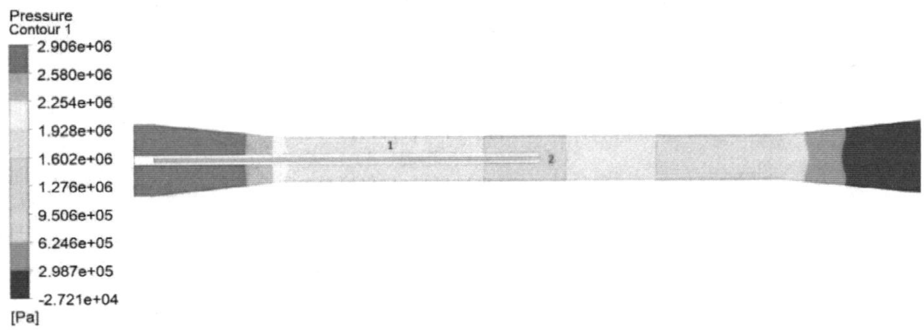

Figure 4. Pressure field for two-channel supply of the working fluid.
1 - external channel; 2 - central small tube.

Next, we should choose which channel will supply gas under high-pressure, and which one will do it with a high temperature. The option considered in the Figure 4 provides for the supply of hot gas through the central tube, while the gas flowing around it is a cooler. However, the external high-pressure flow creates the following problem. Figure 5 shows the flow velocity vectors at the end of a small tube. A change in the direction of movement is observed, which occurs due to a large pressure drop at the outlet of the small central tube, which leads to blocking of the flow [8].

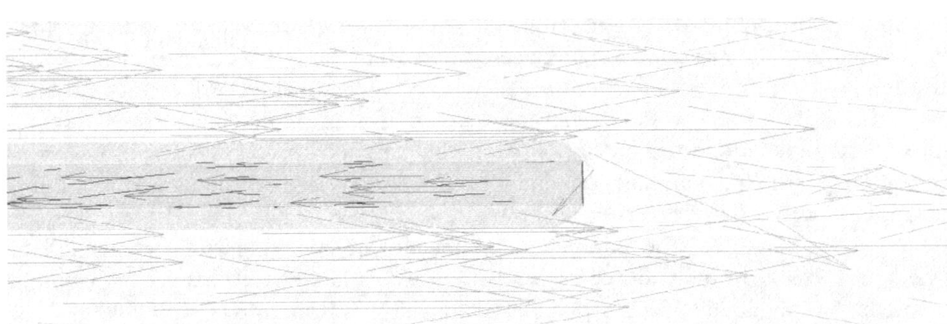

Figure 5. Velocity vectors when flowing out of a small tube.

To avoid the occurrence of reverse currents, it is advisable to interchange the flow of high-pressure and temperature flows. High-pressure gas is supplied through a small central tube, and a high temperature flow is supplied through an external channel (Figure 6).

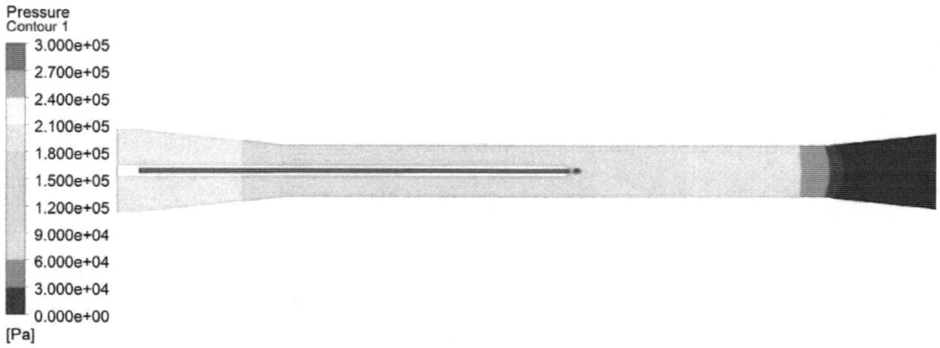

Figure 6. Selected flow diagram of the working fluid. Overpressure field. Upper boundary is limited to 3 atm. Actual pressure is 30 atm.

We consider the options for the location of the powder in the channel. If the powder is supplied through an external channel, its particles will melt and turn into droplets of a liquid substance. The calculation of the trajectory of such droplets and their heat exchange with the surrounding stream is a rather complicated task and it is undesirable. If particles are supplied through the high-pressure channel, then when two flows interact, the particles will be pushed away by the shock wave to the channel periphery, which is also undesirable. It is also difficult to organize the transfer of powder from the dispenser to the high-pressure highway. Without creating high-pressure in the dispenser, the probability of the powder moving in the opposite direction is high.

Low pressure zone that occurs when the hot stream exits from the supplying tube is interesting (Figure 7).

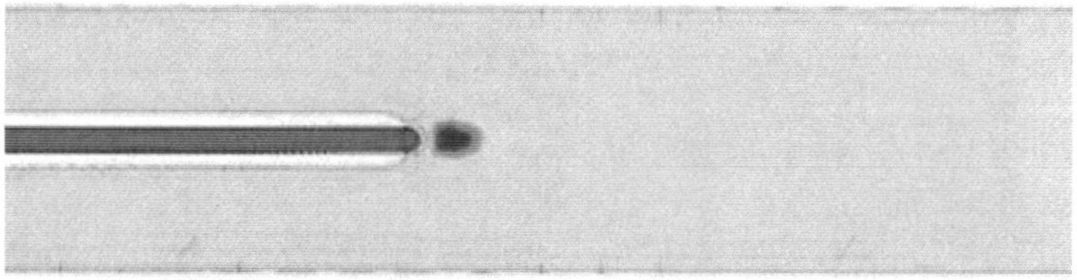

Figure 7. Pressure field during gas outflow from a small central tube.

When the powder is supplied to this area, it is possible to bypass the shock wave and too hot gas stream (Figure 8). Particles are evenly distributed and have the required parameters at the nozzle exit.

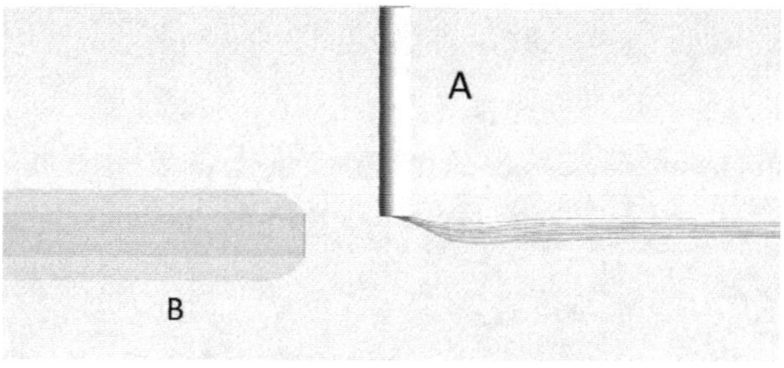

Figure 8. Trajectories of particles when they are supplied through the tube "A" into the main stream; B - small central tube.

4. Conclusion

To ensure the optimal parameters of the heterogeneous flow when using a high-pressure carrier gas during the formation of anticorrosion coatings by the low-temperature gasodynamic method on structural elements, it is advisable to use a rotary type dispenser.

The simulation of the motion of the gas and particle flow in the channel was used to determine the location of the input of the solid phase into the heterogeneous flow while maintaining high quality coating formation.

The type of dispenser that ensures the continuity of the supply of the solid phase at high-pressures of the carrier gas is determined.

The conditions and location of the supply of the solid phase to the heterogeneous high-pressure flow are determined for the formation of a coating with anticorrosive properties on the elements of metal structures using the low-temperature gasodynamic method.

Acknowledgment

This work was financially supported by the Ministry of Education and Science of Russia, agreement No. 05.607.21.0299 of December 3, 2019, unique project identification number: RFMEFI60719X0299.

References

[1] Nikitin P V 2010 *Heterogeneous flows in innovative technologies* (Moscow: Yanus-K)

[2] Nikitin P V, Borisov S A, Dobrovolskiy S V and Glukhovskaya Ju I 2017 Investigation of the mechanism of interaction of a particle with a substrate when a supersonic heterogeneous flow on a flat solid wall *Surf. X-ray, synchr. and neutr. stud.* **3** 100-5

[3] Zemerev E S and Malinin V I 2016 Analysis of the flow of a powder-gas medium in a conical channel with an outlet. Bulletin of Perm National Research Polytechnic University *Aer. Eng.* **47** 154-76

[4] Deeva V S, Romanishin A E and Slobodyan S M 2015 Analysis of dispensers for heterogeneous bulk media flow *Bull. of Alt. Stat. Agr. Un.* **8(130)** 135-9

[5] Zenkov R L, Grinevich G P and Isaev V S 1977 *Hopper devices* (Moscow: Machine building)

[6] Shishkovsky I V 2009 *Laser synthesis of functional gradient mesostructures and bulk products* (Moscow: FIZMATLIT)

[7] Molchanov A M 2013 *Mathematical modeling of gasodynamics and heat and mass transfer problems* (Moscow: Moscow Aviation Institute Publishing House)

[8] Liu X and Miller J E 2013 *System and method for detecting blockage in the channel for supplying solids and carrier gases to the gas stream* Invention # 216.012.78AC, October 27, 2013

IOP Conf. Series: Materials Science and Engineering **927** (2020) 012014 doi:10.1088/1757-899X/927/1/012014

Modification of Gaussian elimination: technique, high resolution seismic model for area of natural accumulation of gases

T A Smaglichenko[1], A V Smaglichenko[2], N V Sokolova[1] and M K Sayankina[1]
[1]*Center for Innovative Oil and Gas Technologies*, Research Oil and Gas Institute of Russian Academy of Sciences, 3 Gubkina str., Moscow, Russia
[2]*Laboratory of Identification of control systems*, V.A. Trapeznikov Institute of Control Sciences of Russian Academy of Sciences,
65 Profsoyuznaya str., Moscow, Russia

E-mail: t.a.smaglichenko@gmail.com

Abstract. The modification of Gaussian elimination is described here as numerical method to determine the physically adequate solution of the linear equations system that is affected by complexity of the data error (measurements, parameterization et. al.). Other modifications and algorithms are mainly established to efficiently overcome calculating issues. In order to avoid influence of errors in the data the proposed technique selects a set of invertible sub-matrixes. Developing the Tikhonov regularization theory the regularized subsequence is determined for each unknown under consideration. Application of the method to local seismicity data derived high resolution areas and revealed that the hydrothermal field of high concentration CH4 and CO2 is surrounded by rock having high seismic velocity. Based on this result we conclude that nature protects the gas flow due to special structure, which is likely presented by dense basalt. Decrypt of satellite images shows that the zone of lowest elevation does not intersect this structure. Particularities of background seismicity are examined for the gas area being studied. Factors learned from nature experience permit us to formulate conditions to select a place for underground gas storage.

1. Introduction

Systems of linear equations are applied in various industry fields to solve engineering tasks including signal processing, analysis of mechanical system, image restoration, robotics and others. Reliable solutions of systems can be very practical to reveal characteristics of objects, to create models of their behaviour. At the same time, it is known that the adequacy of models depends on assumptions that are made. In other words, there is a modelling error. For instance, when solving the problem in a linear approximation there is the linearization error. Setting of inappropriate grid parameters to determine unknowns leads to the parameterization error.

Measurement data that form the known vector of the system may often contain errors having unclear level. A good example of such data would be pixels of astronomical images that are input data for the system to define the linear contribution of different sources. The images depend on atmospheric absorption, involve an error, which is complex combination of noise of receivers and photon noise that exists in the galaxy region [1].

Existing methods to solve the system of algebraic equations are mainly directed to overcome a high complexity of computations including round off errors. In the LSQR algorithm large and sparse systems are processed requiring a minimal memory and time owing to utilizing of the singular value decomposition of large and sparse matrixes with following application of the Lanczos iterative process [2]. In earlier works the sparse matrix processing techniques involved the direct Gaussian elimination method as efficient algorithm, when the computer storage is limited. Taking advantage of the zeros in initial matrix and by reordering of the matrix rows the method is able economically spend virtual memory [3]. Recently developed numerical modified scheme uses the projective reformulation of the Gaussian elimination with purpose to avoid slow division operator, and such calculations can be rapid and convenient for graphics processing unit (GPU) [4]. Thus, the current inversion algorithms control numerical issues. The question arises. Can a method restrain the influence of complex error of a physical experiment during the inversion process? Otherwise this error is included in mathematical calculations and the obtained solution can be significantly distorted. It is known that traditional Tikhonov regularization [5] and adaptive regularizations [6] minimize the regularized functional by smoothing all components of solution under assumption of random expected small value of the error.

In this paper we provide a description of the Modification of Gaussian Elimination (MGE), which mitigate the influence of unclear level of the initial data error to the system solution. New modification has been created taking into account ambiguity and complexity of seismic observations [7, 8]. In the present paper, we investigate the MGE result, which was obtained for the gas area in northern Iceland. Velocity model has been reconstructed by means of inversion of time data from seismic stations.

2. Modified Gaussian elimination involving statistical regularization of solution

Physical tasks result into a solution of system of linear equations $Ax = b + \varepsilon$, where A is a rectangular matrix of dimension $m \times k$, whose elements are determined by numerical distribution of values of known physical parameter in the grid points; $\vec{x}$ is an unknown vector-column, whose k components correspond to values of unknown parameter in the k grids; and $\vec{b}$ is the known vector-column with m components, which contain the error $\vec{\varepsilon}$. In the general case, due to the Gauss scheme the initial invertible matrix is transformed to a triangular form, then step by step the unknowns are eliminated. In practice the initial matrix is often large, over determined. It has the complex structure, which consists of zero elements arbitrary distributed. Let us divide the matrix into cells of smaller sub-matrixes. As an example, figure 1 displays the cell configuration with non-zero sub-matrices A^q and zero sub-matrices 0.

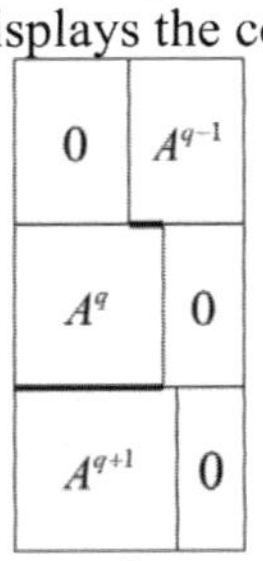

Figure 1. Illustration of the cells structure of the matrix. Bold lines denote common parts of sub-matrices A^{q-1} and A^q or A^q and A^{q+1}, which correspond to the same components of unknown vector-column.

Non-zero sub-matrices correspond to over determined sub-systems of various dimensions. Then a solution of the initial system becomes equivalent to solutions of a set of sub-systems, each of which may be written as:

$$A^q x^q = b^q + \varepsilon^q . \qquad (1)$$

Using the theorem that defines the condition for unique solution of a system of linear equations and selecting rows of sub-matrices that are linearly independent, one can form the basis subsystems $A^p x^p = b^p + \varepsilon^p$. This theorem is known under many names: Kronecker-Capelli-Rouché- Fontené-Frobenius. The statement of uniqueness given in the theorem is equivalent to the situation, when all diagonal elements of resolution matrix for big data are equal to 1.0. Unknown vector of the basis subsystem $\tilde{x}^p$ can be estimated from inversion of the basis sub-system with right part $\tilde{b}^p = b^p + \varepsilon^p$ by using the Gauss- Jordan method: The error of the solution is estimated by using the next formula [9]:

$$\tilde{x}^p - (x^p)^{true} = ((A^p)^{-1} A^p - I)(x^p)^{true} + (A^p)^{-1} \varepsilon^p . \quad (2)$$

Combining all solutions for various basis subsystems, we get a range of values for i - component of unknown vector $\tilde{x}$:

$$\tilde{x}_i = \{x_i^1 \dots x_i^n\}, \quad (3)$$

where n is a number of the basis subsystems participated in determination of the component.

Let estimate the statistical regularity of the solution $\tilde{x}_i$ by means of the search a group of approximately equal values:

$$x_i^1 \approx x_i^2 \dots \approx x_i^{no}, \quad (4)$$

where no is a number of approximately equal found values.

According to the Tikhonov regularization theory the sequence will be regularized if its elements belong to a compact set. A set is called compact, if from its sequence the convergent subsequence can be picked. Hence stability of the system solution can be defined by picking the subsequence of nearly equal values for the given unknown.

Obviously, not all values from the sequence (3) will be in the group presented in (4). We may assume that they involve large error. At each step of the main solution process the group of components for the unknown vector $\tilde{x}$ can be simultaneously found using solutions of various basis subsystems. Then, values of this group are eliminated from the initial system by the direct substitution. Thus, the size of the system is decreased and next one should find and solve new basis subsystems by applying the same procedure.

3. The MGE outcome for the area containing natural reservoir of methane and carbon dioxide

3.1. Reconstructed seismic image

MGE has been utilized to define unknown seismic velocity distribution in blocks of geological medium in Tjornes fracture zone, the northeast Iceland. The experiment has been organized by Prof. W.R.Jacoby (Mainz University, Germany). The temporary seismic stations were deployed in the northeast cost of Iceland from 1986 till 1989. The complexity of the data is caused by the geographical restriction of the position of receivers, which could only be on the shore and register seismic waves from seismic sources in sea. This fact can lead to errors in the hypocenters determining. Beside the mislocation error of sources any seismic data contain: systematic error of station-receiver, random error of picking up the wave arrivals on seismogram. The modeling and parameterization error is added to total error in the data.

In [10] authors showed that LSQR is effective in retrieving of small-sized structures, and severe issues can be, when the model presents large-sized structure. Therefore to test MGE the complex large-sized model was built taking into account all existing information about real features of the zone. Namely, bathymetry, morphological and topographic maps, as well as data on gravitational anomalies and Bouguer anomalies have been used to obtain numerical values of the input vector that imitated the zone structure. Results showed that the inverted model coincides with the initial one in the domain of weak sparseness of the matrix (upper layer 0-5 km of depth) and partly coincides in deeper layers, where the number of zero elements is increased. It was clear that MGE successfully overcomes the parameterization error. Next, we assumed that synthetic data have random errors. The noise vector having normal distribution has been added to the synthetic data and noisy synthetic data were again inverted by utilizing MGE. This revealed that the inverted model coincides with the initial one only along Grimsey lineament. Application of MGE to real data set showed retrieved seismic velocity structure also along this lineament.

Figure 2-a shows the MGE outcome for the Grimsey hydrothermal field with predominant CH4 and CO2 gases. Axes of the figure correspond to longitude and latitude. Segment of the Grimsey lineament is denoted by solid line. Epicenters of local earthquakes correspond to black circles. Note that hydrothermal vents were observed with the manned Swiss underwater vehicle GEO and the German research manned underwater vehicle JAGO [11].

Obtained tomographic image revealed the following. The Grimsey gas field having area of 1.0 x 0.3 km (white square in figure 2-a) is located inside of the high velocity structure of 20.0 x 9.0 km extending to 5 km of depth (dark gray color). The structure is characterized by the found P-wave seismic velocity that is 4.89 *km/s* and presumably corresponds to dense basalt. Since stations- receivers were located from the south and the south-west the structure can be isotropic for these directions owing to particularities of the MGE algebraic reconstruction (see equation 4). Figure 2-b displays the schematic 3D location of the gas field inside of the found high velocity anomalous element of geological medium.

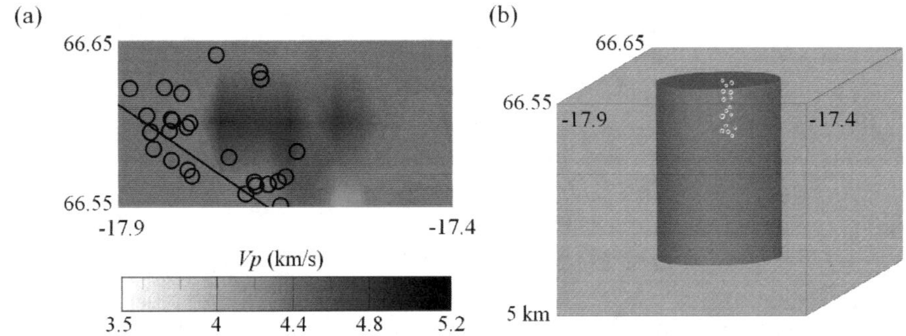

Figure 2. (a) The MGE seismic velocity image in domain of the gas field (b) 3D schematic drawing showing gas inside the high velocity structure.

3.2. The protective area in comparison with the thalweg zone identified from satellite imagery

It is known that landforms created by tectonic movements are very clearly visible in satellite images. For example, faults and volcanic rocks in a deep crust can be well decrypted. It is important to note that landforms having different geological age keep the property of individuality. Elements of the same age are homogeneous surfaces on images. Such elements may be lines of lowest elevation, so called zones of thalwegs, which are independent indicators of maximal degree of dissection of landform within water course. The thalweg zones continuously decrease dynamic stresses associated with the Earth's deformation. Assume that a fault is the main response to stress, then activity of rivers would be blocked. However, often after strong earthquakes a position of the river system remains relatively stable. In the northeastern Iceland the thalweg systems of various ranks were revealed by decrypting of satellite images provided by software "Google Earth Pro" [12]. Relatively independent zones were determined applying the method described in [13]. To decrypt the zones on satellite imagery, we used a visual analysis. An informative map was constructed by choosing the same brightness values to distinguish homogeneous lines of the thalweg. Figure 3 shows the thalweg zone (solid line) located near the area being investigated (rectangular block). The border line of the thalweg that is in immediate vicinity of the block corresponds to a local landform funnel, which indicates the increased risk of earthquakes with small magnitudes at this location. The figure is built in the GMT program by using data from ASTER GDEM v.2 (METI and NASA product).

Epicenters of seismic events with magnitudes $M > 3$ that occurred during the period 1989-2020 [14] are plotted using dots. Thus, the restored thalweg line does not intersect the found block of dense rock that protects the gas field.

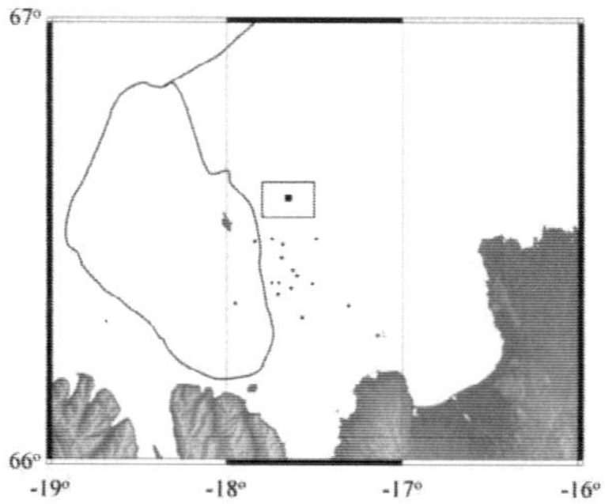

Figure 3. The thalweg zone close to the gas field (small black square).

3.3. Earthquakes depths and "locking depth" for the Tjornes fracture zone

In figures 2a and 3 one can see that none of hypocenters entered inside the reported high velocity structure. At whole, in the depth range from 0 to 5 km the number of seismic events is smaller than their number in the depth range from 5 to 10 km. Figure 4 demonstrates distribution of local earthquakes that occurred during the expedition 1986 - 1989 and were used in this study as initial data set.

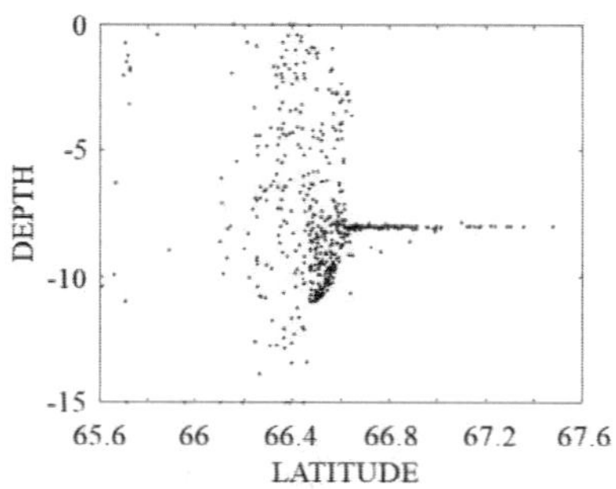

Figure 4. Distribution of seismicity versus depth.

The depth range of seismic energy accumulation can be estimated as 8-12 km. It is deeper than location of the gas area. The limit 8 km best corresponds to the locking depth $6.3^{+1.7}_{-1.2}$ km that was found for Tjornes fracture zone on the base of the surface deformation measurements by using GPS stations and processing the GPS data by standard weighted least-square approach [15]. The GPS data were collected from 2006 till 2010. In assumption of authors [15] the combined model of the fracture zone with a plate boundary has few segments, which "move freely below their locking depth, but are fully locked above". This model is in agreement with the MGE result. High velocity anomalous structure surrounding the gas field can lock the tectonic process above shallow layer 0-5 km of depth.

4. Conclusion

The obtained MGE result and its interpretation have at least important practical meaning for the industry challenge connected with underground gas storages. Note that there is no complete analogy between the considered reservoirs of gas accumulations in sediments covering the sea floor and the underground gas storage on land. At the same time, there is correlation between values of the diffusion coefficients for methane in geothermal underground formations and rocks. In geothermal formations containing methane under high pressure values of the diffusion coefficient vary from 2.37×10^{-9} to 6.02×10^{-9} cm^2/s at the temperature range 26.85 - 226.85 C° [16]. These values are comparable with the methane diffusion coefficients having the order of 10^{-9} cm^2/s for samples of sedimentary rocks (mudstones, quartz sandstones and siltstones, lime stones) at a temperature of 20 C° under natural humidity [17].

Based on the revealed circumstances of the existence of a complex system of gases in nature, we can conclude about conditions that are necessary for the functioning of underground gas storages. Such conditions are: 1) the presence of the protective barrier that is the high velocity seismic structure presented by dense rocks; 2) absence of intersection of the gas area with boundaries of the thalweg zones; 3) if the gas area is in a vicinity of the fault zone, the fault locking depth should be below the gas area.

We suppose that MGE is promising to solve systems in other industries, where it is necessary to overcome the influence of errors, including instrumental measurement errors, errors in parameterization of functions, and others.

Acknowledgments
We wish to thank Prof. Wolfgang R. Jacoby (Mainz University, Germany) for help in promotion of new approaches of seismic tomography, for fruitful discussion and useful ideas. Our thanks go to DAAD foundation for support of part of the research conducted here. The research was carried out within the framework of the state projects № AAAA-A19-119013190038-2, № AAAA-A19-119101690016-9 and № 10.331-17.

References
[1] Lantéri H and Theys C 2005 Restoration of astrophysical images - the case of Poisson data with additive Gaussian noise *EURASIP J. Appl. Signal Process* **15** 2500–2513
[2] Paige C C and Saunders M A 1982 LSQR: an algorithm for sparse linear equations and sparse least squares *ACM Trans. Math. Soft.* **8(1)** 43–71
[3] Ogbuobiri F C, Tinney M F and Walker J W 1970 Sparsity directed decomposition for Gausson Elimination on matrixes *IEEE Trans.* **PAS-89(1)** 141–150
[4] Skala V 2013 Modified Gaussian elimination without division operations *ICNAAM 2013, AIP Conf. Proc.* **1558** 1936-1939
[5] Tikhonov A N 1965 On ill-posed problems of linear algebra and a stable method for their solving *Soviet Mathematics Doklady* **163(3)** 591–594
[6] Anconellia B, Berteroa M, Boccaccia P, Carbilletb M and Lanterib H 2007 Iterative methods for the reconstruction of astronomical images with high dynamic range *Journal of Computational and Applied Mathematics* **198** 321–331
[7] Smaglichenko T A 2011 Modification of Gaussian elimination for the complex seismic data *Numerical Analysis and Applied Mathematics, ICNAAM 2011, AIP Conference Proceedings, American Institute of Physics 978-0-7354-0956-9/S30.00* **1389** 1003–1006
[8] Smaglichenko T A, Horiuchi S and Takai K 2012 A differentiated approach to the seismic tomography problem: Method, testing and application to the Western Nagano fault area (Japan) *International Journal of Applied Earth Observations and Geoinformation* **16** 27–41
[9] Nolet G, Montelli R and Virieux J 1999 Explicit, approximate expressions for the resolution and a posteriori covariance of massive tomographic systems *Geophys. J. Int.* **138** 36–44
[10] Leveque J J, Rivera L and Wittlinger G 1993 On the use of the checker-board test to assess the resolution of tomographic inversions *Geophys. J. Int.* **115** 313–318
[11] Scholten J et al 2000 Hydrothermalism at the Kolbeinsey Ridge, Iceland *Technical Cruise Report PO 253, University of Kiel* 22–43
[12] 2017 *Google Earth Pro* **7.3.0.3832**
[13] Sokolova N V 2019 Influence of hydro and litho dynamic flow on the character of modern erosion division of territories *International Research Journal* **10, 1(88)** 60–64
[14] *U.S.Geological Survey* USGS Earthquake hazard program
[15] Metzger S, Jonsson S and Geirsson H 2011 Locking depth and slip-rate of the Husavik Flatey fault, North Iceland, derived from continuous GPS data 2006–2010 *Geophys. J. Int.* **187** 564–576
[16] Clauser C 2006 Geothermal Energy Renewable Energies *Landolt-Börnstein, Group VIII. Advanced Materials and Technologies* **3** (Heidelberg-Berlin: Springer) pp 480–595
[17] Antonov P L 1970 Results of the investigation of diffusion permeability of sedimentary rocks for hydro- carbon gases *Trudy VNIIYaGG* **8** 51–59

IOP Conf. Series: Materials Science and Engineering **927** (2020) 012015 doi:10.1088/1757-899X/927/1/012015

Stability of the collinear point L_1 in the planar restricted photogravitational three-body problem in the case of equal masses of primaries

Boris S Bardin[1,2] **and Andrey N Avdyushkin**[1]
[1] Moscow Aviation Institute (National Research University), 125993, Russia, Moscow, Volokolamskoe Shosse, 4
[2] Mechanical Engineering Research Institute of the Russian Academy of Sciences, 101990 Russia, Moscow, M. Kharitonyevskiy Pereulok, 4

E-mail: bsbardin@yandex.ru, avdyushkin.a.n@yandex.ru

Abstract. The planar restricted photogravitational three-body problem is considered. The motion of an infinitesimal small body (particle) under the Newtonian gravitational attraction of two bodies (primaries) is studied. It is supposed that primaries having equal masses move in weakly elliptic orbits and act on the particle with radiation pressure repulsive forces. Using the method of normal forms, the nonlinear stability study of the collinear point L_1 is performed and rigorous conclusions on stability in the sense of Lyapunov or formal stability are obtained.

1. Introduction

The planar restricted photogravitational three-body problem deals with the motion of a small body (particle) in the Newtonian gravitational field of two bodies, called primaries. The particle has negligibly-small mass, so it does not affect the motion of the primaries, which move in Keplerian orbits about their barycenter. Besides of gravitational attraction the primaries act on the particle with radiation pressure repulsive forces. It is known that the remarkable Euler solutions (the collinear points) of the classical three-body problem exists as well in the photogravitational three-body problem. The problem of existence and stability of collinear solutions in the photogravitational three-body problem has been considered in papers [1-7]. Most of stability studies have been performed in linear approximation. In this paper, it is supposed that the primaries has equal masses and act on the particle with equal forces of the radiation pressure. In this case, the collinear point L_1 is located in barycenter of the three-body system. We perform nonlinear stability analysis of L_1.

2. Problem statement and equations of motion

To describe the motion of the particle P we introduce a rotating coordinate system $Oxyz$ with the origin in the barycenter O. The axis Ox passes through the primaries P_1 and P_2, axis Oz is perpendicular to the plane of bodies motion and the axis Oy complements the coordinate system to right hand and orthogonal one (see Figure 1).

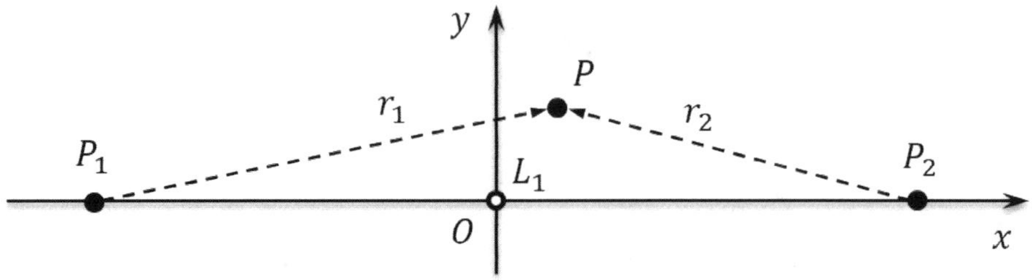

Figure 1. Coordinate system.

In what follows we use coordinates ξ, η, which are introduced by means of formulae

$$x = r\xi, \qquad y = r\eta, \qquad r = \frac{p}{1 + e\cos\nu}, \tag{1}$$

where p and e are parameter and eccentricity of the primaries orbits.

The equations of motions of the particle can be written in the Hamiltonian canonical form with the following Hamiltonian

$$H = \frac{1}{2}\left({p_\xi}^2 + {p_\eta}^2\right) + p_\xi\eta - p_\eta\xi + \frac{(\xi^2 + \eta^2)e\cos\nu}{2(1 + e\cos\nu)} - \frac{Q}{2(1 + e\cos\nu)}\left(\frac{1}{r_1} + \frac{1}{r_2}\right). \tag{2}$$

where Q is the coefficient of mass reduction $Q \in [0; 1]$. In the limiting case $Q = 1$ the radiation pressure forces are absent, that is we have the classical three body problem.

The distances between particle and primaries reads

$$r_1 = \sqrt{\left(\xi + \frac{1}{2}\right)^2 + \eta^2}, \qquad r_2 = \sqrt{\left(\xi - \frac{1}{2}\right)^2 + \eta^2}.$$

The equations of motion admit the trivial solution $\xi = \eta = p_\xi = p_\eta = 0$, which corresponds to the collinear point L_1. Let us expand the Hamiltonian in the power series with respect to ξ, η, p_ξ, p_η.

$$H = H_2 + H_4 + \cdots, \tag{3}$$

where

$$H_2 = \frac{1}{2}\left({p_\xi}^2 + {p_\eta}^2\right) + p_\xi\eta - p_\eta\xi + \frac{e\cos\nu - 16Q}{2(1 + e\cos\nu)}\xi^2 + \frac{e\cos\nu + 8Q}{2(1 + e\cos\nu)}\eta^2, \tag{4}$$

$$H_4 = \frac{-32Q\xi^4 + 96Q\xi^2\eta^2 - 12Q\eta^4}{1 + e\cos\nu}. \tag{5}$$

Supposing that the orbit eccentricity of the primaries is small ($e \ll 1$) we perform nonlinear stability analysis of the trivial solution and obtain rigorous conclusions on stability L_1. To this end we use the method of normal form and apply theorems on stability for Hamiltonian systems with two degrees of freedom.

3. Linear stability analysis

We start the stability study from the analysis of the linear system with Hamiltonian (4). Taking into account that $e \ll 1$ we expand (4) in a convergent series in powers e

$$H_2 = H_2^{(0)} + eH_2^{(1)} + e^2H_2^{(2)} + \cdots. \tag{6}$$

If the orbits of primaries are circular ($e = 0$), then the linear system is autonomous. Its characteristic equation reads

$$\lambda^4 - 8Q\lambda^2 + 2\lambda^2 - 128Q^2 + 8Q + 1 = 0\,. \tag{7}$$

For $Q \notin \left[\frac{1}{9}; \frac{1}{8}\right]$, the equation (7) has roots with nonzero real parts. It means that the trivial solution is unstable. If $Q \in \left(\frac{1}{9}; \frac{1}{8}\right)$, then the equation (7) has two pairs of pure imaginary simple roots: $\pm i\omega_1$, $\pm i\omega_2$, where $\omega_1 = \left(1 - 4Q + 4\sqrt{9Q^2 - Q}\right)^{1/2}$, $\omega_2 = \left(1 - 4Q - 4\sqrt{9Q^2 - Q}\right)^{1/2}$. In this case the trivial solution is stable in linear approximation. To study the stability of the linear system at $e \neq 0$ we perform canonical transformations, which bring Hamiltonian H_2 into the most simple (normal) form.

At first by means of linear autonomous canonical change of variables

$$\xi = a_{11}q_1 + a_{12}q_2, \quad \eta = b_{11}p_1 + b_{12}p_2, \quad p_\xi = b_{21}p_1 + b_{22}p_2, \quad p_\eta = a_{21}q_1 + a_{22}q_2, \tag{8}$$

we bring the Hamiltonian (6) into the following form

$$\widetilde{H}_2 = \frac{1}{2}\omega_1(q_1{}^2 + p_1{}^2) - \frac{1}{2}\omega_2(q_2{}^2 + p_2{}^2) + e\widetilde{H}_2^{(1)} + e^2\widetilde{H}_2^{(2)} + O(e^3), \tag{9}$$

where

$$a_{1i} = \frac{\kappa_i(\omega_i{}^2 - 8Q + 1)}{\omega_i}, \qquad a_{2i} = \frac{\kappa_i(-\omega_i{}^2 - 8Q + 1)}{\omega_i},$$

$$b_{1i} = 2\delta_i\kappa_i, \qquad b_{2i} = \delta_i\kappa_i(\omega_i{}^2 - 8Q - 1), \qquad \delta_1 = 1, \qquad \delta_2 = -1,$$

$$\kappa_i = \left(\frac{\omega_i}{\omega_i{}^4 - 16Q\omega_i{}^2 + 64Q^2 + 2\omega_i{}^2 + 16Q - 3}\right)^{\frac{1}{2}}, \qquad i = 1,2\,.$$

If at least one of the resonance relations holds: $2\omega_1 = n_1$, $2\omega_2 = n_2$ or $\omega_1 - \omega_1 = k$ (where n_1, n_2 and k are some integers), then for $e \neq 0$ in the system with Hamiltonian (9) a parametric resonance leading to instability takes place. The above resonance relations are satisfied at one internal point of the interval $\left[\frac{1}{9};\frac{1}{8}\right]$ and at its borders. In particular, at the point $Q = \frac{5+\sqrt{97}}{128}$, the resonance $2\omega_2 = 1$ is realized, on the left boundary of this interval (where $Q = 1/8$), the resonance relation $\omega_2 = 0$ takes place, and on its right boundary (where $Q = 1/9$) the relation $\omega_1 = \omega_2$ is satisfied. Thus, in the plane of the parameters Q and e the instability regions emanate from points with the coordinates $\left(\frac{1}{8}, 0\right)$, $\left(\frac{5+\sqrt{97}}{128}, 0\right)$, $\left(\frac{1}{9}, 0\right)$. If the eccentricity value is small, then the boundaries of these regions can be obtained analytically in the form of convergent series in powers of e.

$$Q = Q_0 + eQ_1 + e^2Q_2 + \cdots, \tag{10}$$

where one of the above resonance values $\frac{1}{8}$, $\frac{5+\sqrt{97}}{128}$ or $\frac{1}{9}$ should be set as Q_0.

In order to determinate the coefficients Q_j ($j = 1,2,\ldots$) we normalize non-autonomous part of Hamiltonian (9). The form of the normalized Hamiltonian depends on the type of resonance, that is why each resonant case must be considered separately.

To construct the boundary of the instability region emanating from the point $\left(\frac{5+\sqrt{97}}{128}, 0\right)$ we perform the linear canonical change of variables $q_i, p_i \to x_i, y_i$ ($i = 1,2$), which reduces the Hamiltonian (9) to the following normal form [8]

$$K_2 = \frac{1}{2}\omega_1(x_1{}^2 + y_1{}^2) + k_{0200}x_2{}^2 + k_{0101}x_2y_2 + k_{0002}y_2{}^2, \tag{11}$$

where $k_{0200}, k_{0101}, k_{0002}$ are convergent series in powers of e, whose coefficients are expressed in terms of Q_j ($j = 1,2,\ldots$). The above coefficients can be obtained in an explicit form by using, for example, the Deprit-Hori method described in [8,9,10].

The Hamiltonian (11) does not depend explicitly on ν. By taking into account that at the boundary of the instability region, the characteristic equation of a linear system with Hamiltonian (11) has two zero roots, we can calculate the coefficients Q_j ($j = 1,2,\ldots$) of series (10). As a result, we obtain the following expressions for the boundaries of the instability region (parametric resonance region) emanating from the point $\left(\frac{5+\sqrt{97}}{128}, 0\right)$.

$$Q_+ = \frac{5+\sqrt{97}}{128} + \frac{9}{32}\cdot\frac{\sqrt{97}+8}{19\sqrt{97}+97}e + \frac{27}{256}\cdot\frac{444341\sqrt{97}+9550369}{\left(19\sqrt{97}+97\right)^2\left(107\sqrt{97}+97\right)}e^2 + O(e^3),$$
$$Q_- = \frac{5+\sqrt{97}}{128} - \frac{9}{32}\cdot\frac{\sqrt{97}+8}{19\sqrt{97}+97}e + \frac{27}{256}\cdot\frac{444341\sqrt{97}+9550369}{\left(19\sqrt{97}+97\right)^2\left(107\sqrt{97}+97\right)}e^2 + O(e^3). \quad (12)$$

For values of the parameter Q closed to boundary points of the interval $\left(\frac{1}{9};\frac{1}{8}\right)$, the Hamiltonian (6) cannot be reduced to the form (9), but the above method used for constructing the boundaries of the instability regions in the form of series in powers of the small parameter e is also applicable in these cases. In particular, to construct the boundary of the instability region starting from the point $\left(\frac{1}{9},0\right)$, one can perform a linear canonical change of variables $\xi,\eta,p_\xi,p_\eta \to q_1,q_2,p_1,p_2$, which brings Hamiltonian (6) into the form

$$\hat{H}_2 = \frac{1}{2}({q_1}^2+{q_2}^2) + \frac{\sqrt{5}}{3}(q_1p_2 - q_2p_1) + e\hat{H}_2^{(1)} + e^2\hat{H}_2^{(2)} + O(e^3). \quad (13)$$

Then by a linear 2π-periodic in ν change of variables $q_i,p_i \to x_i,y_i$ $(i=1,2)$, the Hamiltonian (13) can be reduced to the following normal form, which does not depend explicitly on ν.

$$K_2 = \frac{1}{2}(1+2k_{2000})({x_1}^2+{x_2}^2) + k_{0020}({y_1}^2+{y_2}^2) + \frac{\sqrt{5}}{3}(x_1y_2 - x_2y_1) + k_{1010}x_1y_1 + k_{0110}x_2y_1. \quad (14)$$

On the basis of an analysis of the roots of the characteristic equation of the canonical system with the normalized Hamiltonian (14) we can obtain an analytic expression for the boundary of the instability region. It has the following form

$$Q_* = \frac{1}{9} - \frac{35}{2178}e^2 + O(e^3). \quad (15)$$

4. Nonlinear stability analysis

Rigorous conclusions on stability for $Q \in \left(\frac{1}{9};\frac{1}{8}\right)$ and $e \ll 1$ can be obtained by using theorem on stability of Hamiltonian systems with two degrees of freedom [11-13]. In order to apply this theorems it is necessary to construct canonical change of variables, which brings the Hamiltonian into the normal form up to terms of the fourth order with respect to canonical variables. The normalization of the quartic part of the Hamiltonian can be performed by means of a canonical near-identity transformation $x_i,y_i \to u_i,v_i$. The generating function for such a transformation has the form

$$S = x_iv_i + S^{(4)}(x_i,v_i,\nu) \quad (i=1,2), \quad (16)$$

where $S^{(4)}(x_i,v_i,\nu)$ is a convergent power series, which starts from terms of order fourth. We have also used the Deprit-Hori method to calculate the coefficients of the generating function (16) and normal form of the Hamiltonian up terms of fourth order.

In nonresonant case the Hamiltonian normal form can differ from ones in resonant cases. In particular, the normalization as well as the stability study in cases of resonances up to fourth order should be performed separately from the general case.

At first we suppose that the frequencies of the linear system do not satisfy a resonant relation up to fourth order. That is, the inequality $n_1\omega_1 + n_2\omega_2 \neq N$ holds, where $n_1$and n_2 are integers satisfying the condition $|n_1| + |n_2| \leq 4$. Then the Hamiltonian normal form reads

$$H = \omega_1r_1 - \omega_2r_2 + c_{20}{r_1}^2 + c_{11}r_1r_2 + c_{02}{r_2}^2 + O\left((r_1+r_2)^{5/2}\right), \quad (17)$$

where the canonical variables φ_i, r_i are introduced by formulae $u_i = \sqrt{2r_i}\sin\varphi_i\,, v_i = \sqrt{2r_i}\cos\varphi_i$.

The most rigorous conclusions on stability can be obtained in the case of circular primaries orbits ($e = 0$). If $\Delta(Q) = c_{02}\omega_1^2 + c_{11}\omega_1\omega_2 + c_{20}\omega_2^2 \neq 0$, then it follows from the Arnold-Moser theorem [11-13] the stability in the sense of Lyapunov for the trivial solution of the system with Hamiltonian (17). In our problem the calculations have shown that

$$\Delta(\mathrm{Q}) = c_{20}{\omega_2}^2 + c_{11}\omega_1\omega_2 + c_{02}{\omega_1}^2 = -\frac{9(18432Q^4 - 2304Q^3 - 136Q^2 + 32Q - 3)}{4(9Q-1)(16Q+1)(8Q-1)} \tag{18}$$

The value Δ is not equal to zero in the interval $\left(\frac{1}{9};\frac{1}{8}\right)$. It can be seen on the plot of the function Δ, which is shown on Fig.2. Thus, if $e = 0$ and $\in \left(\frac{1}{9};\frac{1}{8}\right)$, then in nonresonant case the collinear point L_1 is stable in the sense of Lyapunov.

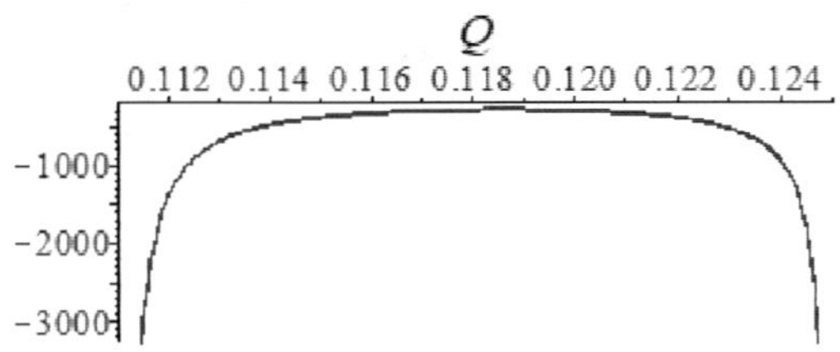

Figure 2. $\Delta(\mathrm{Q}) = c_{20}{\omega_2}^2 + c_{11}\omega_1\omega_2 + c_{02}{\omega_1}^2$.

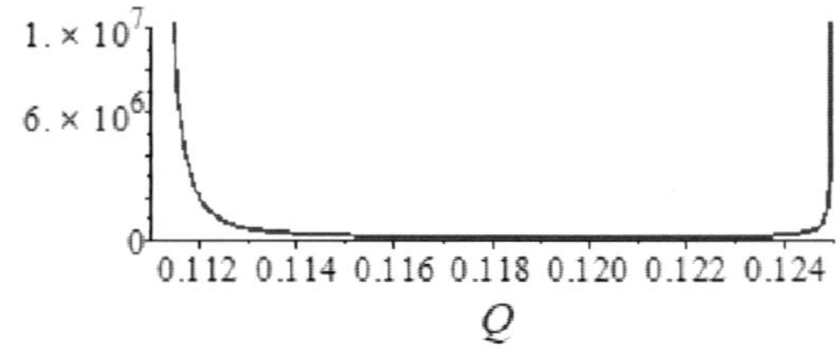

Figure 3. The dependence ${c_{11}}^2 - 4c_{20}c_{02}$ on Q.

If the orbits of the primaries are elliptic ($e \ll 1$), then the Hamiltonian is 2π-periodic function of independent variable ν. In this case we can use Glimm theorem on formal stability [14]. It states that the trivial solution of the system with Hamiltonian (17) is formally stable if the quadratic form $c_{20}{r_1}^2 + c_{11}r_1r_2 + c_{02}{r_2}^2$ is positive (or negative) definite for $r_1 \geq 0,\ r_2 \geq 0$. To show that the above condition of Glimm theorem is fulfilled in our problem we have to analyze the signs of values c_{20}, c_{02} and ${c_{11}}^2 - 4c_{20}c_{02}$. The calculation have shown

$${c_{11}}^2 - 4c_{20}c_{02} = -\frac{81(92160Q^4 - 16384Q^3 + 1920Q^2 - 96Q + 3)}{16(9Q-1)^2(16Q+1)(8Q-1)} + O(e) \tag{19}$$

The expressions for c_{20}, c_{02} are very awkward what is why we do not write down them here. The plots of c_{20}, c_{02} and ${c_{11}}^2 - 4c_{20}c_{02}$ at $e = 0$ are shown on Fig.3-5.

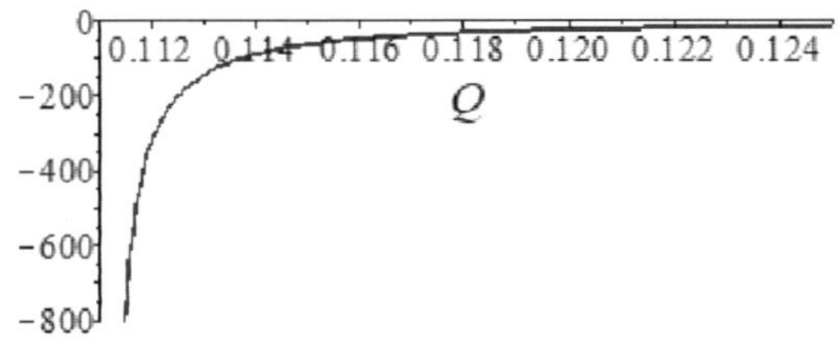

Figure 4. Dependence c_{20} on Q.

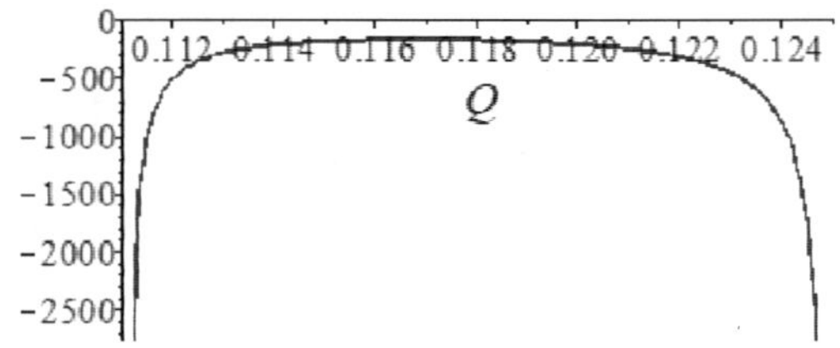

Figure 5. Dependence c_{02} on Q.

Hence, $c_{20} < 0$, $c_{02} < 0$ and ${c_{11}}^2 - 4c_{20}c_{02} > 0$ the condition of Glimm theorem is fulfilled. Thus, for $e \ll 1$ and $Q \in \left(\frac{1}{9};\frac{1}{8}\right)$ the collinear point L_1 is formally stable in nonresonant case.

If a resonance takes place, then the Hamiltonian normal form includes an additional (resonant) term.

The Hamiltonian expansion (3) do not includes terms of order three. That is why at the third order resonances disappear in considered approximation. In our problem the following fourth order resonances are possible: $4\omega_1 = 3$, $4\omega_2 = 1$, $2(\omega_1 - \omega_2) = 1$, $\omega_1 - 3\omega_2 = 0$. In fact, we have to study only the last one, because in the case of other resonances the resonant term is of order $O(e)$ and it cannot affect conclusion on stability obtained for nonresonant case.

In the case of resonance $\omega_1 - 3\omega_2 = 0$ the Hamiltonian normal form reads

$$H = 3\omega_2 r_1 - \omega_2 r_2 + c_{20} r_1^2 + c_{11} r_1 r_2 + c_{02} r_2^2 + B r_2 \sqrt{r_1 r_2} \sin(\varphi_1 + 3\varphi_2) + O\left((r_1 + r_2)^{5/2}\right) \tag{20}$$

In this case the conditions stability obtained in [10] reads: $|3\sqrt{3}B| < |C|$, here $C = c_{20} + 3c_{11} + 9c_{02}$. The calculations have shown that the above stability condition is fulfilled. It yields the stability of L_1 in the sense of Lyapunov at $e = 0$ and formal stability for $e \ll 1$.

5. Conclusions

The results of our study can be summarized as follows. In the case of circular orbits of the primaries, the collinear point L_1 is stable in the sense of Lyapunov for $Q \in \left(\frac{1}{9}; \frac{1}{8}\right)$ and it is unstable for $Q \notin \left[\frac{1}{9}; \frac{1}{8}\right]$. In the case of weakly elliptic orbits of the primaries ($e \ll 1$) the collinear point L_1 is formally stable for $Q \in \left(\frac{1}{9}; \frac{1}{8}\right)$ outside of the parametric resonance region defined by (12) and it is unstable for $Q \notin \left[\frac{1}{9}; \frac{1}{8}\right]$.

Acknowledgments

This work was supported by the grant of the Russian Scientific Foundation (project No. 19-11-00116) at the Moscow Aviation Institute (National Research University).

References

[1] Kunitsyn A L and Tureshbaev A T 1983 The collinear libration points in the photogravitational three-body problem *Sov. Astron. Lett.* **9** 228–229

[2] Kunitsyn A L and Tureshbaev A T 1985 Stability of the coplanar libration points in the photogravitational three-body problem *Sov. Astron. Lett.* **11** 391–392

[3] Kunitsyn A L and Tureshbaev A T 1985 On the collinear libration points in the photo-gravitational three-body problem *Celestial Mechanics* **35** 105–112

[4] Tkhai V N and Zimovshikov A S 2009 The Possible Existence of Cloud-like Clusters of Microparticlesat the Libration Points of a Binary Star *Astronomy Reports* **53** 552–560

[5] Zimovshikov A S and Tkhai V N 2010 Stability diagrams for a heterogeneous ensemble of particles at the collinear libration points of the photogravitational three-body problem *J. Appl. Math. Mech.* **74** 158–163

[6] Tkhai N V 2012 Stability of the collinear libration points of the photogravitational three-body problem with an internal fourth order resonance *J. Appl. Math. Mech.* **76** 441–445

[7] Bardin B S and Avdushkin A N 2018 *Stability analysis of an equilibrium position in the photogravitational Sitnikov problem* – AIP Conference Proceedings **1959**, 040002

[8] Markeev A P 2009 *Linear Hamiltonian systems and certain stability problems of satellite's motion about its center of mass* (Moscow, Izhevsk: Regular and Chaotic Dynamics, Institute of Computer Reasearch)

[9] Giacaglia G E O 1972 *Perturbation Method in Non-Linear Systems* (New York: Springer) 369 p.

[10] Markeev A P 1978 *Libration Points in Celestial Mechanics and Space Dynamics* (Moscow: Nauka) 312 p.

[11] Arnold V I 1961 The Stability of the Equilibrium Position of a Hamiltonian System of Ordinary Differential Equations in the General Elliptic Case *Soviet Math. Dokl.* **2** 247–249

[12] Arnold V I 1963 Small Denominators and Problems of Stability of Motion in Classical and Celestial Mechanics *Russ. Math. Surv.* **18** 85–191

[13] Siegel C L and Moser J K 1971 *Lectures on Celestial Mechanics* (New York: Springer) 290 p.

[14] Glimm J 1964 Formal Stability of Hamiltonian Systems *Communs. Pure Appl. Math.*. **17** 509–526

IOP Conf. Series: Materials Science and Engineering **927** (2020) 012016 doi:10.1088/1757-899X/927/1/012016

On resonant periodic motions close to conical precession of a dynamically symmetric satellite in a weakly elliptic orbit

B S Bardin[1,2] **and A A Antipov**[1]
[1]Moscow Aviation Institute (National Research University), 125993, Russia, Moscow, Volokolamskoe Shosse, 4
[2]Mechanical Engineering Research Institute of the Russian Academy of Sciences, 101990 Russia, Moscow, M. Kharitonyevskiy Pereulok, 4

E-mail: bsbardin@yandex.ru , antipant@mail.ru

Abstract. The motion of a dynamically symmetric rigid body (satellite) about its centre of mass in an elliptical orbit is considered. It is assumed that the eccentricity of the orbit is small and an external resonance caused by gravitational torque takes place. Periodic motions emanating from conical precession of the satellite are studied. Such periodic motions can be constructed in a form of convergent series of fractional powers of the eccentricity. The first terms of these series expansion have been calculated in an explicit form. The bifurcation phenomenon of the above periodic motions is investigated.

1. Introduction

We consider attitude motion of a satellite in the gravitational field. The satellite is modeled by a rigid body, which size is small enough compared with the size of the orbit of its mass center. In this case, it can be assumed that the satellite motion about its center of mass does not affect the motion of the mass center itself. Further, it is assumed that the center of mass of the satellite moves along a given Kepler orbit. If the orbit of the center of mass is circular, then there are partial stationary motions of the satellite, which are regular precessions. There are three types of regular precessions: conical, cylindrical, and hyperboloid [1-3].

In the case of conical precession axis of dynamical symmetry of the satellite is perpendicular to the velocity vector of its center of mass and forms a fixed angle with the normal to the plane of the orbit, while the satellite rotates uniformly around its axis of dynamic symmetry. In an elliptical orbit, conical precession is not possible, but the problem of periodic motions of a satellite close to its conical precession is of interest. If the orbit of the satellite mass center is weakly elliptic, i.e. the orbit eccentricity e is small enough ($e \ll 1$), then using the method of small parameter the periodic motions close to the conical precession can be obtained analytically. For nonresonant case, there exists a unique periodic motion of the above type, which can be constructed in a form of series with respect to integer powers of e [4]. In [5] the case of an external resonance in the motion of the dynamical symmetric satellite was studied. It was shown that depending on values of parameters there exist either one or three periodic motions close to the conical precession. These motions can be represented by series of fractional powers of e.

IOP Conf. Series: Materials Science and Engineering **927** (2020) 012016 doi:10.1088/1757-899X/927/1/012016

In this paper we study the case of the external resonance, which was not considered before. We use method developed in [6-8] to obtain the resonant periodic solutions in a form of series and analyze the bifurcation phenomena.

2. Mathematical Formulation of the Problem

To describe the motion of the satellite about its center of mass, we introduce two orthonormal reference frames:

– The orbital frame $OXYZ$. The axes OX, OY, OZ are directed along the radius vector of the center of mass O, the transversal and normal of the orbit, respectively.

– The principal axes frame $Oxyz$. The axes Ox, Oy, Oz are fixed in the satellite such that the axis Oz is directed along the symmetry axis of the inertia ellipsoid.

We choose three Euler angles ψ, θ, φ as the generalized coordinates defining the orientation of the principal axis frame with respect to the orbital frame. By introducing momenta associated with the coordinates ψ, θ, φ, the equations of motion of the satellite about its mass center can be written in canonical form with the following Hamiltonian [3]

$$H = \frac{p_\varphi^2}{2\alpha(1+e\cos\nu)^2} + \frac{p_\varphi^2 \mathrm{ctg}^2\theta}{2(1+e\cos\nu)^2} + \frac{p_\theta^2}{2(1+e\cos\nu)^2} + \frac{p_\psi^2}{2(1+e\cos\nu)^2\sin^2\theta} - \\ - \frac{p_\varphi p_\psi \cos\theta}{(1+e\cos\nu)^2\sin^2\theta} - p_\psi + \frac{3}{2}(1+e\cos\nu)(\alpha-1)\sin^2\theta\,\sin^2\psi\,, \tag{1}$$

where $p_\psi, p_\theta, p_\varphi$ are momenta dimensioned by the multiplier $\frac{A\omega_0}{(1-e^2)^{\frac{3}{2}}}$, ν is true anomaly, $\alpha = C/A$.

The angle φ is a cyclic coordinate and, hence, $p_\varphi = const$ is an integral of motion.

If the orbit of the center of mass is circular ($e = 0$), then the equations of motion allow a partial solution

$$\theta = \theta^*,\ \psi^* = \frac{\pi}{2};\ p_\theta{}^* = 0;\ p_\psi{}^* = 1 + 3(1-\alpha)\cos^2\theta^*,\ \theta^* \epsilon \left(0; \frac{\pi}{2}\right] \tag{2}$$

describing the satellite's conical precession. On the solution (2) we have $p_\varphi = (4-3\alpha)\cos\theta^*$.

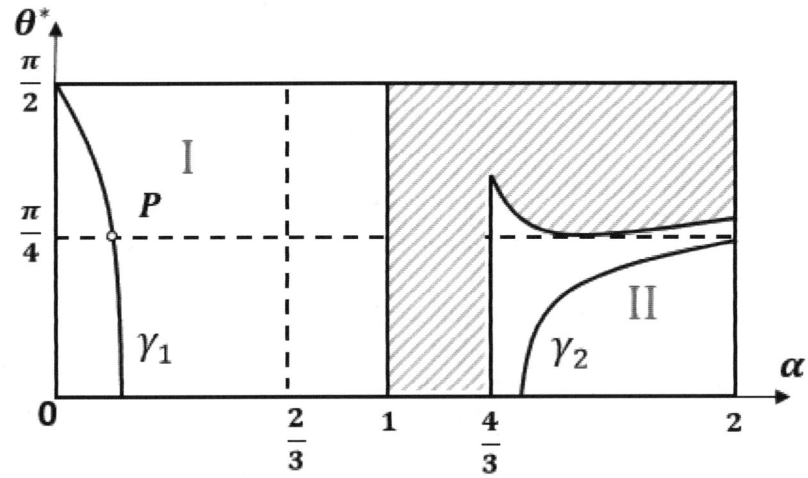

Figure 1. Stability and instability domains.

The stability of conical precession was studied in detail [1-3,9]. Figure 1 shows the results of these studies in the plane of the problem parameters. The instability domain is shaded, and the stability domains I and II are highlighted in white.

The conical precession is not possible in an elliptical orbit ($e \neq 1$). However, if the conic precession is stable (at least in the linear approximation), then 2π -periodic (with respect to ν) motions approaching at $e \to 0$ the conical precession can exist [4,5]. Following the generally accepted terminology, we will further refer to these motions as emanating from the conical precession at small e. As we have already mentioned in nonresonant case, when the frequencies ω_1, ω_2 of a linear system are not equal to integers, such motions have been constructed in a form of power series in e [4]. If at least one of the frequencies is equal to an integer, then the case of external resonance takes place in the system. In our problem the external resonance of the following types: $\omega_i = 1,2,3$ $(i = 1,2)$ is possible. In [5] the case of external resonance, when $\omega_1 = 1$ or $\omega_2 = 1$, was considered and 2π - periodic motions emanating from the conical precession have been constructed in a form of series in $\varepsilon = e^{\frac{1}{3}}$. In the plane of parameters α and θ^*, the above resonance takes place on dashed straight lines.

In what follows, we assume that one of the frequencies of a linear system is equal or close to 2. In Fig. 1, the solid lines γ_1, γ_2 denote the resonant curves, where $\omega_1 = 2$. The purpose of this work is to construct satellite's 2π-periodic motions emanating from the conical precession for the values of parameter in the neighborhood of the resonant curves γ_1, γ_2. We also discuss the phenomenon of bifurcation for such 2π-periodic motions.

3. Constructing the satellite's periodic motions emanating from the conical precession

For small values of the eccentricity, the system of equations of motion with the Hamiltonian (1) is a system close to the so-called the Lyapunov system. The theory of periodic solutions of such systems was developed in [10]. The methods for constructing periodic solutions of these systems were proposed in [10] and developed in [6-8] for resonant cases.

To apply these methods, we introduce the small parameter $\varepsilon = e^{\frac{1}{3}}$ and pass to new canonical variables ξ_i, η_i $(i = 1,2)$ by the following formulas

$$\theta = \theta^* + \varepsilon^2\xi_1, \quad p_\theta = \varepsilon^2\eta_1, \quad \psi = \psi^* + \varepsilon^2\xi_2, \quad p_\psi = p_\psi{}^* + \varepsilon^2\eta_2. \tag{3}$$

where θ^*, ψ^* and $p_\psi{}^*$ are values of canonical variables θ, ψ and p_ψ on conical precession (see formulas (2))

Then, we expand the Hamiltonian of the problem into a series of powers ξ_i, η_i $(i = 1,2)$ and perform the following linear change of variables $\xi_i, \eta_i \to q_i, p_i$ $(i = 1,2)$, which brings the autonomous quadratic part of the Hamiltonian into the most simple (normal) form

$$\begin{bmatrix}\xi_1\\ \xi_2\\ \eta_1\\ \eta_2\end{bmatrix} = N\begin{bmatrix}q_1\\ q_2\\ p_1\\ p_2\end{bmatrix}, \quad N = \begin{bmatrix} A_1 & A_2 & 0 & 0\\ 0 & 0 & \omega_1 B_1 & \omega_2\delta B_2\\ 0 & 0 & \omega_1 A_1 & \omega_2\delta A_2\\ 3(1-\alpha)B_1\sin^2\theta^* & 3(1-\alpha)B_2\sin^2\theta^* & 0 & 0\end{bmatrix}, \tag{4}$$

where $A_i = -\dfrac{\kappa_i(\omega_i^2 + 3\alpha - 3)}{\omega_i}$, $\quad B_i = \kappa_i(3\alpha - 2)\mathrm{ctg}\theta^*$,

$$\kappa_i{}^2 = \frac{\omega_i\delta^{i-1}}{(1 + 9(\alpha - 1)\cos^2\theta^*)\omega_i^2 - 3(\alpha - 1)(9\alpha^2 - 15\alpha + 8)\cos^2\theta^* - 3 + 3\alpha} \quad (i = 1,2)$$

$\delta = 1$ in the stability region I and $\delta = -1$ in the stability region II.

The frequencies of linear oscillations in the vicinity of conical precession read

$$\omega_1 = \left(Q + \frac{1}{2}\sqrt{R}\right)^{\frac{1}{2}} \quad \omega_2 = \left(Q - \frac{1}{2}\sqrt{R}\right)^{\frac{1}{2}}, \tag{5}$$

where

$$Q = -\frac{9}{2}(\alpha - 1)\alpha\sin^2\theta^* + \frac{9}{2}\alpha^2 - \frac{15}{2}\alpha + \frac{7}{2},$$

$$R = 81(\alpha - 1)^2\alpha^2\sin^4\theta^* - 162(\alpha - 1)(27\alpha^3 - 45\alpha^2 + 27\alpha - 8)\sin^2\theta^* + (9\alpha^2 - 15\alpha + 7)^2.$$

The Hamiltonian expansion into a power series in q_i, p_i $(i = 1,2)$ and e has the form

$$H = \frac{1}{2}\omega_1(q_1^2 + p_1^2) + \frac{1}{2}\delta\omega_2(q_2^2 + p_2^2) + \varepsilon^2 H_3^{(0)} + \varepsilon^4 H_4^{(0)} + \varepsilon H_1^{(1)} + \varepsilon^3 H_2^{(1)} + \varepsilon^4 H_1^{(2)} + O(\varepsilon^5), \tag{6}$$

where $H_k^{(m)}$ is a form of degree k, whose coefficients for $m = 0$ are constant values, and for $m > 0$ they are 2π-periodic functions of the true anomaly ν.

The next simplification of the Hamiltonian can be done by means of nonlinear near-identity transformation $q_i, p_i \to x_i, y_i$. The generating function for such a transformation has the form

$$S(q_i, y_i) = q_i y_i + \varepsilon^2 S_3(q_i, y_i) + \varepsilon^2 S_4(q_i, y_i) \quad (i = 1,2), \tag{7}$$

where $S_3(q_i, y_i)$ and $S_4(q_i, y_i)$ are forms of order three and four respectively.

The coefficients of forms $S_3(q_i, y_i)$ and $S_4(q_i, y_i)$ can be chosen by such a way that the new Hamiltonian takes the following normal form

$$\begin{aligned} H = &\frac{1}{2}\omega_1(q_1^2 + p_1^2) + \frac{1}{2}\delta\omega_2(q_2^2 + p_2^2) + \\ &+\varepsilon^4\left(\frac{c_{20}}{4}(q_1^2 + p_1^2)^2 + \frac{c_{11}}{4}(q_1^2 + p_1^2)(q_2^2 + p_2^2) + \frac{c_{02}}{4}(q_2^2 + p_2^2)^2\right) + \\ &+\varepsilon H_1^{(1)} + \varepsilon^3 H_2^{(1)} + \varepsilon^4 H_1^{(2)} + O(\varepsilon^5). \end{aligned} \tag{8}$$

We have used the Deprit-Hori method [3,11] to calculate the coefficients of the generating function (7) and normalized Hamiltonian (8).

Now we describe the algorithm for construction of 2π-periodic motions emanating from the conical precession. This algorithm based on methods developed in [6-8]. Taking into account that ω_1 is close to 2 we put

$$\theta^* = \theta_1(\alpha) + \varepsilon^4\theta_2 + O(\varepsilon^5), \tag{9}$$

where the first term in (8) satisfies the relation $\omega_1(\theta_1^*(\alpha), \alpha) = 2$ and has the following form

$$\theta_1^*(\alpha) = \arcsin 2\,(3\alpha^2 - 5\alpha + 1)^{\frac{1}{2}}(15\alpha^2 - 19\alpha + 4)^{-\frac{1}{2}}. \tag{10}$$

Then, by substituting (9) in (5) we have

$$\omega_1 = 2 + \varepsilon^4\omega_{12} + O(\varepsilon^5), \quad \omega_2 = \omega_{20} + \varepsilon^4\omega_{22} + +O(\varepsilon^5), \tag{11}$$

where

$$\omega_{12} = \frac{3}{2}\frac{\sin\theta_1 \cos\theta_1\,(15\alpha - 4)^2(\alpha - 1)\theta_2^*}{27\alpha^3 - 81\alpha^2 + 9\alpha + 4},$$
$$\omega_{20} = \left(\frac{27\alpha^3 - 81\alpha^2 + 69\alpha - 12}{15\alpha - 4}\right)^{\frac{1}{2}}. \tag{12}$$

The 2π -periodic motions can be constructed in the form of series with a small parameter ε

$$q_i = q_i^{(1)} + \varepsilon q_i^{(2)} + \cdots, \qquad p_i = p_i^{(1)} + \varepsilon p_i^{(2)} + \cdots. \tag{13}$$

Substituting (12) into the equations of canonical system with the Hamiltonian (8) and equating the coefficients with equal powers of ε, we have the following linear equations for $q_i^{(k)}, p_i^{(k)}$ $(i = 1,2;\ k = 1,2, \dots)$.

$$\frac{dq_i^{(k)}}{d\nu} = \omega_i p_i^{(k)} + G_i^{(k)}, \qquad \frac{dp_i^{(k)}}{d\nu} = -\delta^{i-1}\omega_i q_i^{(k)} + F_i^{(k)}. \tag{14}$$

The solutions of the equations (14) has the following form

$$\begin{aligned} q_i^{(k)} &= C_i^{(k)}\sin\omega_i\nu - D_i^{(k)}\cos\omega_i\nu + \varphi_i^{(k)}(\nu), \\ p_i^{(k)} &= C_i^{(k)}\sin\omega_i\nu + D_i^{(k)}\cos\omega_i\nu + \psi_i^{(k)}(\nu), \end{aligned} \tag{15}$$

where

$$\begin{aligned}\varphi_i^{(k)}(\nu) &= \cos\omega_i\nu\int_0^\nu\left(G_i^{(k)}\cos\omega_i\tau+F_i^{(k)}\sin\omega_i\tau\right)d\tau+\\&+\sin\omega_i\nu\int_0^\nu\left(G_i^{(k)}\sin\omega_i\tau-F_i^{(k)}\cos\omega_i\tau\right)d\tau,\\\psi_i^{(k)}(\nu) &= \sin\omega_i\nu\int_0^\nu\left(G_i^{(k)}\cos\omega_i\tau+F_i^{(k)}\sin\omega_i\tau\right)d\tau-\\&-\cos\omega_i\nu\int_0^\nu\left(G_i^{(k)}\sin\omega_i\tau-F_i^{(k)}\cos\omega_i\tau\right)d\tau.\end{aligned} \tag{16}$$

The function $G_i^{(k)}$ and $F_i^{(k)}$ are 2π-periodic with respect to ν and depend on constants $C_i^{(m)}, D_i^{(m)}$ $(i = 1,2;\ m = 1,2,\dots k-1)$. The constants $C_i^{(m)}, D_i^{(m)}$ should be chosen from the following conditions of 2π-periodicity of the solution (15)

$$\begin{aligned}I_i^{(k)} &= \frac{1}{2\pi}\int_0^{2\pi}\left(G_i^{(k)}\cos\omega_i\nu+F_i^{(k)}\sin\omega_i\nu\right)d\nu=0,\\J_i^{(k)} &= \frac{1}{2\pi}\int_0^{2\pi}\left(G_i^{(k)}\sin\omega_i\nu-F_i^{(k)}\cos\omega_i\nu\right)d\nu=0.\end{aligned} \tag{17}$$

From the equation of the first approximations, we have:

$$\begin{aligned}q_1^{(1)} = C_1\sin 2\nu - D_1\cos 2\nu\,,\qquad p_1^{(1)} = C_1\cos 2\nu + D_1\sin 2\nu\,,\\q_2^{(1)} = 0\,,\qquad p_2^{(1)} = 0\,.\end{aligned} \tag{18}$$

The constants C_1 and D_1 can be found form 2π -periodicity of the solution of the fourth order approximation

$$\begin{aligned}I_1^{(4)} &= c_{20}D_1\left({D_1}^2+{C_1}^2\right)+\omega_{12}D_1-\mu_1=0,\\J_1^{(4)} &= c_{20}C_1\left({D_1}^2+{C_1}^2\right)+\omega_{12}C_1+\mu_2=0.\end{aligned} \tag{19}$$

Calculations have shown that in our problem

$$\mu_1=\frac{9A_1\sin\theta_1^*\cos\theta_1^*(1-\alpha)+9(1-\alpha)B_1\sin^2\theta_1^*}{4},\quad \mu_2=0\,. \tag{20}$$

Thus, we have $C_1 = 0$ and D_1 is a root of the equation

$$c_{20}{D_1}^3+\omega_{12}D_1-\mu_1=0\,. \tag{21}$$

Now we can write down the satellite's 2π-periodic motions emanating from the conical precession in the original variables

$$\begin{aligned}\psi &= \frac{\pi}{2}+2B_1D_1\sin(2\nu)\,e^{\frac{2}{3}}+O(e),\\p_\psi &= 1+3(1-\alpha)cos^2\theta_1-3(1-\alpha)B_1\sin^2\theta_1 D_1\cos 2\nu\, e^{\frac{2}{3}}+O(e),\\\theta &= \theta_1-A_1D_1\cos(2\nu)\,e^{\frac{2}{3}}+O(e),\quad p_\theta=2A_1D_1\sin 2\nu\, e^{\frac{2}{3}}+O(e).\end{aligned} \tag{22}$$

Simple analysis of the equation (21) shows that if $\omega_{12} > -\frac{3}{4}\left(4\mu_1\sqrt{c_{20}}\right)^{\frac{2}{3}}$, then the equation (21) has one real root. By taking into account the formulas (10) and (12) we can rewrite the last inequality in terms of parameters α and θ_2

$$\theta_2>\theta_{2*}=-\frac{1}{4}\frac{\left(4\mu_1\sqrt{c_{20}}\right)^{\frac{2}{3}}(27\alpha^3-81\alpha^2+9\alpha+4)\sqrt{(15\alpha-4)(\alpha-1)}\sqrt{15\alpha^2-19\alpha+4}}{\sqrt{(3\alpha+1)\alpha}\sqrt{3\alpha^2-5\alpha+1}(15\alpha-4)^2(\alpha-1)}\,. \tag{23}$$

Thus if the inequality (23) holds, then there exists a unique 2π-periodic motion described by formulas (22).

In the case $\omega_{12} < -\frac{3}{4}\left(4\mu_1\sqrt{c_{20}}\right)^{\frac{2}{3}}$ the equation (21) has three real roots. It means that if the inequality (23) holds with the opposite sign, i.e. $\theta_2 < \theta_{2*}$, then there exist three 2π-periodic motions described by formulas (22).

In the limiting case $\omega_{12} = -\frac{3}{4}\left(4\mu_1\sqrt{c_{20}}\right)^{\frac{2}{3}}$ the equation (21) has one simple root D_{11} and one double real root D_{12} given by formulas.

$$D_{11} = -\frac{1}{2}\left(\frac{4\mu_1}{c_{20}}\right)^{\frac{1}{3}}, \quad D_{12} = \left(\frac{4\mu_1}{c_{20}}\right)^{\frac{1}{3}}. \tag{24}$$

Thus, the phenomenon of bifurcation takes place at $\theta_2 = \theta_{2*}$. That is, if $\theta_2 \to \theta_{2*} - 0$, then two of three 2π-periodic motions approach to each other until they coincide at $\theta_2 = \theta_{2*}$ and disappear when the parameter θ_2 pass through the value θ_{2*}.

4. Conclusions and final remarks

The results of our study can be summarized as is follows. If the frequency of proper oscillations of the dynamically symmetric satellite in a neighbourhood of conical precession is close to double frequency of the external perturbation caused be gravitational torque, then we observe either one or three 2π-periodic motions close to conical precession of a dynamically symmetric satellite. In the case of a weakly elliptic orbit the above motions are represented in the analytical form (22). The bifurcation phenomenon of the above periodic motions takes place at $\theta_2 = \theta_{2*}$.

We note that at $\theta_2 = \frac{\pi}{4}$, $\alpha = \frac{7-\sqrt{33}}{6}$ (point *P* in Figure 1.) the double external resonance takes place ($\omega_1 = 2,\ \omega_2 = 1$). In this case an additional study is necessary to construct the periodic motions.

Acknowledgments
This research was supported by the Russian Foundation for Basic Research (Project № 20-01-00637).

References

[1] Beletsky V V 1975 *Motion of a Satellite about Its Center of Mass in the Gravitational Field* (Moscow: Moscow State University Press) p 380
[2] Chernousko F L 1964 On stability of satellite's regular precession *Journal of Applied Math. and Mech.* **28** 155–157
[3] Markeev A P 2009 *Linear Hamiltonian Systems and Some Problems of Stability of the Satellite Center of Mass* (Moscow–Izhevsk: R&C Dynamics, Institute of Computer Science) 394 p.
[4] Sarychev VA 1965 Asymptotically stable stationary rotations of the satellite *Kosm. Issled.* **3** 667 – 673
[5] Chekhovskaia T N 1986 Resonant periodic motions of an axisymmetric satellite in an elliptical orbit *Kosm. Issled.* **24** 15 – 23
[6] Markeev A P, Chekhovskaya T N On resonant periodic solutions of Hamiltonian systems emanating from equilibrium position *J. Appl. Math. Mech.* **46** 27 – 33
[7] Bardin B S 1999 Bifurcation of periodic solutions of a system close to a Lyapunov system *J. Appl. Math. Mech.* **63** 517 – 525
[8] Bardin B S 2005 *Periodic solutions of systems close to Lyapunov systems* (Moscow: Moscow Aviation Institute Press) p 60
[9] Sokolsky, A.G. 1980 On problem of stability of regular precessions of symmetric satellite *Kosm. Issled.* **18** 698 – 706
[10] Malkin, I. G.: 1956, *Some Problems of the Theory of Nonlinear Oscillations*, (Moscow: Nauka) p 530
[11] Giacaglia G E O 1972 *Perturbation Methods in Non-Linear Systems* (New York: Springer) p 369

IOP Conf. Series: Materials Science and Engineering **927** (2020) 012017 doi:10.1088/1757-899X/927/1/012017

Multi-agent modelling of conflicts with uncertainty

A Karandeev[1], V Osipov[2] and V Baluta[1]
[1]*Plekhanov Russian University of Economics,* 36 Stremmyanny lane, 117997, Moscow, Russia
[2]*Keldysh Institute of Applied Mathematics,* 4 Miusskaya square 4, 125047, Moscow, Russia

E-mail: karalex755@gmail.com, vpo53@mail.ru, vbaluta@yandex.ru

Abstract. In the framework of the paradigm of the multi-agent modelling system, the features of the adaptive behaviour of agents under conditions of competition with increased uncertainty are often considered. Uncertainty is the result of increased dynamics in environmental conditions. Often these changes are the result of the actions of agents, but there are situations when the conditions change on their own. As a consequence, it would be a reasonable solution to investigate not the entire system at a certain point in time, but the dynamics of the behaviour of objects and participants in the phase space of the conflict state. This allows us to reduce the study of agent adaptation processes in a conflict environment to the analysis of the trajectories of achieving the goal in the state space. Given that the main parameters of the trajectory are the coordinates of the agent in the multidimensional state space at the time of the situation assessment, the degree of adaptation to the situation is characterized by the proximity of the current state to the optimal path leading to the achievement of the goal. The rationality of the ways to achieve the target state is evaluated according to the criteria of admissibility of resource costs and the degree of proximity to achieve the goal.

1. Introduction

The development of a rational sequence of actions in a conflict is one of the tasks that is important for the formation of an effective strategy. The complexity of this task increases as the level of uncertainty in a dynamic conflict increases, which is often caused by a lack of resources and time to get enough information for solutions developing.

The best way to analyse situations with uncertainty is to demonstrate emerging phenomena or generate unforeseen patterns is to model and simulate them [1]. The goal of mathematical modelling is to analyse the role and importance of various behavioural strategies in conflict modelling, highlighting the most effective actions in various conditions. Its purpose is to analyse the role and importance of various behavioural strategies in conflict modelling, while highlighting the most effective actions in various conditions, including uncertainty. Conflict modelling is a description of the dynamics of the interaction of the warring parties.

The development of a mathematical model of conflict begins with the formalization of the problem. The statement of the problem includes a description of the source data, accepted conditions and limitations. In any conflict it is possible to highlight certain areas of confrontation, to formulate goals that pursued by different parties, to characterize the state of the parties at the moment, and to describe the actions that lead to a change in the mutual position of the parties. Further, each of the areas of conflict development must be divided into parameters. Some of which may be uncertain at a given time or only general information is known. This parameters could characterize the state of the parties, the scales for setting the values of these parameters and the ranges of their possible changes must be introduced and the actions that change the parameters must be designated. The description of the area of possible changes in the parameters that describe the state of the conflicting parties makes it possible to identify the phase space for the development of the conflict. So, if the conflict model is created in the form of a phase space of opponents' state changes, then the possible paths of conflict development may be explored and the most rational actions in the presented situation may be pre-identified using the methods of mathematical modelling.

The developed research technology is based on numerical modelling of possible ways to actualize an antagonistic conflict. In the simulation model, each of the subjects of the conflict can be represented as an "intelligent" agent. The theory of agents or multi-agent systems [2] is a computer theory which seeks to apprehend the coordination of competing independent process. An agent is thus a computer process [3], which can be considered as autonomous since it is capable of adapting when its environment changes.

It is recognized that due to the specifics of the interaction, the agents are in different ways dependent on each other. For the successful achievement of the set goal each agent must take into account and promptly respond to any change in the situation associated with the opponent's actions, his decisions and actions must be situational. Nevertheless, the key problem is the presence of significant uncertainty in the knowledge of the intentions, plans, and capabilities of the opposing side. Furthermore, the uncertainty is of a dual nature, associated with the limited information on both sides of the conflict regarding each other in terms of both the noted and many other aspects. In view of the above, for analysis in the state space, it is necessary to take into account not only individual ways of achieving goals, but also interdependent ones.

With the optimal actions of each of the agents, the mutual position of their trajectories will tend to Nash equilibrium. That is such a mutual behaviour when any change of one's of the participants behaviour strategy can lead to an increase in his own losses.

The search for an equilibrium system of trajectories is a complex task, for the solution of which various approaches can be applied. In this paper, a co-evolutionary approach is considered - the process of iterative refinement of the optimal trajectories of movement to the target for each of the agents, taking into account changes in the parameters of the medium in the previous step.

2. State graph and price matrix

As mentioned above, each of the subjects of the confrontation can be represented in the form of a complex system that has certain resources and has its own goals. It is worth considering that many types of conflicts, such as social and political, cannot be strictly defined [4]. Because of this, it may be necessary to discard some non-significant parameters. It should be borne in mind that the intervals of state changes may not coincide with the intervals of the estimated time, since they can be considered in relation to individual characteristics and include the ability to instantly switch the parameters characterizing the states (transfer from one degree of readiness to another, inclusion of additional means of response and exposure, change some physical characteristics, replacement of equipment used, etc.).

Any change of the subject's position in the phase space can be associated with a change in some scalar quantity — the transition price, which refers to some conditional costs of this action. Moreover, for each of the parameters, its own transition price is set depending on the degree of significance of this parameter for solving the task, whereby this value itself can be dynamic and can depend on the current state of the subject.

The introduction of scalar parameters of the cost of transition makes it possible to use the principles of optimization in developing decisions on the implementation of actions, assuming that the most rational trajectory in the phase state space will correspond to the trajectory with the minimum cost of all transitions from the initial state to the target one. The opposite will mean increased risks to achieve the goal.

IOP Conf. Series: Materials Science and Engineering **927** (2020) 012017 doi:10.1088/1757-899X/927/1/012017

To interpret the results of operating with the phase space of conflict subjects, the concept of a price matrix [5, 6] is used, an approximate form of which is shown in Figure 1.

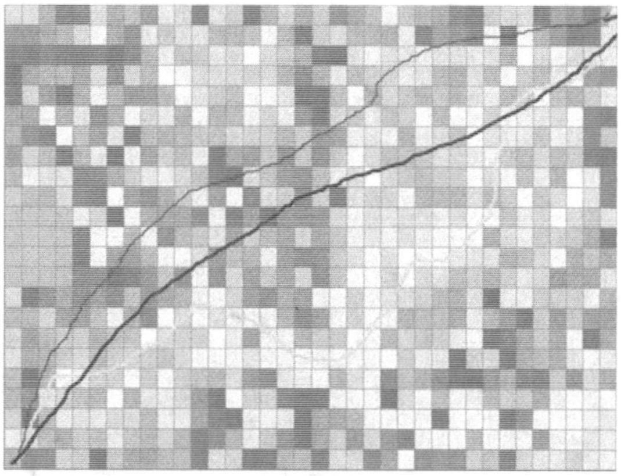

Figure 1. Price matrix example.

The meaning of the presentation through the price matrix allows us to set the cost of changing each of the characteristics that describe the state of the subject. In this case, the cells of the price matrix reflect the costs of the transition from the current state to its adjacent one ("transition price"). By displaying the trajectory of movement from one point to another along the cells of the matrix (lines in figure 1) and summing up the values in the cells that intersect with this line, we can get the total cost of overcoming the path along this trajectory. By comparing different trajectories, one or more optimal trajectories that require less cost for the corresponding transition can be selected. Each of the parameters has its own price scale and its own price matrix, and their combination allows to calculate the costs of changing the subject's states in multidimensional phase space. In computational problems the cell size of the price matrix can be reduced to the size visually displayed on the screen with one pixel, which enables to operate with the concept of a map.

Planning the dynamic behavior of an intelligent agent in a multi-agent environment consists in developing the rules of behavior $\pi(s, a)$ taking into account the evolution of the entire system. After clarifying the general model of the world $Q\pi(s, a)$, the intelligent agent changes its behavior $\pi(s, a)$ in order to choose the most optimal action.

$$a = arg \max_{a^*} Q^* \pi(s,\ a^*). \quad (1)$$

In this case, the choice of rules of behavior of the intelligent agent $\pi(s, a)$ is formed from a certain set of the most valuable possible actions a in the state of s. Thus, the technology for generating decisions of an intelligent agent is implemented as feedback by changing the behavior of $\pi(s, a)$ and the corresponding most valuable actions.

Changing the rules of behavior of an intelligent agent is responsible for the effectiveness of its adaptive actions. If the situation cannot be correlated with the actions of the corresponding intellectual agent, then the expansion of the domain for determining the fuzzy correspondence of the agent is required.

3. An example of solving a model problem

The statement of the problem includes a description of the source data, accepted conditions and limitations. The model of the situation is described by the following models: a model of obstacles to movement (with the allocation of zones that are insurmountable by subjects during movement), a model of obstacles to detect and implement impacts (blocking elements), a model of additional costs for overcoming sections (complexity of the relief), a model of the agents acting on each other (damage methods). In addition, the initial and final (target) positions of the agents are specified.

To simplify the modeling process, it is accepted that the first three models are the same for both sides and are due to the features of the relief on which the interaction of the subjects occurs. Impact models are asymmetric in terms of reachability, while the size of the damage caused by agents varies depending on the distance between them, but the value of the integral of the amount of the damage done over the area is the same for both participants. The initial and target position of the agents is given by intersecting diagonals, the distances between the points are also the same. Moreover, the speed of movement of agents is considered to be equal. Each of the agents is tasked with achieving the target position with the least cost (loss) and at the same time inflicting maximum damage to the other agent.

Obstacle models are represented by a single price matrix, which is shown in Figure 2.

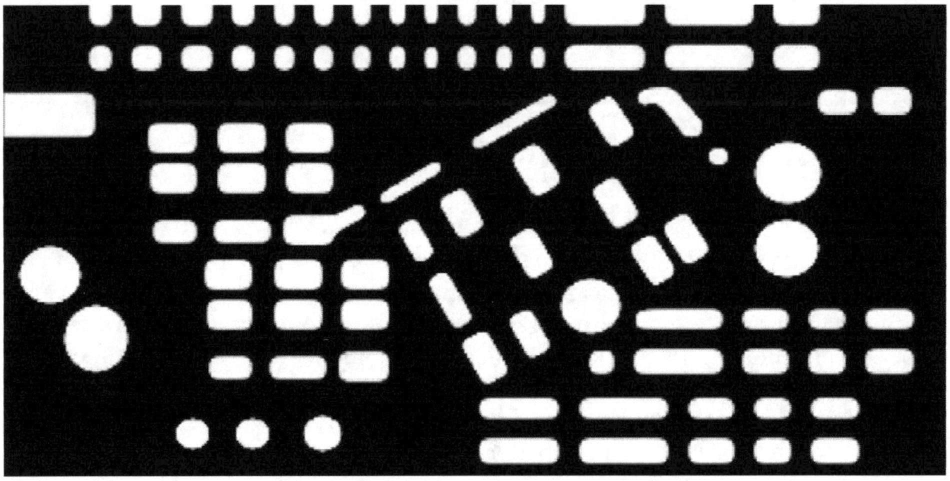

Figure 2. Initial obstacle model.

In the figure below irresistible sections are highlighted. They are also obstacles to means of detection and the effects of subjects on each other. As indicated in the conditions of the problem, a single price matrix is used for both opponents in this example. In dealing with real problems generally it is necessary to use different matrices, since the parties have different possibilities for moving in space. By aggregating the situation formed from the initial data, it is possible to construct a composite price matrix that shows the transition price on the obstacle map corresponding to the cost of overcoming areas of space, as shown in Figure 3.

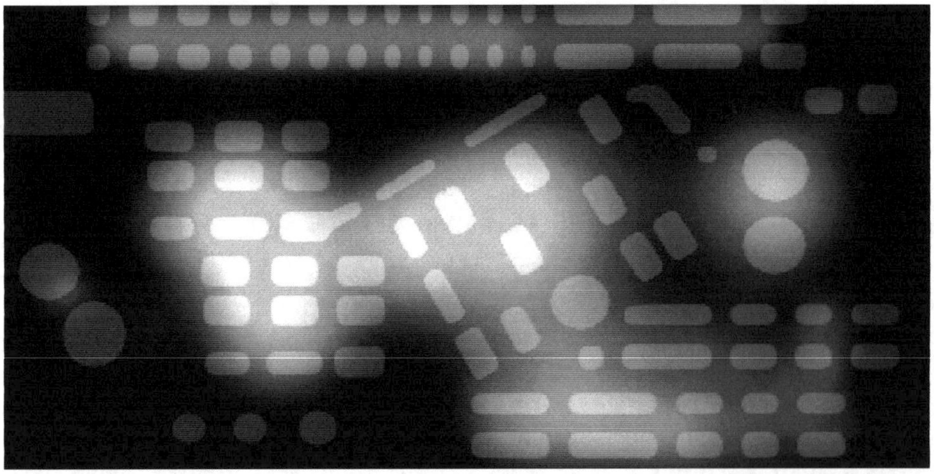

Figure 3. The integrated cost model for overcoming areas of space.

As an example, the transition price matrix on the obstacle map was generated depending on the areas of the absolute obstacle. The areas of absolute obstruction are shown in white, and the costs of moving around the territory are shown with a gray gradient in Figure 3. The lighter the value, the greater the cost of moving around the area. When working with real data, the transition price matrix on the obstacle map can be generated using area elevation maps and height maps, as well as taking into account additional information about the area.

The initial and final positions of the parties are presented in the form of cells on the same price matrix of the situation. These points are shown in Figure 4 in the form of arrows, the beginning of each of which shows the initial position, the end shows the target state, which should be achieved by the corresponding subject.

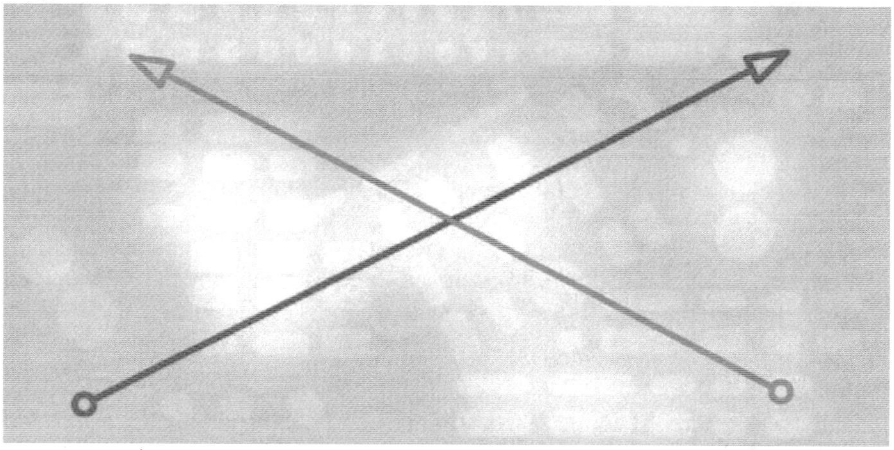

Figure 4. The initial and target position of opponents on the map.

To find the optimal path based on the above factors, the task of finding the optimal path for each of the participants in the conditions of the situation formed by the initial data without taking into account the conflict is solved firstly.

At the next stage a search for the rational trajectory of each of the participants is made, given the possibility of mutual damage. In this example, an asymmetric impact model is adopted. It is assumed that the capabilities of opponents in influencing each other are different. Opponent A has the ability to influence adversary B within the line of sight, and adversary B can influence adversary A along a ballistic trajectory in a certain radius. The manifestation of asymmetry is demonstrated by the circuit in Figure 5.

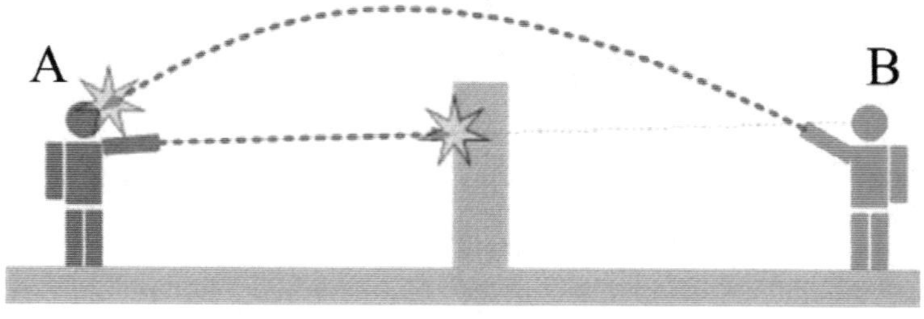

Figure 5. Asymmetric impact model.

The basis of the algorithm is the sequential improvement of the desired solution, broken down by iteration. The solutions obtained as a result of many runs in the form of trajectory bundles that satisfy the above requirements are shown in Figure 6, which shows the results of solving the problem of finding optimal interconnected trajectories for two opposing agents with an asymmetric pattern of actions on each other acting in the same conditions.

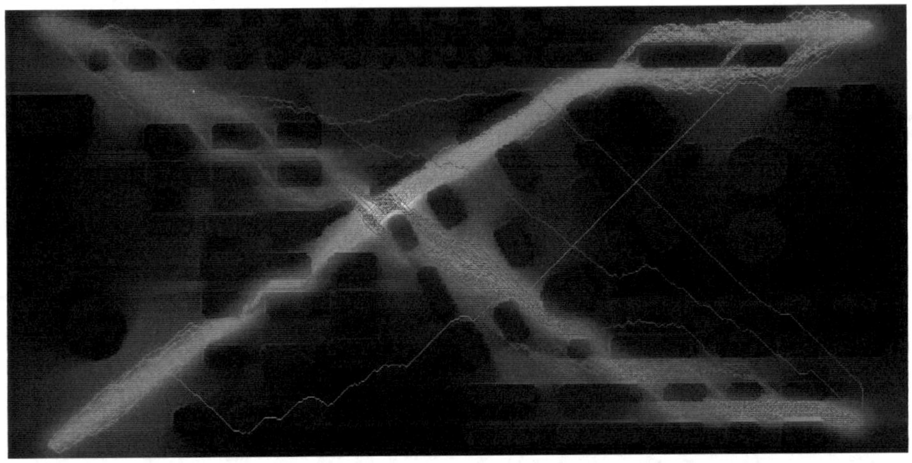

Figure 6. Solving the problem of finding optimal interconnected trajectories.

The effect of the interdependence of the trajectories is clearly shown in Figure 6. Taking into account the algorithm used to search for rational trajectories, that is, to search for the relative position at each moment of time with minimal losses to achieve the target state of each of the opponents in situations of conflict, it can be argued that the result reflects the Nash equilibrium for this task: any change in its trajectory by any of the participants will lead to an increase in its costs to achieve the target state.

Acknowledgments

The reported study was funded by RFBR according to the research project No. 18-00-00012 (18-00-00011) KOMFI.

References

[1] Klügl F, Oechslein C, Puppe F and Dornhaus A, 2014 Multi-Agent Modelling in Comparison to Standard Modelling. *Simulation News Europe* p 3-9

[2] Bousquet F, Barreteau O, Le C, Mullon C and Weber J, An environmental modelling approach. The use of multi-agent simulations, 1999.

[3] Ferber J, La kénétique: dés systems multi-agents à une science de l'interaction, 1994.

[4] Petukhov A, Malhanov A, Sandalov V and Petukhov V, *Mathematical Modeling of the Ethno-social Conflicts by Non-linear Dynamics*. Proceedings of the 7th International Conference on Simulation and Modeling Methodologies, Technologies and Applications, 2017

[5] Alexandrov V, Baluta V, Varykhanov S, Karandeev A and Rodnin Y 2018 Simulation of antagonistic conflicts in the third-order cybernetics paradigm. *Preprint* doi.org/10.20948/prepr-2018-86.

[6] Baluta V, Osipov V, Chetverushkin B and Yakovenko Y 2019 *Conceptual issues of model representation of conflicts*. Proceedings of the International Scientific Conference. (Nizhny Novgorod: NGASU, NICEFTI) p 27

IOP Conf. Series: Materials Science and Engineering **927** (2020) 012018 doi:10.1088/1757-899X/927/1/012018

Longitudinal-transverse ("breathing") vibration modes of orthotropic plates with and without account of its interaction with the surrounding acoustic media

V N Paimushin[1,2] **and R K Gazizullin**[1]

[1] Department of Structural Strength, Kazan National Research Technical University named after A.N. Tupolev, 10 K. Marx Street, Kazan 420111, Russian Federation

[2] Department of Computational Technologies and Computer Modeling, Kazan Federal University, 18 Kremlyovskaya Street, Kazan 420008, Russian

E-mail: vpaimushin@mail.ru

Abstract. Based on the previously obtained improved equations of motion, the problem of free cylindrical vibration of a hingedly supported plate surrounded by acoustic media is considered. Plate motion equations are obtained taking into account the internal friction of the plate material according to the Kelvin – Voigt model. The motion of acoustic media is described by two-dimensional generalized Helmholtz wave equations, compiled taking into account the internal dissipation of energy by introducing into consideration the complex sound velocity according to Skudrzyk. Based on the obtained solution, a study of resonant nonclassical shearless longitudinal-transverse ("breathing") forms of free vibrations has been carried out at various values of the physical-mechanical properties of the plate material and the parameter characterizing energy dissipation in the acoustic media surrounding the plate.

1. Introduction

A study of the stationary and non-stationary interactions of solid deformable bodies with surrounding acoustic media has been the subject of a fairly large number of works [1–5, etc.]. It should be emphasized that the deformation models of thin-walled structural elements (rods, plates and shells), which are correct and justified for the formulation of deformation mechanics problems under static types of loading, may turn out to be completely incorrect and insufficiently accurate for setting problems in dynamic deformation processes. To substantiate this statement, it seems appropriate to consider the mechanism of sound waves formation that occurs during the process of concrete pile driving into the ground at building sites. As is known, upon longitudinal impact on the pile face end unsteady longitudinal waves of stresses and displacements are formed. If they are described by the classical motion equations of the straight rod theory based on the hypotheses of plane sections, then the strains and displacements formed in the transverse directions turn out to be connected with axial strains and displacements only through Poisson's ratios. Such dependencies, as is known, take place in the classical theory of rods together with the assumption that the normal stresses in the transverse directions are equal to zero at all points of the cross section. Studies conducted, in particular [6], show that, in fact, straight rods, in addition to bending forms, must have longitudinal-transverse ("breathing" [7]) forms of free vibrations. The implementation of the latter is always accompanied by a distortion of the flat cross-sectional shape, which does not remain flat under the dynamic deformation process. These forms of dynamic deformations due to displacements in the transverse directions of the side

IOP Conf. Series: Materials Science and Engineering **927** (2020) 012018 doi:10.1088/1757-899X/927/1/012018

faces of the piles form sound waves propagating in the transverse directions to the surrounding air (acoustic medium), which a person hears even at a sufficiently large distance from the pile driven into the ground.

The described example convincingly shows that in case of end face impact on a straight rod or the same loading on plates and shells, as well as at their forced vibrations on average frequencies the degree of accuracy of the classical and majority of refined versions of the motion equations of rods, plates and shells theory (especially those made of composite materials having small values of elastic moduli in the transverse directions and shear moduli) are insufficient for the formulation of sound wave emission problems. Therefore it is necessary to develop such refined models and construct such motion equations that would have the necessary degree of accuracy and richness for the formulation of the corresponding problems of acoustoelasticity.

In connection with the foregoing, in this paper we consider the simplest problem of the class described above, associated with the study of the longitudinal-transverse forms of free cylindrical vibrations of orthotropic plates, taking into account its interaction with acoustic media surrounded on both sides. In this formulation refined equations of motion [6, 8] are used compiled taking into account the internal friction of the plate material according to the Kelvin–Voigt model. The motion of acoustic media is described by the two-dimensional generalized Helmholtz wave equations [9], compiled taking into account the internal energy dissipation by introducing into consideration the complex sound velocity according to Skudrzyk [10].

2. Statement and solution of the problem

Consider a plate of orthotropic material having dimensions a and $2h$ along the axes $x_1 = x$ and $x_3 = z$. Small free vibrations of the plate will be described by motion equations based on approximations of displacements in the directions of the axes x and z by functions ($\lambda_1 = \pi/(2h), -h \le z \le h$)

$$u = U - zW' + \psi \sin\lambda_1 z + \varphi' \frac{\cos\lambda_1 z}{\lambda_1} - p_3^{a\prime} \frac{\sin\lambda_1 z}{\lambda_1^2 E_3} - p_3^{c\prime} \frac{\cos 2\lambda_1 z}{4\lambda_1^2 E_3},$$
$$w = W + \varphi \sin\lambda_1 z + p_3^a \frac{\cos\lambda_1 z}{\lambda_1 E_3} - p_3^c \frac{\sin 2\lambda_1 z}{2\lambda_1 E_3}. \tag{1}$$

In equations (1) unknown functions U and φ describe longitudinal-transverse (“breathing”) forms, and functions W, ψ describe bending forms of motion. p_3^c and p_3^a are symmetric (in-phase) and antisymmetric (antiphase) components of the external surface load, equal to $p_3^c = (p_3^- + p_3^+)/2$, $p_a^c = (p_3^- - p_3^+)/2$.

Based on the results of [6,8] and the adopted approximation (1), it is possible to compose the equations of cylindrical motion of the plate bend. Of these equations, those that describe its longitudinal-transverse modes of vibration can be represented as

$$\tilde{E}_1^*\left(hU'' + \frac{\varphi'''}{\lambda_1^2} + \nu_{31}\varphi'\right) + \Omega^2\left(hU + \frac{\varphi'}{\lambda_1^2}\right) = 0,$$
$$2\left[\tilde{E}_1^*\left(U''' + \nu_{31}\lambda_1^2 U'\right) + \Omega^2 U'\right] + h\left[\tilde{E}_1^*\left(\varphi^{IV} + 2\nu_{31}\lambda_1^2\varphi'' + e_3\lambda_1^4\varphi\right) - \Omega^2\left(\lambda_1^2\varphi - \varphi''\right)\right] - \tag{2}$$
$$-\frac{1}{6\tilde{E}_3}\left[\tilde{E}_1^*\left(\frac{p_3^{cIV}}{\lambda_1^2} + 5\nu_{31}p_3^{c''} + 4e_3\lambda_1^2 p_3^c\left(4 - 3\nu_{13}\nu_{31}\right)\right) - \Omega^2\left(4p_3^c - \frac{p_3^{c''}}{\lambda_1^2}\right)\right] = 0.$$

where

$$\tilde{E}_1^* = \tilde{E}_1/(1-\nu_{13}\nu_{31}),\ e_3 = \tilde{E}_3/\tilde{E}_1,\ \Omega^2 = \rho\omega^2,\ \tilde{E}_j = E_j(1+i\delta_j/\pi),\ j=1,3,$$

and through δ_j the logarithmic decrements of oscillations of the plate material are indicated.

Let us assume that the plate is surrounded on both sides by acoustic media "1" and "2", occupying half-spaces V_1 and V_2 bounded by the plane $z=0$. As a result of harmonic vibrations of the plate in the spaces V_k $(k=1,2)$, sound waves with pressures p_1 and p_2 will be formed. To determine the pressures $p_3^- = p_1(z=0)$, $p_3^+ = p_2(z=0)$ applied to the boundary planes of the plate, it is necessary to find solutions of the generalized Helmholtz wave equations [9]

$$\tilde{c}^2\left(\Phi_{k,xx}+\Phi_{k,zz}\right)-\ddot{\Phi}_k = 0;\ k=1,2, \tag{3}$$

These equations are written relative to the potentials of velocities Φ_1 and Φ_2 with account of energy dissipation in acoustic media [9] by introducing into consideration the complex sound velocities $\tilde{c} = c(1+i\eta)$ by Skudrzyk [10] (i is an imaginary unit). The pressures and velocity components in half-spaces V_1, V_2 through functions Φ_k are expressed by dependencies (ρ is density of media)

$$p_k = -\rho\dot{\Phi}_k,\quad v_x^k = \Phi_{k,x},\quad v_z^k = \Phi_{k,z};\ k=1,2, \tag{4}$$

In the case of continuous interaction of acoustic media with the plate the following conditions must be satisfied (by virtue $W=0$ of the problem under consideration)

$$-\dot{\varphi} = v_z^1\big|_{z=0},\quad \dot{\varphi} = v_z^2\big|_{z=0}. \tag{5}$$

In accordance with the assumption adopted above on the monoharmonic form of oscillations the representations $\Phi_k = \tilde{\Phi}_k e^{i\omega\tau}$ are used for functions Φ_k. For a hingedly supported plate at the $x=0$ and $x=a$ edges, the solutions of equations (2) will be

$$U = \tilde{u}_m e^{i\omega\tau}\cos I_m x,\ \varphi = \tilde{\varphi}_m e^{i\omega\tau}\sin I_m x,\ I_m = m\pi/a,\ m=1,2,\dots. \tag{6}$$

By virtue of (6), solutions of equations (3) must be sought in the following form

$$\Phi_k = \tilde{\Phi}_k^m \sin I_m x\cdot e^{i\omega\tau}. \tag{7}$$

After substituting representations (7) into equations (2) with respect to one-dimensional functions $\tilde{\Phi}_k^m(z)$, we arrive at the equations

$$\tilde{\Phi}_{k,zz}^m - \left(æ^m\right)^2\tilde{\Phi}_k^m = 0, \tag{8}$$

where the quantities $æ^m$ determined by the formulas $æ^m = \sqrt{I_m^2 - k^{*2}}$, $k^{*2} = \omega^2/c^{*2}$ generally are complex numbers.

We will take solutions of equations (8) in the form [9, etc.] (B_1^m and A_2^m are integration constants)

$$\tilde{\Phi}_1^m = B_1^m e^{æ^m z},\quad \tilde{\Phi}_2^m = A_2^m e^{-æ^m z}. \tag{9}$$

corresponding to the propagation of the sound waves emitted by plate.

Using (9) and in accordance with relations (4) and (7), at $\left(\tilde{æ}_k^m\right)^2 > 0$ we can obtain the expressions

$$v_z^{1m}\big|_{z=0} = æ^m B_1^m \sin I_m x,\quad v_z^{2m}\big|_{z=0} = -æ^m A_2^m \sin I_m x, \tag{10}$$

$$p_1^m\Big|_{z=0} = -i\rho\omega B_1^m \sin I_m x,\ \ p_2^m\Big|_{z=0} = -i\rho\omega A_2^m \sin I_m x. \tag{11}$$

Let us note that at $\left(æ^m\right)^2 < 0$ in expressions (9) - (11) the quantity $æ^m$ should be replaced by $i\tilde{æ}^m$, where $\left(\tilde{æ}^m\right)^2 = \left|\left(æ^m\right)^2\right|$. Considering that modern research systems (for example, Matlab) provide the ability to perform operations with complex numbers, all the transformations below we give only for the case $\left(æ^m\right)^2 > 0$. The case $\left(æ^m\right)^2 < 0$ is taken into account automatically in the software implementation of solution as the quantity $æ^m$ can be complex.

Submitting functions (6) and (7) further to conditions (5) and using the composed expressions (10), we obtain

$$-i\omega\tilde{\varphi}_m \sin I_m x = æ^m B_1^m \sin I_m x,\ \ i\omega\tilde{\varphi}_m \sin I_m x = -æ^m A_2^m \sin I_m x. \tag{12}$$

These conditions will be satisfied if the following equalities are satisfied

$$B_1^m = -i\frac{\omega}{æ^m}\tilde{\varphi}_m,\ \ A_2^m = -i\frac{\omega}{æ^m}\tilde{\varphi}_m. \tag{13}$$

When they are substituted into relations (11), we obtain the dependences for calculating the pressures of sound waves emitted by the plate at the points of its boundary planes

$$p_k^m\Big|_{z=0} = \frac{-\rho\omega^2}{æ^m}\tilde{\varphi}_m \sin I_m x,\ p_3^c = p_k^m\Big|_{z=0}. \tag{14}$$

Therefore, in the case under consideration

$$p_3^c = \frac{p_3^- + p_3^+}{2} = -\frac{\rho\omega^2}{æ^m}\tilde{\varphi}_m \sin I_m x. \tag{15}$$

Let us note that for calculating the pressure values of sound waves p_1^m and p_2^m in media V_1 and V_2 at the found integration constants B_1^m and A_2^m the following relations are valid

$$p_1^m = -i\omega\rho B_1^m e^{æ^m z} \sin I_m x,\ \ p_2^m = -i\omega\rho A_2^m e^{-æ^m z} \sin I_m x. \tag{16}$$

Substituting further in the equations of motion representations (6) and relation (16), we arrive at a system of homogeneous algebraic equations of the following form

$$\left(\begin{bmatrix} a_{11} & a_{12} \\ a_{21} & a_{22} \end{bmatrix} - \Omega^2 \begin{bmatrix} b_{11} & b_{12} \\ b_{21} & b_{22} \end{bmatrix}\right)\begin{Bmatrix} \tilde{u}_m \\ \tilde{\varphi}_m \end{Bmatrix} = 0, \tag{17}$$

where

$$a_{11} = -E_1^* I_m^2 h,\ a_{12} = E_1^*\left(\nu_{31} I_m - \frac{I_m^3}{\lambda_1^2}\right),\ a_{21} = 2E_1^*\left(I_m^3 - \nu_{31}\lambda_1^2 I_m\right),$$

$$a_{22} = E_1^* h\left(I_m^4 - 2\nu_{31}\lambda_1^2 I_m^2 + e_3\lambda_1^4\right) - \frac{E_1^* K}{6E_3}\left(\frac{I_m^4}{\lambda_1^2} - 5\nu_{31} I_m^2 + 4e_3\lambda_1^2\left(4 - 3\nu_{13}\nu_{31}\right)\right),$$

$$b_{11} = -h,\ b_{12} = -\frac{I_m}{\lambda_1^2},\ b_{21} = 2I_m,\ b_{22} = h\left(\lambda_1^2 + I_m^2\right) - \frac{K}{6E_3}\left(4 + \frac{I_m^2}{\lambda_1^4}\right),\ K = -\frac{\rho\omega^2}{æ^m}.$$

Solving the obtained problem (17) the eigenfrequencies and their corresponding longitudinal-transverse vibrational modes of the plate can be determined taking into account the internal energy dissipation in the plate material as well as its aerodynamic interaction with the surrounding acoustic media. Note that in the case $K=0$ the eigenvalue problem (17) becomes linear and its solution determines the eigenfrequencies and vibration modes of the plate, taking into account only the internal energy dissipation in the plate material. To find a solution to the nonlinear problem in the case $K \neq 0$ an iterative algorithm has been developed. According to it at the first step of the iteration, the problem is solved in the zeroth approximation at $K=0$. In this case, its eigenvalues $\Omega_{(k)}^2 = \rho\omega_{(k)}^2$ (k is iteration number) are determined. Then we substitute this eigenvalues in (17). Further, at each step, a linear eigenvalue problem is solved with respect to $\Omega_{(k)}^2$ using the Matlab built-in *eig* function. Iterations are performed until the error $\Delta = \left|\left(\Omega_{(k)}^2 - \Omega_{(k-1)}^2\right)/\Omega_{(k)}^2\right|$ becomes less than the specified value.

Let us note that at $K=0$ and $\varphi \equiv 0$ the first equation of system (17) with accuracy $E_1 \approx E_1^*$ turns into the classical equation $E_1 U'' + \Omega^2 U = 0$ of longitudinal (axial) plate vibrations. Its solution of the form (6) for determination of $\Omega_{cl}^2 = \rho\omega_{cl}^2$ leads to the formula $\Omega_{cl}^2 = E_1 I_m^2$.

3. The results of the calculations and their analysis

On the basis of constructed solutions, there were calculations carried out for plates of geometrical sizes $a = 1\,\mathrm{m}$, $2h = 0.1\,\mathrm{m}$ for two variants of the physical and mechanical characteristics of the material. In the first variant, it is believed that the material is isotropic and has the following stiffness, inertial and damping characteristics

$$E_1 = E_3 = 40\,\mathrm{GPa},\ \nu_{13} = 0.3,\ \rho = 1500\,\mathrm{kg/m^3},\ \delta_1 = \delta_3 = 0.012\,.$$

In the second variant, it is assumed that the rod is made of composite material with the following mechanical characteristics and logarithmic decrements of vibrations

$$E_1 = 40\ \mathrm{GPa},\ E_3 = 5.9\ \mathrm{GPa},\ \nu_{13} = 0.3,\ \rho = 1500\ \mathrm{kg/m^3},\ \delta_1 = \delta_3 = 0.012\,.$$

In both cases, it is believed that the plate is surrounded on both sides by air having the parameters $\rho = 1.225\ \mathrm{kg/m^3}$ and $c = 340\,\mathrm{m/s}$. Tables 1 and 2 present the results of determining the frequencies of natural longitudinal vibrations $f_{cl} = \omega_{cl}/(2\pi)$ obtained for each vibration mode m for the first and second variant of plate material characteristics respectively. The tables also show the real parts of the longitudinal-transverse frequencies $f_{1,2} = \omega_{1,2}/(2\pi)$ of natural vibrations (determined by the solution of equations (17), having two roots for each given m), the corresponding parameters $\xi_{1,2}$ that characterize the imaginary parts of the complex vibration frequencies $\tilde{f}_{1,2} = f_{1,2}(1 + i\xi_{1,2})$, and the relations $\tilde{\varphi}_{1,2}/\tilde{u}_{1,2}$ characterizing the form of natural vibrations. The analysis of the results showed that account the aerodynamic interaction of the plate with the surrounding acoustic media does not affect the determined frequencies $f_{1,2}$, parameters $\xi_{1,2}$ and $\tilde{\varphi}_{1,2}/\tilde{u}_{1,2}$. Therefore, the results in the tables are given only for the case with account of the internal energy dissipation in the plate material.

Analyzing the results, one can see that the lowest determined frequencies correspond to practically pure longitudinal modes of vibration only for isotropic plates at low values m, because in this cases $\tilde{\varphi}/\tilde{u} \approx 0$. It should be noted that, at $\tilde{\varphi}/\tilde{u} \neq 0$ the determined eigenfrequencies of the oscillations (f_1 or f_2) substantially coincide with the frequencies f_{cl} in a rather wide range of changes in the wave number m. A change in the material parameter $e_3 = \tilde{E}_3/\tilde{E}_1$ leads to a noticeable change in the values of the real parts of the natural frequencies. For small e_3 taking place for plates made of fibrous composite materials the main ones are longitudinal-transverse ("breathing") forms of natural vibrations, characterized by distortion of a flat cross-sectional shape. It is important to note that with the geometrical and physical-mechanical parameters of the plate under consideration these frequencies lie in the range of audible sound frequencies, and precisely these forms of vibrations emit the sound waves in the surrounding media.

Table 1. Results obtained for isotropic material ($E_3 = 40$ GPa)

m	f_{cl}, Hz	f_1, Hz	ξ_1	$\tilde{\varphi}_1/\tilde{u}_1$	f_2, Hz	ξ_2	$\tilde{\varphi}_2/\tilde{u}_2$
1	2581.99	2604.99	0.00191	0.039	27035.92	0.00191	-12.049
2	5163.98	5203.254	0.00191	0.081	26948.82	0.00191	-5.971
3	7745.97	7786.648	0.00191	0.129	26820.93	0.00191	-3.918
4	10327.96	10343.61	0.00191	0.192	26679.50	0.00191	-2.867
5	12909.94	12855.67	0.00191	0.278	26566.27	0.00191	-2.211
6	15491.93	15290.6	0.00191	0.413	26543.23	0.00191	-1.747
7	18073.92	17589.82	0.00191	0.654	26704.07	0.00191	-1.385
8	20655.91	27188.64	0.00191	-1.083	19652.02	0.00191	1.184
9	23237.90	28164.16	0.00191	-0.831	21349.51	0.00191	2.923
10	25819.89	29704.87	0.00191	-0.637	22645.58	0.00191	3.610

Table 2. The results obtained for the composite material ($E_3 = 5.9$ GPa)

m	f_{cl}, Hz	f_1, Hz	ξ_1	$\tilde{\varphi}_1/\tilde{u}_1$	f_2, Hz	ξ_2	$\tilde{\varphi}_2/\tilde{u}_2$
1	2581.99	2878.62	0.00191	0.27461	16083.46	0.00191	-5.029
2	5163.98	5373.711	0.00191	0.655653	16813.02	0.00191	-2.314
3	7745.97	7114.734	0.00191	1.31994	18349.19	0.00191	-1.375
4	10327.96	20783.50	0.00191	-0.9261	8022.566	0.00191	2.705
5	12909.94	23882.17	0.00191	-0.68411	8423.61	0.00191	7.160
6	15491.93	27368.49	0.00191	-0.54018	8749.92	0.00191	-54.182
7	18073.92	31071.32	0.00191	-0.44653	9348.32	0.00191	-7.091
8	20655.91	34900.55	0.00191	-0.38108	10431.55	0.00191	-4.095
9	23237.90	38808.35	0.00191	-0.33278	12066.49	0.00191	-2.978
10	25819.89	42767.91	0.00191	-0.29563	14215.32	0.00191	-2.381

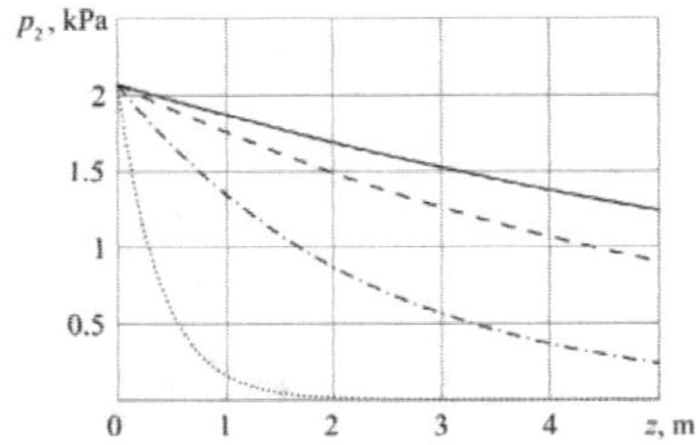

Figure 1. Change in sound pressure p_2 with increasing distance from the plate (at $f = 2878.62$ Hz)

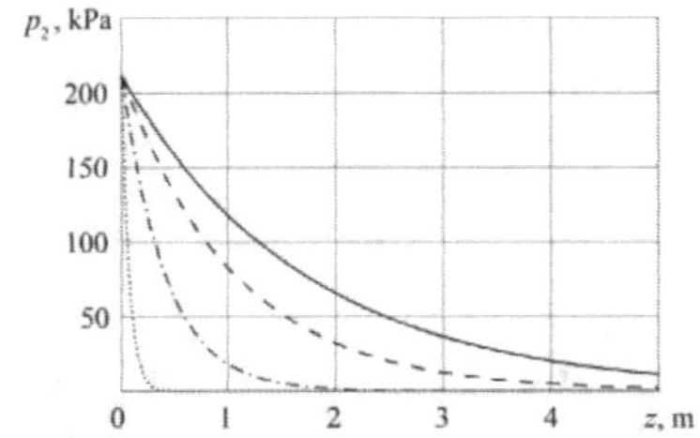

Figure 2. Change in sound pressure p_2 with increasing distance from the plate (at $f = 16083.46$ Hz)

We also analyzed the graphs of changes in the pressure of sound waves emitted by the plate in resonant modes of vibration along the studied longitudinal-transverse forms in the surrounding acoustic media. Fig. 1 and 2 show graphs of changes in the amplitude value of sound pressure p_2 along the coordinate z at $f_1 = 2878.62$ Hz (Fig. 1) and $f_2 = 16083.46$ Hz (Fig. 2) found for the composite plate considered above and at a given displacement $\tilde{u} = 1$ mm. The solid line indicates results obtained at $\eta = 0$, the dashed line, at $\eta = 0.005$, the dash-dotted line, at $\eta = 0.01$, the dotted line, at $\eta = 0.05$. It can be seen that the use of the generalized Helmholtz wave equation with introducing into consideration the complex sound velocities by Skudrzyk to take into account energy dissipation allows dramatically refine the laws of decreasing amplitudes of the sound wave pressures with increase of the distance from the plate.

Acknowledgment
This research was supported by a grant of the Russian Science Foundation (project No. 19-19-00058).

References

[1] Pretlove A J 1965 *J. Sound Vib.* **2(3)** 197
[2] Pretlove A J 1966 *J. Sound Vib.* **3(3)** 252
[3] Jeyaraj P, Ganesan N and Padmanabhan, C 2009 *J. Sound Vib.* **320(1)** 322
[4] Li X, Yu K, Han J, Song H and Zhao R 2016 *Int. J. Mech. Sci.* **199** 370
[5] D'Alessandro V, Petrone G, Franco F and De Rosa S 2013 *J. Sandw. Struct. Mater.* **15 (5)** 541
[6] Paimushin V N, Polyakova T V 2011 *J. Appl. Math. Mech.* **75(1)** 49
[7] Paimushin V N, Polyakova T V 2019 *Mech. Compos. Mater.* **55(5)** 575
[8] Paimushin V N, Polyakova T V, Polyakova N V and Gazizullin R K 2020 *Russ. Math.* **64(5)** 56
[9] Paimushin V N, Tarlakovskii D V, Firsov V A and Gazizullin R K 2020 *Z. Angew. Math. Mech.* e201900102
[10] Skudrzyk E 1971 *The foundations of acoustics* (Wien, New York: Springer–Verlag)

IOP Conf. Series: Materials Science and Engineering **927** (2020) 012019 doi:10.1088/1757-899X/927/1/012019

Through calculation method of fatigue damage

I S Nikitin, N G Burago, A D Nikitin and B A Stratula

Institute for computer aided design of RAS, 2-ya Brestskaya st., 123056, Moscow, Russia

E-mail: i_nikitin@list.ru

Abstract. A kinetic model of damage development under cyclic loading is proposed to describe the process of fatigue failure. To determine the coefficients of the kinetic equation of damage, the well-known criterion of multiaxial fatigue failure is used. A numerical method for calculating crack-like zones up to macro fracture is proposed. The results of fatigue experiments for various geometries and asymmetry coefficients were reproduced; the operability of the model and calculation algorithm were verified.

1. Introduction

In recent decades, entire classes of criteria have been constructed that relate the number of cycles before the initiation of fatigue damage (microcracks) with the amplitudes and maximum values in the cycle (or average) that characterize the uniform stress-strain state of the working part of the specimen in a fatigue test. A large number of stress-based criteria are based on a direct generalization of the S-N Wöhler-type curves described by Basquin-type relations [1], and based upon the results of fatigue tests. Reviews on this topic are given in [2-4].

In this paper we study the processes of fatigue damage zones development using the damage theory approach dating back to [5, 6]. In the application to the cyclic loading and fatigue failure problems this approach was applied in [7, 8]. A model for the development of fatigue failure based on the evolutionary equation for the damage function is proposed. The model parameters are determined for various fatigue failure modes – low-cycle and high-cycle fatigue (LCF, HCF).

The following scheme of the amplitude fatigue curve is used. Up to value $N \sim 10^3$ the regime of repeated-static loading with amplitude slightly differing from the static tensile strength σ_B is realized. Further the fatigue curve (Wöhler curve) describes the modes of the LCF-HCF up to $N \sim 10^7$ with an asymptotic exit to the fatigue limit σ_u.

It should be noted that at present the idea of an explicit division of the classic Wöhler branch into two parts – in fact, the LCF and the HCF. The boundary of this transition region is determined not by the value of N, but by the value of the loading amplitude equal to the yield strength of the material σ_T [9], since this changes the physical mechanism of fatigue failure. In addition, the boundary of the repeated-static range $N \sim 10^3$ is rather arbitrary. It is also specified in [9] depending on the strength and plastic characteristics of the material. However, in this paper we will keep the suggestion of the proposed model of damage development based on the scheme described above.

In order to match the model with the well-known criteria for multiaxial fatigue failure, a stress-based criterion has been selected that describes the fatigue failure associated with the normal crack microcracks development. This is a modification of the Smith-Watson-Topper (SWT) criterion [10]

described in [11], in which the amplitudes of maximum tensile stresses play a decisive role in the development of fatigue damage.

2. Kinetic equation for damage in LCF-HCF mode

The criterion of multiaxial fatigue failure in the LCF-HCF mode with the development of normal crack microcracks [11] (stress-based SWT) corresponding to the left branch of the bimodal fatigue curve (figure 1) is:

$$\sqrt{\langle \sigma_{1_{\max}} \rangle \Delta\sigma_1 / 2} = \sigma_u + \sigma_L N^{-\beta_{LH}} \tag{1}$$

where σ_1 is the largest principal stress, $\Delta\sigma_1$ is the spread of the largest principal stress per cycle, $\Delta\sigma_1 / 2$ is its amplitude. From the condition of repeated-static fracture up to values of $N \sim 10^3$ by the method [4] it is possible to obtain the value $\sigma_L = 10^{3\beta_{LH}}(\sigma_B - \sigma_u)$. According to the chosen criterion only tensile stresses lead to failure, so it includes the value $\langle \sigma_{1_{\max}} \rangle = \sigma_{1_{\max}} H(\sigma_{1_{\max}})$. In these formulas σ_B is the static tensile strength of the material, σ_u is the classic fatigue limit of the material during a reverse cycle (asymmetry coefficient of the cycle $R = -1$), β_{LH} is power index of the left branch of the bimodal fatigue curve.

From the fatigue fracture criterion we obtain the number of cycles before fracture at uniform stressed state:

$$N_{LH} = 10^3 \left[(\sigma_B - \sigma_u) / \langle \sigma_{LH} - \sigma_u \rangle \right]^{1/\beta_{LH}}, \quad \sigma_{LH} = \sqrt{\langle \sigma_{1_{\max}} \rangle \Delta\sigma_1 / 2} \tag{2}$$

In order to describe the process of fatigue damage development in the LCF-HCF mode, a damage function $0 \le \psi(N) \le 1$ is introduced, which describes the process of gradual cyclic material failure. When $\psi = 1$ a material particle is considered completely destroyed. Its Lame modules become equal to zero. The damage function ψ as a function on the number of loading cycles for the LCF-HCF mode is described by the kinetic equation:

$$\partial\psi / \partial N = B_{LH} \psi^{\gamma} / (1 - \psi^{\alpha})$$

where α and $0 < \gamma < 1$ are the model parameters that determine the rate of fatigue damage development. The choice of the denominator in this two-parameter equation, which sets the infinitely large growth rate of the zone of complete failure at $\psi \to 1$, is determined by the known experimental data on the kinetic growth curves of fatigue cracks, which have a vertical asymptote and reflects the fact of their explosive, uncontrolled growth at the last stage of macro fracture.

An equation for damage of a similar type was previously considered in [8], its numerous parameters and coefficients were determined indirectly from the results of uniaxial fatigue tests. In our case, the coefficient B_{LH} is determined by the procedure that is clearly associated with the selected criterion for multiaxial fatigue failure of one type or another. It has the following form.

The number of cycles to complete failure N_{LH} at $\psi = 1$ is from the equation for damage for a uniform stress state:

$$\left[\psi^{1-\gamma} / (1-\gamma) - \psi^{(1+\alpha-\gamma)} / (1+\alpha-\gamma) \right]\Big|_0^1 = B_{LH} N\Big|_0^{N_{LH}}, \quad N_{LH} = \alpha / (1+\alpha-\gamma) / (1-\gamma) / B_{LH} \tag{3}$$

By equating the values N_{LH} from the fracture criterion (2) and from the solution of the equation for damage (3), we obtain the expression for the coefficient B_{LH}:

$$B_{LH} = 10^{-3}\left[\langle\sigma_{LH} - \sigma_u\rangle / (\sigma_B - \sigma_u)\right]^{1/\beta_{LH}} \alpha / (1+\alpha-\gamma)/(1-\gamma)$$

where the value σ_{LH} is determined by the selected mechanism of fatigue failure and the corresponding multiaxial criterion (1).

When $\sigma_{LH} \le \sigma_u$ fatigue failure does not occur, when $\sigma_{LH} \ge \sigma_B$ it occurs instantly.

3. Fatigue damage development calculation algorithm

The Ansys software was used to calculate the loading cycle of a deformable specimen, supplemented by a code to calculate the damage equation and changes of elasticity modulus.

To integrate the equation $d\psi/dN = B_{LH}\psi^{\gamma}/(1-\psi^{\alpha})$, the damage function approximation was applied at the *k*-th node of the computational grid for given discrete values ψ_k^n at moments N^n and sought ψ_k^{n+1} at moments N^{n+1}.

To calculate the damage equation, the value $\alpha = 1-\gamma$ was chosen for which by analytic integration an explicit expression for $\psi_k^{n+1}(\psi_k^n, \Delta N^n)$ can be obtained:

$$\left[\psi^{1-\gamma}/(1-\gamma) - \psi^{2(1-\gamma)}/2/(1-\gamma)\right]_{\psi_k^n}^{\psi_k^{n+1}} = B\,N\Big|_{N_n}^{N_{n+1}}$$

With the denotations $(\psi_k^{n+1})^{1-\gamma} = x$, $q = 2(1-\gamma)B\Delta N^n + (\psi_k^n)^{1-\gamma} - 2(\psi_k^n)^{2(1-\gamma)}$ and $\Delta N^n = N^{n+1} - N^n$ the equation transforms to $x^2 - 2x + q = 0$ and its valid root $x = 1-\sqrt{1-q} < 1$.The analytical expression for the increment of damage on the increment of the number of cycles ΔN^n is derived from it:

$$\psi_k^{n+1} = \left(1-\sqrt{1-\left[2(1-\gamma)B\Delta N^n + (\psi_k^n)^{1-\gamma} - 2(\psi_k^n)^{2(1-\gamma)}\right]}\right)^{1/(1-\gamma)}$$

Increment value ΔN^n defined as follows.

Based on the stress state calculation data, the coefficient B_{LH} is calculated for each node. After that, for each node, the following values are calculated

$$\Delta\tilde{N}_k^n = \left[\psi^{1-\gamma}/(1-\gamma) - \psi^{2(1-\gamma)}/2/(1-\gamma)\right]_{\psi_k^n}^{1} / B_{LH}$$

corresponding to the number of cycles at which in the node *k* from its current level of damage and equivalent stress complete destruction will be achieved (damage is equal to 1).If the damage level in the considered node is less than the threshold ψ_0 (threshold $\psi_0 = 0.95$ is selected), then the value for this node $\Delta\tilde{N}_k^n$ is multiplied by a factor of 0.5. Otherwise, it is multiplied by a factor of 1. Thus, the step of incrementing the number of cycles for a given node is $\Delta N_k^n = 0.5(1+H(\psi_k^n - 0.95))\Delta\tilde{N}_k^n$. Of all the ΔN_k^n values the smallest one is selected. The increment of the number of loading cycles for the calculation of the entire specimen is $\Delta N^n = \min_k \Delta N_k^n$. For each node, based on its current level of damage and equivalent voltage, a new level of damage is found taking into account the calculated increment ΔN^n.

All elements are sorted out, for each of them the most damaged node is searched and according to its damage the mechanical properties of the element are adjusted:

$$\lambda(\psi_k^n) = \lambda_0(1-\kappa\psi_k^n), \ \ \mu(\psi_k^n) = \mu_0(1-\kappa\psi_k^n)$$

Those elements that belong to nodes with damage $\psi = 1$ are removed from the calculation area and form a localized zone (crack-like) of completely destroyed material. The calculation ends when the boundaries of a completely damaged region exit to the specimen surface (macro destruction) or the evolution of this region stops.

4. Calculation results

To determine parameters of the proposed model and verify its performance, one of the fatigue experiments described in [8] was performed numerically. From the condition of matching the experimental and calculated fatigue curves for a specimen of certain geometry for a given loading amplitude and cycle asymmetry the numerical coefficients were found. Using the obtained values, the experimental results on specimens of a different geometry and asymmetry coefficients were reproduced and calculation algorithm operability was confirmed.

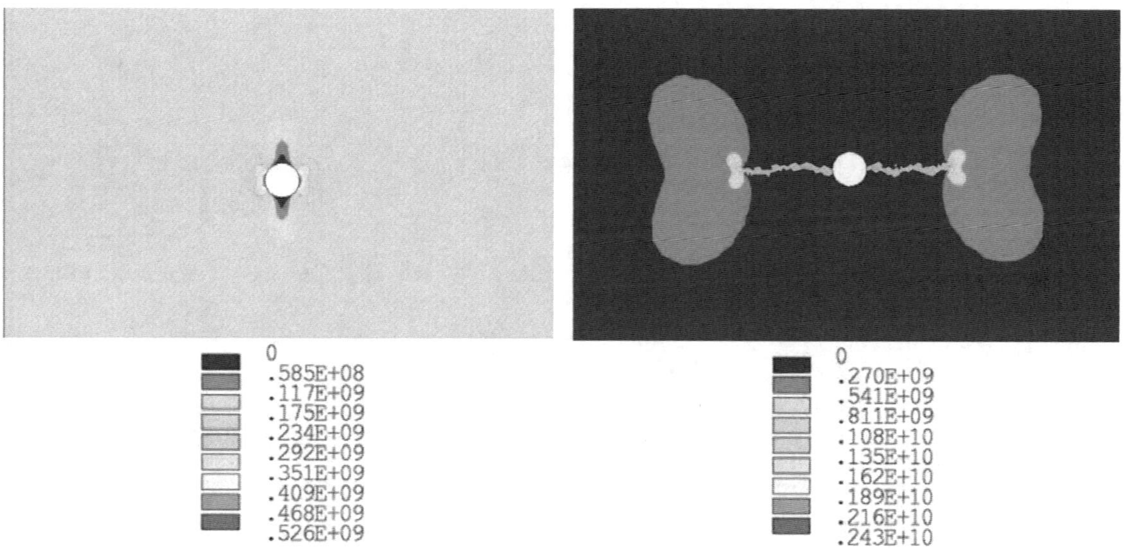

a) b)

Figure 1. Specimen with a hole at R = -1:
a) emergence of a "quasi-crack", b) growth of a "quasi-crack".

Initial tests were conducted on a plate 100x25x1.57 mm in size with 1.56 mm diameter through hole in the ones center. Ratification tests were conducted on a V-notched specimen that has 15 mm width without a notch, thickness of 1.7 mm, the notch depth of 1.32 mm, the V-notch angle of 60 degrees and the notch radius of 0.675 mm. The cyclic loading of the upper and lower boundaries of the specimen with an amplitude of 0.096 mm with the development of damage zones up to macroscopic destruction was simulated and matched with the results from [8]. In the center of the plate there is a through hole with diameter of 1.56 mm.

Plate material – titanium alloy with strength and fatigue parameters = 1135 MPa, = 330 MPa, = 0.31. Elasticity modulus of intact alloy are = 77 GPa, = 44 GPa. Figures 1 and 3 show the lines of the effective stress level for the specimen with a hole (figure 1) and for the specimen with a notch (figure 3) in two states – before the fatigue quasi-crack initiation and at the moment when it has passed approximately halfway to macro-destruction.

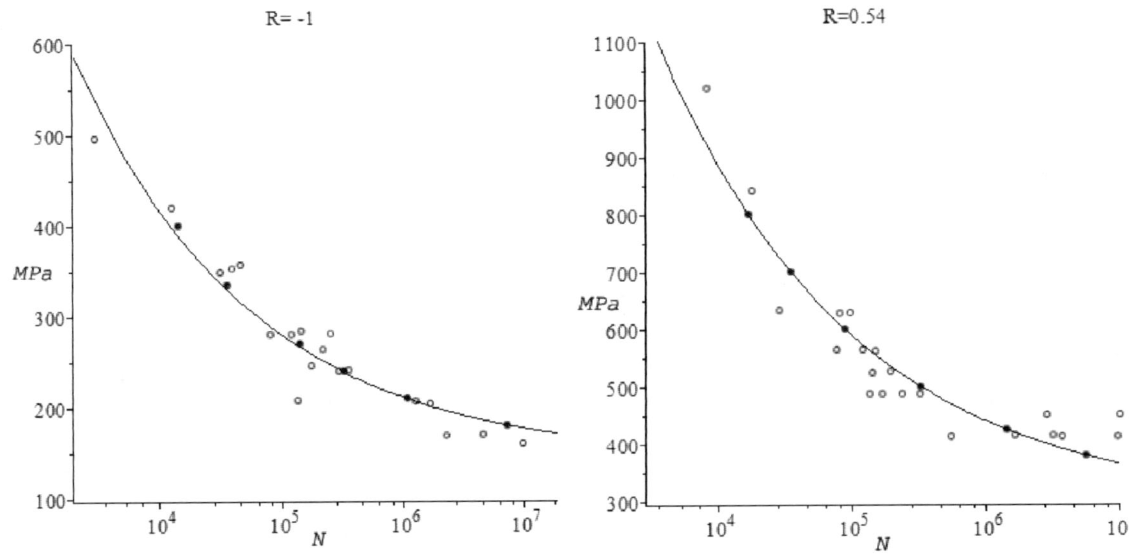

a) b)

Figure 2. Fatigue curves for the specimen with a hole: ○ - real test points, • - calculating points.

In figures 2 and 4 the results of real and computational experiments on constructing fatigue curves for specimens with a hole and a side notch are presented. Both real and calculated points represent the moment of a crack initiation. The curves in the figures approximate the experimental points. The calculations presented in figure 2-b almost exactly fit the approximation curve for the values of the model parameters. Utilizing these parameters, the fatigue curves presented in figure 2-a (specimen with a hole, R = -1) and in figure 4 (notched specimen, R = -0.5and 0.1. At figure 2-b, the calibration series, the relative error is 0. The average relative errors at figures 2-a, 4-a and 4-b are 1%, 7% and 6% respectively. The obtained satisfactory quality reproduction of real fatigue experiments indicates the efficiency and prospects of the model and calculation algorithm. The considered model represents the development of the damage model in case of cyclic loads, presented in [12] for the description of damages during dynamic loading.

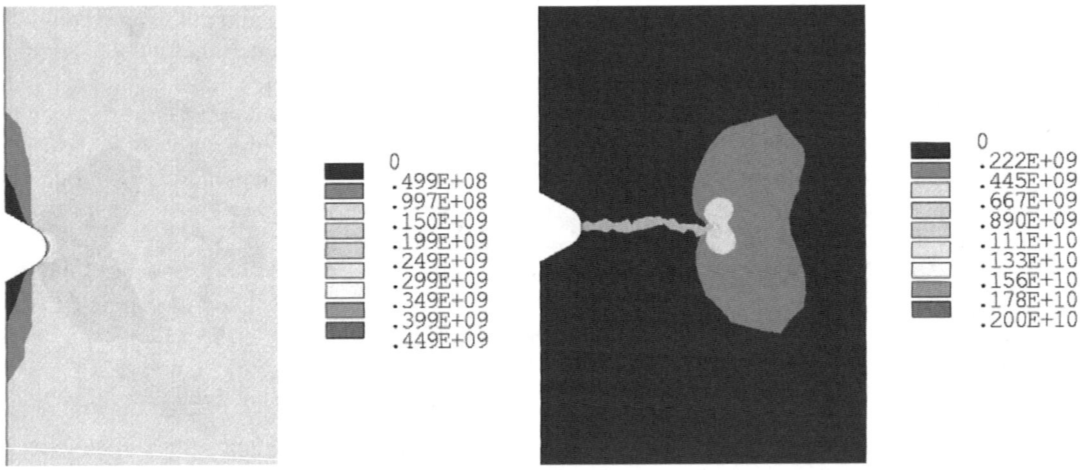

a) b)

Figure 3.V-notched specimen at R = -0.5:
a) emergence of a "quasi-crack", b) growth of a "quasi-crack".

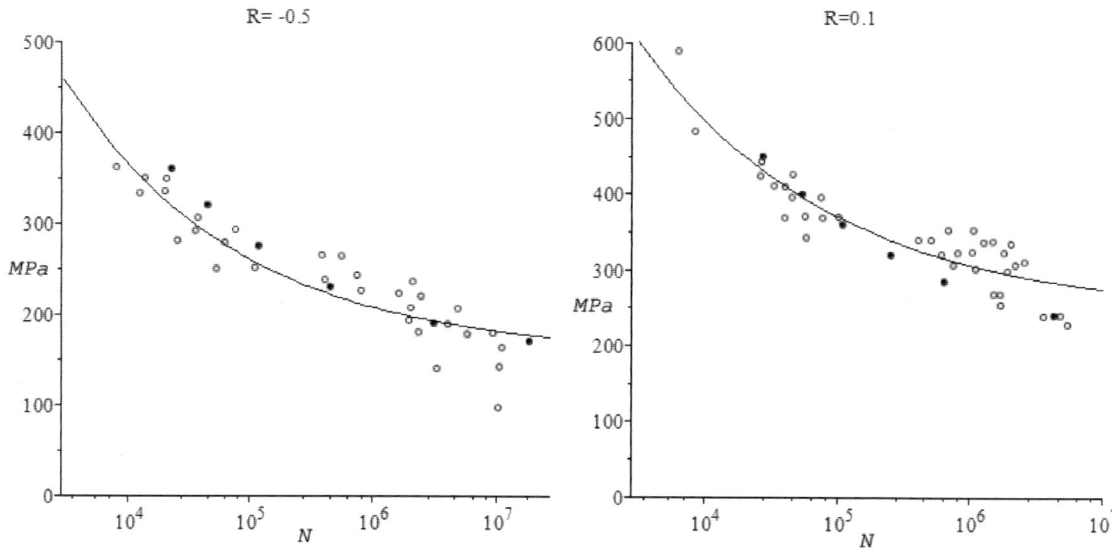

a) b)

Figure 4. Fatigue curves for V-notched specimen: ○ - real test points, • - calculating points.

5. Conclusions

A kinetic model of cyclic loading damage development is proposed to describe the fatigue fracture process development. To determine the coefficients of the kinetic equation of damage, the known SWT criterion of multiaxial fatigue fracture was used. A numerical method for calculating crack-like zones up to macrofracture is proposed. The model parameters are determined from the condition of matching the experimental and calculated fatigue curve for a specimen of a certain geometry at a given load amplitude and cycle asymmetry coefficient. Using the obtained values, the results of experiments on specimens of a different geometry and asymmetry coefficients were reproduced and the model and calculation algorithm performance was confirmed.

Acknowledgments

The study was carried out with a grant from the Russian Science Foundation (project No. 19-19-00705).

References

[1] Basquin O H 1910 The exponential law of endurance tests*Proc of the American society for testing and material***10** 625–30
[2] Meggiolaro MA Miranda AC de Castro J 2007 Comparison among fatigue life prediction methods and stress-strain models under multiaxial loading*Proceedings of 19th Int. Congress of Mech. Eng*
[3] Karolczuk A Papuga J Palin-Luc T 2020 Progress in fatigue life calculation by implementing life-dependent material parameters in multiaxial fatigue criteria*Int. J. of fatigue***134**1-13
[4] Bourago NG Zhuravlev AB Nikitin IS 2011 Models of multiaxial fatigue fracture and service life estimation of structural elements*Mechanics of Solids***46** 828–38
[5] Kachanov L M 1958 On the time of failure under creep conditions*Izv. AN SSSR OTN***8** 26–31
[6] Rabotnov J N 1959 On the mechanism of prolonged destruction. Strength issues of materials and structures*AN SSSR OTN* 5–7
[7] Lemaitre J Chaboche JL 1994 *Mechanics of solid materials* (Cambridge: Cambridge University Press) p 582
[8] Marmi AK Habraken AM Duchene L 2009 Multiaxial fatigue damage modeling at macro scale

of Ti6Al4V alloy*Int. J. of fatigue***31** 2031–40

[9] Shanyavskiy AA Soldatenkov AP 2019 The fatigue limit of metals as a characteristic of the multimodal fatigue life distribution for structural materials*Procedia Structural Integrity***23** 63–8

[10] Smith RN Watson P Topper TH 1970 A stress-strain parameter for the fatigue of metals*J. of Materials***5** 767–78

[11] Gates N Fatemi A 2016 Multiaxial variable amplitude fatigue life analysis including notch effects*Int. J. of fatigue***91** 337–51

[12] Burago NG Nikitin IS Nikitin AD Stratula BA 2019 Algorithms for calculation damage processes*Frattura ed Integrità Strutturale***49** 212–24

IOP Conf. Series: Materials Science and Engineering **927** (2020) 012020 doi:10.1088/1757-899X/927/1/012020

Simulation of seismic responses from the 3D non-linear model of the Bazhenov formation

I S Nikitin[1], V I Golubev[2], A V Ekimenko[3] and M B Anosova[1]
[1] Department of Mathematical Simulation, Institute of Computer Aided Design of RAS, 19/18 2-nd Brestskaya str., Moscow, 123056, Russia
[2] Laboratory of Applied Numerical Geophysics, Moscow Institute of Physics and Technology, 9 Institutskiy per., Dolgoprudny, Moscow Region, 141701, Russia
[3] Seismic Data Analysis Department, Gazprom Neft Science & Technology Centre, Moika river embankment 75-79, Saint Petersburg, 190000, Russia

E-mail: w.golubev@mail.ru

Abstract. The article is devoted to the investigation of seismic responses from probable oil-bearing formations. The three-dimensional geological model of the Bazhenov formation situated at Western Siberia is constructed. It is based on the elastic two-dimensional cross-section with curvilinear boundaries between different layers. For this purpose, logging and seismic data were processed jointly. To describe correctly the internal structure of mudstones the novel continuum model of media with interlayer slippage and delamination was utilized. The grid-characteristic method was used to solve numerically the govern system of the linear elasticity. The non-linearity of the model is incorporated with the appropriate explicit-implicit approximation procedure. For increasing the reliability of obtained synthetic seismograms the presence of the static rock pressure was taken into account. Reasonable computational time was achieved with the help of the parallelization based on the Message Passing Interface (MPI) technology. To analyze the structure of seismic responses wave patterns for the model with different parameters were presented. The influence of fractured layer on registered seismic traces were estimated. The proposed numerical approach may be used for simulation of complex geological objects.

1. Introduction

The seismic survey process is the most popular method for planning of oil and gas deposit exploration. Positions of fractured clusters are associated with the producing formation. In the case of the complex medium structure, it is crucial to improve the reliability of obtained estimations. In this direction, the numerical simulation of synthetic responses from geological objects with known properties has a great potential.

Different approaches have been developed to describe the dynamic behavior of fractured geological media to date. The homogenization method proposed by Schoenberg [1] and Hudson [2] can be used. Unfortunately, this method is not well suitable for the arbitrary crack orientation distribution. It is possible to solve explicitly acoustic equations inside each crack; however, it spends a huge amount of computational resources [3, 4]. The other method is to set explicitly the crack geometry and use appropriate correctors [5]. It allows to improve significantly the robustness of the computational algorithm. It should be noticed, that the most of these articles deal with simplest contact conditions.

However, the discrete variant of the slippage theory [6] allows us to formulate continuum models with nonlinear (dry friction or viscous-plastic types) slip conditions [7, 8].

In the paper we demonstrate the application of this model for simulation of Bazhenov formation in 3D. Firstly, the geological model is described, and the procedure of properties setting is explained. Secondly, the grid-characteristic method for solving hyperbolic equations and the non-linear continuum model are formulated. Finally, results of numerical simulations are presented, and synthetic seismograms are analyzed.

2. Geological model

In this work the 3D seismic-geological model was constructed. Elastic properties and geometries of layers reflect the main features of geological structures of Western Siberia. At Figure 1 a fragment of this model is represented. The fall of layers in the model is gentle and less than 5 degrees. The P-wave velocity is based on the results of the acoustic logging. The S-wave velocity and the density are obtained using empirical relationships. For calculating the S-wave velocity the Castagna's formula [9] is used

$$c_s = 0.58 * c_p - 0.078,$$

and for calculating the density the Gardner's formula [9] is used

$$\rho = 1.74 * c_p^{0.25}.$$

These dependences were obtained as a result of a statistical analysis of measurements on rock images and are widely used in the seismic exploration.

The geological object of an attention is the low-speed layer corresponding to the Bazhenov formation. The description of traditional reservoirs is possible using an elastic model. In this paper we made an attempt to describe the non-traditional Bazhenov formation as a medium with micro-layering using more complicated non-linear continuum model. This approach is justified and applicable because of the lithological composition (clay mud) and properties (leafiness) of the formation.

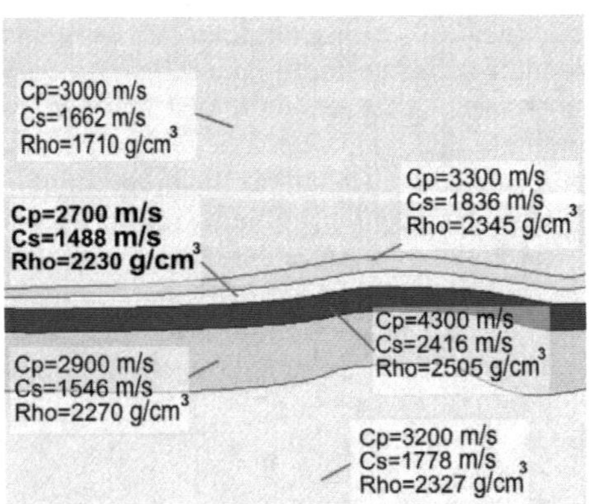

Figure 1. The geological model of the Bazhenov formation. The 2D slice along the OYZ plane is presented.

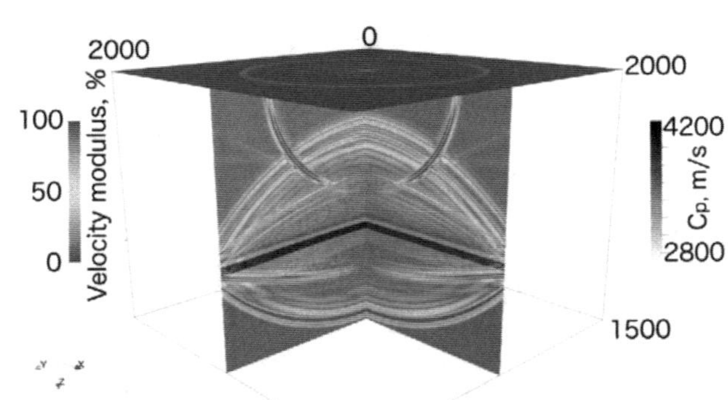

Figure 2. The wave pattern after interactions with curvilinear layer boundaries is depicted. Geological model is put under it for better representation. Complex superposition of reflected and transmitted waves is registered.

3. Elastic medium and numerical method

The grid-characteristic method [10] used in this paper was previously successfully applied for describing media with different rheology [11-14]. The stress state of the deformable body, in the case of small deformations, can be represented by three independent components of the velocity vector $\mathbf{v}(\mathbf{r}, t)$ and six independent components of the stress tensor $\boldsymbol{\sigma}(\mathbf{r}, t)$. All of them can be combined in a single unknown vector as

$$\mathbf{u} = \{v_x, v_y, v_z, \sigma_{xx}, \sigma_{xy}, \sigma_{xz}, \sigma_{yy}, \sigma_{yz}, \sigma_{zz}\}^T.$$

The govern system of partial differential equations is

$$\mathbf{u}'_t + \mathbf{A}_x\mathbf{u}'_x + \mathbf{A}_y\mathbf{u}'_y + \mathbf{A}_z\mathbf{u}'_z = \mathbf{0},$$

where $\mathbf{A}_x, \mathbf{A}_y, \mathbf{A}_z$ are matrices. Three one-dimensional systems can be produced by eliminating derivatives along both other directions in the form of

$$\mathbf{u}'_t + \mathbf{A}_\xi\mathbf{u}'_\xi = \mathbf{0}, \quad \xi = x, y, z.$$

The one-dimensional systems above are hyperbolic and matrix $\mathbf{A}_\xi$ is hyperbolic and have a full set of eigenvalues $\boldsymbol{\Lambda}_\xi$ and eigenvectors $\boldsymbol{\Omega}_\xi$. Due to this fact the equation above can be rewritten as

$$\mathbf{u}'_t + \boldsymbol{\Omega}_\xi\boldsymbol{\Lambda}_\xi\boldsymbol{\Omega}_\xi^{-1}\mathbf{u}'_\xi = \mathbf{0}, \quad \xi = x, y, z.$$

$\boldsymbol{\Lambda}_\xi$ is the diagonal matrix with $\lambda_\xi = \{\pm c_p, \pm c_s, \pm c_s, 0,0,0\}$, and $c_p = \sqrt{\frac{\lambda+2\mu}{\rho}}$, $c_s = \sqrt{\frac{\mu}{\rho}}$, where λ and μ are Lame parameters. The value of c_p can be estimated from the well-logging data and other parameters are directly calculated according to the procedure described at Section 2.

With the substitution $\mathbf{w} = \boldsymbol{\Omega}_\xi^{-1}\mathbf{u}$ we obtain independent linear transport equations:

$$\mathbf{w}'_t + \boldsymbol{\Lambda}_\xi\mathbf{w}'_\xi = \mathbf{0}, \quad \xi = x, y, z.$$

Its analytical solution is

$$\mathbf{w}(\xi, t + \Delta t) = \mathbf{w}(\xi - \lambda_\xi \mathrm{sign}(\lambda_\xi), t).$$

The interpolation problem encountered was solved in this work on the constant step grid with the third order of the spatial approximation. To eliminate non-physical oscillations the common grid-characteristic monotonization procedure was used. After a single step along ξ axis values of $\mathbf{u}$ are calculated with the inverse transformation. The sequential systems solution along three Cartesian axes used to solve the initial three-dimensional system of equations.

To obtain correct values of the total stress tensor in the geological medium the static pressure should be added. In the case of elastic half-space (*Z*-axis goes up), it is equal to

$$\sigma_{zz}^0 = \rho g z, \quad \sigma_{xx}^0 = \sigma_{yy}^0 = \frac{\lambda}{\lambda+2\mu}\sigma_{zz}^0.$$

4. Continuum non-linear model

The internal structure of mudstones can be described like a set of elastic horizontal layers with the possibility of slippage and delamination. These effects can be taken into account with the distributed slippage velocity vector $\boldsymbol{\gamma}$ and the delamination velocity vector $\boldsymbol{\omega}$. We treat them as continuous functions of coordinates and time. The plane of the layer is set by the unit normal vector $\mathbf{n} = (0,0,1)$. The total strain tensor $\mathbf{e}$ consists of elastic part $\mathbf{e}^e$ and inelastic part $\mathbf{e}^i = \mathbf{e}^\omega + \mathbf{e}^\gamma$. Relations for its calculations are [6]

$$\mathbf{e}^\gamma = (\mathbf{n}\otimes\boldsymbol{\gamma} + \boldsymbol{\gamma}\otimes\mathbf{n})/2, \quad \boldsymbol{\gamma}\cdot\mathbf{n} = 0,$$
$$\mathbf{e}^\omega = (\mathbf{n}\otimes\boldsymbol{\omega} + \boldsymbol{\omega}\otimes\mathbf{n})/2, \quad \boldsymbol{\omega} = \omega\mathbf{n},$$
$$\mathbf{e} = \mathbf{e}^e + \mathbf{e}^\omega + \mathbf{e}^\gamma,$$
$$\mathbf{e} = (\nabla\mathbf{v} + \nabla\mathbf{v}^T)/2.$$

The real fractured medium demonstrates the non-linear behavior described in our model with two regimes. In the sliding mode ($\sigma_{zz} + \sigma_{zz}^0 < 0$):

$$\boldsymbol{\gamma} = \mathbf{0}, \quad \omega = 0, \quad \text{if } |\boldsymbol{\tau}| \le q|\sigma_{zz} + \sigma_{zz}^0|,$$
$$\boldsymbol{\tau} = q|\sigma_{zz} + \sigma_{zz}^0|\left(\frac{\boldsymbol{\gamma}}{|\boldsymbol{\gamma}|} + \eta\boldsymbol{\gamma}\right), \quad \omega = 0, \quad \text{if } |\tau| > q|\sigma_{zz} + \sigma_{zz}^0|,$$

where $\boldsymbol{\tau}$ is the shear stress, q is the dry friction coefficient, η is the viscosity coefficient.

Both conditions may be generalized as

$$\boldsymbol{\gamma} = \frac{\boldsymbol{\tau}}{\eta|\tau|}\left\langle\frac{|\tau|}{q|\sigma_{zz} + \sigma_{zz}^0|} - 1\right\rangle, \quad \omega = 0,$$

where $\langle F\rangle = FH(F)$ and $H(x)$ is the Heaviside function.

In the delamination mode ($\Omega > 0$):

$$\boldsymbol{\tau} = 0, \quad \sigma_{zz} + \sigma_{zz}^0 = 0, \quad \dot{\Omega} = \omega,$$

where Ω is the distributed displacement jump along the interlayer boundary.

With the procedure of implicit-explicit approximation the non-linearity of the model can be expressed effectively with the correction step. If we denote the results of elastic step with the superscript $(\sigma_{ij}^{e}, i, j = x, y, z)$, we obtain the final formulas as

$$\sigma_{xz}^{n+1} = \mathrm{sign}(\sigma_{xz}^{n+1e}) q |\sigma_{zz}^{n+1e} + \rho g z| \frac{1+\delta|\sigma_{xz}^{n+1e}|}{1+\delta q|\sigma_{zz}^{n+1e} + \rho g z|},$$

$$\sigma_{yz}^{n+1} = \mathrm{sign}(\sigma_{yz}^{n+1e}) q |\sigma_{zz}^{n+1e} + \rho g z| \frac{1+\delta|\sigma_{yz}^{n+1e}|}{1+\delta q|\sigma_{zz}^{n+1e} + \rho g z|},$$

where $\delta = \eta/(\mu\Delta t)$ and Δt is the time step. The conditions for applying this correction is

$$\sigma_{zz}^{n+1e} + \rho g z < 0 \text{ and } \sqrt{(\sigma_{xz}^{n+1e})^2 + \left(\sigma_{yz}^{n+1e}\right)^2} > q|\sigma_{zz}^{n+1e} + \rho g z|.$$

5. Simulation results

The central part of the geological Bazhenov formation model was covered with the squared grid. The total mesh size was 2000x2000x1500 m with 2 m between neighbouring nodes. In each of them we stored material properties (Lame parameters and density). According to the Courant stability condition the time step was set to 0.4 ms. One run included 3000 steps and covered 1.2 s of physical time. To support meshes with more than 750 million nodes the numerical algorithm was paralyzed with the MPI technology. Utilizing 100 cores of the NRC "Kurchatov Institute" a single run took 10 hours. The Ricker point source at the centre of the top boundary was applied. To simulate the real shot-gather systems three components of the velocity vector were saved at the day surface on the grid with the 20 m step.

The three-dimensional distribution of the velocity modulus is presented at Figure 2. The synthetic seismograms obtained are shown at Figure 3. The wavefield is not noisy, spurious reflections from edges of the computational domain have a low intensity and do not complicate the interpretation of the signal. The target reflection is easily recognized as the negative phase from 630 to 675 ms.

The analysis of simulation results may be used to identify some wavefield features that make it possible to confirm the presence of fractures in the medium. It is useful for the interpretation of the field data. In this article a traditional analysis of amplitude variation with the offset (AVO) is performed. Figure 4 shows the two-dimensional distribution of amplitudes corresponding to the reflection from the Bazhenov formation. The mean square of amplitudes obtained for the entire negative reflection phase is depicted. In general, distributions for open and closed fractures are similar. However, the last one demonstrates large amplitudes and a more intense amplitude decrease with distance from the source.

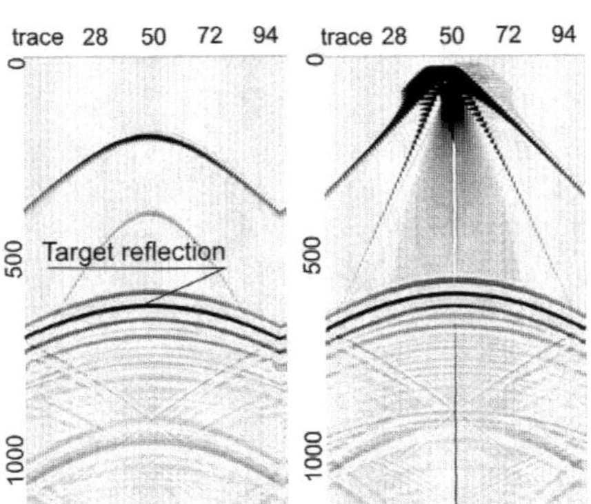

Figure 3. Synthetic seismograms: far from the source (left) and near to the source (right).

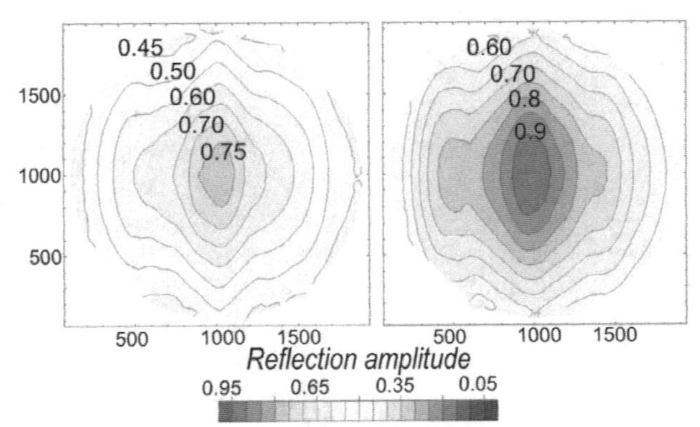

Figure 4. The response amplitude based on the vertical velocity component for closed fractures (left) and open fractures (right).

6. Conclusions

In this paper dynamic processes in the complex geological model were considered. Based on the logging and seismic data the three-dimensional multi-layered model was constructed. To take into account the internal structure of Western Siberia curvilinear layer boundaries was simulated with the usage of cubic meshes. To describe the non-linear behaviour of the leafiness Bazhenov formation the continuum model with possible interlayer slippage and delamination was utilized. To obtain the numerical solution the grid-characteristic method with the 3rd accuracy order was used. To achieve the reasonable computational time the MPI-parallelized algorithm was applied. The comparison of seismic responses from open and closed horizontal fractures shows a more intense amplitude decrease with the distance increase from the source for the last one. This feature may be used during the interpretation of the field data. The described approach can be applied for other oil deposits with layered structures.

Acknowledgments
This work has been carried out using computing resources of the federal collective usage center Complex for Simulation and Data Processing for Mega-science Facilities at NRC "Kurchatov Institute", http://ckp.nrcki.ru/. This work was carried out with the financial support of the Russian Science Foundation, project no. 19-71-10060. We are very grateful for an anonymous reviewer for valuable comments.

References

[1] Coates R and Schoenberg M 1995 Finite-difference modeling of faults and fractures *Geophysics* **60(5)** 1514-26

[2] Hudson J A, Pointer T and Liu E 2008 Effective-medium theories for fluid-saturated materials with aligned cracks *Geophysical Prospecting* **49(5)** 509-522

[3] Frehner M 2014 Krauklis wave initiation in fluid-filled fractures by seismic body waves *Geophysics* **79** T27-T35

[4] Caspari E, Novikov M, Lisitsa V, Barbosa N D, Quintal B, Rubino J G and Holliger K 2019 Attenuation mechanisms in fractured fluid-saturated porous rocks: a numerical modelling study *Geophysical Prospecting* **67(4)** 935-955

[5] Khokhlov N and Stognii P 2020 Novel approach to modeling the seismic waves in the areas with complex fractured geological structures *Minerals* **10** 122–139

[6] Nikitin I S 2008 Dynamic models of layered and block media with slip, friction and separation *Mechanics of Solids* **43(4)** 652-661

[7] Nikitin I S, Burago N G, Golubev V I and Nikitin A D 2019 Mathematical modeling of the dynamics of layered and block media with nonlinear contact conditions on supercomputers *J. Phys.: Conf. Ser.* **1392** 012057

[8] Nikitin I S, Burago N G, Golubev V I, Nikitin A D 2020 Methods for Calculating the Dynamics of Layered and Block Media with Nonlinear Contact Conditions *Advances in Theory and Practice of Computational Mechanics. Smart Innovation, Systems and Technologies* **173** 171–183

[9] Mavko G, Mukerji T and Dvorkin J 2009 *The Rock Physics Handbook. Second Edition Tools for Seismic Analysis of Porous Media* (Cambridge University Press)

[10] Khokhlov N I, Golubev V I 2019 On the Class of Compact Grid-Characteristic Schemes. *GCM50 2018. Smart Innovation, Systems and Technologies* **133** 64–77

[11] Beklemysheva K A, Vasyukov A V, Golubev V I and Zhuravlev Yu I 2018 On the Estimation of Seismic Resistance of Modern Composite Oil Pipeline Elements *Dokl. Math.* **97** 184–187

[12] Golubev V I, Voinov O Y and Zhuravlev Y I 2017 On seismic imaging of fractured geological media *Dokl. Math.* **96** 514–516

[13] Golubev V I, Shevchenko A V and Petrov I B 2019 Taking into Account Fluid Saturation of Bottom Sediments in Marine Seismic Survey *Dokl. Math.* **100** 488–490

IOP Conf. Series: Materials Science and Engineering **927** (2020) 012020 doi:10.1088/1757-899X/927/1/012020

[14] Golubev V I 2019 The Usage of Grid-Characteristic Method in Seismic Migration Problems *GCM50 2018. Smart Innovation, Systems and Technologies* **133** 143–155

IOP Conf. Series: Materials Science and Engineering **927** (2020) 012021 doi:10.1088/1757-899X/927/1/012021

On orbital stability of resonant periodic motions originating from hyperboloidal precession of a dynamically symmetric satellite

E A Sukhov

Department of Mechatronics and Theoretical Mechanics, Moscow Aviation Institute, Moscow, Russia.

E-mail: Sukhov.george@gmail.com

Abstract. Periodic motions arising from Hyperboloidal precession of a dynamically symmetric rigid-body satellite on a circular orbit are studied in this work. There exist two types of these motions categorized by frequencies ω_1 and ω_2 of the linearized system: short-periodic motions with period close to $2\pi/\omega_2$ and long-periodic ones with period close to $2\pi/\omega_1$ ($\omega_2 > \omega_1$). These motions describe oscillations of the satellite's principal axis in the neighbourhood Hyperboloidal precession and can be obtained as small-parameter power series of the oscillation amplitude. It has been shown before that in a resonant case there exist more than one family of long-periodic motions. In this work, families of long-periodic motions originating from Hyperboloidal precession in case of fourth order resonance are computed and analysed.

1. Problem formulation

Motion of a rigid-body satellite about its center of mass O on a circular orbit in central Newtonian gravitational field is considered. To describe this motion, an orbital reference frame $OXYZ$ and a fixed reference frame $Oxyz$ are introduced as shown in figure 1. The axes OX, OY and OZ of the orbital frame are aligned with transversal and normal vectors to the satellite's orbit and with its center of mass' radius-vector $\vec{R}$, respectively. The axes of the fixed frame $Oxyz$ are pointed along the satellite's principal axes with corresponding moments of inertia being J_1, J_2 and J_3. Relative position of these frames is described by Euler angles ψ, θ, φ.

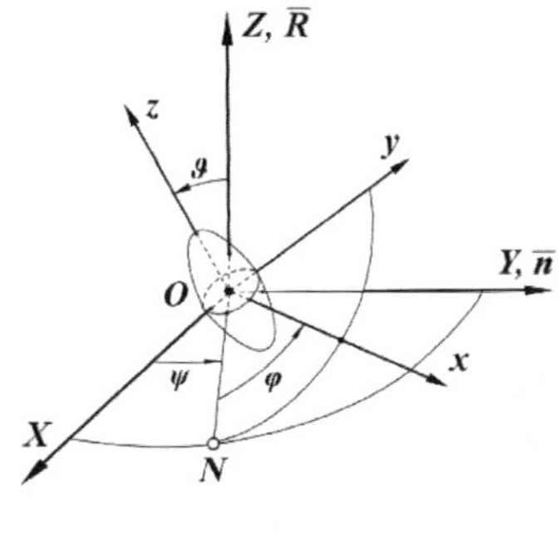

Figure 1. The satellite's reference frames.

Following [1, 2] ψ, θ, φ are taken as generalized coordinates and corresponding generalized impulses $p_\psi, p_\theta, p_\varphi$ are introduced. It is then assumed that the satellite is dynamically symmetric, i. e. $J_1 = J_2$, which leads to φ being a cyclic coordinate so its respective impulse retains constant value

$$p_\varphi = \frac{J_3}{J_1}\frac{r_0}{\omega_0} = \gamma \tag{1}$$

where $r_0 = \dot{\psi}\cos\theta - \omega_0 \cos\psi \sin\theta$ and ω_0 is the angular velocity of the radius-vector $\vec{R}$. In this case the satellite's motion is described by the following canonical equations

$$\frac{d\psi}{dv} = \frac{\partial H}{\partial p_\psi}, \; \frac{d\theta}{dv} = \frac{\partial H}{\partial p_\theta}, \; \frac{dp_\psi}{dv} = -\frac{\partial H}{\partial \psi}, \; \frac{dp_\theta}{dv} = -\frac{\partial H}{\partial \theta} \tag{2}$$

with Hamiltonian

$$H = \frac{p_\psi^2}{2\sin^2\theta} + \frac{p_\theta^2}{2} - \left(\frac{\gamma\cos\theta}{\sin^2\theta} + \frac{\cos\psi}{\tan\theta}\right)p_\psi - p_\psi \sin\psi + \frac{\gamma^2}{2\tan^2\theta} + \frac{\gamma\cos\psi}{\sin\theta} + \frac{1}{2}\delta\cos^2\theta. \tag{3}$$

where $\nu = \omega_0 t$,

$$\delta = 3\left(\frac{J_3}{J_1} - 1\right), \quad -3 < \delta < 3 \tag{4}$$

and γ is given by (1).

Equations (2) possess a particular solution

$$\theta_0 = \frac{\pi}{2}, \cos\psi_0 = -\gamma, p_{\theta 0} = 0, p_{\psi 0} = \delta \sin\theta_0 \cos\theta_0$$

known as Hyperboloidal precession [3, 4]. In this case, the satellite's principal axis Oz rotates at a constant rate relative to the absolute space and describes a hyperboloid as shown on figure 2.

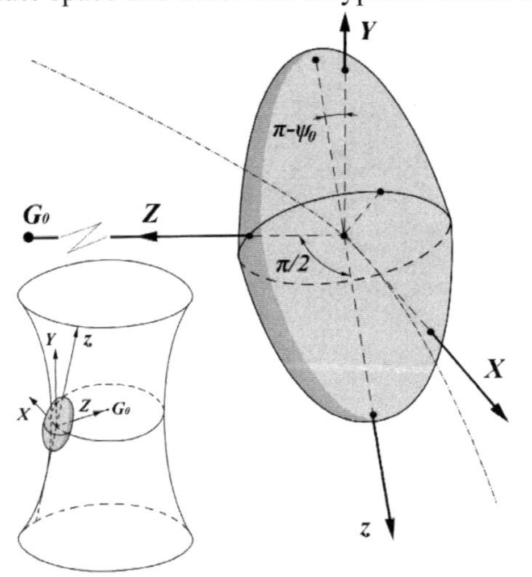

Figure 2. Hyperboloidal precession of a satellite.

Hyperboloidal precession exists if $|\gamma| \leq 1$ and is Lyapunov stable if $\delta > 0$ [5]. In the neighborhood of Lyapunov-stable Hyperboloidal precession there exist two types of periodic motions: short-periodic motions with period close to $\frac{2\pi}{\omega_2}$ and long-periodic motions with period close to $\frac{2\pi}{\omega_1}$ where $\omega_{1,2}$ are frequencies of the linearized system given by

$$\omega_{1,2} = \sqrt{\frac{\delta+1}{2} \mp \sqrt{\frac{(\delta-1)^2 + 4\gamma^2\delta}{4}}}. \tag{5}$$

These periodic motions describe oscillations of the satellite's axis Oz in the neighborhood of Hyperboloidal precession (figure 3). If $\omega_2 \neq N\omega_1$ these motions can be obtained as small parameter power series of oscillation amplitude following Lyapunov's method. In the non-resonant case said motions have been obtained and studied both analytically and numerically in works [6-8].

In this work, the case of fourth order resonance ($\omega_2 = 3\omega_1$) is taken into consideration. In a resonant case it's not possible to analytically compute the periodic motions using Luapunov's method so they are obtained using an approach developed in [9, 10] and previously applied in [11, 12]. It is further shown that in case of fourth-order resonance there always exist three families of long-periodic motions originating from Hyperboloidal precession with two of them being orbitally stable and one being orbitally unstable.

2. Analytical computation of periodic motions

Assuming that the frequencies ω_1 and ω_2 of the linearized system are in a resonant relationship $\omega_2 = 3\,\omega_1$ and the amplitude of the oscillations is small, canonical univalent transformation $\psi, \theta, p_\psi, p_\theta \to q_1, q_2, p_1, p_2$ is applied to the initial system (2). It transforms the initial system's Hamiltonian (3) to the normal form

$$K = \frac{1}{2}\omega_1(q_1^2 + p_1^2) + \frac{1}{2}\omega_2(q_2^2 + p_2^2) + \frac{1}{4}c_{20}(q_1^2 + p_1^2)^2 + \frac{1}{4}a_{11}(q_1^2 + p_1^2)(q_2^2 + p_2^2) + \frac{1}{4}a_{02}(q_2^2 + p_2^2)^2 - \frac{B_{31}}{4}[q_1^2(q_1q_2 + 3p_2p_1) - p_1^2(p_1p_2 + 3q_1q_2)] + O_6. \quad (6)$$

where a_{10}, a_{11}, a_{02} are constant coefficients given by the following formulae [12]

$$a_{11} = \frac{1}{560\omega_1^6}(678\omega_1^6 + 13\omega_1^4 - 20\omega_1^2 + 1),$$
$$a_{20} = \frac{1}{320\omega_1^6}(118\omega_1^6 - 319\omega_1^4 + 60\omega_1^2 - 3),$$
$$a_{02} = -\frac{1}{6720\omega_1^6}(906\omega_1^6 - 983\omega_1^4 - 220\omega_1^2 + 11)$$

and

$$B_{31} = \frac{1}{120}\frac{(1 - 15\omega_1^2 + 56\omega_1^4)\sqrt{30\omega_1^2 - 27\omega_1^4 - 3}}{\omega_1^6}$$

is a resonant coefficient. O_6 are terms of order 6 and higher.

New time $\tau = \omega_1 v$ is then introduced and a scaling canonical transformation

$$q_i = \sqrt{\frac{\varepsilon\omega_1}{B_{31}}}x_i, \qquad p_i = \sqrt{\frac{\varepsilon\omega_1}{B_{31}}}y_i \quad (i = 1,2)$$

with valence $n = \frac{B_{31}}{\varepsilon\omega_1}$ is performed. It transforms Hamiltonian (6) to form

$$K^{II} = \frac{1}{2}\omega_1(x_1^2 + y_1^2) + \frac{1}{2}\omega_2(x_2^2 + y_2^2) + \frac{1}{4}\varepsilon c_{20}(x_1^2 + y_1^2)^2 + \frac{1}{4}\varepsilon c_{11}(x_1^2 + y_1^2)(x_2^2 + y_2^2) + \frac{1}{4}\varepsilon c_{02}(x_2^2 + y_2^2)^2 - \varepsilon[x_1^2(x_1x_2 + 3y_1y_2) - y_1^2(y_1y_2 + 3x_1x_2)] + O_6 \quad (7)$$

where ε is a scaling coefficient and

$$c_{20} = \frac{a_{20}}{B_{31}}, c_{02} = \frac{a_{02}}{B_{31}}, c_{11} = \frac{a_{11}}{B_{31}}.$$

Omitting terms of order higher than four in (7) and introducing polar canonical variables

$$x_i = \sqrt{2r_i}\sin\phi_i, y_i = \sqrt{2r_i}\cos\phi_i$$

gives a truncated canonical system with Hamiltonian

$$K^{III} = r_1 + 3r_2 + \varepsilon\{c_{20}r_1^2 + c_{11}r_1r_2 + c_{02}r_2^2 + r_1^{\frac{3}{2}}\sqrt{r_2}\cos(\phi_1 - 3\phi_2)\}$$

and first integral $3r_1 + r_2 = const$. Applying another canonical transformation given by

$$\psi = \phi_1, \theta = \phi_2 - 3\phi_1, J = 3r_2 + r_1, R = r_2$$

the truncated system assumes the following form

$$\frac{d\theta}{d\tau} = \varepsilon\left\{\frac{1}{2}\frac{\sqrt{J-3R}(J-12R)\cos\theta}{\sqrt{R}} + \sigma J + \mu R\right\}, \quad (8)$$
$$\frac{dR}{d\tau} = \varepsilon\sqrt{R}(J - 12R)^{\frac{3}{2}}\sin\theta.$$

with Hamiltonian $F = J + \varepsilon\{c_{20}J^2 + \sigma JR + \mu R^2 + \varepsilon\sqrt{R}(J - 12R)^{\frac{3}{2}}\cos\theta\}$ and a first integral $J = J_0 = const$. The constant parameters μ and σ depend on the initial problem's parameters γ, δ and are given by the following formulae

$$\sigma = c_{11} - 6c_{20}, \quad \mu = c_{02} + 9c_{20} - 3c_{11}.$$

The system (8) possesses equilibria that correspond to periodic motions in the initial Hamiltonian system (2). These equilibria are given by the expression

$$\theta = \frac{\pi}{2}\left[1 - sign\big((12x_* - 1)(\sigma + 2\mu x_*)\big)\right], \qquad R = x_* J_0$$

where x_* is a Real-value root of a cubic equation

$$(432 + 16\mu)x^3 + (16\mu\sigma - 216)x^2 + (4\sigma^2 + 27)x - 1 = 0. \tag{9}$$

It has been shown in [10] that depending on the values of μ and σ the initial system may possess one or three families of long-periodic motions in case of fourth-order resonance. The curves c_1 and c_2 derived from the equation (9) separate the parameter space of μ and σ into three domains S_1, S_2 and S_3 (figure 3). Domains S_1 and S_2 contain values that correspond to the case of one family and domain S_3 contains values corresponding to the case of three families of long-periodic motions. The dash line corresponds to the case of spherically-symmetric satellite ($\delta = 0$).

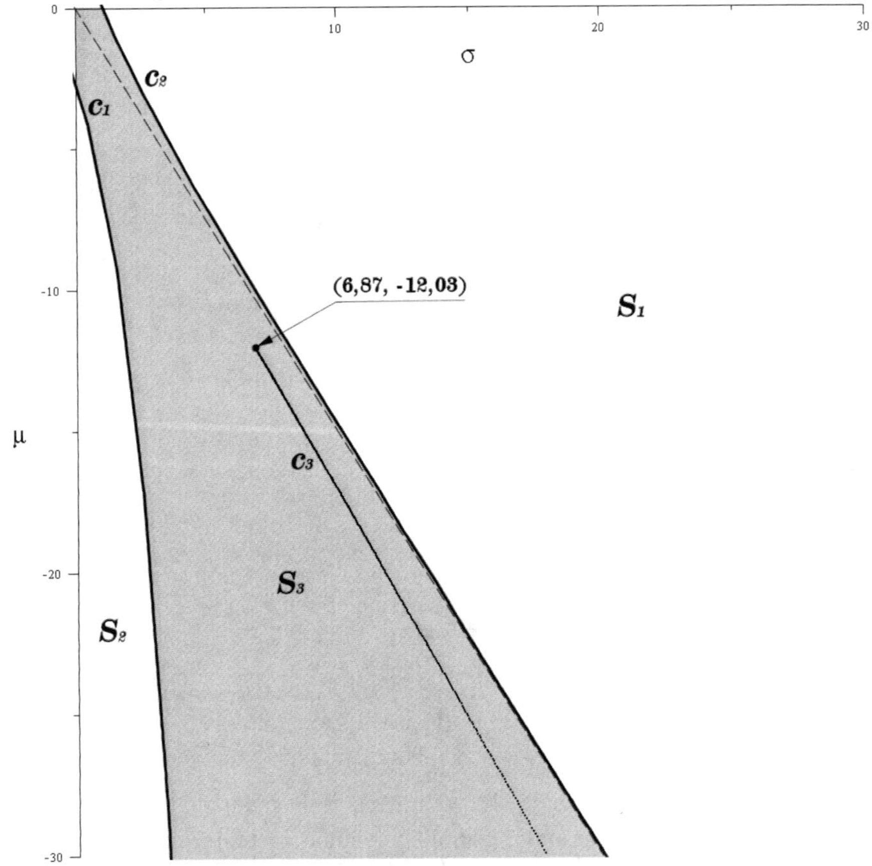

Figure 3. Domains with one (S_1, S_2) and three (S_3) families of long periodic motions in μ-σ parameter space.

In case of Hyperboloidal precession the resonant values of μ and σ corresponding to the case of fourth-order resonance are given by curve c_3. Thus, in case of strict fourth order resonance there always exist three families of long-periodic motions in the neighborhood of Hyperboloidal precession. These families are denoted as $\Gamma_1, \Gamma_2, \Gamma_3$ [11, 12].

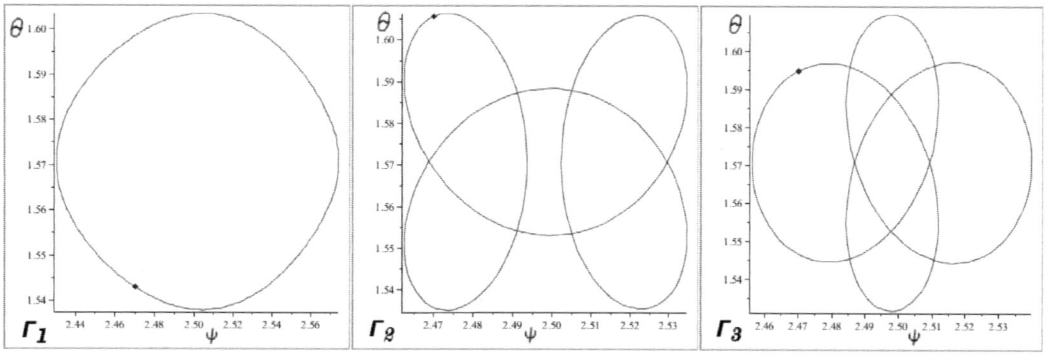

Figure 4. Resonant long-periodic motions originating from Hyperboloidal precession.

Figure 4 shows trajectories of long-periodic motions belonging to families $\Gamma_1, \Gamma_2, \Gamma_3$ in a $\psi - \theta$ plane in case $\gamma = 0.8, \delta = 1.0, \Delta h = 0.001$ where Δh is deviation of the initial system's energy constant h from its value on Hyperboloidal precession.

3. Orbital stability of long-periodic motions

Analytical study carried out in [10] shows that in domain S_3 there are two families of orbitally stable long-periodic motions and one family of orbitally unstable long-periodic motions. To verify this result in case of Hyperboloidal precession Poincare maps have been computed along the fourth order resonance curve c_3. Figure 4 shows a Poincare map for the parameter values of $\gamma = 0.8, \delta = 1.0, \Delta h = 0.001$. The points marked Γ_i $(i = 1..3)$ correspond to orbitally stable long-periodic motions belonging to families Γ_1 and Γ_2 and an orbitally unstable long-periodic motion of the family Γ_3. The point Γ_s corresponds to an orbitally stable short-periodic motions. Computation shows that pattern seen on figure 5 is retained for all values of γ and δ along the resonance curve c_3.

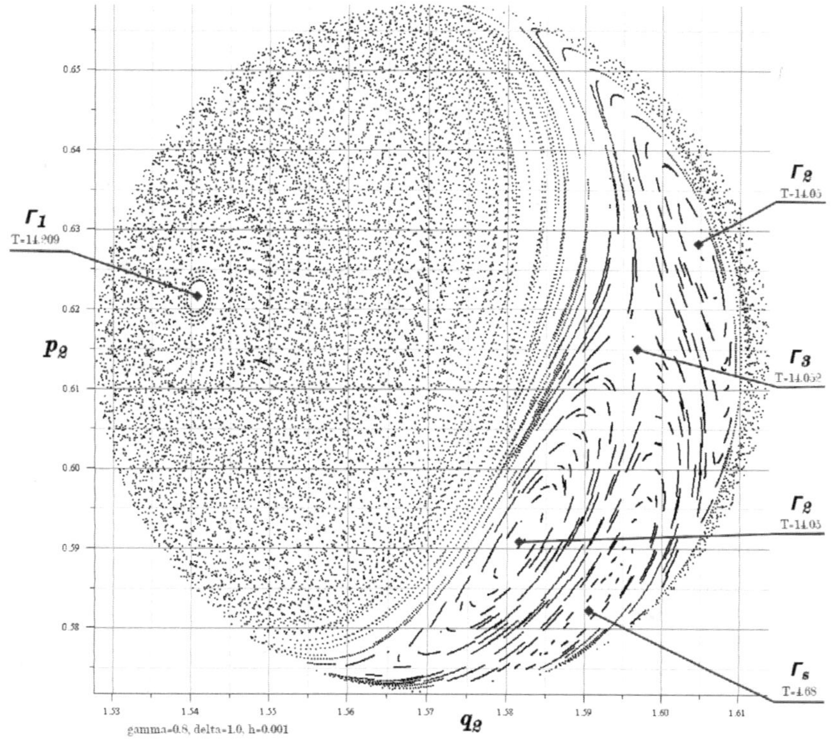

Figure 5. Poincare map showing short- and long- periodic motions originating from Hyperboloidal precession.

IOP Conf. Series: Materials Science and Engineering **927** (2020) 012021 doi:10.1088/1757-899X/927/1/012021

4. Conclusion

For the case of fourth-order resonance, families of long-periodic motions originating from Hyperboloidal precession have been constructed. It has been shown that for all admissible values of problem's parameters there are always thee of these families, two of which are orbitally stable and one is orbitally unstable. Poincare maps have been used to illustrate and verify this result.

Acknowledgement

The study was supported by a grant from the Russian Science Foundation (No. 19-11-00116) at the Moscow Aviation Institute (National Research University).

References

[1] Beletskiy V V 1975 Motion of a satellite relative to its center of mass in gravitational field (Moscow: MSU press)

[2] Markeev A P 2009 Linear Hamiltonian systems and certain stability problems of satellite's motion about its center of mass (Moscow, Izhevsk: Regular and Chaotic Dynamics, Institute of Computer Reasearch)

[3] Kondurar V T 1959 Particular solutions to the general problem of rotational-translational motion of a spheroid in gravitational field of a ball *Astronomical Journal* **5** 890–901

[4] Duboshin G N On rotational motion of artifitial celestial bodies *Bulletins of USSR AS ITA* **7** 511–520

[5] Chernousko F L 1964 On stability of satellite's regular precession *Journal of Applied Math. and Mech.* **28** 155–157

[6] Sokolskiy A G, Khovanskiy S A 1979 Periodic motions close to Hyperboloidal precession of a satellite on circular orbit *Cosmic Research* **17** 208–217

[7] Sukhov E A, Bardin B S 2016 Analytical and numerical computation of symmetric satellite's family of periodic motions originating from its Hyperboloidal precession *Engineering Journal: Science and Innovation* **53**

[8] Sukhov E A, Bardin B S 2017 Analytical and numerical computation and stability analysis of periodic motions of a symmetric satellite *Engineering Journal: Science and Innovation* **11**

[9] Schmidt D S 1974 Periodic solutions near a resonant equilibrium of a Hamiltonian system *Celestial Mechanics* **9**

[10] Bardin B S, Chekin A M 2009 On non-linear oscillations of a Hamiltonian system in case of 3:1 resonance *Journal of Applied Math. and Mech.* **73**

[11] Sukhov E 2018 Analytical and numerical computation ans study of long-periodic motions originating from hyperboloidal precession of a symmetric satellite *Proc. of the 8th Polyakov Readings St. Petersburg*

[12] Sukhov E A 2019 Bifurcation analysis of periodic motions originating from regular precessions of a dynamically symmetric satellite *Russian Journal of Nonlinear dynamics* 4 593–609

IOP Conf. Series: Materials Science and Engineering **927** (2020) 012022 doi:10.1088/1757-899X/927/1/012022

Nonlinear flutter response of cylindrical shell in thermal field and supersonic gas flow

G Y Baghdasaryan[1,2], M A Mikilyan[1,2], I A Vardanyan[1,2], A V Panteleev[3] and N S Severina[3]

[1] Russian-Armenian (Slavonic) University, Department of Mathematics and Mathematical Modeling, 132 Hovsep Emin str., 0051 Yerevan, Armenia

[2] Institute of Mechanics, National Academy of Sciences of Armenia, 24B Marshal Baghramyan ave., 0019 Yerevan, Armenia

[3] Moscow Aviation Institute (National Research University), 4, Volokolamskoe Shosse, 125993, Moscow, Russia.

E-mail: marine.mikilyan@rau.am

Abstract. The work is devoted to the investigation of flutter oscillations and stability of closed cylindrical shell in supersonic gas flow and placed in an inhomogeneous temperature field. It is assumed that supersonic gas flows on the outside of the shell with an unperturbed velocity U, directed parallel to the cylinder generatrix. Under the action of an inhomogeneous temperature field, the shell bulges out; this deformed state is accepted as unperturbed, and the stability of this state is studied. The main nonlinear equations and relations describing the behavior of the examined system are derived. The formulated boundary value problem is solved using Galerkin method. The joint influence of the flow and the temperature field on the relation between the amplitude and frequency of nonlinear oscillations of a cylindrical shell is studied. The critical velocity values are calculated from the corresponding linear system and are given in tables. The numerical results show that: a) the surrounding flow significantly affects the nature of the investigated relation; b) it is impossible to excite steady-state flutter oscillations (the silence zone) up to a certain value of the oscillations frequency; c) the dependence of the amplitude on the oscillations frequency can be either multi-valued or single-valued.

1. Introduction

An important problem in many branches of engineering industries is to increase the reliability and extend the service life of structures interacting with the gas flow. When designing aviation and space structures, one of the most important issues is the stability of the skin of aircraft during its interaction with a supersonic gas flow and the undamped elastic vibrations generated by this interaction – the panel flutter.

Among the problems of panel flutter of thin bodies, the issues of vibrations and stability of cylindrical shells are significant. The experiments [1] showed that the amplitude of flutter oscillations is comparable to the thickness of the shell; therefore, to predict accurately the amplitude of flutter, nonlinear theory of shells should be used. Huge work was done in [2], where the author showed that when solving nonlinear problems, steady flutter oscillations may occur at any stage (subcritical, postcritical) of the perturbed state. In this case, at subcritical velocity, either only soft excitation of flutter oscillations takes place, or only hard.

The main solution method in studies [3], devoted to nonlinear flutter, is the Bubnov–Galerkin method when constructing an approximate solution by representing it in the form of Fourier series in eigenfunctions. Particular attention is paid to the number of retained terms in the Fourier series. Similar questions were studied in detail in [4] where in addition to considering a satisfactory number of modes in the Fourier expansion; also the influence of the shell's edges fixing type is studied. In the study of flutter in both linear and nonlinear formulations, a significant contribution was made by the representatives of the Armenian scientific school [5–9].

Since then, investigation of panel flutter has not stopped, but so far many of the significant aspects of the flutter phenomenon have not been sufficiently studied. There is a large number of works devoted to flutter, but little attention has been paid to the relation between amplitude and velocity taking into account the quadratic aerodynamic nonlinearity. The influence of the surrounding flow on the amplitude-frequency relation has been studied even less. Also, the effect of inhomogeneity of the temperature field on flutter characteristics has not been fully studied, while the task of studying this relations and effects is very relevant.

This work is devoted to investigation of flutter oscillations of a closed cylindrical shell in a supersonic gas flow in the presence of an inhomogeneous temperature field.

2. The problem statement

Let us consider a thin isotropic closed cylindrical shell of constant thickness h and length a in a stationary temperature field T. The shell is assigned to cylindrical coordinates (r,φ,x), the coordinate lines x and φ coincide with the curvature lines of the middle surface of the shell (x – along the generatrix, φ – along the arc of the cross section). There is a supersonic gas stream on the outside of the shell with an unperturbed velocity U directed along the axis $0x$. The axisymmetric stability issues of this aero-thermo-elastic system are investigated, as well as the influence of the temperature field on the relation between the amplitude and frequency of nonlinear flutter oscillations of this system. The investigation is based on the following well-known assumptions:

a) the Kirchhoff – Love hypothesis of non-deformable normal [10]:

$$u_x = u-(r-R)\frac{\partial w}{\partial x},$$

$$u_\varphi = \left(1+\frac{r-R}{R}\right)v-(r-R)\frac{\partial w}{\partial \varphi},$$

$$u_r = w(x,\varphi,t),$$

b) the basic assumptions of the theory of flexible plates, assuming that normal displacements are comparable with the thickness of the plate [11];

c) the "law of plane sections" for determining the aerodynamic pressure [12]:

$$p = p_\infty\left(1+\frac{æ-1}{2}\frac{\text{v}_3}{a_\infty}\right)^{\frac{2æ}{æ-1}}, \tag{1}$$

where p is the gas pressure on the surface of the shell, a_∞ is the speed of sound in the unperturbed gas $\left(a_\infty^2 = æp_\infty\rho_\infty^{-1}\right)$, p_∞ and ρ_∞ are the gas pressure and density in the unperturbed state, æ is the polytropic exponent, v_3 is the normal component of the speed of the shell's surface points;

d) the linear law of temperature change over the thickness of the shell [2]

$T = T_0(x,\varphi)+(R-r)\Theta(x,\varphi)$, where $T_0 = \dfrac{T^+ + T^-}{2}$, $\Theta = \dfrac{k(T^+ - T^-)}{kh-2\lambda}$, λ is the thermal conductivity of the shell material, k is the heat transfer coefficient;

e) the Neumann hypothesis on the absence of displacements from the temperature changes [13].

For simplicity and clarity it is assumed that from the front surfaces of the shell $(r = R \pm h/2)$ there occurs a heat exchange with the medium according to the Newton–Richmann law (constant temperatures T^+ and T^- are maintained on the surfaces), and the side surfaces ($x = 0$ and $x = a$) are heat-insulated.

Under the action of a stationary temperature field inhomogeneous over the thickness $(\Theta \neq 0)$, the shell buckles (with a deflection $w_T(x)$ and a longitudinal displacement $u_T(x)$) and, as a result, an additional aeroelastic pressure appears. The specified buckled state is accepted as unperturbed [9], and its stability under the influence of the temperature field and the pressure of the flowing gas stream is studied.

For simplicity and clarity, the axisymmetric problem is considered in this work. Similarly to [9], first the characteristics of the unperturbed state are determined.

The following equations and relations are derived for thermoelastic stresses and displacements (in account of the formula (1))

$$\frac{d^2u_T}{dx^2} + \frac{\mu}{R}\frac{dw_T}{dx} = 0, \tag{2}$$

$$D\left[\frac{d^4w_T}{dx^4} + \frac{12}{Rh^2}\left(\mu\frac{du_T}{dx} + \frac{w_T}{R}\right)\right] + æp_\infty M\frac{dw_T}{dx} = 0, \tag{3}$$

for the internal forces of the unperturbed state

$$T_{11}^0 = \int_{R-h/2}^{R+h/2} \sigma_{11}^0 dr = \frac{Eh}{1-\mu^2}\left[\frac{du_T}{dx} + \mu\frac{w_T}{R} - \alpha(1+\mu)T_0\right],$$

$$T_{22}^0 = \int_{R-h/2}^{R+h/2} \sigma_{22}^0 dr = \frac{Eh}{1-\mu^2}\left[\mu\frac{\partial u_T}{\partial x} + \frac{w_T}{R} - \alpha(1+\mu)T_0\right]. \tag{4}$$

Here $M = Ua_\infty^{-1}$ is the Mach number for the unperturbed flow, $D = Eh^3/12(1-\mu^2)$, E is the elastic modulus, μ is the Poisson's ratio, α is the coefficient of linear thermal expansion of the shell material.

The solutions of equations (2) and (3) must satisfy the conditions for the edges fixing type at $x = 0$ and $x = a$. Two types of boundary conditions are considered:

hinge supported edges, free to move along the axis $0x$

$$w_T = 0, \quad \frac{d^2w_T}{dx^2} + \alpha(1+\mu)\Theta = 0 \;\; at \;\; x = 0, x = a, \tag{5}$$

$$\frac{du_T}{dx} - \alpha(1+\mu)T_0 = 0 \;\; at \;\; x = 0, x = a; \tag{6}$$

hinge supported and fixed edges

$$w_T = 0, \quad \frac{d^2w_T}{dx^2} + \alpha(1+\mu)\Theta = 0 \;\; at \;\; x = 0, x = a, \tag{7}$$

$$u_T = 0 \;\; at \;\; x = 0, x = a. \tag{8}$$

Finding a solution u_T of equation (2) satisfying the conditions (6) and (8), substituting it into (3) and expressions (4), we obtain the following equation for determining w_T :

$$D\frac{d^4w_T}{dx^4} + \frac{12}{Rh^2}\left(\frac{1-\mu^2}{R}w_T + \frac{\delta\mu^2}{Ra}\int_0^a w_T(x)dx + (1-\delta)\alpha\mu(1+\mu)T_0\right) + æp_\infty M\frac{dw_T}{dx} = 0 \tag{9}$$

with the boundary conditions (5) or (7), where

$$\delta=\begin{cases}0, & \text{when the shell's edges move freely,}\\ 1, & \text{when the shell's edges are fixed.}\end{cases}$$

By finding the solution of equation (9), based on the general theory of thermoelasticity of isotropic bodies, the Kirchhoff–Love hypothesis, similarly to [9], the following nonlinear equation is obtained with respect to the displacement $w(x,t)$ of shell points

$$D\left(\frac{\partial^4 w}{\partial x^4}+\frac{12}{Rh^2}\left(\frac{w}{R}+\mu\frac{dw_T}{dx}\frac{\partial w}{\partial x}+\frac{\mu^2}{a}\int_0^a\frac{w}{R}dx+\frac{\mu}{a}\int_0^a\frac{dw_T}{dx}\frac{\partial w}{\partial x}dx\right)\right)+\frac{Eh}{1-\mu}\alpha T_0\frac{\partial^2 w}{\partial x^2}+$$

$$+\frac{Eh}{2a\left(1-\mu^2\right)}\left\{\frac{d^2w_T}{dx^2}\int_0^a\left(\frac{\partial w}{\partial x}\right)^2dx_1+\frac{\partial^2 w}{\partial x^2}\int_0^a\left[2\frac{dw_T}{dx}\frac{\partial w}{\partial x}+\left(\frac{\partial w}{\partial x}\right)^2\right]dx_1\right\}+\rho h\frac{\partial^2 w}{\partial t^2}+\left(\rho h\varepsilon+\frac{æp_\infty}{a_\infty}\right)\frac{\partial w}{\partial t}+ \quad (10)$$

$$+æp_\infty M\frac{\partial w}{\partial x}+\frac{æ(æ+1)}{4}p_\infty M^2\left[2\frac{dw_T}{dx}+\frac{\partial w}{\partial x}\right]\frac{\partial w}{\partial x}+\frac{æ(æ+1)}{12}p_\infty M^3\left[3\frac{dw_T}{dx}+\frac{\partial w}{\partial x}\right]\left(\frac{\partial w}{\partial x}\right)^2=0.$$

The solutions of equation (10) must satisfy the following condition of hinge support:

$$w=0,\ \frac{\partial^2 w}{\partial x^2}=0 \quad \text{for} \quad x=0,\ x=a. \quad (11)$$

3. Solution of the problem

The obtained nonlinear problem (10)–(11) will be solved approximately by representing its solution in the form

$$w(x,t)=f_1(t)\sin\lambda_1 x+f_2(t)\sin\lambda_2 x\ \left(\lambda_i=i\pi/a\right). \quad (12)$$

Substituting representation (12), which satisfies the boundary conditions (11), into equation (10) and solving it by the Bubnov–Galerkin method, a nonlinear system of ordinary differential equations with respect to the unknown dimensionless functions $x_i=f_i(t)/h$ is obtained. Thus, the stability problem of the hydroelastic system under consideration is reduced to the study of the behavior of solutions of a system of nonlinear ordinary differential equations depending on the value of the velocity of the flowing gas stream (on the value of the parameter ν) and temperature (on the values T, Θ).

3.1. Solution of the linear problem

The solution of the non-linear problem is usually preceded by the analysis of the corresponding linear problem. The linear system corresponding to the obtained non-linear system has the form

$$\frac{d^2x_1}{d\tau^2}+\chi\frac{dx_1}{d\tau}-\frac{2}{3}K\nu x_2+\frac{1}{\pi^2\Delta}\left\{\bar{T}_0+\left(\frac{96}{\pi^2}\mu^2+12\left(1-\mu^2\right)+\pi^4\xi^2\eta^2\right)\xi^2\beta+\bar{\Theta}\left[(1+æ)K\nu^2-\right.\right.$$

$$-96\mu\beta\xi^3\eta+\beta\xi^3\eta^3\left\langle 4608\mu\eta\left(3\left(1-\mu^2\right)+4\pi^4\xi^2\eta^2\right)\left(24\pi^2\left(1-\mu^2\right)+\left(288-\pi^4\right)\mu^2+3\pi^6\xi^2\eta^2\right)+\right.$$

$$+K\nu^2\frac{\pi^2}{\xi^7\beta^3}\left\{-96(1+æ)\left(12\left(1-\mu^2\right)+\pi^2\mu^2\right)\left(3\left(1-\mu^2\right)+4\pi^4\xi^2\eta^2\right)\xi^4\beta^2+\right.$$

$$\left.\left.\left.+\beta K\left(48\left(32-\pi^2\right)\beta\mu\eta\xi^3-K\nu^2(1+æ)\pi^2\right)\right\}\right]\right\}x_1+$$

$$+\frac{2}{5}K\nu\bar{\Theta}\frac{\xi\eta\left(96\left(4-7\mu^2\right)-4\pi^2\left(9-17\mu^2\right)-3\pi^6\xi^2\eta^2\right)\left(48\mu\beta\xi^3\eta+7(1+æ)K\nu^2\right)}{\pi^2\xi^3\eta\Delta}x_2=0,$$

$$\frac{d^2x_2}{d\tau^2}+\chi\frac{dx_2}{d\tau}+\frac{2}{3}K\nu x_1-$$
$$-\frac{2}{5}K\nu\bar{\Theta}\frac{\xi\eta\left(96\left(4-7\mu^2\right)-4\pi^2\left(9-17\mu^2\right)-3\pi^6\xi^2\eta^2\right)\left(48\mu\beta\xi^3\eta+7(1+æ)K\nu^2\right)}{\pi^2\xi^3\eta\Delta}x_1+ \quad (13)$$
$$+4\left\{\bar{T}_0+\left(3\left(1-\mu^2\right)+4\pi^4\xi^2\eta^2\right)\xi^2\beta+4\bar{\Theta}\left[(1+æ)K\nu^2-6\mu\beta\xi^3\eta+\frac{1}{20\pi^2\xi^4\Delta}\left(96\beta\xi^4\left(12\left(1-\mu^2\right)+\right.\right.\right.\right.$$
$$\left.\left.\left.\left.+\pi^2\mu^2\right)\left(3\left(1-\mu^2\right)+4\pi^4\xi^2\eta^2\right)+\pi^2K^2\nu^2\right)\left(96\mu\beta\xi^3\eta-(1+æ)K\nu^2\right)\right]\right\}x_2=0,$$

where

$$\xi=\frac{R}{a},\quad \eta=\frac{h}{a},\quad \beta=\frac{\left(1+\pi^2\xi^2\right)^2}{12\pi^4\left(1-\mu^2\right)\xi^6+\left(1+\pi^2\xi^2\right)^4\eta^2},$$

$$\Delta=9\xi^4\beta^2\left(3\left(1-\mu^2\right)+4\pi^4\xi^2\eta^2\right)\left(\frac{96}{\pi^2}\mu^2+12\left(1-\mu^2\right)+\pi^4\xi^2\eta^2\right)+\pi^2K^2\nu^2.$$

Representing the solution of the system (13) in the form

$$x_1=y_1e^{p\tau},\quad x_2=y_2e^{p\tau},$$

one can obtain the characteristic equation with respect to p. Coefficients of the noted characteristic equation depend on the speed parameter. The value of the parameter $\nu=\nu_{cr}$ at which two of the characteristic indicators become purely imaginary, and the rest still lie in the left half-plane, is critical, it corresponds to the critical speed of the panel flutter in the linear formulation of this problem. Taking $\nu=\nu_{cr}$, from the characteristic equation, one can find the value θ_{cr} of the oscillations frequency of the shell for a linear flutter. Based on the characteristic equation the following data are obtained for flutter critical speed at several values of parameters of thermal field.

Table 1. The value of unidimensional critical velocity depending on a constant temperature field (^0C).

h/a \ T_0	−500	−300	−100	−50	0	5
1/70	8.2821	4.9792	1.6764	0.8506	0.0249	0.0577
1/100	3.9496	2.3312	0.7128	0.3082	0.0964	0.1369
1/200	0.9678	0.5632	0.1586	0.0575	0.0437	0.0538
1/500	0.1540	0.0892	0.0245	0.0083	0.0079	0.0095

Tables 1 and 2 show that: a) at a fixed temperature, the thinner the shell, the lower the critical velocity of the flow, b) for a fixed shell thickness, the critical velocity of the flow as a function of the temperature field has a minimum point.

Table 2. The value of the critical velocity depending on an inhomogeneous temperature field.

h/a \ Θ	−1000	−500	−100	100	500	1000
1/70	0.0354	0.0302	0.0260	0.0238	0.0196	0.0143
1/100	0.0919	0.0942	0.0960	0.0968	0.0986	0.1008
1/200	0.0426	0.0432	0.0436	0.0438	0.0442	0.0447
1/500	0.0077	0.0078	0.0078	0.0079	0.0080	0.0081

Validation of the present results. To check the reliability of the approach in this paper, the parameters from [2] are used: $K = 30.2,\ æ = 1.4,\ \chi = 0.05$. The flutter critical speed in this book was calculated and obtained $\nu_* = 0.11$. For the same data, in absence of thermal field, the following value $\nu_* = 0.1098$ for the critical speed is obtained. Comparing with the result presented in [2], it can be seen that the good agreement is observed. In other hand, for $R = 200h,\ æ = 1.4,\ \chi = 0.05$, in absence of thermal field [14] it was obtained $\nu_* = 0.0376$, while as one can see from the Table 1 of this paper, we have found $\nu_* = 0.0437$. Similar to the above-brought comparison, the brought results have a good agreement, also.

3.2. *Solution of the nonlinear problem*

The obtained nonlinear system is solved by the harmonic balance method, i.e. an approximate periodic solution of the obtained system of ordinary differential equations is sought in the form

$$x_1 = A_1 \cos\theta\tau + B_1 \sin\theta\tau + C_1 + ...,$$
$$x_2 = A_2 \cos\theta\tau + B_2 \sin\theta\tau + C_2 + ... \quad (14)$$

Here A_i, B_i, C_i и θ $(i = 1,2)$ are the unknown constants and the terms with high harmonics are neglected. Solution (14) is substituted into the system, then the coefficients of the free term, $\cos\theta\tau$ and $\sin\theta\tau$ are equated to zero (the terms containing high harmonics are neglected). The resulting system of nonlinear algebraic equations is quite cumbersome to study and is not given here.

The specified system of nonlinear algebraic equations is solved numerically for the following initial parameters: $E = 7.3\cdot 10^{10}$ N/m^2; $\mu = 0.34$; ρ_0=2.79·10^{3}kg/m^3 (Duralumin), $æ$=1.4; $\rho_\infty = 1.29 kg/m^3$; $a_\infty = 340.29 m/s$ (air). The dependence of the amplitude A of steady-state flutter oscillations on the parameters θ and ν for the values $a = 200h$ and $R = 2a$ is investigated. Here are some interesting results in our opinion.

3.2.1. *Influence of an inhomogeneous temperature field*

For small values of the flow velocity, the inhomogeneous temperature field has the following effect on the dependence of the amplitude on the frequency of flutter oscillations, shown in Figure 1. This figure indicates the existence of bifurcation points.

An increase in the flow velocity leads to a qualitative change in the nature of this dependence and to occurrence of a silence zone. In this case, there is a frequency variation interval $[\theta_*, \theta^*]$ where it is possible to excite flutter oscillations and the function $A(\theta)$ has a maximum point, when $\theta^* < \theta < \bar{\theta}$ it is impossible to excite steady flutter oscillations, and when $\theta = \bar{\theta}$ the amplitude value jumps to a certain final value, after which the function $A(\theta)$ is two-valued, one branch of which decreases, and the second increases (Figure 2).

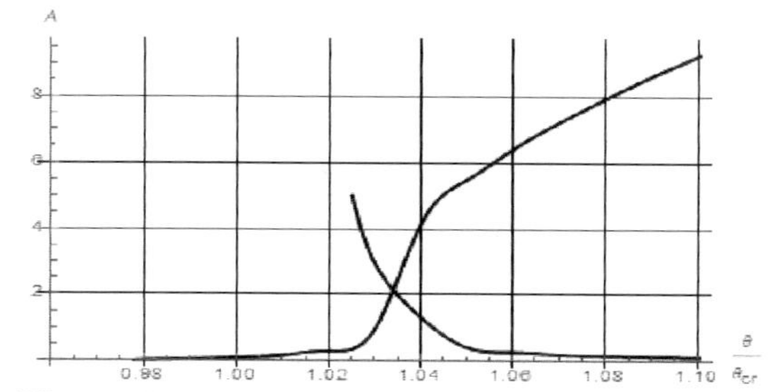

Figure 1. The nature of the amplitude frequency relation at $\nu = 0.1\nu_{cr}, \Theta = -100,\ T_0 = 0$.

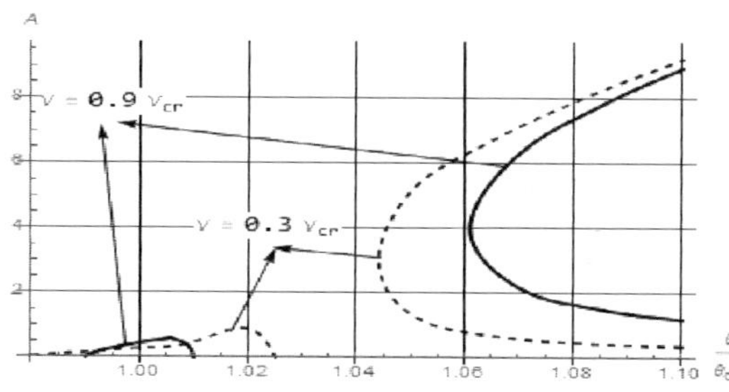

Figure 2. The nature of the amplitude-frequency relation at $\Theta = 0,\ T_0 = -50$.

A further increase in the flow velocity leads to a decrease in the length of the interval $[\theta_*, \theta^*]$.

3.2.2. Influence of a constant temperature field

A constant temperature field significantly changes the above picture of the amplitude-frequency relation. Namely:

At negative temperatures, there is such a certain value θ_* of the parameter θ $(\theta_* < \theta_{cr})$ that for $\theta \in [0, \theta_*]$ it is impossible to excite flutter oscillations (the silence zone). Starting with θ_* the function $A(\theta)$ is a single-valued monotonically increasing function up to a certain value θ^*, after which the function $A(\theta)$ is two-valued (Figure 3).

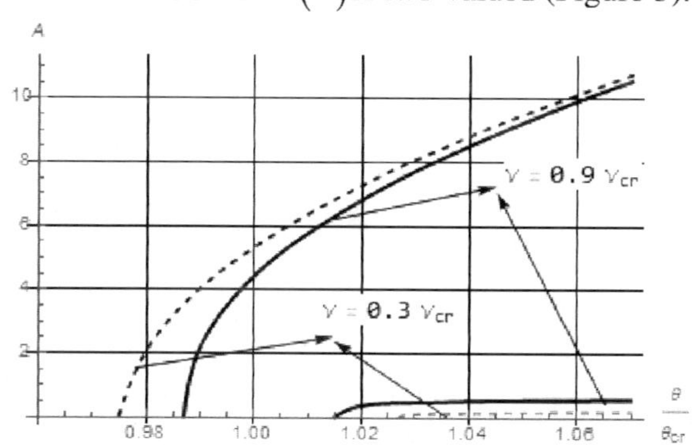

Figure 3. The nature of the amplitude-frequency relation at $\Theta = 0,\ T_0 = -50$.

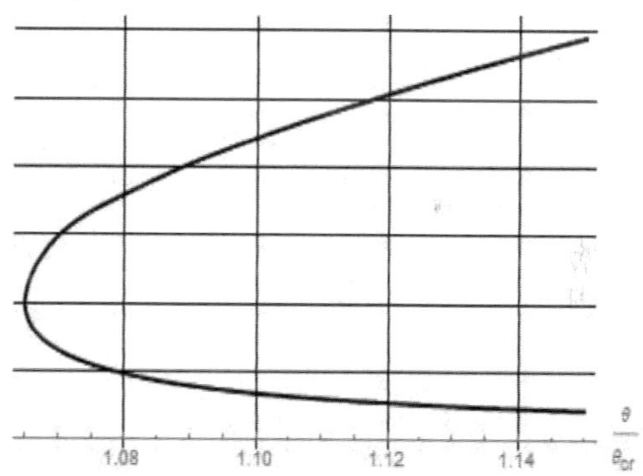

Figure 4. The nature of the amplitude-frequency relation at $\nu = 0.9\nu_{cr},\ \Theta = 0,\ T_0 = 5$.

The indicated relation remains the same with an increase in the velocity of the flowing stream, while the length of the interval $[\theta_*, \theta^*]$ decreases.

At positive temperatures, the dependence shown in Figure 4 takes place.

3.2.3. Joint influence of the temperature field components.

In this case when $\nu < \nu_{cr}$ the nature of the function $A(\theta)$ in the main is similar to those shown in Figs. 3 and 4. For $\nu = \nu_{cr}$ and negative values of the components of the temperature field, a new picture appears, which is shown in Figure 5.

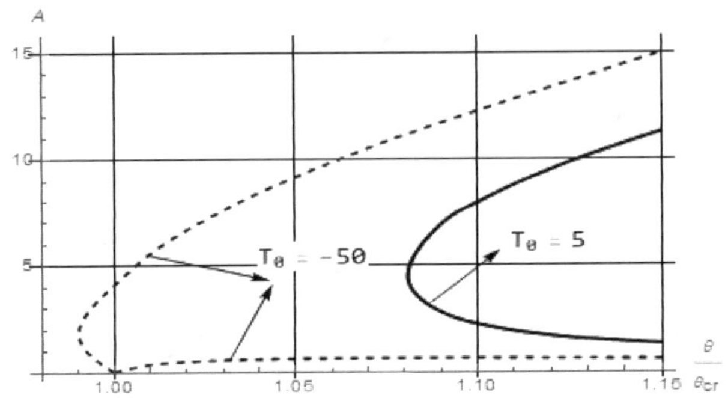

Figure 5. The nature of the amplitude-frequency relation at $\nu = \nu_{cr}$, $\Theta = -100$.

4. Conclusion

The numerical solutions, tables and graphs show that: a) the flow significantly affects the nature of the investigated relation; b) it is impossible to excite steady-state flutter oscillations (the silence zone) up to a certain value of the oscillation frequency; c) the dependence of the amplitude on the oscillation frequency can be both single-valued and multi-valued; d) the value of the critical velocity is a monotonously decreasing function of the relative thickness of the shell and has a minimum point depending on the value of the temperature field.

References

[1] Awrejcewicz J and Krysko V A 2003 *Nonclassical Thermoelastic Problems in Nonlinear Dynamics of Shells. Applications of the Bubnov-Galerkin and Finite Difference Numerical Methods* (Berlin: Springer)

[2] Bolotin V V 1963 *Non-conservative problems of the theory of elastic stability* (New York: Macmillan)

[3] Dowell E H 1975 *Aeroelasticity of plates and shells* (Leyden: Noordhoff)

[4] Amabili M, Pellicano F 2001 *AIAA J.* **39(4)** 564–573

[5] Baghdasaryan G Y, Mikilyan M A, Vardanyan I A and Panteleev A V 2020 *J. Phys.: Conf. Ser.* **1474** id 012007

[6] Bagdasaryan G Y, Mikilyan M A and Vardanyan I A 2019 Mechanics, *Proc. NAS RA* **72**(1) 10–34

[7] Baghdasaryan G, Mikilyan M *Effects of Magnetoelastic Interactions in Conductive Plates and Shells* (Springer International Publishing)

[8] Baghdasaryan G Y, Mikilyan M A 2015 *Int. J. of Non-Linear Mechanics* **77** 51–60

[9] Baghdasaryan G Y, Mikilyan M A, Vardanyan I A and Marzocca P 2019 *TAML* **9**(5) 285–288

[10] Vlasov V Z 1964 *General theory of shells and its application to the engineering (English translation)* NASA-TT F-99

[11] Volmir A S 1972 *The nonlinear dynamics of plates and shells* (Department of the Air Force Foreign Technology Division, Wright-Patterson AFB Dayton, OH United States 45433)

[12] Ashley H, Zartarian C 1956 *Journ. Aeronaut. Sci.* **23**(6) 1109–1118

[13] Novatskiy V 1962 *Issues of thermoelasticity* (Moscow: Acad. of Sci. of USSR)

[14] Shulman Y 1959 *Some dynamic and aeroelastic problems of plate and shell structures* (USA: MIT)

IOP Conf. Series: Materials Science and Engineering **927** (2020) 012023 doi:10.1088/1757-899X/927/1/012023

Benchmark analysis of novel multi-agent optimization algorithm using linear regulators for agents motion control

M Karane and A Panteleev

Moscow Aviation Institute, 4 Volokolamskoe shosse, 125993, Moscow, Russia

E-mail: mmarselina@mail.ru

Abstract. The aim of this paper is to develop algorithm and software for a novel multi-agent constrained global optimization method. In multi-agent algorithms, the solution search area is populated with agents. Some transformations occur over them, which lead to a global extremum. It was proposed to take as a basis the use of linear regulators to control the movement of agents group. At each stage of the algorithm, a search is made by closed loop agents control with various types of criteria. Using criteria of various types, a more detailed study of the solution search area and finding a global extremum is provided. This is the main feature and novelty of the algorithm under consideration. The method efficiency is studied on a standard set of test functions of two variables with a complex structure of level lines using the software developed. Efficiency analysis is carried out for Schweffel function and multi-function from the benchmark set. During the study, statistical characteristics were calculated, according to which the most suitable parameters were selected. The results obtained indicate that the algorithm successfully copes with such problems and can be used, for example, for problems in the theory of optimal control of dynamic systems.

1. Introduction

Over the past few decades, new numerical algorithms have been actively developed to solve a variety of optimization problems. In almost every field of science and technology, numerical algorithms are required. Metaheuristic optimization algorithms are one of the types of effective numerical algorithms. They contain several heuristic procedures governed by upper level methods. For example, they can describe physical and mathematical processes, or even those that occur in the animal world. The metaheuristic algorithms developed can be divided into four groups: evolutionary methods [1–2], swarm intelligence methods [3–4], methods that simulate physical processes, and multi-start methods. Multi-agent algorithms belong to the class of metaheuristic algorithms. A review of the developed multi-agent methods is given in [5]. These algorithms are being developed due to the fact that many classical methods turn out to be unsuitable and cannot cope with the solution of an optimization problem due to its complexity. The objective function can be multi-extreme, have a large number of variables, many local extrema [6–8]. This all greatly complicates the search for a solution, but such tasks are becoming more demanded by practice. Therefore, the use of multi-agent algorithms is extremely relevant and useful in many industries [9–16] and allows one to solve successfully problems both in optimization, optimal control, and, for example, in computer vision, which is now very popular.

IOP Conf. Series: Materials Science and Engineering **927** (2020) 012023 doi:10.1088/1757-899X/927/1/012023

In this paper, we propose a new multi-agent optimization algorithm using linear regulators for agents motion control. The purpose of this study is the formation of algorithm and software of multi-agent algorithm, its testing on a standard set of test functions.

The novelty of the described approach lies in the fact that to find the optimal solution of the problem, the movement of agents is described by a system of differential equations. The task is to find optimal control of a linear dynamic system with full feedback. The type of optimal or suboptimal control and the equations used are known for each type of criterion considered.

The scope of applicability of multi-agent algorithms is quite wide. Accordingly, the problems requiring solution have many differences. Therefore, it is impossible to develop one universal algorithm for solving all existing problems. In this regard, it is necessary to develop this area and develop new algorithms that can more effectively cope with various optimization problems, and perhaps take into account the specifics of the practical engineering tasks.

2. Methods and materials

In this section optimization problem of function and strategy of multi-agent method are considered.

2.1. Optimization problem

Given the objective function $f(x)=f(x_1,x_2,...,x_n)$, we defined the set of admissible solutions $D \subseteq R^n$.

It is required to find the constrained global maximum of a function $f(x)$ on set D, i.e. such a point $x^* \in D$, that

$$f(x^*)=\max_{x\in D} f(x), \tag{1}$$

where $x=(x_1,x_2,...,x_n)^T$, $D=\{x \mid x_i \in [a_i,b_i],\ i=1,2,...,n\}$. Function $f(x)$ can be multi-extremal.

2.2. Solution search strategy

It is required to generate a population of NP agents on a set of admissible solutions D using a uniform distribution law.

The search for an extremumof problem (1) is realized in a given number of passes $P_{\max}$. In the next pass, all agents move under the action of appropriate control for a certain number of iterations.

Let suppose the equation of the agent motion of the form

$$\frac{dx}{dt}=v,\ x(t_0)=x_0,$$

$$\frac{dv}{dt}=u,\ v(t_0)=v_0, \tag{2}$$

where $x-n$ is dimensional vector of agent position, $v-n$ is dimensional vector of agent velocity, t – time, t_0 – initial time in the next passage, x_0 – initial position, v_0 – initial velocity, $u-n$ is dimensional vector of agent control. As $t_0=0$ let $v_0=0$ (0 – zero n - dimensional vector).

Denote $X=\begin{pmatrix} x \\ v \end{pmatrix}$ – the extended state vector of the agent and rewrite equation (2) in the form

$$\frac{d\begin{pmatrix} x \\ v \end{pmatrix}}{dt}=\begin{pmatrix} O_n & E_n \\ O_n & O_n \end{pmatrix}\begin{pmatrix} x \\ v \end{pmatrix}+\begin{pmatrix} O_n \\ E_n \end{pmatrix}u,$$

or

$$\frac{dX}{dt} = AX(t) + Bu(t),\ X(t_0) = X_0 = \begin{pmatrix} x_0 \\ v_0 \end{pmatrix}, \tag{3}$$

where O_n, E_n – zero and unit matrices of order n, $A = \begin{pmatrix} O_n & E_n \\ O_n & O_n \end{pmatrix}$, $B = \begin{pmatrix} O_n \\ E_n \end{pmatrix}$ – matrices of size $(2n \times 2n)$, $(2n \times n)$ respectively; no restrictions are imposed on the control vector, i.e. $u \in R^n$.

In the initial population ($k = 0$), as well as at the end of each k-th pass, determine the position of the leader among the agents of the population and the corresponding maximum value of the objective function: $x^{best,k}, f(x^{best,k})$. During the next pass, he does not change his position:

$$\frac{dx^{best}}{dt} = 0,\ x^{best}(t_k) = x^{best,k},$$

$$t \in [t_k, t_{k+1}), k = 0,\dots,P_{\max} - 1, \tag{4}$$

$$\frac{dv^{best}}{dt} = 0,\ v^{best}(t_k) = 0,$$

where the control of leader $u^{best} = 0$, or $\frac{dX^{best}}{dt} = 0$ with leader state vector $X^{best} = (x^{best}, v^{best})^T$.

The remaining $(NP - 1)$ agents are divided into 4 equal groups:

- agents using the minimization criterion of agents movement to the current leader in a finite time interval;
- agents using the minimization criterion of agents movement to the current leader in an infinite time interval;
- agents using the minimization semi-defined criterion (so called criterion of the generalized work) control agents for a finite time interval;
- agents using functional increment minimization of agents movement to the current leader at the current moment(locally optimal approach).

For all agents of each group, positions and velocity vectors are different. However, we can find and apply the same feedback control law determined by the relation for a linear optimal controller, the gain matrix of which is found from the minimization condition of the quadratic control quality criterion. It characterizes the nature of the approximation agent to agent leader at the current iteration, as well as the intensity of the control signal applied.

Introduce the deviation from the leader $\Delta X = X - X^{best}$, whose change is described by the equation (subtracting equation (4) from (3)):

$$\frac{d\Delta X}{dt} = A\Delta X(t) + Bu(t),\ \Delta X(t_k) = \begin{pmatrix} x^k - x^{best,k} \\ v^k \end{pmatrix},\ t \in [t_k, t_{k+1}],\ k = 0,\dots,P_{\max} - 1,$$

where x^k, v^k – the agent position and velocity at the end of the previous pass.

Movement of the first group of agents. Quality criterion for controlling the agents trajectories of the first group is $I = \frac{1}{2}\int_{t_k}^{t_{k+1}} \left[\Delta X^T(t) S(t) \Delta X(t) + u^T(t) Q(t) u(t) \right] dt + \frac{1}{2}\Delta X^T(t_{k+1}) \Lambda \Delta X(t_{k+1})$, where Λ – non-negative definite symmetric matrices of size $(n \times n)$, $Q(t)$ – positive definite symmetric matrix $(n \times n)$.

For any initial states, optimal feedback control $u^*(t, \Delta X)$ has the form: $u^*(t, \Delta X) = -Q^{-1}(t) B^T(t) P(t) \Delta X = -F(t)\Delta X$, where matrix coefficients of the gain of the linear optimal regulator $F(t) = Q^{-1}(t) B^T(t) P(t)$, $P(t)$ – symmetric matrix of sizes $(n \times n)$, satisfying the Riccati differential equation $\dot{P}(t) = -A^T(t) P(t) - P(t) A(t) + P(t) B(t) Q^{-1}(t) B^T(t) P(t) - S(t)$,

$P(t_{k+1})=\Lambda$. Here $t\in[t_k,t_{k+1}]$, $t_0=0$, value $t_{k+1}=t_k+NMAX\cdot h$, where $NMAX$ – given number of iterations, h – integration step (4). To simplify the solution, you can let everywhere $A(t)=A, B(t)=B, Q(t)=Q, S(t)=S$, because the model (3), (4) are linear stationary.

Movement of the second group of agents. The quality criterion for controlling the trajectories of agents of the second group is $I=\frac{1}{2}\int\limits_{t_k}^{+\infty}\left[\Delta X^T(t)S\,\Delta X(t)+u^T(t)Qu(t)\right]dt$, where S – non-negative definite symmetric matrix of sizes $(n\times n)$, Q – positive definite symmetric matrix $(n\times n)$.

For any initial states, optimal feedback control $u^*(\Delta X)$ has the form: $u^*(\Delta X)=-Q^{-1}B^TP\Delta X=-F\Delta X$, where matrix coefficients of the gain of the linear optimal regulator $F=Q^{-1}B^TP$, a P – positive definite symmetric matrix satisfying the Riccati algebraic equation: $-A^TP-PA+PBQ^{-1}B^TP-S=0$.

The solution to this equation satisfying the Sylvester criterion is the only one.

Movement of the third group of agents. The quality criterion for controlling the agents trajectories of the third group can be written:

$$I^{\text{o.p.}}=\frac{1}{2}\int\limits_{t_k}^{t_{k+1}}\left[\Delta X^T(t)S(t)\Delta X(t)+u^T(t)Q(t)u(t)+\Delta X^T(t)P(t)B(t)Q^{-1}(t)B^T(t)P(t)\Delta X(t)\right]dt+$$

$$+\frac{1}{2}\Delta X^T(t_{k+1})\Lambda\,\Delta X(t_{k+1}),$$

where Λ – non-negative definite symmetric matrices of sizes $(n\times n)$, $Q(t)$ – positive definite symmetric matrix $(n\times n)$.

For any initial states, optimal feedback control with $u^{\text{o.p.}}(t,\Delta X)$ has the form: $u^{\text{o.p.}}(t,\Delta X)=-Q^{-1}(t)B^T(t)P(t)\Delta X=-F(t)\Delta X$, where matrix coefficients of the gain of the linear regulator $F(t)=Q^{-1}(t)B^T(t)P(t)$, a $P(t)$ – symmetric matrix satisfying the linear differential equation: $\dot{P}(t)=-A^T(t)P(t)-P(t)A(t)-S(t)$, $P(t_{k+1})=\Lambda$.

Movement of the fourth group of agents (for all group agents, locally optimal control is applied). The quality criterion for controlling the trajectories of agents of the fourth group is $I^{loc}=\frac{1}{2}\int\limits_{t_k}^{t}\left[\Delta X^T(\tau)S(\tau)\Delta X(\tau)+u^T(\tau)Q(\tau)u(\tau)\right]d\tau+\frac{1}{2}\Delta X^T(t)\Lambda(t)\Delta X(t)$, where $\Lambda(t)$ – non-negative definite symmetric matrices of sizes $(n\times n)$, $Q(t)$ – positive definite symmetric matrix $(n\times n)$.

For any initial states, the locally optimal feedback control $u^{loc}(t,x)$ has the form: $u^{loc}(t,\Delta X)=-Q^{-1}(t)B^T(t)\Lambda(t)\Delta X=-F(t)\Delta X$, where matrix coefficients of the gain of the linear regulator $F(t)=Q^{-1}(t)B^T(t)\Lambda(t)$.

After performing $NMAX$ iterations, each agent presents the best result obtained during the movement. This completes the next passage.

Then, among all the agents of the population, the leader is selected again and a new division into groups is made. The passage repeat process ends when the maximum number of passes is reached.

3. Results and discussion. Efficiency analysis

Based on the described algorithm, we developed a software that allows finding the extrema of functions of many variables. We used the Microsoft Visual Studio development environment, the C# programming language for creating software. On the main form, the user can select the objective function, the set of admissible solutions and algorithm parameters. The result of the program is the minimum point (x^*,y^*) of the objective function and the value of the function at this point $f(x^*,y^*)$.

The software also demonstrates the exact solution if it is known and the deviation of this solution from the one found. The program depicts the level lines of the selected function, shows the position of group of agents on the last iteration, and separately highlights the best agent.

Using software, we studied the effectiveness of the algorithm on a standard set of test functions [6–8] of two variables with a complex structure of level lines. The article provides an analysis of the effectiveness of two functions from the set: Schwefel function and multi-function. For demonstration, we selected functions with a different number of extrema, a range of feasible solutions.

The analysis of the effectiveness of a multi-agent optimization algorithm using linear regulators to control the movement of agents is carried out as follows.We will perform a series of 100 solutions of the considered problem with the same parameter values (Tables 1–4) [6]. For the sample obtained $\{f^1, f^2, ..., f^{100}\}$ we calculate:

- the average value of the deviation: $\overline{\Delta f} = \frac{1}{100}\sum_{i=1}^{100} \Delta f_i$ where $\Delta f_i = f(x^*) - f^i$ – the deviation of the value f^i from the maximum;
- $\Delta f_{best} = \min_i \Delta f_i$ – the smallest deviation value;
- standard deviation: $\bar{\sigma}_f = \sqrt{\bar{S}_{100}}$, where $\bar{S}_{100} = \frac{1}{100}\sum_{i=1}^{100}(\Delta f^i - \overline{\Delta f})^2$;
- n_{succ} –the number of successes (success is considered to be the best point in the ε- neighborhood of the exact solution, $\varepsilon = \frac{\max_{i=1,...,n} |b_i - a_i|}{1000}$).

Table 1. Definition of Schwefel function [6-8].

Objective function	Set of admissible solutions	Minimum point	Function value
$f(x,y) = x\sin(\sqrt{\lvert x\rvert}) + y\sin(\sqrt{\lvert y\rvert})$	$x, y \in [-500; 500]$	$(420.9687; 420.9687)^T$	$f(x^*, y^*) = 837.9657$

Table 2. Influence of method parameters($NMAX = 50$, $h = 0.0001$). Schwefel function [6-8].

Method parameters				$\overline{\Delta f}$	Δf_{best}	$\bar{\sigma}_f$	n_{succ}
NP	P_{max}	k_s	k_l				
5001	10	0.1	1.0	0.000385	0.000025	0.001561	79
2001	10	0.1	1.0	0.003127	0.000025	0.009733	55
1601	10	0.1	0.5	0.005495	0.000025	0.021443	99
1601	10	0.1	1.5	0.005399	0.000025	0.021717	99
1601	15	0.1	1.5	1.188174	0.000025	11.784093	99
1601	10	0.1	0.5	1.187471	0.000025	11.784456	99
1601	10	1.0	1.0	2.370793	0.000026	16.581389	98
1601	10	3.0	1.0	2.475812	0.000042	3.999449	27

Table 3. Definition of Multi-function [6–8].

Objective function	Set of admissible solutions	Minimum points	Function value

$f(x,y)=x\sin(4\pi x)-$ $-y\sin(4\pi y+\pi)+1$	$x,y\in[-2;2]$	$(1.6288;1.6288)^T$ $(-1.6288;-1.6288)^T$ $(-1.6288;1.6288)^T$ $(1.6288;-1.6288)^T$	$f(x^*,y^*)=4.2539$

Table 4. Influence of method parameters ($h=0.0001$). Multi-function [6–8].

Method parameters					$\overline{\Delta f}$	Δf_{best}	$\overline{\sigma}_f$	n_{succ}
NP	*NMAX*	$P_{\max}$	k_s	k_l				
501	20	10	0.01	5.00	0.062945	0.000001	0.176148	81
501	80	20	0.01	5.00	0.029248	0	0.104200	82
501	80	10	10	5.00	0.029720	0	0.101900	79
501	80	10	1	0.01	0.006591	0	0.051225	93
501	80	10	1	0.50	0.006981	0	0.050309	93
1201	80	10	1	0.10	0.000031	0	0.000058	100
501	80	15	1	0.10	0.005267	0	0.049655	98
801	80	10	1	0.10	0,000904	0	0,008349	99

The analysis of the effectiveness of a multi-agent optimization algorithm using linear regulators for agent motion control is obtained. The algorithm run with various parameters in order to find out what parameter values should be specified to achieve the best result. This can help in solving problems that do not have a predetermined answer.

Analyzing the results obtained in Tables 1–4, we can say which set of parameters is optimal for each task and give general recommendations for their selection:

- the size of population *NP* determines the number of calculations of the objective function at each iteration. The value $NP-1$ must be divisible by four. For a task with a large range of admissible solutions, it is recommended to take a larger value of the parameter *NP*. Recommended parameter value $NP\in[501;1601]$;
- the number of iterations *NMAX* during the passage determines how long the search for new solutions per passage will continue. As the parameter *NMAX* increases, the accuracy of the solution increases. Recommended values for the considered set of standard functions depending on the complexity of the function $NMAX\in[20;80]$;
- the maximum number of passes $P_{\max}$ determines how long the search for new solutions will continue. As the number of passes increases, the accuracy of the solution increases. Recommended value for this parameter $P_{\max}\in[10;30]$;
- coefficient k_s for matrix S, defining the criterion of quality of agent trajectories control. Recommended value for this parameter $k_s\in[0.1;1]$;
- coefficient k_l for matrix Λ, defining the criterion of quality of agent trajectories control. Recommended value for this parameter $k_l\in[1;5]$;
- step of integrating Riccati differential equations h. Recommended value for this parameter $h=0.0001$.

As a result, Table 5 shows the best parameters for each function under study. These parameters were chosen on the basis of the least computational cost and so that the probability of getting into the vicinity of the solution tends to unity. The time for which the program receives a result with the specified parameters is also indicated.All calculations were carried out on the Intel Core i7-4710HQ processor with a clock frequency of 2.5 GHz.

Table 5. Efficiency analysis result.

Objective function	The best combination of parameters	Solution time, sec
Schwefell function (Tables 1–2)	$NP = 1601,\ NMAX = 50,\ P_{max} = 10,\ k_s = 0.1,\ k_l = 1.5,\ h = 0.0001.$	12.82
Multi-function (Tables 3–4)	$NP = 801,\ NMAX = 80,\ P_{max} = 10,\ k_s = 1,\ k_l = 0.1,\ h = 0.0001.$	7.49

Using the recommendations on the selection of parameters described above and the optimal sets of parameters for each function, shown in table 5, you can search for solutions to more complex problems and apply ready-made specified sets of parameters. For example, for those tasks those have a large dimension, or whose solution is not known.

4. Conclusion

A novel multi-agent optimization algorithm using linear regulators for agent motion control is proposed. At each step of the algorithm, an optimal feedback control is sought. It is required to find such a control that a minimum of the following criteria is achieved: a criterion for the quality of agent control at a finite period of time; a criterion for the quality of agent control at an infinite period of time; a semi-defined quality criterion (criterion for generalized work) of agent control at a finite period of time; an increment minimization of the criterion of agent control quality (locally optimal approach). Using criteria of various types, a more detailed study of the set of admissible solution and finding a global extremum is provided. The aim was fulfilled - a novel multi-agent optimization algorithm using linear regulators for agent motion control was developed, and software was created that allows finding the solution to optimization problems using the proposed algorithm. The software allows you to find the constrained global extremum of functions of many variables with a complex structure of level surfaces. The software also allows you to depict graphs of function level lines, select the most suitable algorithm parameters, and observe the operation of the algorithm in a step-by-step mode.

The effectiveness of the proposed algorithm on a standard set of test functions of two variables with a complex structure of level lines. For demonstration, two functions from the set were selected: Schwefel function and multi-function. Efficiency analysis consists in the fact that for each function a series of 100 runs of the solution to the problem with the same parameters was carried out. As a result, the statistical characteristics were calculated from the sample: the average value of the deviation, the smallest deviation value, standard deviation and the number of hits in the vicinity of optimal solution. Based on the obtained characteristics, it was concluded that for what parameters the algorithm gives the best result for each function. The recommendations for the algorithm parameters selection were suggested.

Based on these results, we can conclude that the proposed multi-agent optimization algorithm using linear regulators for agent motion feedback control can be successfully applied to solve various problems and give results close to optimal.

References

[1] Kordestani J, Ahmadi A and Meybodi M 2014 *Applied intelligence* **41**(4) 1150–69
[2] Firouzjaee A, Kordestani J and Meybodi M 2017 *Engineering Optimization* **49**(4) 597–616
[3] Blum C and Li X 2008 *Swarm Intelligence in Optimization* (Berlin: Springer)
[4] Zambrano-Bigiarini M, Clerc M and Rojas R 2013 *IEEE Congress on Evolutionary Computation (CEC)* 2337–44
[5] Beheshti Z, Shamsuddin S 2013 *Int. J. Adv. Soft Comput. Appl.* **5**(1) 1–35
[6] Herrera F, Lozano M and Molina D 2010 *Test suite for the special issue of soft computing on scalability of evolutionary algorithms and other metaheuristics for large scale continuous optimization problems* (Technical Report. University of Granada)
[7] Mishra S 2007 *Some new test functions for global optimization and performance of repulsive*

IOP Conf. Series: Materials Science and Engineering **927** (2020) 012023 doi:10.1088/1757-899X/927/1/012023

particle swarm method (Munich Personal RePEc Archive MPRA)
[8] Neumaier A Personal page, set of test functions // URL: http://www.mat.univie.ac.at/~neum/glopt/test_results.html.
[9] Panovskiy V and Panteleev A 2017 *J. Comput. Syst. Sci. Int.* **56**(1) 52–63
[10] Panteleev A and Pis'mennaya V 2018 *J. Comput. Syst. Sci. Int.* **57**(1) 25–36
[11] L.M. Rere, M.I. Fanany and A. Arymurthy 2016 *Metaheuristic Algorithms for Convolution Neural Network* (Computational Intelligence and Neuroscience)
[12] Petukhov V, Ivanyukhin A and Wook W 2019 *Cosmic Research* **57** 188–203
[13] Starchenko A 2019 *Cosmic Research* **57** 289–300
[14] Kondratieva L 2019 *AIP Conf. Proc.* **2181** id 020002
[15] Karane M and Panteleev A 2019 *AIP Conf. Proc.* **2181** id 020028
[16] Rybakov K 2019 *Gyroscopy and Navigation* **10**(1) 27–34

IOP Conf. Series: Materials Science and Engineering **927** (2020) 012024 doi:10.1088/1757-899X/927/1/012024

Application of bio-inspired methods for optimal open loop discrete time deterministic control problems

A V Panteleev and Y V Stanovskaya
Moscow Aviation Institute, 4 Volokolamskoe shosse, 125993, Moscow, Russia

E-mail: avpanteleev@inbox.ru

Abstract. The article proposed the novel Sparrow colony optimization (SCO) algorithm to solve the optimal open loop control problem for nonlinear discrete deterministic dynamic systems described by difference equations. Numerical algorithm and software in Python programming language are implemented. Model examples demonstrating the effectiveness of the developed software are solved. Comparison of the considered algorithm results with a Dragonfly optimization algorithm at the solution of model problems is made. The recommended parameters values of the proposed algorithm are suggested.

1. Introduction

Application of optimization methods in technical tasks leads to necessity of development of new approaches by reason of use of nonlinear objective functions with complex structure of level surfaces and constraints of different types. One of the modern trends in optimization theory is meta-heuristic algorithms and their modifications, including bio-inspired optimization methods, such as particle swarm optimization algorithm, ant colony method, wolf pack method. Detailed reviews and comparison of described types of methods are given in [1-6].

The paper deals with the method of Sparrow colony optimization algorithm for a constrained global extremum and its application to the problem of finding the optimal open loop control of discrete deterministic dynamic systems. Different meta-heuristic methods are successfully used for solving quite a wide class of optimal control problems [7-11]. To solve the main problem of this article – the search of optimal open loop control [11] – a Python software was created and used to solve two model open loop discrete time deterministic control problems.

2. Methods and materials

The behavior of a nonlinear deterministic discrete model of a control object is described by a difference equation:

$$x(t+1) = f(t, x(t), u(t)), \tag{1}$$

where $t \in T = [0,1,\dots,N-1]$ – discrete time, $x \in R^n$ – state vector, $u \in U(t) \subseteq R^q$ – control vector, $U(t)$ – set of admissible control values for each time moment t representing a direct product of segments $[a_i(t), b_i(t)]$, $i = 1,2,\dots,q$; $f(t,x,u) = (f_1(t,x,u), f_2(t,x,u),\dots, f_n(t,x,u))^T$ – continuous vector function; N – number of discrete time steps. The initial condition is given in the form:

$$x(0) = x_0. \tag{2}$$

It is assumed that only discrete time information t is used in control law, so the open loop control is considered. The set of allowable processes $D(0,x_0)$ – is a set of pairs $d=(x(\cdot),u(\cdot))$, where $x(\cdot)=\{x_0,x(1),...,x(N)\}$ – trajectory, $u(\cdot)=\{u(0),u(1),...,u(N-1)\}$ – admissible control, satisfying the state equation (1) and initial condition (2). The cost functional on the set $D(0,x_0)$ is defined $I(d)=\sum_{i=0}^{N-1} f^0(t,x(t),u(t))+F(x(N))$ or $I(d)=F(x(0),...,x(N);u(0),...,u(N-1))$, where $f^0(t,x,u)$, $F(x)$, $F(x(\cdot),u(\cdot))$ – continuous functions.

It is required to find such a pair $d^*=(x^*(\cdot),u^*(\cdot))\in D(0,x_0)$, that satisfies:

$$I(d^*)=\min_{d\in D(t_0,x_0)} I(d).$$

The problem of finding the optimal program control for discrete deterministic systems can be solved with the method of Sparrow colony optimization.

The principle of the Sparrow colony optimization [9] algorithm is the following: sparrows, united in packs, obey the commands of the pack leader, having some freedom to choose the method of food search. What distinguishes them from other birds and animals is their special unity, coordination of collective actions, intensity of found food source utilization before its disappearance, clear and friendly execution of commands common for the members of the pack.

In the first iteration, an initial population with NP members is generated on a set of acceptable solutions using a uniform distribution law. For each sparrow (solution), the value $f(x)$ of the target function is calculated. The best value of the function is associated with the position of the pack leader. It does not participate in the food search procedure (finding the function extremum) at the current iteration, but only waits for its results to be processed and further stored in the memory matrix of M size, where M is the maximum number of records. The best result achieved at each k-th iteration up to its final filling is recorded in the matrix. The number M determines the number of iterations performed by the algorithm in one loop. When M records are reached, the memory is cleared and the best solution from the records obtained at a given pass is placed in a special $Pool$ set, which stores the best results of the passes.

A new leader position is generated randomly by Levi distribution:

$$x^{1,k}=x^{1,k,search}+\frac{\alpha}{k+1}\circ Levy(\lambda),$$

where $\circ$ – component product of vectors (by Hadamard), α – is a value of a leader's step.

The position of the other members of the pack is modeled with the help of uniform distribution on the parallelepiped, the center of which is determined by the position of the pack leader, and the lengths of the sides are equal $r^k(b_i-a_i)$ $i=1,...,n$, where a_i and b_i – the left and right borders of the allowed solutions area respectively. The trajectory of each individual, in the course of which the search at the current iteration is realized, is described by solving the stochastic differential equation:

$$dx=f(t,x^{j,k}(t))dt+\sigma(t,x^{j,k}(t))dW+dq\ ,\ x^{j,k}(0)=x^{j,k},$$

where W – the standard Wiener random process, dq – is a Poisson's component of the form:

$$dq=\sum_p \theta_p\delta(t-\tau_p)dt,$$

$\delta(t)$ – asymmetric delta function, τ_p – leap moments. At random moments of time τ_p the position of the sparrow receives random increments θ_p, that form a Poisson's flow of events of a given intensity μ. The solution of the equation determines the sparrow trajectories that implement the diffusion search procedure with jumps.

The drift vector $f(t,x^{j,k}(t))$ is determined by the expression: $f(t,x^{j,k}(t))=c_1r_1[x^{best}-x^{j,k}(t)]$.

The diffusion matrix $\sigma(t, x^{j,k}(t))$ takes into account information about the best solution obtained by the individual for all iterations, about the best solution in the vicinity of the current solution, determined by the radius ρ: $\sigma(t, x^{j,k}(t)) = c_2 r_2 [x^{j,best} - x^{j,k}(t)] + c_3 r_3 [x^{j,local} - x^{j,k}(t)]$ where c_1, c_2, c_3 – influence coefficients, depending on the number of iterations; r_1, r_2, r_3 – random parameters, uniformly distributed on a segment $[0;1]$. Usually c_2 characterizes the process of forgetting about the search history.

3. Experimental part

Algorithm for searching the optimal open loop control using the method of Sparrow colony optimization is implemented in Python language in the Qt environment. The user can set the parameters of the algorithm to study their influence on the work of the algorithm. At the output, the user gets an approximate solution of the problem and its deviation from the exact solution, as well as diagrams of optimal control and corresponding trajectories. This allows you to determine the accuracy of the algorithm's work and analyze how close the obtained results are to the expected ones. The algorithm of Sparrow colony optimization has been tested on a set of test functions of many variables with the already known solution, which made it possible to verify the accuracy of its work. As for the speed of program execution, which solves the problem of searching for optimal open loop control using the described method, the number of iterations in the algorithm increases greatly. To get an acceptable accuracy the program execution takes from 20 to 50 minutes.

4. Results and discussion

SCO algorithm is proposed to finding an optimal open loop control of discrete deterministic systems and theirs trajectories. The recommended values of its parameters were formed.

4.1. Example 1.

Schwefel function: $f(x;y) = x \cdot \sin(\sqrt{|x|}) + y \cdot \sin(\sqrt{|y|})$. The admissible values set D: $x \in [-500;500]$, $y \in [-500;500]$. Algorithm parameters for the function: $NP = 20$, $M = 20$, $L = 15$, $T = 15$, $h = 50$, $\mu = 0.7$, $\alpha = 1$, $\gamma = 0.75$, $\eta = 0.89$, $\lambda_{Levi} = 1.4$, $c_1 = 0.5$, $c_2 = 0.6$, $c_3 = 0.5$, $\varepsilon = 10^{-5}$.

The algorithm's test results: approximate solution $(x^*; y^*) = (420.972; 420.9664)$; value of the objective function: $f(x^*; y^*) = 837.96577$; exact solution: $(x; y) = (420.9687; 420.9687)$; value of the function accorded to exact solution: $f^{exact}(x; y) = 873.9658$; deviation of the approximate solution from exact: $\Delta = f^{exact} - f = 2.74 \cdot 10^{-5}$; algorithm's runtime: $t_{work} = 2.82$ seconds.

The recommended values for the SCO algorithm are as follows: the maximum number of entries in the memory matrix must be at least 5, so that the algorithm retains its meaning and strategy of finding a solution. Recommended value $M \in [5;30]$. The number of records M may be greater, but the number of generated members of pack NP should be increased so that the solution does not go deep into the local extremum. Parameters involved in modeling the flock trajectory: $\alpha \in [0;1]$, $\gamma \in [0;1]$, $\eta \in [0;1]$, $\lambda \in [0;1]$, $c_1 \in [0;1]$, $c_2 \in [0;1]$, $c_3 \in [0;1]$, $\mu \in [0;1]$.

4.2. Example 2. The plant model:

$$x_1(t+1) = x_2(t), \; x_2(t+1) = 2x_2(t) - x_1(t) + \frac{1}{N^2}u(t),$$

where $x \in R^2$, N – number of discrete steps. The initial conditions are given: $x_1(0) = 0$, $x_2(0) = 0$. The cost functional defined on a set of admissible solutions is:

$$I = x_1(N) - \frac{1}{2N}\sum_{t=0}^{N-1} u^2(t).$$

It is required to find a maximum value of the cost functional and an optimal process $(x^*(\cdot),u^*(\cdot))$.

The known exact solution:

$$\max I = \frac{1}{3} - \frac{3N-1}{6N^2} - \frac{1}{2N^3}\sum_{t=0}^{N-1} t^2 .$$

With the following parameters of SCO algorithm: $NP = 2500$, $M = 25$, $L = 15$, $T = 15$, $\mu = 0.2$, we receive the optimal open loop control and trajectories accordingly. The results obtained: $I^* = 0.134827$ – approximate functional value, $\Delta = I^{exact} - I^* = 0.021972$ – deviation from the exact solution, $\Delta / I^{exact} = 0.14013$ - relative accuracy of algorithm, $I^{exact} = 0.1568$ – exact functional value, $t_{work} = 1191$ seconds – algorithm runtime. The graphs of the obtained optimal open loop control and corresponding trajectories are shown in the figure 1.

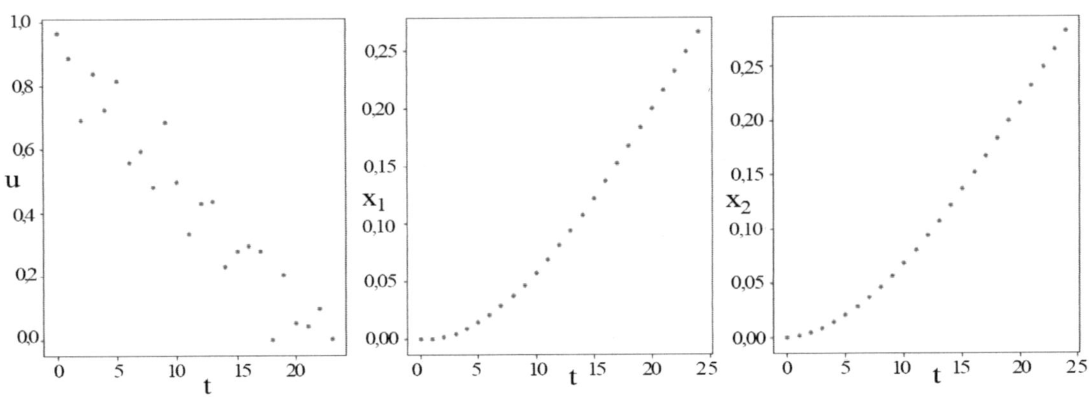

Figure 1. Optimal control and corresponding trajectories in example 2.

Two algorithms were compared – the DA [8] and SCO algorithm in solving the discrete problem with different number of discrete steps N. At close values of deviations of the approximate solutions from the exact ones in both algorithms, we get different operating time. SCO takes $t_{work} = 21664$ seconds to get the accuracy of $\Delta = 0.02058$, while DA handles the work faster, though a little less accurately. In Table 1, you should pay attention to the column with Δ / I^{exact} values, which shows the relative accuracy of the algorithm. For the SCO algorithm these values are less than for DA, from here we can conclude that SCO works more accurately. Decision, which algorithm is better, depends on the requirements of the task: if time is more important than accuracy, then the best algorithm can be considered DA and vice versa: if you need an accurate solution without time constraints, the best - the algorithm SCO.

Table 1. Comparison of DA and SCO in solving the discrete time control example 2

Method	N	NP	I^*	I^{exact}	Δ	Δ / I^{exact}	t_{work}
DA	5	2000	0.055795	0.1200	0.064205	0.53505	2331
SCO	5	900	0.055999	0.1200	0.064000	0.53333	4206
DA	10	2000	0.095889	0.1425	0.046610	0.32709	2304
SCO	10	900	0.101629	0.1425	0.040870	0.28681	4671
DA	25	2000	0.119512	0.1568	0.037280	0.23776	2202
SCO	25	2000	0.136218	0.1568	0.020580	0.13125	21664

4.3. Example 3. Luus-Tassone problem [11].

The plant model:

$$x_1(t+1) = \frac{x_1(t)}{1+0,001u_1(t)(3+u_2(t))};$$

$$x_2(t+1) = \frac{x_2(t)+u_1(t)x_1(t+1)}{1+u_1(t)(1+u_2(t))};$$

$$x_3(t+1) = \frac{x_3(t)}{1+0,001u_2(t)(1+u_3(t))},$$

where $x \in R^3$, $t = 0,...,N-1$, $N = 100$. Initial state vector: $x(0) = (2,5,7)^T$.

Control constraints: $0 \le u_1(t) \le 4$, $0 \le u_2(t) \le 4$, $0 \le u_3(t) \le 0,5$.

Cost functional:

$$I = x_1^2(N) + x_2^2(N) + x_3^2(N) + \left[\left(\sum_{k=1}^{N} x_1^2(k-1) + x_2^2(k-1) + 2u_3^2(k-1)\right)\left(\sum_{k=1}^{N} x_3^2(k-1) + 2u_1^2(k-1) + 2u_2^2(k-1)\right)\right]^{\frac{1}{2}}.$$

With the following parameters of SCO algorithm: $NP = 100$, $M = 50$, $L = 15$, $T = 15$, $h = 50$, $\mu = 0.7$, $\alpha = 1$, $\gamma = 0.75$, $\eta = 0.89$, $\lambda_{Levi} = 1.4$, $c_1 = 0.5$, $c_2 = 0.6$, $c_3 = 0.5$, $\varepsilon = 10^{-5}$, $h = 5$, we get optimal open loop control and the corresponding trajectories shown in figure 2. Algorithm work results: $I^* = 341.04623$ – approximate functional value, $\Delta = I^{exact} - I^* = -82.7069$ – deviation from the exact solution, $I^{exact} = 257.3393$ – exact functional value. $t_{work} = 554$ – algorithm runtime.

Figure 2 shows charts of the obtained optimal open loop control and corresponding trajectories.

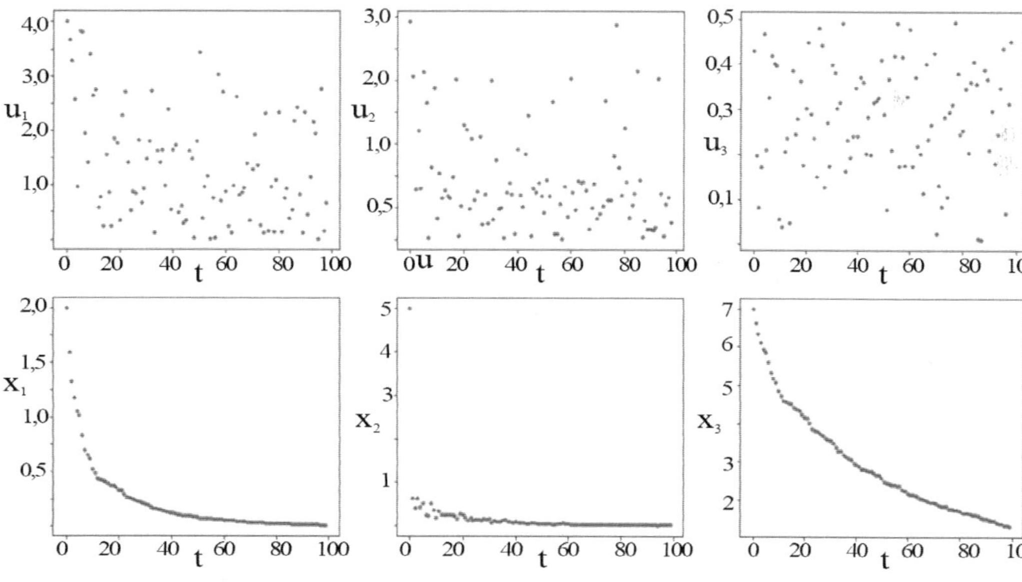

Figure 2. Optimal control and corresponding trajectories in Luus-Tassone problem.

Two algorithms were compared – the DA[8] and SCO algorithm in solving the Luus-Tassone problem. In this example, as in the previous ones, the decision as to which algorithm should be considered the best depends on what is the priority: the accuracy of the solution or the time of the algorithm runtime. The results are presented in Table 2.

Table 2. Comparison of DA and SCO in solving the Luus-Tassone problem

Method	*N*	*NP*	I^*	I^{exact}	Δ	Δ / I^{exact}	t_{work}
DA	5	50	139.2620	-	-	-	15
SCO	5	30	137.3700	-	-	-	18
DA	10	150	180.9700	-	-	-	25
SCO	10	40	176.7960	-	-	-	17
DA	100	1000	353.8831	268.3393	-95.5438	-0.35606	475
SCO	100	100	341.0460	268.3393	-82.7069	-0.30822	554

5. Conclusion

The paper describes the algorithm of Sparrow colony optimization and the solution of the problem of finding the optimal open loop control of nonlinear discrete deterministic systems. For the method of simulation of sparrow colony software is implemented and tested on a set of test functions with the known solutions. Solution of two model discrete optimal control problems using the described method is given: with control constraints and Luus-Tassone example. The results of the described algorithm's work are compared with the Dragonfly algorithm while solving the optimization task and model tasks of searching for optimal open loop control of discrete deterministic systems. The comparison has shown that the method of Sparrow colony optimization works more accurately, than the Dragonfly algorithm, but it takes more time to work. As mentioned earlier, when choosing an algorithm for solving a problem, it is necessary to decide what is more important: to get the most accurate result without strict time limits or to find a solution as quickly as possible. Also the recommended values for the algorithm of Sparrow colony optimization are given.

References

[1] Chen J and Mahfouf M 2009 Artificial immune systems as a bio-inspired optimization techniques and its engineering application in *Handbook of Research on Artificial Immune Systems and Natural Computing: Applying Complex Adaptive Technologies*(IGI Global)
[2] Yang X, Cui Z, Xiao R,Gandomi A and Karamanoglu M 2013 *Swarm Intelligence and Bio-Inspired Computation: Theory and Applications* (London: Elsevier)
[3] Ding Y, Chen L and Hao K 2018 *Bio-Inspired Collaborative Intelligent Control and Optimization* (Berlin: Springer)
[4] Floudas C and Pardalos P 2009 *Encyclopedia of Optimization* (Berlin: Springer)
[5] *Bioinspired Optimization Methods and Their Applications* 2018 Proc. 8th Int. Conf. BIOMA Paris, France, May 2018 ed. Korošec P, Melab N and Talbi E–G 10835 (Berlin.: Springer)
[6] Mirjalili S 2016 *The Neural Comput. and Appl.* **27**(4) 1053–73
[7] Panovskiy V and Panteleev A 2017 *J. Comput. Syst. Sci. Int.* **56**(1) 52–63
[8] Panteleev A and Stanovskaya Y 2019 *AIP Conf. Proc.* **2181**(1) id020027
[9] Panteleev A and Stanovskaya Y 2019 *Inform. and Telecommun. Technol.* **43** 35–42
[10] Rudenko E 2018 *J. Comput. Syst. Sci. Int.* **57**(1) 43–62
[11] Luus R 2000 *Iterative Dynamic Programming* (London: Chapman & Hall/CRC Press)
[12] Herrera F, Lozano M and Molina D 2010 *Test suite for the special issue of soft computing on scalability of evolutionary algorithms and other metaheuristics for large scale continuous optimization problems* Techn. Report Univ. of Granada
[13] Mishra S 2007 *Some new test functions for global optimization and performance of repulsive particle swarm method* (Munich Personal RePEc Archive MPRA)

The mini-batch adaptive method of random search (MAMRS) for parameters optimization in the tracking control problem

A V Panteleev and A V Lobanov

Moscow Aviation Institute, 4Volokolamskoe shosse, 125993,Moscow, Russia

E-mail: avpanteleev@inbox.ru

Abstract. The tracking control problem for nonlinear dynamic system linear in control and signal tracked is considered. It is proposed to formalize the tracking problem as a problem of controlling a bunch of trajectories emanating from a given set of possible initial states. The quality of a single trajectory control is described by the value of the integral criterion, which depends on the square of the tracking error and the intensity of the control used. The control is sought from the condition of the minimum value of the average criterion defined on the set of the bunch trajectories. When choosing a control structure, the well-known result was used that provides internal stability and the desired quality of the output control. The unknown parameters of the control law within the selected structure are found using the new mini-batch adaptive method of random search (MAMRS). This new optimization algorithm combines the ideas of the adaptive random search and stochastic gradient descent methods. The examples of solving the tracking problem as a result of parametric optimization of controlling a bunch of trajectories are presented.

1. Introduction

In applied problems of synthesis of complex control systems, tracking tasks occupy a special place [1–4]. There are many known methods for solving such problems, providing asymptotic stability, the desired quality of transients, etc. For particular case of dynamic systems linear in control and signal tracked the fulfillment of two conditions can be achieved [2]. One of them is internal stability (in the absence of a tracked parameter, the system is asymptotically stable). Another one is asymptotic decrease in the tracking error to zero for all initial conditions from a neighborhood of the coordinate origin. Its application gives rise to some structure of the control law with an unknown constant matrix. The elements of the regulator matrix are determined from known conditions. This article proposes to use the specified control structure for this class of dynamical systems, but use a different procedure for selecting matrix elements. The tracking quality criterion takes the form of an integral of the weighted sum of the squared tracking error and the norm of the regulator matrix. For a set of initial states in a neighborhood of the origin, determine the value of the functional that describes the average value of the quality control criterion for controlling an individual trajectory along a given set. To solve the problem of finding the regulator matrix, it is proposed to use a new optimization method developed by the authors for the case when the objective function can be represented as the sum of some functions. This method is metaheuristic [5, 6] and uses the idea of forming a mini-batch of realizations of a fixed volume at each iteration and its use in the adaptive random search procedure. Thereby, the idea of gradient optimization methods used in machine learning is generalized, for example, the stochastic gradient descent method and the mini-batch gradient descent method [7].

The purpose of study is to develop a new optimization method and its application in the problems of the trajectory bunchcontrolof dynamic systems.

An algorithm is developed for parametric control optimization of the bunch trajectories emanating from a given set of initial states [8–12]. Its application is demonstrated by solving a three-dimensional model example of controlling a chaotic system in which it is required to track the desired value of the selected coordinate. An analysis of the obtained comparison results with the trajectories obtained in [1] showed the effectiveness of the algorithm proposed.

2. Statement of the problem

The behavior of a nonlinear continuous deterministic model of the control object is described by the ordinary differential equation

$$\dot{x}(t) = f(x(t)) + g(x(t))\,u(t) + p(x(t))\,w(t), \tag{1}$$

where t – time, $t \in T = [t_0; t_1]$, moments t_0 and t_1 are given; x– state vector, $x \in R^n$; u – control vector, $u \in R^q$; $w \in R^m$ – tracked signal, $f(x) \in R^{n\times 1}$, $g(x) \in R^{n\times q}$, $p(x) \in R^{n\times m}$ – continuously differentiable functions. The tracked signal is given by the equation $\dot{w}(t) = s(w(t))$, $w(t_0) = w_0$, and the tracking error is given by the relation $e(t) = h(x(t)) - q(w(t))$, where $s(w) \in R^{m\times 1}$, $h(x), q(w)$ – continuously differentiable functions. It is assumed that $f(0) = 0, g(0) = 0, p(x) = 0, s(0) = 0, h(0) = 0, q(0) = 0$. The initial conditions are given by a compact set Ω of positive measure with a piecewise smooth boundary:

$$x(t_0) = x_0 \in \Omega \subset R^n,$$

where the set Ω characterizes the uncertainty of the initial conditions, and the vectors x_0, w_0 belong to a neighborhood of the origin in R^n and R^m respectively.

It is assumed that the control uses information about vectors x, w, i.e. $u(t) = \boldsymbol{u}\big(x(t), w(t)\big)$, where $\boldsymbol{u}\big(x, w\big)$ – continuously differentiable function.

In [2], the nonlinear control problem that ensures the fulfillment of two properties of a closed loop system is solved: internal stability (at $w \equiv 0$ the system is asymptotically stable); output control (for all initial states x_0, w_0, close enough to the origin, the error $e(t) \to 0$ at $t \to \infty$). Moreover, assumptions are made about the Lyapunov stability of the system $\dot{w}(t) = s(w(t))$, as well as the asymptotic stability of the linearized system obtained as a result of linearization of equation (1) in a neighborhood of the origin. It is shown that the control has the form

$$u(x, w) = \varphi(w) + K[x - \pi(w)], \tag{2}$$

where $\varphi(w), \pi(w)$ – functions satisfying conditions

$$\frac{\partial \pi}{\partial w} s(w) = f(\pi(w)) + g(\pi(w))\varphi(w) + p(\pi(w))w, \quad h(\pi(w)) - q(w) = 0,$$

and $\varphi(0) = 0, \pi(0) = 0$, K – constant gain matrix $(q \times n)$ such that the eigenvalues of the matrix $A + BK$ belong to the left hand half-plane $\left(A = \left.\frac{\partial f}{\partial x}\right|_{x=0}, B = \left.\frac{\partial g}{\partial x}\right|_{x=0} \right)$.

The paper considers the problem of finding the matrix K^* in the framework of a pre-fixed control structure (2) from the condition of minimizing the quality criterion for controlling a bunch of trajectories emanating from the set Ω. We define the cost functional for the individual trajectory control that characterizes the integral tracking error and control intensity:

$$I\left(x_0, K\right)=\int_{t_0}^{t_1}[\lambda_1 e^2(t)+\lambda_2\|K\|^2]dt, \tag{3}$$

where λ_1, λ_2 – specified weights. To each admissible control (2) and the set Ω we associate a bunch [8] of trajectories of the system of equations (1): $X\left(t, K\right)=\bigcup\left\{x\left(t, K, x_0\right) \mid x_0 \in \Omega\right\}$, that is, the union of the solutions of the system of equations (1) for all possible initial states (4).

It is proposed to evaluate the quality of control of the trajectory bundle by the value of the functional

$$J\left[K\right]=\int_{\Omega} I\left(x_0, K\right)dx_0 \;/\; \text{mes}\,\Omega, \tag{4}$$

where mes Ω – measure of set Ω.

It is required to find a matrix K^* such that:

$$J\left[K^*\right]=\min_K J\left[K\right]. \tag{5}$$

The resulting control will be suboptimal on average, since the parameters of a fixed control structure (2) are best selected, solving the problem of minimizing the average value of the functional defined on individual trajectories.

3. Solution Search Strategy

The proposed approach to the solution consists in the transition from problem (5) to the problem of finite-dimensional optimization, that is, the problem of finding the best values of the indefinite matrix elements K that define the control structure of the control object model (1).

Further we will use the following assumptions.

1. The set of initial states Ω is a parallelepiped defined by the direct product of segments $\left[\alpha_i;\beta_i\right], i=\overline{1,n}$, i.e. $\Omega=\left[\alpha_1;\beta_1\right]\times\cdots\times\left[\alpha_n;\beta_n\right]$. With the value of a step Δx_i line segment $\left[\alpha_i;\beta_i\right], i=\overline{1,n}$, is divided into N_i segments, and the parallelepiped Ω is divided into $N=N_1\cdots N_n$ elementary disjoint subsets Ω_k, $k=\overline{1,N}$. In each elementary subset Ω_k, an initial state $x_0^{\,k}$ (the center of the parallelepiped Ω_k) is specified.

2. Functions $\varphi(w), \pi(w)$ in (2) are considered to be found.

3. To determine the approximate value of criterion (4) in the general case, it follows:

a) find a solution $x^k(t)$ to the system of equations (1) with a control of the form (2) and the initial condition $x(t_0)=x_0^k$, $w(t_0)=w_0$, $k=\overline{1,N}$. To do this, apply one of the methods of numerical solution of ordinary differential equations. Find tracking error $e^k(t)=h(x^k(t))-q(w(t))$.

b) calculate the value of the functional (3): $I\left(x_0^{\,k}, K\right), k=\overline{1,N}$. The value of the integral term of the functional can be calculated in paragraph a, if we add the following equation to the system (1)

$$\dot{x}_{n+1}^k(t)=\lambda_1[e^k(t)]^2+\lambda_2\|K\|^2,\; x_{n+1}^k(t_0)=0.$$

c) find the approximate value of criterion (4), setting

$$J\left[K\right]\cong(1/N)\sum_{k=1}^{N} I\left(x_0^k, K\right). \tag{6}$$

4. In order to reduce the amount of calculations when finding the value of criterion (4), instead of N trajectories starting from a given set of initial states Ω, it is proposed to use only the part of them

that forms the mini-batch. In this case, one could set the mini-batch size d $(1 \le d \le N)$ and a suitable amount of d pairwise mismatching initial states x_0^k is generated, where k - random number belonging to the set $\{1,...,N\}$. Instead of calculating the value of (6), it is proposed to find the value of criterion (9) on a mini-batch of size d :

$$J_d[K] = (1/d)\sum_{k=1}^{d}\left(\int_{t_0}^{t_1}[\lambda_1 e^2(t) + \lambda_2 \|K\|^2]dt\right) \to \min_K . \quad (7)$$

It is required to find a matrix K^* on which the objective function $J_d[K]$ reaches a minimum value.Note that for $d = N$ the equality $J_d[K] = J[K]$ is valid.

The following options for specifying a mini-batch are possible: fix the number d ; consistently increase the number d , i.e. $d = 1,2,...,N$ (this method is used in the technique described below to compare the efficiency of the method for various mini-batch sizes); change the mini-batch size dynamically depending on the achievement of a given estimation accuracy, determined by the value of the criterion (4).

If the mini-batch size and the total number of iterations of the applied optimization algorithm are fixed, then the result of applying the algorithm can change with each new start due to the random nature of the choice of the initial states x_0^k included in the mini-batch. To obtain a more stable result, it is proposed to perform a given number of starts $S_{\max}$. Then the average value of criterion (7) and the standard deviation, i.e. $m_d = S_d^s / S_{\max}$; $\sigma_{m_d} = \left(\frac{1}{S_{\max}-1}\sum_{s=1}^{S_{\max}}\left[J_d^s - m_d\right]^2\right)^{1/2}$, are calculated. Here we use the notations $\hat{K}_d^s$ and $J_d^s = J_d^s(\hat{K}_d^s)$ – vector of estimates and the value of the objective function (7) obtained as a result of the s -th start ($s = 1,...,S_{\max}$) with a fixed mini-batch size d . Upon completion of $S_{\max}$ starts the pass ends.

Since the indices m_d and σ_{m_d} change during the implementation of independent passes, we can re-average over the given number of passes $B_{\max}$ and find the mathematical expectation and the standard deviation of the mean values of criterion (7): $\bar{m}_d = \sum_{b=1}^{B_{\max}} m_d^b / B_{\max}$; $\sigma_{\bar{m}_d} = \left(\frac{1}{B_{\max}-1}\sum_{b=1}^{B_{\max}}\left[m_d^b - \bar{m}_d\right]^2\right)^{1/2}$, where m_d^b – value m_d after the b – th pass ($b = 1,...,B_{\max}$). As a result, one can find the best estimate of $\hat{K}_d^*$ after $B_{\max}$ passes, each of which implements $S_{\max}$ starts of the minimization algorithm, and consider it as the final search result. To analyze the resulting estimation accuracy, the value of the objective function (6) is found.

5. We form from the elements of matrix K of size $(q \times n)$ a column vector K of size $(qn \times 1)$ by sequentially adding matrix rows to it, starting from the first. The matrix and its corresponding column for the sake of abbreviation are denoted by the same symbol K .

4. The mini-batch adaptive method of random search (MAMRS)

Denote: J_d^s – the minimum value of the function after the s – th run; $\hat{K}_d^s$ – best parameter vector column after startup; d – the size of the mini-batch.

*Step 0.*Set: $d = 1$ – initial mini-batch size (in the general case, one can start with any value of $1 \le d \le N$); λ_1, λ_2 – weights; $S_{\max}$ – maximum number of starts; $B_{\max}$ – maximum number of passes; $\alpha = 1.618$ – expansion coefficient; $\beta = 0.618$ – compression coefficient; M – the maximum number of failed tests at the current iteration; $t_0 = 1$ – initial step size (can be set any value $t_0 > R$), R – minimum step size, L – The maximum number of iterations in the startup procedure.

*Step 1.*Put: $b = 1$ (number of passes counter); $P_d = 0$ (initial value of the sum of the average values of the objective function).

*Step 2.*Put: $s = 1$ (counter of the number of starts); $J_d^1 = 10^8 \div 10^{10}$; $S_d = 0$ (initial value of the sum of the values of the objective function).

Step 3. Define the initial approximation of the matrix K and the corresponding column vector $K_d^{s,0}$. Put $l = 0, j = 1$.

Step 4. Get a random vector $\xi^j = \left(\xi_1^{\ j},\ldots,\xi_{qn}^{\ j}\right)^T$, where $\xi_i^{\ j}$ – a random variable uniformly distributed on the interval $[-1, 1]$.

Step 5. Calculate: $y^j = K_d^{s,l} + t_l \dfrac{\xi^j}{\|\xi^j\|}$.

*Step 6.*Generate mini-batch of size d. To do this, generate d pairwise mismatched sets of $q_1 \in N_1,\ldots,q_n \in N_n$ values defining the initial states x_0^k with numbers $k = q_1 \cdot \ldots \cdot q_n \in \{1,\ldots,N\}$, or corresponding to the tuple $< q_1,\ldots,q_n >$.

Check the fulfillment of the conditions:

a) if $J_d(y^j) < J_d(K_d^{s,l})$, a successful step. Put $z^j = K_d^{s,l} + \alpha\left(y^j - K_d^{s,l}\right)$. Determine if current direction $y^j - K_d^{s,l}$ is successful: if $J_d(z^j) < J_d(K_d^{s,l})$, successful search direction. Put $K_d^{s,l+1} = z^j$, $t_{l+1} = \alpha t_l$, $l = l + 1$ and check the termination condition. If $l < L$, put $j = 1$ and go to step 4. If $l = L$, search complete: $\hat{K}_d^s = K_d^{s,l}$, go to step 8; if $J_d(z^j) \ge J_d(K_d^{s,l})$, search direction is unsuccessful, go to step 7;

b) if $J_d(y^j) \ge J_d(K_d^{s,l})$, unsuccessful step and go to step 7.

Step 7. Estimate the number of failed steps from the current point:

a) if $j < M$, put $j = j + 1$ and go to step 4;

b) if $j = M$, check the termination condition: if $t_l \le R$, process finish: $\hat{K}_d^s = K_d^{s,l}$, $J_d^s = J_d(K_d^{s,l})$, go to step 8; if $t_l > R$, put $t_l = \beta t_l$, $j = 1$ and go to step 4.

Step 8. Check the improvement in the value of the objective function as a result of the s-th run: if $J_d(K_d^{s,l}) < J_d^s$, put $J_d^s = J_d(K_d^{s,l})$, $\hat{K}_d^s = K_d^{s,l}$ and go to step 9; if $J_d(K_d^{s,l}) \ge J_d^s$ go to step 9;

Step 9. Calculate $S_d = S_d + J_d^s$ and verify that the conditions for the end of the number of starts are met:if $s < S_{\max}$, put $s = s + 1$ and go to step 3; if $s = S_{\max}$, put $\hat{K}_d = \hat{K}_d^s$ - the best solution during the b-th pass for a given k; calculate m_d, σ_{m_d} and go to step 10.

*Step 10**. Put $P_d = P_d + m_d$, $m_d^b = m_d$ and check the condition for completing a given number of passes: if $b < B_{\max}$, put $b = b + 1$ and go to step 2; if $b = B_{\max}$, calculate: $\bar{m}_d$, $\sigma_{\bar{m}_d}$.

*Step 11**. Check the condition for completing studies of the effect of the mini-batch size: if $k < N$, put $k = k + 1, s = 1$ and go to step 1; if $k = N$, go to step 12.

Step 12. As a result, find the best estimate of $\hat{K}_d^*$ after $B_{\max}$ passes and indicators $\bar{m}_d$, $\sigma_{\bar{m}_d}$ for each value of the mini-batch size d. To analyze the resulting estimation accuracy, find the value $J(\hat{K}_d^*)$ of the estimation accuracy criterion (4): $J(\hat{K}_d^*) = (1/N)\sum_{k=1}^{N}\left(\int_{t_0}^{t_1}[\lambda_1 e^2(t) + \lambda_2 \left\|\hat{K}_d^*\right\|^2]dt\right)$.

Steps 10 and 11 are performed if necessary.
The flow chart of MAMRS is shown in the figure 1.

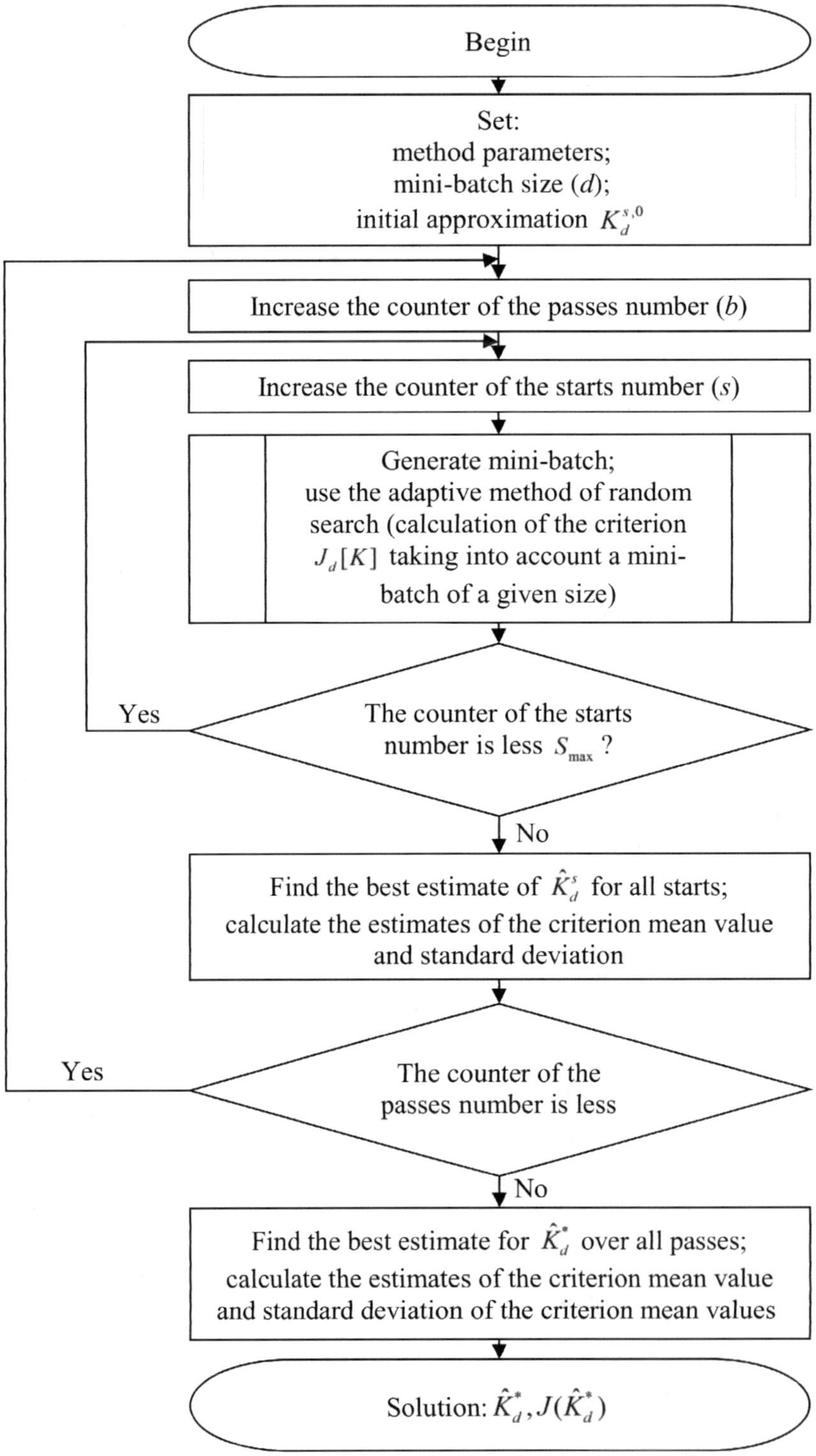

Figure 1. MAMRS flow-chart

5. Numerical example

The behavior of a nonlinear continuous deterministic model of the control object is described by the system [1]

$$\dot{x}_1(t) = a(x_2 - x_1),\ \dot{x}_2(t) = cx_1 - x_1x_3 + u,\ \dot{x}_3(t) = -bx_3 + x_1x_2,\ \dot{w}(t) = 0,\ w(0) = 2,$$

where $a = 10; b = 8/3; c = 16;\ t \in T = [0;4];\ x \in R^3;\ u \in R, n = 3, q = 1$.

The set of possible initial states $\Omega : |x_{10}| \le 20, |x_{20}| \le 30, |x_{30}| \le 24$.

We consider three cases of tracking the coordinates of the state vector.

Case (A): Constant tracking problem for x_1. Equation (3) has the form $e(t) = x_1(t) - w(t)$. It is required to find a control ensuring the fulfillment of condition $x_1(t) = 2$, i.e. tracking a given coordinate value x_1.

In [1], using the theory from [2], elements of the control law (2) were obtained:

$$\pi_1(w) = w, \pi_2(w) = w, \pi_3(w) = \frac{w^2}{b}, \varphi(w) = -cw + \frac{w^3}{b},\quad K = [-20, 9; 4; 0].$$

Table 1 presents the solution depending on mini-batch size. In all tests $\Omega_k : N = 19425$ – number of elementary subsets, $N_1 = 37, N_2 = 21, N_3 = 25$, $t_0 = 0$, $t_1 = 4$, $\lambda_1 = 1$, $\lambda_2 = 0$, $L = 300$ – maximum number of iterations, $M = 3$ – maximum number of failed tests at the current iteration, $R = 8 \cdot 10^{-6}$ – minimum step size, $S_{\max} = 100$ – maximum number of starts, $K_d^{s,0} = (-25, 5, -5)^T$ – initial approximation.

Table 1. Result of the solution using the mini-batch adaptive method of random search

d	K_1	K_2	K_3	$J[K]$	m_d	σ_{m_d}
1	-25.48860	-1.18872	-9.18406	10.62927	22.31498	6.51919
5	-26.43757	-1.68196	-2.96053	14.85412	22.64488	5.15948
10	-27.07250	3.81421	-2.94559	7.53309	20.35456	3.90334
50	-29.46386	-12.76871	-1.62380	5.97487	6.86996	0.70619
100	-30.60633	-5.37748	-0.33078	5.76164	6.75285	0.70501
500	-33.52515	-7.97133	1.54822	5.60249	5.89001	0.26432

For comparison with the known solution [1], Figure 2 shows a graph of tracking a given value of the coordinate $x_1 = 2$ under initial conditions $x_1(0) = 18, x_2(0) = 10, x_3(0) = 12$. In the control law (6), the matrix $K = [-33.52515; -7.97133; 1.54822]$ is used, which was found by MAMSP with the mini-batch size $d = 500$ that is less then maximum possible size $d = 19425$.

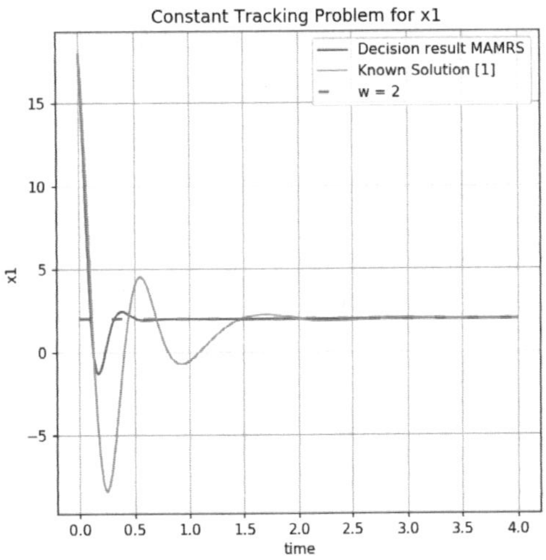

Figure 2. Comparison of methods in case A.

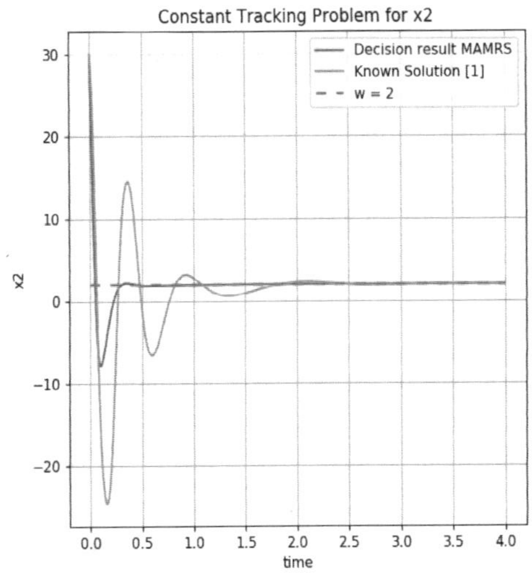

Figure 3. Comparison of methods in case B.

Case (B): Constant tracking problem for x_2. Equation (3) has the form $e(t) = x_2(t) - w(t)$. It is required to find a control ensuring the fulfillment of condition $x_2(t) = 2$, i.e. tracking a given coordinate value x_2. In [1], elements of the control law (2) were obtained:

$$\pi_1(w) = w, \pi_2(w) = w, \pi_3(w) = \frac{w^2}{b}, \varphi(w) = -cw + \frac{w^3}{b}, \; K = [-20, 9; 4; 0].$$

Table 2 presents the solution depending on the mini-batch size. In all tests $\Omega_k : N = 32513$ – number of elementary subsets, $N_1 = 41, N_2 = 61, N_3 = 13$, $t_0 = 0$, $t_1 = 4$, $\lambda_1 = 1$, $\lambda_2 = 0$, L = 300 – maximum number of iterations, M = 3 – maximum number of failed tests at the current iteration, $R = 8 \cdot 10^{-6}$ – minimum step size, $S_{\max} = 100$ – maximum number of starts, $K_d^{s,0} = (-25, 5, -5)^T$ – initial approximation.

Table 2. Result of the solution using the mini-batch adaptive method of random search.

d	K_1	K_2	K_3	$J[K]$	m_d	σ_{m_d}
1	-21.33687	-7.58870	-2.32745	20.38419	28.43374	4.216244
5	-20.45931	-11.22378	-1.59738	16.76418	18.79866	1.254322
10	-20.80403	-12.18129	1.12886	16.38702	17.74631	1.203542
50	-19.69187	-11.17048	-1.04255	16.94842	17.71211	1.022311
100	-20.25975	-11.60871	-0.30464	16.54648	17.52009	0.914954
500	-22.43584	-13.79402	0.53847	15.82680	16.83972	0.619967

For comparison with the known solution [1], Figure 3 shows a graph of tracking a given value of the coordinate $x_2 = 2$ under initial conditions $x_1(0) = 20, x_2(0) = 30, x_3(0) = 6$. In the control law (2), the matrix $K = [-22.43584; -13.79402; 0.53847]$, is used, which was found by MAMSP with the mini-batch size d = 500.

Case (C): Constant tracking problem for x_3. Equation (3) has the form $e(t) = x_3(t) - w(t)$. It is required to find a control ensuring the fulfillment of condition $x_3(t) = 2$, i.e. tracking a given coordinate value x_3. In [1], elements of the control law (2) were obtained:

$$\pi_1(w) = \sqrt{bw} = \pi_2(w), \pi_3(w) = w, \varphi(w) = \sqrt{bw}(w - c), \; K = [-20, 9; 4; 0].$$

IOP Conf. Series: Materials Science and Engineering **927** (2020) 012025 doi:10.1088/1757-899X/927/1/012025

Table 3 presents the solution depending on the mini-batch size. In all tests $\Omega_k : N = 20825$ – number of elementary subsets, $N_1 = 25, N_2 = 17, N_3 = 49$, $t_0 = 0$, $t_1 = 4$, $\lambda_1 = 1$, $\lambda_2 = 0$, $L = 300$ – maximum number of iterations, $M = 3$ – maximum number of failed tests at the current iteration, $R = 8 \cdot 10^{-6}$ – minimum step size, $S_{\max} = 100$ – maximum number of starts, $K_k^{s,0} = (-25, 5, -5)^T$ – initial approximation.

Table 3. Result of the solution using the mini-batchadaptive method of random search.

d	K_1	K_2	K_3	$J[K]$	m_d	σ_{m_d}
1	-30.74191	-8.33088	-9.95571	24.83133	27.97781	8.164195
5	-30.68321	-7.89112	-9.53304	24.80551	25.17827	0.560775
10	-30.00224	-8.93992	-10.02461	24.86136	25.13478	0.361106
50	-30.19378	-7.56466	-8.77771	24.80905	24.95795	0.162247
100	-30.00534	-9.13703	-9.84967	24.85348	24.90627	0.093409
500	-30.37724	-8.44442	-9.91808	24.83875	24.90092	0.050236

6. Conclusion

A strategy and an algorithm have been developed for an approximate solution to the problem of finding a suboptimal average control of one class of nonlinear continuous deterministic dynamical systems under conditions of uncertainty in the initial conditions. The developed algorithm uses a new mini-batch adaptive method of random search to solve the problem of parametric optimization of the specially selected structure of control law. The proposed approach is tested on a model example of a certain coordinates tracking of a chaotic process.

References

[1] Sundarapandian V 2011 *ISRN Appl. Math.* **2011**Article ID 983136
[2] Byrnes C, Isidori A 1990 *IEEE Trans. Autom. Contr.* **35** 131–40
[3] Pan L, Xu D, Zhou W 2010 *J. of Inform. and Comput. Sci.* **5** 117–24
[4] Sundarapandian V 2011 *Int. J. of Comput. Inf Syst.* **2** 27–31
[5] Floudas C, Pardalos P 2009 *Encyclopedia of Optimization* (New York: Springer)
[6] Gendreau M 2010 *Handbook of Metaheuristics* (New York: Springer)
[7] Ruder S 2017 *An Overview of Gradient Descent Optimization Algorithms*
[8] Kurzhanski A, Varaiya P 2014 *Dynamics and Control of Trajectory Tubes* (Basel: Birkhauser)
[9] Panovskiy V, Panteleev A 2017 *J. Comput. Syst. Sci. Int.* **56**(1) 52–63
[10] Panteleev A, Pis'mennaya V 2018 *J.Comput. Syst. Sci. Int.* **57**(1) 25–36
[11] Bortakovskii A 2016 *J. Comput. Syst. Sci. Int.* **55**(1) 1–20
[12] Bortakovskii A, Nemychenkov G 2017 *J. Comput. Syst. Sci. Int.* **56**(6) 914–29

IOP Conf. Series: Materials Science and Engineering **927** (2020) 012026 doi:10.1088/1757-899X/927/1/012026

Approximate methods for H-infinity output feedback control

A V Panteleev and A A Yakovleva
Moscow Aviation Institute, 4 Volokolamskoe shosse, 125993, Moscow, Russia

E-mail: ayakovleva982@gmail.com

Abstract. The problem of finding H^∞ – regulators in the linear output control problem is considered. Sufficient H^∞ – control conditions for closed loop control law depending on the state vector are formulated. Two approaches for approximate synthesis of the output feedback control systems are proposed. The first is associated with finding a pseudo-solution of the linear model of measuring system as well as with its use in control, the second one – with the synthesis of a state observer generating an estimate of the state vector and its application in control. The step by step algorithm is presented. It is shown that both of suggested methods provide the desired quality of transient processes and asymptotic stability in the model example. The approximate methods could be applied in autopilot design problems for aircrafts and rotorcrafts.

1. Introduction

Problems and methods for finding control laws have already become classic [1–11]. However, it is still difficult to achieve a solution to the problem of synthesis of regulators with incomplete state information. Using the technique described in [1, 3–4], new sufficient H^∞ – control conditions are formulated. Firstly the control laws depending of the full state vector are found. Since the full state vector is generally unavailable, two approximate approaches are proposed.

The purpose of this study is to design output feedback controller using the H^∞ approach. To achieve this, we have to formulate and prove the sufficient conditions theorem, to find controller in conditions of incomplete information and propose two approximate methods which could be applied to the different control problems.

An example of solving the model stabilization control problem is given, illustrating the effectiveness of the proposed approaches for different applied control problems [5–7, 10–11].

2. Methods and materials

2.1. Statement of the Problem

The mathematical model of the plant is

$$\dot{x}(t) = Ax(t) + B_1 w(t) + B_2 u(t), \quad x(0) = 0, \tag{1}$$

and model of a measuring system is

$$y(t) = Cx(t), \tag{2}$$

where $x \in R^n$ – state vector, $u \in R^q$ – control input, $w \in R^p$ – disturbance, $y \in R^m$ – measured output, $t \in [0,\infty)$ – time. Matrices $A \in R^{n\times n}$, $B_1 \in R^{n\times p}$, $B_2 \in R^{n\times q}$, $C \in R^{m\times n}$ are given. It is assumed that $w(.) \in L_2[0,\infty)$, $u(.) \in L_2[0,\infty)$; $m \le n$, $\operatorname{rg} C = m$; the system is fully controlled and observable, i.e. $\operatorname{rg}(B \;\; AB_2 \; \ldots \; A^{n-1}B_2) = n$, $\operatorname{rg}(C^T \;\; A^T C^T \ldots (A^T)^{n-1} C^T) = n$.

Let us denote $\|z(t)\|^2 = y(t)^T Sy(t) + u^T(t)Qu(t)$ as performance output, where Q – positive definite symmetric matrix, S – non-negative definite symmetric matrix. It is required to ensure (if possible) the fulfillment of the inequality

$$\frac{\int_0^\infty \|z(t)\|^2 dt}{\int_0^\infty \|w(t)\|^2 dt} = \frac{\int_0^\infty [y(t)^T Sy(t) + u^T(t)Qu(t)]dt}{\int_0^\infty \|w(t)\|^2 dt} = \frac{\int_0^\infty [x^T(t)C^T S\,C\,x(t) + u^T(t)Qu(t)]dt}{\int_0^\infty w^T(t)w(t)dt} \le \gamma^2, \quad (3)$$

where $\gamma > 0$ – a given non-negative number, as well as the asymptotic stability of a closed loop system. It is preferred to find the minimum value of γ at which these properties are still valid, which can be achieved by minimizing the value of the fraction numerator while maximizing the denominator. In other words, the quality functional must satisfy the condition

$$I(u,w) = \int_0^\infty \left[\|z(t)\|^2 - \gamma^2 \|w(t)\|^2 \right] dt = \int_0^\infty \left[x^T(t)C^T S\,C\,x(t) + u^T(t)Qu(t) - \gamma^2 w^T(t)w(t) \right] dt \le 0,$$

which will be fulfilled when the output vector tends to zero, minimizing control costs under the worst influence of disturbances possible.

2.2. Synthesis of full state vector H^∞ – control

Let there be a function $V(x) \in C^1$ and the construction

$$R(x,u,w) = \left(\frac{\partial V}{\partial x}\right)^T (Ax + B_1 w + B_2 u) + x^T C^T S\,Cx + u^T Qu - \gamma^2 w^T w, \quad (4)$$

where $\frac{\partial V}{\partial x} = \left(\frac{\partial V(x)}{\partial x_1}, \ldots, \frac{\partial V(x)}{\partial x_n}\right)^T$.

The following sufficient condition of H^∞ – control was proved.

Theorem. If a function $V(x) \in C^1$ exists, satisfying the condition $V(0) = 0$ and

$$R(x, u^*(x), w^*(x)) = \min_u \max_w R(x,u,w) = 0, \quad (5)$$

where

$$u^*(x) = -\frac{1}{2} Q^{-1} B_2^T \frac{\partial V}{\partial x} = -Q^{-1} B_2^T K_2 x, \; w^*(x) = \frac{1}{2\gamma^2} B_1^T \frac{\partial V}{\partial x} = \frac{1}{\gamma^2} B_1^T K_2 x, \quad (6)$$

matrix $K_2 \ge 0$ satisfies the Riccati algebraic equation

$$K_2 A + A^T K_2 - K_2 [B_2 Q^{-1} B_2^T - \frac{1}{\gamma^2} B_1 B_1^T] K_2 + C^T S\,C = 0 \quad (7)$$

and a condition $\sigma(A + [\frac{1}{\gamma^2} B_1 B_1^T - B_2 Q^{-1} B_2^T] K_2) \subset C^-$, where σ – matrix spectrum, C^- – open left half-plane of the complex plane, then inequality (3) is true.

If the energy of disturbances acting on the system is limited, i.e. $\int_0^\infty \|w(t)\|^2 dt \le 1$, then $\int_0^\infty \|z(t)\|^2 dt \le \gamma^2$.

2.3. Synthesis of Output H^∞ – control

If $m = n$ and the matrix C is non-degenerate, then the state vector could be found directly from the output vector: $x = C^{-1} y$. Then output control $u^*(y) = -Q^{-1} B_2^T K_2 C^{-1} y$.

If $m \le n$, $\operatorname{rg} C = m$, we can apply two approaches. The first one is related to finding a pseudo-solution $\tilde{x}$ of system $y = Cx$ and its application in control law, the second - to the synthesis of the state observer that develops the estimate $\hat{x}$ of the state vector, and using estimates in control.

First approach. Find system pseudo-solution $y = Cx$ using a pseudo-inverse matrix $C^{-1} = C^T (CC^T)^{-1}$: $\tilde{x} = C^{-1} y$, i.e. column with the smallest modulus among all columns minimizing $|Cx - y|$, where the column modulus is $|x| = \sqrt{x_1^2 + ... + x_n^2}$. Then the output control has the form

$$u^*(y) = u^*(\tilde{x}) = -Q^{-1} B_2^T K_2 C^T (CC^T)^{-1} y. \tag{8}$$

Second approach. Synthesize an asymptotic full order observer

$$\frac{d\hat{x}}{dt} = A\hat{x}(t) + B_1 w(t) + B_2 u(t) + K[y(t) - C\hat{x}(t)], \quad \hat{x}(0) = x_0^*, \tag{9}$$

where K is the gain observer matrix with a size of $(n \times m)$, x_0^* – column containing the a priori information about the initial state. Matrix K is selected with respect to condition $\sigma(A - KC) \subset C^-$, providing that the error $\varepsilon(t) = x(t) - \hat{x}(t)$ asymptotically tends to zero. Then the control law has the form

$$u^*(y_0^t) = u^*(\hat{x}) = -Q^{-1} B_2^T K_2 \hat{x}, \tag{10}$$

where $y_0^t = \{ y(\tau), 0 \le \tau \le t\}$ is an accumulated information about the output measurement results.

2.4. Algorithm of Output H^∞ – control

Step 1. Set parameter $\gamma > 0$. Find a solution to the Riccati algebraic equation

$$K_2 A + A^T K_2 - K_2 [B_2 Q^{-1} B_2^T - \frac{1}{\gamma^2} B_1 B_1^T] K_2 + C^T S C = 0,$$

at fixed γ, satisfying conditions $K_2 \ge 0$ and $\sigma(A + [\frac{1}{\gamma^2} B_1 B_1^T - B_2 Q^{-1} B_2^T] K_2) \subset C^-$.

Sequentially reducing γ, we find the minimum value under which all conditions remain fulfilled.

Step 2. Find the control of a plant of the form (8) or (10) and the law of change of the disturbing effect $w^*(x) = \frac{1}{\gamma^2} B_1^T K_2 x$.

Step 3. Find output control law and closed-loop system trajectories.

Case 1. Control system with arbitrary disturbances satisfying the condition $\int_0^\infty \|w(t)\|^2 \, dt \le 1$:

$$\dot{x} = Ax(t) + B_1 w(t) + B_2 u(t), \quad x(0) = 0, \quad y(t) = Cx(t),$$

first approach: $u(t) = -Q^{-1} B_2^T K_2 C^T (CC^T)^{-1} y(t)$;

second approach: $u(t) = -Q^{-1} B_2^T K_2 \hat{x}(t)$,

$$\frac{d\hat{x}}{dt} = A\hat{x}(t) + B_1 w(t) + B_2 u(t) + K[y(t) - C\hat{x}(t)], \quad \hat{x}(0) = x_0^* = 0.$$

Case 2. Control system with the worst perturbations $w(t) = w^*(x(t)) = \frac{1}{\gamma^2} B_1^T K_2 x(t)$:

$$\dot{x} = Ax(t) + B_1 w(t) + B_2 u(t), \quad x(0) = 0, \quad y(t) = Cx(t),$$

first approach: $u(t) = -Q^{-1} B_2^T K_2 C^T (CC^T)^{-1} y(t)$;

second approach: $u(t) = -Q^{-1} B_2^T K_2 \hat{x}(t)$,

$$\frac{d\hat{x}}{dt} = A\hat{x}(t) + B_1 w(t) + B_2 u(t) + K[y(t) - C\hat{x}(t)],\ \ \hat{x}(0) = x_0^* = 0.$$

Verify the asymptotic stability of the closed loop system for each case.

3. Results and discussion

Based on the outlined algorithms, software was developed to find the parameters of synthesized controllers, to simulate processes in dynamic systems for a given period of time, to obtain information about the quality of transient processes and analyse the asymptotic stability of a closed loop system. MATLAB as a computing environment was used. Following is a model example, for which an analytical solution is obtained. Numerical experiment was carried out using the data of the model example.

3.1. Model example

$$\dot{x}_1 = x_2,\ x_1(0) = x_{10},\ \ \dot{x}_2 = w + u,\ x_2(0) = x_{20},$$

$$y = x_1,$$

$$\int_0^\infty \|z(t)\|^2 dt = \int_0^\infty [y^T(t) S\, y(t) + u^T(t) Q u(t)]\, dt.$$

Establish the connection with the problem statement: $A = \begin{pmatrix} 0 & 1 \\ 0 & 0 \end{pmatrix}, B_1 = B_2 = \begin{pmatrix} 0 \\ 1 \end{pmatrix},\ C = \begin{pmatrix} 1 & 0 \end{pmatrix}$.

Step 1. Compose the Riccati algebraic equation

$$\begin{pmatrix} K_{11} & K_{12} \\ K_{12} & K_{22} \end{pmatrix}\begin{pmatrix} 0 & 1 \\ 0 & 0 \end{pmatrix} + \begin{pmatrix} 0 & 0 \\ 1 & 0 \end{pmatrix}\begin{pmatrix} K_{11} & K_{12} \\ K_{12} & K_{22} \end{pmatrix} -$$

$$-\begin{pmatrix} K_{11} & K_{12} \\ K_{12} & K_{22} \end{pmatrix}\left[\begin{pmatrix} 0 \\ 1 \end{pmatrix}\cdot Q^{-1}\cdot(0\ \ 1) - \frac{1}{\gamma^2}\begin{pmatrix} 0 \\ 1 \end{pmatrix}(0\ \ 1)\right]\begin{pmatrix} K_{11} & K_{12} \\ K_{12} & K_{22} \end{pmatrix} + \begin{pmatrix} 1 \\ 0 \end{pmatrix}\cdot S\cdot(1\ \ 0) = \begin{pmatrix} 0 & 0 \\ 0 & 0 \end{pmatrix}$$

or

$$\begin{pmatrix} 0 & K_{11} \\ 0 & K_{12} \end{pmatrix} + \begin{pmatrix} 0 & 0 \\ K_{11} & K_{12} \end{pmatrix} - \begin{pmatrix} (Q^{-1}-\gamma^{-2})K_{12}^2 & (Q^{-1}-\gamma^{-2})K_{12}K_{22} \\ (Q^{-1}-\gamma^{-2})K_{12}K_{22} & (Q^{-1}-\gamma^{-2})K_{22}^2 \end{pmatrix} + \begin{pmatrix} S & 0 \\ 0 & 0 \end{pmatrix} = \begin{pmatrix} 0 & 0 \\ 0 & 0 \end{pmatrix}$$

or in coordinate form

$$-(Q^{-1}-\gamma^{-2})K_{12}^2 + S = 0,\ K_{11} - (Q^{-1}-\gamma^{-2})K_{12}K_{22} = 0,\ 2K_{12} - (Q^{-1}-\gamma^{-2})K_{22}^2 = 0.$$

Solution:

$$K_{11} = \pm\sqrt{\pm 2S\sqrt{\frac{S}{(Q^{-1}-\gamma^{-2})}}};\ K_{12} = \pm\sqrt{\frac{S}{(Q^{-1}-\gamma^{-2})}};\ K_{22} = \pm\sqrt{\frac{\pm 2\sqrt{\frac{S}{(Q^{-1}-\gamma^{-2})}}}{(Q^{-1}-\gamma^{-2})}};$$

In this case, for a solution to exist, the condition $\gamma^2 > Q$ must be satisfied. Condition $K_2 > 0$ is satisfied by

$$K_{11} = \sqrt{2\cdot S\sqrt{\frac{S}{(Q^{-1}-\gamma^{-2})}}};\ K_{12} = \sqrt{\frac{S}{(Q^{-1}-\gamma^{-2})}};\ K_{22} = \sqrt{\frac{2\sqrt{S}}{\sqrt{(Q^{-1}-\gamma^{-2})^3}}};$$

because $K_{11} = \sqrt{2\cdot S\sqrt{\frac{S}{(Q^{-1}-\gamma^{-2})}}} > 0, K_{11}K_{22} - K_{12}{}^2 = \frac{S}{Q^{-1}-\gamma^{-2}} > 0.$

Check if the stability condition $\sigma(A + [\frac{1}{\gamma^2}B_1B_1{}^T - B_2Q^{-1}B_2{}^T]K_2) \subset C^-$ is satisfied:

$$A+[\frac{1}{\gamma^2}B_1B_1^T-B_2Q^{-1}B_2^T]K_2=\begin{pmatrix}0&1\\0&0\end{pmatrix}-(Q^{-1}-\gamma^{-2})\begin{pmatrix}0&0\\0&1\end{pmatrix}\begin{pmatrix}K_{11}&K_{12}\\K_{12}&K_{22}\end{pmatrix}=$$

$$=\begin{pmatrix}0&1\\-(Q^{-1}-\gamma^{-2})K_{12}&-(Q^{-1}-\gamma^{-2})K_{22}\end{pmatrix}.$$

The characteristic equation has the form

$$\begin{vmatrix}-\lambda&1\\-(Q^{-1}-\gamma^{-2})K_{12}&-(Q^{-1}-\gamma^{-2})K_{22}-\lambda\end{vmatrix}=0 \text{ or } \lambda^2+(Q^{-1}-\gamma^{-2})K_{22}\lambda+(Q^{-1}-\gamma^{-2})K_{12}=0.$$

Therefore

$$\lambda_{1,2}=-\frac{(Q^{-1}-\gamma^{-2})K_{22}}{2}\pm\sqrt{\frac{(Q^{-1}-\gamma^{-2})^2K_{22}^{\ 2}}{4}-(Q^{-1}-\gamma^{-2})K_{12}}.$$

Since $K_{22}>0, K_{12}>0$, then $\mathrm{Re}\,\lambda_1<0, \mathrm{Re}\,\lambda_2<0$, i.e. a closed loop system is asymptotically stable.

Step 2. The feedback control looks like

$$u^*(x)=-Q^{-1}B_2^TK_2x=-Q^{-1}\begin{pmatrix}0&1\end{pmatrix}\begin{pmatrix}K_{11}&K_{12}\\K_{12}&K_{22}\end{pmatrix}\begin{pmatrix}x_1\\x_2\end{pmatrix}=-Q^{-1}K_{12}x_1-Q^{-1}K_{22}x_2,$$

$$w^*(x)=\frac{1}{\gamma^2}B_1^TK_2x=\frac{1}{\gamma^2}\begin{pmatrix}0&1\end{pmatrix}\begin{pmatrix}K_{11}&K_{12}\\K_{12}&K_{22}\end{pmatrix}\begin{pmatrix}x_1\\x_2\end{pmatrix}=\frac{1}{\gamma^2}(K_{12}x_1+K_{22}x_2).$$

Step 3. Find the output control:

first approach: $u(t)=-Q^{-1}B_2^TK_2C^T(CC^T)^{-1}y(t)$ or

$$u(t)=-Q^{-1}\begin{pmatrix}0&1\end{pmatrix}\begin{pmatrix}K_{11}&K_{12}\\K_{12}&K_{22}\end{pmatrix}\begin{pmatrix}1\\0\end{pmatrix}\left[\begin{pmatrix}1&0\end{pmatrix}\begin{pmatrix}1\\0\end{pmatrix}\right]^{-1}y(t)=-Q^{-1}K_{12}y(t);$$

second approach: $u(t)=-Q^{-1}B_2^TK_2\hat{x}(t)=-Q^{-1}K_{12}\hat{x}_1-Q^{-1}K_{22}\hat{x}_2$,

$$\frac{d\hat{x}}{dt}=\begin{pmatrix}0&1\\0&0\end{pmatrix}\hat{x}(t)+\begin{pmatrix}0\\1\end{pmatrix}w(t)+\begin{pmatrix}0\\1\end{pmatrix}u(t)+\underbrace{\begin{pmatrix}K_1\\K_2\end{pmatrix}}_{K}[y(t)-\begin{pmatrix}1&0\end{pmatrix}\hat{x}(t)],\ \hat{x}(0)=\begin{pmatrix}x_{10}\\x_{20}\end{pmatrix}.$$

Matrix $A-KC=\begin{pmatrix}0&1\\0&0\end{pmatrix}-\begin{pmatrix}K_1\\K_2\end{pmatrix}\begin{pmatrix}1&0\end{pmatrix}=\begin{pmatrix}-K_1&1\\-K_2&0\end{pmatrix}$, characteristic equation $|A-KC-\lambda E|=$ $=\begin{vmatrix}-K_1-\lambda&1\\-K_2&-\lambda\end{vmatrix}=0$ or $\lambda^2+K_1\lambda+K_2=0$. Let the choice of matrix K be desirable to provide a double root λ_{giv} of the characteristic equation, which will have the form $(\lambda-\lambda_{giv})^2=\lambda^2-2\lambda\lambda_{giv}+\lambda_{giv}^2=0$. Equating the coefficients at the same degrees of λ, we get $K_1=-2\lambda_{giv}, K_2=\lambda_{giv}^2$. For example, with $\lambda_{giv}=-3$ we have $K_1=6, K_2=9$.

3.2. Simulation Results

3.2.1. Simulation Settings. For the correct computation of the output feedback control, it is necessary to select Q and S. During a search for suitable elements, we changed their design several times. After all we selected the design matrices as in Table 1, which contains parameters of numerical simulations. In addition, we choose function given in Table 1 to describe disturbance, which affects the state vector.

Table 1. Numerical simulation parameters.

Initial conditions $(x_{10}, x_{20})^T$	S	Q	$T=[t_0;t_1]$	**Disturbance**
$(0.1, 1)$	1000	0.001	[0;10000]	$w(t)=\begin{cases}0, & \text{if } t<0 \text{ or } t>t_w \\ 1, & 0\le t\le t_w,\ t_w=1\end{cases}$

Table 2. Calculations with different γ values.

γ	$\sigma(K_2)$	$\sigma(A+[\frac{1}{\gamma^2}B_1B_1^T - B_2Q^{-1}B_2^T]K_2)$
0.2	$\begin{pmatrix}0.0228\\45.0281\end{pmatrix}$	$\begin{pmatrix}-22.2196 + 22.2196i\\-22.2196 - 22.2196i\end{pmatrix}$
0.1	$\begin{pmatrix}0.0242\\45.9392\end{pmatrix}$	$\begin{pmatrix}-21.7794 + 21.7794i\\-21.7794 - 21.7794i\end{pmatrix}$
0.05	$\begin{pmatrix}0.0328\\50.8461\end{pmatrix}$	$\begin{pmatrix}-19.6799 + 19.6799i\\-19.6799 - 19.6799i\end{pmatrix}$
0.04	$\begin{pmatrix}0.0466\\57.1955\end{pmatrix}$	$\begin{pmatrix}-17.4982 + 17.4982i\\-17.4982 - 17.4982i\end{pmatrix}$
0.03	$\begin{pmatrix}54.6900 - 54.8548i\\0.2466 - 0.0818i\end{pmatrix}$	$\begin{pmatrix}-18.2574i\\18.2574i\end{pmatrix}$

3.2.2. Numerical Simulations. For one of the described cases, namely the case of control system with arbitrary disturbances, we will perform solution of the model example with the different γ parameter values (Table 2). As we can see from the results of series of calculations from Table 2, parameter $\gamma = 0.04$ is the minimum value of γ, which satisfies (3) and still guarantees the preservation of the property of asymptotic stability of a closed loop system. For each case condition $K_2 \ge 0$ is satisfied. As previously mentioned, we consider case 1, which is a case of control system with arbitrary disturbances satisfying the condition $\int_0^\infty \|w(t)\|^2\, dt \le 1$:

$$\dot{x} = Ax(t) + B_1 w(t) + B_2 u(t),\ \ x(0)=0,\ \ y(t)=Cx(t),$$

using first approach: $u(t) = -Q^{-1}B_2^T K_2 C^T (CC^T)^{-1} y(t)$ or second approach: $u(t) = -Q^{-1}B_2^T K_2 \hat{x}(t)$,

$$\frac{d\hat{x}}{dt} = A\hat{x}(t) + B_1 w(t) + B_2 u(t) + K[y(t) - C\hat{x}(t)],\ \ \hat{x}(0) = x_0^* = 0.$$

For the second approach we set the gain observer matrix $K = \begin{bmatrix}100\\100\end{bmatrix}$, which is selected with respect to condition $\sigma(A-KC) \subset C^-$, i.e. the solution of characteristic equation $|A - KC - \lambda E| = 0$ in this case is $\lambda = \begin{bmatrix}-0.5013\\-199.4987\end{bmatrix}$.

Following are plots of the transient processes for the first approach (Fig. 1–2). The plots of the transient processes for the second approach look similar.

IOP Conf. Series: Materials Science and Engineering **927** (2020) 012026 doi:10.1088/1757-899X/927/1/012026

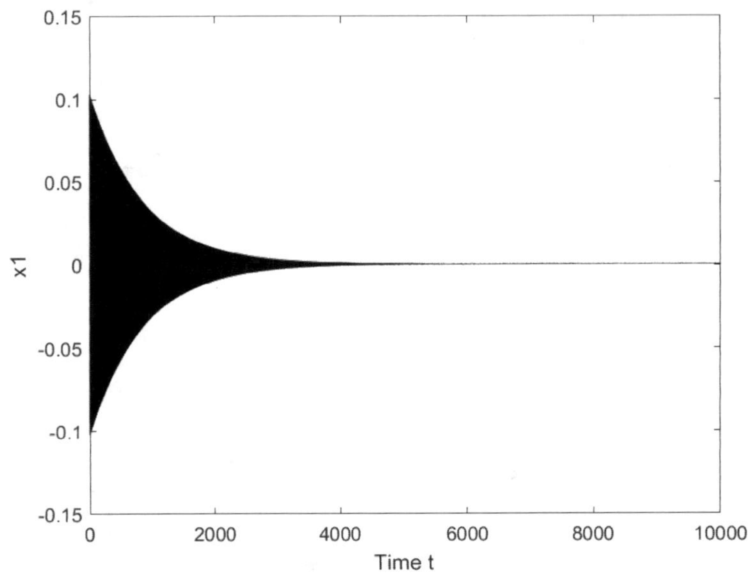

Figure 1. State response $x_1(t)$ for the first approach with $\gamma = 0.04$.

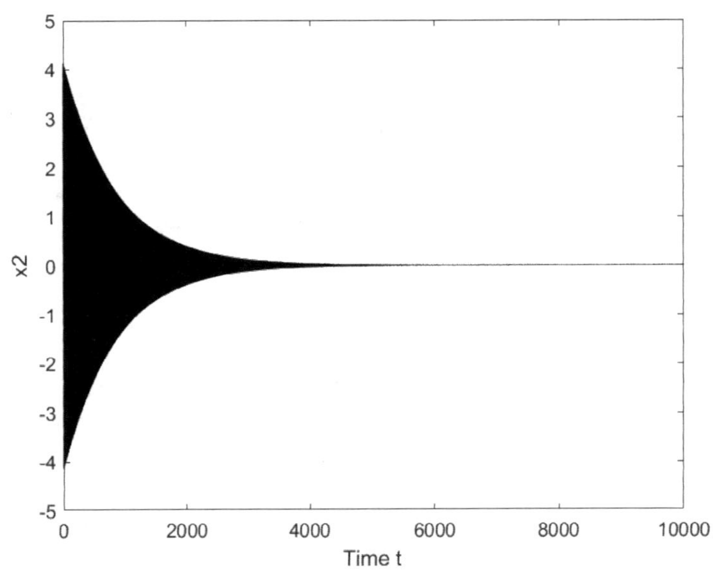

Figure 2. State response $x_2(t)$ for the first approach with $\gamma = 0.04$.

Analysis of transient processes plots indicates that problems of synthesis of suboptimal H^{∞} – controller gives good results. A rather large influence is exerted by the matrix Q, S elements in the quality criteria. The coordinates of state vector asymptotically converge to zero, which indicates the robustness of the system and a correct selection of parameters, for which the system retains its property of stability for any given disturbances. However, time for which the system converges to zero indicates a need to investigate the parameters' influence on the properties of the synthesized closed loop system. Generally, both of the approaches produce generally comparable results observed with the processes change over time, which allows to make a conclusion about them both being a viable option in further research.

IOP Conf. Series: Materials Science and Engineering **927** (2020) 012026 doi:10.1088/1757-899X/927/1/012026

4. Summary

In this study, we designed output feedback controllers using the H^{∞} control. Based on the results obtained we can state that suggested algorithms of linear controllers synthesis provide the desired quality of transient processes and asymptotic stability of closed loop systems. Using H^{∞} – control makes it possible to effectively neutralize the negative influence of limited disturbances on the system. The model example of H^{∞} – approach is solved. The described method successfully deals with the problem, demonstrating good results. The methods proposed could be useful for autopilot design problems for aircrafts and rotorcrafts.

References

[1] Koobloch H W, Isidori A and Flockerzi D 1993 *Topics in control theory* (Basel: Springer)
[2] Doyle J, Francis B and Tannenbaum A 2009 *Feedback Control Theory* (New York: Dover Publications)
[3] Skogestad S and Postlethwaite I 2005 *Multivariable Feedback Control: Analysis and Design* (New York: John Wiley and sons)
[4] Green M and Limebeer D J N 2012 *Linear Robast Control* (New York: Dover Publications)
[5] Gadewadikar J, Lewis F L and Abu-Khalaf M 2006 *J. Guid. Control Dyn.* **29**(4)
[6] Davtyan L G and Panteleev A V 2019 *J. Comput. Syst. Sci. Int.* **58**(3) 360–373
[7] Bortakovskii A S 2018 *J. Comput. Syst. Sci. Int.* **57**(4) 543–560
[8] Rybakov K A 2019 *Gyroscopy and Navigation* **10**(1) 27–34
[9] Rudenko E A 2018 *J. Comput. Syst. Sci. Int.* **57**(4) 505–528
[10] Panteleev A V and Pis'mennaya V A 2018 *J. Comput. Syst. Sci. Int.* **57**(1) 25–36
[11] Ivanyukhin A V and Petukhov V G 2019 *Cosmic Research* **57** 378–388

IOP Conf. Series: Materials Science and Engineering **927** (2020) 012027 doi:10.1088/1757-899X/927/1/012027

Experimental study on viscous fingering in Hele-Shaw cell under acoustic impact

E A Marfin[1,2], S V Garaeva[2] and A A Abdrashitov[1]
[1] Institute of Power Engineering and Advanced Technologies, FRC Kazan Scientific Center, Russian Academy of Sciences, 2/31 Lobachevsky Street, Kazan, Russia
[2] Kazan Federal University, Institute of Physics, 16a Kremlyovskaya Street, Kazan, Russia

E-mail: marfin_ea@mail.ru

Abstract. Viscous fingering shaped as curvature of the fluid displacement fronts are usually formed in porous media and Hele-Shaw cell when a liquid with a high viscosity is displaced by a liquid with a lower viscosity. Such interfacial instability is undesirable in many displacement processes, and mainly in oil production. We present the results of physical modeling of the displacement of viscous oil by air in Hele-Shaw cell. As a method of suppressing undesirable interphase instability, the effect of elastic vibrations on the process is proposed. The experiment was performed in stationary conditions and under acoustic action on the displacement process. In the absence of superimposed vibrations, the growth rate of viscous finger is 2.15 times higher than the speed of liquid flow at the cell border. The imposition of acoustic oscillations can both accelerate and slow down the growth rate of viscous fingers depending on their frequency. The paper discusses the peculiarities of the development of viscous fingering. The results are compared with the calculated data.

1. Introduction

In Russia, oil production is not efficient enough. The average oil recovery factor is about 30%. In most fields, produced oil is heavily waterlogged. The main reason for this situation seems to be the fact that oil is extracted from a porous reservoir by displacing it with water, which is a less viscous liquid than oil. From a physical point of view, this process can be described as the Saffman-Taylor instability [1]. With this displacement, the interphase between the two liquids is deformed with formation of folds. These folds do not grow at the same rate, forming viscous fingers. Such "fingers" of water in the reservoir, reaching the producing wells, form a breakthrough, which significantly increases the water content of the extracted oil. At the same time, areas of immobile oil that are not involved in filtration are formed in the reservoir itself, surrounded by water. The solution to the problem, according to oilmen, lies in the use of various gels instead of water. The gels are more viscous liquids than oil. However, this approach has significant environmental risks related to possible contamination of groundwater and surrounding areas.

A large number of papers have been devoted to the simulation of Saffman-Taylor instability. The development of computer technology has allowed us to expand the field of solving problems with moving borders. The paper [2] studies the stability of a system with two propagating fronts in a Hele-Shaw cell, where the viscosity increases monotonously from the eigenvalue to the viscosity of the external liquid. The critical parameter is the ratio of viscosities of liquids. The minimum value, below which the system is stable at any flow rate, is determined. It is proposed to use the stability criterion for choosing the viscosity of a displacing liquid in relation to the problem of pumping liquid into an oil well. In the paper [3], the authors identified three flow modes that characterize the development of viscous fingers and the conditions for their fusion. In the paper [4], the process of formation and growth of viscous fingers in inhomogeneous porous media is numerically investigated.

Various ways of suppressing undesirable instability are suggested. Thus, the paper [5] proposes a gradual reduction in the pore size of a porous medium. In the work [6], the inhomogeneity of the permeability was studied. In this study, three different models of inhomogeneity are considered: an exponential decrease in permeability in the transverse direction, an exponential decrease and an increase in the longitudinal direction. In the paper [7], the influence of cracks in a porous medium on the formation and growth of viscous fingers were studied. In many works, for example, in [8-10], the displacement of oil by various liquids with non-Newtonian properties is studied. In the paper [11], the influence of perturbations of the initial interface of liquids in the Hele-Shaw cell is investigated. The authors show that the period and amplitude of these perturbations affect the growth rate of viscous fingers.

The presented brief overview of the state in the field of Saffman-Taylor instability shows that various groups of scientists have been actively studying this phenomenon in various manifestations. At the same time, studies on the management of such instability by external acoustic treatment, as well as similar results, have not been found in the world literature. The main idea of this paper is that the phenomenon of Saffman-Taylor instability, often observed in the Hele-Shaw cell, can be influenced by the exposure to elastic oscillations. Moreover, the field experience of acoustic treatment of the reservoir indicates an increase in oil production and a decrease in water cut [12-13].

2. Experiment

A series of experimental studies was conducted on the displacement of oil by air in the Hele-Shaw cell. A Hele-Shaw cell is a device consisting of two closely located parallel plates containing a thin layer of viscous liquid, which is mainly used to simulate the flow of a liquid in a porous medium. To avoid the influence of gravity, the model was placed horizontally.

2.1. Materials

The viscous phase in the experiment was mineral oil of the I-20A brand (GOST 20799-88). The rheological properties of the oil were measured using a vibration viscometer SV-10 (A&D). The experiment was conducted in room conditions at a temperature of 23 ° C. At this temperature, the oil had a density of 0.860 g/cm^3, viscosity - 45.4 mPa·s, and the surface tension coefficient - 0.1472 N/m.

2.2. Experimental setup

For laboratory research, an experimental stand was created, which included the actual Hele-Shaw cells and a reservoir model, a liquid supply system and pressure measurement system, a lighting and video recording system, and a system for generating elastic oscillations. Hele-Shaw cells are made of two organic glasses with a thickness of 4 mm and a size of 200x100 mm. The gap between the glasses is 0.8 mm. On the sides, the cell is hermetically closed, and the inlet and outlet are connected through a slotted hole with pipes that supply liquids. The liquid supply system is based on syringe pump. The system for keeping a record of liquid displacement process in the cell is based on a digital video camera and a smartphone which provides video recording with a resolution of 1920 x 1080 pixels. The semi-transparent horizontal surface, on which the Hele-Shaw cell is placed, and the panel with LED illumination placed under it provide a contrasting image of the observed process. Acoustic treatment was carried out by means of created emitters from piezoelectric rings, working and back plates. Four emitters were placed under the Hele-Shaw cell and used to influence the displacement process with sound. The AKIP-3408/2 electric oscillator was used as the master generator. Measurement of the pressure difference at the cell boundaries is performed using a digital pressure sensor indicating the pressure. The experimental setup is shown in figure 1.

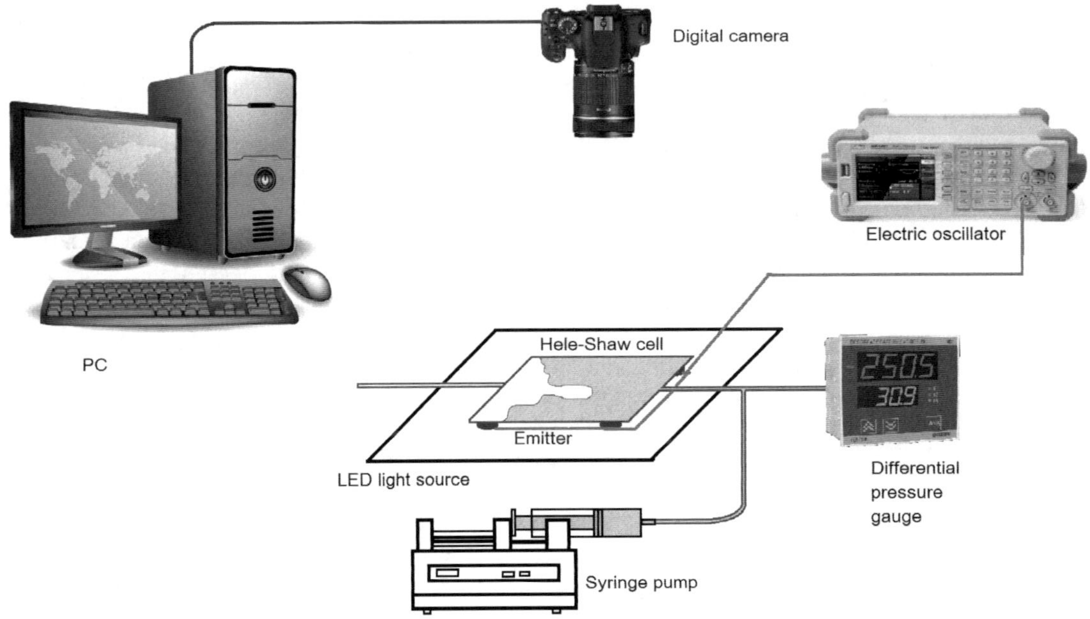

Figure 1. Experimental setup.

2.3. Experimental method

The study of the fluid displacement process in Hele-Shaw cell on an experimental installation is based on the processing of a video recording of the experiment. The recorded video file was opened in the VegasPro video processing program from where pictures of Hele-Shaw cell were extracted with the observed border for various moments of time. The pictures also displayed the pressure and recorded the time. Video was recorded at a resolution of 1920x1080 pixels at 30 frames per second. The duration of one experiment was about 1 minute, which corresponds to about 1000 frames.

The tests were divided into two groups: the one without acoustic treatment and the one with acoustic treatment at different frequencies. For each group, three different values of liquid flow rate were set. The experiments were repeated several times for each condition. The purpose of these experiments was to study the effect of superimposed vibrations on the formation of viscous fingers.

The following sequence was used for each test:

1) Filling Hele-Shaw cell with oil.

2) Turning on the video recording; the piston starts moving in the syringe pump at a fixed speed.

3) The process stops when reaching the summit of a viscous finger at Hele-Shaw cell borders. Video recording is over.

4) Repetition of pp. 1-3, including different values of the volumetric velocity.

5) Repetition of pp. 1-4 under conditions of superimposed vibrations.

Subsequent processing of video recordings was carried out on a computer. The graph editor processed footage, where the boundaries of the liquid interface were selected using filters. Then the position of the vertex of the maximum viscous finger was measured for each moment of time.

3. Results and discussion

At the beginning of the experiment, small perturbations of the interface are observed. After about 10 seconds, it is possible to recognize prominent three or four folds that begin to grow forming viscous fingers. Approximately at a distance of 10% of the cell length, there is an accelerated growth of the large "viscous finger", while the growth of others slows down significantly. The average growth rate of the "viscous finger" is more than twice the initial speed of the interface. This result is well described in previous studies, for example, in [1]. Figure 2 shows the change of the liquid interface in time steps of 5 seconds.

IOP Conf. Series: Materials Science and Engineering **927** (2020) 012027 doi:10.1088/1757-899X/927/1/012027

When elastic vibrations are applied, the picture of the process endures some changes (see figure 2 b). There is an uneven curvature of the liquid interface. Nevertheless, a single large viscous finger is formed, which grows in the same way as in the absence of superimposed vibrations. As viscous fingers grow, the following feature is manifested. The side surface of the viscous finger also oscillates and side protrusions can appear, which forms a complex shape of the viscous finger. With a certain combination of factors, it is possible to "bud" and grow new fingers from the main one.

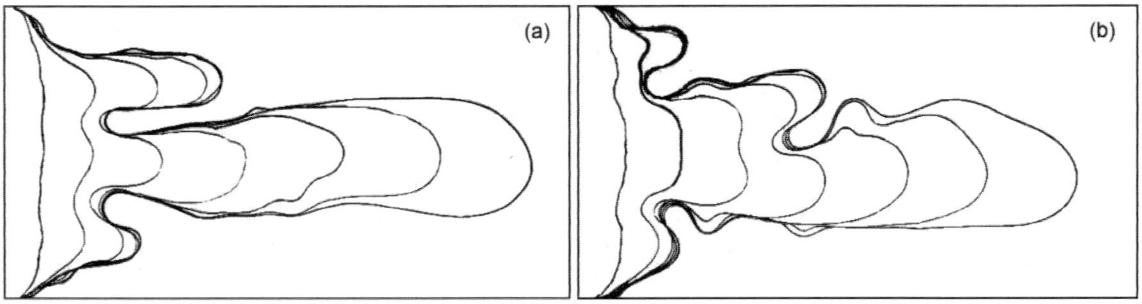

Figure 2. Comparison of the development of viscous fingerings in a Hele-Shaw cell: (a) without acoustic impact, (b) with acoustic impact at a frequency of $f = 45$ kHz.

Experiments conducted at various frequencies revealed that the growth rate of viscous fingers can vary with the frequency of superimposed oscillations. Figure 3 presents the dependence of the relative growth rate of the viscous finger on the frequency. On the graph, the value of the growth rate of the viscous U_{vf} finger was divided by the rate of fluid displacement from the Hele-Shaw u_x cell. It can be seen that in the absence of acoustic treatment, the growth rate of the viscous finger is approximately 2.18 times greater than the displacement rate. Initially, there is a tendency to increase the growth rate of the viscous finger, with a maximum value at a frequency of 22 kHz. Then the relative speed decreases to a minimum at 45 kHz.

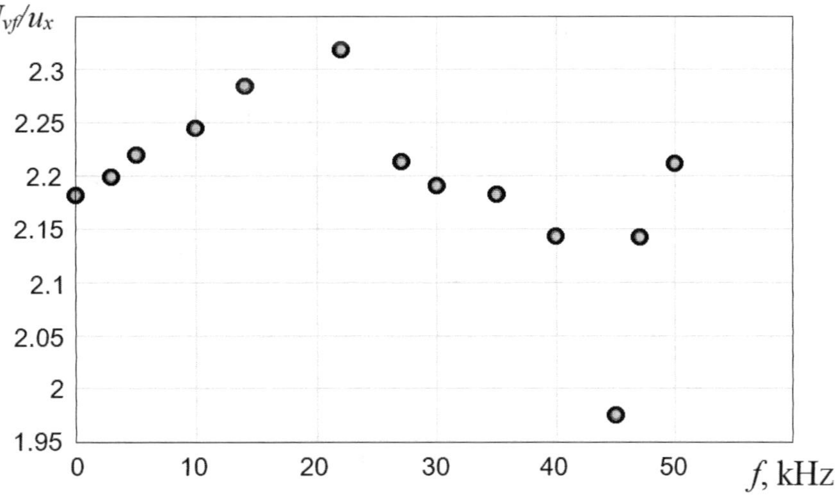

Figure 3. Dependence of the relative growth rate of the viscous finger on the frequency.

As the reason for the change in the growth rate of "viscous fingers", it is most likely that the oil viscosity changes in the field of elastic oscillations. Indeed, the study of oil before and after exposure to ultrasound on a viscometer revealed the effect of reducing its viscosity.

4. Conclusion

The paper studies the influence of acoustic treatment on the formation of viscous fingers in Hele-Shaw cell during the displacement of a viscous liquid. Experiments in stationary conditions confirmed the known data. When the viscous oil is displaced by a less viscous liquid, the initially smooth liquid interface curves, forming several folds that increase in size. Approximately at a distance of 10% of the cell length, there is an accelerated growth of the large "viscous finger", while the growth of others slows down significantly. The average growth rate of a "viscous finger" is more than twice the initial speed of the interface. When elastic vibrations are applied, the picture of the process endures some changes. So, with an increase in frequency to 22 kHz, the average growth rate increases by 6.4%, then decreases to a minimum at 45 kHz and then increases again. The maximum reduction in the speed of the "viscous finger" was 9.6%.

References

[1] Saffman P G and Taylor G I 1958 *Proc. Roy. Soc. London*. A245 **1242** 311
[2] Beeson-Jones T H and Woods A W 2015 *J. Fluid Mech.* **782** 127
[3] Nijjer J S, Hewitt D R and Neufeld J A 2018 *J. Fluid Mech.* **837** 520
[4] Christou K, Radünz W C, Lashore B, de Oliveira F B S and Gomes J L M A 2019 *Adv. Water Resour.* **130** 46
[5] Rabbani H S, Or D, Liu Y, Lai C-Y, Lu N B, Datta S S, Stone H A and Shokri N 2018 *Proc. Natl. Acad. Sci. U. S. A.* **115** 4833
[6] Shoghi M R and Norouzi M 2015 *Rheol. Acta* **54** 973
[7] Kargozarfard Z, Riazi M and Ayatollahi S 2019 *Petrol. Sci.* **16** 105
[8] Eslami A and Taghavi S M 2019 *J. Non-Newton. Fluid.* **264** 25
[9] Norouzi M, Yazdi A A, Birjandi A K 2018 *J. Non-Newton. Fluid.* **260** 109
[10] Shokri H, Kayhani M H and Norouzi M 2019 *J. Eng. Mech.* **145** 04019098
[11] Marfin E A, Garaeva S V, Ermolaev D V 2019 **J**. *Phys. Conf. Ser.* **1158** 032028
[12] Marfin E A, Kravtsov Y I, Abdrashitov A A, Gataullin R N and Galimzyanova A R 2015 *Petrol. Sci. Technol.* **33** 1526
[13] Marfin E A, Kravtsov Ya I, Abdrashitov A A and Gataullin R N 2014 *Georesursy [Georesources]* **57** 14

IOP Conf. Series: Materials Science and Engineering **927** (2020) 012028 doi:10.1088/1757-899X/927/1/012028

Trajectory Optimization of a Lunar Direct Landing from Halo-Orbits

A V Ivanyukhin[1,2] , E D Korolevskiy[2] and A E Starchenko[1]
[1]Research Institute of Applied Mechanics and Electrodynamics of the Moscow Aviation Institute (RIAME MAI), Leningradskoye shosse 5, 125080, POB 43, Moscow, Russian Federation
[2]Peoples Friendship University of Russia (RUDN University), ulitsa Miklukho-Maklaya 6, 117198, Moscow, Russian Federation

E-mail: ivanyukhin.a@yandex.ru

Abstract. This paper illustrates the problem of optimizing the trajectory of a direct landing from a halo orbit to the lunar surface with given coordinates. The study is considered in circular restricted three body problem. The descent operation consists of two stages: change in velocity for descent from orbit (the descent is considered as an instantaneous change in velocity of a spacecraft) and the main stage of braking, which is considered as a task of optimal performance of the spacecraft with finite thrust. The formulation and solution of the optimal control problem is presented, as well as numerical examples illustrating the accessibility of various regions of the lunar surface and examples of landing trajectories.

1. Introduction

In the beginning of practical astronautics, the achievement and exploration of the Moon were one of the highest priorities. In that period a huge amount of new knowledge on a flight mechanics were acquired, which made it possible to implement successfully the flights to the Moon with and without the man. Moreover, researches in that time were carried out in formulation of the two-body problem, which did not fully consider the specific features of the dynamics in the Earth-Moon system. Soon after the completion of the Apollo and Soviet lunar programs, the interest in the study of near-moon orbits had decreased for a rather long period of time.

The current renewal of interest to the Moon is associated with a significant expansion of the planned research programs, which has led to the need for reconsideration of new options for targeted near-moon orbits and new schemes for launching, stabilizing and supporting spacecraft take-off and landing spacecraft operations.

One of the promising options for near-moon basing orbits are the orbits around the libration points L1 and L2 of the Earth–Moon system. So halo-orbits around the libration point L2 should be reasonably used to place communication satellites in order to provide radio communication of the landing spacecraft on the far side of the Moon with the earth, which has already been implemented by the Queqiao spacecraft of China [2]. Nearly rectilinear halo orbits (NRHO) are planned to be used as the base orbit of the international lunar orbital station [3]. Such base orbit has several advantages over a near-lunar orbit. For example, start window trajectories for returning to Earth are constantly open from this kind of orbits. However, these orbits are significantly different from the well-studied

IOP Conf. Series: Materials Science and Engineering **927** (2020) 012028 doi:10.1088/1757-899X/927/1/012028

Keplerian type orbits and were not considered as part of the programs implemented in the 60s – 70s of the last century. The matter of which have motivated interest in this research.

2. Circular restricted three-body problem (CR3BP)

In order to understand distinctive features of this problem, let us consider the mathematical formulation of the circular restricted three-body problem. Since the libration points and halo orbits are the consequence of the exact solution of CR3BP [1].

Three material points (bodies) P_1, P_2 and P_3, move under the influence of forces of mutual attraction. The bodies P_1 and P_2 have masses m_1 and m_2 ($m_1 > m_2$), and the mass of body P_3 is negligible and does not affect the movement of P_1 and P_2. That is, the bodies P_1 and P_2 move around their common center of mass, and in that case as they move in circular orbits, so the problem is called the circular restricted three-body problem. Let us write the motion of the third body in the dimensionless synodic barycentric coordinate system. Let the unit of length be the distance between P_1 and P_2, the unit of mass is the sum of m_1 and m_2, and let us choose the unit of time so that the period of rotation of these points around their center of mass equal 2π. Let us direct the X axis along the vector, directed from P_1 to P_2, the Z axis along the vector of the kinetic moment of these points, and the Y axis complements the reference frame so that it meets the right hand rule. Then the motion equations of P_3 could be written in the form of [1]:

$$\begin{cases} \ddot{x} - 2\dot{y} = U_x \\ \ddot{y} + 2\dot{x} = U_y, \\ \ddot{z} = U_z \end{cases} \quad U = \frac{x^2 + y^2}{2} + \frac{1-\mu}{r_1} + \frac{\mu}{r_2}, \quad \mu = \frac{m_2}{m_1 + m_2}. \tag{1}$$

The libration points are stationary solutions of system (1). In addition, the system of equations has one first integral found by Jacobi, which defines the curves of zero velocity ($v = 0$), limiting the range of possible motions of the point P_3:

$$v^2 = 2U - J, \tag{2}$$

where J – Jacobi constant, v – vector of velocity.

Since the system (1) is autonomous, and has its first integral and a stationary solution, Lyapunov theorem on the existence of periodic solutions could be applied here: flat and vertical Lyapunov orbits. A numerical analysis of the solutions of system (1) shows that it has the following solutions in the vicinity of collinear libration points [4]: periodic halo orbits (for which the periods in the XY plane and perpendicular to it coincide) and quasiperiodic orbits for which the periods in the XY plane and perpendicular to it do not match. In the latter case, two types of orbits are distinguished: quasigalo-orbits (if the periods are equal on average) and Lissajous orbits (if the periods are rationally comparable).

Figure 1 shows a family of northern halo orbits near point L1 and L2 of the Earth-Moon system. The family of southern halo orbits is absolutely symmetrical to the one given above and differs only in the sign at the Z coordinate. Studies of the dynamics near the halo orbits show that the family of halo orbits has islands of increased stability, in particular resonant orbits, the period of which is rationally commensurate with the period of rotation of the Moon around the Earth. Exactly this kind of orbits get the greatest interest nowadays [3,7].

In particular, as noted earlier, NRHO orbits are of a great practical interest, one of which is a halo orbit near L2 with a resonance of 9:2, which could be chosen to accommodate the lunar orbital station in the future. This orbit has the following parameters: $R_p = 2032.2$ km, $R_a = 68126.7$ km, $T = 6.1$ day, $J = 1.4478$ km^2/sec^2. In further calculations, we assume that the spacecraft is located on that orbit.

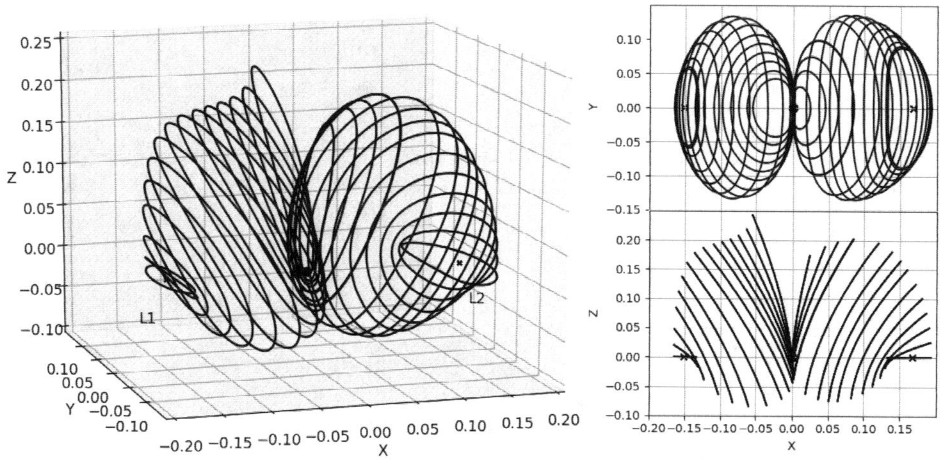

Figure 1. A family of northern halo-orbits near libration points L1 and L2.

3. Trajectory Optimization of a Lunar Direct Landing in the CR3BP model

This section is focused on the problem of optimization of direct landing from a halo orbit onto the lunar surface at a point with given coordinates. Traditionally the direct landing task consists of two stages: descent from orbit and the main braking stage [5-9]. The first sets the direction of descent from orbit, usually it is not consuming a lot of energy and therefore it will be considered as an instantaneous change in speed. The last maneuver requires basic energy costs and usually has a noticeable radial component, in connection with which it cannot be accurately estimated in a pulse setting. Thus, its optimization has the greatest effect on total energy costs.

Optimization of the main breaking will be considered in the following form. The movement occurs within the CR3BP with a finite thrust and described by the equations

$$\frac{d\mathbf{r}}{dt}=\mathbf{v},\quad \frac{d\mathbf{v}}{dt}=\mathbf{U}_r+\begin{pmatrix}2v_y\\-2v_x\\0\end{pmatrix}+\frac{T}{m}\mathbf{e},\quad \frac{dm}{dt}=-\delta\frac{T}{c},, \tag{3}$$

where $\mathbf{r}$ = (x, y, z) is the radius of the spacecraft position vector, $\mathbf{v}$ = (v_x, v_y, v_z) is the spacecraft velocity vector, m is the spacecraft mass, t is time, T is the thrust vector module, c is the outflow velocity, δ is the switching function (engine operation control: on / off), $\mathbf{e}$ is the unit thrust direction vector.

The initial conditions are determined by the position and velocity vectors of the spacecraft corresponding to its motion along a passive trajectory, which is a solution to system (1) and is dependent on the leaving point from the halo orbit, the velocity momentum for leaving the halo orbit $\mathbf{\Delta v}$, the time of passive motion $\mathbf{\Delta t}$, and the initial spacecraft mass:

$$\mathbf{r}(t_0)=\mathbf{r}(t_H,\Delta\mathbf{v},\Delta t),\quad \mathbf{v}(t_0)=\mathbf{v}(t_H,\Delta\mathbf{v},\Delta t),\quad m(t_0)=m_0e^{-\frac{\Delta v}{c}}, \tag{4}$$

The halo-orbit leaving point is given by the time period t_H elapsed after the passage of the apolune. The finite boundary conditions correspond to the landing on the lunar surface with the given selenographic coordinates: longitude $\lambda\in[-\pi;\pi]$ and latitude $\varphi\in[-\pi/2;\pi/2]$, delayed from the direction to the Earth. In a rotating coordinate system, relative to the Moon, they have the form:

$$\mathbf{r}(t_k)=\mathbf{r}_k(\lambda,\varphi)=\begin{pmatrix}-R_{Moon}\cos(\lambda)\cos(\varphi)\\R_{Moon}\sin(\lambda)\cos(\varphi)\\R_{Moon}\sin(\varphi)\end{pmatrix},\quad \mathbf{v}(t_k)=\begin{pmatrix}0\\0\\0\end{pmatrix}, \tag{5}$$

where R_{Moon} is the radius of the Moon.

The velocity impulse for the descent from the halo orbit and the time of passive motion are not specified. It is crucial to minimize engine runtime during main braking:

$$t_k - t_0 \to \min . \quad (6)$$

Using the Pontryagin's maximum principle, we can reduce this optimal control problem to a boundary-value problem for equations (4) with boundary conditions (5) and (6), supplemented with differential equations for conjugate variables, optimal control expressions and transversality conditions:

$$\frac{d\mathbf{p}_r}{dt} = -\mathbf{U}_{rr}\mathbf{p}_v, \quad \frac{d\mathbf{p}_v}{dt} = -\mathbf{p}_r + \begin{pmatrix} -2v_y \\ 2v_x \\ 0 \end{pmatrix}, \quad \frac{dp_m}{dt} = \delta \frac{T}{m^2}|\mathbf{p}_v|, \quad (7)$$

$$\mathbf{e} = \frac{\mathbf{p}_v}{|\mathbf{p}_v|}, \quad \delta \equiv 1, \quad (8)$$

$$\frac{m_0}{c} e^{-\frac{\Delta v}{c}} p_m(t_0) \frac{\Delta \mathbf{v}}{\Delta v} - \left(\frac{\partial \mathbf{r}_0(t_H, \Delta \mathbf{v}, \Delta t)}{\partial \Delta \mathbf{v}} \right) \mathbf{p}_r(t_0) - \left(\frac{\partial \mathbf{v}_0(t_H, \Delta \mathbf{v}, \Delta t)}{\partial \Delta \mathbf{v}} \right) \mathbf{p}_v(t_0) = 0,$$

$$-\left(\frac{\partial \mathbf{r}_0(t_H, \Delta \mathbf{v}, \Delta t)}{\partial \Delta t} \right)^T \mathbf{p}_r(t_0) - \left(\frac{\partial \mathbf{v}_0(t_H, \Delta \mathbf{v}, \Delta t)}{\partial \Delta t} \right)^T \mathbf{p}_v(t_0) = 0, \quad (9)$$

$$p_m(t_k) = 0, \quad H(t) = \mathbf{v}^T \mathbf{p}_r + \mathbf{U}_r^T \mathbf{p}_v + 2(p_{vx} v_y - p_{vy} v_x) + \frac{T}{m}|\mathbf{p}_v| - \frac{T}{c} p_m = 1.$$

This boundary value problem with 12 unknowns $\Delta \mathbf{v}, \Delta t, \mathbf{p}_r(t_0), \mathbf{p}_v(t_0), p_m(t_0), t_k$. The time of departure from the halo-orbit and the coordinates of the Moon landing t_H, λ, φ are considered to be set by (external) parameters of the task.

To study the dependence of direct landing costs on parameter values t_H, λ, φ, we chose 4 times of descent from the halo orbit t_H, equally spaced from each other (by 1/5 of the halo orbit period), we will call them 1, 2, 3, 4.

For each of them, the problem in the impulse formulation was solved: a two-point boundary value problem with velocity pulses at the beginning and end of the trajectory with optimizing the landing duration. The solutions obtained in this way are displayed on the isolines of the total velocity impulse figures 2-5. The marker in these figures shows the minimum, and the dashed line shows the projection of the 9:2 L2 halo-orbit onto the Moon.

These solutions can be used to select the optimal landing point, since it is obvious that the costs at the main stage of braking are proportional to the second velocity impulse, while the first impulse is small and mainly serves to "aim" at the landing point.

For the local minimum point of the obtained isolines, the boundary-value problem with finite thrust was solved at the main stage of braking (4)-(6), (8)-(10). During numerical calculations, the following propulsion system parameters were selected: thrust 2016 N, specific impulse of thrust 321 s, initial mass of the spacecraft in halo-orbit was assumed to be 7000 kg.

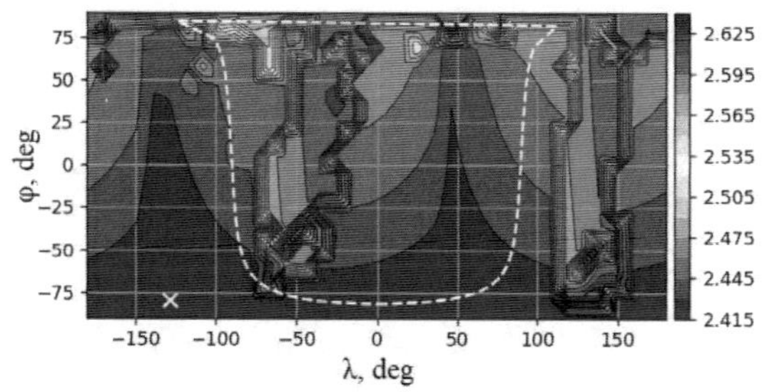

Figure 2. Isolines of characteristic velocity(km/sec) as leaving the halo orbit at point 1 (Moon's apocenter)

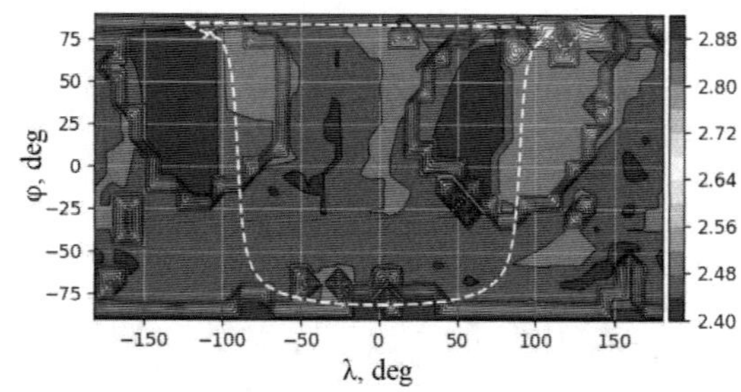

Figure 3 Isolines of characteristic velocity (km/sec) as leaving the halo orbit at point 2

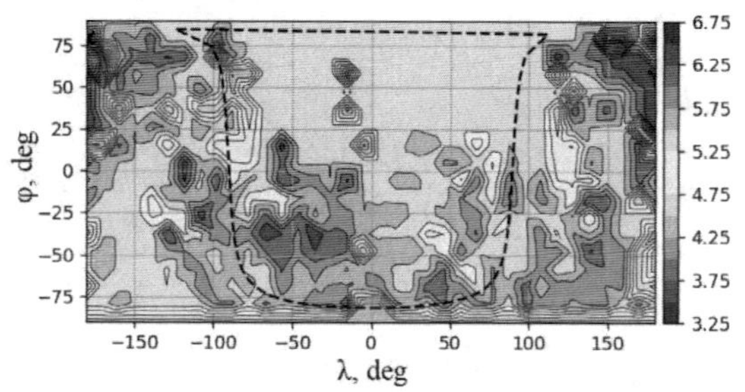

Figure 4. Isolines of characteristic velocity (km/sec) as leaving the halo orbit at point 3 (Moon's pericenter)

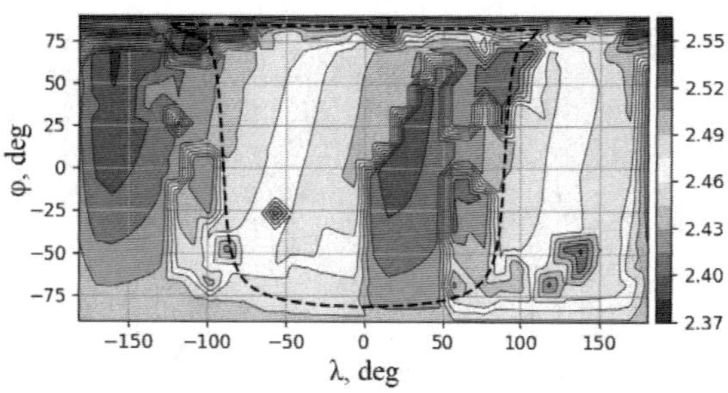

Figure 5. Isolines of characteristic velocity (km/sec) as leaving the halo orbit at point 4

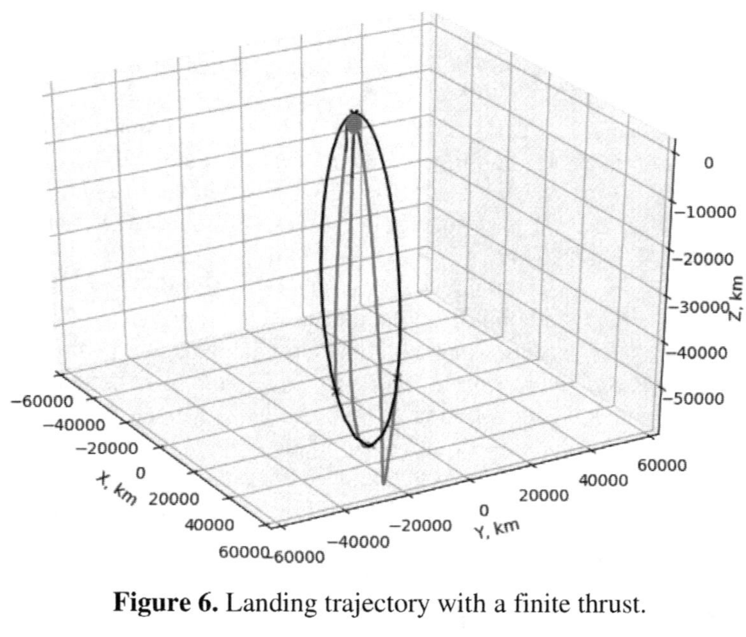

Figure 6. Landing trajectory with a finite thrust.

The results of the solution with impulse and finite thrust are shown in Table 1. It can be seen that taking into account the finiteness of the thrust at the main stage of braking leads to significant changes in the relative final mass. Obviously, this is due to large gravitational losses at this stage. The resulting trajectories are shown in figure 6.

Table 1. Comparison of solutions with impulse and finite thrust.

Leaving point	(λ, φ), deg.	Landing time for an impulse solution, day	Relative final mass for impulse solution	Landing time with a finite thrust, day	Relative final mass for solution with finite thrust
1	(-130,-80)	1.530	0.451	3.294	0.187
2	(-108, 80)	1.568	0.453	0.876	0.297
3	(-180, 80)	0.026	0.350	-	-
4	(138, 90)	5.808	0.456	5.888	0.398

4. Conclusion

The problem of a direct landing on the lunar surface from an almost rectilinear halo orbit at the libration point L2 is article's primary focus. It is shown that the time of launch from the halo orbit makes it possible to reach most of the lunar surface with total characteristic velocity expenses of 2.4 km/sec in the impulse setting. The only exceptions are the regions in the vicinity of the poles and the region of the lunar surface located near the projection of a halo-orbit onto it. In this case, impulse setting can only serve to compare the costs of landing in different regions of the surface, and to obtain the values of the fuel costs per landing, it is necessary to switch to a model with finite thrust.

Acknowledgments

This study was supported by the grant of the Government of the Russian Federation allocated from the federal budget for state support of scientific research conducted under the guidance of leading scientists in Russian educational institutions of higher education, research institutions and state research centers of the Russian Federation (Contest 7, Resolution No.220 of the Government of the Russian Federation of April 9, 2010). Agreement No. 075-15-2019-1894 of December 3, 2019.

References

[1] Szebehely V 1967 *Theory of orbits: the restricted problem of three bodies* (New York and London: Academic Press)

[2] Wu W, Tang Y, Zhang L and Qiao D 2018 Design of communication relay mission for supporting lunar-farside soft landing *Sci. China Inf. Sci.* **61** 040305

[3] Davis D C, Bhatt S A, Howell K C, et al. 2017 Orbit Maintenance and Navigation of Human Spacecraft at Cislunar Near Rectilinear Halo Orbits *27th AAS/AIAA Space Flight Mechanics Meeting*, Paper AAS 17-269, 20 p.

[4] Folta D C, Pavlak T A, Haapala A F et al. 2014 Earth-Moon Libration Point Orbit Stationkeeping: Theory, Modeling, and Operations *Acta Astronautica* **94** 421–33.

[5] Cho D-H, Jeong B, Lee D and Bang H 2009 Optimal Perilune Altitude of Lunar Landing Trajectory *International Journal of Aeronautical and Space Sciences* **10** (1) 67-74

[6] Bury L and McMahon J W 2020 Landing Trajectories to Moons from the Unstable Invariant Manifolds of Periodic Libration Point Orbits *AIAA Scitech 2020 Forum* p. 2181.

[7] Trofimov S, Shirobokov M, Tselousova A and Ovchinnikov M 2018 Transfers between near-rectilinear halo orbits and the moon *69th International Astronautical Congress* IAC-18.C1.8.10.

[8] Cheng Y, Gómez G, Masdemont J J and Yuan J 2017 Study of the transfer between libration point orbits and lunar orbits in Earth–Moon system. *Celestial Mechanics and Dynamical Astronomy* **128(4)** 409–33

[9] Cao P, He B and Li H 2017 Analysis of direct transfer trajectories from LL2 halo orbits to LLOs *Astrophysics and Space Science* **362(9)** 153

Experimental study for stochastic pulse component of emission from stationary plasma thrusters in the radio-frequency range

N A Vazhenin, A P Plokhikh and G A Popov
Research Institute of Applied Mechanics and Electrodynamics of the Moscow Aviation Institute (RIAME MAI)
Leningradskoye shosse 5, 125080, POB 43, Moscow, Russia

E-mail: riame@sokol.ru

Abstract. This paper is a sequel to the authors' publications devoted to the investigation of characteristics of self emission from electric propulsion thrusters. We consider the possibility of using the applications of the analytical signal theory and of the Fourier transform with a sliding window to study the spectral-temporal structure of the stochastic pulse component of the emission from stationary plasma thrusters in the radio-frequency range.

1. Introduction

A number of experimental works revealed that generally the electromagnetic emission from stationary plasma thrusters is non-Gaussian, has a pronounced component in the form of a random sequence of broadband radio pulses and a noise component being a sum of the SPT thermal radiation and noise of the measuring (reception) equipment [1-3]. An example of the pulse train envelope for the Hall thruster SPT-100 is shown in figure 1 [3].

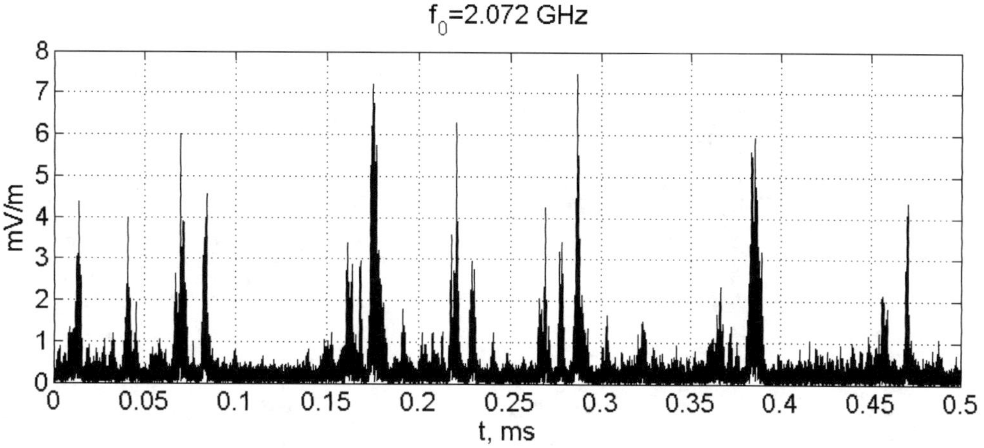

Figure 1. Signal waveform at the detector output for EP of SPT-100 type

2. Investigation problem statement

Despite the studies fulfilled, detailed information on the spectral-temporal structure of the stochastic pulse component of SPT emission is not available in the publications. This is due to the fact that typical measuring instruments (oscilloscopes, spectrum analyzers) do not allow us to obtain in an explicit form such important interference parameters as instantaneous phase and frequency.

We have stated and solved the problem of studying the spectral-temporal structure of the stochastic pulse component of SPT emission in the radio-frequency band using various methods of processing sample functions (as applied to the SPT-100 laboratory model), obtained with the help of a broadband digital oscilloscope connected to the output of the pick-up antenna mounted at a distance of 1 m from the operating thruster and overlapping the specified frequency range. The problem geometry meets the conditions described in [2].

Examples of 2 sample functions of single radio pulses are shown in figures 2, 3. We can see that in the time domain the pulses have a complex multimodal structure and different duration. Within the investigated data arrays it lies in the range from 0.5 to 1 μs. The sample function spectra obtained using the usual Fourier transform are presented in figures 4, 5. We can see that their spectral components are grouped in the frequency range of 0.5-1.5 GHz.

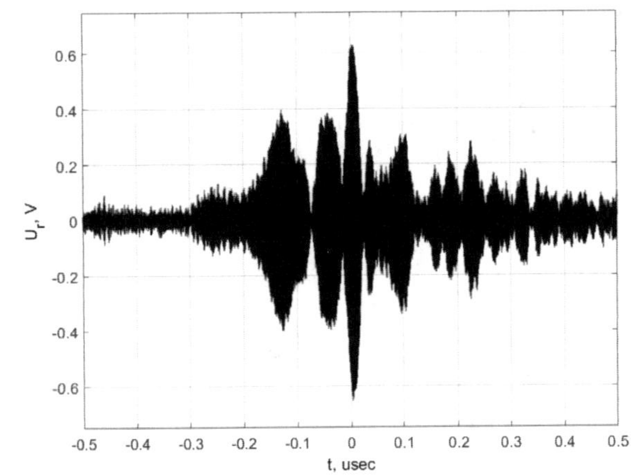

Figure 2. Sample function No. 1 of the single pulse of emission from SPT-100 model

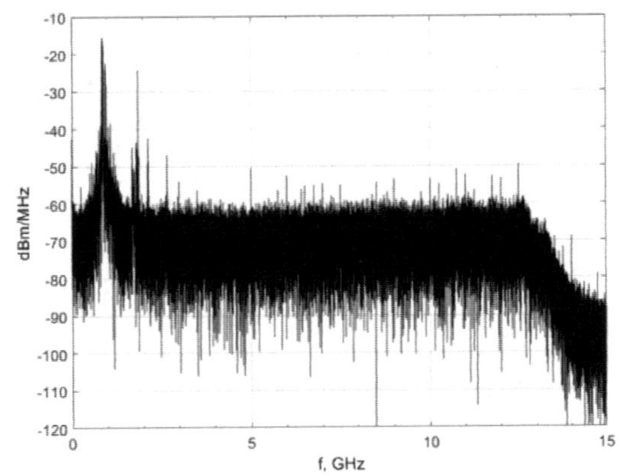

Figure 4. Spectrum of the sample function No. 1

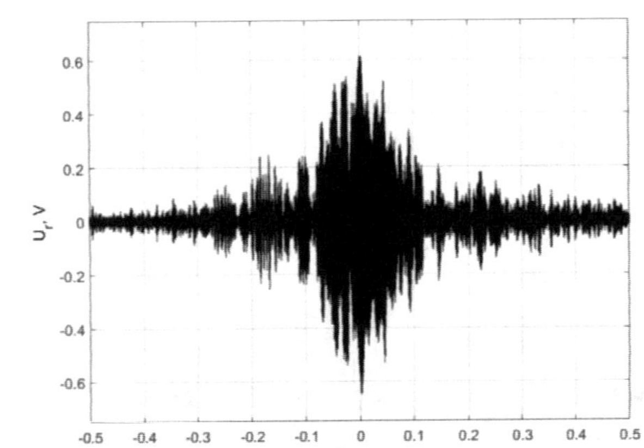

Figure 3. Sample function No. 2 of the single pulse of emission from SPT-100 model

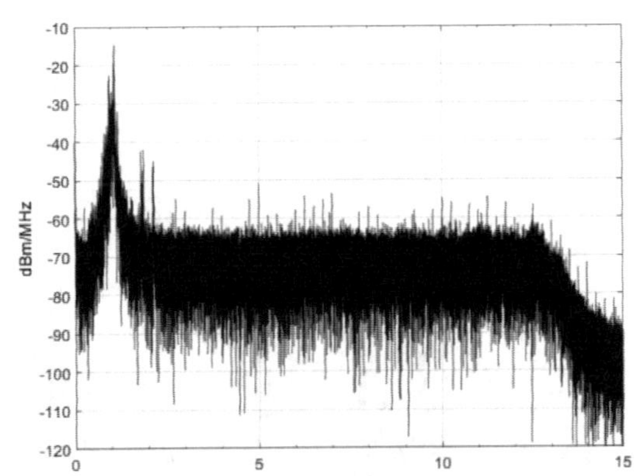

Figure 5. Spectrum of the sample function No. 2

At the same time, the question of phase and frequency changes in time within pulses remains open.

IOP Conf. Series: Materials Science and Engineering **927** (2020) 012029 doi:10.1088/1757-899X/927/1/012029

3. Application of the theory of analytical signals

In order to obtain new knowledge about the spectral-temporal structure of the stochastic pulse component of emission, let us use the results of the theory of analytical signals and write down the general expression for such signals [4]:

$$\dot{s}_a(t) = s(t) + js_H(t)$$

Here: $s(t)$ is the real sample function of the signal obtained by test, $s_H(t) = \frac{1}{\pi} p \int_{-\infty}^{+\infty} \frac{s(\xi)}{t-\xi} d\xi$ is the direct Hilbert transformation of an actual signal $s(t)$, p is the principal value of the Cauchy integral at $\xi = t$. The function $s_H(t)$ is defined by Hilbert transformation of initial function $s(t)$.

The analytical signal depends on the actual argument, is unambiguous and differentiable. On the complex plane it is displayed by a vector, the module and phase angle of which are defined by the argument, and the projection of the signal onto the real axis for any value of the argument is equal to the value of the original signal s(t). The analytical signal does not carry any new information as it is obtained by linear transformation from the initial signal and represents its alternative mathematical model.

However, such mathematical representation allows us to determine the time dependencies of amplitude envelope, current phase of the signal, and instantaneous frequency of the signal spectrum. The results of calculation of these functional dependencies for 2 sample functions are presented in figure 6, where the time dependencies of amplitude and phase are plotted within ±π (the first and the second graphs from above). According to the analysis, there is a strong dependence of instantaneous phase values on the signal-to-noise ratio, which does not allow to reliably estimate the instantaneous frequency of the signal. Therefore, to reduce the influence of additive noise component of the signal, the phase smoothing in a sliding window of 100 ns and transition from instantaneous frequency to the average one in the specified averaging interval was carried out. More complicated methods described in [5], for example, can be used also.

The results of calculations of the total phase and the average frequency as applied to the pulses (figures 2, 3) are shown in figure 6 (the third and the fourth graphs from above).

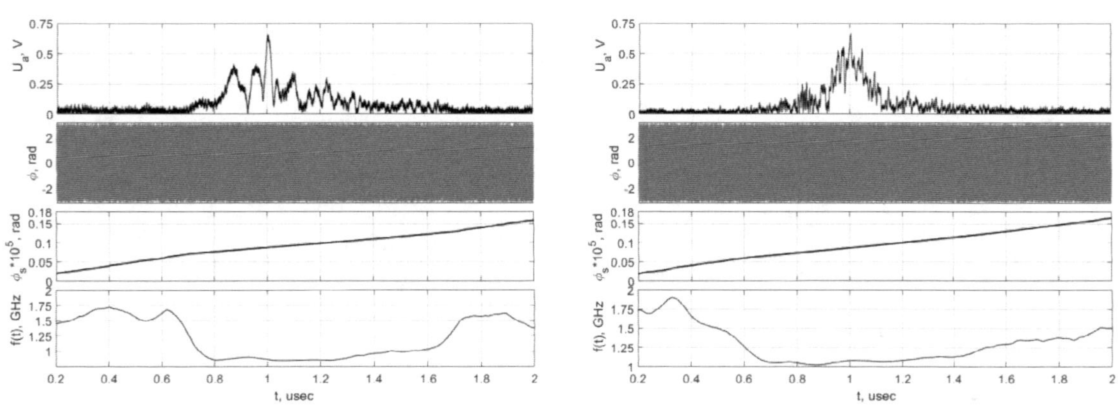

a) Sample function No. 1 b) Sample function No. 2

Figure 6. Amplitude, phase and frequency characteristics of the pulse in the time domain

It is obvious that with such test results during the entire multimodal pulse a considerable variation of the average signal frequency is observed. At the same time, variations in the average frequency of the signal spectrum without taking into account the current spectrum form are relatively rough estimates of the internal spectral-temporal characteristics of pulses generated by the SPT.

IOP Conf. Series: Materials Science and Engineering **927** (2020) 012029 doi:10.1088/1757-899X/927/1/012029

The next step in the analysis was the use of spectrograms based on the Fourier transform with a sliding window to reveal the finer spectral-temporal characteristics of the signal.

4. Analysis for spectrograms of the processes under study

Such spectrograms for the considered pulses are shown in figure 7. For clarity, they show curves corresponding to the maximum value of the spectrum in each time section of the spectrogram.

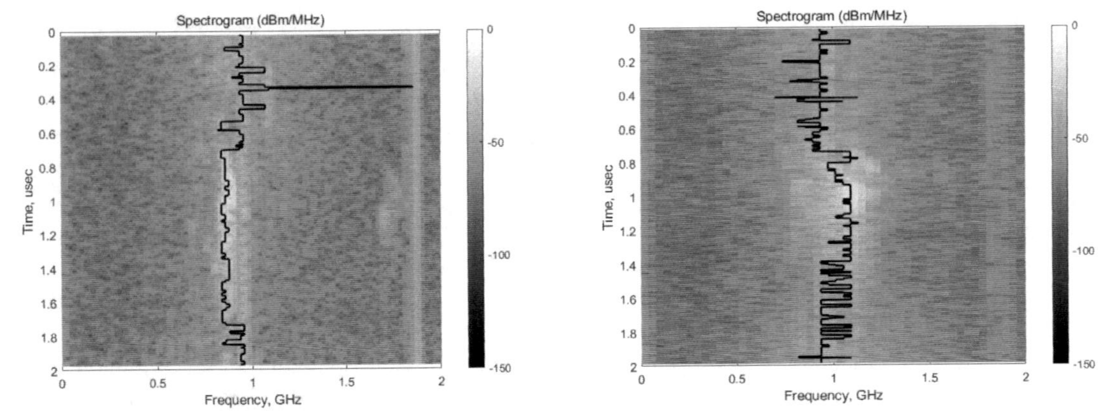

a) Sample function No. 1 b) Sample function No. 2

Figure 7. Spectrogram of emission from SPT-100 model

It can be seen that in this case within the pulse the maximum of the local spectrum is shifted relative to the same maximum for the "grass" by about 100 MHz.

For illustrative purposes, figure 8 shows the spectra for time sections 1, 1.2, 1.6 and 2 μs, corresponding to the spectrograms in figure 7.

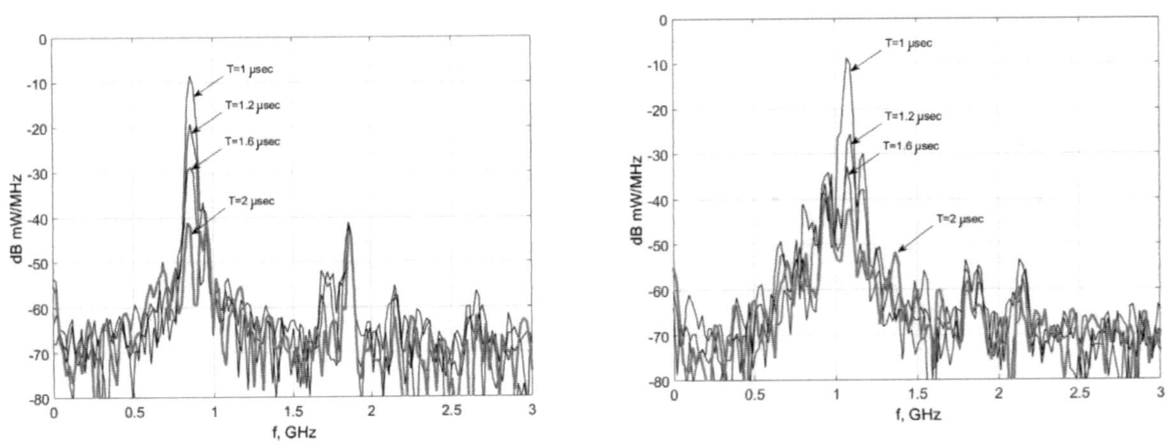

a) Sample function No. 1 b) Sample function No. 2

Figure 8. Local spectra of the SPT-100 emission

It is obvious, that for the sample function of radio pulse No. 1 all local spectra within the pulse are located at the frequency of about 0.9 GHz, which corresponds to the results of spectrogram in figure 6. As for the sample function No. 2, the qualitative results coincide with the sample function No. 1; however, the maximum amplitude of the local spectrum is not at 0.89 GHz, but at 1.07 GHz (figure 8). Taking into account the results obtained for other recorded sample functions of the processes, we can conclude that for the considered mode of operation the stochastic pulse component of SPT-100 emission has a random value of the local spectrum frequency corresponding to the maximum of the amplitude spectrum, the mean-square spread of which is approximately 100 MHz.

The study of the "fine" spectral-temporal structure of SPT radio-frequency emission made it possible to reveal new features inside the pulse frequency-temporal characteristics of SPT radio-frequency emission, which should be taken into account when developing models of radio channels and evaluating the reliability of data transmission in space communication radio systems.

5. Conclusions

1. The available methods of analysis for spectral and temporal characteristics of emission from stationary plasma thrusters based on experimental data are considered. It is shown that direct use of typical measuring devices allows obtaining data only on the average energy spectrum without taking into account information on the instantaneous phase and frequency.
2. It is shown that the most accurate method of processing is the use of spectrograms based on the Fourier transform with a sliding window, which allow to reveal more "fine" spectral and temporal characteristics of the signal with a high degree of reliability.
3. For the SPT 100 model, a full cycle of research for the stochastic pulse component of emission in the radio-frequency range was completed, which showed that while modeling the frequencies corresponding to the maximums of local spectra of independent pulses can be considered within the limits of the pulse as the constant random variables, the statistical characteristics of which should be determined experimentally for each type of SPT and the mode of its operation.

References

[1] Beiting E, Garrett M, Pollard J, Pezet B and Gouvernayre P 2005 Spectral Characteristics of Radiated Emission from SPT-100 Hall Thrusters, *Proc. 29th International Electric Propulsion Conference* (Princeton University), IEPC-2005-221

[2] Beiting E, Garrett M, Pollard J, Pezet B and Gouvernayre P 2005 Temporal Characteristics of Radiated Emission from SPT-100 Hall Thrusters in the L, S, and C Bands, *Proc. 29th International Electric Propulsion Conference* (Princeton University), IEPC-2005-222

[3] Arbatskiy V M, Baranov S V, Vazhenin N A, Plokhikh A P and Popov G A 2017 Peculiarities of determination of electromagnetic interference emission from electric propulsion thrusters under ground conditions *Procedia Engineering* **185** 97-104

[4] Vainshtein L A, Vakman D E 1983 *Frequency separation in the wave oscillations theory* (Moscow, Nauka)

[5] Chugai K N, Kosachev I M and Rybakov K A 2020 Approximate Filtering Methods in Continuous-Time Stochastic Systems. In: Jain L, Favorskaya M, Nikitin I and Reviznikov D (eds) *Advances in Theory and Practice of Computational Mechanics. Smart Innovation, Systems and Technologies* **173** 351-371 (Springer, Singapore)

IOP Conf. Series: Materials Science and Engineering **927** (2020) 012030 doi:10.1088/1757-899X/927/1/012030

Colloid thrusters. Development status and prospects for application

V V Nigmatzyanov, V A Obukhov, G A Popov and V V Svotina
RIAME MAI, 5 Leningradskoe shosse, 125080, POB 43, Moscow, Russia

E-mail: deutscher@mail.ru

Abstract. The article examines the principle of operation, the basic laws of electrostatic spraying of fluid, which constitutes the working process of colloid thrusters (CT). The features of two competing technologies for producing charged colloids based on the design of capillary type sprayers and the design of porous sprayers are considered. The experience of CT development and application is considered.

1. Introduction

The development of micro-satellite and nano-satellite technology is one of the trends in the space technology development for the coming years. The spacecraft of the indicated size should be equipped with engines with the level of power consumption from several units to ten Watts and thrust from several units to 200 μN. Colloid thrusters (CT), based on electrostatic spraying of an electrically conductive fluid, are one of the acceptable engine options for spacecraft of such sizes.

The CT idea belongs to H. Obert, who in the 30s of the twentieth century proposed using charged fluid droplets accelerated in an electrostatic field to create thrust [1, 2, 3]. H. Preston-Thomas suggested that accelerated beams of colloid charged particles can be used to obtain thrust in space [4]. J.J. Barre presented a theoretical work on engines, which contained calculations for the parameters of flows of liquid and solid colloidal particles [5]. In 1958, an experimental and theoretical study of a charged colloid particle thruster was conducted at the Astronautical Laboratory of the Aerojet-General Corporation [6]. In the 60s and 70s of the past Century and later the CT studies were conducted in the USA [7], and in the USSR [8, 9, 10]. Certain research on CT was continued at the Moscow Aviation Institute [8, 9, 10]. Until the early 2000s, the interest in CT around the world was negligible. The European program LISA (Laser Interferometry Space Antenna) [11] for the registration of gravitational waves gave rise to the CT development at a new stage. The spacecraft being the interferometer parts should compensate of any gravitational perturbations that interfere with the registration of gravitational waves. In the mission completed within the framework of the European LISA Pathfinder (LPF) program, flight tests of the BET-300-P CT developed by Busek Co. were successfully conducted under the contract with NASA Jet Propulsion Laboratory (JPL) [12]. (A laboratory model of the thruster was developed and delivered to NASA JPL in 2008).

2. The principle of CT operation

The CT (figure 1) contains the following elements: a charged particle emitter, which is a combination of a number of sprayers — sources of charged particles made, for example, in the form of capillary tubes; an electrostatic, as a rule a two-electrode system of charged particle acceleration formed by the

IOP Conf. Series: Materials Science and Engineering **927** (2020) 012030 doi:10.1088/1757-899X/927/1/012030

extracting and accelerating electrodes with holes made so that individual cells of acceleration of charged particles are formed; a propellant storage and feed system; a neutralizer of a charged particle beam at the thruster exit; and a power processing unit including power supply sources for electrodes. The emitter is maintained under the positive potential of several *kV* relative to the extracting electrode, due to which a strong (above 10^9 $V{\cdot}m^{-1}$) electric field is formed at the end of the sprayers. The propellant in the form of an ion-conducting fluid is supplied through the capillaries. At the ends of the capillaries, in the region of strong electric field, the fluid is sprayed into charged droplets. At the thruster exit, the current and space charge of the flow of charged droplets accelerated in the electrostatic system is neutralized by particles of the opposite sign from the neutralizer. The exiting quasi-neutral flow generates thrust.

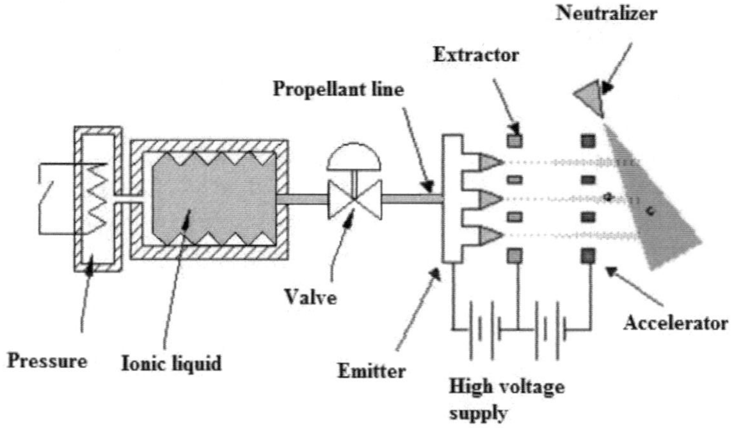

Figure 1. CT principal diagram.

3. The CT basic physics

The features of electrostatic spraying will become clearer if at least approximate relationships are obtained between the required value of the electric field strength *E*, the physical properties of the liquid, and the main parameters of the charged particles: radius *r* characterizing the dispersion fineness, and specific charge of droplets q/m. These relations are determined by the conditions for the stable existence of charged droplets.

Charged droplets retain such macroscopic properties of a substance as: density ρ, surface tension α, liquid permittivity ε_r, if they contain more than 10^3 atoms or molecules, which in order of magnitude corresponds to droplet size of 10^{-9} m. A charged droplet is stable when the Relay condition is fulfilled: $P_e \le P_\alpha$, where $P_e = 0.5\varepsilon_0 E^2 = q^2\left(2\varepsilon_0 r^4\right)^{-1}$ - is the pressure of electrostatic forces; ε_0 is vacuum permeability; $P_\alpha = 2\alpha r^{-1}$ - capillary pressure.

Table 1 shows *r* and *E* on the surface of the droplet as functions of q/m, calculated for glycerol. The solutions of KI and NaI in glycerol were considered and studied as the propellant for CT in the 60s and 70s of the past century [7-9]. In addition, the exhaust velocity of charged droplets accelerated by a given potential difference U_0 is also given.

Table 1. Calculated values.

Specific charge (kCal kg^{-1})	Droplet radius, r (nm)	Exhaust velocity (at U_0 3 kV/5 kV) (ms^{-1})	Electrostatic field strength, E (Vm^{-1})
0.5	30	1700 / 2200	5×10^9
1	20	2450 / 3200	6.5×10^9
5	7	5470 / 7000	10^{10}

It follows from the above data that at relatively moderate values of the accelerating potential of *3-5 kV,* the exhaust velocity of more than $5000\ m{\cdot}s^{-1}$ can only be obtained for droplets with a specific charge of more than $5\ kCal\ kg^{-1}$. In this case, $r = 7{\cdot}10^{-9}\ m$, and $E = 10^{10}\ V{\cdot}m^{-1}$. At such electric field strength, ion field emission is possible, which can cause a drop electric charge to be lost. Based on this, for example, for glycerol, as follows from Table 1, the specific charge of $5\ kCal\ kg^{-1}$ can be considered as a limit from the point of view of droplet stability in both mechanical and charge terms. The above assessment can be extended to other fluids, which are considered as a CT propellant. The general properties of these liquids are such that having high electrical conductivity ε_r they are good solvents.

4. Main regularities of electrostatic spraying

Let us consider the technology, in which liquid is supplied by a thin (0.2-0.4 mm in diameter) metal capillary. The experiments [9] were carried out at vacuum of $1.33{\cdot}10^{-3}\ Pa$ according to the capillary-plane scheme at a distance between the electrodes of 8-12 cm. The circuit of high-voltage power supply source (up to 20 kV) comprised resistors $R_1 = 30.5\ MOhm$, $R_2 = 650\ kOhm$ and R_3; from the latter the signal was output to the oscilloscope. The optical system provided 22.4 magnification when projected onto a 9x12 cm photographic plate. The backlight was produced by two lamps in a pulse of *0.1-0.2 s* in duration and power of up to *200 W*. The modes of KJ solution spraying in glycerol were studied. The experimental design is shown in figure 2.

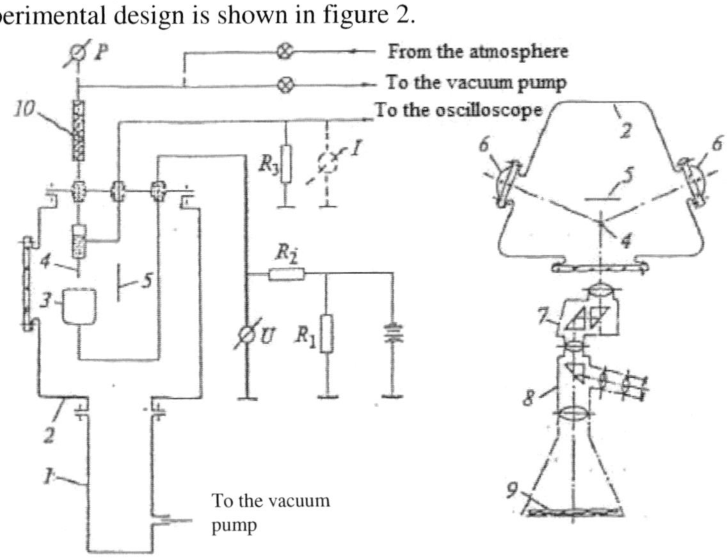

Figure 2. Scheme of the experimental setup [9]. 1-vacuum pump, 2-vacuum chamber, 3-collector, 4-capillary, 5-blackened screen, 6-pulse backlights, 7, 8 - optical system elements, photographic plate, 10-insulator.

The process of charged droplets formation proceeds as follows. Under the action of a field near the end of the capillary, the surface of the outflowing fluid becomes unstable. A microscopic jet is formed, that disintegrates into charged multi-molecular drops. At the same liquid flow rate, as the potential difference increases, the meniscus is pressed against the end of the capillary, and then two or more jets may appear on the edge of the capillary instead of one jet.

Taylor [13] showed that from the mathematical point of view the only stable stationary form of meniscus of the liquid with high electrical conductivity is a cone with a half-angle at the apex of 49.3^0. The cone-like menisci of liquid from which the formation of the thin spray occurs are called as the “Taylor cones”, although their shape may differ from the canonical one.

The change in the spray pattern can be seen in microphotographs of figure 3. In the single-jet spraying mode, the current is weakly dependent on voltage. This feature has been experimentally discovered for a number of liquids with high electrical conductivity and fairly high viscosity at low flow rates [14]. Dependence of electric current on physical properties and flow rate $I_p = F(Q, K, \alpha, \varepsilon_0\varepsilon_r)$ can be determined by resorting to dimensional analysis, from where the formula follows:

$$I_p = C(QK\alpha)^{1/2} \tag{1}$$

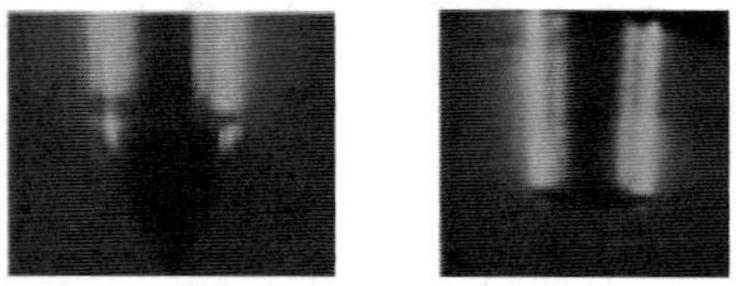

Figure 3. Microphotography of a charged jet: a single-jet mode is on the left and a two-jet mode is on the right

Here K is the fluid electrical conductivity (St·m^{-1}), Q is the volumetric flow rate (m^3 s^{-1}), C is the constant, which is determined from experiment for each fluid. It is assumed in [15] based on experimental data, that C can be represented as a function of relative liquid permeability:

$$I_p = f_1(\varepsilon_r)(QK\alpha)^{1/2} \tag{2}$$

The average specific charge of particles can be defined from (2):

$$\frac{q}{m} = \frac{I_p}{\rho Q} = \frac{f_1}{\rho}\left(\frac{K\alpha}{Q}\right)^{1/2} \tag{3}$$

The single-jet spraying mode is limited by the flow rate (from the bottom) and the potential applied to the sprayer (from the top). This implies the need to obtain a sufficiently high specific charge of *5 kCal kg^{-1}* at moderate values of the potential (*3-5 kV*) (ref. to table 1). It follows from (3) that this can be obtained with a sufficiently high electrical conductivity of the liquid and low flow rate. Let us consider what determines the minimum flow rate, at which we will implement a single-jet spraying mode. From experiments it follows:

$$Q_{\min} = f_2 \frac{\alpha \varepsilon_0 \varepsilon_r}{\rho K} \tag{4}$$

By substituting (4) in (3) we obtain the maximum estimate for the specific charge of droplets:

$$\mathbf{max}\left(\frac{q}{m}\right) = f_3 \frac{K}{\sqrt{\rho \varepsilon_0 \varepsilon_r}} \tag{5}$$

From (5), a strong dependence of specific charge on the electrical conductivity of the fluid follows. This defines the choice of CT propellant, the physical properties of which are determined by high polarity; such fluids are good solvents of salts with a covalent bond. Table 2 shows the fluids currently being studied, their properties, and the dependence of the specific charge of particles on fluid flow rate. The data presented are taken selectively from the experimental data given in [16].

Table 2. Properties of working fluids and spray characteristics.

Substance	Composition, salt concentration,% by weight	Dencity (g m^{-3})	Electric conductivity, (St m^{-1})	Flow rate (μg s^{-1})	Specific charge (kCal kg^{-1})
F3	Formamid+11.8 %	1.234	0.93 (at t=19 ^{0}C)	0.69	0.76
F3	NaJ		0.3 (at 21 ^{0}C)	0.33	1.27
F3				0.11	2.27
F3				**0.026**	**11.62**
F4	T			0.56	0.91
F4	Formamid+16.8 %	1.289	1.22 (at t=19 ^{0}C)	0.17	2.0
F4	NaJ			0.079	4.76
F4				**0.045**	**6.25**

As it is obvious from data of Table 2, to obtain high specific charge of droplets in a single-jet mode, it is necessary to use liquids with high electrical conductivity $K > 1 Stm^{-1}$ at the flow rate less than

0.05 μg s^{-1}, which is in qualitative agreement with the above conclusions. So, the substitution of the following values for a 16.8% NaI solution in formamide in (6): $K = 1.289\, Stm^{-1}, \rho = 1.13 \cdot 10^3 kgm^{-3}$, $\mathcal{E}_0 = 8.85 \cdot 10^{-11} Fm^{-1}$, $\mathcal{E}_r = 111$ gives for q/m the order of magnitude of 10^3 Ckg^{-1}, which is observed in the experiment at low flow rates (see table 2).

5. Development of CT based on capillary technology

The experimental data presented in table 2 allow us to conclude that the simplest capillary technology has not lost its relevance. The Busek BET-300-P CT was developed under the ESA LISA Pathfinder program based just on such a technology. The thruster passed flight tests in 2016 successfully [17]. The thruster meets the technical requirements of the LISA project (see Table 3).

Table 3. Parameters of the thruster BET-300-P developed by Busek Co [18]

Parameter	Estimated	Demonstrated
Thrust, μN	4–30	4.35–35.8
Thrust control accuracy (μN)	less than 0.1	0.08
Working liquid	NaI solution in Formamid	
Specific impulse (s)		150–200
Mass (kg)	less than 2	achieved
Power (W)	less than 10	achieved
Lifetime (hour)	40000–50000	3478

With such value of electrical conductivity, the specific charge of particles was about 1.3 kC kg^{-1}, and the specific impulse was 150–200 s. Later, when working under the ST7 program to increase the specific impulse, the company managed to achieve significant progress. The Busek BET-300-P thruster used a working fluid with higher electrical conductivity. The average specific charge of the particles was 35 kC kg^{-1} [26]. Flight tests of the Busek BET-300-P as part of the Lisa Pathfinder SC of the European Space Agency and the works under the ST7 program represent a significant success in terms of CT technology demonstration.

6. Trends in CT technology development

The development of CT technology proceeds in the field of creating atomizers of various types, the design of which largely determines the nature of the process of charged particles formation. The specific charge of the droplets is limited by the magnitude of the electric field, at which the droplet starts to loose charge due to ionic emission. However, it is possible to obtain not only charged droplets, but also the charged clusters of complex structure with specific charge of 10–30 kC kg^{-1}, which ensures the achievement of a specific thrust impulse of more than 500 s at moderate accelerating potentials.

The use of a porous atomizer along with the Emi-Im ionic conductive liquid can give a corresponding effect. Studies of a thruster with porous atomizer [19] showed that it can have a high specific impulse with relatively high thrust. A feature of physical processes is the emission of "solvated" ions, that is, of cluster ions formed in Emi-Im.

The composition of the cluster is difficult to predict. The initial beam contains single ions ("monomers"), clusters formed by a single ion with an attached pair anion-cation (dimers), and clusters formed by the attachment of two pairs anion-cation to a single ion (trimmers). In addition, after its extraction from the surface, the cluster is extremely unstable and dissociates into lighter ions and neutral components with a characteristic lifetime of less than a microsecond. Fragmentation, if it occurs at the stage of beam formation and acceleration, leads to the thruster performance degradation. In [20] the processes of solvated ions' fragmentation in Emi-Im were calculated. The spray conditions are: beam current - *526 nA;* accelerating voltage - *1618 V;* 50% of clusters with a dimer structure with a specific charge of *20 kC*·kg^{-1}; 50% are the monomer clusters with a specific charge of 38 kC·kg^{-1}. According to the calculations, the neutral fraction should appear after the passage of approximately *0.3 mm* from the end of the atomizer. In the cited work, an emitter with an interelectrode gap of *0.4 mm* was used. With such gap, most of the clusters do not manage to loose their charge, and the fraction of neutral particles flowing out with the ion beam is still not large. Thruster efficiency may be high.

7. Conclusion

Currently, the most advanced technology for creating CT is the technology with capillary atomizers. By now, the first CT models that have passed space flight tests have been developed. The CT performances achieved with this technology are the thrust of *1-200 μN*, specific impulse of *250-300 s* when using a working fluid in the form of a sodium iodide solution in formamide with an electrical conductivity of $0.5\ Ohm{\cdot}m^{-1}$.

CE activity will develop both in research and in the creation of new technologies designed for large-scale serial production of micro- and nano-spacecraft.

Acknowledgements

The work was supported by The Ministry of Science and Higher Education of the Russian Federation in framework of "Goszadanie" grant number FSFF-2020-0014.

References

[1] Oberth H 1923 *Die Rakete zu den Planetenraumen* (Munich and Berlin: Oldenbourg)
[2] Oberth H 1929 *Wege zur Raumschiffahrt* (Munich and Berlin: Oldenbourg)
[3] Stuhlinger E 1964 *Ion Propulsion for Space Flight* (NY: McGraw-Hill Book Company) p 373
[4] Preston-Thomas H 1952 *J. of the British Interplanetary Society* **11** 173
[5] Barre J 1957 Essai de Contribution a la Propulsion Ionique *VIII Int. Astr.Congr. (Barcelona)*
[6] Ehricke K 1957 Comparisons of Propulsion Systems: Solar Heating, Arc-Thermodynamics and Arc-Magnetohydrodynamics *Advanced Propulsion Systems Symposium of the US Office of Scientific Research (Los Angeles, USA. - 11-13 December 1957)* pp 71-104
[7] Schultz R and Branson L 1959 The Colloid Rocket: Progress Toward a Charged-Liquid-Colloid Propulsion System *Proc. of 2 Symposium on Adv. Prop. Conc. (Boston, MA, USA)* p 54
[8] Shtyrlin A 1995 IEPC-95-103 *24th International Electric Propulsion Conference (Russia, Moscow 19-23 September 1995)* p 692
[9] Stirlin A 2011 Neutralizing of colloidal electroroket thruster charged particles beam *Aerospace MAI Journal* **18** 1 p 54
[10] Stirlin A Charged-particle beam acceleration and energy efficiency of colloid electric propulsion thruster *Aerospace MAI Journal* **16** 5 p 111
[11] Website ESA LISA http://sci.esa.int/lisa
[12] Gamero-Castaño M and Hruby V 2001 IEPC-01-282 *27th International Electric Propulsion Conference (USA, CA, Pasadena 15-19 October 2001)*
[13] Taylor G 1964 *Proc. Royal Soc.* A **280** 1382 p 383
[14] Gafian-Calvo A, Davila J and Barrero A 1997 *J. Aerosol Sci* **28** 2 p 249
[15] Fernandez de la Mora J and Loscertales I 1994 *J. of Fluid Mechanics* **260** p 155
[16] Gamero-Castaño M, Hruby V, Falkos P and Shenoy S 2001 IEPC-01-281 282 *27th International Electric Propulsion Conference (USA, CA, Pasadena 15-19 October 2001)*
[17] Meghami P, O-Donelli J, Hsu O, Ziemmer J and Dunn C 2017 Drag-Free Performance of the FT7 Disturbance Reduction System Flight Experiment on the Lisa Pathfinder *10 Int. ESA Conf. on Guidance, Navigation & Control Systems (Salzburg, Austria)*
[18] Courtney D, Wood Z, Grey S and Model J 2019 IEPC-C-2019-A-788 *36th International Electric Propulsion Conference (Vienna, Austria. 15-20 September 2019)*
[19] Ma C and Ryan C 2019 IEPC-2019-A223 *36th Int. Electric Propulsion Conf.(Vienna, Austria)*
[20] Petro E, Miller C, Schmidt J and Lozano P 2019 IEPC-2019-830 *36th International Electric Propulsion Conference (Vienna, Austria. 15-20 September 2019)*

IOP Conf. Series: Materials Science and Engineering **927** (2020) 012031 doi:10.1088/1757-899X/927/1/012031

On the exoplanet obliquity variations in planetary four – body problem with evolving satellite's orbit

P S Krasil'nikov
Moscow Aviation Institute (National Research University), 4 Volokolamskoe shosse, Moscow 125993, Russian Federation

E-mail: krasil06@rambler.ru

Abstract. The evolution of the obliquity (the axial tilt) of a hypothetical exoEarth under the influence of gravitational moments from a star, a planet's satellite in evolving orbit, and an outer exoplanet Evolution was investigated analytically. The exoEarth is assumed to be rigid, axially symmetric and almost spherical. The case where the orbits of two planets and satellite are given Keplerian ellipses while the satellite's ellipse evolves on the lunar type was considered. The canonical equations describing the oscillations of spin axis of exoEarth were obtained. These equations contain one parameter (the longitude of the ascending node of the satellite's orbit) slowly changing in time. Using the theory of the adiabatic invariant, approximate formulas for the precession angle and obliquity in time are obtained.

1. Introduction

More than 3000 planets outside our Solar system have been identified to date. Whether any of the exoplanets are hospitable to life depends on a number of factors. They involve the planet's obliquity, i.e., the orientation of spin axis of a planet relative to its orbit [1,2]. A modest steady tilt causes seasons and assures a smooth temperature distribution over the entire globe. Stable climate is beneficial for advanced life to develop. The studies of exoplanets' climate indicate that even small variations of obliquity can result in drastic variations of the climate; similarly, both extremely high and very low obliquities suggest that the planet is inhabitable. The Earth's obliquity varies between 22.1° and 24.5° and its orbital eccentricity between 0 and just 0.06, but even such slight periodic changes are sufficient to drive the Earth into glacial-interglacial cycles with periods of 3000–4000 years [3,4].

In this paper, we obtain asymptotic formulas describing exoplanet obliquity variations of exoEarth under the influence of gravitational moments from a star, a planet's satellite and an outer exoplanet, while the satellite's orbit evolves on the lunar type: longitude Ω_2 of the ascending node and the argument of the periapsis ϖ_2 change slowly over time. ExoEarth is an axially symmetric rigid body. Other planets and star are assumed to be point masses.

There are *two classes of the non-resonant problem*. In the first class, the orbits of celestial bodies are the quasiperiodic functions in time with base frequencies $(\omega_1,\ldots,\omega_N)$ that are of the order of one. The trajectories of celestial bodies can also depend on time in a complex way [5]. Then the Fourier series with the specified frequencies is considered as an approximating series. This class of problems has been studied both in the three-body problem and in the N-body problem [6-9]. We have studied the non-resonant rotations of the exoEarth under the torque due to these bodies. Since the frequencies of the orbital motions are of the order of one, the averaged Hamiltonian has three first integrals in involution, so by the Liouville theorem, there exists a complete set of first integrals, and hence the problem can be solved analytically for an arbitrary number of bodies in the system. It was shown that the topology of the averaged rotations of the exoEarth for different types of celestial body orbits is the same for any $N \geq 3$ [6,7]. For planetary N-body problem, the exoEarth's angular momentum describes the spoiled precession around some axis making up a small obliquity. The observe librational motions of angular momentum near the equator of the planet are also observed (see Figure 1).

Note that further study of the first class of problems is also possible using models of deformable bodies [10,11]. The averaging method is also efficiently used in the problems of rotational artificial satellite motions. Out of the entire set of papers on this subject matter, let us mention only some of them [12,13].

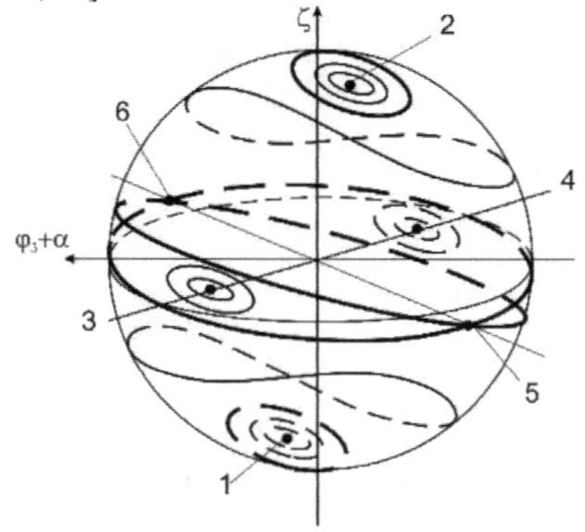

Figure 1. Trajectories of exoEarth's angular momentum on the celestial sphere

In the *second class of problem*, some orbits of celestial bodies have the small base frequencies ω_j. The second class of problem is less studied although many articles describe the rotations of celestial body with small base orbital frequencies (for example, [14-20]). Note that Cassini's laws belong to this class of problem.

Let us consider the problem of exoEarth's rotation under the torque due to star, a planet's satellite and an external planet (exoJupiter) in a general sky-mechanical setup. It has advantages: research results can be applied both to the planets of the solar system, and to exoplanets. Using this setup, you can study the general laws of planetary rotation, avoiding specific facts that often interfere with analysis.

2. Hamiltonian of problem

Let m_1 be the star mass, m_2 the satellite mass, m_3 the outer planet mass, m the mass of exoEarth. Suppose $m_1 \gg m, m_2, m_3$; then we have planetary system here. Denote by *OXYZ* an inertial reference frame whose origin is a star, the *OXY* plane coincides with the exoEarth orbit plane, the axis *OX* is directed to the perigee of the orbit. Denote by $Mx_1y_1z_1$ the coordinate system, which origin is the center of mass of the exoEarth and axes are parallel to those of the *OXYZ*, and by *Mxyz* the coordinate system with the same origin and coordinate axes coinciding with the exoEarth's principal axes. Assume that *Mz* is the axis associated with the maximum moment of inertia.

Let $L, I_2, I_3, l, \varphi_2, \varphi_3$ be Depri-Andoyer canonical variables [21], which define the attitude of *Mxyz* w.r.t. axes $Mx_1y_1z_1$. I_2 is a value of exoEarth's angular momentum. The Hamiltonian of the problem is

$$H = \frac{I_2^2 - L^2}{2A} + \frac{L^2}{2C} + \sum_{i=1}^{3} U_i, \quad U_i = \frac{3\mu_i}{2R_i^3}(C - A)\gamma_i^2, \quad R_i = \sqrt{(X - X_i)^2 + (Y - Y_i)^2 + (Z - Z_i)^2} \qquad (1)$$

Here *A*, *C* are the equatorial and axial moments of inertia of the exoEarth, $\mu_i = fm_i$, f is the universal gravitational constant, R_i is the distance between the exoEarth and mass point m_i, γ_i is the cosine of the angle between the radius vector $\mathbf{R}_i$ and the axis Mz.

IOP Conf. Series: Materials Science and Engineering **927** (2020) 012031 doi:10.1088/1757-899X/927/1/012031

Let us consider the case where the orbits of two planets and satellite are given Keplerian ellipses while the satellite's ellipse evolves on the lunar type. The absolute coordinates of the exoEarth are

$$X = \frac{a(1-e^2)}{1+e\cos\nu}\sin\nu,\ Y = \frac{a(1-e^2)}{1+e\cos\nu}\cos\nu,\ Z = 0$$

where a, e, ν are a semi-major axis, eccentricity and true anomaly of the exoEarth orbit. The relative coordinates of the planet's satellite are described by the formulas

$$\begin{gathered} X - X_2 = [\cos\Omega_2\cos\varpi_2 - \sin\Omega_2\cos i_2\sin\varpi_2]\xi_2 - [\cos\Omega_2\sin\varpi_2 + \\ + \sin\Omega_2\cos i_2\cos\varpi_2]\eta_2 + \sin\Omega_2\sin i_2\zeta_2 \\ Z - Z_2 = \sin i_2\sin\varpi_2\xi_2 + \sin i_2\cos\varpi_2\eta_2 + \cos i_2\zeta_2 \\ Y - Y_2 = [\sin\Omega_2\cos\varpi_2 + \cos\Omega_2\cos i_2\sin\varpi_2]\xi_2 + [-\sin\Omega_2\sin\varpi_2 + \\ + \cos\Omega_2\cos i_2\cos\varpi_2]\eta_2 - \cos\Omega_2\sin i_2\zeta_2 \\ \xi_2 = \frac{a_2(1-e_2^2)}{1+e_2\cos\nu_2}\sin\nu_2,\ \eta_2 = \frac{a_2(1-e_2^2)}{1+e_2\cos\nu_2}\cos\nu_2,\ \zeta_2 = 0 \end{gathered} \tag{2}$$

Here

$$\Omega_2 = \Omega_{20} + \varepsilon_1 t,\ \varpi_2 = \varpi_{20} + \varepsilon_2 t, \quad \varepsilon_1, \varepsilon_2 \ll 1$$

is the evolving longitude of the Earth's satellite orbit and the argument of the pericenter, i_2 is the satellite orbital inclination to the OXY plane, $\xi_2, \eta_2 \zeta_2$ are satellite coordinates in the perigee coordinate system $M\xi_2, \eta_2, \zeta_2$. The absolute coordinates X_3, Y_3, Z_3 of the third celestial body (exoJupiter) are calculated by formulas (2) with the replacement of index 2 by 3.

3. The averaging

Averaging Hamiltonian (1) over fast variables $\theta = \omega t$, $\theta_2 = \omega_2 t$, $\theta_3 = \omega_3 t$ and φ_2, we get

$$\bar{H} = \frac{I_2^2}{2J_0} + \varepsilon\bar{H}_1,\ C = J_0 + \varepsilon C_1,\ A = J_0 + \varepsilon A_1$$

where

$$\bar{H}_1 = F(I_2, L, l)\left(\frac{I_2^2}{2J_0^2} + \frac{9}{4}G(I_2, I_3, \varphi_3)\right),\ F = -(C_1 - A_1)\left(\frac{2}{3} - \sin^2\delta_2\right),\ I_2 = \text{const}$$

$$G = \left(-D_1(\tau)\sin^2\varphi_3 - D_2(\tau)\cos^2\varphi_3 + D_3(\tau) + D_4(\tau)\sin 2\varphi_3\right)\sin^2\delta_1 - \sin 2\delta_1\left(D_5(\tau)\sin\varphi_3 - D_6(\tau)\cos\varphi_3\right)$$

Note that $\cos\delta_1 = I_3 / I_2$, where δ_1 is the obliquity. Functionals $D_j(\tau)$ slowly change over time $\tau = \varepsilon_1 t$:

$$\begin{gathered} D_1(\tau) = \frac{\mu_1(1-e^2)^{3/2}}{2p^3} + \Xi\left(\cos^2(\Omega_{20}+\tau)\sin^2 i_2 + \cos^2 i_2\right) + \frac{\mu_3}{2\pi}\int_0^{2\pi}\frac{(X-X_3)^2}{R_3^5}\Sigma d\nu d\nu_3, \\ D_2(\tau) = \frac{\mu_1(1-e^2)^{3/2}}{2p^3} + \Xi\left(-\cos^2(\Omega_{20}+\tau)\sin^2 i_2 + 1\right) + \frac{\mu_3}{2\pi}\int_0^{2\pi}\frac{(Y-Y_3)^2}{R_3^5}\Sigma d\nu d\nu_3, \\ D_3(\tau) = \Xi\sin^2 i_2 + \frac{\mu_3}{2\pi}\int_0^{2\pi}\frac{(Z-Z_3)^2}{R_3^5}\Sigma d\nu d\nu_3, \\ D_4(\tau) = \Xi\sin(\Omega_{20}+\tau)\cos(\Omega_{20}+\tau)\sin^2 i_2 + \frac{\mu_3}{2\pi}\int_0^{2\pi}\frac{(X-X_3)(Y-Y_3)}{R_3^5}\Sigma d\nu d\nu_3, \end{gathered} \tag{3}$$

$$D_5(\tau) = -\Xi\sin(\Omega_{20}+\tau)\cos i_2 \sin i_2 + \frac{\mu_3}{2\pi}\int_0^{2\pi}\frac{(X-X_3)(Z-Z_3)}{R_3^5}\Sigma dv dv_3,$$

$$D_6(\tau) = \Xi\cos(\Omega_{20}+\tau)\cos i_2 \sin i_2 + \frac{\mu_3}{2\pi}\int_0^{2\pi}\frac{(Y-Y_3)(Z-Z_3)}{R_3^5}\Sigma dv dv_3$$

$$\Sigma = \frac{(1-e^2)^{3/2}(1-e_3^2)^{3/2}}{(1+e\cos v)^2(1+e_3\cos v_3)^2},\quad \Xi = \frac{\mu_2}{2a_2^3(1-e_2^2)^{3/2}}$$

Note that all terms in the expressions for $D_j(\tau)$ are constants except $\sin(\Omega_{20}+\tau)$, $\cos(\Omega_{20}+\tau)$. The averaged equations in I_3 and precession angle φ_3 are of the following form:

$$\dot{\varphi}_3 = \Lambda\frac{\partial G}{\partial I_3},\quad \dot{I}_3 = -\Lambda\frac{\partial G}{\partial \varphi_3},\quad \Lambda = \frac{3}{4}(A-C)\quad (\delta_2 \approx 0) \tag{4}$$

Let us simplify equations (4) considering I_3 as a series of small parameters μ_2, μ_3 and expanding function G in a power series in these parameters:

$$I_3(\varphi_3) = I_{31} + I_3^*(\varphi_3) + O\left(\mu_2^2, \mu_3^2\right),\ I_{31} = \text{const},\ I_3^* = \mu_2 I_{32}(\varphi_3) + \mu_3 I_{33}(\varphi_3)$$

$$\tilde{G}(I_3^*, \varphi_3, \tau) = KI_3^* + \left(-D_1^*(\tau)\sin^2\varphi_3 - D_2^*(\tau)\cos^2\varphi_3 + D_3(\tau) + D_4(\tau)\sin 2\varphi_3\right)\sin^2\delta_{11} - \tag{5}$$

$$\sin 2\delta_{11}\left(D_5(\tau)\sin\varphi_3 - D_6(\tau)\cos\varphi_3\right) + O\left(\mu_2^2, \mu_3^2\right),\ K = \frac{\mu_1\cos\delta_{11}}{I_2 a^3(1-e^2)^{3/2a}}$$

Here $\cos\delta_{11} = I_{31}/I_2$, where δ_{11} is a constant obliquity in the exoEarth-star system, parameters D_1^*, D_2^* coincide with D_1, D_2 up to the first terms. Everywhere below, we keep the terms linear in μ_2, μ_3 in formulas (5).

So, equation (4) has one degree of freedom and one slowly evolving parameter τ. To study these equations, we use the theory of the adiabatic invariant.

4. Adiabatic invariant

Let's introduce the action-angle variables in equations (4). Suppose $t' = \Lambda t$ is a new time; then G is a Hamiltonian of equations (4). In the future, let us keep the old designations t for time. We introduce a superposition of canonical maps $(I_3, \varphi_3) \to (I_3^*, \varphi_3)$, $(I_3^*, \varphi_3) \to (I, w)$

$$I_3 = I_{31} + I_3^*,\ \varphi_3 = \varphi_3,\quad I_3^* = \frac{\partial S}{\partial I_3},\ w = \frac{\partial S}{\partial I} \tag{6}$$

such that the new Hamiltonian H depends on variable I only: $H = H(I)$. The generating function S satisfies Hamilton-Jacobi type equation:

$$K\frac{\partial S}{\partial I_3} + U(\varphi_3, \tau) + \frac{\partial S}{\partial t} = H(I),$$

$$U = \left(-D_1^*(\tau)\sin^2\varphi_3 - D_2^*(\tau)\cos^2\varphi_3 + D_3(\tau) + D_4(\tau)\sin 2\varphi_3\right)\sin^2\delta_{11} -$$

$$\sin 2\delta_{11}\left(D_5(\tau)\sin\varphi_3 - D_6(\tau)\cos\varphi_3\right)$$

Calculations show that S consists of two terms

$$S = S_0(I, \varphi_3) + S_1(\varphi_3, \tau)$$

Assuming $H = I$, we get

$$S_0 = K^{-1}I\varphi_3 - K^{-1}\int_0^{\varphi_3} U\big|_{\tau=0}\,d\varphi_3,\quad S_1 = b_1(\tau)\cos\varphi_3 + b_2(\tau)\sin\varphi_3 + b_3(\tau)\cos 2\varphi_3 + b_4(\tau)\sin 2\varphi_3,$$

where parameters $b_j(\tau)$ satisfy integrable linear differential equations

$$\begin{cases} \varepsilon_1 \dfrac{db_1}{d\tau} = -Kb_2 - \sin 2\delta_{11}\Delta_6(\tau) \\ \varepsilon_1 \dfrac{db_2}{d\tau} = Kb_1 + \sin 2\delta_{11}\Delta_5(\tau) \end{cases} \quad \begin{cases} \varepsilon_1 \dfrac{db_3}{d\tau} = -2Kb_4 - \sin^2 \delta_{11} \dfrac{\Delta_1(\tau) - \Delta_2(\tau)}{2} \\ \varepsilon_1 \dfrac{db_4}{d\tau} = 2Kb_3 - \sin^2 \delta_{11}\Delta_4(\tau) \end{cases} \tag{7}$$

Here $\Delta_j(\tau) = D_j^*(\tau) - D_j^*(0)\ (j = 1,2)$, $\Delta_i(\tau) = D_i(\tau) - D_i(0)\ (i = 3,\ldots 6)$. We will not give an explicit form of the solutions of these equations. Note that the problem of integrating equations (7) is similar to the problem [22, 23]. From equalities (6) it follows that

$$\begin{aligned} I_3^* = K^{-1}I - K^{-1}U(\varphi_3,\tau)\big|_{\tau=0} + \big(-b_1(\tau)\sin\varphi_3 + b_2(\tau)\cos\varphi_3 - \\ - 2b_3(\tau)\sin(2\varphi_3) + 2b_4(\tau)\cos(2\varphi_3)\big),\ \varphi_3 = Kw + \varphi_{30} \end{aligned} \tag{8}$$

In the new variables, equations (4) are of the following form:

$$\dot{I} = 0,\ \dot{w} = 1$$

So, $I = \text{const}$, therefore I is the adiabatic invariant of original equations with Hamilton function (1). Then the obliquity δ_1 of exoEarth is calculated by the formula

$$\delta_1(t) = \arccos\left(\frac{I_{31} + I_3^*}{I_2}\right) \approx \delta_{11} - \frac{I_3^*}{C\omega_r \sin\delta_{11}} \tag{9}$$

where ω_r is the angular speed of exoEarth's rotation in inertial space, δ_{11} is a constant obliquity in the exoEarth-star system.

5. Conclusion

It has been shown that non-resonant problems of the celestial body rotations are divided into two classes. In the first class, the orbits of celestial bodies are the quasiperiodic functions in time with base frequencies of the order of unity. In the second class, some orbits of celestial bodies have the small base frequencies. These classes differ in research methods and rotation topology. In this paper, we have studied the problem of second class: the evolution of obliquity of exoEarth under the influence of gravitational moments from a star, a planet's satellite in lunar type evolving orbit, and an outer exoplanet. It was shown that equations describing oscillations of exoEarth's spin axis take the form (4). Using the transformation to action-angle variables, the explicit formulas (7) - (9) describing obliquity as a function in time was obtained.

Acknowledgments

Our research was partially financed by the grant 18-01-00820 from the Russian Foundation for Basic Research.

References

[1] Armstrong J C, Barnes R, Domagal-Goldman S, Breiner J, Quinn T R and Meadows V C 2014 Effects of extreme obliquity variations on the habitability of Exoplanets *Astrobiology*. **14** 277-291

[2] Cowan N B, Voigt A and Abbot D S 2012 Thermal phases of Earth-like planets: estimating thermal inertia from eccentricity, obliquity, and diurnal forcing *Astrophys. J.* **757** 80

[3] Kasting J F and Catling D 2003 Evolution of a habitable planet *Annu. Rev. Astron. Astrophys*. **41** 429-463

[4] Williams D M and Pollard D 2003 Extraordinary climates of Earth-like planets: threedimensional climate simulations at extreme obliquity. *Int. J. Astrobiol*. **2** 1–19

[5] Ivanyukhin A V and Petukhov V G 2019 Low-energy sub-optimal low-thrust trajectories to libration points and halo-orbits *Cosm. Res.* **57 (5)** 378–388

[6] Markeev A P and Krasilnikov P S 1981. On motion of a satellite relative to the center of mass in the elliptic restricted three body problem. *Cosm. Res.* **19** 178–190
[7] Krasilnikov P S and Zaharova E E 1993 Non-resonant rotation of a satellite relative to the center of mass on a quasi-periodic orbit in the restricted N body problem. *Cosm. Res.* **31** 11–21
[8] Krasilnikov P S and Podvigina O M 2018. On evolution of the planet's obliquity in a nonresonant planetary system (in Russian). *Bulletin of Udmurt University. Mathematics, Mechanics, Computer Science.* **28** 549–564
[9] Podvigina O M and Krasilnikov P S 2020 Evolution of the obliquity of an exoplanet: A non-resonant case *Icarus*. **335** 113371
[10] Zlenko A A (2015) A celestial-mechanical model for the tidal evolution of the Earth–Moon system treated as a double planet *Astronomy reports* **59** 1 72
[11] Zlenko A 2019 The perturbing potential and the torques in one three-body problem *J. Phys.: Conf. Ser.* **1301** 012022
[12] Tikhonov A A 2005 *Cosmic Res.* **43** 107
[13] Aleksandrov A and Tikhonov A A 2019 *Bulletin of St. Petersburg University Mathematics. Mechanics. Astronomy* **6(64)** 271-281
[14] Colombo G 1966 Cassini's Second and Third Laws. *Astron. J.* **71(9)** 891
[15] Ward W R 1975 Tidal friction and generalized Cassini's laws in the solar system. *Astron. J.* **80** 64-70
[16] Beletskii V V 1972 Resonance rotation of celestial bodies and Cassini's laws *Celestial Mechanics*. **6(3)** 356-378
[17] Lidov M L and Neustadt A I 1973 The method of canonical transformations in problems of the rotation of celestial bodies and Cassini's laws *Preprint No 9. Keldysh Institute of Applied Mathematics* (in Russian)
[18] Peal S J 2006 The proximity of Mercury's spin to Cassini state 1 from adiabatic invariance *Icarus* **181** 338-347
[19] Tisserand F 1891 *Traité de mécanique célest*. T. II. Paris.
[20] Smart W M 1953 *Celestial Mechanics* Longmans, Green and Co, London-New York-Toronto.
[21] Andoyer H 1923 *Cours de Mécanique Céleste*, Vol. 1, Gauthier-Villars
[22] Kuznetsov E , Leonov S and Tsapko K 2019 On the exact solution of a singularly perturbed aerodynamic problem *AIP Conference Proceedings* **2181** 020014
[23] Krylov S, Perepelkin V and Filippova A 2019 Short-term prediction of universal time variations dUT1 *AIP Conference Proceedings* **2181** 020023

IOP Conf. Series: Materials Science and Engineering **927** (2020) 012032 doi:10.1088/1757-899X/927/1/012032

A new numerical approach for solving initial value problems with exponential growth integral curves

E B Kuznetsov, S S Leonov and E D Tsapko
IT and Applied Mathematics Department, Moscow Aviation Institute, 4 Volokolamskoe shosse, Moscow 125993, Russia

E-mail: zapkokaty@gmail.com

Abstract. Some physical processes raised in aerodynamics, hydrodynamics, and theory of catalytic reactions are modeled by stiff initial value problems. This paper considers a class of stiff problems such as singularly perturbed problems for ordinary differential equations. Singularly perturbed problems have regions of rapid change in the solution, which are called contrast structures. Most often, boundary layer exhibits in such problems, but several pieces of research devoted to inner layers too. There are many numerical and asymptotic methods for solving singularly perturbed initial value problems with contrast structures. However, it is still an open field for researches because the universal method does not exist. Problems with an exponential growth rate of integral curves have a particular interest for researchers because most of the methods for solving singularly perturbed problems cannot handle these. We are going to discuss methods for solving singularly perturbed initial value problems with contrast structures. Also, we are proposing a new technique based on the method of solution continuation for problems with the exponential growth of integral curves. This technique allows explicit numerical methods to get the solution of the considered problem when it seems impossible for them. The efficiency of the new approach will be shown on the test initial value problem.

1. Introduction

We consider a singularly perturbed n-th order differential equation with initial conditionsof the form

$$\begin{cases} \varepsilon \dfrac{d^n y}{dt^n} = f\left(t, y, y', \dots, y^{(n-1)}\right), \quad \varepsilon > 0, \quad t \in [t_0, T], \\ y(t_0) = y_0, \;\; y'(t_0) = y_1, \dots, \;\; y^{(n-1)}(t_0) = y_{n-1}. \end{cases} \tag{1}$$

Singularly perturbed problems belong to a class of stiff problems, which are characterized by a stiffness parameter $1/\varepsilon$, where ε is a small parameter. Such problems arise in various fields of mathematical and engineering sciences, e.g., aerodynamics, hydrodynamics, chemical kinetics, the theory of catalytic reactions, differential geometry. In practice, the function $f\left(t, y, y', \dots, y^{(n-1)}\right)$ is nonlinear, so the equation does not have an exact analytical solution. Therefore, a researcher has to use other techniques, such as numerical and asymptotic methods.

The small parameter ε affects the problem in such a way that the solution varies rapidly in some regions of the problem domain and slowly in other parts. It is a well-known fact that the singularly perturbed problem possesses boundary or interior layer, i.e., regions of rapid change in the solutionnear the endpoints or some interior points. The interior layers are also called contrast structures [1, 2]. Hence, the numerical solving of singularly perturbed differential equations gives rise to computational difficulties. Another technique is to use asymptotic methods, but complex analytical calculations are required.

The method of solution continuation with respect to the best argument (the best parameterization) is a very efficient method for solving singularly perturbed problems [3, 4]. It allows us to get a reasonably accurate solution and a less computational time compared to other traditional methods. However, if the integral curve of the problem has an exponential growth rate, this method also becomes inefficient even for a mildly stiff problem.

The purpose of the paper is to propose a new technique for solving singularly perturbed initial value problems,especially with the exponential growth rate of integral curves — the technique bases on the method of solution continuation. The efficiency of the proposed technique is shown in the test exponential problem.

2. Methodology

For convenience, we transform the Cauchy problem (1) to normal form:

$$\begin{cases} \dfrac{dy_1}{dt} = f_1(t, y_1, \dots, y_n), \quad y_1(t_0) = y_1, \\ \dots \\ \dfrac{dy_{n-1}}{dt} = f_{n-1}(t, y_1, \dots, y_n), \quad y_{n-1}(t_0) = y_{n-1}, \\ \dfrac{dy_n}{dt} = \dfrac{1}{\varepsilon} \cdot \tilde{f}_n(t, y_1, \dots, y_n) = f_n(t, y_1, \dots, y_n), \quad y_n(t_0) = y_n, \\ \varepsilon > 0, \quad t \in [t_0, T], \end{cases} \tag{2}$$

It is possible to apply different numerical explicit or implicit schemes to the system of equations (2). However, in some cases, when the stiffness parameter is a very large value, they are not able to obtain the approximate solution in the inner regions at all. The method of solution continuation with respect to various arguments, including the best argument,can increase the efficiency of these numerical techniques.

2.1. The method of the best parameterization

The main idea of the method of the best parameterization is to introduce new parameter λ, measured tangentially along the integral curve:

$$d\lambda^2 = dy_1^2 + dy_2^2 + \dots + dy_n^2 + dt^2. \tag{3}$$

It allows to transform the system of equations (2) as follows:

$$\begin{cases} \dfrac{dy_i}{d\lambda} = \dfrac{f_i(t, y_1, \dots, y_n)}{Q(t, y_1, \dots, y_n)}, \quad \dfrac{dt}{d\lambda} = \dfrac{1}{Q(t, y_1, \dots, y_n)}, \quad i = 1, \dots, n, \\ y_i(0) = y_i, \quad t(0) = t_0, \end{cases} \tag{4}$$

where $Q(t, y_1, \dots, y_n) = \sqrt{1 + f_1^2(t, y_1, \dots, y_n) + \dots + f_n^2(t, y_1, \dots, y_n)}$ and λ is now an argument of the problem.

If the integral curves of the problem (2) have a power-law growth rate, this transformation can significantly improve the accuracy of the approximate solution. However, in the case of an exponential-law growth rate, this technique can be inefficient as well.

2.2. Modification of the best parameterization method

We propose a modified best argument of the form:

$$d\mu^2 = dy_1^2 + dy_2^2 + \dots + dy_n^2 + \exp(2\alpha t) \cdot dt^2, \tag{5}$$

where $\boldsymbol{\alpha}$ is the chosen value such that it extremely exceeds the stiffness parameter of the problem (2).Then the transformed system is

$$\begin{cases} \dfrac{dy_i}{d\mu} = \dfrac{f_i(t, y_1, \dots, y_n) \cdot \exp(-\alpha t)}{P(t, y_1, \dots, y_n)}, \dfrac{dt}{d\mu} = \dfrac{\exp(-\alpha t)}{P(t, y_1, \dots, y_n)}, \quad i = 1, \dots, n, \\ y_i(0) = y_i, \quad t(0) = t_0, \end{cases} \tag{6}$$

where $P(t, y_1, \dots, y_n) = \sqrt{1 + e^{-2\alpha t} f_1^2(t, y_1, \dots, y_n) + \dots + e^{-2\alpha t} f_n^2(t, y_1, \dots, y_n)}$.

If the problem (2) does not have limiting or essential singular points, the proposed modified argument (5) allows the weakeningof theproblem stiffness parameter. Otherwise, this transformation is not effective, but we can modify the argument as necessary.

3. The test problem

We will consider the test exponential problem proposed by Kalitkin N N and Belov A A [5]:

$$\frac{du}{dt} = -\xi(t) \cdot u \cdot (u^2 - a^2), \quad u(0) = u_0, \quad t\epsilon[0, 2\pi]. \tag{7}$$

The problem (7) has an exact solution

$$u(t) = \frac{a u_0}{\sqrt{u_0^2 + (a^2 - u_0^2) \cdot \exp(-2a^2 \Xi(t))}},$$

where $\xi(t) = \xi_0 \cos t$, $\Xi(t) = \xi_0 \sin t$, $u_0 = 0.5$, $a = \pi$. The exact solution ishelpfulin estimating the efficiency of the technique. In figure 1 the exact solution is shown for $\xi_0 = 1$ and $\xi_0 = 50$.Thus, the larger the parameter ξ_0, the faster the solution changes in the regions of contrast structures.

3.1. The test problem transformed to the best argument

The system transformedto the best argument (3) has the form:

$$\begin{cases} \dfrac{du}{d\lambda} = \dfrac{-\xi(t) \cdot u \cdot (u^2 - a^2)}{\sqrt{1 + \xi^2(t) \cdot u^2 \cdot (u^2 - a^2)^2}}, u(0) = u_0, \\ \dfrac{dt}{d\lambda} = \dfrac{1}{\sqrt{1 + \xi^2(t) \cdot u^2 \cdot (u^2 - a^2)^2}}, \quad t(0) = t_0 \end{cases} \tag{8}$$

Let us immediately present the problem with the modified argument to compare their efficiency.

3.2. The test problem transformed to the modified best argument

Using the argument (5) we obtain the transformed system of the form:

$$\begin{cases} \dfrac{du}{d\mu} = \dfrac{-\xi(t) \cdot u \cdot (u^2 - a^2) \cdot \exp(-\alpha t)}{\sqrt{1 + \xi^2(t) \cdot u^2 \cdot (u^2 - a^2)^2 \cdot \exp(-2\alpha t)}}, u(0) = u_0, \\ \dfrac{dt}{d\mu} = \dfrac{1}{\sqrt{1 + \xi^2(t) \cdot u^2 \cdot (u^2 - a^2)^2 \cdot \exp(-2\alpha t)}}, \quad t(0) = t_0 \end{cases} \tag{9}$$

Table 1 presents the average absoluteerror of the numerical solution $\boldsymbol{\varepsilon_{avg}}$ and computational time $\boldsymbol{t_c}$ for problems (7), (8), and (9). The problems were solved using the Euler method with a variable step size varies by the Runge rule [6] with the local tolerance parameter $\boldsymbol{\theta}$. Thus, according to the Runge rule, step size decreases in the regions of contrast structures and increases otherwise. We consider here the initial step size $\boldsymbol{h_0 = 10^{-5}}$ and parameter $\boldsymbol{\alpha = 100}$ for the modified best argument. It was not able to obtain the numerical solution for the problem (7) and (8) for $\boldsymbol{\xi_0 = 10, 50}$ at all. The best parameterization is ineffective both in terms of computational time, obtained accuracy, and ability to obtain a solution to the original problem (7) when it is not possible by traditional methods. However, the modified best argument allows a higher accuracy solution to be obtained with the same level of the local tolerance, as well as for larger stiffness parameters.

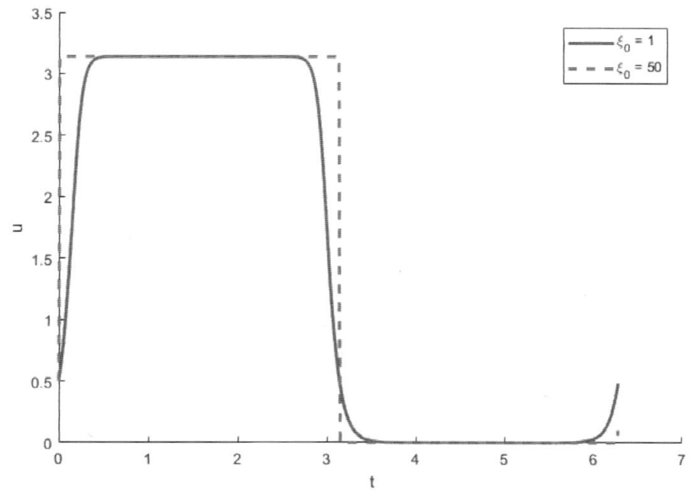

Figure 1. The exact solution to the problem (7) for $\xi_0 = 1$ and $\xi_0 = 50$.

Table 1. The average absolute error of the numerical solution and the computational time for problems (7), (8), and (9), explicit Euler method with variable step size varies by the Runge rule with the local tolerance θ, the initial step size $h_0 = 10^{-5}$, and parameter $\alpha = 100$ for the modified best argument.

ξ_0	θ	The original problem		The best argument		The best modified argument	
		ε_{avg}	t_c, s	ε_{avg}	t_c, s	ε_{avg}	t_c, s
1	10^{-4}	0.0072	0.01	0.020	0.02	$1.1 \cdot 10^{-4}$	0.33
	10^{-5}	0.0200	0.05	0.020	0.03	$2 \cdot 10^{-5}$	0.80
	10^{-6}	0.0060	0.11	0.010	0.18	$3.6 \cdot 10^{-6}$	2.50
	10^{-7}	$8.5 \cdot 10^{-4}$	0.30	0.002	0.40	$6.3 \cdot 10^{-7}$	8.20
	10^{-8}	$1.4 \cdot 10^{-4}$	0.80	$2.3 \cdot 10^{-4}$	1.25	$1.1 \cdot 10^{-7}$	25
10	10^{-6}	-	-	-	-	$7.9 \cdot 10^{-4}$	2.3
	10^{-7}	-	-	-	-	$8.5 \cdot 10^{-4}$	7.2
	10^{-8}	-	-	-	-	$2.1 \cdot 10^{-4}$	22
50	10^{-6}	-	-	-	-	0.006	2
	10^{-7}	-	-	-	-	0.003	6.6
	10^{-8}	-	-	-	-	$1.6 \cdot 10^{-4}$	22

Although, the solution of the problem (8) by the variable step method was not very successful, using the fourth-order Runge-Kutta method with constant step size showedan interesting result.Namely, as Table 2 shows, it was able to obtain solutions for $\boldsymbol{\xi_0 = 10, 50}$. In turn, the modified parameter due to the presence of an exponent does not work well with constant step size numerical methods. However,it is still better in the computation time and error than the classic best parameter with constant step size.

Table 2. The computational error and the computational time for the problem (8), explicit fourth-order Runge-Kutta method with the constant step size

$\boldsymbol{\xi_0}$	$\boldsymbol{h}$	The best argument, Runge-Kutta method with the constant step size

		ε_{avg}	t_c, s
1	10^{-3}	$7.2\cdot10^{-5}$	1.25
	10^{-4}	$2.3\cdot10^{-6}$	12
	10^{-5}	$7.2\cdot10^{-8}$	29
10	10^{-3}	0.02	1
	10^{-4}	0.006	9
	10^{-5}	$8.2\cdot10^{-5}$	127
50	10^{-3}	0.02	1
	10^{-4}	0.003	12.5
	10^{-5}	0.002	92
	10^{-6}	$2.9\cdot10^{-4}$	1564

All computations are carried out on a personal computer with CPU Intel Core i5-8400 2.8GHz, RAM DDR4 8GB (x2), graphics card NVIDIA GeForce GTX 1070 Ti 8GB on Windows 10 x64. Program code has been written in MATLAB.

4. Conclusion

A test initial value problem with contrast structures was considered, the stiffness of which depended on the constant value ξ_0. The Euler method with a variable step size was applied to the solving of the original problem and problems transformed by the best classical and modified arguments. It was shown that even with a small stiffness parameter such as $\xi_0 = 10, 50$, only the transition to the best modified argument made it possible to obtain a numerical solution to the problem. Thus, the best modified argument shows its effectiveness in solving problems with contrast structures in which integral curves have an exponential growth rate. It provides getting a sufficiently accurate solution in cases where the classic best argument and traditional methods cannot. The disadvantage is the need to select the parameter. This can be a severe problem when solving complex problems because the empirical selection of this parameter can take much time. Now, there is no strict criterion for choosing this parameter. This issue will be considered in the further works of the author.

Acknowledgments

This paper is based on research carried out with the financial support of the grant of the Russian Science Foundation (project 18-19-00474).

References

[1] Chang K W and Howes F A 1984 *Nonlinear singular perturbation phenomena: theory and application* (New York: Springer) 187 p

[2] Vasil'eva A B, Butuzov V F and Nefedov N N 1998 Contrast structures in singularly perturbed problems *Fundamental and Applied Mathematics* **4** 799–851

[3] Shalashilin V I and Kuznetsov E B 2003 *Parametric continuation and optimal parametrization in applied mathematics and mechanics* (Dordrecht/Boston/London: Kluwer Academic Publishers) p 228

[4] Kuznetsov E, Leonov S, Tarkhov D, Tsapko E and Babintseva E 2020 Arc Length and Multilayer Methods for Solving Initial Value Problems for Differential Equations with Contrast Structures *Modern Information Technology and IT Education. SITITO 2018. Communications in Computer and Information Science* **1201** 335-351

[5] Belov A A and Kalitkin N N 2017 Features of calculating contrast structures in the Cauchy problem *Math. Models Comput. Simul.* **9** 281–291

[6] Hairer E, Nørsett S. and Wanner G 1993 *Solving Ordinary Differential Equations I. Nonstiff Problems* (Springer-Verlag Berlin Heidelberg) p 528

IOP Conf. Series: Materials Science and Engineering **927** (2020) 012033 doi:10.1088/1757-899X/927/1/012033

The problem of forecasting Earth pole trajectory when changing the average parameters of its motion

V V Perepelkin[1], D S Rumyantsev[2] and Wai Yan Soe[1]

[1] Moscow Aviation Institute, 4 Volokolamskoe shosse, Moscow 125993, Russia
[2] V A Trapeznikov Institute of control sciences of RAS, 65 Profsoyuznaya street, Moscow 117997, Russia

E-mail: vadimkin1@yandex.ru

Abstract. The change of the Earth pole motion modes based on few-parameter two-frequency model has been investigated. It is shown that the effect of the change in the average frequency during the period of the Chandler and annual harmonics modulation leads to the change in the average parameters of the two-frequency model of the Earth pole oscillations. A method is proposed for detecting this effect by analyzing the phase variations of the observed Earth pole's modulation.

1. Introduction

The study of the fundamental astrometric problem of predicting the motion of the Earth's pole [1] is of theoretical interest and is one of the fundamental issues of satellite navigation [2, 3]. High-precision orbit prediction of spacecraft [4, 5] is one of the most important tasks. Accounting in the equations of motion of various perturbing factors [3, 6] is necessary to build a forecast. A high-precision spacecraft navigation require determination of the Earth's orientation in inertial space and its rotation parameters [1], which should be predicted. The Earth pole coordinates are the most difficult parameters for predictions, because the Earth pole motion is affected by non-stationary perturbations of various physical nature [7].

The polar oscillation process is not completely regular. To predict its position it is often proposed to use numerical methods, for example, ones based on regression models [8, 9]. However, this approach takes regular effects into account only partially, within the approximation order, which negatively affects the long-term forecast accuracy, and affects the model autonomy, i.e. the model performance without adjusting its parameters.

The numerical-analytical approach can be based on a deterministic model of the Earth pole motion obtained by means of equations of the deformed Earth rotation about its center of mass.

Due to the relatively slow change in the oscillations characteristics, the parameters in this model are assumed to be quasi-constant during the modulation period of the main harmonics (Chandler and annual) [10-12]. Then the observed process could be characterized by a few average parameters. By correcting the quasi-constant parameters of the model, this approach can be refined.

This numerical-analytical approach has a significant advantage. That is the ability to analytically take into account significant changes in the Earth pole oscillation process. For example, when analyzing the observation data from the International Earth Rotation and Reference Systems Service (IERS) [1] it is possible to detect the phenomenon of Earth pole oscillatory mode change, connected with change of the average frequency value of its motion about the center - the average position after excluding the trend component. Changing the mode of the Earth pole motion could significant deterioration in the accuracy of the numerical forecast [13]. Until new motion statistics accumulate, the accuracy of the forecast will remain low. A positive effect could be brought by reducing the time interval for estimating model parameters. However, a more efficient solution would be to change the structural properties of the model after such mode switch while maintaining the interval length of the observation data processing. It should be noted that a decrease in accuracy characteristics will be observed in any forecasting model if this effect is not taken into account. In this regard, there is a problem of identifying the moment of the mode change in the Earth pole motion as well as correcting the forecasting model while maintaining the few-parameters property.

In the article, within a simple two-component model of the Earth pole motion with quasi-constant coefficients, the dependence between the average motion frequency (derivative of the polar angle) and the ratio of the Chandler and annual harmonics amplitudes are investigated. An expression is obtained for the phase variation of the Earth pole motion after the transition from one mode to another. This expression is a modification of a model structural property. A method of identifying the average frequency variation is proposed by processing the IERS observation data. A condition for the algorithmic determination of such transition is found.

2. Phase variation of the Earth pole two-frequency motion

It is known [14, 15] that the motion of the Earth pole is the sum of the two quasistationary oscillations with variable amplitudes - Chandler and annual ones. The Earth pole coordinates are functions of time τ and depend on the amplitudes a_{ch}, a_h of the Chandler and annual harmonics:

$$x_p = x_p(a_{ch}, a_h, \alpha_{ch}(\tau), \alpha_h(\tau)),$$
$$y_p = y_p(a_{ch}, a_h, \alpha_{ch}(\tau), \alpha_h(\tau)). \quad (1)$$

Here, the phases $\alpha_{ch}(\tau)$, $\alpha_h(\tau)$ of the Chandler and annual oscillations are linear functions of time.

The observational data analysis allows us to assume that a_{ch}, a_h are quasi-constant during the modulation period. Then system (1) is a four-parameter two-frequency model of the main pole motion (without the trend component) [14]:

$$x_p = a_{ch}\cos\alpha_{ch}(\tau) + a_h\cos\alpha_h(\tau),$$
$$y_p = a_{ch}\sin\alpha_{ch}(\tau) + a_h\sin\alpha_h(\tau), \quad (2)$$
$$\alpha_{ch}(\tau) = 2\pi N\tau + \alpha_{ch}^0, \quad \alpha_h(\tau) = 2\pi\tau + \alpha_h^0.$$

The value of the Chandler frequency N is assumed to be a constant and equals to 0.843 cycles per year, τ is the time measured in years.

The coefficients a_{ch}, a_h in (2) have the meaning of the average amplitudes of the Chandler and annual harmonics and can be obtained by averaging over the modulation period. Changing the averaging interval of the process leads to a variation in the coefficients. When varying the values of quasi-constant amplitudes, the instantaneous frequency $\dot{\psi}$ of the Earth pole circulation around the "middle" point will turn out to be dependent on the amplitude ratio $k = a_h / a_{ch}$ and phase difference $\gamma = \alpha_{ch}(\tau) - \alpha_h(\tau)$:

$$\dot{\psi} = 2\pi \frac{Na_{ch}^2 + (N+1)a_{ch}a_h\cos\left(2\pi(1-N)t + \gamma\right) + a_h^2}{a_{ch}^2 + 2a_{ch}a_h\cos\left(2\pi(1-N)t + \gamma\right) + a_h^2} \quad (3)$$

In the stationary oscillations model with constant coefficients, when varying the ratio of the Chandler and annual harmonics' amplitudes, the average frequency $\langle\dot{\psi}\rangle_T$ during the modulation period $T = 2\pi / (\dot{\alpha}_h - \dot{\alpha}_{ch})$ will jump from $\dot{\alpha}_{ch}$ to $\dot{\alpha}_h$ (if the amplitude of the Chandler component has become less than the annual amplitude) or vice versa, immediately passing through their average value when the amplitudes are equal:

$$\langle \dot{\psi} \rangle_T = \begin{cases} \dot{\alpha}_h, & \text{if } k > 1 \\ \dfrac{\dot{\alpha}_{ch} + \dot{\alpha}_h}{2}, & \text{if } k = 1 \\ \dot{\alpha}_{ch}, & \text{if } k < 1 \end{cases} \tag{4}$$

Let such a transition happen at the moment τ^*. We denote by ψ the polar angle of the pole when $k = k_1$ before the mode change, and by $\tilde{\psi}$ - the polar angle when $k = k_2$ after the mode change. Then, provided that $k_1 k_2 = 1$, up to the additive constant, we get the equation:

$$\tilde{\psi} = -\psi + \alpha_{ch} + \alpha_h. \tag{5}$$

This property allows us to find the polar angle variation $\delta\tilde{\psi} = \tilde{\psi} - \langle \tilde{\psi} \rangle_T$ after changing the oscillation mode of the Earth pole at $k_1 k_2 = 1$:

$$\delta\tilde{\psi} = -\delta\psi. \tag{6}$$

Due to the model approximation by one with quasi-constant coefficients (1) to the observed Earth pole trajectory, the coefficient $k = a_h / a_{ch}$ during the approximation and forecast intervals is assumed to be a constant. This allows us to write down the condition for the periodicity of variations $\delta\psi$ before and after the change in the oscillation mode:

$$\delta\psi(t+T) = \delta\psi(t), \tag{7}$$
$$\delta\tilde{\psi}(t+T) = \delta\tilde{\psi}(t).$$

Within such an approximation, the forecast before and after the mode change turns out to be of acceptable accuracy [12].

The obtained condition (6) also retains the accuracy of the Earth pole motion forecast within the few-parameter two-frequency model after the oscillatory mode change, since the approximations (7) and (6) have the same accuracy, and the amplitude of the resulting pole oscillatory process is not sensitive to such changes.

As an example, let us cite the results of approximating the oscillations of the coordinates of the Earth's pole on an interval approximately in the middle of which there is a change in the amplitude ratio. For example, let's take the time interval 2000–2016. So, in the interval from 2006 to 2012 approximately k = 1 can be assumed, and until 2006 and from 2012, k <1 and k> 1, respectively. It is essential to improve the accuracy of approximation while maintaining the same number of model parameters. Since it is not yet possible to predict a change in the ratio of amplitudes, just as it is impossible to determine it in real time, this effect is always detected late and it will fall into the approximation interval for a considerable time, which will negatively affect the forecast. In this case, the approximation of the modified model will be more accurate. Figure 1 compares the approximations of the pole oscillations based on the two-frequency model (2) and using a modified model taking into account dependence (5). Approximation of the modified model is significantly better at the ends of the interval. Such an approximation will be more suitable for prediction.

3. Condition of Earth pole oscillatory mode change

The condition for using the equation (6) is that the Earth pole oscillation mode change. The problem of determining the exact moment of transition of the pole motion to another mode may be related to the problem of forecasting, filtering and approximating. To give a confident forecast and to determine in advance the moment of amplitudes equality of the Chandler and annual harmonics (when there is a change of modes in Earth pole motion) turns out to be difficult. At this stage of the research, it is sufficient to determine the fact of the change with a slight delay. For this purpose, by solving the filtration problem, it is possible to estimate amplitude values of Chandler and annual harmonics.

However, amplitudes are non-observed values, and this fact results in a significant error in their estimates at the current time. But there is another way, which will be discussed below.

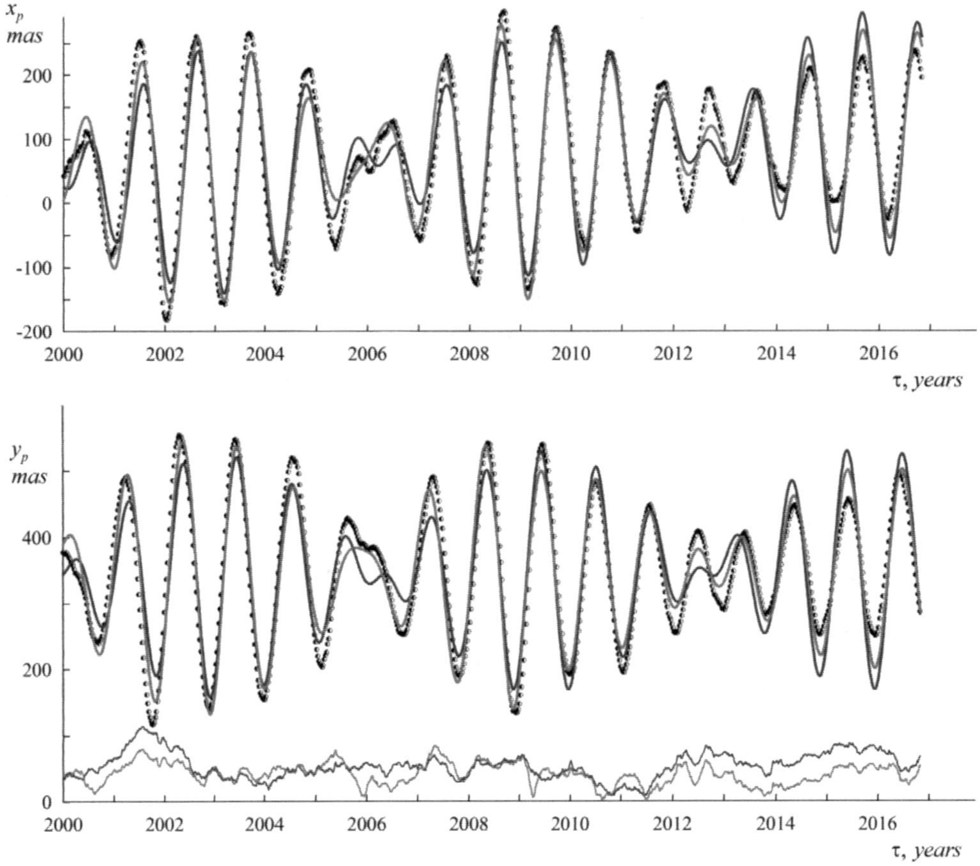

Figure 1. Approximations of fluctuations in the coordinates of the Earth's pole, constructed using a two-frequency model (blue line) and a modified model taking into account the found structural property (red line); The blue and red lines below are the corresponding residuals.

The change in amplitude ratio affects the dynamics of the polar angle and, in particular, the average frequency. The polar angle is also a non-observed value due to the presence of the long-period trend of the middle pole. But the error that appears in process of removing the trend component affects the dynamics of the angle not so significantly. Therefore, it is possible to determine the actual change in the amplitude ratio by evaluating the angle dynamics by knowing the functional relationship (3). The estimate of the non-averaged angle ψ is the most important practical problem, because the average value $\langle\dot{\psi}\rangle_T$ is sensitive to the change in the amplitude ratio with a large lag.

This task includes estimating the derivatives – frequency $\dot{\psi}$ and angular acceleration $\ddot{\psi}$, as well as derivatives of higher orders. Moreover, the higher the order of the estimated derivative is, the shorter the delay interval of transition identification becomes. In a real situation when corresponding to the measurements it is advisable to estimate the angle ψ and the first two derivatives $\dot{\psi}$, $\ddot{\psi}$.

From the relation (4) follows the condition for instantaneous angle value for when the pole oscillation mode changes:

$$\dot{\psi} = \left\langle \dot{\psi} \right\rangle_T = \frac{\dot{\alpha}_{ch} + \dot{\alpha}_h}{2}, \qquad (8)$$
$$\ddot{\psi} = 0.$$

Condition (8) is valid for two-frequency motion of the Earth pole, and in the case of observed motion it can be used not for instantaneous values, but for average angle, frequency and angular acceleration at the selected averaging interval $T_1 < T$. Thus, the angle variations can be approximated by a second order polynomial over the interval T_1. The choice for averaging interval duration of the observational data is a compromise between the quality of the smoothing and the reduction of the transition identification delay. The delay can be reduced by shrinking the smoothing interval, but only to a certain value, since in this case the spread of values $\dot{\psi}$, $\ddot{\psi}$ increases relative to functions $\dot{\psi}(\tau)$, $\ddot{\psi}(\tau)$ of the two-frequency model.

It seems appropriate to determine the maximum deviation $\Delta_{\max}$ of the angular velocity variation $\delta\dot{\psi} = \dot{\psi} - \left\langle \dot{\psi} \right\rangle_T$ from the function $\delta\dot{\psi}(\tau)$ of the two-frequency model by the approximating polynomial. Then the condition for the transition identifying is to exceed the deviation $\Delta_{\max}$. Namely, the condition that establishes the transition takes the following form

$$\pm\left(\left\langle \dot{\psi} \right\rangle_{T_1} - \frac{\dot{\alpha}_{ch} + \dot{\alpha}_h}{2} \right) > \Delta_{\max}, \qquad (9)$$

where the plus or minus sign is determined by the coefficient $k = a_h / a_{ch}$ increase or decrease when passing through the unit value.

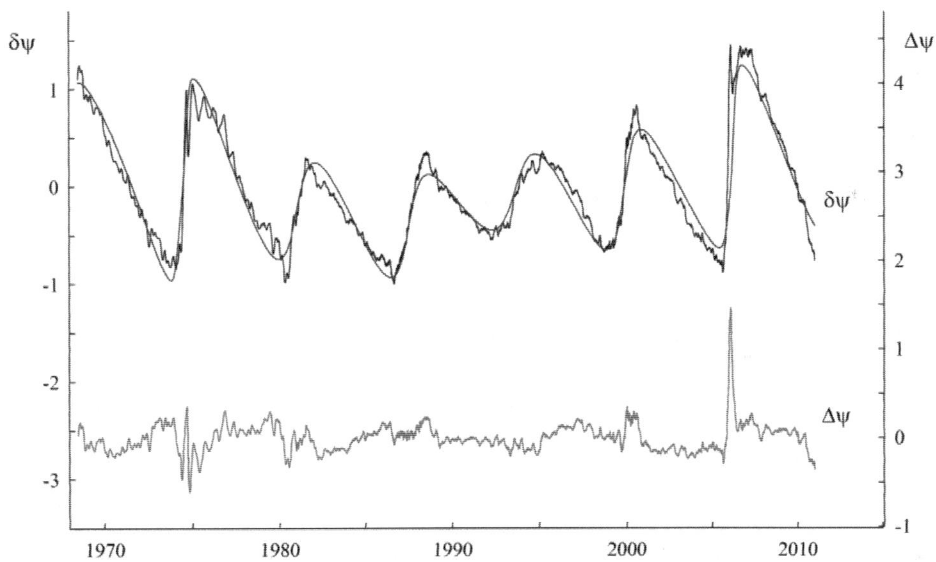

Figure 2. The polar angle variations $\delta\psi$ for the Earth pole observed motion (black line) and according to the two-frequency model with variable amplitudes (blue line), measured in radians. The red line is a residual.

Comparison between the polar angle variation graphs of the observed Earth pole motion and its variation based on the two-frequency model with variable amplitudes, as well as their residual, was provided in figure 2. It should be noted that ψ is a nonlinear function with respect to the coordinates of the pole and, according to (3), its sensitivity to high-frequency oscillations is inversely proportional to the amplitude of the total motion of the pole around the central point. Therefore, high-frequency fluctuations lead to bursts with a minimum amplitude of its motion, which is clearly visible on the residual graph in figure 2. By approximating the resulting residual the value Δ_{max} of the maximum deviation $\dot{\psi}$ can be estimated.

A more reliable estimate Δ_{max} can be obtained by increasing statistics, shifting the residual relative to the main oscillation $\delta\psi$. Since the angle is calculated from observations, it is possible to calculate the residuals in the coordinates ($\Delta x, \Delta y$) and convert them to the residual $\Delta\psi$. Moreover, the residuals ($\Delta x, \Delta y$) obtained from observational data should be replicated before calculations so that the last element of residual is followed by its first element.

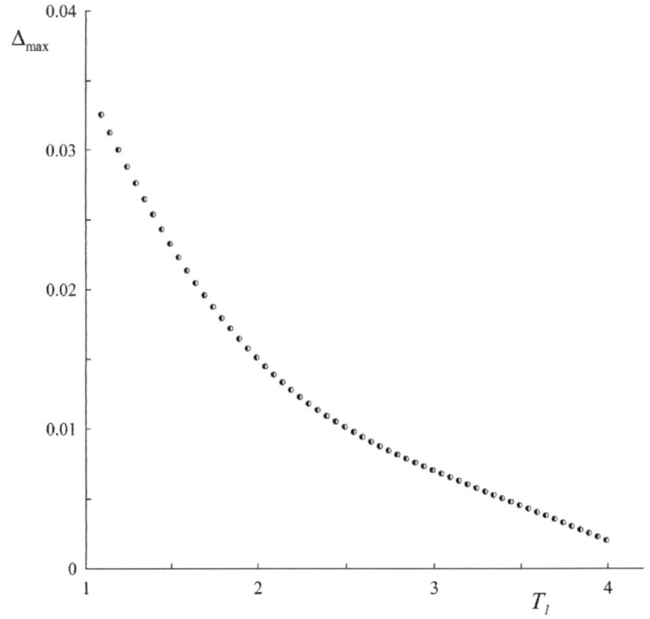

Figure 3. The dependence of the estimated deviation Δ_{max} on the averaging interval T_l

The graph in figure 3 shows the dependence of the deviation Δ_{max} measured in cycles per year on the length of the averaging interval measured over the years. The numerical experiment also shows that changing the averaging interval T_l affects the delay in determining the transition rather weakly - within a few months.

4. Conclusion

The obtained condition (9) of the Earth pole motion transition from one mode to another depends on the averaging interval. Condition (9) in numerical experiments on the test examples allows to find optimal duration of T_l averaging interval to minimize delay of the transition detection. In this case, for approximation, we have limited ourselves to only a first-order polynomial, i.e. an estimate of angular position and frequency. But to reduce the delay in determining the transition, the approximation can be refined by taking a second-order polynomial, and condition (9) should be supplemented by a similar one, that follows from the second equation (8). These conditions must be considered together. In this case, the deviation Δ_{max} will have a different value.

Once the transition to another motion mode is identified, the relation (6) should be used. Using (6) and Earth pole motion equations (2) it is easy to find changes in quasi-constant parameters after transition.

The above method allows one to extend the approximation of the Earth pole two-frequency motion model to the observational data on the whole time interval.

Acknowledgments
The research was performed as a part of realization of the base part of state assignment of Ministry of Education and Science of the Russian Federation (project FSFF-2020-0013).

References

[1] International Earth Rotation and Reference Systems Service - IERS Annual Reports, http://www.iers.org.
[2] Markov Yu G, Mikhaylov M V, Lar'kov I I et al. 2015 *Cosmic Res.* **53** 143–154
[3] Markov Yu G, Mikhailov MV, Perepelkin V V et al. 2016 *Cosmic Res.* **54** 155–163
[4] Petukhov V G 2019 *Cosmic Res.* **57** 351–363
[5] Ivanyukhin A V, Petukhov V G 2019 *Cosmic Res.* **57** 378–388
[6] Starchenko A E *Cosmic Res.* **57** 289–300
[7] Munk H, MacDonald G 1960 *The rotation of the Earth* (University Press) 323
[8] Kalarus M, Schuh H, Kosek W at el. 2010 *Journal of Geodesy* **84** (10) 587-596
[9] Markov Yu G, Perepelkin V V, Rykhlova L V, Filippova A S 2018 *Astron. Rep.* **62** (4) 299-308
[10] Markov Yu G, Perepelkin V V, Filippova A S 2017 *Dokl. Phys.* **62** (6) 318-322
[11] Klimov D M, Akulenko L D, Kumakshev S A 2013 *Dokl. Phys.* **58** (11) 505–509
[12] Kumakshev S A 2018 *Mech. Solids.* **53** (2) 159–163
[13] Barkin M Yu, Krylov S S, Perepelkin V V 2019 *IOP Conf. Series: Journal of Physics: Conf. Series* **1301** 012005
[14] Klimov D M, Akulenko L D, Kumakshev S A 2014 *Dokl.Phys.* **59** (10) 472–475
[15] Bondarenko V V, Krylov S S, Perepelkin V V 2018 *IOP Conf. Series: Materials Science and Engineering* **468** 012016

Features of aggregation of fuzzy expert judgments

V Sudakov[1] and T Sivakova[2]
[1] *Moscow Aviation Institute*, 4, Volokolamskoe shosse, 125993, Moscow, Russia
[2] *Plekhanov Russian University of Economics*, 36 Stremyanny lane, 117997, Moscow, Russia

E-mail: sivakova15@mail.ru

Abstract. The paper discusses methods of aggregating fuzzy judgments of experts based on fuzzy weighted sums and fuzzy areas of preference. The study is based on soft calculations. The approach based on fuzzy weighted summation allows you to take into account the degree of trust in the expert and his qualifications. The fundamental nature of the study is determined by the fact that the proposed approaches are invariant with respect to the subject area, it will allow you to effectively solve the problems of supporting the adoption of group decisions in case of fuzzy preferences. The authors developed algorithmic and software methods on the portal http://ws-dss.com.

1. Introduction

Expert assessment is a dynamically developing area of the decision support system (DSS). The use of decisions made on the basis of expert judgments is quite in demand. As a rule, experts' responses are considered as numbers, attributing strictly defined numerical values to the opinion, which are then processed using methods of applied statistics. In the case of such processing, the obtained data can significantly differ from reality [1]. In practice, decisions based on expert judgment are often non-numerical, making it difficult to process the information received. As a rule, the error in evaluating objects is also not taken into account.

To obtain more "accurate" results of fuzzy judgments, it is advisable to use the apparatus of "soft" calculations, the theory of fuzzy sets.

2. Literature review

In decision support tasks, expert judgment is used sufficiently long. A large number of studies by Russian and foreign authors are devoted to this [2-5].

Various quantitative and qualitative methods are used to process responses received from experts. The choice of a particular method depends on the complexity of the problem being solved, and the form in which the opinions of experts are presented. All these methods have their own advantages and disadvantages, but most often they are used to process discrete estimates.

In the practical use of decision-making methods, issues of increasing the consistency of judgments are quite difficult for experts, since the presence of inaccuracies leading to low consistency is primarily due to the "human factor". If the consistency is violated, it may turn out that the resulting vector estimates characterizing the properties of the objects being compared can significantly differ from the values obtained as a result of the "ideal experiment". And this, in turn, leads to incorrect decisions made on the basis of poorly coordinated source data.

According to the authors of the proposed study, in order to increase the correctness of decisions, an urgent scientific task is the creation of methods and algorithms for solving decision support problems based on fuzzy judgments of a group of experts.

In decision support problems, the apparatus of fuzzy sets has been used for a long time. A large number of studies by V. D. Nogin [6], A.N. Borisov [7], Orlov [8] and other scientists in Russia and abroad [9,10]. Typically, authors try to use the classical approaches of making decision theory in their estimates, which conception can give an incorrect result.

In the article of Nogin V.D. [11] solves the problem of multicriteria optimization of a numerical vector function on a fuzzy set. The membership function of a fuzzy feasible set is attached to the initial set of criteria, which allows the original multicriteria optimization problem to be interpreted as the task of finding a suitable compromise (Pareto-optimal) solution for an extended set of criteria. It is assumed that there is only fuzzy information about the preferences of the decision maker (DM) in the form of information quanta.

In an article by Podinovsky [12], methods for analyzing decisions in conditions of uncertainty are proposed, for the implementation of which utility functions and quantitative probabilities are not required. They are based on the ideas and results of the theory of the importance of criteria in multicriteria decision-making problems in the conditions of certainty. The qualitative probability and the partial relationships of preference and indifference generated by it on a variety of strategies are taken into account. Computational methods for constructing such relationships are considered.

In the work of Cosco [13], the author proposes to use the soft computing apparatus for calculating both the values of factors and the degree of their influence on each other. This method takes into account the degree of confidence of the decision maker.

Thus, one of the demanded is the class of problems in which experts express fuzzy judgments about a finite set of objects of assessment. This work is dedicated to this. Typically, objects are evaluated according to a given set of criteria, or an expert evaluates the alternative as a whole. To bring together several fuzzy primary criteria into a single integral fuzzy assessment, an original procedure based on fuzzy areas of preference is proposed. In complex decision support tasks, often enough it is necessary to take into account the opinions of various specialists. Therefore, the urgent task is to reduce fuzzy estimates of many experts into a single assessment. Experts can make fuzzy judgments both about one criterion, if they have similar specialization, and about different criteria, if they evaluate various aspects of the object of examination. The simplest solution to this problem is to defazzify estimates using standard approaches, for example, based on the center of gravity method.

If a fuzzy estimate of X with a membership function is given $\mu(x)$, then the defuzzified estimate of X^{α} is found by the formula:

$$X^{\alpha} = \frac{\int x\mu(x)dx}{\int \mu(x)dx}. \quad (1)$$

After defuzzification, you can use a wide range of common methods for processing expert judgments [11]. However, this approach loses significant information. A fuzzy assessment allows you to reflect not only the "value" of the criterion value for the object of examination, but also the degree of confidence of the expert in this assessment.

3. Calculation

3.1. *Fuzzy Weighted Amount*

The simplest option for a general assessment is to find the average of fuzzy estimates. The membership function of the general assessment will take the form:

$$\mu(x^*) = \sup_{\substack{x_1, x_2, \ldots x_n: \\ \frac{1}{n}\sum_{i=1}^{n} x_i = x^*}} (\min_i \mu_i(x_i)) \quad , \tag{2}$$

where $\mu_i(x_i)$ is the membership function of the assessment of the *i*-expert, *n* is the number of experts. The calculation of the final grade occurs according to the standard formula for the weighted amount:

$$X^* = \sum_{i=1}^{n} W_i X_i \quad , \tag{3}$$

where is X_i the fuzzy assessment of the alternative, and W_i is the fuzzy competence of the expert, X^* is the final assessment of the alternative.

The rules of summation and the product of fuzzy numbers are carried out on the basis of the principle of communication. Membership function corresponding to the operation:

$$\mu(y^*) = \sup_{\substack{y_1, y_2, \ldots y_n: \\ \eta(y_1, y_2, \ldots y_n) = y^*}} (\Theta_i \mu_i(y_i)) \quad . \tag{4}$$

Where η is the operation to be applied (in the case of calculation $W_i X_i$ it is the product, and for calculation $\sum_{i=1}^{n} W_i X_i$ it is the sum), y_i - the values to which the required operation is applied, $\mu_i(y_i)$ - the membership function of fuzzy values, $\mu(y^*)$ - the membership function for the result of applying the operation η. Θ is an intersection operation for membership functions.

The calculation by formula (2) is good because it does not require the involvement of a decision maker (DM), on the other hand, if a DM has certain preferences or decision rules on the competencies of experts, then it is more expedient to calculate according to formula (3).

3.2. Fuzzy Rank Comparison

Let us get fuzzy estimates of several objects. Instead of the often used defuzzification, this paper proposes a transition to pairwise comparisons of fuzzy ranks. Further, we assume that the larger the value of the rating, the better the object.

Consider two objects: A_i and A_j.. Between them, you can define a fuzzy binary relation $A_i \geq A_j$. It says that the object A_i is no worse than the object A_j. The degree of certainty that two objects are in a relationship $\geq$ is specified by a number $r_{ij} \in [0,1]$. Then we represent the binary relation matrix in the form:

$$R = \begin{bmatrix} 1 & r_{12} & r_{13} & & r_{1m} \\ r_{21} & 1 & r_{23} & & r_{2m} \\ r_{31} & r_{32} & 1 & & r_{3m} \\ & & & \ldots & \\ r_{m1} & r_{m2} & r_{m3} & & 1 \end{bmatrix} . \tag{5}$$

Elements of this matrix are found by analyzing the membership functions corresponding to the objects A_i and A_j:

$$r_{ij} = \sup_{\substack{x_i, x_j: \\ x_i \geq x_j}} [\min \mu_i(x_i), \mu_j(x_j)] \quad (6)$$

The final ranks of objects can be found using the Saati pair comparison method [14]. The Saati method involves the use of a multiplicative matrix: they show how many times one object is preferable to another, i.e. but in our case this is not so. Having performed the simplest arithmetic transformations, one can achieve the fulfillment of the required condition:

$$R^* = \left\| r_{ij}^* = \frac{r_{ij}}{r_{ji}} \right\| . \quad (7)$$

Further from equation (8):

$$(\lambda I - R^*)u = 0 \quad (8)$$

we find the maximum eigenvalue $\lambda_{\max}^*$ and the corresponding vector of influence factors u^*. The element u^* will be the final rank of the A_i. alternative. The value $\lambda_{\max}^*$ allows you to determine the inconsistency of the fuzzy relationship of preference C:

$$C = \frac{\lambda_{\max}^* - m}{m - 1} . \quad (9)$$

3.3. Average fuzzy ranking

Formulas (2) and (3), designed to aggregate expert judgments, have the disadvantage that the result of their application can be far from the opinion of most experts. This happens when a significantly high or substantially low score of one expert can greatly bias the final score.

To eliminate these shortcomings, the following algorithm is proposed:

- According to formula (6), the elements of the matrices R_k are calculated for each k -th expert.
- By solving the optimization problem, a "median" fuzzy ratio R^v is found, which is "close" to the opinion of most experts.
- Elements of R^v are substituted into formula (7) instead of elements of R, and thus the final estimates of the objects are found.

Consider the second step of this algorithm in more detail.

Let the fuzzy ratio R^v be represented as the following matrix:

$$R^v = \begin{bmatrix} 1 & r_{12}^v & r_{13}^v & & r_{1m}^v \\ r_{21}^v & 1 & r_{23}^v & & r_{2m}^v \\ r_{31}^v & r_{32}^v & 1 & & r_{3m}^v \\ & & & \cdots & \\ r_{m1}^v & r_{m2}^v & r_{m3}^v & & 1 \end{bmatrix} . \quad (10)$$

The fuzzy ratio for the k-th expert is presented in the form of the following matrix:

$$R_k = \begin{bmatrix} 1 & r_{12}^k & r_{13}^k & & r_{1m}^k \\ r_{21}^k & 1 & r_{23}^k & & r_{2m}^k \\ r_{31}^k & r_{32}^k & 1 & & r_{3m}^k \\ & & & \dots & \\ r_{m1}^k & r_{m2}^k & r_{m3}^k & & 1 \end{bmatrix}. \quad (11)$$

As a metric for the distance between the ratios R^v and R_k, we use the Manhattan distance [9]. It is calculated by the formula:

$$d(R^v, R_k) = \sum_{i=1}^{m} \sum_{\substack{j=1 \\ j \neq i}}^{m} \left| r_{ij}^k - r_{ij}^v \right|. \quad (12)$$

The sum of the distances from all relations R_k to R^v defines the objective function of the optimization problem:

$$\min_{r_{ij}^v} \sum_{k=1}^{n} d(R^v, R_k) = \min_{r_{ij}^v} \sum_{k=1}^{n} \sum_{i=1}^{m} \sum_{\substack{j=1 \\ j \neq i}}^{m} \left| r_{ij}^k - r_{ij}^v \right|. \quad (13)$$

Such a statement of the problem is close to the Kemeny median for the case of "clear" rankings.

The initial setting of the Kemeni median is difficult to solve due to binary optimization variables, whereas in this setting the restriction is much milder: $r_{ij}^v \in [0,1]$.

The objective function (12) is nonlinear, but the problem can easily be transformed to a linear programming problem. We introduce two new non-negative real variables: y_{ij}^+ and y_{ij}^-. For them, the restriction must be met:

$$y_{ij}^+ - y_{ij}^- = r_{ij}^k - r_{ij}^v. \quad (14)$$

Since the variables y_{ij}^+ и y_{ij}^- are non-negative, you can get rid of the module:

$$\left| r_{ij}^k - r_{ij}^v \right| = \left| y_{ij}^+ - y_{ij}^- \right| = y_{ij}^+ + y_{ij}^-. \quad (15)$$

As a result, the converted task will be written as:

$$\min_{\substack{r_{ij}^v, \\ y_{ij}^+, y_{ij}^-}} \sum_{k=1}^{n} \sum_{i=1}^{m} \sum_{\substack{j=1 \\ j \neq i}}^{m} \left(y_{ij}^+ + y_{ij}^- \right)$$

$$y_{ij}^+ - y_{ij}^- + r_{ij}^v = r_{ij}^k \quad . \quad (16)$$

$$y_{ij}^+ \geq 0, y_{ij}^- \geq 0, 0 \leq r_{ij}^v \leq 1$$

4. Conclusion

A set of methods has been developed for solving problems of the decision support system in a fuzzy information environment based on fuzzy expert judgments. The proposed methods for aggregating fuzzy expert judgments allow you to save information about the fuzziness and uncertainty of expert judgments until the very last stage of decision-making. Fuzzy ranking based on expert judgment allows us to assess the degree of consistency of the results. Formalization of the linear programming problem allows us to efficiently find an analogue of the Kemeny median for fuzzy judgments.

Acknowledgments
The reported study was funded by RFBR according to the research project No 19-01-00520.

References

[1] Orlov A 2018 *Methods of making management decisions* (Moscow: Knorus)
[2] Saaty T and Shang J 2011 An innovative ordersof-magnitude approach to AHP-based mutli-criteria decision making: Prioritizing divergent intangible humane acts *European. J. of Oper. Reseacher* **214(3)** 703–15
[3] Saaty T and Vargas L 2017 Origin of Neural Firing and Synthesis in Making Comparisons *European J.l of Pure and Applied Mathematics* **10(4)** 602-13
[4] Podinovski V V 2014 Potential optimality in multicriterial optimization *Comp. Mathem. and Math. Physics.* **54(1)** 429 - 38
[5] Podinovskaya O V and Podinovski V V 2017 Criteria importance theory for multicriterial decision making problems with a hierarchical structure *European. J. of Oper. Reseacher* **258** 983 – 92
[6] Nogin V D 2016 *Narrowing the Pareto set: An axiomatic approach* (Moscow: Fizmatlit)
[7] Borisov V V,. Fedulov A S and Zernov M M 2014 *Fundamentals of the theory of fuzzy sets* (Moscow: Hot Line – Telecom)
[8] Orlov A I 2011 *Organizational and economic modeling, textbook for university students: at 3 o'clock .: Part 2. Expert assessments* (Moscow: Publishing House of MSTU. N.E. Bauman)
[9] Mardani A J and Zavadskas E K 2015 *Expert Systems with Applications* **42 (8)** 4126-48
[10] Rezaei J, Nispeling T, Sarkis J and Tavasszy L 2016 *J. of Cleaner Production.* **135** 577-88
[11] Nogin V D 2019 Multi-criteria selection based on fuzzy information *Artif. Intelligence and Dec. Making* **2** 50-61
[12] Podinovsky V V 2020 Analysis of decisions in conditions of uncertainty with non-numerical preferences and probabilities *Management Problems* **1** 48-58
[13] Kosko B 2018 *Int. J. of Intelligent Systems* **33(8)** 1573-1623
[14] Saaty T L 2016 The Analytic Hierarchy and Analytic Network Processes for the Measurement of Intangible Criteria and for Decision-Making. *In Multiple Criteria Decision Analysis: State of the Art Surveys; Greco, S., Ehrgott, M., Figueira, J.R., Eds .; International Series in Operations Research & Management Science* (New York: Springer) pp 363-419
[15] Sudakov V 2020 Web Services for Decision Support Systems Retrieved from https://ws-dss.com/?locale=en

Using GRU based deep neural network for intrusion detection in software-defined networks

I I Kurochkin[1] and S S Volkov[2,3]
[1]Institute for Information Transmission Problems of Russian Academy of Sciences, Moscow, Russia
[2]Peoples' Friendship University of Russia, Moscow, Russia
[3]Federal Research Center "Computer Science and Control" RAS, Moscow, Russia

E-mail: qurochkin@gmail.com

Abstract. This paper considers the possibility of using machine learning methods in solving the problem of intrusion detection in software-defined networks (SDN). The work is devoted to the research and development of a network attack classifier, which is a core of the intrusion detection systems. To evaluate the methods, an existing data set was used, which includes network traffic records with a several different network attack scenarios. A comparison of machine learning methods implementing neural networks on a selected data set is presented. Based on the results, it can be concluded that the task of intrusion detection in software-defined networks can be successfully solved using deep neural networks.

1. Introduction

In today's world, more and more devices require an Internet connection. Fast growth in connectivity demand has challenged traditional network architectures. To meet current requirements, SDN (Software-Defined Network) was proposed [1,2], in which the control plane functionality is separated (abstracted) from the underlying packet forwarding layer. SDN enables network programming and allows you to dynamically change thread policies. SDNs can provide new capabilities by addressing the tasks of increasing channel throughput, simplifying network management, redistributing load, and increasing network scalability.

Since SDN technology provides network virtualization capabilities, separates control and forwarding planes, implements logically centralized management and opens up network capabilities for higher-level applications, it is especially suitable for implementing networks of large data centres.

Since the SDN architecture assumes a completely different approach to implementing a network infrastructure, it also brings new security issues to the network. Now the main target of attacks is the central controller, as it is a key component in managing the entire SDN infrastructure. To protect the controller and the network as a whole, Intrusion Prevention System (IPS) is used, which is used to detect and prevent unauthorized actions in the information system. The core of the IPS-system is the analysis subsystem. It is responsible for the quality of detection of unauthorized actions in the network.

The main goal of the research is to check the possibility of using machine learning methods in solving the problem of intrusion detection in SDN. The task of intrusion detection in the network is reduced to the classification of traffic passing through it. The easiest way is to divide traffic into two categories: malicious and normal. A more advanced way is to try to identify a specific type of threat among malicious traffic.

The process of analysing network traffic can be divided into two stages:

- Feature extraction based on all available data for each connection;
- Traffic classification based on selected features.

The second stage is considered in more detail in this paper.

2. Related works

One of the most important conditions for the smooth operation of the network is security. There are many studies on security in SDN. The paper [4] describes the main vulnerabilities of the SDN technology, as well as the OpenFlow protocol. Several examples of using vulnerabilities to disrupt the system are given.

Due to the centralized control of SDN networks, the most serious threat to them is DoS (Denial of Service) attacks, which can lead to exhaustion of network resources. Many researchers pay special attention to this type of attack because it is widespread and have a high degree of danger to SDN. The paper [5] presents trends and characteristics of DDoS attacks in cloud computing, and also provides a comprehensive overview of the protection mechanisms against DDoS attacks using SDN. In addition, various options for launching DDoS attacks on SDN, as well as methods to counteract DDoS attacks were investigated. This work allows you to understand how to take full advantage of SDN to protect against DDoS attacks in cloud computing environments. Paper [6] is also devoted to identifying network attacks in SDN. The authors consider a specific SYN flood Distributed Denial of Service (DDoS) attack. For some classifiers, there are cases in which these attacks cannot be recognized. The authors give some recommendations on how to increase the resistance of classifiers to demonstrated attacks. In the paper [7], the authors identify the problems that they face in protecting the SDN data plane, one of the least studied but most important components of this technology. They argued the importance of security SDN data planes, surveyed the latest advances in data plane development and analyzed the security implications of stateful data planes. In the work [8], the authors propose an effective solution for preventing DDoS attacks in an SDN cloud environment. They present a new hybrid machine learning model based on the support vector algorithm and self-organizing cartographic algorithms to improve traffic classification. In the paper [9], the authors describe various basic functions of SDN, which make it a suitable network technology for cloud computing. Also, the work presents flow-table sharing approach to protect SDN-based clouds from flow table overloading DDoS attacks. This approach uses the flow table of other OpenFlow switches on the network to protect the switch table from overload. The approach increases the resilience of the cloud system to DDoS attacks with minimal involvement of the SDN controller.

To detect intrusions in SDN, deep neural networks can be used. In [10], the authors use GRU-RNN (Gated Recurrent Unit Recurrent Neural Network). The proposed approach has been tested using the NSL-KDD dataset. To solve this problem, you can also use a multilayer perceptron. So, for example, in [11], the authors used a model based on a neural network that was trained on the NSL-KDD data set. Even simple neural networks (for example, a single-layer perceptron) can solve the problem of intrusion detection in software-configured networks [12]. Model trained on the same NSL-KDD dataset.

The review shows that the problem of intrusion detection in SDN can be solved by using machine learning methods. This work is devoted to the research and development of a network attack classifier, which is a core of the intrusion detection systems. To implement and test it, an existing data set was used.

3. Datasets

To implement methods of detecting network attacks, an existing data set was used, which includes network traffic records with a several different network attack scenarios. There are currently several prepared datasets [3]:

- CSE-CIC-IDS2018 on AWS;
- Intrusion Detection Evaluation Dataset (CICIDS2017);
- Intrusion detection evaluation dataset (ISCXIDS2012);
- NSL-KDD dataset;
- DDoS Evaluation Dataset (CICDDoS2019);
- CIC DoS dataset (2017).

The most popular dataset is the NSL-KDD dataset. Various experiments in many studies were carried out on it. But its main problem is that the data presented in it can be considered obsolete. NSL-KDD dataset was assembled in 2009. The traffic presented in it no longer corresponds to today's realities. For our research, it was decided to use the most voluminous and relatively new data set - CSE-CIC-IDS2018.

CSE-CIC-IDS2018 dataset includes several different attack scenarios: Brute force, Botnet, DoS, DDoS, Web-attack, infiltration of the network from inside. Some of them are divided into subgroups. Total attack classes are 15:

- Benign (normal traffic) – records corresponding to the permissible actions of users / services on the network.
- Bot (botnet) – computer network consisting of a number of hosts with bots (autonomous software) running. Most often, a bot as part of a botnet is a program that is covertly installed on the victim's device and allows an attacker to perform certain actions using the resources of an infected computer.
- Brute Force – Complete enumeration (or the "brute force" method) - breaking the password by enumerating all possible keys. The dataset contains several variations of this attack: SSH-Bruteforce; Brute Force –Web; Brute Force –XSS; FTP-BruteForce.
- DoS/DDOS – a cyber-attack in which the perpetrator seeks to make a machine or network resource unavailable to its intended users by temporarily or indefinitely disrupting services of a host connected to the Internet. Denial of service is typically accomplished by flooding the targeted machine or resource with superfluous requests in an attempt to overload systems and prevent some or all legitimate requests from being fulfilled. The dataset contains several variations of this attack: DDOS attack-LOIC-UDP; DoS attacks-GoldenEye; DoS attacks-SlowHTTPTest; DDoS attacks-LOIC-HTTP; DoS attacks-Slowloris; DoS attacks-Hulk; DDOS attack-HOIC.
- Infiltration – Infiltration can be accomplished by directly breaching a network, or by infecting a host, which is then joined to a private network. Infiltration attacks can be grouped into two basic categories: Machine-based attacks (Attackers look for known vulnerabilities on internet-facing systems, as well as any misconfiguration of network perimeter controls, exposed applications, or other oversights that allow for an initial entryway into a corporate environment.), and human-based attacks (Social engineering, or tricking humans by taking advantage of their trusting nature etc.).
- SQL Injection – a code injection technique, used to attack data-driven applications

The attacking infrastructure includes 50 machines and the victim organization has 5 departments and includes 420 machines and 30 servers. The dataset includes the captures network traffic and system logs of each machine, along with 80 features extracted from the captured traffic using CICFlowMeter-V3.

The authors of the dataset offer a processed version created specifically for working with machine learning algorithms. This set consists of several CSV files. They contain a set of 80 features, as well as labels indicating the class of traffic. The following are all traffic classes in the selected data set and the number of their records

4. Neural network architecture

The neural network has following architecture:

- Input layer;
- GRU (Gated Recurrent Units) [14] layer - Takes input values of current traffic features;
- GRU layer;
- Self-Attention layer [15];
- GRU layer. Return sequences = False;
- Fully-connected hidden layer. Activation function is ReLU [13];
- Fully-connected hidden layer. Activation function is ReLU;
- Output layer – 7 neurons. Each neuron corresponds to the type of traffic. Activation function is SoftMax.

Using a deep recurrent neural network [10] can improve classification results compared to direct propagation neural networks and deep convolutional neural networks.

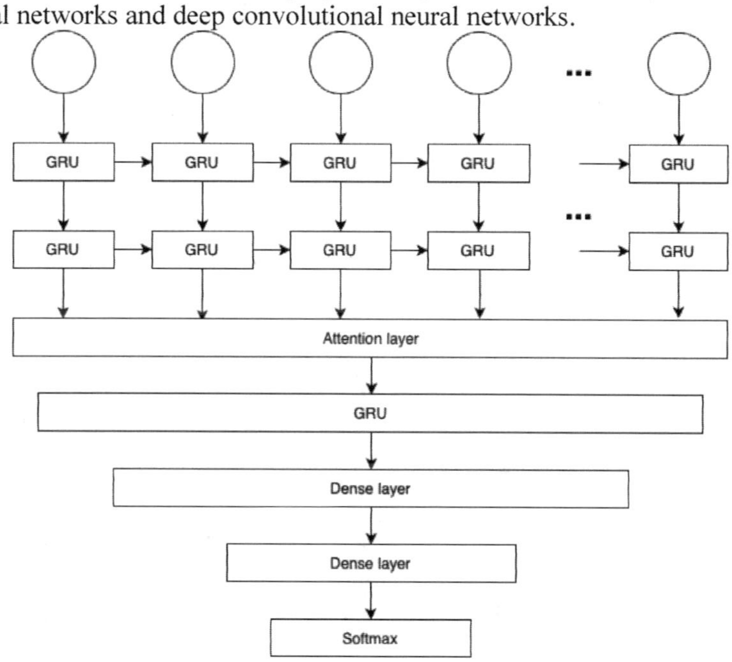

Figure 1. Neural network with GRU layers.

5. Data pre-processing

The presented dataset contains too detailed grouping of attack types. It makes sense to combine some of subgroups, since they will have similar characteristics. Thus, we obtain 7 groups: Benign, Bot, Brute Force, DDOS, DOS, Infiltration, Web-attacks.

This data set is not balanced. It has too many benign traffic records compared to the rest. We can also observe a small number of records corresponding to web attacks. It is necessary to balance at least these two classes for better model training. First of all, we decided to reduce the number of benign traffic records to 2 million. Then the records for each traffic category were randomly divided into 3 groups: 70% (train) / 15% (validation) / 15% (test).

For Web attack classes, we artificially increased the number of records for the training set. SMOTE (Synthetic Minority Oversampling Technique) [16] algorithm was used. This algorithm allows you to generate a number of artificial examples that would be similar to those in the minority class but did not duplicate them. Web attack class expanded to 100/150/200/250 thousands of records. We managed to achieve the highest quality of training with 200 thousand records. Oversampling was not performed for test and validation samples. As a result, we obtain the following distribution of records:

Table 1. Dataset classes list.

Class\Set	Train	Validation	Test
Benign	1 400 000	300 000	300 000
Brute Force	266 664	57 142	57 143
Web-attacks	200 000	139	140
DDOS	884 753	189 589	189 591
DOS	458 010	98 145	98 145
Bot	200 333	42 928	42 930
Infiltration	113 353	24 290	24 291

6. Results

The experiments were carried out on the following hardware and software configuration:

- Intel(R) Xeon(R) CPU E5-2683 v4 @ 2.10GHz;
- 4x NVIDIA Tesla K80;
- 512 GB RAM;
- Python 3.6.9 programming language.

We used the Python programming language, as well as libraries for this language that implement deep neural networks (Keras, TensorFlow). Pandas library was used for data pre-processing. Imbalanced-learn library was used for SMOTE algorithm.

Model was trained on pre-processed data. Figure 2 shows the confusion matrix for the presented model in 7 classes. Table 2 shows the result of the model on a testing set. We can see a low F-measure for two classes. For a web-attack class, this is caused by a small number of class instances. The classifier completed its main task (131 out of 140 attacks were identified, with only 5 of the 10 remaining being identified as benign traffic). The Infiltration class has a more difficult situation. This type of traffic is very similar to benign traffic, and it turned out to be difficult to detect even with the help of deep neural networks. Based on this, we can conclude that it is necessary to take additional measures to identify attacks of this type in order to introduce such a model into an intrusion detection system

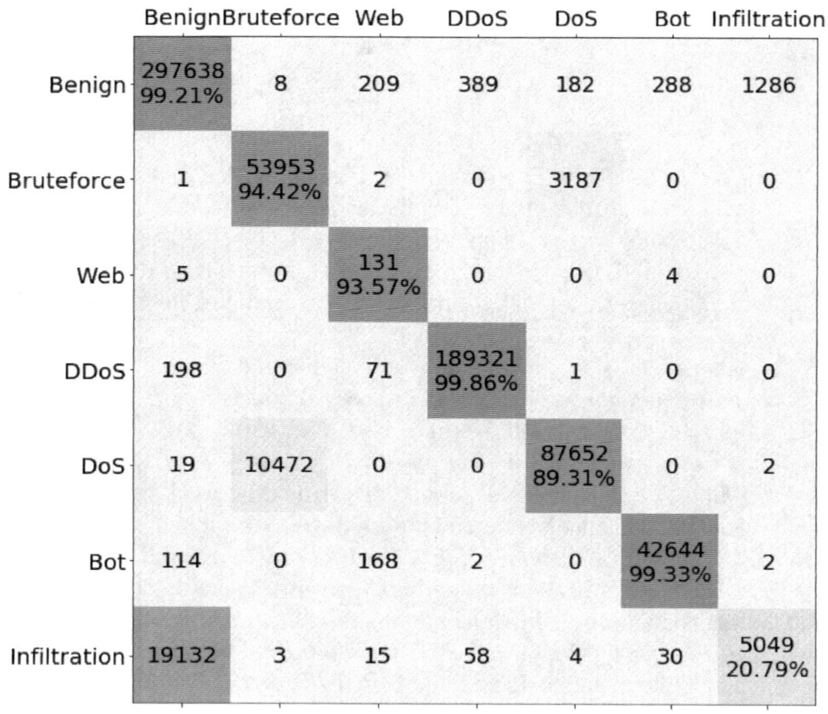

Figure 2. 7 classes confusion matrix (GRU based NN).

Table 2. Classification result on test data.

Class\Set	Precision	Recall	F1-score	Records count
Benign	0.94	0.99	0.96	300 000
Brute force	0.84	0.94	0.89	57 143
Web	0.23	0.94	0.36	140
DDOS	1.00	1.00	1.00	189 591
dos	0.96	0.89	0.93	98 145
Bot	0.99	0.99	0.99	42 930
Infiltration	0.80	0.22	0.34	24 291

7. Conclusions

Based on the results of the study, it can be concluded that the task of intrusions detection in software-defined networks can be successfully solved using deep neural networks. This approach showed a fairly good result on one of the latest network attack datasets. The specificity of SDN allows you to use the security methods other than the methods inherent in networks with traditional architecture. Because the OpenFlow allows you to use the network itself as a sensor. This can increase the efficiency of such classifiers because additional features will be considered. Machine learning techniques can also be used to analyse traffic in general in SDN. This will increase the efficiency of network resource allocation.

Further work will be devoted to the analysis of intrusion detection methods and the research and development of an intrusion detection system (IDS), which is used to detect unauthorized actions in software-defined networks using machine learning methods.

Acknowledgments
The reported study was funded by RFBR according to the research project No. 18-29-03264.

References

[1] Casado M, Garfinkel T, Akella A, Freedman M J, Boneh D, McKeown N and Shenker S 2006 SANE: A Protection Architecture for Enterprise Networks. *15-th Usenix Security Symposium. Vancouver, Canada* **49** pp 50

[2] McKeown N, Anderson T, Balakrishnan H, Parulkar G, Peterson L, Rexford J, Shenker S and Turner J 2008 OpenFlow: enabling innovation in campus networks *ACM SIGCOMM Comput. Commun. Rev.* **38(1)** 69-74

[3] Datasets. Research. Canadian Institute of cybersecurity. *https://www.unb.ca/cic/datasets/index.html* (Accessed 20 May 2020)

[4] Dacier M C, König H, Cwalinski R, Kargl F and Dietrich S 2017 Security challenges and opportunities of software-defined networking *IEEE Security & Privacy* **15(2)** 96-100

[5] Yan Q, Yu F R, Gong Q and Li J 2016 Software-defined networking (SDN) and distributed denial of service (DDOS) attacks in cloud computing environments: A survey, some research issues, and challenges *IEEE Commun. Surv. Tutorials* **18(1)** 602-622

[6] Aiken J and Scott-Hayward S 2019 Investigating Adversarial Attacks against Network Intrusion Detection Systems in SDNs *IEEE Conference on Network Function Virtualization and Software Defined Networks, NFV-SDN 2019 - Proceedings* pp 1-7

[7] Shaghaghi A, Kaafar M A, Buyya R and Jha S 2019 Software-Defined Network (SDN) data plane security: Issues, solutions, and future directions *Handbook of Computer Networks and Cyber Security: Principles and Paradigms* pp 341-387

[8] Phan T V. and Park M 2019 Efficient distributed denial-of-service attack defense in SDN-based cloud *IEEE Access* **7** 18701-14

[9] Bhushan K and Gupta B B 2019 Distributed denial of service (DDoS) attack mitigation in software defined network (SDN)-based cloud computing environment. *Journal of Ambient Intelligence and Humanized Computing* **10(5)** 1985-97

[10] Tang T A, Mhamdi L, McLernon D, Zaidi S A R and Ghogho M 2018 Deep Recurrent Neural Network for Intrusion Detection in SDN-based Networks *4th IEEE Conference on Network Softwarization and Workshops (NetSoft 2018)* pp 202-206

[11] Tang T A, Mhamdi L, McLernon D, Zaidi S A R and Ghogho M 2016 Deep learning approach for Network Intrusion Detection in Software Defined Networking *Proceedings - 2016 International Conference on Wireless Networks and Mobile Communications, WINCOM 2016: Green Communications and Networking* pp 258-263

[12] Abubakar A and Pranggono B 2017 Machine learning based intrusion detection system for software defined networks *Proceedings - 2017 7th International Conference on Emerging Security Technologies (EST 2017)* pp 138-143

[13] Nair V and Hinton G E 2010 Rectified linear units improve Restricted Boltzmann machines *ICML 2010 - Proceedings, 27th International Conference on Machine Learning* pp 807-814

[14] Cho K, Van Merriënboer B, Gulcehre C, Bahdanau D, Bougares F, Schwenk H and Bengio Y 2014 Learning phrase representations using RNN encoder-decoder for statistical machine translation *EMNLP 2014 - 2014 Conference on Empirical Methods in Natural Language Processing, Proceedings of the Conference*

[15] Keras Self-Attention *https://pypi.org/project/keras-self-attention* (Accessed 20 May 2020)

[16] Chawla N V., Bowyer K W, Hall L O and Kegelmeyer W P 2002 SMOTE: Synthetic minority over-sampling technique *J. Artif. Intell. Res.* **16** 321-357

IOP Conf. Series: Materials Science and Engineering **927** (2020) 012036 doi:10.1088/1757-899X/927/1/012036

Estimation of the contribution of geophysical perturbations to the Earth pole oscillatory process at the precession frequency of the lunar orbit

S S Krylov, V V Perepelkin and A S Filippova
Moscow aviation institute, 4, Volokolamskoe shosse, 125993, Moscow, Russia

E-mail: vadimkin1@yandex.ru

Abstract. The contribution of the main geophysical perturbations (atmospheric and oceanic) to the oscillatory process in phase with the precession of the lunar orbit is investigated. The solutions for the differential equations of Earth pole motion are obtained, taking into account the atmospheric angular moment and the oceanic angular momentum, and the combination harmonics associated with the oscillatory process under consideration are identified. It is shown that the harmonics found are partially caused by fluctuations in the mobile media of the atmosphere and ocean.

1. Introduction

The Earth pole motion is a multi-frequency dynamic process that occurs under the influence of external and internal forces - both potential (gravitational forces and centrifugal forces [1]), and non-potential, associated with the movement of masses in geophysical media and affecting the dynamic characteristics of the deformable Earth rotation [2, 3].

Modern high-precision observations give a possibility for detecting irregularities in the Earth rotation with small amplitudes. Analyzing the observational data of the oscillatory Earth pole motion from the International Earth Rotation and Reference Systems Service (IERS) [2] a suggestion could be made that there is a complex dynamic process that occurs in the Earth-Moon-Sun system, which is not explained in the theory of Earth rotation.

Earlier in [4], a transformation of the Earth pole coordinates was found, the result of which shows the connection between some variations of the Earth pole oscillatory process and the Moon's orbit precession. To be more precise, a transformation of the Earth coordinate system by using the rotating and shifting operations to a new system, where the pole motion is synchronized with the precession of the lunar orbit, could be found.

In this regard, the studies aimed at establishing the celestial-mechanical and geophysical causes of such behavior of the Chandler and annual oscillatory components are of interest.

In the article, the numerical integration of the Earth pole equations of motion is carried out, using the data of the NCEP/NCAR [5] atmospheric circulation and NASA/JPL [6] data of the oceanic angular momentum. As a result of integration, the Earth's pole motion is obtained, taking into account the main geophysical perturbations. The contribution of atmospheric and oceanic perturbations to the oscillatory process, synchronous with the lunar orbit precession, is determined. Combinational harmonics are found, obtained as a combination of the fundamental harmonics of the pole oscillations and harmonic with the lunar orbit precession frequency.

2. Differential equations of the Earth pole motion taking into account geophysical perturbations

The lunar orbit spatial motion consists of a series of rotations around intersecting axes. These rotations lead to the cyclical motion of its nodes and perigee. In addition, the derivatives of the orbit parameters are nonzero and the values are not permanents, being a subject to small variations. A consequence of the lunar orbit precession and the associated cyclical change in the longitude of the ascending node with a period of 18.61 years is a change in the orbit plane inclination relative to the Earth equator. The inclination of the lunar orbit to the Earth's equator varies from 18.3° to 28.58°. In this case, the intersection point of the lunar orbit with the equator oscillates along the equator near its average position, which coincides with the point of the vernal equinox. Unlike the node (the intersection point of the the lunar orbit and the ecliptic on the celestial sphere), which makes a complete revolution, the intersection point between the lunar orbit and the Earth equator oscillates from -13.2° to 13.2°.

It was shown in [4, 7-10] that one can find a Earth pole coordinates transformation that illustrate the in-phase mode between the Earth pole oscillatory process with combination frequencies 1 ± 0.05373, 0.843 ± 0.05373 cycle per year and precession of the lunar orbit. Namely, the Earth pole oscillatory motion minus the Chandler (or annual - depending on the values of the Chandler and annual harmonics amplitudes) and six-year cycles occurs in phase with the oscillations of the lunar orbit intersection point with the equator that happens along the equator. This property requires a more detailed analysis and study of the causes of such oscillations. In particular, it is of interest to establish the contribution of geophysical (atmospheric and oceanic) perturbations to these oscillations.

To do this, it is necessary to determine the Earth pole motion by integrating the differential equations of its motion taking into account geophysical perturbations. Differential equations of the Earth pole oscillations can be obtained from the Euler-Liouville equations of the deformable Earth rotational motion [11-13]:

$$J\dot{\omega}+\omega\times J\omega=\mathrm{M}-\dot{J}\omega,\quad \omega=(p,q,r)^{\mathrm{T}},\quad J=J^{*}+\delta J,\quad J^{*}=\mathrm{const},$$
$$J^{*}=\mathrm{diag}(A^{*},B^{*},C^{*}),\quad \delta J=\delta J(t),\quad \|\delta J\|\ll\|J^{*}\| \tag{1}$$
$$\mathrm{M}=\mathrm{M}^{S}+\mathrm{M}^{L}-\dot{\mathrm{h}}-\omega\times\mathrm{h}.$$

Here ω is the angular velocity vector in the coordinate system relative to the Earth. The axes of the chosen coordinate system approximately coincide with the main central axes of inertia J^{*} of the Earth "frozen" shape, that take into account the "equatorial bulge". It is assumed that small variations of the inertia tensor δJ include various harmonic components, which are caused by the regular disturbing effect of gravitational diurnal tides from the Sun and the Moon and, possibly, other, for example, annual, monthly, etc. M^{S} and M^{L} - are disturbing moments from the Sun and Moon, respectively, a h - is the vector of the total angular momentum of geophysical fluids (atmospheric and oceanic).

Due to the smallness of p, q ($p,q\ll r$), up to a first order approximation in p, q, the system of differential equations for them takes the form:

$$\dot{p}+N_{p}q=\dot{j}_{qr}^{0}+\mu_{p},\quad p(t_0)=p_0,$$
$$\dot{q}-N_{q}p=-\dot{j}_{pr}^{0}+\mu_{q},\quad q(t_0)=q_0,$$
$$\delta\dot{r}=\mu_{r},\quad r(t_0)=r_0, \tag{2}$$
$$N_{p}=\frac{C^{*}-B^{*}}{A^{*}}r_0,\quad N_{q}=\frac{C^{*}-A^{*}}{B^{*}}r_0,\quad r_0=7.29\cdot10^{-5}\ \text{rad/s},$$

where $N=\sqrt{N_pN_q}\cong0.84\div0.85$ cycles per year is the average Chandler frequency, and $j_{pr}^{0}=\left\langle J_{pr}r_0^2\right\rangle_{\varphi}$, $j_{qr}^{0}=\left\langle J_{qr}r_0^2\right\rangle_{\varphi}$ are tidal "bulges" averaged by diurnal rotation, including dissipative terms due to the pole tide. Values μ_p, μ_q have the meaning of specific moments of forces and characterize the total perturbations leading to the observed motion of the Earth pole. They are expressed through geophysical perturbations χ_x, χ_y, published by IERS.

To determine the contribution of geophysical perturbations to the studied oscillatory process, the system of differential equations for the Earth pole motion is written in standard form taking into account the perturbing functions χ_x, χ_y [14]. For numerical integration atmospheric and oceanic perturbations were taken into account both separately and jointly. Integration was carried out by the Runge-Kutta 4th order method.

Figures 1-2 display the taken into account perturbations χ_x, χ_y from the atmosphere and the ocean. In Figure 3 the joint atmospheric and oceanic perturbations are shown.

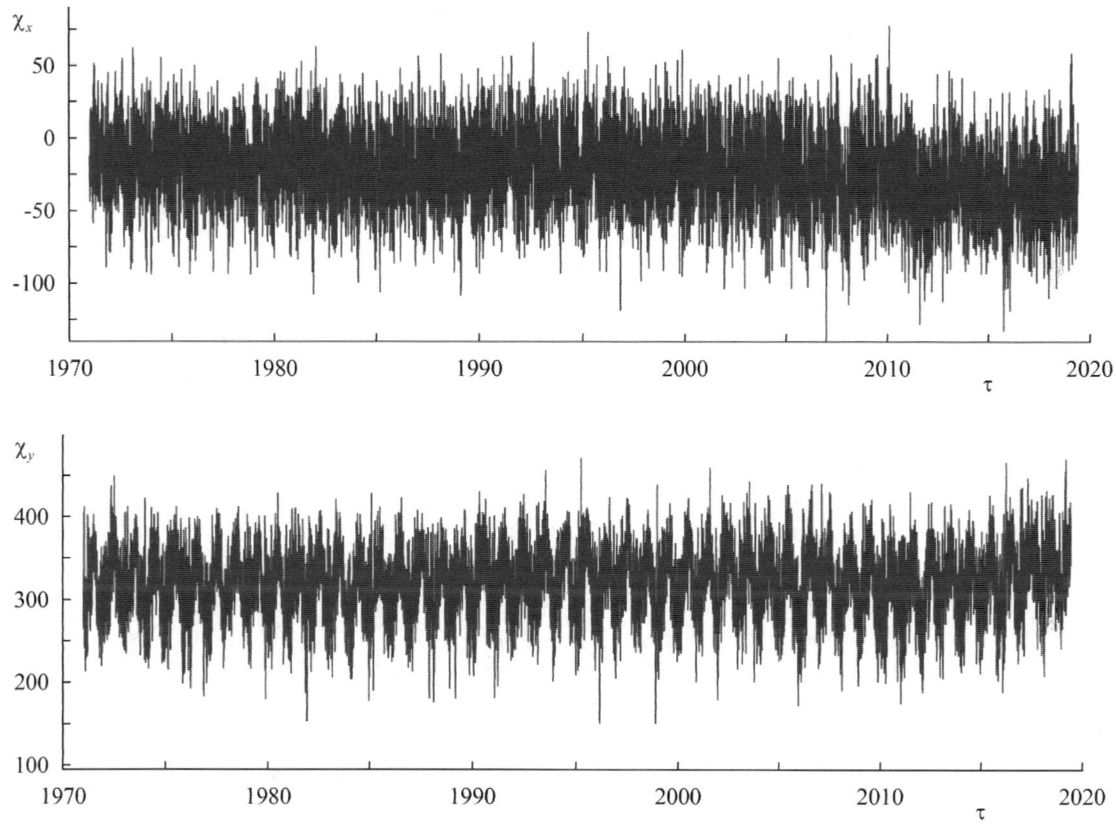

Figure 1. Perturbations χ_x, χ_y from the atmosphere

IOP Conf. Series: Materials Science and Engineering **927** (2020) 012036 doi:10.1088/1757-899X/927/1/012036

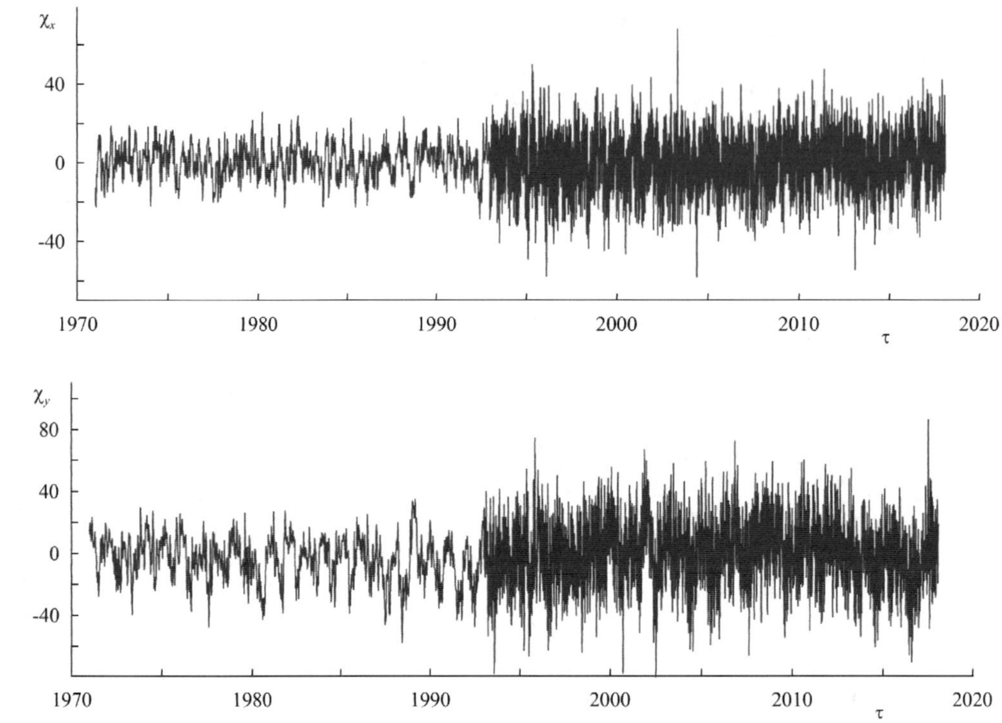

Figure 2. Perturbations χ_x , χ_y from the ocean

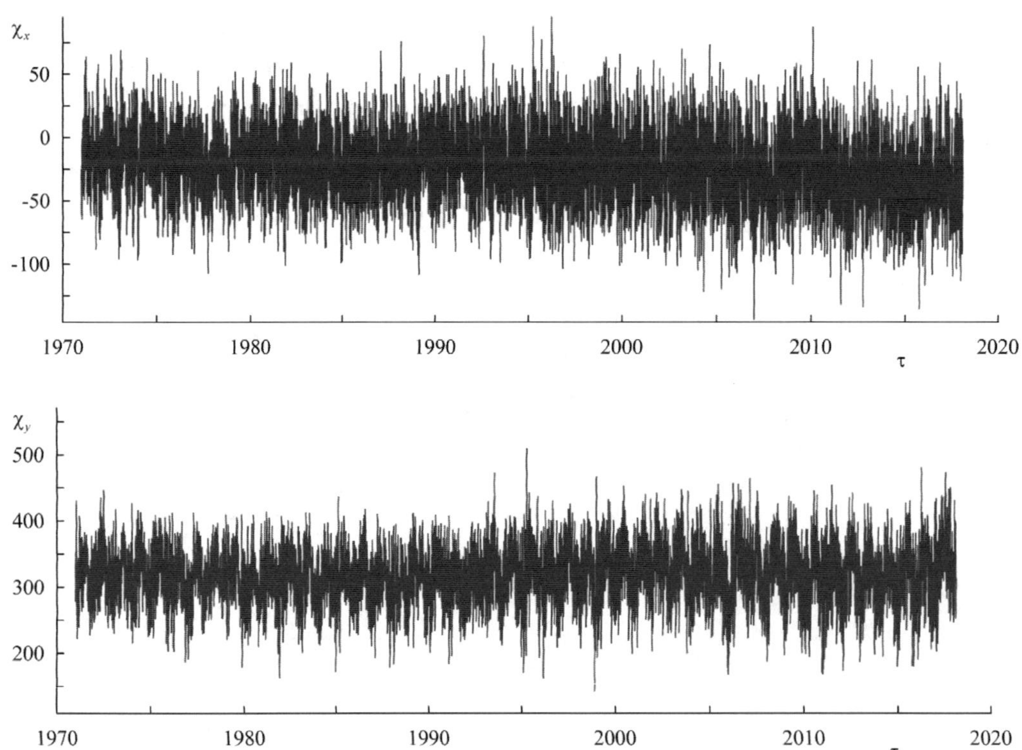

Figure 3. Joint perturbations χ_x , χ_y from the atmosphere and the ocean

3. The contribution of geophysical perturbations in synchronizing the Earth pole motion and the lunar orbit precession

To perform a prior assessment of the geophysical perturbations contribution to the oscillatory process that is in phase with the lunar orbit precession a comparison between the amplitude spectra of geodetic, atmospheric, and oceanic perturbations χ_y is given in Figure 4. It shows that the studied oscillatory process with combination frequencies 1 ± 0.05373 cycle per year does not determine completely by the atmospheric and oceanic perturbations. There will be differences between the sums of these harmonics extracted from the calculated and observed Earth pole trajectories both in the amplitude and in the phase shift of the amplitude modulation. The same conclusions are true for perturbations χ_x. Its graph is less readable due to overlapping spectral peaks and therefore it is not given.

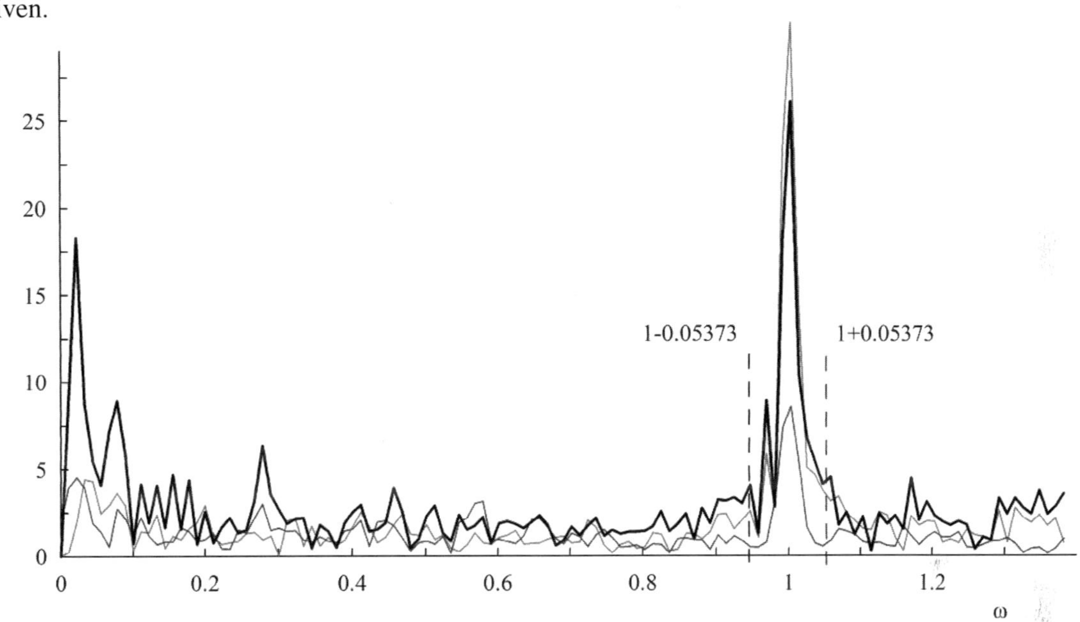

Figure 4. Comparison of the amplitude spectra of geodetic perturbations χ_y and perturbations from the atmosphere (blue line) and the ocean (red line)

As a result of numerically solving the differential equations of the Earth pole motion, the trajectories of the pole are obtained for three cases of perturbations. The perturbing functions were tabulated according to the IERS data.

For example, Figure 5 shows the oscillations in the calculated Earth pole motion, taking into account the joint perturbations from the atmosphere and the ocean in comparison with the oscillations of its observed motion.

To extract the combinational harmonics with frequencies 1 ± 0.05373 cycle per year from calculated and observed Earth pole oscillations the procedure proposed in the work [4] was applied. Using the transformations of the Earth pole coordinates, i.e. eliminating the two cycles - with the Chandler and six-year periods, it is possible to obtain a pole oscillation that is in phase with the lunar orbit precession. Then, centering the resulting trajectory, the inverse transformation is performed. As a result, we obtain the desired oscillatory process from the harmonics combination. In fig. 6 the comparison between the considering oscillations, which are taken from the observed and calculated trajectories of the Earth pole (taking into account atmospheric and oceanic perturbations) is shown. The obtained result confirms the preliminary conclusions made using the spectral analysis. The process turned out to be with a significantly lower amplitude and with a shifted modulation phase, which indicates the possibility of a more complex physical nature of these oscillations and indicates the incompleteness of the perturbations taken into account.

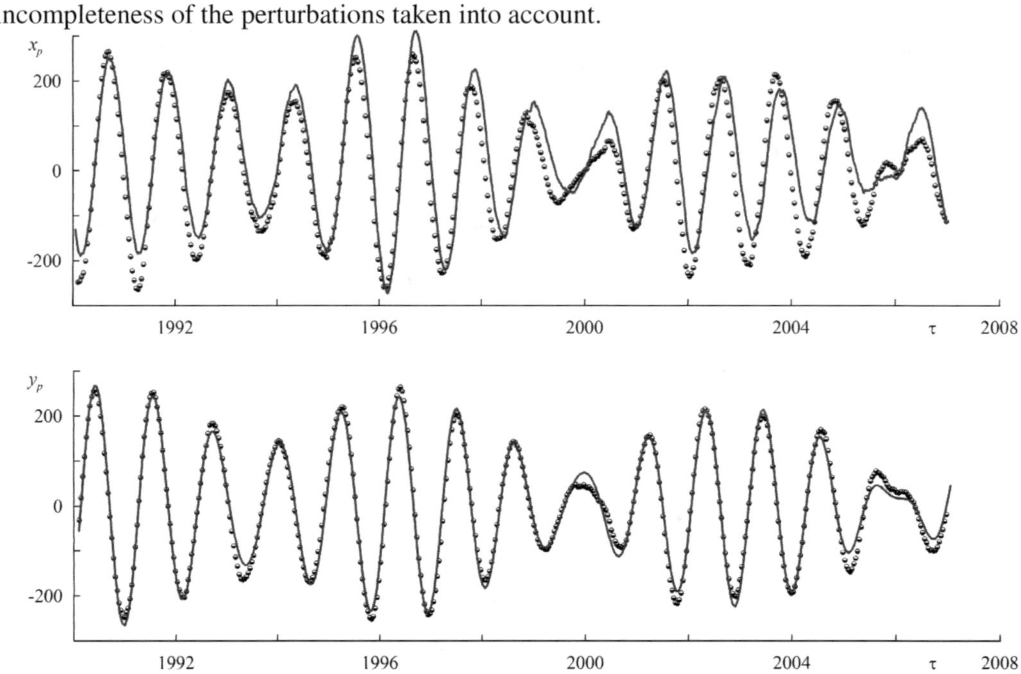

Figure 5. Earth pole oscillations according to the IERS observational data (discrete data) in comparison with the calculated oscillations caused by the perturbations of the atmosphere and the ocean

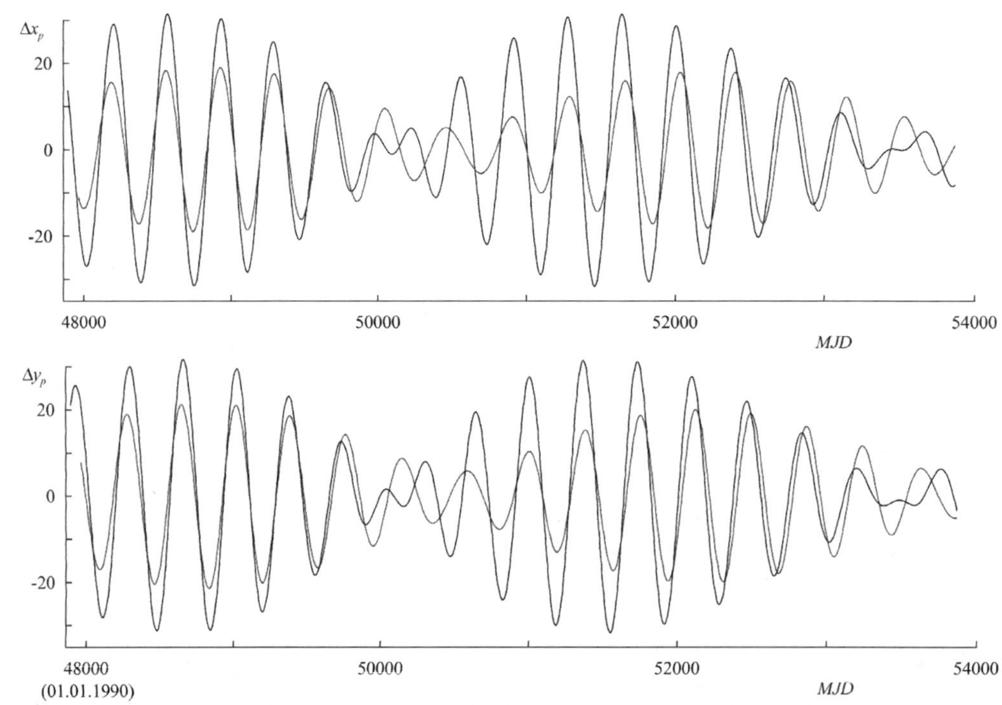

Figure 6. The studied oscillatory process of the Earth pole, extracted from the observational data (black line) and Earth pole motion, calculated taking into account the perturbations from the atmosphere and the ocean (blue line)

4. Conclusion

Thus, it was established in the work that more than 50% of the energy of the considered oscillatory process, that is in phase with the lunar orbit precession can be caused by perturbations in the atmosphere and the ocean. Oscillations with frequencies 1 ± 0.05373 cycle per year, caused by the atmospheric and oceanic perturbations coincide in phase with the corresponding oscillations of the observed pole motion. However, the amplitude ratios of these harmonics in the observed motion and the motion calculated when taking into account geophysical perturbations turn out to be different. This leads to a phase shift of their amplitude modulation.

At least a third of this process is not explained by the perturbations taken into account in the work. Since the perturbations from other geophysical fluids have a much smaller effect on the Earth pole motion, various Earth's media can be affected by such oscillations. Then this process is more global in nature, and its excitation in the said media can be caused mostly not by the internal perturbations but by the disturbances external to the Earth.

Acknowledgments

This work was carried out within the state task no. 9.7555.2017/BCh.

References

[1] Zlenko A A 2019 *J. Phys.: Conf. Ser.* **1301** 012022
[2] International Earth Rotation and Reference Systems Service - IERS Annual Reports, http://www.iers.org.
[3] Munk H and MacDonald G 1960 *The rotation of the Earth* (University Press) 323
[4] Markov Yu G, Perepelkin V V, Filippova A S and Rykhlova L V 2019 *Astron. Rep.* **63** (3) pp 238-247
[5] Zhou Y H, Salstein D A and Chen J L 2006 *J. Geophys. Res.* **111** D12108
[6] IERS Special Bureau for the Oceans, https://euler.jpl.nasa.gov/sbo/sbo_data.html
[7] Markov Yu G, Perepelkin V V and Filippova A S 2017 *Dokl. Phys.* **62** (6) pp 318-322
[8] Barkin M Yu, Markov Yu G, Perepelkin V V and Filippova A S 2018 *IOP Conf. Series: Materials Science and Engineering* **468** 012006
[9] Krylov S S, Perepelkin V V and Filippova A S 2019 *Advances in Theory and Practice of Computational Mechanics: Smart Innovation, Systems and Technologies* 173 eds. Jain L C et al. (Springer Nature Singapore) pp 315-331
[10] Perepelkin V V 2020 *Mech. Solids.* **54** (4) pp 584-588
[11] Klimov D M, Akulenko L D and Kumakshev S A 2013 *Dokl. Phys.* **58** (11) pp 505–509
[12] Kumakshev S A 2018 *Mech. Solids.* **53** (2) pp 159–163
[13] Klimov D M, Akulenko L D and Kumakshev S A 2014 *Dokl.Phys.* **59** (10) pp 472–475
[14] Schubert G 2007 *Treatise on Geophysics* vol. 3 (Amsterdam: Elsevier)

IOP Conf. Series: Materials Science and Engineering **927** (2020) 012037 doi:10.1088/1757-899X/927/1/012037

A Godunov-type method for a multi-temperature plasma with the tabulated equation of state

Alexey G Aksenov

Institute for Computer Aided Design, Russian Academy of Sciences, 2-nd Brestskaya, 19/18, 123056 Moscow, Russia

E-mail: aksenov@icad.org.ru

Abstract. A multi-temperature code for a multi-component gas dynamic is considered. The velocities of components with nonzero mass are assumed to be identical to each other. The gas dynamic part is a Godunov-type method based on the efficient approximate solution of the Riemann problem operating with all components of the gas mixture. The method assumes the arbitrary table equation of state, but the system of the hydrodynamic equations should be hyperbolic. An arbitrary equation of state (EOS) can contain domains with a negative square of the sound speed $c^2 \equiv (dP/d\rho)_s$ at phase transitions. In that domains the gas enthalpy should be corrected to provide the nonnegative sound speed square. This work contains the test of the method on a strong shock wave in hydrogen plasma.

1. Introduction

A special interest in physics is concerned with problems for a multi-component gas of different substances α described by a set of densities $\rho_\alpha(\mathbf{r},t) \equiv c_\alpha(\mathbf{r},t)\rho(\mathbf{r},t)$, where c_α are concentrations, $\rho\epsilon_\alpha(\mathbf{r},t)$ are internal energy densities, ϵ_α is specific energy. All massive particles (atoms, ions, nucleons) have identical velocities $\mathbf{v}(\mathbf{r},t)$ and temperatures, while the "massless" fast particles from viewpoint of the total density ρ (like electrons, photons) have their own temperatures. The equations of state are $P = \sum_\alpha P(\rho,\epsilon_\alpha)$, $\epsilon_\alpha = \epsilon_\alpha(\rho,T_\alpha)$. The components can exchange energy, can transfer energy by heat conduction not associated with the velocity of the massive particles, and can participate in reactions. Such problems arise in inertial thermonuclear fusion [1], laser ablation experiments [2], [3], and astrophysics [4]. This is an intermediate case between the description based on the Boltzmann equations for the one particle distribution function $f_\alpha(\mathbf{r},\mathbf{p},t)$ and classical single component gas dynamics. The basic mathematical problem concerns the gas dynamic part because of the discontinuities in the solution.

Efficient Riemann problem solvers for such hydrodynamic equations are constructed, see e.g. [5, 6, 7] for some special case of equation of state (EOS). Below an original method based on the Riemann problem solver for the multi-temperature non-equilibrium gas [8, 9, 10] is briefly described. The method was applied within the plasma physics for the inertial heavy ion fusion [1], [11], for the laser ablation [8], and useful in astrophysical tasks with hydrodynamic and the radiation transfer [12, 13]. Each component in the mixture has its own density and specific internal energy. In the local model for the EOS proposed, it is assumed that the entropy variations in neighboring mesh cells are small at the evaluation of the dimensionless coefficients EOS from the pressure jump across the discontinuity. In the case of an arbitrarily large pressure jump, the model yields physically reasonable results. However, the algorithm for solving the Riemann problem should be tested for correctness. It was assumed that the Riemann solver deals with small pressure jumps, but in real cases the pressure jumps are not small in nearest cells of the computational grid. The simplest multi-temperature shock wave (SW) structure in plasma was considered in [14]. This solution is a suitable test for the technique developed. previous This work prevents some disadvantages of the strong SW test from the view point of the physics in the last publications [8, 9, 10]. Also the test demonstrate some mistakes of the solution by Shafranov [14].

The only alternative code in the literature is CASTRO [15], which uses a different approximation for the Riemann problem solver. Other numerous gas dynamical models for mixtures with the separate descriptions of the gas dynamics of the matter and "massless" fast particles are very limited by the computational power in the explicit multidimensional tasks. In an optically dense, opaque region, the number of time steps for the computations of the energy exchange between fast particles of various energies will be determined by the opacity for the fastest reaction. In a joint treatment of the matter and fast particles, the transport only weakly affects the thermal distribution, and reactions computed using an implicit scheme in individual steps does not influence the number of time steps required for the gas dynamical transport.

2. Formulation of the problem and the numerical method

The system in the fixed Euler coordinates is: the mass transfer equations for the components

$$\frac{\partial \rho_\alpha}{\partial t} + \mathrm{div}\rho_\alpha \mathbf{v} = \rho \dot{c}_\alpha, \tag{1}$$

the momentum conservation law

$$\frac{\partial \rho \mathbf{v}}{\partial t} + \mathrm{Div}\Pi = 0, \tag{2}$$

and the energy density equations

$$\frac{\partial \rho E_\alpha}{\partial t} + \mathrm{div}(\rho E_\alpha + P_\alpha)\mathbf{v} + \mathbf{v}(c_\alpha \mathrm{grad}P - \mathrm{grad}P_\alpha) = \mathrm{div}(\kappa \mathrm{grad}T_\alpha) + \rho Q_\alpha, \tag{3}$$

where the energy densities $E_\alpha = \epsilon_\alpha + c_\alpha \mathbf{v}^2/2$, the tensor $\Pi_{ij} = \rho v_i v_j + P\delta_{ij}$, and the equation of state $P = \sum_\alpha P_\alpha(\rho, \mathbf{c}, \epsilon_\alpha)$ with specific energies $\epsilon_\alpha(\rho, \mathbf{c}, T_\alpha)$. The kinetic coefficients $\dot{c}_\alpha$, κ_α, and Q_α depend on $\rho, \mathbf{c}$, and $\mathbf{T}$. The problem is computed by applying dimensional splitting. The heat conduction equations are solved using central difference approximations. As a result, the system of partial differential equations (PDEs) is reduced to an ordinary differential equations (ODEs) system for $\dot{\epsilon}_{\alpha,i}$. The ODEs system is solved by applying the implicit Gear's method [16]. To describe the kinetics of reactions, the ODEs system for $\dot{\rho}_\alpha$ and $\dot{\epsilon}_\alpha$ is solved in each grid cell also by Gear's method. The transport of "massless" particles in both transparent and opaque cases can be described in the frame of diffusion with flux limiters.

The system in Euler variables involves the term $\mathbf{v}(c_\alpha \mathrm{grad}P - \mathrm{grad}P_\alpha)$, which is different from the divergence of a flux. This term requires a special treatment at discontinuities. The same non-conservative term is involved in the Lagrangian coordinates ($m = \int_{r0}^{r} \rho(\xi, t)d\xi, t$) only when the temperatures of the components are different. Generally speaking the term $\mathrm{grad}P_\alpha$ at a discontinuity is not defined. In the classical problem of SW in a hydrogen plasma, one has only three conservation laws and, due to heat conduction, a piecewise smooth temperature, for which a differential equation can be used instead of a conservation law.

The hydrodynamic part of the code is based on a high order explicit Godunov scheme for single-temperature single-component gas dynamics [8, 17]. A local model for the EOS simplifies the solution of the Riemann problem and makes it possible to obtain fluxes and partial pressures

IOP Conf. Series: Materials Science and Engineering **927** (2020) 012037 doi:10.1088/1757-899X/927/1/012037

of the components in any flow region with discontinuities. Following [18], such model for a multi-component gas is constructed so as it holds strictly in the case of weak discontinuities. The increment of the specific entropy s across SW is a quantity of the third order of smallness with respect to the pressure jump: $\mathcal{O}([P]^3)$. Neglecting the entropy variation behind the SW, one computes the dimensionless coefficients $\gamma_\alpha \equiv P_\alpha \tau/\epsilon_\alpha + 1$, as functions of the state ahead of the SW and the total pressure behind the SW. The local model for the EOS is used to solve the Riemann problem with constant values specified initially to the left and to the right of $x/t = 0$. On the left and right of the contact discontinuity, the concentrations remain constant, the EOS is independent of c_α. Using the relation $d\epsilon_\alpha(s_\alpha, \tau) = T_\alpha ds_\alpha - P_\alpha d\tau$ and the assumption $ds_\alpha = 0$, one can write

$$\frac{d\gamma_\alpha}{dP_\alpha} = \frac{\tau}{\epsilon_\alpha} + \frac{P_\alpha}{\epsilon_\alpha}\frac{d\tau}{dP_\alpha} - \frac{P_\alpha \tau}{\epsilon_\alpha^2}\frac{d\epsilon_\alpha}{d\tau}\frac{d\tau}{dP_\alpha} = \frac{\tau}{\epsilon_\alpha}\left(1 - \frac{\gamma_\alpha}{\Gamma_\alpha}\right) = (\gamma_\alpha - 1)\left(1 - \frac{\gamma_\alpha}{\Gamma_\alpha}\right)\frac{1}{P_\alpha}, \tag{4}$$

i.e., one obtains an explicit dependence of the increment of γ_α on the increment of the pressure of a component. The partial pressure increment depends explicitly on the total pressure increment:

$$dP_\alpha = \frac{C_\alpha^2}{C^2} dP, \tag{5}$$

where the squared Lagrangian speed of sound of a component is $C_\alpha^2 \equiv -dP_\alpha/d\tau = (\partial P_\alpha/\partial \epsilon_\alpha)P_\alpha - \partial P_\alpha/\partial \tau$, and total pressure is $P(\epsilon, \tau) = \sum_\alpha P_\alpha(\epsilon_\alpha, \tau)$. In the computations it is convenient to use the fraction of the specific energy of a component $\gamma_\alpha^\epsilon \equiv \epsilon_\alpha/\epsilon$:

$$d\gamma_\alpha^\epsilon = \frac{\epsilon d\epsilon_\alpha - \epsilon_\alpha d\epsilon}{\epsilon^2} = \frac{(\epsilon P_\alpha - \epsilon_\alpha P)dP}{\epsilon^2 C^2} = \frac{(\epsilon P_\alpha - \epsilon_\alpha P)\tau dP}{\Gamma P \epsilon^2} = \gamma_\alpha^\epsilon \frac{\gamma_\alpha - \gamma}{\Gamma}\frac{dP}{P}. \tag{6}$$

With notations $\Gamma_\alpha \equiv C_\alpha^2 \tau/P_\alpha$, $\Gamma \equiv C^2\tau/P$, and $d\epsilon_\alpha = -P_\alpha d\tau = -P_\alpha(d\tau/dP_\alpha)dP_\alpha = P_\alpha dP/C^2$ the increment of the dimensionless variable $\gamma \equiv \sum_\alpha \gamma_\alpha \epsilon_\alpha / \sum_\alpha \epsilon_\alpha$ can be evaluated (see [8])

$$d\gamma = (\gamma - 1)\left(1 - \frac{\gamma}{\Gamma}\right)\frac{dP}{P}, \tag{7}$$

as expected for a single component (4). A key point of the method and the main difference of the proposed gasdynamic approach for multi-component multi-temperature gases from [17, 18] lie in formula (6) and in the derivation of formula (7) for a multi-component gas. The assumption that the entropy variation is negligibly small is used to compute only the variations in the dimensionless coefficients. This assumption is not used in the computation of other parameters behind the SW.

Firstly, the local model for the EOS proposed resolves the uncertainty occurring when the specific internal energy and the pressure of a mixture component behind the SW (rarefaction wave) are computed from known values of $\gamma_\alpha^\epsilon \equiv \epsilon_\alpha/\epsilon$ behind the wave. Secondly, the local EOS with the a priory known dimensionless coefficients as the functions of the full pressure behind the wave reduce to the Riemann problem solver to the case of one temperature gas with EOS of the ideal gas [8, 10, 18]. When the energy equations for all components are added, the resulting difference scheme is conservative if the normalization $\sum_\alpha P_{\alpha *} = P_*$ is used.

3. Shock wave structure in hydrogen plasma

As a test, an SW structure in hydrogen arising in hydrogen plasma in a tube at rest with a piston moving with the constant velocity into the gas is considering. If the hydrogen is completely ionized, the system involves protons and electrons. The "massless" electrons transfer heat by conduction. Also electrons exchange energy with the protons. The temperatures of protons and electrons near the SW are different. The kinetic coefficients are specified as by Shafranov [14] $\rho\chi_{ie}(T_i - T_e) = (3m_e/m_p)(n_e/\tau_e)(kT_e - kT_i)$, where $\tau_e = 3\sqrt{m_e}(kT_e)^{3/2}/(4\sqrt{2\pi}\lambda q^4 n_e)$; the Coulomb logarithm $\lambda = \frac{1}{2}\ln\left(\frac{kT_eT_i}{T_e+T_i}\right)^3 \Big/ (q^6 n_e)$, $n_e = n_i = \rho/m_p$, and the heat flux in electron heat conduction is $-\kappa_e\nabla(kT_e)$ with thermal conductivity

$$\kappa_e = 3.16 n_e kT_e\tau_e/m_e. \tag{8}$$

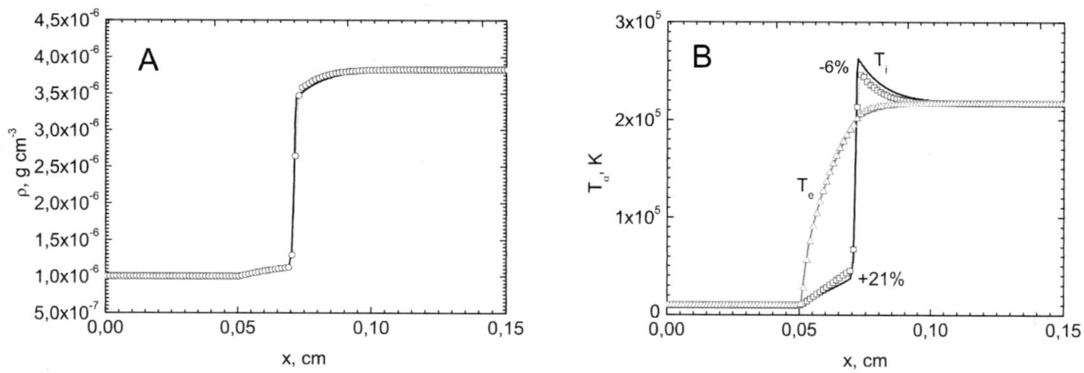

Figure 1. Density profiles (circles) (A), proton temperatures profiles (squares), electron temperature profiles (triangles) (B), and comparison with exact solution of Shafranov problem (solid curves) for strong SW for $M = 8$ at time moment $7.2\cdot 10^{-7}$ s . Numbers near SW on the right plot show relative accuracy of numerical solution.

An ideal monoatomic gas for protons and electrons with $\Gamma_{i,e} = 5/3$ is used in EOS. The initial hydrogen pressure is selected approximately equal to the atmospheric pressure. The initial temperature is chosen so as to achieve the ionization. The hydrogen is half ionized when the initial temperature is 10^4 K and the density is 10^{-6} g$\cdot$cm^{-3}. The Saha ionization equation for hydrogen $n_e n_p/n_H = (m_e kT)^{3/2}/(2\pi\hbar)^3 g_e \exp(-I/(kT))$, with the ionization potential $I = 13.6$ eV provides the ionized hydrogen at rather low temperature $kT \ll I$.

The electroneutrality condition is assumed to hold, the concentrations and velocities of the electrons and protons everywhere coincide, the jump of charge on the SW is neglected. Shafranov [14] worked with relations on discontinuities and solved a system of ODEs from both sides of the discontinuity. Due to the piecewise-smooth temperature of electrons $(-\partial T_e/\partial x)_1 > (-\partial T_e/\partial x)_2$ (Fig. 1) the heat fluxes are different on an SW, because the independence of the conduction coefficient, κ_e (8) from the concentration. Shafranov [14] does not take into account the heat conduction on SW. Only PDEs can operates with the different heat fluxes. The easiest way is to solve the system of PDEs in Lagrangian coordinates for density, momentum, and specific energies of protons and electrons on fine computational grid, see details in [10]. To exclude discontinuities, we need to introduce the viscosity protons passing from the hyperbolic system of equations to a parabolic one. In the present calculations we are not interesting by the fine structure of the SW due to finite protons viscosity, and artificially reduced the physical protons viscosity. The SW width is larger than the heat conduction transfer region on the factor $\sqrt{m_e/m_p}$.

The velocity of the gas injected at the right boundary (the velocity a moving piston) is chosen so as to obtain a strong stationary SW on which the density jump near to $\rho_2/\rho_1 = 4$. In the previous publication [10] we selected the piston velocity to $-4 \cdot 10^7$ cm $\cdot$ s^{-1} and obtained the pressure $5.37 \cdot 10^8$ din $\cdot$ cm^{-2} and the temperature $8.11 \cdot 10^5$ K behind SW. The black body radiation pressure with such temperature is $\frac{4}{3}\frac{\pi^2}{60(\hbar c)^3}(kT)^4 = 1.2 \cdot 10^9$ dyn$\cdot$cm^{-2} and exceeds the gas pressure. Therefore, for a correct physical formulation of the problem, one need to introduce photons and take into account their energy transfer and energy exchange with electrons or, alternatively, the velocity on the right boundary and the SW intensity have to be reduced. The estimations give for the comptonization of photons on electrons the time scale $(\sigma_{\mathrm{T}} n_e c)^{-1} \sim 10^{-4}$ s and the free path of electrons $(\sigma_{\mathrm{T}} n_e)^{-1} \sim 10^6$ cm. This means, the solution from [10] is quasi stationary at considered scales $\sim 10^{-7}$ s of time and space ~ 10 cm.

In present calculations we prefer to reduce the velocity of the piston to value $-2 \cdot 10^7$ cm$\cdot$s^{-1} to have the strongly correct physical task about the steady SW stricture in hydrogen plasma. Behind the SW, the gas velocity is $-1 \cdot 10^7$ cm $\cdot$ s^{-1}, the pressure is $1.37 \cdot 10^8$ din $\cdot$ cm^{-2}, and the temperature is $2.17 \cdot 10^5$ K. The density behind the SW is $3.83 \cdot 10^{-6}$ g $\cdot$ cm^{-3} and its Mach number is 8.

The numerical solution of the problem is shown in Fig. 1. A numerical grid contains 8000 intervals on the region 10 cm. In the direction of motion of the gas, a stationary SW is formed, propagating relative to the unperturbed gas. The profiles of all quantities near the SW shift at a constant velocity and remain unchanged. The plasma is in a non-equilibrium state near the SW, but equilibrium is established at some distance behind the SW front. Jumps in the proton density and temperature are observed on the SW. Due to electron heat conduction, the electron temperature T_e is continuous and piecewise smooth. Fig. 1 demonstrates good agrement of the numerical solution with the "exact" solution obtained in Lagrangian coordinates.

4. Conclusions

Proposed approximate Riemann problem solver yields qualitatively and also quantitative good results despite the assumption for the entropy jump smallness used in computation of the dimensionless coefficients of the gases.

References

[1] Basko M M, Churazov M D and Aksenov A G 2002 *Laser and Particle Beams* **20** 411–414
[2] Anisimov S I, Zhakhovskiĭ V V, Inogamov N A, Nishihara K, Petrov Y V and Khokhlov V A 2006 *JETP* **103** 183–197
[3] Fortov V E, Hoffmann D H and Sharkov B Y 2008 *Physics Uspekhi* **51** 109–131
[4] Bruenn S W 1985 *ApJS* **58** 771–841
[5] Pelanti M and Shyue K M 2014 *Journal of Computational Physics* **259** 331–357
[6] Zhukov V T, Zabrodin A V and Feodoritova O B 1993 *Zhurnal Vychislitelnoi Matematiki i Matematicheskoi Fiziki* **33** 1240–1250
[7] Miller G H and Puckett E G 1996 *Journal of Computational Physics* **128** 134–164
[8] Aksenov A G 2015 *Computational Mathematics and Mathematical Physics* **55** 1752–1769
[9] Vereshchagin G and Aksenov A 2017 *Relativistic Kinetic Theory With Applications in Astrophysics and Cosmology* (Cambridge University Press)
[10] Aksenov A G, Chechetkin V M and Tishkin V F 2019 *Mathematical Models and Computer Simulations* **11** 360–373
[11] Aksenov A G and Churazov M D 2003 *Laser and Particle Beams* **21** 81–84
[12] Aksenov A G and Chechetkin V M 2018 *Astronomy Reports* **62** 834–839
[13] Aksenov A G and Chechetkin V M 2019 *Astronomy Reports* **63** 900–909
[14] Shafranov V D 1957 *Sov. Phys. JETP* **5** 1183–1188
[15] Dolence J C, Burrows A and Zhang W 2015 *Astroph. J.* **800** 10 (*Preprint* `ArXiv: 1403.6115`)
[16] Gear C W 1971 *Numerical initial value problems in ordinary differential equations* (Prentice-Hall, Inc. Englewood Cliffs, New Jersey)
[17] Colella P and Woodward P R 1984 *Journal of Computational Physics* **54** 174–201
[18] Colella P and Glaz H M 1985 *Journal of Computational Physics* **59** 264–289

Packaging and unfolding of shells using highly elastic joints

V M Pestrenin[1,2], I V Pestrenina[1], S V Rusakov[1], V G Gilev[2] and G V Buzmakova[1]
[1] Department of Mechanics and Mathematics, Perm state national research university, Bukireva street 15, Perm, Russia
[2] Department of Physics, Perm state national research university, Bukireva street 15, Perm, Russia

E-mail: pestreninvm@mail.ru

Abstract. A computational-experimental method for studying the packaging and unfolding of shell structures made of composite material is proposed. The construction consists of rigid elements interconnected by elastic hinges made of uncured prepreg. This material is elastic and therefore it can store an elastic energy during the structure packaging, and later a stored energy will facilitate unfolding. The hinge's material is modelled by a hyper elastic isotropic medium, the Neo-Hookean potential is applied. The parameters of the potential are determined by the computational-experimental method. The viscous and viscoelastic characteristics of the binder are determined by the rheological method with a Physica MCR 501 rotational rheometer. The calculated effective properties of the hinge's material are used to demonstrate the packaging of the cylindrical shell. Three different packaging methods for a cylindrical shell with six prepreg hinges are considered. The article studies the dependence of the load required for packaging on the type of packaging and the curing degree of the binder.

1. Introduction

Space exploration involves the creation of large-sized shell structures on the Earth's orbit. These can be, for example, greenhouses, warehouses, energy storage devices, reflectors of telescopes and antennas, etc. In the transportation state, such structures should be packed as compact as possible on the ground with followed unfolding in space.

The possibility of manufacturing a structure from a composite material such as a prepreg was considered in [1,2]. In space, it pneumatically unfolds from a packed state, takes the necessary shape and becomes rigid due to the binder polymerisation. The mechanical state of the shell loaded by internal pressure was described in earlier author's articles.

The prepreg polymerisation in space under the action of the solar radiation energy was studied in [3]. In this publication, packaging and pneumatic unfolding of large-sized shells made of composite elastic elements fastened with ideal hinges where considered. It was shown that this method of manufacturing and unfolding of structures leads to undesirable stresses in its elastic elements. In [4], a different approach was proposed for the manufacture of packaging structures with elastic joints from a curing prepreg. In the elastic joints of such structures, elastic energy is accumulated during packaging in order to be used in further self-unfolding. The hardening of the hinge as a result of polymerization binds makes it rigid, suitable for use.

In this paper, we demonstrate the results of studying of packaging shells with hinges made from a curable prepreg in different hardness states using numerical-experimental method.

2. Sample preparation for the experiment

The construction that is shown in Figure 1 was used for an experimental studying of the hinges mechanical properties (curable prepreg) [5]. It was made from the preform-hose with carbon thread twill weaving. The "soft preform-hose" was pulled onto a solid form with the required geometric parameters, then the workpiece faces were partially soaked with a binder to obtain a semi-finished product in the combination of soaked flat sections separated by not soaked ribs. Impregnated sections become cured; ribs between them become elastic hinges. You can see cured elements – 1 and elastic sections (hinges) – 2 in figure 1b.

3. Experimental devices

The curing time and viscous and viscoelastic characteristics of the binder were determined by the rheological method. The measurements were performed by a Physica MCR 501 rotary rheometer. The cone-plate system was used in all measurements. The cone diameter d = 25 mm, the angle α = 1°. Such geometry ensures uniformity of the shear rate gradient in the measuring range.

To maintain the temperature regime, a special H-PTD 200 temperature device based on the Peltier effect was used. During the experiment, about 0.07 ml of the solution was placed on the working surface of the rheometer plate. The thickness of the solution layer along its outer radius was 0.047 mm, which ensured the fast establishment of the operating temperature even under heat release conditions caused by the polymerization reaction.

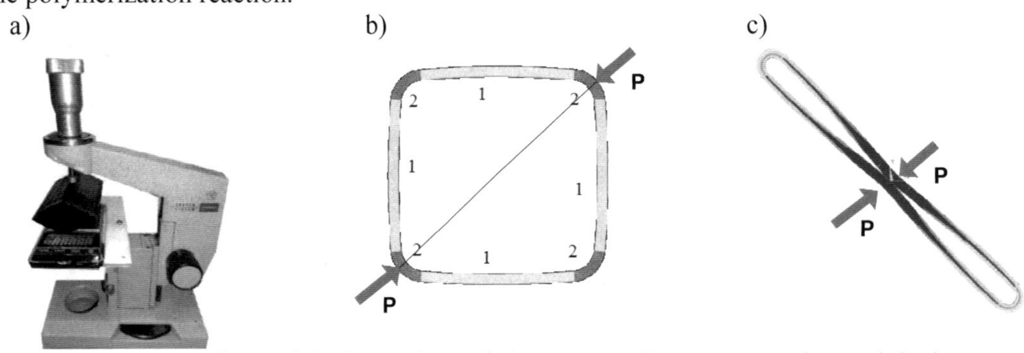

Figure 1. Experimental device and sample's cross-section: a – experimental device; b – section of a four-hinged carbon-fiber shell in an equilibrium state; c – the section in a deformed state under load

4. Computational-experimental method for studying the deformation of shell structures with hardening hinges

The effective properties of the cured elements were determined by the methods of mechanics of composite materials. When assigning mechanical characteristics to hinges, the following has taken into account. The binder at the initial stage of curing exhibits the properties of a viscous body, which mainly resists shear. The reinforcing elements (fibers) of the joints at the considered stage of binder curing undergo large displacements but small strains so we suppose they are nonlinearly elastic. These properties of the hinge material (prepreg) modelled by a hyperelastic isotropic medium with Neo Hookean potential

$$w(\varepsilon_{ij}) = \frac{\mu}{2}(I_1 - 3) + \frac{K}{2}(J-1)^2, \tag{1}$$

where μ – is the initial shear modulus, K – is the bulk modulus of elasticity, I_1 – is the first invariant of the Cauchy-Green strain tensor, $J = dV / dV_0$ – is the ratio of elementary volume in the current state to the initial one. The Poisson's coefficient does not change and is equal $\nu = 0.48$. The bulk module K corresponding to the current degree of prepreg curing is determined through μ and ν of linear elasticity formulas.

The parameter μ in the potential representation (1) is determined in the numerical solution and two parallel experiments. In the first experiment, a four-hinged cylindrical box-shaped shell subjected to mechanical tests (figure 1a). Here we fix value P at the position, shown on figure 1c. In a second experiment a dynamic shear modulus is determined at the same time. For this purpose a rotational rheometer is used. A compound of both results is shown in Figure 2.

The numerical solution of the same problem with the current degree of binder curing is carried out by the method of successive approximations. It is built from the initial state (figure 1b) by successive loading with force P until full preload (figure 1c). The initial state is already tensed. A natural state (unstressed and unstrained) of this construction is a plate with inclusions modelling hinges. Gluing the plate's ends into a ring we obtain the desired initial stress state.

The parameter μ is determined from the condition that the calculated load P coincides with its experimental value. The calculation results for the dependence the effective parameter μ on the current binder shear modulus are presented in Figure 3.

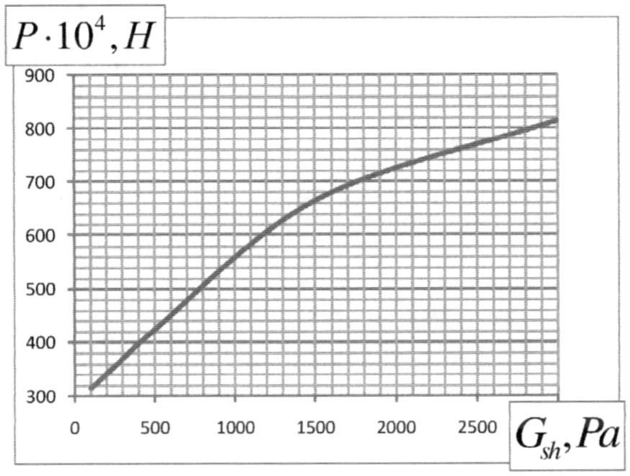

Figure 2. Dependence of the value of the full preload on the dynamic shear modulus of the binder

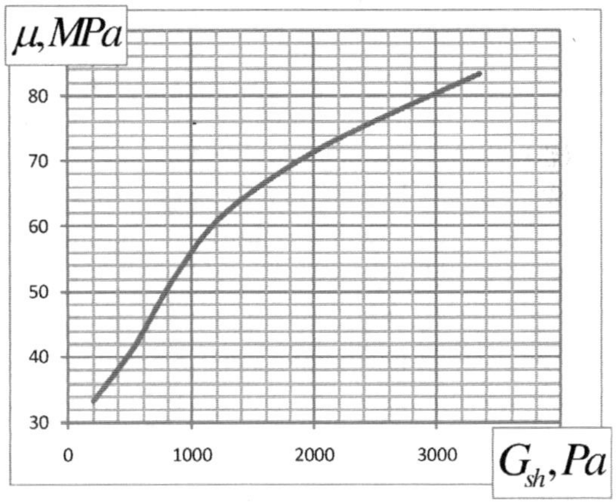

Figure 3. Dependence of the effective parameter μ on the dynamic shear modulus of the binder

5. Packaging of a cylindrical shell with six prepreg hinges

A cylindrical shell with six hinges is considered. The cross section of the shell is shown in Figure 4. The natural state of the hinges (unstressed and unstrained) is straight, therefore, in the state shown here, the shell elements are at the stress state.

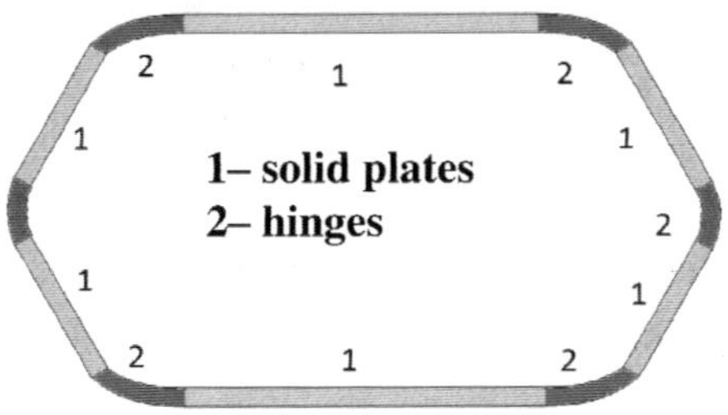

Figure 4. Cross section of six-hinged shell

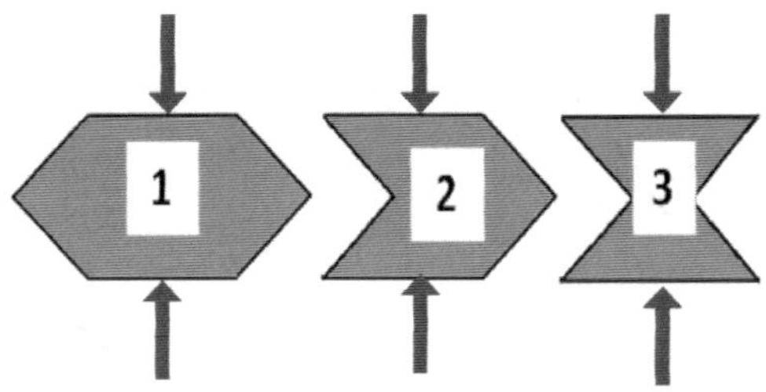

Figure 5. Schemes of packaging of six-hinged shell

The packaging process of the shell is the bringing the product from state at figure 4 to any states shown schematically in Figure 5. To solve the problem, the results given in the previous paragraph are used. In Figure 6, numerical solutions are compared with experimental data for all shell packaging variants shown in Figure 5.

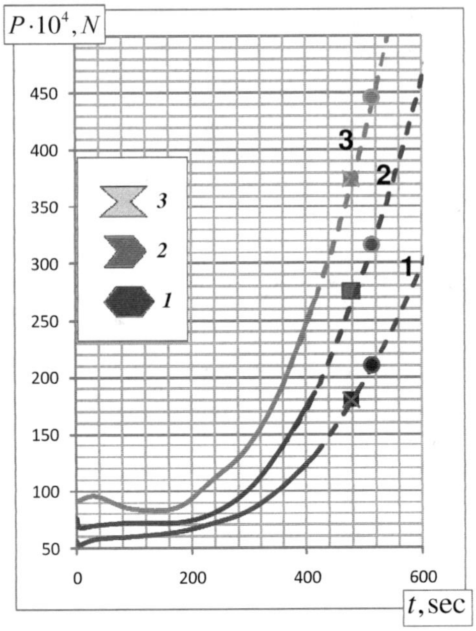

Figure 6. Dependence of the load providing full packing of the six-hinged shell in three versions: 1– in the first variant; 2– in the second; 3 - in the third.
The solid lines correspond to the experimentally obtained data; the dashed lines correspond to their extrapolations. Numerical solutions for all three packaging cases were obtained for two values of the binder curing degree. The squares indicate the calculation results for the current dynamic binder shear modulus of 4948 Pa, and the circles for 12520 Pa.

In figure 6 one can see that numerical estimates of the load required for packaging the structure are consistent with experimental data in all cases.

Note that the required load depends both on the curing degree of the binder and on the packaging option. In particular, the packaging of the structure in the first variant is carried out with a half-load in comparison with the third variant.

6. Conclusion

The results of the study are following:

1) the calculation-experimental method is proposed for determining the effective mechanical characteristics of a hyperelastic prepreg with a current degree of cure;

2) the method for packaging and unfolding of shell composite structures containing hinges from a prepreg is studied;

3) an assessment of the efforts required for three different packaging options for a cylindrical shell with six prepreg hinges is given. The dependence of such efforts on the type of packaging and the degree of curing of the binder was studied.

The calculation-experimental method proposed in this work may be useful in constructing of packaging and unfolding big-size shell structures for space applications.

References

[1] Kondyurin A, Lauke B, Kondyurina I and Orba E 2004 Creation of biological module for self-regulating ecological system by the way of polymerization of composite materials in free space *Advances in Space Research* **34** 1585–91

[2] Kondyurina I, Kondyurin A, Lauke B, Figiel Ł, Vogel R and Reuter U 2006 Polymerisation of composite materials in space environment for development of a Moon base *Advances in Space Research* **37** 109–15.

[3] Kondyurin A, Rusakov S V, Pestrenin V M and Pestrenina I V 2018 Curing of large prepreg shell in solar synchronous Low Earth Orbit:Precession flight regimes *Acta astronavtica* **151** 342–47 DOI: 10.1016/j.actaastro.2018.06.029

[4] Gilev V G, Rusakov S V, Pestrenin V M and Pestrenina I V 2018 Estimation of the cylindrical composite shell stiffness at the initial stage of curing during deployment by internal pressure *PNRPU Mechanics Bulletin* **1** 93–9 DOI : 10.15593/perm.mech/2018.1.08
[5] Gilev V G and Rusakov S V 2019 A method of manufacturing a self-expanding bulk product from composite material *Patent for invention 2697463, 08.14.2019*

Effect of ablation at the inlet of submerged nozzle on discharge characteristics

A N Kirillova and A N Sabirzyanov
Kazan National Research Technical University named after A N Tupolev – KAI, 15 Karla Marksa str, Kazan 420111, Russia

E-mail: ansabirzyanov@kai.ru

Abstract. The effect of ablation on the discharge characteristics of solid rocket engines with submerged nozzles is examined. State-of-the-art computational fluid dynamics software package ANSYS Fluent is employed for simulation. The ablation is considered from the standpoint of the effect caused by uniform blowing of degradation products on the boundary layer developing over the entrance length of the submerged nozzle. Different masses of injected substance are considered. The impact of injection on velocity profiles is estimated. The discharge coefficient is affected by the non-isothermal boundary layer when degraded thermal protection material is injected and chemical reactions are neglected. The effect of injection on the profiles of velocity and temperature, and hence, on discharge characteristics is revealed.

1. Introduction

The effect of ablation of thermal protection materials covering the submerged part of a solid rocket engine nozzle on the gas dynamics is integral to estimation of losses in the engines. Due to complex nature of the interaction between high-temperature flows and thermal protection materials, it is reasonable to consider the components of this interaction separately and employ some simplifying assumptions. The boundary layer developing on the submerged part of nozzles features a wide diversity of processes, including the interaction between high-temperature combustion products and thermal protection materials (TPM), their heating, change of properties, degradation and ablation, physicochemical interaction between degradation products, scaffold material and oncoming flow, absorption and emission of heat due to the interaction, direct injection of degradation products, etc. The influence of TPM degradation on the discharge characteristics of engines can be estimated in terms of energy balance, according to which it is assumed that when combustion products interact with TPM, the supplied total heat flux is spent on heating of external and internal layers of material, surface radiation, melting and other physicochemical transformations [1].

Composite materials used for thermal protection in rocket engineering generally include a matrix filled with a reinforcement material of lower degradation temperature. Consequently, the properties of ablation products and the moment when the blowing of ablation products into main flow starts are determined by the reinforcement material reaching its degradation temperature. The research into modification of properties of thermal protection materials and composition of pyrolysis products [2, 3] provides the initial information on interaction of flows in the near-wall layer, energy expenditures on heating and degradation, and contribution of physicochemical interaction on the surface. Some considerations on the current state of knowledge about near-wall flows of chemically reacting

IOP Conf. Series: Materials Science and Engineering **927** (2020) 012039 doi:10.1088/1757-899X/927/1/012039

substances are provided in [4, 5]. Papers [1, 6-8] describe the mathematical models of TPM destruction and implementation of these models.

Injection of TPM degradation products makes the largest contribution to the gas dynamics of outflow. When the mass flow rate of ablation products is high enough, they can force the main flow away from the wall and affect the outflow process significantly. The main advantage of blowing is the reduction of supplied heat flux [1, 9]. Blowing along the whole converging part of the nozzle can distort the fields of velocity and temperature. Authors [9] note that the gradients of velocity and temperature near the wall are reduced when the blowing is performed. Thus, the flow temperature variation within the boundary layer implies that there is an extremum if chemical reactions are taken into account. The velocity profile exhibits a delayed transition to the core flow velocity due to combustion products being forced away from the wall and degradation products being added to the flow [4]. These processes modify the velocity profile in the nozzle throat, which plays a key role in the studies of discharge characteristics of nozzles. Ablation, including the one occurring in the nozzle throat, has a profound effect on intraballistic parameters of engines [8]. And while ablation of material in the throat leads to pressure reduction in the combustion chamber and requires compensation for the loss, the contribution from e.g. sublimation of inhibitory coating benefits the intraballistic parameters [10] and the discharge complex in particular.

The present paper focuses on estimation of the limit to which the ablation from the inlet part of the submerged nozzle affects the velocity profile and discharge characteristics by injecting the degradation products into the main flow.

2. Physical and mathematical models of the research. Acknowledgement

The injection of TPM degradation products from the surface of the inlet part of submerged nozzle was arranged in conditions of high-temperature impact under the following assumptions: degradation is quasi steady, ablation rate is constant, and no liquid film is formed during gasification of degradation products. Shrinkage and destruction of thermal protection material during filtration of pyrolysis gases were neglected, and hence geometry did not change during the simulation. Thermal equilibrium was assumed inside the pores, so the temperature of ablating wall was equal to the temperature of the supplied gas. Uniform blowing of gas was arranged in simulation similar to the main flow from the surface of the submerged nozzle. Combustion and ablation products were considered homogeneous, and the mass balance was a difference between the inflow from the charge surface and from the submerged nozzle part and the outflow from the nozzle.

The effect on gas dynamics in the engine was estimated using the variation of the discharge coefficient of the engine calculated as:

$$\mu = \dot{m}\left(A(k)p_0F_{\text{M}}/\sqrt{RT_0}\right)^{-1}, \tag{1}$$

where $\dot{m}$ is the actual flow rate specified on the charge surface and ablating surface; p_o is the stagnation pressure obtained from simulation; T_o, k and R are the stagnation temperature, ratio of heat capacities and equilibrium value of the gas constant in the nozzle throat, respectively; F_{M} is the throat area; $A(k) = \sqrt{k}[2/(k+1)]^{(k+1)/[2(k-1)]}$.

The comparison was made between the cases with and without blowing (different percentage of the flow rate from the inlet part of the submerged nozzle was considered). *ANSYS Fluent* software was employed for numerical simulation. Quasi-steady axisymmetric approximation with ideal gas assumption were considered. Navier-Stokes equations were used to simulate the turbulent flows of viscous compressible fluid. Two-equation RNG k-ε turbulence model was employed for closure. This model is sensitive to rapid deformation of flows and variable curvature of streamlined bodies. The grid size was approximately 10^5 cells. At the inlet to the submerged part of the nozzle, y^+ varied from 10 to 20. Coupled solver and second-order discretization scheme were used in simulation. The applicability of this approach is confirmed in [10].

The geometry under consideration is demonstrated in figure.1. The temperature of combustion products supplied from the charge surface was 2932 K, and the temperature of degradation products from the ablating surface equaled 2000 K. The required flow rate was provided by the pressure value of

4 MPa. Specific mass arrival from the surface of ablating TPM was 0, 0.01, 0.02, 0.03, 0.1, 0.2, 0.3, 0.6, and 1 kg/m^{2}s. The assigned specific mass arrival corresponds to the percentage of the flow rate from the charge surface: 0, 0.0007, 0.0014, 0.002, 0.007, 0.014, 0.021, 0.041, and 0.069 %. The following simulation cases were considered: case V1 – no blowing of degradation products at the inlet to the submerged part of the nozzle, adiabatic wall parameters, constant wall temperature T_w = 2000 K; V2 – uniform blowing at the inlet from the surfaces with the radius R_3 (contour B) and from ellipsoidal part R_2/R_4 over contour A according to the specified mass arrival; V3 – same as V2 but without blowing from the part intended for the throat insert. The submergence was $\bar{L}_{sub} = L_{sub}/L_{ch} = 0.1$.

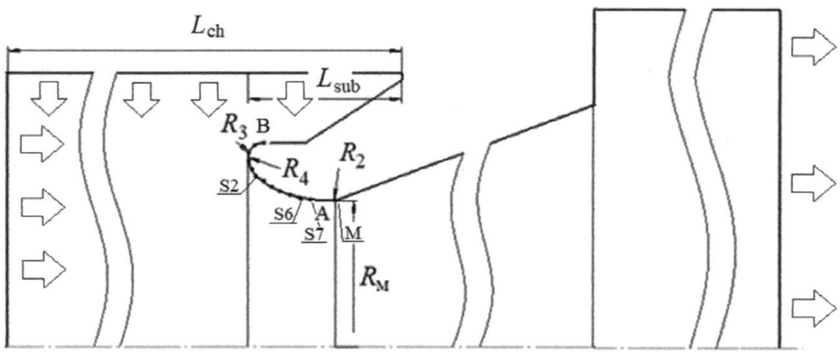

Figure 1. Computational model: R_M – radius of the nozzle throat; L_{ch} – charge length; L_{sub} – submerged length; R_2 – radius of the throat inlet, R_4 – stretch radius of the inlet part, R_3 – radius of the outer edge of the submerged part.

3. The discussion of the results

For the considered simulation cases, figure.2 (a, b) demonstrates the profiles of velocity according to the locations specified in figure.1. The curve names include the case number, the value of specific mass arrival (if any) and the location of the profile. E.g., V2-0.03-M curve was obtained for the simulation case V2 and specific mass arrival of 0.03 kg/m^{2}s in the nozzle throat (Figure 2.a). The results for case V1 are given under the assumption of adiabatic wall. The calculation procedure of the discharge coefficients for submerged nozzles under adiabatic assumption without blowing was tested in [10]. The mass arrival from the ablating coating to the submerged nozzle inlet promotes flattening of the velocity profile, which should be taken into account when optimizing the geometry of the submerged part of the nozzle [10]. Over the distance between the nozzle inlet and the throat, the profiles are progressively more extended near the wall, and the gradient of velocity variation is reduced there due to additional arrival of mass. Authors [11] note that turbulent regime is typical of the boundary layer at the inlet to the submerged nozzle and further downstream. However, for the accelerating flows [12], the relative reduction in heat and mass transfer coefficients is associated with the growing thickness of viscous sublayer and reduction of fluctuations in the turbulent flow. Therefore, the expected effect of uniform injection of compressible flow can lead to laminarization of the boundary layer, while the non-uniformity can result in turbulization. Case V2 demonstrates the effect of thickening boundary layer leading to inferior discharge characteristics. Meanwhile, in the case V3, despite the growth of the boundary layer from S2 to S6, S7 and further downstream to M, it "sticks" to the wall (imitating a noninflammable insert) in the nozzle throat, making the velocity profile uniform and contributing to the increase of the discharge coefficient. This effect corresponds to the curtain zone after the permeable section. In the region where ablation products are injected, the displacement thickness of the boundary layer grows, the profiles of velocity and temperature rearrange. The rearrangement of the temperature profile is caused by uneven temperature distribution resulting from temperature difference between the injected and main flows. Around the insert, the near-wall layer rearranges compared to the flow with blowing because of the changes in boundary conditions. The discrepancy of V2 and V3 profiles resulting from rearrangement of density and velocity in the boundary layer in the noninflammable zone should be noted as well as their dominant effect on the pattern of the upstream flow, which corresponds to larger difference in the discharge coefficient for V2 cases than for V3 and V1 cases.

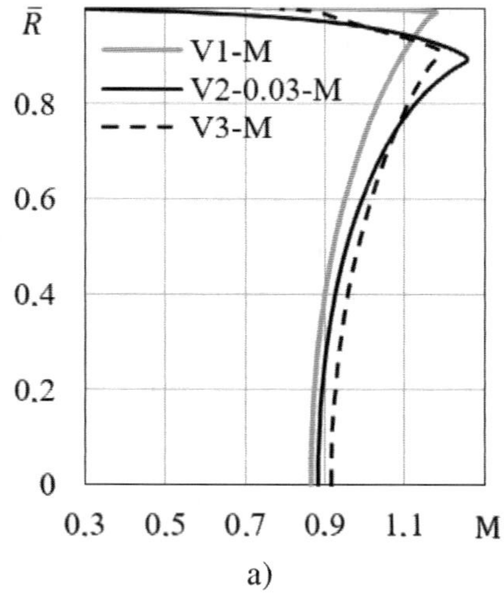

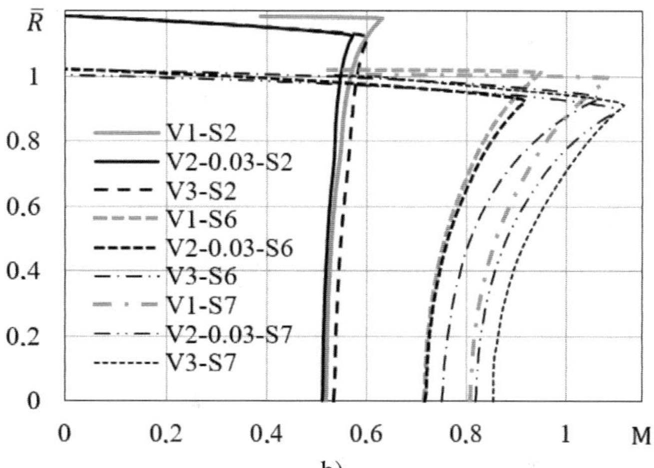

Figure 2. Profiles of normalized velocity.

The effect of non-uniform temperature on estimation of the discharge coefficient is reflected in taking into account the variation of temperature in the throat when the injection is performed (Figure 3). Figure 4 shows the behavior of the discharge coefficient depending on the specific discharge of the degradation products for two cases of discharge coefficient calculation (1): with and without taking into account the changes in temperature and k during the injection. The difference between these two cases can be as high as 0.9% and the pattern of curves does not change, which demonstrates the impact of non-isothermal flow on the discharge coefficient. To illustrate the effect of constant wall temperature (T_w = 2000 K), Figure 3 is supplemented with the temperature profile for the case V1 (V1-M-2000). This case does not promote significant changes in the discharge coefficient, which allowed us to make satisfactory comparisons for the case V1 with the adiabatic wall. For V2 cases, when the specific arrival of gas from the surface of submerged part of the nozzle changes, the variation of discharge coefficient is negligible. This contradiction is probably associated with the small mass of injected products uniformly pressed by the dominant combustion products.

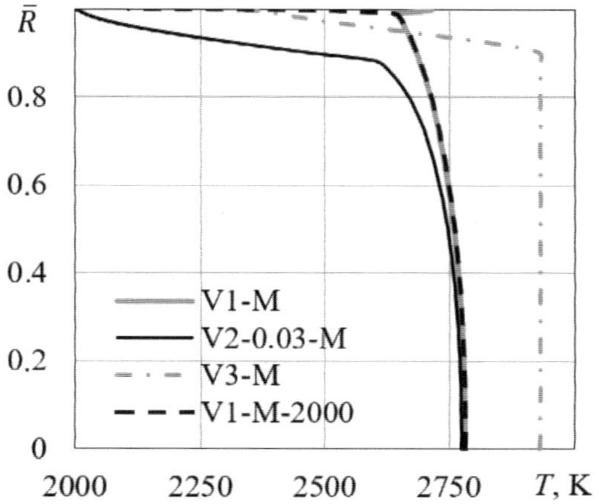

Figure 3. Temperature profiles in the nozzle throat.

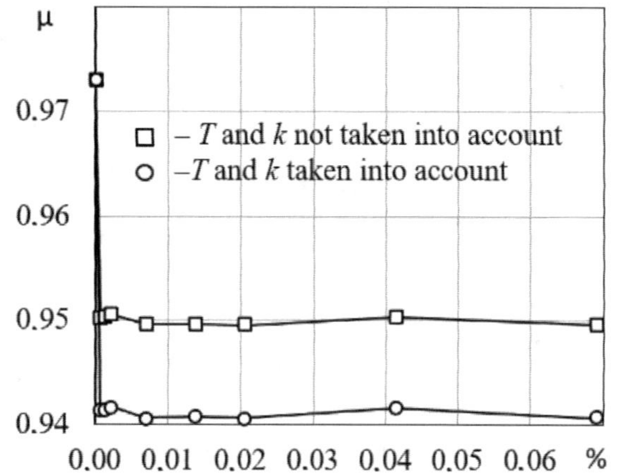

Figure 4. Discharge coefficient depending on the normalized discharge from ablating surface.

4. Conclusions

Our numerical results comply with the general idea of the boundary layer being forced away by the injected products of ablation, which results in lower gradients of velocity and temperature, and hence, thicker boundary layer. Based on our numerical experiments, we made the following conclusions. The paper provides qualitative assessment of the effect caused by blowing from the surface of the submerged nozzle on the discharge characteristics. The temperature of injected ablation products affects the discharge coefficient. The effect of non-isothermal boundary layer on the discharge coefficient when TPM degradation products are injected is estimated to lie within 1% if chemical reactions are neglected. Thermal protection materials applied to the submerged part of the nozzle define the temperature profile, and hence, the displacement thickness of the boundary layer, which should be taken into account when choosing the inlet geometry of the submerged nozzle. Apparently, if blowing of degradation products affecting the discharge coefficient is taken into consideration, the submergence of the nozzle does not influence the discharge coefficient. Improved discharge characteristics are observed in the case with impermeable zone in the throat if compared to the case with blowing over the whole submerged surface. However, the impact depends on whether the blowing is performed but not on the specific mass arrival.

Acknowledgement

The study was financially supported by the Russian Foundation for Basic Research (grant no.19-38-90277).

References

[1] Polezhaev Yu V and Yurevich F B 1976 *Thermal protection* Edited by A V Lykov (Moscow: Energia)
[2] Gribanov V M, Slobodchikov S S, Potapenko A I and Ulyanenkov R V 2012 Computational-experimental method for estimation of energy expenditure for physicochemical transformations in polymeric materials *Issues in Electromechanics* **130** 51-57
[3] Tovstonog V A 2007 Experimental studies and analysis into mechanisms of thermal breakdown of reinforced plastics heated by radiation *Herald of N.E.Bauman MSTU. Mechanical Engineering series* **2** 15-34
[4] Lukashov V V, Terekhov V V and Terekhov V I 2015 Near-wall flows of chemically reacting substances. State of the art review *Physics of combustion and explosion.* **51 2** 23-36
[5] Volchkov E P, Terekhov V I and Terekhov V V 2004 Flow structure, heat and mass transfer in boundary layers with injection of reacting substances (review) *Physics of combustion and explosion.* **40 1** 3-20
[6] Erokhin B T, Abramova N V and Gusev S V 2014 Mathematical model of TPM ablation *Herald of MSUII.* **55** 56-63
[7] Frolov G A 2003 Energy intensity of the main process of heat absorption during aerodynamic heating of thermal protection of rocket and space equipment. 1. Limit energy intensity of internal processes of heat absorption at thermal breakdown of material / *Space Science and Technology* **9** 2/3 58-67
[8] Zhang J, Jackson T L, Najjar F and Buckmaster J 2009 *High-Fidelity Multiphysics Simulations of Erosion in SRM Nozzles* 45th AIAA/ASME/SAE/ASEE Joint Propulsion Conference and Exhibit. (Denver, Colorado)
[9] Pancratov B M, Polezhaev Yu V and Rudko A K 1975 *Interaction between materials and gas flows.* Edited by V S Zuev (Moscow: Mashinostroenie) 224
[10] Sabirzyanov A N and Kirillova A N 2018 Multi-factor nature of the effect caused by nozzle submergence on the discharge coefficient *Herald of Almaz-Antey Corporation* 1 43–50 (in Russian)
[11] Shishkov A A, Panin S D and Rumyantsev B V 1988 *Operating cycles in solid rocket engines: Handbook* (Moscow: Mashinostroenie) 240

IOP Conf. Series: Materials Science and Engineering **927** (2020) 012039 doi:10.1088/1757-899X/927/1/012039

[12] Gubertov A M, Mironov V V, Borisov V N et al. 2004 *Gasdynamic and thermophysical processes in solid rocket propulsion* Edited by Academican A S Koroteev (Moscow: Mashinostroenie) 512

IOP Conf. Series: Materials Science and Engineering **927** (2020) 012040 doi:10.1088/1757-899X/927/1/012040

Development and application of combined methods to expand the potential for predicting the defining characteristics of composites

E L Gusev[1,2] , V N Bakulin [3] and V D Chernykh[4]
[1]Institute of Oil and Gas Problems of the Siberian Branch of the RAS, Yakutsk, Russia
[2]Institute of Mathematics and Informatics of the North-Eastern Federal University, Yakutsk, Russia
[3]Institute of Applied Mechanics of the RAS, Moscow, Russia
[4]Institute of Physics and Technology, North-Eastern Federal University, Yakutsk, Russia

E-mail: elgusev@mail.ru

Abstract. The issue of promising ways to expand potential opportunities for improving the effectiveness of methods for predicting the remaining life, durability, and reliability of composite materials and structures under the influence of extreme environmental factors is considered. Modifications of methods for searching for the global extremum of multiparametric functions have been developed and studied, taking into account the structural peculiarities of the efficiency criteria used in forecasting the remaining life and durability of structures made of composite and polymer materials. One of the promising ways to increase the potential opportunities has been analyzed. It is connected with the improvement of methods for constructing optimal directions for searching for the absolute minimum of efficiency indicators related to solving the problems of predicting the determining characteristics of polymer composites under the influence of extreme environmental factors and operational loads.

1. Introduction

Polymer materials (PM) and polymer composite materials (PCM) have a wide variety of properties and are widely used in all spheres of human life, including in extreme conditions [1-9]. PM and PCM are considerably used in such modern areas of technology as aviation, space technology, shipbuilding, oil and gas industry, etc. When choosing the best composite material for a specific use in industry, construction, and transport, the complex problem of multiple choice arises. The problem is to choose from a large number of possible options the composition that provides the most effective combination of the necessary properties. At the same time, one of the most important advantages of a composite material is its ability to maintain its performance as long as possible, that is, to resist aging – a combination of physicochemical reversible and irreversible transformations under the influence of aggressive external factors (temperature, humidity, solar radiation, mechanical stresses, etc. [2].

The combination of the required material properties, acceptable cost, as well as the ability to resist aging are the basis for making a decision on the use of a composite material. At the same time, such PM and PCM dominate, which are able to function without replacement for 20-30 years, and in necessary cases for more than 50 years. In accordance with this, the need to justify the long-term performance of PM and PCM increases the relevance of climate tests [10-13]. Currently, the main problem is that there are no universal programs for conducting climate tests. Usually, in practice, programs are used taking into account only two factors: material composition and operating conditions. One of the most common approaches is to find the best option from a set of similar materials. In this case, comparative tests of candidate materials are carried out using accelerated methods in climate chambers for modes with increased aggressiveness, and as a result of the tests, the most resistant material is selected. The advantage of this approach is that it is operational. The disadvantage of this approach is that there is a need to adjust the obtained conclusions when switching from excessively aggressive test modes in climate chambers to real operating conditions, where the mechanisms and kinetics of aging may differ significantly [14].

2. Materials and methods

Models describing the dependence of changes in the defining characteristics of R polymer composites under the influence of climate factors in General can be presented in the form of functional dependencies of the following type:

$$R(u_1,u_2,...,u_n;t)=R_0+\Phi(u_1,u_2,...,u_n;t). \tag{1}$$

Here R is the studied defining characteristic of the composite, R_0 is the value of the defining characteristic of the composite in the initial state; $u_1,u_2,...,u_n$ - composite parameters that reflect the influence of extreme environmental factors.

In General, polymer, composite materials, and composite structures can be affected in various combinations simultaneously by several different F_j factors, related to hardening processes, solar radiation, moisture saturation, ultraviolet radiation, extreme factors, and operating loads of both cyclic and non-cyclic nature, etc. Then, assuming that various physical factors F_j have an effect on the polymer composite that is independent of the influence of other factors, and the changes caused in the composite are summed up, it can be assumed that in the generalized model (1) describing the simultaneous effect of several factors p on the composite, the function $\Phi(u_1,u_2,...,u_n;t)$ can be represented as:

$$\Phi(u_1,u_2,...,u_n;t)=\sum_{j=1}^{p}F_j(u_1,u_2,...,u_n;t) \tag{2}$$

In the refined variational formulation, the problems of predicting the defining characteristics of complex composites can be formulated as follows. On the basis of short-term tests conducted on the time interval of retrospection [0, Tmin], it is necessary to find the optimal parameters $u_1^*,u_2^*,...,u_n^*$ of the developed multiparametric forecasting model $R(u_1,u_2,...,u_n;t)$, that deliver a global minimum to the standard error criterion (3) on a set of multiparametric forecasting models $\{R(u_1,u_2,...,u_n;t)\}$, that allow to implement the forecast with an accuracy not exceeding the pre-set maximum permissible accuracy of the forecast $\gamma_R^{\max}$ (4):

$$S^2(u)=\frac{1}{m}\sum_{i=1}^{m}\left[R(u_1,u_2,...,u_n;t_i)-\tilde{R}_i\right]^2\Rightarrow\min_{u\in U}, \tag{3}$$

$$\max_{T_{\min}\le t\le T_{\max}}\left|R(u_1,u_2,...,u_n;\ t)-R^*(t)\right|\le\gamma_R^{\max}, \tag{4}$$

$$(u_1,u_2,...,u_n)\in U,$$

The optimal solution of the prediction problem in the variational setting (3)-(4) should deliver the global minimum of the function S (u). Based on the conducted computational experiments, it was found that the efficiency indicator has a complex multi-extreme structure. The complex structure of the efficiency indicator (3) established on the basis of computational experiments is a significant obstacle to building an optimal solution to the forecast problem. In accordance with this, research has been conducted to develop effective methods for constructing optimal solutions within the framework of refined variational statements of forecast problems (3)-(4).

In [6], we proposed a method for finding the extremum of multiparametric functions with an optimal choice of parameters for solving problems of predicting changes in the strength of polymer composites, associated with the modification of gradient methods. However, the effectiveness of this method depends significantly on the successful choice of the search direction for the minimum performance indicator, which is usually possible only for performance indicators that have a fairly simple structure. For performance indicators that have a complex structure, one of the most promising ways to increase potential opportunities is to improve methods for building optimal search directions for the absolute minimum of performance indicators.

In [7], modifications of the methods of possible directions were developed to determine the optimal search directions. The structure of modification of methods of possible directions includes such additional elements that increase their effectiveness as global optimization of the efficiency indicator along the selected conditionally optimal search direction $\vec{l}_p^{\,*}$; construction of the next (p+1)-th approximation to the globally optimal solution associated with the expansion of the efficiency index J (u) in the Taylor series in the vicinity of the p-th approximation along the conditionally optimal search direction $\vec{l}_p^{\,*}$:

$$J(u) = J(u^p) + \frac{\partial J(u^p)}{\partial \vec{l}_p^{\,*}}\left(u-u^p\right) + \frac{1}{2}\left(u-u^p\right)^T \frac{\partial^2 J(u^p)}{\partial \vec{l}_p^{\,*} \partial \vec{l}_p^{\,*}}\left(u-u^p\right) + \ldots \tag{5}$$

$$\frac{\partial J(u^p)}{\partial \vec{l}} = \left(\nabla J(u^p), \vec{\tau}\right) = \sum_{i=1}^{n} \frac{\partial J(u^p)}{\partial u_i}\alpha_i = \sum_{i=1}^{n} \frac{\partial J(u^p)}{\partial u_i}\cos\varphi_i, \tag{6}$$

$$\frac{\partial^2 J(u^p)}{\partial \vec{l}^{\,2}} = \sum_{i=1}^{n}\sum_{k=1}^{n} \frac{\partial^2 J(u^p)}{\partial u_i \partial u_k}\alpha_i\alpha_k = \vec{\tau}^T \nabla^2 J(u^p)\vec{\tau}, \quad p = 0, 1, 2, \ldots$$

In these notations $\varphi_1, \varphi_2, \ldots, \varphi_n$ the angles, formed by the unit vector $\vec{l}$, that sets the search direction with coordinate axes $Ou_1, Ou_2, \ldots, Ou_n$. Also, in [7], we considered an approach to constructing optimal solutions to forecast problems within the framework of refined variational statements of forecasting problems of the form (3)-(4), when mutually conjugate directions are chosen as promising search directions.

3. Results and discussion

Comparative computational experiments were conducted to evaluate the effectiveness of the developed methods and their modifications for constructing optimal solutions to forecast problems in the framework of variational statements (3)-(4) with a complicated structure of significantly nonlinear performance indicators of the form (3). It is established that the developed methods allow us to build sufficiently effective solutions to forecast problems, if the initial approximation to the optimal solution of the forecast problem in the variational problem (3)-(4) is relatively well chosen.

For variational statements of forecasting problems (3)-(4), a study of the design features of performance indicators with regard to their structure of the form (1)-(2) was conducted to build effective approximations to optimal solutions of forecasting problems. Based on the study of the design features of performance indicators of the form (3)-(4), a methodology for constructing effective approximations to optimal solutions of forecasting problems was developed for the first time. The developed methodology largely uses the additivity property of the generalized efficiency indicator, represented as (1)-(2). The developed methodology is based on the immersion of the original variational problem (3)-(4) in a parametric family of problems for the integer parameter τ, which has a physical meaning of the number of destabilizing factors affecting the composite (τ=1,2,.., p). As a result, we obtain a parametric family of variational problems (3)-(4) that depends on the parameter τ and, respectively , a parametric family of performance indicators $\{S_\tau(u)\}$, that depends on the integer parameter τ (τ=1,2,..,p). The efficiency indicator in a one-parameter variational problem corresponding to the value of the parameter τ can be represented as:

$$S_\tau^2(u)=\frac{1}{m}\sum_{i=1}^{m}\left[\sum_{j=1}^{\tau}F_j\left(u_1,u_2,...,u_n;t_i\right)-\Delta R_i\right]^2, \quad (\tau=1,2,...,p). \tag{7}$$

In these notations: $\Delta R_i=\tilde{R}_i-R_0$. At the first step, the variational problem of a parametric family corresponding to the value of the parameter τ=1 is solved:

$$S_1^2\left(u^{1*}\right)=\min_{u\in U}S_1^2\left(u\right) \tag{8}$$

At the step with the number τ, the variational problem of the parametric family corresponding to the value of the parameter τ is solved:

$$S_\tau^2\left(u^{\tau*}\right)=\min_{u\in U}S_\tau^2\left(u\right), \quad (\tau=2,3,...,p). \tag{9}$$

The initial approximation to the optimum solution of the problem of the parametric family corresponding to the value of the parameter τ, taken to be the optimal solution of the variational problem parametric family of (7) corresponding to the parameter value (τ-1): $u^{\tau-1*}$. The solution to the variational problem parametric family of (7) corresponding to the value значению τ=p - u^{p*} is taken as the initial approximation to the optimal solution of the original variational problem (3)-(4). Further, to solve the initial variational problem (3)-(4), the methods described earlier [7-9] are applied. The conducted comparative computational experiments have shown the high efficiency of the developed approach for solving the problems of high-precision forecasting of the defining characteristics of composite materials (CM), polymer composite materials (PCM) and structures under the influence of extreme environmental factors.

4. Conclusion

Conducted research and developed modifications to existing methods of finding the minimum of multivariable functionsж; comparative estimation of their efficiency; development of recommendations on the optimal layout and efficient design modifications in accordance with the specific features of the performance indicators significantly expand the potential in improving the efficiency of methods of forecasting of a residual resource and durability of composite materials and structures when exposed to extreme environmental factors.

References

[1] Kablov E N, Lebedev M P , Startsev O V and Golikov N I 2013 *Proceedings of the VI*

Eurasian Symposium on the problems of strength of materials and machines for cold climate regions (Yakutsk, Russia) **1** 5-7

[2] Kablov E N, Startsev O V, Krotov A S and Kirillov V N 2011 *Russian Metallurgy (Metally) 2011* **10** 993–1000

[3] Gusev E L and Bakulin V N 2018 *Doklady physics* **63** 213-17

[4] Gusev E. L 2016 *Material Physics and Mechanics* **26** 70-2

[5] Gusev E L and Bakulin V N 2018 *Doklady physics* **63** 388-92

[6] Gusev E L and Babenko F I 2015 Proceedings «Materials for technical devices and structures used in the Arctic» (Moscow: All-Russian Institute of aviation materials) 132-41

[7] Gusev E. L., Bakulin V. N., Chernykh V. D. Development of combined search methods for efficiency indicator extreme in variation statement of forecasting tasks for determine characteristics of composite materials // Journal of Physics: Conference Series. , Supercomputer Technologies in Mathematical Modelling, 19–21 June 2019, Moscow, Russian Federation 2019. Vol. 1392. 012008.

[8] Gusev, E.L., Ivanova, M.A. Development and application of combined methods for solving the prediction problems of defining characteristics of composites // Mathematical Notes of NEFU, 2019, Vol. 29, No. 1, pp. 60-69. DOI: 10.25587/SVFU.2019.101.27247

[9] Gusev E.L., Bakulin V.N. The use of generalized models in the variational formulation of the prediction tasks defining characteristics of composite materials//Journal of Physics: Conference Series, 2020, v. 1431, p. 012017

[10] Bulmanis V N, Yartsev V A and Krivonos V V 1987 *Mechanics of composite materials* **5** 915-20

[11] *Ageing of composites 2008* ed R Martin (Woodhead Publishing Limited: Cambridje) 544 p

[12] *Long-term durability of polymeric matrix composites* 2012 ed K V Pochiraju, G P Tandon and GA Schoeppner (Springer) 677 p

[13] *Service Life Prediction of Polymers and Plastics Exposed to Outdoor Weathering* 2017 Eds C C White, K M White and L E Pickett (William Andrew Publ) 342 p

[14] Startsev O V , Anikhovskaya L I, Litvinov A A and Krotov AS *Increasing the reliability of predicting the properties of polymer composites in hygrothermal aging* 2009 *Doklady Chemistry* **428** 228-32

IOP Conf. Series: Materials Science and Engineering **927** (2020) 012041 doi:10.1088/1757-899X/927/1/012041

Methodology for processing expert judgments based on fuzzy weighted summation

V Sudakov[1] and T Sivakova[2]
[1] *Moscow Aviation Institute*, 4, Volokolamskoe shosse, 125993, Moscow, Russia
[2] *Plekhanov Russian University of Economics*, 36 Stremyanny lane, 117997, Moscow, Russia

E-mail: sudakov@ws-dss.com

Abstract. The paper is devoted to the development and research of a new technique for fuzzy ranking of alternatives. The authors of this approach suggest dividing the criteria space into fuzzy preference areas. A criterion for the proximity of fuzzy preference ratings is also formulated. The scientific novelty of the proposed approach lies in the proposed procedure for clarifying the level of preference based on fuzzy weighted summation, for alternatives with close fuzzy levels of preference. The solution of the problem in this statement makes it possible to increase the distinguishability of alternatives by preference. Two approaches to the construction of membership functions of fuzzy preferences were investigated: the first method - choosing the type of membership function and then entering the parameters of the membership function; the second method is entering the characteristic points of the membership function and constructing a piecewise polynomial approximation. After entering preferences and presetting, this method allows ranking an arbitrary number of alternatives in a mode close to real time. In cases of high dimensionality of the vector criterion and optimized parameters of the domain model, as well as in the presence of complex algorithmic rules, the solution of such problems is of high scientific importance. In the public domain at http://ws-dss.com one can use this method - fuzzy ranking based on preference areas. The proposed approach was implemented (tested) on the problem of selecting an electronic pilot tablet.

1. Introduction

Expert assessment is a dynamically developing area of the decision support system (DSS). The use of decisions made on the basis of expert judgments is quite in demand. As a rule, experts' responses are considered as numbers, attributing strictly defined numerical values to the opinion, which are then processed using methods of applied statistics. In the case of such processing, the obtained data can significantly differ from reality [1]. In practice, decisions based on expert judgment are often non-numerical, making it difficult to process the information received. As a rule, the error in evaluating objects is also not taken into account.

To obtain more "accurate" results of fuzzy judgments, it is advisable to use the apparatus of "soft" calculations, the theory of fuzzy sets.

2. Literature review

Over the past years, methods of multi-criteria analysis of alternatives have been actively developed and applied. The use of computer technology and computer networks can greatly simplify the receipt of expert opinions of a wide range of specialists in various fields of activity. An expert often cannot unambiguously determine the relationship of preference between two objects and give a clear assessment of the object of examination. Interval assessments of various characteristics of the objects of expertise are more "accurate", they are closer to the real situation. Very important information from an expert can be an assessment of the degree of his confidence that the assessment takes on one or another value. For such estimates, the well-known apparatus of the theory of fuzzy sets, proposed by Lotfi Zadeh [1,2], is well suited.

Despite the fact that the theory of fuzzy sets has been used in decision-making for a long time, its popularity is only growing, which is reflected in the works of authors in Russia and abroad [3–9]. In most cases, the authors try to use the classical approaches of decision theory, "mechanically" replacing clear numbers with fuzzy ones. Issues of optimization of solutions with fuzzy described sets of feasible solutions are investigated in detail in the work of Orlovsky S.A. [6]. The article uses fuzzy selection methods based on methods of fuzzy qualitative importance of criteria [7]. In the works of A.I. Orlova analyzed fuzzy approaches to expert evaluation [8].

The prerequisites for this work were the early studies of the authors [9] in the field of individual functions of fuzzy preferences built on the basis of fuzzy domains

Let us dwell on the early work of the authors of [9], where it was proposed to introduce fuzzy scales for all criteria, consisting of fuzzy gradations (ranges of values with fuzzy boundaries). The intersection of fuzzy gradations within the framework of one criterion is allowed taking into account restrictions on the minimum and maximum degree of belonging to an arbitrary value. A decision maker (decision maker) evaluates some of the combinations of fuzzy criteria values in a fuzzy preference scale. Next, the completeness of the preference model is checked: for all points of the criterion space, the level of preferences should not be lower than the given. This model allows you to automatically evaluate an arbitrary number of alternatives. For all areas of preferences, the method allows you to build membership functions of alternatives. Clear and / or fuzzy criteria values are input to the multicriteria assessment model. To obtain a final ranking of alternatives, one should: either defuzzify the fuzzy ranks of the alternatives, or determine the fuzzy dominance ratio for all pairs of alternatives.

In this paper, we propose a new method for determining fuzzy judgments of decision makers, combining the approaches used to date with fuzzy automatic control based on expert judgments [10], and the idea of dividing the criteria space into areas proposed by the authors in a combined decision support method [9]. Let us pass to the mathematical formulation of the problem.

3. Formalization of the method and solution algorithm

Suppose that for the case of p criteria and k specific evaluated alternatives, we obtain k membership functions corresponding to all alternatives (1):

$$z_t = h_t(y, x_1^t, x_2^t, \ldots, x_k^t) \tag{1}$$

where t is the number of the alternative ($t = 1 \ldots k$).

To assess the proximity of two membership functions h_1 and h_2, metrics of various function spaces are used. For example (2) or (3):

$$\|h_1 - h_2\| = \sup|h_1(y) - h_2(y)| \tag{2}$$

or

$$\|h_1 - h_2\| = \sqrt[2]{\int (h_1(y) - h_2(y))^2 dy} \tag{3}$$

Since the values of the criteria in general terms also correspond to the membership function, it makes sense to use the rule: "If the value of at least one criterion is significantly different, then the ranks of the alternatives should differ significantly." The lexical term "significantly different" can be expressed as the following membership function (4):

$$\mu(\|h_1 - h_2\|) = \begin{cases} 0, \ \|h_1 - h_2\| \le \alpha, \\ \dfrac{\|h_1 - h_2\| - \alpha}{\beta - \alpha}, \ \alpha < \|h_1 - h_2\| < \beta, \\ 1, \ \|h_1(y) - h_2(y)\| \ge \beta. \end{cases} \tag{4}$$

The constants α and β are selected based on the specifics of the problem being solved (for many practical problems, a good approximation is $\alpha = 0.3$ and $\beta = 0.7$. To check the above rule for the closeness of two alternatives t_1 and t_2, one should calculate the degree of confidence in the validity of the premise using formula (5):

$$c(t_1,t_2) = \min_i \mu\left(\left\|h_{t_1}(x_i^{t_1}) - h_{t_2}(x_i^{t_2})\right\|\right) \tag{5}$$

For the corollary, membership should be calculated by the formula (6):

$$m(t_1,t_2) = \mu\left(\left\|h_{t_1}(y^{t_1}) - h_{t_2}(y^{t_2})\right\|\right) \tag{6}$$

And if there exist t_1, t_2 such that $c>m$, then it is necessary to clarify the preference function using quantitative fuzzy methods.

The most common quantitative fuzzy method is fuzzy weighted summation. For all criteria, fuzzy weights w_i with membership function $\mu_i(w)$ are assigned, and for the criteria values, the membership function is also given: $\lambda_i(x)$.

Weighted sum for the case of ordinary (clear) numbers (7):

$$s = S(x_1, x_2, \ldots, x_n) = \sum_{i=1}^{k} w_i x_i \tag{7}$$

It can be upgraded for the case of fuzzy numbers using the principle of generalization. For the components of the sum, the membership function is calculated by the formula (8):

$$f_i(g_i) = \sup_{g=w_i x_i} \min(\mu_i(w_i), \lambda_i(x_i)) \tag{8}$$

And the membership function of the weighted sum will take the form (9):

$$r(s) = \sup_{s=\sum_{i=1}^{k} g_i} \min_i (f_i(g_i)) \tag{9}$$

The integrated assessment of the alternative in the proposed method consists of two components of a fuzzy preference level y and fuzzy result of weighted summation s. The integral rank of the alternative $I = y + s$ for the fuzzy case is also found using the generalization principle. Membership Function (10):

$$\Phi(\mathrm{I}) = \sup_{I=y+s} \min(p(y), r(s)) \tag{10}$$

The use of this integral criterion makes it possible to increase the distinguishability of the ranks of alternatives.

4. Discusion

Using fuzzy preference areas with a fuzzy weighted summation procedure on a single criteria space allows us to talk about the originality of the method.

When designing user interfaces for working with fuzzy propositions, the most important is the input and editing interface of the membership function. Two approaches are proposed for construction:

1. Two-stage procedure: choosing the type of membership function and then entering the parameters of the membership function.
2. Pointwise procedure: input of characteristic points of the membership function and then whether the points are connected by polynomials of order n.

The first procedure can cause difficulties for an expert with poor mathematical training, although in some cases it is faster than the second approach. The second approach is more universal and allows you to set an arbitrary curve with the required degree of accuracy. Typically, the points are connected by linear segments, since the required accuracy at the initial stage of entering preferences is not too high.

Figure 1 shows a fragment for entering the membership function using the second approach. The user can use the mouse to select arbitrary points on which the function is built and, if necessary, change the position of the characteristic points by moving the mouse.

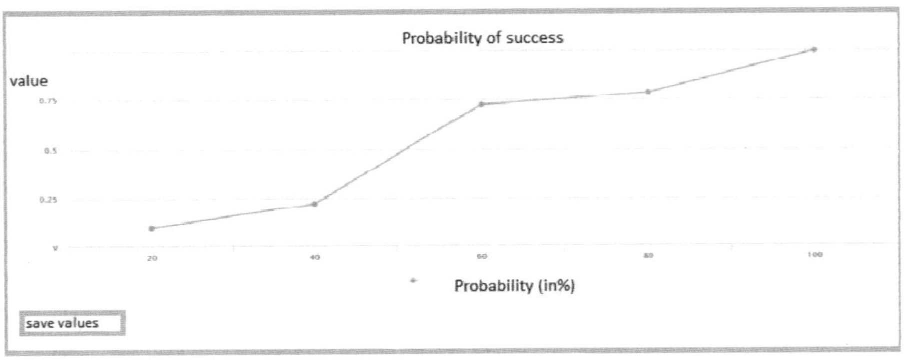

Figure 1. Fragment of input and editing membership function.

Each membership function must have a unique symbolic identifier (name) that is understandable to the user and reflects its physical meaning. The decision support system should maintain a database of such membership functions. Further, when setting both the weights and gradations of the scales of criteria, the user must have access to the choice:

1. Select the membership function from the directory.
2. Create a new membership function.
3. Indicate a "clear" value.

The proposed scheme allows you to reduce the number of membership functions that you must enter to the user in the interface. For example, a fuzzy number of "about 0.5" can be used to indicate the weights of several criteria, and to indicate their fuzzy values.

5. Application

This method of fuzzy ranking of alternatives was applied to the problem of choosing a tablet computer for an aircraft pilot [11]. Electronic Flight Pads (EFB) have been widely used in aviation as pilot assistance devices, since they improve the quality and speed of decision-making on take-off and landing, are convenient in use, have low weight, and have been developed and are constantly being improved for them specialized software. The use of EFB is certified by aviation services in most developed countries. There are many different models of tablet computers currently on the market. This leads to the fact that at the stage of creating an EFB implementation plan, a non-trivial task arises of choosing the most effective and economically affordable tablet computer model.

Many tablets are represented by a large number of models of famous manufacturers based on the operating systems iOS, Windows and Android. If the airline carries out commercial air transportation, it is permissible to use only models certified by the aviation authorities. Certified models include iPad and Window Surface Pro. Window Surface Pro is used as EFB only by some large airlines, and iPad is used by most airlines in the world, including many Russian ones. Therefore, when developing a tablet selection model, iPad models presented on the commercial market were analyzed. The value system, which is reflected in the corresponding membership function for fuzzy areas of preferences and in the corresponding fuzzy estimates of alternatives, is based on the following judgments:

1. The space of the cockpit is very limited, so working with large tablets is difficult.
2. A comfortable perception of information is determined by the size of the screen. There may be difficulty in perceiving information if the tablet screen is less than 9 inches. A comfortable perception of information is provided by medium screens (9-11 inches) and large screens (more than 12 inches).
3. Significant importance is attached to the presence of 3G/LTE Internet on the tablet, as well as the built-in GPS/GLONASS module.
4. The preferred amount of memory is determined by whether it is planned to take photos and videos from a tablet, whether multiple backups of documents are required, and whether audio and video materials are required.
5. The novelty of the model directly determines the period of use of the tablet. With a high probability, a tablet released more than 3 years ago will require replacement in 2-3 years.

6. Conclusion

The proposed approach based on fuzzy weighted summation allows not only ranking an arbitrary number of alternatives in automatic mode, but also improving the quality of decisions made by experts. On the portal http://ws-dss.com [12] in the public domain one can get acquainted with the software implementation of this technique. This portal allows one to publish console software implementations of mathematical models in C ++, R, Ruby languages on the Internet, and also makes it possible to connect other web services and organize a sequence of calls to a group of mathematical models with the transfer of input/output parameters between them.

The rule is formalized in the work: "If the value of at least one criterion is significantly different, then the ranks of the alternatives should differ significantly". To improve the quality of decisions made, rules and ratios have been developed to assess the completeness of the task of fuzzy judgments.

Acknowledgments

The reported study was funded by RFBR according to the research project No 19-01-00520.

References

[1] Zadeh L A 1965 Fuzzy sets *Inform. and control* **8** 338-53
[2] Bellman R, Zade L 1976 Decision making in vague conditions *Issues of analysis and decision-making procedures* (Moscow: Mir) pp 172 - 215
[3] Aggarwal H, Arora H D and Kumar V 2019 Decision Making Problem As An Application Of Fuzzy Sets *International J. of Scientif & Technol. Research* **8(11)**
[4] Rezaei J, Nispeling T, Sarkis J and Tavasszy L 2016 *J. of Cleaner Production* **135** 577–88
[5] Puchkov A Y 2017 *Software Product, Systems and Algorithm* **4**
[6] Orlovsky S A 1981 *Decision making problems with fuzzy initial information* (Moscow.: Nauka)
[7] Hyun J and Kim S 2019 J. of Korean Institute of Communic. and Information Sci. **44** 936–42
[8] Orlov A 2018 *Methods of making management decisions* (Moscow: Knorus)
[9] Osipov V P and Sudakov V A 2017 Multicriteria decision analysis for fuzzy preference areas KIAM Preprints named after M V Keldysh **6**
[10] Ding Z, Gong W, Li S and Wu Z 2018 A Review of Complexity Simulation in Construction Waste Management. *Sustainability* **10** 1-13

[11] Dutov A V, Nesterov V A , Sudakov V A and Sypalo K I 2018 Fuzzy Preference Domains and Their Use for Selecting an Electronic Flight Bag for Flight Crews *J. of Comp.r and Syst. Sciences Intern.* **57(2)** 230–238
[12] Sudakov V 2020 Web Services for Decision Support Systems Retrieved from https://ws-dss.com/?locale=en

Structure of a viscoplastic fluid flow in a pipe with a given obstruction shape

I Ryltsev and G Shrager
Tomsk State University, 36 Lenin Ave., Tomsk 634050, Russia

E-mail: ryltsev_i@ftf.tsu.ru

Abstract: In this paper, we present a study of the axisymmetric steady-state laminar flow of a viscoplastic fluid in a pipe with variable cross section. Mathematical formulation of the problem is written in terms of stream function and vorticity variables in a cylindrical coordinate system. Rheological behavior of the medium is described by the Shvedov-Bingham model. Coordinate transformation is carried out to implement a numerical algorithm. The problem is solved numerically using the finite-difference method and the false transient method. This paper demonstrates the effect of Reynolds number, Bingham number, and geometric parameters of a flow domain on the flow structure.

1. Introduction

Flow of a rheologically complex fluid in pipes with variable cross section is widespread in technological processes of various industries. In particular, such channels encounter as structural elements of the heat exchangers and lubrication systems. In biomechanics, such geometry is used to simulate the flow in stenosed blood vessels.

There is a large number of experimental and numerical works considering the flow in pipes with variable cross section, where the Newtonian fluid is used as a studied medium [1–5]. In these works, the impact of the Reynolds number and geometric parameters on the kinematic and dynamic characteristics of the flow is shown on the basis of the obtained dependency diagrams. In [6], the authors study the Newtonian fluid flow in the case of asymmetric obstruction. In [7–8], an analysis of the flow in a channel with two successive stenoses is carried out, and the effect of the second stenosis on the distribution of flow characteristics is demonstrated.

When modeling a real fluid flow such as blood, lubricants or fuel suspensions, it is necessary to take into account rheological properties of the medium. In [9], the power-law model and Casson's model are used to describe blood rheology. Impact of non-Newtonian properties of the medium on the flow resistance is assessed. Study presented in [10] demonstrates changes in the structure of the Ostwald – de Waele fluid flow in a pipe with variable cross section depending on the main parameters of the problem. In [11], the flow in a large artery is studied numerically, where a viscoelastic fluid is used as a medium. Flow of five non-Newtonian models in a vessel is considered in [12], where the rheological law constants for blood are determined experimentally. Stress level on the vessel wall is obtained and analyzed for various rheological models. Author of [13] uses the Herschel-Bulkley, Carreau, and Ostwald-de Waele rheological models to assess the need of taking into account of non-Newtonian properties when modeling the flow in coronary arteries. Results of numerical simulation of a steady-state fluid flow are presented in [14], where rheological properties of the fluid are described by the Herschel-Bulkley model. The use of this model allows considering of the following models: Newton, Shvedov-Bingham, and Ostwald-de Waele. In [15], a mathematical model is proposed for a stationary blood flow through stenosed artery.

Purpose of this work is to study the effect of non-Newtonian properties of the liquid medium on the flow structure in a pipe with variable cross section using the Shvedov-Bingham model.

2. Formulation of the Problem

Steady-state laminar flow of an incompressible viscoplastic fluid in a pipe with variable cross section is considered. Flow region is shown in figure 1.

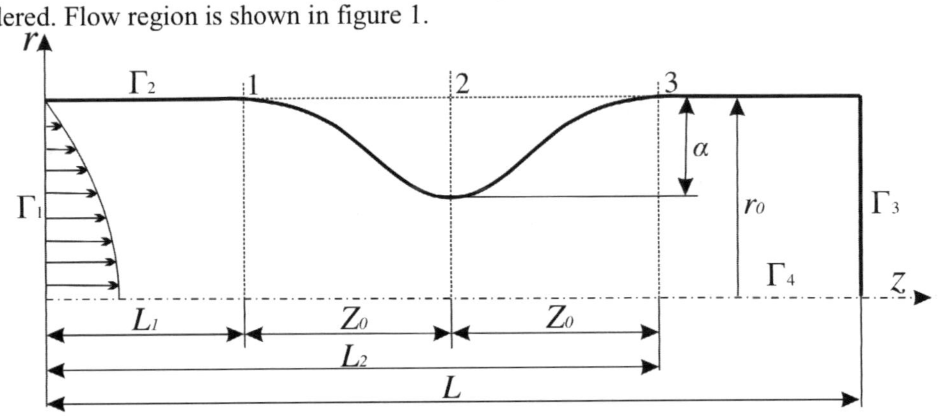

Figure 1. Flow region.

Mathematical description of the flow is based on the equations in terms of stream function and vorticity which are written in dimensionless form in a cylindrical coordinate system as follows [16]

$$\frac{\partial(u\omega)}{\partial z}+\frac{\partial(v\omega)}{\partial r}=\frac{B}{\mathrm{Re}}\left(\nabla^2\omega-\frac{\omega}{r^2}\right)+\frac{S}{\mathrm{Re}}, \tag{1}$$

$$\nabla^2\psi-\frac{2}{r}\frac{\partial\psi}{\partial r}=-r\omega, \tag{2}$$

$$\nabla^2=\frac{\partial^2}{\partial z^2}+\frac{\partial^2}{\partial r^2}+\frac{1}{r}\frac{\partial}{\partial r},$$

$$S=2\frac{\partial^2 B}{\partial z\partial r}\left(\frac{\partial v}{\partial r}-\frac{\partial u}{\partial z}\right)+2\frac{\partial B}{\partial z}\frac{\partial\omega}{\partial z}+2\frac{\partial B}{\partial r}\frac{\partial\omega}{\partial r}+\left(\frac{\partial^2 B}{\partial z^2}-\frac{\partial^2 B}{\partial r^2}\right)\left(\frac{\partial u}{\partial r}+\frac{\partial v}{\partial z}\right)+\frac{\omega}{r}\frac{\partial B}{\partial r},$$

where dimensionless velocity components and the vorticity are determined by formulas

$$u=\frac{1}{r}\frac{\partial\psi}{\partial r},\quad v=-\frac{1}{r}\frac{\partial\psi}{\partial z},\quad \omega=\frac{\partial v}{\partial z}-\frac{\partial u}{\partial r}.$$

Rheological properties of the medium are described by the Shvedov-Bingham equation [17, 18]. Corresponding formula for dimensionless apparent viscosity is written as

$$B=(\mathrm{Bn}+A)/A.$$

Here, u and v are the axial and radial components of the velocity vector, respectively, $\mathrm{Re}=(\rho U r_0)/\mu_0$ is the Reynolds number, $\mathrm{Bn}=\tau_0 r_0/U\mu_0$ is the Bingham number, ρ is the density, U is the average velocity, μ_0 is the rheological parameter, τ_0 is the yield stress, $A=\left(2e_{ij}e_{ji}\right)^{\frac{1}{2}}$, and e_{ij} are the components of strain rate tensor. Length, velocity, and stress are scaled with the pipe radius r_0, the average velocity U, and the value of $U\mu_0/r_0$, respectively.

In the section Γ_1, the velocity profile corresponds to the steady-state Shvedov-Bingham fluid flow in a cylindrical pipe [19]. On the boundary Γ_2, no-slip conditions are used; on the symmetry axis Γ_4, symmetry conditions are satisfied; in the outlet section Γ_3, "soft" boundary conditions are assigned. Thus, the boundary conditions in terms of stream function – vorticity are written as [16]

$$\Gamma_1 : \psi = \int_0^r urdr,\ \omega = -\frac{\partial u}{\partial r},$$

$$\Gamma_2 : \frac{\partial \psi}{\partial n} = 0,\ \omega = \frac{1}{r}\frac{\partial^2 \psi}{\partial n^2},$$

$$\Gamma_3 : \frac{\partial \psi}{\partial z} = 0,\ \frac{\partial \omega}{\partial z} = 0,$$

$$\Gamma_4 : \psi = 0,\ \omega = 0,$$

where n is a normal to the curved boundary Γ_2, which is defined by the following function:

$$f(z) = \begin{cases} 1 - \dfrac{\alpha}{2}\left(1 - \cos\dfrac{\pi(z - L_l)}{Z_0}\right), & z \in [L_l; L_2]; \\ 1, \quad z \in [0; L_l) \cup (L_2; L]. \end{cases}$$

Values of L_1 and L_2 are set assuming that the obstruction has no effect on the flow behavior in vicinity of the boundaries Γ_1 and Γ_3.

3. Computational Method

The problem is solved using the finite-difference method. False transient method is used to obtain the fields of stream function and vorticity. This method implies substitution of the stationary problem with a non-stationary one. Thus, "fictitious" time derivatives of the unknown functions ψ, ω are introduced into equations (1) and (2), and the calculations continue until the introduced derivatives are zeroed [20]. Physical flow region with a curved boundary $f(z)$ is transformed into a rectangular one by introducing new coordinates $\xi = z,\ \eta = r / f(z)$.

In the new coordinate system (ξ, η) equations (1) and (2) with introduced time derivatives take the following form:

$$\begin{aligned} &\frac{\partial \omega}{\partial t} + \frac{\partial(u\omega)}{\partial \xi} - g\frac{\partial(u\omega)}{\partial \eta} + \frac{1}{f}\frac{\partial(v\omega)}{\partial \eta} = \\ &= \frac{B}{\mathrm{Re}}\left(\frac{\partial^2 \omega}{\partial \xi^2} + H\frac{\partial \omega}{\partial \eta} - 2g\frac{\partial^2 \omega}{\partial \xi \partial \eta} + G\frac{\partial^2 \omega}{\partial \eta^2} + \frac{1}{f^2 \eta}\frac{\partial \omega}{\partial \eta} - \frac{\omega}{f^2 \eta^2}\right) + \frac{S}{\mathrm{Re}}, \end{aligned} \tag{3}$$

$$\frac{\partial \psi}{\partial t} + \frac{\partial^2 \psi}{\partial \xi^2} + \left(H - \frac{1}{f^2 \eta}\right)\frac{\partial \psi}{\partial \eta} - 2g\frac{\partial^2 \psi}{\partial \xi \partial \eta} + G\frac{\partial^2 \psi}{\partial \eta^2} = -f\eta\omega, \tag{4}$$

where $G = \dfrac{1}{f^2} + \left(\eta\dfrac{f'}{f}\right)^2,\ H = \eta\left(2\dfrac{f'^2}{f^2} - \dfrac{f''}{f}\right),\ g = \eta\dfrac{f'}{f},$

$$S = S_1 + S_2 + S_3 + S_4,$$

$$S_1 = 2\left(\frac{1}{f}\frac{\partial^2 B}{\partial\xi\partial\eta} - \frac{g}{f\eta}\frac{\partial B}{\partial\eta} - \frac{g}{f}\frac{\partial^2 B}{\partial\eta^2}\right)\left(\frac{1}{f}\frac{\partial v}{\partial\eta} - \frac{\partial u}{\partial\xi} + g\frac{\partial u}{\partial\eta}\right),$$

$$S_2 = 2\left(\frac{\partial B}{\partial\xi} - g\frac{\partial B}{\partial\eta}\right)\left(\frac{\partial\omega}{\partial\xi} - g\frac{\partial\omega}{\partial\eta}\right) + 2\frac{1}{f^2}\frac{\partial B}{\partial\eta}\frac{\partial\omega}{\partial\eta},$$

$$S_3 = \left(\frac{\partial^2 B}{\partial\xi^2} + H\frac{\partial B}{\partial\eta} - 2g\frac{\partial^2 B}{\partial\xi\partial\eta} + \left(g^2 - \frac{1}{f^2}\right)\frac{\partial^2 B}{\partial\eta^2}\right)\left(\frac{1}{f}\frac{\partial u}{\partial\eta} + \frac{\partial v}{\partial\xi} - g\frac{\partial v}{\partial\eta}\right),$$

$$S_4 = \frac{\omega}{f^2\eta}\frac{\partial B}{\partial\eta}.$$

Difference representation of the equations in transformed coordinate system is obtained using an explicit difference scheme. In governing equations, the convective terms are approximated by upwind differences. To use a through computation of flow with unyielded regions, a regularization of the initial rheological equation is carried out [21]:

$$B = \frac{\mathrm{Bn} + (\sqrt{A^2 + \varepsilon^2})}{(\sqrt{A^2 + \varepsilon^2})},$$

Inequality $BA \leq \mathrm{Bn}$ is referred to as a condition for separating unyielded regions out of flow. It represents a dimensionless analogue of condition for tracking the regions, where the stress level is less than yield stress. To verify the approximation convergence of the algorithm, series of calculations are performed on a sequence of grids. Figure 2 shows the velocity distributions in section *2* illustrated in figure 1. Table 1 demonstrates values of the axial velocity calculated on the symmetry axis in the output section of the pipe at various grid steps *h*. It also contains the velocity values calculated analytically. Presented data demonstrate approximation convergence of the algorithm. Further results are obtained using the grid step of 1/40.

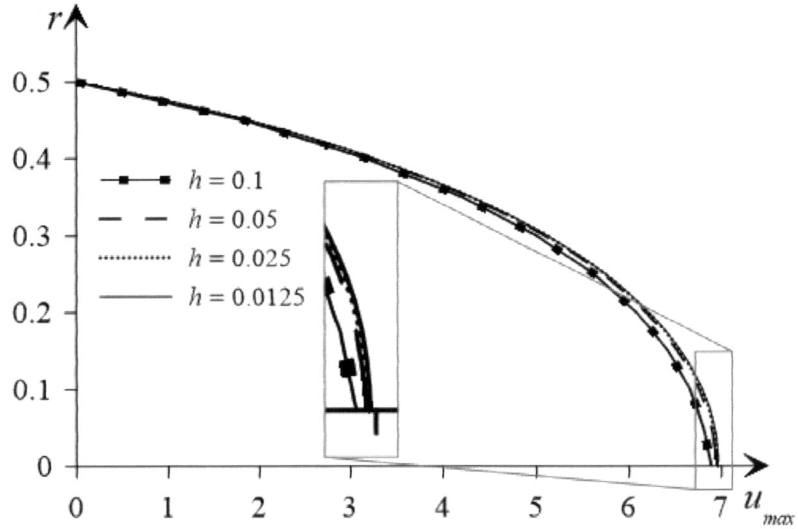

Figure 2. Axial velocity distribution in cross section *2* (Re = 1, Bn = 1, $\alpha = 0.5$, $Z_0 = 1$).

Table 1. Maximum axial velocity at the pipe outlet (Re = 1, Bn = 1, $\alpha = 0.5$, $Z_0 = 1$).

h	u_{max}
0.1	4.2544
0.05	4.3366
0.025	4.3336
0.0125	4.3322
Analytical solution	4.3311

4. Results and Discussion

Figure 3 shows distribution of the streamlines and unyielded regions at various Reynolds numbers for a given Bingham number and geometric parameters α, Z_0. According to the presented flow patterns, in both cases, the fluid flow includes both one-dimensional flow zones occurring at a sufficient distance upstream and downstream of the obstruction, and a two-dimensional flow zone in the vicinity of the obstruction. Flow structure is characterized by the presence of unyielded regions in the one-dimensional flow area. In addition, with an increase in the Reynolds number a dead zone occurs downstream of the obstruction. The unyielded regions and dead zone are both blackened, while the viscous flow zone is painted white. In vicinity of the obstruction, the unyielded region in central part of the flow disappears as stress level in two-dimensional flow zone increases with respect to that in the one-dimensional flow zone, which is typical for a pipe of uniform radius. Increase in the Reynolds number leads to a relative increase in the inertia force of the flow, which, in turn, causes an upsize of the dead zone downstream of the obstruction and a displacement of the unyielded region towards the outlet section.

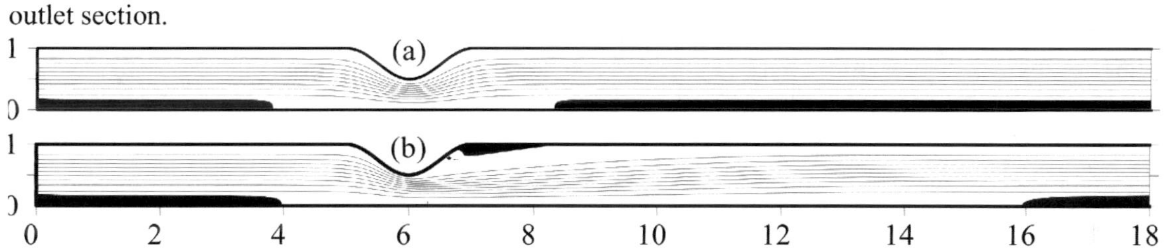

Figure 3. Flow structure (Bn = 1, $\alpha = 0.5$, $Z_0 = 1$): (a) Re = 1 and (b) Re = 25.

Figure 4 shows location of the unyielded regions at various Bn values. As observed, the rise of the Bingham number widens the unyielded regions, while the size of the dead zone downstream of the obstruction varies insignificantly. The higher the Bingham number, the shorter the distance between unyielded regions and obstruction.

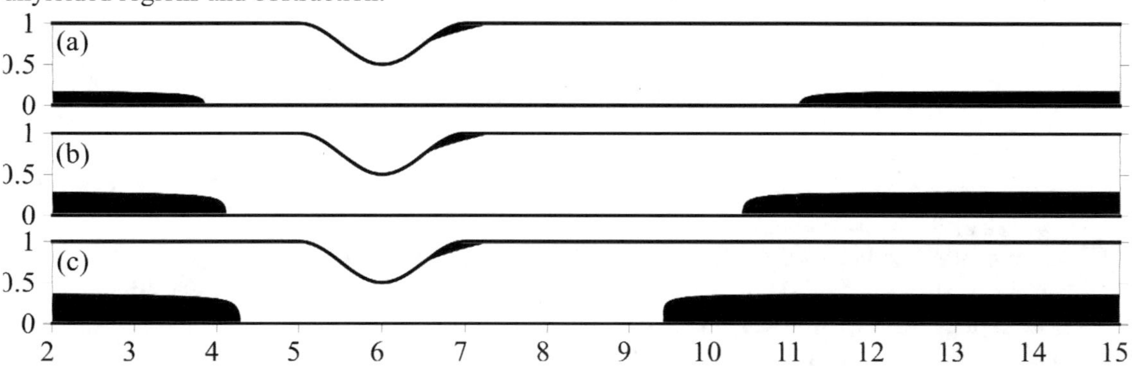

Figure 4. Flow structure (Re = 10, $\alpha = 0.5$, $Z_0 = 1$): (a) Bn = 1, (b) Bn = 2, and (c) Bn = 3.

Pipe overlap depth effect on the flow structure is shown in figure 5. At a low value of α, a small unyielded region occurs on the symmetry axis right under the obstruction (figure 5 *a*). However, the unyielded region disappears once the depth of pipe overlap increases. A dead zone is formed downstream of the obstruction.

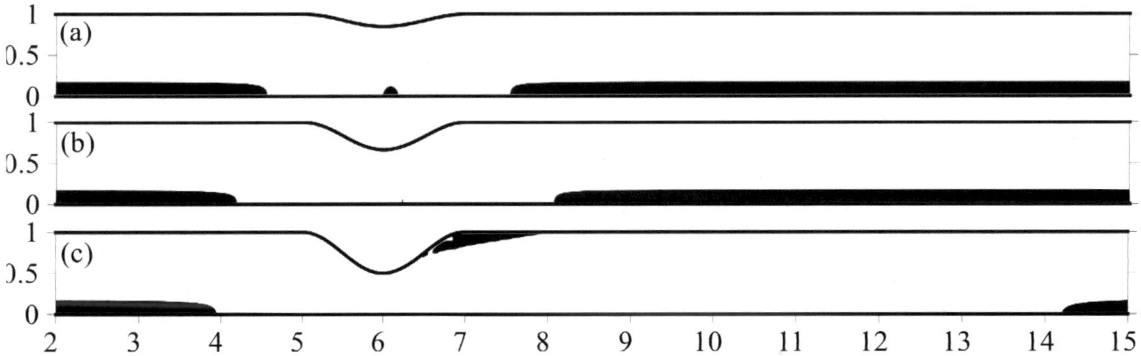

Figure 5. Flow structure (Re = 20, Bn = 1, Z_0 = 1): (a) α = 0.15, (b) α = 0.333, and (c) α = 0.5.

5. Conclusion

As a result of implemented work, the mathematical formulation of the problem of non-Newtonian fluid flow in a pipe with variable cross section is formulated and solved numerically. Flow patterns of the Bingham fluid, which are characterized by the presence of unyielded regions and dead zone in addition to viscous flow zones, have been obtained. It is revealed that with an increase in the Reynolds number, a dead zone occurs downstream of the obstruction. Increasing Bingham parameter contributes to an upsize of the unyielded regions and to a decrease in the two-dimensional flow zone in the vicinity of the obstruction. An increase in the depth of the pipe overlap leads to the destruction of the unyielded region on the symmetry axis under the obstruction and to the formation of a dead zone downstream of the obstruction.

Acknowledgement
This research is implemented at the expanses of the Russian Science Foundation (project no. 18-19-00021).

References

[1] Lee J S and Fung Y C 1970 *J. Appl. Mech.* **37** 9–16
[2] Forrester J H and Yang D F 1970 *J. Biomech.* **3** 297-305
[3] Deshpande M D, Giddens D P and Mabon R F 1976 *J. Biomech.* **9** 165–74
[4] MacDonald D A 1979 *J. Biomech.* **12** 13–20
[5] Lee T S 1994 *J. Fluids Eng. Trans. ASME* 1994 **116** 66–71
[6] Mandal D K, Manna N K and Chakrabarti S 2011 *J. Mech. Sci. Technol.* 2011 **25** 1933–47
[7] Lee T S Liao and W Low H T 2003 *Int. J. Numer. Meth. Fl.* **42** 717–40
[8] Huang H, Lee T S and Shu C 2006 *Int. J. Numer. Method. H.* **16** 185–203
[9] Shukla J B, Parihar R S and Rao B R P 1980 *Bull. Math. Biol.* **42** 283–94
[10] Ryltsev I A, Ryltseva K E and Shrager G R 2020 *Tomsk State University Journal of Mathematics and Mechanics* **63** 125–38 (in Russian)
[11] Leuprecht A and Perktold K 2001 *Comput. Method. Biomec.* **4** 149–63
[12] Jahangiri M, Saghafian M and Sadeghi M R 2017 *J. Mech. Sci. Technol.* **31** 1003–13
[13] Manimaran R 2011 *World Acad. Sc.i Eng. Technol.* **73** 957–62
[14] Tu C Deville M 1996 *J. Biomech.* **29**. 899–908
[15] Prakash O, Makinde O D, Singh S P, Jain N and Kumar D 2015 *Int. J. Biomath.* **8** 1550010

[16] Roache P J 1982 *Computational Fluid Dynamics* (Hermosa: Albuquerque)
[17] Schwedoff T 1889 *J. Phys. Théorique Appliquée* **8** 341–59
[18] Bingham E C 1922 *Fluidity and Plasticity* (New-York: McGraw-Hill)
[19] Borzenko E I 2019 *Simulation of Non-isothermal Flows of a Complex Fluid When Filling Planar or Axisymmetric Channels.* Dissertation (Tomsk: Tomsk State University)
[20] Godunov S K and Ryabenkiy V S 1987 *Difference Schemes* (North-Holland: Elsevier Science Ltd)
[21] Bercovier M and Engelman M 1980 *J. Comput. Phys.* **36** 313–26

IOP Conf. Series: Materials Science and Engineering **927** (2020) 012043 doi:10.1088/1757-899X/927/1/012043

Stability of the flow with an over-compressed detonation wave in a duct of variable cross section

A J Egoryan and Kh F Valiyev

Central Institute of Aviation Motors named after P. I. Baranov, 111116, Russia, Moscow, Aviamotornaya street, 2.

E-mail: arche_hl@mail.ru

Abstract. The stability of stationary flows of an ideal gas in ducts of variable cross section with an over-compressed detonation wave is studied. The combustible mixture before the wave and the combustion products after it are perfect gases. A numerical-analytical method has been developed to analyse the stability of stationary flows with over-compressed detonation waves. The calculations performed, demonstrating its effectiveness, revealed examples of stable flows with over-compressed detonation wave in ducts of variable cross section.

1. Introduction

According to the model developed by Zel'dovich, Neumann, and Döring (ZND model) [1-3], the structure of the detonation wave (DW) includes an intense shock wave (discontinuity surface) and a "reaction zone" of finite width adjacent to it. The actual thickness of the shock wave is several mean free paths of molecules, in contrast to the reaction zone, which is several tens of times larger. Chemical reactions, the rate of which is highly dependent on temperature, begin due to the heating of the combustible mixture behind the shock wave and end when the combustion products reach thermodynamic equilibrium. The detonation adiabats obtained within the ZND model are completely described by the thermodynamic parameters of the mixture in the initial (cold, metastable) and final equilibrium states and not depending on the kinetics of the combustion process. On detonation adiabats, of particular interest are the points that correspond to the DW, named after the names of their discoverers: DW of Chapman - Jouguet (DW_{CJ}). An important feature of DW_{CJ} is that the gas velocity normal to the detonation wave behind it is equal to the speed of sound. Due to this, the DW_{CJ} is self-sustaining, because rarefaction waves, which are coming from the "detonation wave initiation point", cannot overtaken and weaken it.

The cases of the collapse of the stationary detonation wave structures expected within the ZND model experimentally discovered in the mid-twentieth century caused a torrent of studies of their structural stability. In this direction, real kinetics is replaced by one reaction with a certain activation energy and subsequent instantaneous or extended heat release over time. In this approximation, the condition of structural stability leads to constraints on the activation energy. At present, the determination of the one-dimensional structural stability of a detonation wave in any mixture with the known kinetics of its combustion is possible in the process of numerical solution of the relatively simple non-stationary formulation of a nonequilibrium one-dimensional ideal gas flow. If such a solution reaches a stationary structure, the detonation wave in the considered fuel mixture is structurally stable. If the transition to the stationary structure does not occur, then the DW in such a mixture is structurally unstable.

Along with the structural stability of the DW, it is of interest to the stability of stationary flows in ducts of variable area with structurally stable and therefore considered as the surface of the DW break. Stationary flows in ducts of variable area with a close to straight Chapman-Jouguet DW are always unstable [4]. The interest in such flows is a consequence of their attraction as an element of ramjet engines with combustion in a stationary DW. The subject of this study is the analysis of the stability of stationary flows with a compressed DW in ducts of variable area. Moreover, the DW itself is in the expanding part, according to the results of studies [5-7].

2. Analysis of the stability of stationary flows with a compressed DW

Further, stationary and non-stationary flows are described by equations in a one-dimensional approximation with a detonation wave (flat discontinuity surface) that is normal to the duct axis. The combustible mixture before the detonation wave and the combustion products behind it are perfect gases with different constant and specific heat capacities before and behind the wave and additive constants in the formulas for internal energy.

The stability analysis of stationary flows with over-compressed detonation wave and subsonic flows adjacent to them develops the approaches used in [8-14] for flows with a breakdown shock wave. The differences are in the research tools that are now used are in sequentially running programs for the numerical solution of stationary and non-stationary equations, which describe the flows under study in a one-dimensional approximation. The stationary flow calculation program finds the distribution of parameters over the entire duct for given duct shape, parameters determining the flow condition (in particular, supersonic Mach number at the inlet), constants defining the combustible mixture and combustion products and the stationary section of the over-compressed detonation wave (DW). Typical ducts had a diverging-expanding inlet section ("air intake") with a supersonic Mach number in a minimum section ($M_m > 1$). In a certain section x_{DW} of the expanding section of the air intake with the Mach number $M_1 > 1$, an over-compressed DW with a given stationary value of the Mach number $M_2 < 1$ behind it. To the right of the DW, the divergent duct passes into a small cylindrical section, and then there is a sudden narrowing part with a stationary Mach number $M_m = 1$ in its minimum section. An expanding supersonic section may be located after a sudden narrowing part of the duct. Since the perturbations of the flow in it cannot get into the area before the sudden narrowing, the flow behind it does not affect the stability in front of it. And therefore it was not calculated numerically.

In the subsonic flow adjacent to the over-compressed DW, different initial (at time $t = 0$) perturbations are specified. For $M_0 = 5.39$, $M_1 = 2.52$, $M_2 = 0.8$, the contours of the flat duct with a four shock wave in air intake and a Laval nozzle with a sudden narrowing are shown in figure 1*a*. In figure 1*b* shows the stationary distributions of p and M on the x axis for $x < 2$ and the average cross section (from one-dimensional calculation) for $2 \leq x \leq 3.75$. In a one-dimensional approximation in the minimum nozzle section M = 1.

The stability of the presented flow was calculated numerically. At $t = 0$, the same relative perturbations p, ρ, and u were introduced in the subsonic duct (the pressure perturbation Δp is shown in figure 1*b*). The finite difference scheme used is a modified Godunov scheme with a higher approximation order in time and space. The calculation is carried out on a moving and uniform (on each time layer) calculation grid. The moving DW is allocated in such a way that its velocity D, coordinate $x_1(t)$ and all parameters behind the wave on each new layer are determined by three conservation laws, the equation $dx_1 / dt = D$ and the condition of consistency for the C^- – characteristic coming to the DW on the right . Since the flow to the left of the DW is unperturbed and stationary, all its parameters are known functions x_1.

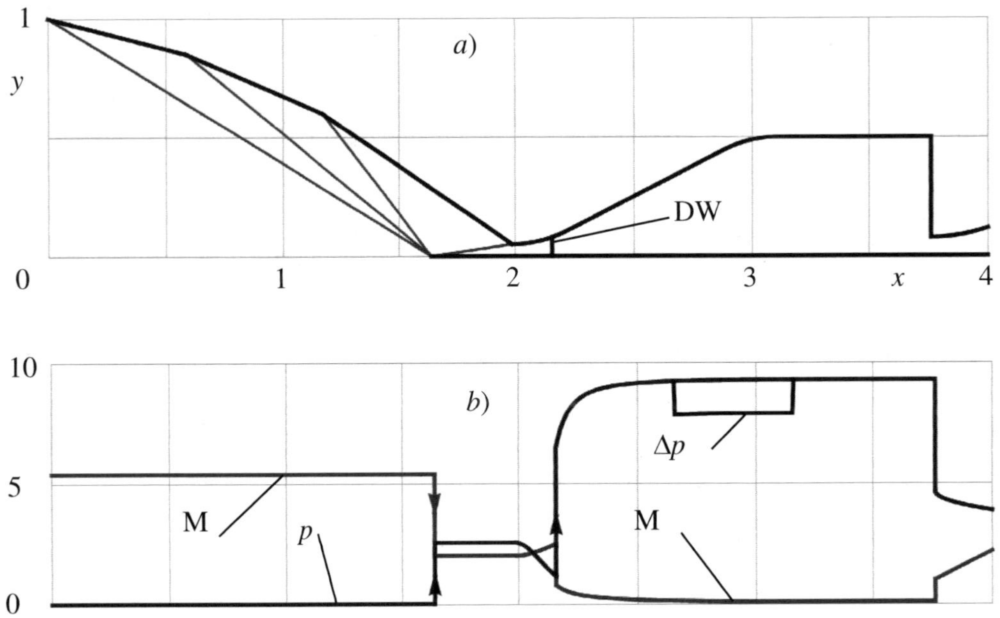

Figure 1. The flow in the duct with over-compressed DW (*a*) and the distribution of pressure and Mach number in it (*b*).

In figure 2, the solid line shows the distribution of p, referred to the pressure of the stationary flow in the cylindrical section, for time $t \leq 5$. The dashed line shows the pressure in the stationary solution. Time t is divided by L / a_0, where $4L$ is slightly less than the length of the duct shown in figure 1*a* and a_0 is the speed of sound at the entrance to the duct. In this example, the initial perturbation, (an extended region in which pressure, density and velocity is reduced by 15% of the parameters in the stationary solution) splits into two regions of reduced pressure, the left region goes to the front of the detonation wave, as shown in figure 2*a*. The arrows in the figure show the direction of movement of the areas at a given time. In this case, the pressure perturbation increases (figure 2*b*) in the narrowing duct and causes the growth of M_2 to 1. Therefore, the detonation wave becomes the Chapman-Jouguet wave, while the front accelerates and moves to the minimum cross section. The pressure behind the DW gradually increases as the perturbation is absorbed, and becomes higher than the pressure in this section of the stationary solution (figure 2*c*). The DW front accelerates in the opposite direction. Meanwhile, the right-hand side of the reduced pressure is reflected from the nozzle with a sudden narrowing, and arrives at the front of the DW (figure 2*d*), increasing the Mach number to 1. If the front did not reach the throat during the passage of the perturbation, then it will again reverse. The pressure perturbations after absorption of the second extended region of reduced pressure are already much smaller, and then the DW front position returns to stationary.

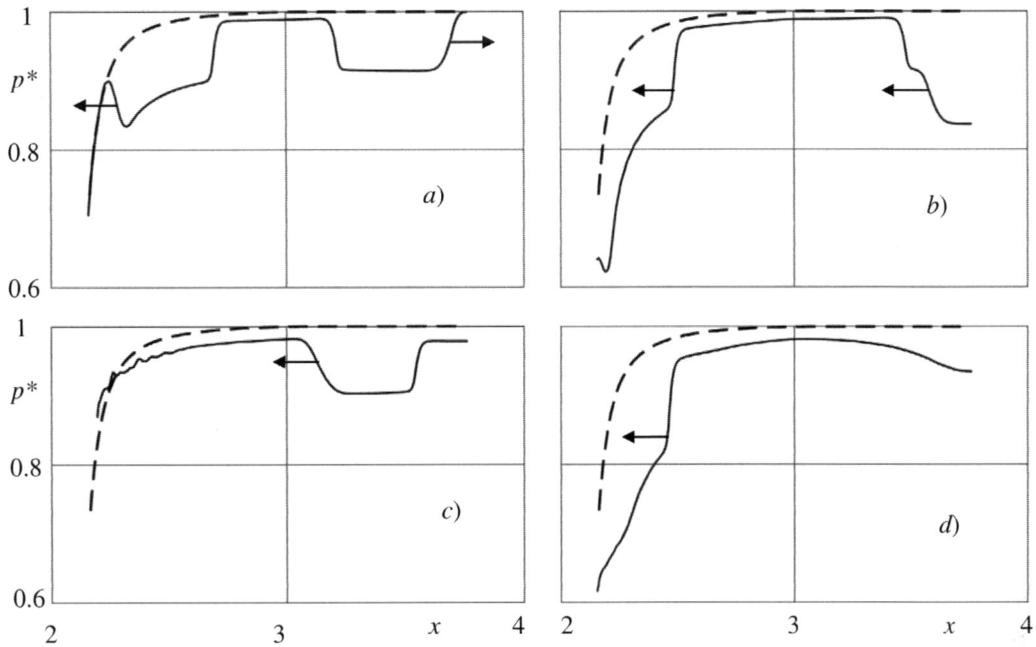

Figure 2. Pressure disturbances between the DW and the minimum nozzle section at $t = 1$ (*a*), 1.5 (*b*), 2.5 (*c*), 5 (*d*).

Also, in numerical studies of the stability of flows with a normal [5] or close to the normal [6] over-compressed DW, the reflection of disturbances from the output sections of the ducts was less intense than from sudden narrowing of the nozzle of the engine-like duct in figure 1.

The resulting evolution of the Mach number for DW M_2 and the coordinates of DW x_{DW} are shown by solid lines in figure 3. For the first time, M_2 increases strongly at the time moment $t \approx 1.5$, when the introduced perturbation arrives at the DW front. The front of the detonation wave begins to move, and it is maximally deflected by the moment $t \approx 2.3$. The second time a significant increase in M_2 occurs at the moment $t \approx 6$ of the arrival of the perturbation to the front, reflected from a sudden narrowing part of the duct. Because of this, the front deviates to $x_{DW} \approx 2.195$. As the calculations show, in the case when the nozzle narrowing is not sudden, but smooth, then the amplitude M_2 and with it the amplitude x_{DW} (due to the reflected disturbance) are smaller. After the arrival of the reflected initial perturbation, M_2 gradually returns to its unperturbed value of 0.8. And x_{DW} gradually moves to the initial value in a stationary flow without disturbances.

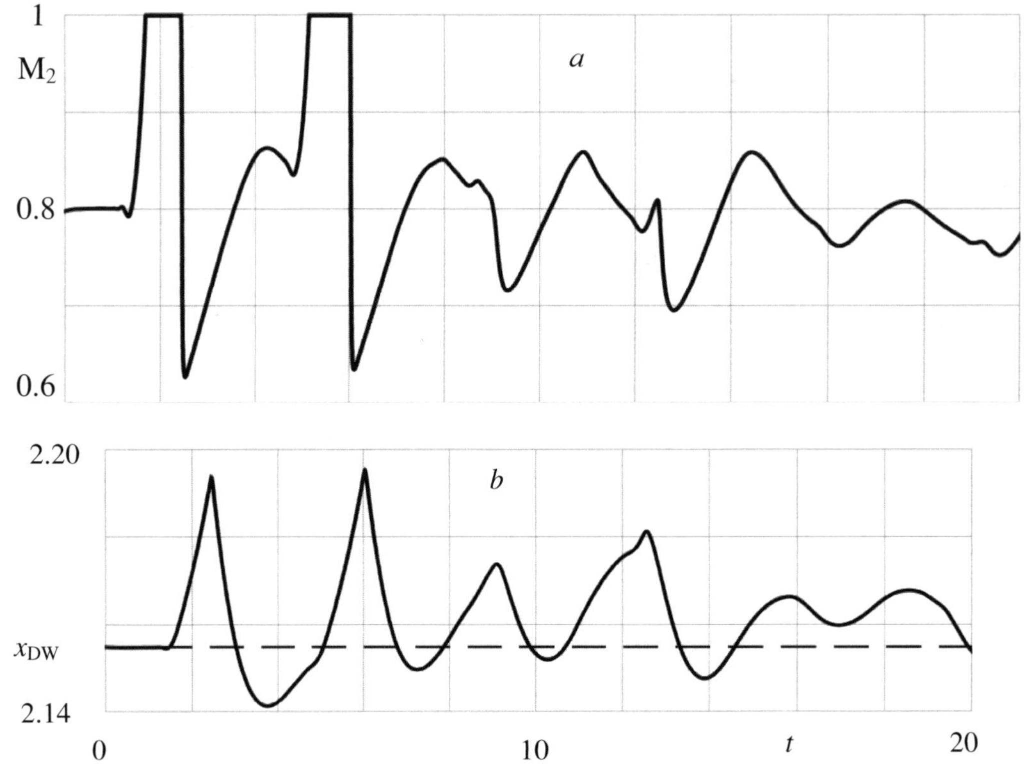

Figure 3. The evolution of M_2 (*a*) and x_{DW} (*b*).

The calculations showed the efficiency and effectiveness of the developed in [15] approach and created programs. We have identified examples of stable flows with over-compressed DW in engine-like ducts of the type described above.

3. Conclusion

The developed approach allows us to study non-stationary one-dimensional flows in ducts of variable cross area with an over-compressed DW at relatively large initial perturbations and analyze the dynamics of perturbation development. This provides additional opportunities for the construction of engines with combustion in a stable stationary over-compressed DW.

The created software package [15] was supplemented with new tools for analyzing flows. It has also been tested for calculating flows with large Mach numbers M_0.

Acknowledgments

Authors are grateful to A.N. Kraiko for leading the scientific work. This work was supported by RFBR grants 20-01-00100 and 18-31-20059

References

[1] Zeldovich Ya B 1940 On the theory of detonation wave propagation in gaseous systems *Zh. Eksp Teor Fiz* **10** (5) 542–569
[2] von Neumann J 1942 Progress report on theory of detonation waves *Office of Scientific Research and Development* Report No. 549
[3] Doring W 1943 Uber der Detonation-vorgang in Gasen *Ann. Phys* **43** 421–436

[4] Kraiko A N 2019 The instability of steady-state flows with a Chapman–Jouguet detonation wave in a duct of variable cross section *Fluid Dynamics* **54** (7) 887–897

[5] Levin V A Manuilovich I S and Markov V V 2010 Excitation and disruption of detonation in gases *Eng Phys J* **83** (6) 1174–1201

[6] Zhuravskaya T A and Levin V A 2016 Stability of gas mixture flow with a stabilized detonation wave in a plane duct with a constriction *Fluid Dynamics*. **51** (4) 544–551

[7] Gidaspov V Y Kononov D S 2020 On the stability of a detonation wave in a channel of variable cross section with supersonic input and output flows *Smart Innovation, Systems and Technologies* **173** 91-106

[8] Babitskii A B 1963 Reflection of Small Disturbances from the Surface of Shock Wave and the Stability of a Shock in a Duct of Variable Cross Section *Tsentr. Nauchn.-Issled. Inst. Mashinostroen., Korolev* [in Russian]

[9] Grin' V T Kraiko A N and Tilliaeva N I 1975 Investigation of the stability of perfect gas flow in a quasi-cylindrical duct *J. Appl. Math. Mech.* **39** (3) 449–460

[10] Grin' V T Kraiko A N and Tilliaeva N I 1976 On the stability of flow of perfect gas in a duct with a closing compression shock and simultaneous reflection of acoustic and entropy waves from the outlet cross section *J. Appl. Math. Mech.* **40** (3) 426–435

[11] Kraiko A N and Shironosov V A 1976 Investigation of flow stability in a duct with a closing shock at transonic flow velocity *J. Appl. Math. Mech.* **40** (4) 533–541

[12] Grin' V T Kraiko A N Tilliaeva N I and Shironosov V A 1977 Analysis of one-dimensional flow stability in a duct with arbitrary variation of stationary flow parameters between the closing shock cross section and duct outlet *J. Appl. Math. Mech.* **41** (4) 651–659

[13] Grin' V T Kraiko A N Tilliaeva N I and Shironosov V A 1982 Investigation of stability of ideal gas flow in ducts with a closing shock wave *Tr. Tsentral. Inst. Aviats. Motorostroen* **1020** 6–21

[14] Galin G Ya and Kulikovskii A G 1975 On the stability of flows appearing at the disintegration of an arbitrary discontinuity *J. Appl. Math. Mech.* **39** (1) 85–92

[15] Valiyev Kh F Kraiko A N and Tillyayeva N I 2020 Stability of one-dimensional steady flows with Detonation Wave in a duct of variable cross-sectional area *Computational mathematics and mathematical physics* **60** (4) 697–710

Fluid flow modeling in solid rock fractures

I N Cherednichenko
Scientific Research Institute for System Analysis, 36/1Nakhimovsky pr., Moscow, Russia

E-mail: inch@jscc.ru

Abstract. This paper considers the solution of the problem of modeling fluid flows occurring while filling solid rock fractures with fluid under pressure. For the mathematical modeling of the solid rock, real data of the rock core sample density was used, obtained in the imaging process of Synchrotron radiation X-ray tomographic microscopy. A calculation method of computational grid cell effective volume, which takes into account rock permeability parameter and is based on tomographic data, is described. To simulate the fluid behavior the modified finite volume method, widely used in hydrodynamics, was applied. Its modification is described for calculating the medium with a variable permeability. The results of numerical experiments are presented.

1. Introduction

Synchrotron radiation X-ray tomographic microscopy is now actively used for non-destructive studies of the internal structure of various rock core samples. Scanning results are widely available on the websites of many geology-oriented scientific teams [1]. Modern tomographs used in geology have high spatial resolution and allow obtaining sufficiently detailed data about the structure of the studied samples.

The result of such studies is a set of values for the optical density of radiation absorption, usually presented in the form of 3D-maps for rock cells, which size is determined by the resolution of the tomograph.

Based on the data obtained from the tomograph, it is possible to create a function of the rock density depending on the coordinates of the elementary volume in the studied sample. But, since the size of the elementary volume is defined by the resolution of the tomograph, it is impossible to obtain additional data on the boundaries of various inclusions and fractures of the studied sample. In a detailed analysis of the optical density function of the rock, three categories of elementary cells can be distinguished. The first type is cells with zero absorption of radiation, which can be interpreted as fractures in the rock. The second type is cells with maximum absorption, which can be considered as parts of the rock without any fractures. The third type of cells corresponds to partially filled rock areas with fractures.

Therefore, the volume that the fluid is able to fill within each cell of the rock sample can be determined by this 3D density -map.

2. Modeling the parameters of the computational domain

In the correspondence with the results of tomographic studies of the rock sample, a computational domain is created, covered with a 3D mesh with a fixed cell size. The cell geometric size corresponds to the spatial resolution of the tomograph. For each cell, the optical density Ro_{xyz} has its own value determined from the initial data.

The ratio of the value Ro_{xyz} in a particular cell to the maximum value of the optical density of the sample determines the porosity of the cell, and therefore can be used to calculate the effective free volume of the cell $V_e$which it is possible to fill with fluid.

$$V_e = V_0\left(1 - \frac{Ro - Ro_{min}}{Ro_{max} - Ro_{min}}\right), Ro > 0$$
$$V_e = V_0, Ro = 0,$$

where $V_0 = \Delta x \Delta y \Delta z$, indexes *x, y, z* are omitted. The cell effective volume V_e will be used in the difference scheme of the finite volume method.

3. The Finite Volume Method

While modeling fluid's behavior, we will accept the following assumptions:

- the fluid is perfect,
- the fluid is incompressible,
- the fluid in the cell cannot occupy a volume greater than the effective volume
- while the entire available volume is not occupied by the fluid, the fluid only flows into the cell.

The behavior of an incompressible fluid is described by the system of Navier-Stokes equations, supplemented by conservation laws.

$$\begin{cases} \oint_{\partial\Omega}(\rho dx - \rho u dt) = 0 \\ \oint_{\partial\Omega}(\rho u dx - (P - \rho u^2)dt) = 0 \\ \oint_{\partial\Omega}(E dx - u(P + E)dt) = 0 \end{cases}$$

Various numerical methods can be used to solve this system of equations numerically, but in this case, conservative methods based on local balances and the laws of conservation of all physical parameters of the computational cell are most suitable. One such method is the finite volume method (FVM).

The main idea of this method is the transition of the initial equations from the integral form to the differential one, which allows one to construct difference schemes for local cell parameters [2], [3].

The classical scheme for the realization of the finite volume method was used, divided into Euler and Lagrangian steps.

The finite volume method is widely known and implemented in such hydrodynamic packages as FLUENT, CFX, STAR – CCM [4], [5], etc., which are used in calculations for solving a wide range of problems. However, the model of the computational domain of solid rock described above, taking into account the permeability of the medium, required to modify the standard difference scheme. The main innovation of the FVM developed by the author is that, depending on the input data of the medium, the effective volume of the computational cell changes although its geometric dimensions remain constant. Such modification of the method allows one to take into account the porosity of the medium at each point of the computational domain. The essential aspects of the realization of the modified difference scheme are given below.

4. The Euler step

In the classical realization, it is assumed that at the Euler step the fluid flow is frozen and the instantaneous values of the flow parameters at the cell boundaries are calculated. The modifications made were due to the need to take into account the permeability of the medium and the homogeneity of the liquid phase of the injected fluid.

The fluid flow velocities at the boundaries were calculated standardly, using approximation by values in neighboring cells:

$$\Delta u = \frac{\Delta t(P_{+0.5} - P_{-0.5})}{\rho(\Delta x + (u_{+0.5} + u_{-0.5}) \cdot 0.5 \cdot \Delta t - \mu \Delta t \Delta x)},$$

μ - viscosity, ρ - fluid density, Δt - time step. Values of variables with fractional indices refer to approximated values at the cell boundaries.

But to calculate the mass flows across the boundaries, the already modified formula was used:

$$\Delta M = \rho \cdot \Delta u \cdot \Delta t \cdot (\Delta x_e \cdot \Delta x_e) \text{, where } \Delta x_e = (V_e)^{1/3}.$$

In this formula, the cell effective volume is used, taking into account the permeability of the medium and determining the changing area of the boundaries $(\Delta x_e \cdot \Delta x_e)$ of each cell.

In addition, significant changes in the realization of the difference scheme relate to the conditions of filling a cell with fluid.

The first condition is that a stream cannot flow from a cell in which there is no fluid yet.

$$\Delta M = \rho \cdot \Delta u \cdot \Delta t \cdot (\Delta x_e \cdot \Delta x_e) \cdot I(M_{\pm 1} > 0)$$

The second condition is that the volume of fluid entering the cell cannot be greater than the effective volume of this cell:

$$\Delta u = \Delta u \cdot I(M > V_e \cdot \rho).$$

Here I - is the indicator [6], equal to 1 when the condition is met and equal to 0 in the opposite case.

In the next step, the impulse balance for the fluid in a cell is calculated:

$$p = \Delta u \cdot \Delta M + M \cdot u.$$

After that, the intermediate mass balance for a cell is calculated:

$$\widetilde{M} = \Delta M + M.$$

At this point Euler step is considered completed, the fluid flow is thawed and the transition to the Lagrangian step is performed.

5. The Lagrangian step

At this step, the fluid moves and the calculation of the final balances of local masses, velocities and pressures in the cells is performed. The pressure in a cell is calculated in proportion to the ratio of the volumes available and occupied by the fluid:

$$P = \begin{cases} P_b + \frac{(P_i - P_b) \cdot \widetilde{M}}{V_e \cdot \rho}, & V_e > 0 \\ P_f, \ V_e = 0 \end{cases}$$

P_b - background air pressure, P_i - pressure of the fluid supplied to the rock sample.

The final fluid velocity in a cell is calculated according to the law of momentum and mass conservation.

6. Determination of boundary conditions

IOP Conf. Series: Materials Science and Engineering **927** (2020) 012044 doi:10.1088/1757-899X/927/1/012044

To simulate conditions on the boundary of the computational domain of the rock sample through which fluid flows in, a layer of cells filled with liquid is set under constant external pressure P_i. On the lateral faces of the domain, a layer of cells with an effective volume of 0 is specified to simulate the effect of fluid non-leakage. On the boundary of the domain, where fluid flows out, a layer of cells with free leakage is defined.

The time calculation step is chosen so that it does not exceed the value obtained from the assessment of stability conditions.

Like all explicit schemes, the finite volume method is sensitive to the choice of the time step integration. Assessment of stability is consistent with the physics of the process. From the equation of dependence of speed on pressure difference:

$$\Delta u = -\frac{\Delta P \Delta t}{\rho(\Delta x + u\Delta t)},$$

one can see that when $\Delta x < u\Delta t$ the denominator value is 0 or less than zero. The physical meaning of this is that the fluid flows a distance greater than the geometric size of a cell during the calculation time step and the integration is not correct. From here we get an estimate of the maximum possible time step: $\Delta t > \Delta x/u$ or $\Delta t < \Delta x \cdot \rho/\Delta P$ which, as it turned out, is quite true.

7. Results of a numerical experiment

The program was developed in C ++, with the parallelization of calculations in one of the coordinates using the technology of displacing multitasking. The division into computational domains for each flow occurred along the *z* coordinate. At each step - Euler and Lagrange, 8 parallel flows were launched with synchronization of the completion of the calculations for each step. This technology allows one to realize the benefits of multi-core parallel data processing by modern processors.

After calculating 5 time steps - to control the results obtained, the procedure of constructing streamlines was carried out and the obtained result was visualized on the monitor screen.

The calculation was carried out for data of tomographic studies of rock samples with a resolution of 200x200x200 cells.

The density of fluid is 1000 kg / m3, which corresponds to ordinary water.

The pressure of the injected fluid at the inlet face of the test sample in different calculations was from 3 to 7 atmospheres. The figures below correspond to calculations for a fluid pressure of 5 atmospheres.

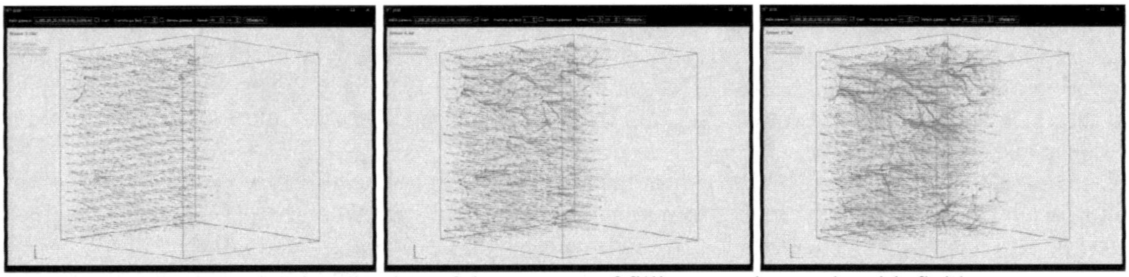

Figure 1. Visualization of the process of filling a rock sample with fluid

Figure 1 clearly shows the formation of fluid flows in the fractures of the studied rock sample.

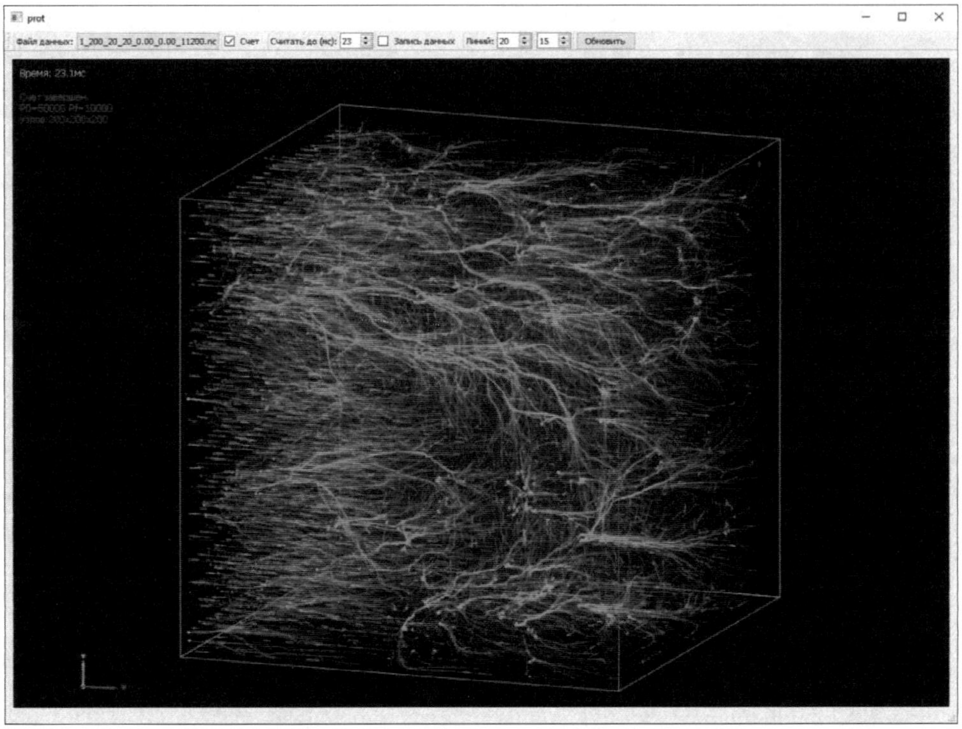

Figure 2. A steady picture of fluid flows

Figure 2 displays the steady-state results of the calculations and gives a visual representation of the filling of the fractures with fluid of the studied rock sample.

It should be noted that when using modern processors, the calculation and visualization of results occur interactively, in almost real time. When switching to higher resolutions of the computational grid (500x500x500 cells and higher), you will need to recompile the program to the x64 runtime or even switch to MPI with a multiprocessor implementation.

8. Conclusion

The result of the project is a software package that allows you to simulate the process of filling rock samples with fluid under pressure. The main advantages include the ability to observe the current process of forming fluid flows interactively in 3D display with a choice of the scale and point of camera position. In the calculations, the results of Synchrotron radiation X-ray tomographic microscopy studies of real rock samples were used as initial data.

The scientific achievements of this work include the concept of cell effective volume, which takes into account rock porosity according to tomographic data, the realization of the modified difference scheme for the FVM and numerical experiments paired with visualization of calculation results.

The created software package allows conducting numerical experiments with existing data obtained as a result of tomography of geological samples. Additional advantages include the fact that this approach allows multiple studies under various initial conditions, which do not destroy the samples

Acknowledgments

The project was completed as part of the work on the state assignment for basic research (The State Science Program 14) on the topic "34.9. Virtual environment systems: technologies, methods and algorithms of mathematical modeling and visualization" (0065-2019-0012).

IOP Conf. Series: Materials Science and Engineering **927** (2020) 012044 doi:10.1088/1757-899X/927/1/012044

References

[1] Rock Physics http://www.rockphysics.org (Internet)

[2] Toro E F 2009 *Riemann Solvers and Numerical Methods for Fluid Dynamics* (Berlin Heidelberg :Springer-Verlag)

[3] Bertolazzi E, Manzini G 2004 *A cell-centered second-order accurate finite volume method for convection-diffusion problems on unstructured meshes Mathematical Models and Methods in Applied Sciences*, **14(8)** 1235–60

[4] Ansys Fluent https://www.ansys.com/products/fluids/ansys-fluent (Internet)

[5] Simcenter STAR-CCM https://www.plm.automation.siemens.com (Internet)

[6] Folland G B 2013 *Real Analysis: Modern Techniques and Their Applications* (John Wiley & Sons)

IOP Conf. Series: Materials Science and Engineering **927** (2020) 012045 doi:10.1088/1757-899X/927/1/012045

Effect of viscous dissipation on the flow structure and pressure losses for a viscoplastic fluid flow through a sudden pipe contraction

K Ryltseva, E Borzenko and G Shrager
Tomsk State University, 36 Lenin Ave., Tomsk 634050, Russia

E-mail: kiraworkst@gmail.com

Abstract. A viscoplastic fluid flow through an axisymmetric sudden pipe contraction is studied numerically in a laminar regime under non-isothermal conditions. Governing equations are written in terms of stream function – vorticity – temperature variables. A modified Schwedoff - Bingham rheological model is used to describe characteristics of the fluid. The problem is solved by the finite-difference method based on the alternative directions scheme. Simulation of the process involves taking into account viscous dissipation effects and temperature dependence of viscosity. Distributions of flow characteristics and flow structures are shown and compared with those obtained in the case when viscous dissipation is neglected. A local resistance coefficient is calculated and presented as a function of the Brinkman number.

1. Introduction

Nowadays, particular attention is paid to flows of polymer melts and solutions due to their frequent application in various industries. These materials usually exhibit both viscous and plastic properties and, therefore, are called viscoplastic fluids, which are characterized by “yield stress”. Yield stress is defined as a critical stress value below which fluids do not flow. Yield stress concept remains an ambiguous issue in rheology, however, it is believed to be a favorable notion for structured fluid flow modeling [1]. Viscoplastic fluid flow is characterized by a specific structure formation, which includes both yielded and unyielded regions. The former occur in the regions where the stress level exceeds yield stress; the latter occur in the regions where the stress level is below yield stress. To analyze the structure of a viscoplastic fluid flow, the extent and shape of yielded/unyielded regions are studied depending on problem parameters [2]. Pioneering works dealing with viscoplastic fluid flows considered simple geometry of the flow domain (straight planar channels, circular pipes, eccentric annuli, etc.) [3], [4]. In course of time, complex geometries, such as sudden pipe expansions or contractions, have become an engineering interest [5]. The authors of [6] have performed a numerical simulation of a Bingham plastic flow through a sudden pipe contraction under dominant inertia effects. It is shown that yield number had no effect on the centerline velocity evolution in contraction plane vicinity. The paper [7] presents an experimental investigation of the ice slurry flow through sudden contractions and expansions. The Bingham model is used to describe the rheology of the slurry. Special attention is paid to calculations of local pressure loss coefficients. The effect of yield stress and geometric parameters on a viscoplastic fluid flow is analyzed on the basis of finite-element numerical simulation in [8]. Variation in the flow structure, namely, extension of moving and rigid static zones, and a vortex size are given in relation to governing parameters.

The majority of works consider viscoplastic fluid flows under isothermal conditions. From a practical point of view, it is more important to study non-isothermal flows, since heat generation is observed in real industrial processes. Thermal effects may cause undesirable changes in the material or dangerous irreversible processes, such as inflammation or explosion. A comprehensive review on convective flow and heat transfer in a channel with backward- and forward-facing steps is given in [9]. Heat transfer phenomenon is analyzed for a Bingham plastic flow through sudden expansions and contractions in [10]. The obtained results show strong impact of the yield number on the Nusselt number distribution for a flow through a sudden expansion comparing to that of the contraction case. Another parametric study of a non-isothermal viscoplastic fluid flow through an axisymmetric sudden expansion is presented in [11]. This work is aimed to assess the effect of dimensionless parameters on the flow regime (recirculating and non-recirculating). Calculations of pressure losses in a Bingham fluid flow through abrupt contractions and expansions are performed in [12]. It has been discovered that the effect of the yield stress number on the local friction coefficient shows different behavior for the considered pipe geometries.

Detailed information about viscoplastic fluid flow structure and its changes with variation in main parameters is very important from a technological point of view. Moreover, to design industrial processes, it is also necessary to know the values of pressure losses. The aim of this work is to study the effect of viscous dissipation on the flow kinematics, flow structure, and pressure losses for a viscoplastic fluid flow through a sudden pipe contraction.

2. Formulation of the Problem

A steady-state viscoplastic fluid flow through a 2:1 sudden pipe contraction is considered under non-isothermal conditions with allowance for viscous dissipation and temperature-dependent viscosity. Schematic representation of an axisymmetric computational domain is illustrated in figure 1.

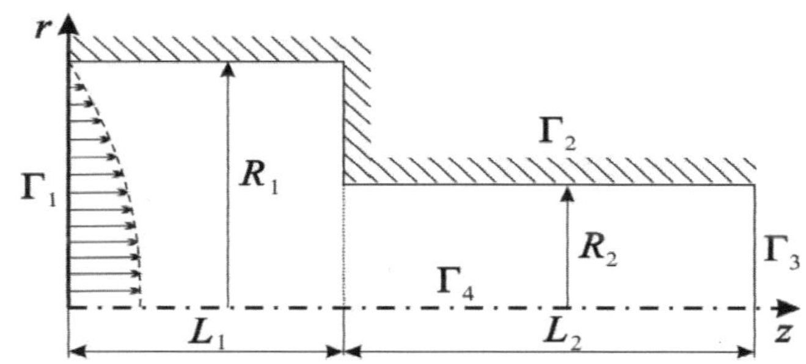

Figure 1. Schematic of the pipe configuration.

Mathematical formulation of the problem includes dimensionless equations written in terms of stream function (ψ), vorticity (ω), and temperature (θ):

$$\frac{\partial(v\omega)}{\partial r}+\frac{\partial(u\omega)}{\partial z}=\frac{2\cdot B}{\mathrm{Re}}\left(\nabla^2\omega-\frac{\omega}{r^2}\right)+\frac{2\cdot S}{\mathrm{Re}}, \tag{1}$$

$$\nabla^2\psi-\frac{2}{r}\frac{\partial\psi}{\partial r}=-r\omega, \tag{2}$$

$$\frac{\partial(v\theta)}{\partial r}+\frac{\partial(u\theta)}{\partial z}=\frac{2}{\mathrm{Pe}}\left(\nabla^2\theta+A^2B\cdot\mathrm{Br}\right)-\frac{v\theta}{r}, \tag{3}$$

where the source term is represented as

$$S=2\frac{\partial^2 B}{\partial r\partial z}\left(\frac{\partial v}{\partial r}-\frac{\partial u}{\partial z}\right)+2\frac{\partial B}{\partial r}\frac{\partial\omega}{\partial r}+2\frac{\partial B}{\partial z}\frac{\partial\omega}{\partial z}+\left(\frac{\partial^2 B}{\partial z^2}-\frac{\partial^2 B}{\partial r^2}\right)\left(\frac{\partial v}{\partial z}+\frac{\partial u}{\partial r}\right)+\frac{\partial B}{\partial r}\frac{\omega}{r}.$$

The stream function and vorticity are defined as follows

$$u = \frac{1}{r}\frac{\partial \psi}{\partial r},\ v = -\frac{1}{r}\frac{\partial \psi}{\partial z},\qquad \omega = \frac{\partial v}{\partial z} - \frac{\partial u}{\partial r}.$$

Viscoplastic fluid rheology is determined using a modified Schwedoff-Bingham model [13, 14]. The dimensionless apparent viscosity is written as a function of temperature

$$B = \left(0.5\mathrm{Bn}\cdot\exp(-\beta\theta) + A\cdot\exp(-\theta)\right)/A.$$

In the governing equations, v and u are the radial and axial components of the velocity vector, respectively, $A = \left[2(\partial u/\partial z)^2 + 2(\partial v/\partial r)^2 + 2(v/r)^2 + (\partial u/\partial r + \partial v/\partial z)^2\right]^{1/2}$ is the intensity of the rate of strain. The following dimensionless criteria are used: $\mathrm{Re} = \rho UD/\mu_1$ is the Reynolds number, $\mathrm{Pe} = c\rho UD/\lambda$ is the Peclet number, $\mathrm{Bn} = \tau_1 D/(U\mu_1)$ is the Bingham number, $\mathrm{Br} = U^2\mu_1\beta_2/\lambda$ is the Brinkman number, where ρ is the density, c is the heat capacity, λ is the thermal conductivity, $D = 2R_2$ is the downstream pipe diameter, $\tau_1 = \tau_0\exp[-\beta_1(T_1 - T_0)]$ and $\mu_1 = \mu_0\exp[-\beta_2(T_1 - T_0)]$ are the yield stress and consistency index at T_1, β_1 and β_2 are coefficients in the temperature dependencies of τ_1 and μ_1, respectively, $\beta = \beta_1/\beta_2$, T and T_1 are dimensional temperatures of the fluid in the flow and on the solid wall, respectively, T_0 is the reference temperature. The properties of the fluid (ρ, c, λ) are assumed to be constant. The space, velocity, and stress are scaled with the downstream pipe radius R_2, the average velocity in the downstream pipe U, and the value of $U\mu_1/R_2$, respectively. The dimensionless temperature is given by $\theta = \beta_2(T - T_1)$.

Specification of boundary conditions at the pipe inlet implies a calculation of the axial velocity and temperature profiles corresponding to a steady one-dimensional non-isothermal flow of a Schwedoff-Bingham fluid in an infinite pipe. Description of this procedure is presented in [15]. The obtained velocity distribution allows calculating vorticity and stream function. The calculated temperature, stream function, and vorticity determine inlet boundary conditions. At the pipe outlet, derivatives of the main variables with respect to z are set to zero. To eliminate geometric singularity effect on the flow behavior near inlet and outlet sections, these boundaries are remote from the contraction plane ($L_1 / R_2 = 20$, $L_2 / R_2 = 120$). On the solid wall, no-slip boundary conditions are assigned, and the dimensionless temperature of the fluid equals to zero. Along the symmetry axis, symmetry conditions are satisfied. Mathematical representation of the boundary conditions is written as follows:

$$\Gamma_1: u = f_1(r),\quad \psi = \int_0^r urdr,\quad \omega = -\frac{\partial u}{\partial r},\quad \theta = f_2(r),\quad 0 \le r \le \frac{R_1}{R_2},\quad z = 0;$$

$$\Gamma_2: \psi = const,\quad \omega = -\frac{1}{r}\frac{\partial^2 \psi}{\partial r^2},\quad \theta = 0,\quad r = \frac{R_1}{R_2},\quad 0 \le z \le \frac{L_1}{R_2},$$

$$\psi = const,\quad \omega = -\frac{1}{r}\frac{\partial^2 \psi}{\partial z^2},\quad \theta = 0,\quad 1 \le r \le \frac{R_1}{R_2},\quad z = \frac{L_1}{R_2},$$

$$\psi = const,\quad \omega = -\frac{\partial^2 \psi}{\partial r^2},\quad \theta = 0,\quad r = 1,\quad \frac{L_1}{R_2} \le z \le \frac{L_1}{R_2} + \frac{L_2}{R_2};$$

$$\Gamma_3: \frac{\partial \psi}{\partial z} = 0,\quad \frac{\partial \omega}{\partial z} = 0,\quad \frac{\partial \theta}{\partial z} = 0,\quad 0 \le r \le 1,\quad z = \frac{L_1}{R_2} + \frac{L_2}{R_2};$$

$$\Gamma_4: \psi = 0,\quad \omega = 0,\quad \frac{\partial \theta}{\partial r} = 0,\quad r = 0,\quad 0 \le z \le \frac{L_1}{R_2} + \frac{L_2}{R_2}.$$

3. Computational Method

The formulated problem of a non-isothermal viscoplastic fluid flow through an axisymmetric sudden pipe contraction is solved numerically by the method, which implies addition of derivatives of the unknown functions (ψ, ω, θ) with respect to "time". Computational process is reduced to a step-by-step approximation of the unsteady problem solution to the solution of the initial (steady) problem [16]. A finite-difference method based on the alternative directions scheme is used to discretize the governing equations [17]. A resulting system of the obtained algebraic equations is solved by a tridiagonal matrix algorithm [16].

The rheological equation is regularized to provide a through computation of the viscoplastic fluid flow without explicit separation of unyielded regions. A temperature-dependent apparent viscosity corresponding to a modified rheological model is determined by formula [18]

$$B=\left(0.5\mathrm{Bn}\cdot\exp(-\beta\theta)+\exp(-\theta)\left(A^2+\varepsilon^2\right)^{1/2}\right)/\left(A^2+\varepsilon^2\right)^{1/2},$$

where a regularization parameter is taken as $\varepsilon = 0.0125\mathrm{Bn}$.

Approximation convergence is tested on a sequence of square grids at Re = 1, Pe = 10, Br = 1, Bn = 6 in order to verify the developed numerical algorithm. Maximum values of the axial velocity and temperature are calculated in the contraction plane and at the pipe outlet at various grid steps (h) and presented in table 1. The results demonstrate convergence of the algorithm with mesh refinement. Further calculations are performed using a grid step of 0.025.

Table 1. Approximation convergence.

h	$u_{\max}$ (contraction)	$\theta_{\max}$ (contraction)	$u_{\max}$ (exit)	$\theta_{\max}$ (exit)
0.1	1.9481	1.0644	2.5062	1.3977
0.05	1.9435	1.0958	2.3725	1.3711
0.025	1.9451	1.1126	2.4091	1.3911
0.0125	1.9460	1.1208	2.4341	1.4028

4. Results and Discussion

To assess the effect of viscous dissipation, the flow characteristics are considered at varying Brinkman number, which is known as the ratio of the heat resulting from viscous dissipation to the heat transported by conduction. Figure 2 demonstrates temperature fields at various Br. At low Brinkman numbers, a temperature rise is insignificant and mostly observed in the area downstream of the contraction plane (figure 2 (a)). With an increase in Br, the temperature rise becomes larger in both parts of the pipe with a sudden contraction. A highly heated area is located in the vicinity of the pipe outlet. Such temperature changes primarily affect the viscosity field and flow structure.

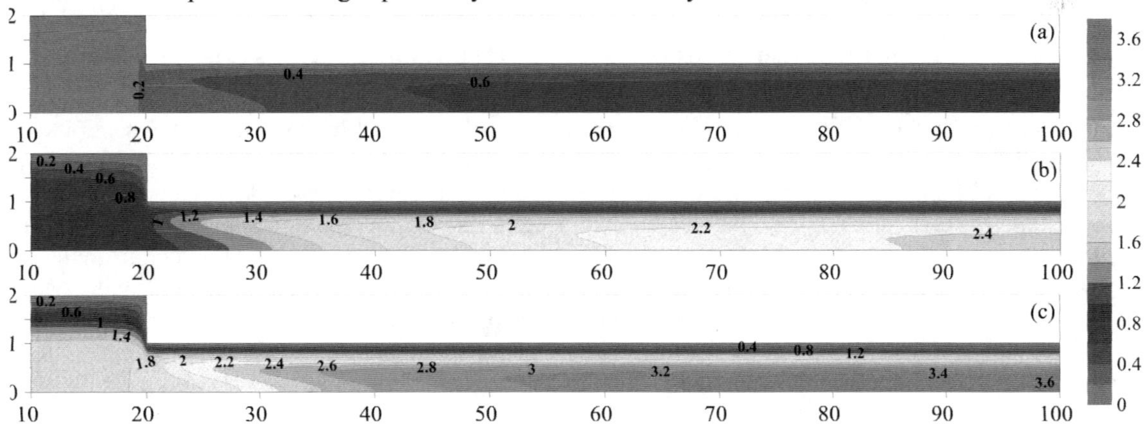

Figure 2. Temperature contours at Re = 1, Pe = 100, Bn = 6: (a) Br = 0.5, (b) Br = 2, and (c) Br = 4.

Flow kinematic characteristics also undergo considerable changes with Brinkman number. The higher the value of this parameter, the faster the fluid flows through the pipe (figure 3 (a)-(c)). At high Br, it is necessary to consider longer downstream pipe in order to attain a fully developed one-dimensional flow. When viscous dissipation is neglected (Br = 0), the flow structure includes a dead zone in the inner corner (shaded black), a yielded region (unshaded), and unyielded regions (shaded black) upstream and downstream of the contraction plane (figure 3 (d)). At Br = 2 and Br = 4, the unyielded regions grow narrow (figure 3 (e),(f)). Notable that at Br > 0, the downstream unyielded region is not uniform in *z*-direction. Moreover, additional dead zones appear in both parts of the pipe along the wall. These rigid layers tend to expand with increasing Brinkman number. In the vicinity of contraction plane, the flow is characterized by large deformation rates and, consequently, by low viscosity. Hence, there is no complete "blockage" of the downstream pipe entrance.

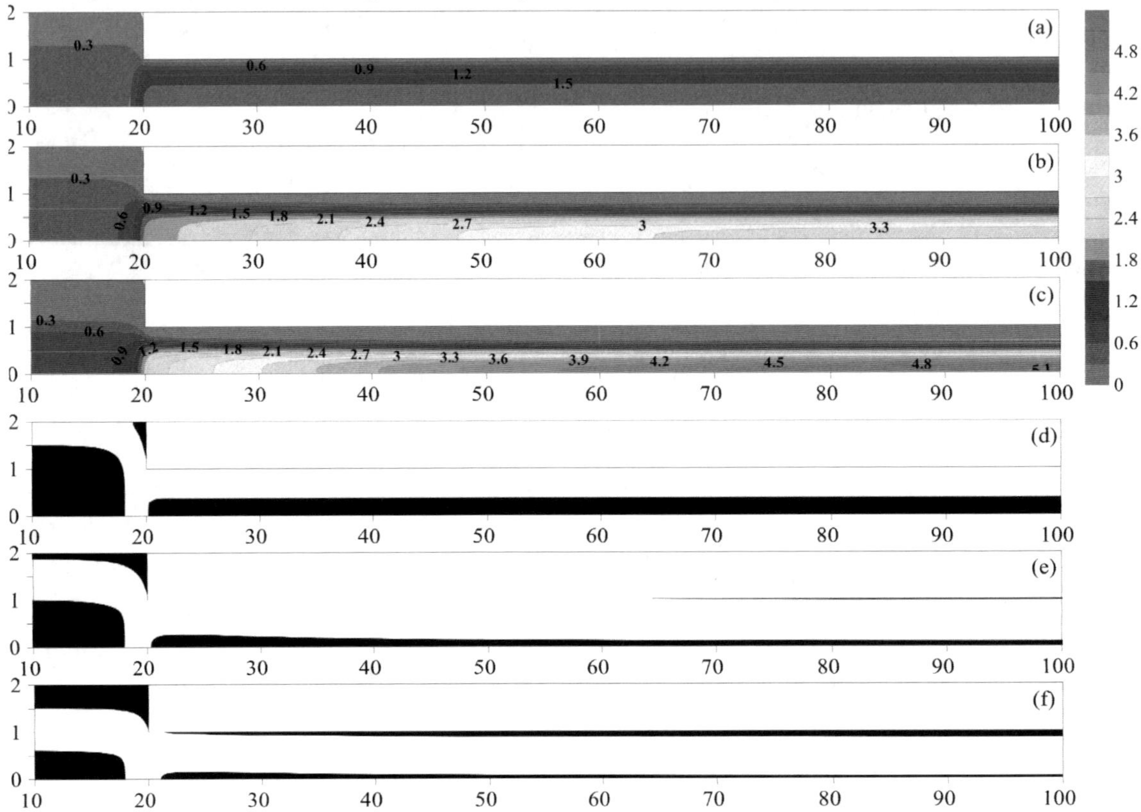

Figure 3. Axial velocity distributions (a) – (c) and flow structures (d) – (f) at Re = 1, Pe = 100, Bn = 6: (a), (d) Br = 0; (b), (e) Br = 2; and (c), (f) Br = 4.

In figure 4 (a) - (b), the temperature and viscosity profiles at the inlet and outlet pipe sections are presented for a Brinkman number equal to either 0.5 or 4. Comparing these two cases and leaving out of account differences in the temperature values, which are higher for Br = 4, the behavior of the temperature profiles are similar. As for viscosity profiles, they are completely different in the considered cases. The main difference is observed near pipe wall, where the rigid layers are revealed at Br = 4. In these regions, the viscosity increases dramatically and attains its maximal values, which is not typical for the cases at low Brinkman numbers, particularly, at Br = 0.5.

Figure 4 (c) demonstrates dependency diagram for a local resistance coefficient representing energy losses due to a sudden pipe contraction [19]. The coefficient is calculated by formula

$$C_{\text{loc}} = \Delta p_{\text{loc}} / 0.5\rho U^2 + \alpha_1 \left(R_2 / R_1 \right)^4 - \alpha_2 .$$

Here, Δp_{loc} is the pressure loss associated with the energy, which is required to alter velocity profile as the fluid enters downstream pipe after flowing through a contraction plane; α_1 and α_2 are kinetic energy correction factors, which account for difference between velocity profiles of fully developed flows in the upstream and downstream pipes. The plot of C_{loc}(Br) exhibit intricate behavior, which implies an increase in the energy losses in the range of Br = [0, 1.5] and a decrease at higher Brinkman numbers.

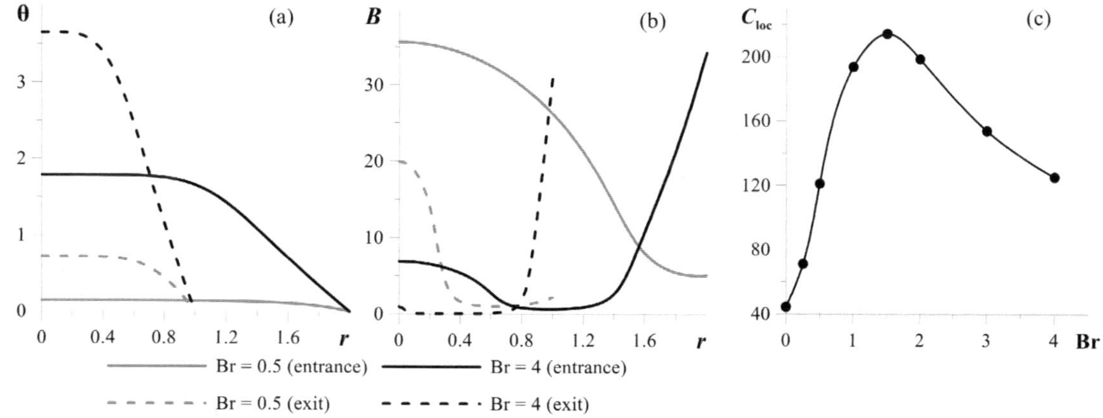

Figure 4. Profiles of the (a) temperature and (b) apparent viscosity along the pipe entrance and exit; (c) local resistance coefficient as a function of the Brinkman number at Re = 1, Pe = 100, Bn = 6.

5. Conclusion

A non-isothermal Bingham fluid flow through a sudden pipe contraction is simulated numerically using the finite-difference method. Based on the computed results, the temperature contours, velocity fields, and flow structures are obtained at various Brinkman numbers. An increase in this parameter leads to a significant heat-up in the entire flow region resulting in a temperature rise and in a higher flow velocity. Comparison of the flow structures for the cases with and without viscous dissipation effects shows that in the former case, alongside with unyielded regions and dead zone, which are typical for the latter case, additional rigid layers appear along the pipe wall. These layers become wider, while the unyielded regions grow narrow with increasing Brinkman number. Dependency of the local resistance coefficient on the Brinkman number is found to have non-monotonic behavior.

Acknowledgement

This research is implemented at the expanses of the Russian Science Foundation (project no. 18-19-00021).

References

[1] Balmforth N J, Frigaard I A and Ovarlez G 2014 *Annu. Rev. Fluid Mech.* **46** 121–46
[2] Abdali S S, Mitsoulis E and Markatos N C 1992 *J. Rheol.* **36** 389–407
[3] Weymann H D, Chuang M C and Ross R A 1973 *Phys. Fluids* **16** 775–83
[4] Thomas D G 1963 *Ind. Eng. Chem.* **55** 18–29
[5] Mitsoulis E and Tsamopoulos J 2017 *Rheol. Acta* **56** 231–58
[6] Vradis G C and Hammad K J 1996 *Can. J. Chem. Eng.* **74** 457–63
[7] Mika L 2012 *Int. J. Refrig.* **35** 1697–708
[8] Jay P, Magnin A and Piau J M 2002 *J. Fluids Eng.* **124** 700–5
[9] Kherbeet A Sh, et al. 2016 *Int. Commun. Heat Mass Transf.* **76** 237–44
[10] Vradis G C and Hammad K J 1995 *Numer. Heat Transf.* A **28** 339–53
[11] Hammad K J 2017 *Int. J. Heat Mass Transf.* **106** 829–40
[12] Kfuri S L D, Soares E J, Thompson R L and Siqueira R N 2017 *J. Fluids Eng.* **139** 021203 1–8
[13] Schwedoff T 1889 *J. Phys. Théorique Appliquée* **8** 341–59
[14] Bingham E C 1922 *Fluidity and Plasticity* (New-York: McGraw-Hill)
[15] Borzenko E I, Ryltseva K E and Shrager G R 2019 *IOP Conf. Ser. Mater. Sci. Eng.* **581** 1–9
[16] Godunov S K and Ryabenkiy V S 1987 *Difference Schemes* (North-Holland: Elsevier Sci. Ltd.)
[17] Samarskiy A A 1971 *Introduction to the Theory of Difference Schemes* (Moscow: Nauka)

IOP Conf. Series: Materials Science and Engineering **927** (2020) 012045 doi:10.1088/1757-899X/927/1/012045

[18] Bercovier M and Engelman M 1980 *J. Comput. Phys.* **36** 313–26
[19] Idelchik I E 1966 *Handbook of Hydraulic Resistance* (Jerusalem: Israel Prog. for Sci. Translat.)

IOP Conf. Series: Materials Science and Engineering **927** (2020) 012046 doi:10.1088/1757-899X/927/1/012046

Hierarchical database management model with uncertain parameters

T I Korotkova and A A Mokhov
Moscow Aviation Institute (National Research University), Volokolamskoe shosse, 4, Moscow 125993, Russia

E-mail: tatyanamail11@yandex.ru

Abstract. Decision making in large technical, biological, economic and other systems always requires the creation of a complex information system. This implies the need to create sophisticated software for the safe storage of information, its updating and management. At the same time, there should be a convenient and easy to use interface that provides the user with all the necessary functions without unnecessary actions. In conditions of uncertainty, the volume of information becomes even greater. A central coordination system must exist to avoid data inconsistencies and redundancies and to increase the reliability of information storage in the database management system (DBMS). In this article, we propose using a database with a hierarchical structure. In such databases, it is possible to quickly update information at each level separately, without accumulating excessive amounts of information for the entire system.

1. Problem statement

DBMS integrates many technologies and is sufficiently complex hierarchical system [1]. This integration of technologies is necessary to provide the ideal conditions to solve the problems of managing large amounts of data. An important task of a database is maintaining the integrity of stored information. Under conditions of uncertainty, the amount of information increases significantly [2–5]. Solving the coordination problem for the whole system exacerbates the problem of inconsistency and redundancy of information that accumulates in local DBMSs. [6,7]. Therefore, there is a need to update information in the database [8-10]. Development of a hierarchical approach to updating databases under conditions of uncertainty eliminates the problem [8,11].

The difficult task of decision-making under uncertainty can be decomposed into a number of local control problems with uncertain parameters. The uncertainty vectors in local problems can both reduce the dimension and keep the same as in the global problem. In the latter case, the control problem based on the guaranteed result principle of the maxmin or minmax type is significantly complicated. Due to this, the volume of redundant information increases. When the conditions for coordination are met, there also appears an accumulation of redundant information which must be updated on time. The challenge is to identify identical blocks of information in the various local objectives of each level. After identification, in this case, it is possible to reduce the total amount of redundant information.

2. Mathematical justification of the rules of decomposition

The database of a complex information system has a hierarchical structure. This is due to the fact that a complex system is composed of subsystems, like modules. Therefore, there is always a coordinating upper level.

The general system model is defined by the mapping: $P: U \to X$, where

U – the set of control vectors u from r -dimensional Euclidean space E^r ,

X – the set of state vectors of the system x of the l -dimensional Euclidean space E^l .

In a complex system of dimension r and l spaces E^r , and E^l are large enough.

The total system model is composed from models of the subsystems. Principles of conformity is defined in the rules of decomposition. And performance indicator G information system is also according to certain rules can be decomposed and linked to the performance indicator of the subsystems.

Note that in a complex system efficiency criterion burdened by dependence on uncontrolled vectors y . It is known that the vector y belongs to some set Y . Thus $G: U \times X \times Y \to R$.

To be specific, the performance indicator should be minimized. It does not limit the generality of the issues and principles of drafting the management model of the database.

In accordance with the principle of the guaranteed result the task of managing the complex information system is:

Find the optimal pair (u^0, x^0) and such y^0 that provide

$$\min_{(u,x)\in Z} \max_{y\in Y} G(u,x,y) = G(u^0(y^0), x^0(y^0), y^0) \tag{1}$$

$$Z = \{(u,x) : u \in U \subset E^r, x \in X \subset E^l, x = P(u)\} \tag{2}$$

Three rules for decomposing a complex system into subsystems:

1) $E^r = \prod_{i=1}^{m} E^{ri}, E^l = \prod_{i=1}^{m} E^{li}, \sum_{i=1}^{m} ri = r, \sum_{i=1}^{m} li = l, X = \prod_{i=1}^{m}, X^i, U = \prod_{i=1}^{m} U^i$.

2) $$x = P(u) = \begin{pmatrix} P_1(u) \\ \vdots \\ P_m(u) \end{pmatrix} = \begin{pmatrix} P_1(u_1, \lambda_1) \\ \vdots \\ P_m(u_m, \lambda_m) \end{pmatrix} \begin{pmatrix} P_1(u_1, \lambda_1) \\ \vdots \\ P_m(u_m, \lambda_m) \end{pmatrix}_{\substack{\lambda i = K_i(u) \\ i=\overline{1,m}}} = \begin{pmatrix} x_1 \\ \vdots \\ x_m \end{pmatrix},$$

where u_i – control vector of the i -th subsystems, $u_i \in U^i$;

x_i – state vector of the i -th subsystems, $x_i \in X^i$;

λ_i – generalized binders the variable that determines the relationship of the i -th subsystem with other subsystems;

$K_i(u)$ – the function that defines the relationship of the i -th subsystem with other subsystems, and $K_i(u) = \sum_{j=1}^{m} K_{ij}(u_j)$.

3) $G(u,x,y)=\sum_{i=1}^{m}G_i(u,x,y)=\sum_{i=1}^{m}G_i(u_i,x_i,\lambda_i,y)_{\substack{\lambda_i=K_i(u)\\ i=\overline{1,m}}}$.

3. The formalization of the decomposition principle with uncertainty

Use obtained by the third rule, the expression for the minimized function. Then the problem (1)-(2) is equivalent to the problem

$$\min_{(u_i,x_i,\lambda_i)\in\bigcup Z_i}\ \max_{y\in Y}\sum_{i=1}^{m}G_i(u_i,x_i,\lambda_i,y)=\sum_{i=1}^{m}G_i(u_i^0(y^0),x_i^0(y^0),\lambda_i^0(y^0),y^0) \tag{3}$$

$$Z_i=\{u_i,x_i,\lambda_i,y):u_i\in U^i\subset E^{ri},x_i\in X^i\subset E^{li},x_i=P_i(u_i,\lambda_i)\} \tag{4}$$

under conditions

$$\lambda_i=K_i(x),\ i=\overline{1,m} \tag{5}$$

The corresponding Lagrangian, after some transformations has the form:

$$L=\sum_{i=1}^{m}G_i(u_i,x_i,\lambda_i,y)+\sum_{i=1}^{m}\beta_i(\lambda_i-K_i(u))$$

From the conditions of rule 2), we obtain

$$L=\sum_{i=1}^{m}G_i(u_i,x_i,\lambda_i,y)+\sum_{i=1}^{m}\beta_i\left(\lambda_i-\sum_{j=1}^{m}K_{ij}(u_j)\right)=\sum_{i=1}^{m}G_i(u_i,x_i,\lambda_i,y)+\sum_{i=1}^{m}\beta_i\lambda_i-\sum_{i=1}^{m}\sum_{j=1}^{m}\beta_iK_{ij}(u_j),$$

In the last term, let us reassign indexes i and j. Then the Lagrange function after the permutation of the signs of the sum has the form

$$L=\sum_{i=1}^{m}G_i(u_i,x_i,\lambda_i,y)+\sum_{i=1}^{m}\beta_i\lambda_i-\sum_{i=1}^{m}\sum_{j=1}^{m}\beta_jK_{ji}(u_i),$$

This manipulation of indices allows you to decompose the Lagrange function:

$$L=\sum_{i=1}^{m}\left[G_i(u_i,x_i,\lambda_i,y)+\beta_i\lambda_i-\sum_{j=1}^{m}\beta_jK_{jj}(u_j)\right],$$

As a result, we have:

$$L=\sum_{i=1}^{m}G_i(u_i,x_i,\lambda_i,y)+\sum_{i=1}^{m}\beta_i(\lambda_i-K_i(u))=\sum_{i=1}^{m}G_{i\beta}(u_i,x_i,\lambda_i,y,\beta),$$

where $G_{i\beta}(u_i,x_i,\lambda_i,y,\beta)=G_i(u_i,x_i,\lambda_i,y)+\lambda_i\beta_i-\sum_{j=1}^{m}\beta_jK_{ji}(u_i),\ i=\overline{1,m}$.

Therefore, the task (3) – (5) is equivalent to the task

$$\min_{\substack{(u_i,x_i,\lambda_i)\in Z_i \\ i=\overline{1,m}}} \max_{y\in Y} \sum_{i=1}^{m} G_{i\beta}(u_i,x_i,\lambda_i,y,\beta) = \sum_{i=1}^{m} G_{i\beta}(u_i^0(y^0,\beta),x_i^0(y^0,\beta),\lambda_i^0(y^0,\beta),y^0,\beta).$$

Moreover, $\lambda_i^0\left(y^0,\beta\right)=K_i\left(u_i^0(y^0,\beta)\right),\ i=\overline{1,m}$, where $u^0\left(y,\beta\right)=(u_1^0\left(y^0,\beta\right),\ldots,u_m^0\left(y^0,\beta\right))$.

It is known that the maximum of the sum is less than or equal to the sum of the maxima. Therefore, the maximization can be performed for each item separately without violating warranty of the result. The minimum amount is greater than or equal to the sum of minimums. But the terms have their local variables, so the minimum of the sum is equal to the sum of the minimums. Therefore, the problem (1)-(2) in the minmax result can be decomposed into local tasks.

$$\min_{(u_i,x_i,\lambda_i)\in Z_i} \max_{y\in Y} G_{i\beta}(u_i,x_i,\lambda_i,y,\beta) = G_{i\beta}(u_i^0(y^0,\beta),x_i^0(y^0,\beta),\lambda_i^0(y^0,\beta),y^0,\beta) \tag{6}$$

These tasks are related to the coordination problem of the 2nd level:

$$\lambda_i^0\left(y^0,\beta\right)=K_i\left(u_i^0\left(y,\beta\right)\right),\ i=\overline{1,m} \tag{7}$$

4. Hierarchical system of accumulation and updating of information in the database

It is known that the hierarchical data model is a model with a relation of subordination between the data. The relation of objects of the hierarchical model is determined by the following rules: *(A subordinate B)* $\Rightarrow$*(B not subordinate A); (A subordinate B)* & *(A subordinate C)*$\Rightarrow$*(B identically C)*, where A, B, C – objects. The hierarchical model is implemented by a tree structure such as an inverted tree. Hierarchical model objects are represented by tree nodes. Descriptions of these nodes are entries called segments. Segments can consist of simple data elements of various types, as well as more complex ones.

A parent record corresponds to an arbitrary number of instances of subordinate (local) records of each type. Here, in the local subsystems, uncertainty manifests itself. A record instance of each type is identified by a key unique to this type. A database is a collection of such trees. The model provides for a lot of navigation operations on the base structure, a segment search operation. Access to data is through a chain of ancestors. There are data manipulation operations: insert, modification and deletion of data. The insert allows for both horizontal and vertical node. Deletion is performed using the natural cascade method, that is, when a record is deleted, all its subordinates are also deleted. In the model the concept of current state is supported. The current position of the segment pointer after the last navigation operation is fixed. Further movement can be made relative to this position.

General view of the hierarchical system for collecting and updating information is presented in the form of a block diagram in figure 1.

5. A few words about software

Software implementation for solving the tasks on the local server requires the use of a high-level programming language (PL) libraries GUI. One will also need a DBMS for hierarchical type of database and framework, which will help to connect database with the program. The popular C # language is suitable as an optimal language solution. For a hierarchical database, a typical representative (the most famous and common) is IBM Information Management System (IMS).

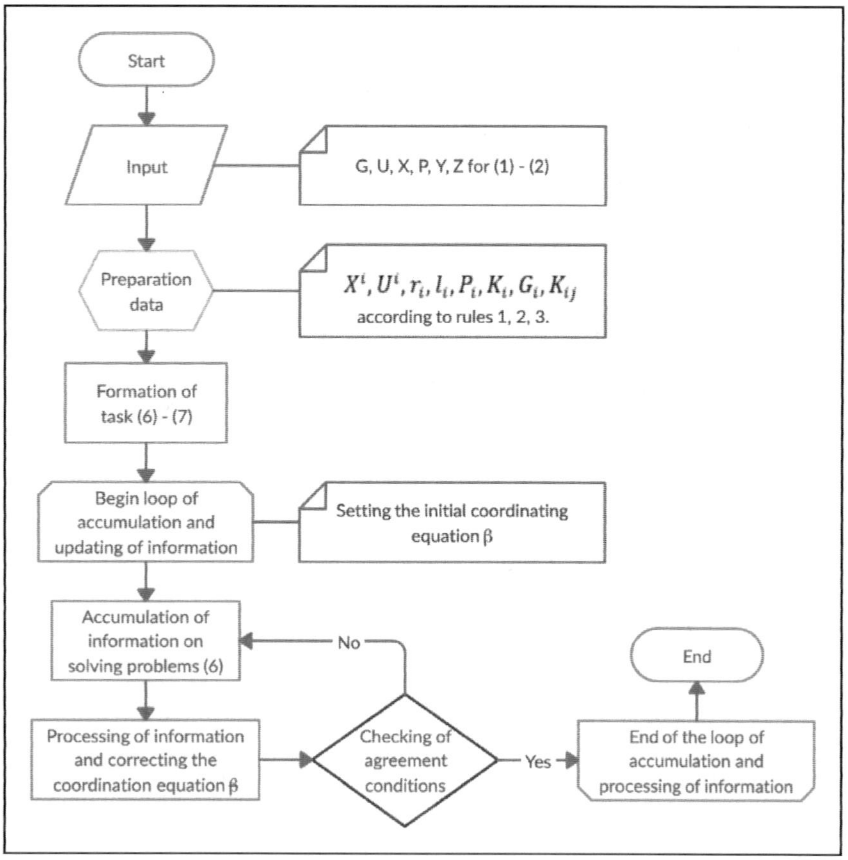

Figure 1. The block diagram of the hierarchical system for the accumulation and updating of information.

References

[1] Korotkova T I and Popova V I 2017 Multicriteria hierarchical iterative interactive algorithm for organizing operational modes of large heat supply systems *J. of Phys.: Conf. Series.* **891** 012164

[2] Panteleev A V and Rybakov K A 2018 Continuous optimal stochastic control systems with incomplete feedback: Approximate synthesis *Automation and Remote Control* **79** 103–116

[3] Davtyan L G, Panteleev A V 2019 Method of parametric optimization of nonlinear continuous systems of joint estimation and control *Journal of Computer and Systems Sciences International* **58** 360–373 doi: 10.1134/S1064230719030079

[4] Panovskiy V N and Panteleev A V 2017 Meta-heuristic interval methods of search of optimal in average control of nonlinear determinate systems with incomplete information about its parameters, *Journal of Computer and Systems Sciences International* **56** 52–63

[5] Panteleev A V, Panovskiy V N and Korotkova T I 2016 Interval explosion search algorithm and its application to hypersonic aircraft modelling and motion optimization *Bulletin of the South Ural State University. Ser. Mathematical Modelling, Programming & Computer Software* (Bulletin SUSU MMCS) **9 3** 55–67

[6] Korotkova T I, Luneva S Yu, Alekseev N S and Vinogradov V I 2018 Minimizing redundant information in a hierarchical database *Information and telecommunication technologies* **39** 11–19

[7] Panteleev A V, Korotkova T I and Alekseychuk A 2017 Usage of the Software System based on Web Conferences for Distance Online Learning *Proceedings of 11th IEEE Inter. Conf. Appl. of Info. and Commun. Tech.* AICT2017 (Moscow) pp 331–335
[8] Korotkova T I, Kuzmina Y 2019 Decision algorithm with uncertainties in a multi-purpose flight safety system *AIP Conference Proceedings* **2181**
[9] Bortakovskii A S, Nemychenkov G I 2019 Optimal in the mean control of deterministic switchable systems given discrete inexact measurements *Journal of Computer and Systems Sciences International* **58** 50–74
[10] Smerchinskaya S O, Yashina N P 2019 Preference levels for clusters of alternatives *International Journal of Modeling, Simulation, and Scientific Computing* **10 4** 1950019
[11] Karane M, Panteleev A V 2019 Hybrid multi-agent optimization method of interpolation search, *AIP Conference Proceedings* **2181** 020028

Improvement of the Landau-Teller model for CO_2 on the basis of the Chapman–Enskog method

E Kustova, A Lukasheva and M Mekhonoshina
Saint-Petersburg State University, 7/9 Universitetskaya nab., St. Petersburg, 199034 Russia
E-mail: m.mekhonoshina@mail.spbu.ru

Abstract. Vibrational relaxation of carbon dioxide in the three-temperature kinetic-theory approach is studied. The focus is on the evaluation of the vibrational energy production terms in the coupled symmetric-bending and asymmetric modes. Using the generalized Chapman–Enskog method, the energy production terms are expressed as functions of the gas temperature, vibrational temperatures of CO_2 modes and probabilities of corresponding transitions. It is shown that a modification of the Landau–Teller formula can be obtained from the rigorous theoretical expressions using several limiting assumptions. Whereas VT relaxation in the bending mode can be described by the Landau–Teller approximation with a satisfactory accuracy, its application to the inter-mode exchange requires introducing relaxation times depending on the vibrational temperatures of the modes involved to the transitions.

1. Introduction

Non-equilibrium kinetics in polyatomic gases is of importance for modeling hypersonic flows, planetary atmospheres, solving environmental problems. Coupled vibrational-chemical kinetics of carbon dioxide may affect gas parameters on the surface of a spacecraft when it enters the Mars or Venus atmosphere. While the kinetic theory of diatomic species received much attention during last decades [1, 2], it is much less developed for polyatomic molecules. This is due to the presence of several vibrational modes and, therefore, multiple relaxation channels.

Under strongly non-equilibrium conditions, state-to-state simulations provide the best accuracy for the prediction of flow variables and heat transfer. However, for polyatomic molecules, implementation of full state-to-state models is extremely challenging since the set of master equations includes several thousands of equations for the vibrational state populations [3, 4]. For real space vehicle geometries, such a description is not suitable, and reliable multi-temperature models have to be used. Two-temperature models [5, 6], while being computationally efficient, cannot describe the inter-mode vibrational energy exchange, which may considerably alter the flow dynamics. More sophisticated multi-temperature transport and relaxation models taking into account different temperatures of CO_2 vibrational modes are developed in [7, 8, 9]. In the above references, the vibrational energy relaxation terms are either calculated by averaging the state-to-state ones or are written in the form of the generalized Landau–Teller formula [10]. Using the state-resolved production terms is not efficient from the computational point of view. On the other hand, the limitations of the Landau–Teller equations for polyatomic gases have not been assessed until the present time. For diatomic species, such a study is carried out in [11, 12], and it is shown that the original Landau–Teller formula is valid only for inviscid gas flows under weak deviations from equilibrium.

IOP Conf. Series: Materials Science and Engineering **927** (2020) 012047 doi:10.1088/1757-899X/927/1/012047

The objective of this study is to develop a kinetic-theory algorithm for calculating the relaxation terms associated with different intra-mode and inter-mode vibrational energy transitions in carbon dioxide in the three-temperature approach and to compare the results with those obtained on the basis of the Landau–Teller approach under various conditions. For the sake of simplicity, we consider a single-component CO_2 gas without chemical reactions.

2. Theoretical model

2.1. Processes in CO_2

CO_2 is a linear triatomic molecule characterized by three types of vibrations: symmetric stretching mode with the frequency ν_1, doubly degenerated bending mode (ν_2) and asymmetric stretching mode (ν_3). The possible channels of vibrational relaxation are the intra-mode vibrational-vibrational (VV_m) and vibrational-translational (VT_m) transitions as well as the inter-mode VV_{m-k} vibrational energy exchange (m, k correspond to different modes).

In the three-temperature model, the vibrational transitions proceeding in the gas-dynamic time scale are VT exchange in the bending mode (VT_2), three-quantum VV transition between bending and asymmetric stretching mode (VV_{2-3}), and VV exchange resulting in the energy redistribution between all modes (VV_{1-2-3}). All other internal energy transitions are assumed to be fast [8]. Thus, in the absence of chemical reactions, the following time relation holds

$$\tau_{\mathrm{el}} < \tau_{\mathrm{rot}} < \tau_{\mathrm{VV}} \sim \tau_{\mathrm{VV}_{1-2}} \ll \tau_{\mathrm{VV}_{1-2-3}} \sim \tau_{\mathrm{VV}_{2-3}} \sim \tau_{\mathrm{VT}_2} \sim \theta. \tag{1}$$

Here, τ_{VV} are characteristic times for intra-mode resonant VV transitions, τ_{VT_2}, $\tau_{\mathrm{VV}_{1-2}}$, $\tau_{\mathrm{VV}_{2-3}}$, $\tau_{\mathrm{VV}_{1-2-3}}$ are the times of corresponding transitions. In this model, fast VV_{1-2} exchange leads to the equilibration of symmetric and bending vibrations with the same vibrational temperature T_{12}. The asymmetric mode has its own vibrational temperature T_3, and equilibrium is attained through slow inter-mode VV_{2-3}, VV_{1-2-3} and VT_2 relaxation.

2.2. Governing equations

Governing equations for the macroscopic variables are obtained starting from the Boltzmann equation using the Chapman–Enskog method and time relation (1). The set of governing equations includes equations of conservation of mass, momentum and energy [2] coupled to the relaxation equations of vibrational energy in different modes:

$$\rho\frac{dE_{12}}{dt} + \nabla\cdot\mathbf{q}_{12} = R_{12}, \tag{2}$$

$$\rho\frac{dE_3}{dt} + \nabla\cdot\mathbf{q}_3 = R_3, \tag{3}$$

where ρ is the density, E_{12} and E_3 are the specific energies of the combined (symmetric–bending) and asymmetric modes, respectively, $\mathbf{q}_{12}$, $\mathbf{q}_3$ are the vibrational energy fluxes in the combined symmetric-bending and asymmetric modes [13]; R_{12} and R_3 are the production terms specified by VT_2 and slow inter-mode VV exchanges [9]:

$$R_{12} = \sum_{i_1,i_2,i_3} (i_1\varepsilon_{100} + i_2\varepsilon_{010}) \sum_j \int \left(J^{\mathrm{VT}_2}_{i_1,i_2,i_3,j} + J^{\mathrm{VV}_{2-3}}_{i_1,i_2,i_3,j} + J^{\mathrm{VV}_{1-2-3}}_{i_1,i_2,i_3,j} \right) d\mathbf{u}, \tag{4}$$

$$R_3 = \sum_{i_1,i_2,i_3} i_3\varepsilon_{001} \sum_j \int \left(J^{\mathrm{VT}_2}_{i_1,i_2,i_3,j} + J^{\mathrm{VV}_{2-3}}_{i_1,i_2,i_3,j} + J^{\mathrm{VV}_{1-2-3}}_{i_1,i_2,i_3,j} \right) d\mathbf{u}, \tag{5}$$

here $\mathbf{u}$ is the molecular velocitiy, j corresponds to the rotational state, $\varepsilon_{i_1 i_2 i_3}$ is the vibrational energy, i_1, i_2, i_3 are quantum numbers of CO_2 vibrational modes, the numbers i_m may vary in the range $i_m = 0, ..., L_m$. The last vibrational level L_m in each mode is determined by the dissociation threshold and is calculated for fixed lowest levels of two other modes. The collision integral operators $J^{\mathrm{VT}_2}_{i_1,i_2,i_3,j}$, $J^{\mathrm{VV}_{2-3}}_{i_1,i_2,i_3,j}$, $J^{\mathrm{VV}_{1-2-3}}_{i_1,i_2,i_3,j}$ are given in [9].

2.3. Relaxation terms

In the zero-order approximation, the distribution function is found in the form [13]:

$$f^{(0)}_{i_1,i_2,i_3,j} = \left(\frac{m}{2\pi kT}\right)^{3/2} \frac{n(2j+1)(i_2+1)}{Z_{\mathrm{rot}}(T)Z_{12}(T_{12})Z_3(T_3)} \exp\left(-\frac{mc^2}{2kT} - \frac{\varepsilon_j}{kT} - \frac{i_1\varepsilon_{100}}{kT_{12}} - \frac{i_2\varepsilon_{010}}{kT_{12}} - \frac{i_3\varepsilon_{001}}{kT_3}\right), \tag{6}$$

here n is the number density, k is the Boltzmann constant, T is the gas temperature, T_{12}, T_3 are vibrational temperatures of the combined and asymmetric modes, m is the mass, $Z_{12}(T_{12})$, $Z_3(T_3)$ are the vibrational partition functions for the corresponding modes [13] obtained by summation over all vibrational states below the dissociation threshold, $Z_{\mathrm{rot}}(T)$ is the rotational partition function, ε_j is the rotational energy, $\mathbf{c} = \mathbf{u} - \mathbf{v}$ is the peculiar velocity.

Let us consider in detail the vibrational energy production terms $R^{VT_2}_{12}$ and $R^{VV_{2-3}}_{12}$ in the combined mode. For the asymmetric mode, the production terms are derived similarly. To obtain the expressions for the relaxation terms we use the microscopic energy conservation in a collision resulting in the corresponding energy transition. For VT_2 transitions, both rotational energy and vibrational energy of the partner are not changed, therefore

$$\frac{mc^2}{2} + \frac{mc_1^2}{2} - \frac{mc'^2}{2} - \frac{mc_1'^2}{2} = \varepsilon_{i_1,i_2+1,i_3} - \varepsilon_{i_1,i_2,i_3} = \varepsilon_{010}. \tag{7}$$

Here, the subscript '1' corresponds to the velocity of a partner in the collision, primes denote the particle velocities after collision. Substituting equation (6) into definition of R^{VT2}_{12} (4) and using the above relations yields the following expression

$$R^{\mathrm{VT}_2}_{12} = \sum_i (\varepsilon_{i_1 00} + \varepsilon_{0 i_2 0}) \sum_j \int J^{\mathrm{VT}_2}_{i_1,i_2,i_3,j} d\mathbf{u} = \frac{1}{2}\left\langle \Delta\varepsilon_{\mathrm{VT}_2}\left(\exp\left[\frac{1}{kT}\varepsilon_{010}\left(1 - \frac{T}{T_{12}}\right)\right] - 1\right)\right\rangle_{\mathrm{VT}_2}. \tag{8}$$

In equation (8), $\Delta\varepsilon_{\mathrm{VT}_2} = \pm\varepsilon_{010}$ is the energy variation during VT_2 transitions (constant for harmonic oscillators), $\langle \dots \rangle_{\mathrm{VT}_2}$ is the averaging operator which is introduced as follows:

$$\begin{aligned} \langle F \rangle_{\mathrm{VT}_2} &= \left(\frac{kT}{\pi m}\right)^{1/2} \frac{8n^2}{Z_{12}Z_3} \sum_{i_1,i_2,i_3} (i_2+1) \exp\left(-\frac{i_1\varepsilon_{100}}{kT_{12}} - \frac{i_2\varepsilon_{010}}{kT_{12}} - \frac{i_3\varepsilon_{001}}{kT_3}\right) \times \\ &\times \int F_{i_1,i_2,i_3} \exp\left(-g_0^2\right) g_0^3 \sigma_{\mathrm{VT}_2} d^2\Omega dg_0, \end{aligned} \tag{9}$$

g_0 is the dimensionless relative velocity, σ_{VT_2} is the cross section of VT_2 exchange. We express the cross section as $\sigma_{\mathrm{VT}_2} = P_{\mathrm{VT}_2}\sigma_{\mathrm{VSS}}$, where σ_{VSS} is the cross section of the elastic collision given by the variable soft sphere (VSS) model [14], P_{VT_2} is the transition probability calculated using the SSH theory [15].

Similarly, for VV_{2-3} transitions, the energy conservation takes on the form

$$\frac{mc^2}{2} + \frac{mc_1^2}{2} - \frac{mc'^2}{2} - \frac{mc_1'^2}{2} = \varepsilon_{i_1,i_2+3,i_3-1} - \varepsilon_{i_1,i_2,i_3} = 3\varepsilon_{010} - \varepsilon_{001} \tag{10}$$

and the relaxation term is given by

$$R_{12}^{\mathrm{VV}_{2-3}} = \sum_i (\varepsilon_{i_1 00} + \varepsilon_{0 i_2 0}) \sum_j \int J_{i_1,i_2,i_3,j}^{\mathrm{VV}_{2-3}} d\mathbf{u} = $$
$$= \frac{1}{2} \langle \Delta\varepsilon_{VV_{2-3}} \left(\exp\left[\frac{1}{kT} \left(3\varepsilon_{010} \left(1 - \frac{T}{T_{12}}\right) - \varepsilon_{001}\left(1 - \frac{T}{T_3}\right)\right)\right] - 1\right) \rangle_{\mathrm{VV}_{2-3}}, \quad (11)$$

here $\Delta\varepsilon_{\mathrm{VV}_{2-3}} = \pm(3\varepsilon_{010} - \varepsilon_{001})$ is the energy variation during VV_{2-3} transitions, and the averaging operator is obtained from equation (9) replacing σ_{VT_2} by $\sigma_{\mathrm{VV}_{2-3}}$. The cross section is specified by the corresponding transition probability, $\sigma_{\mathrm{VV}_{2-3}} = P_{\mathrm{VV}_{2-3}} \sigma_{\mathrm{VSS}}$.

Formulas (8) and (11) for the relaxation terms in the three-temperature approximation are further referred to as those given by rigorous theory.

In computational fluid dynamics (CFD) the relaxation terms are commonly calculated using the Landau–Teller model [10]. For carbon dioxide, the modified Landau–Teller formulas are often used [8]:

$$R_{12} = R_{12}^{\mathrm{VT}_2} + R_{12}^{\mathrm{VV}_{2-3}} + R_{12}^{\mathrm{VV}_{1-2-3}} = \rho \sum_\gamma \frac{E_{12}(T) - E_{12}(T_{12})}{\tau_\gamma}, \quad \gamma = VT_2, VV_{2-3}, VV_{1-2-3}, \quad (12)$$

$$R_3 = R_3^{\mathrm{VV}_{2-3}} + R_3^{\mathrm{VV}_{1-2-3}} = \rho \sum_\gamma \frac{E_3(T) - E_3(T_3)}{\tau_\gamma}, \quad \gamma = VV_{2-3}, VV_{1-2-3}, \quad (13)$$

τ_γ are corresponding relaxation times. One of the main objectives of the present study is to estimate the limits of applicability of the Landau-Teller model for carbon dioxide.

Along with the traditional Landau–Teller model, which is basically applicable only for weak deviations from thermal equilibrium, its modification suitable for strongly non-equilibrium flows can be considered. For diatomic species, such a modification was derived and assessed in [11, 12]. Let us generalize it to the case of carbon dioxide.

If we assume

$$\left|1 - \frac{T}{T_{12}}\right| \ll 1$$

then

$$\exp\left[\frac{1}{kT}\varepsilon_{010}\left(1 - \frac{T}{T_{12}}\right)\right] - 1 \approx \frac{\varepsilon_{010}}{kT}\frac{T - T_{12}}{T_{12}}.$$

Introducing the relaxation time of VT_2 exchange by

$$\frac{1}{\tau_{\mathrm{VT}_2}(T, T_{12})} = \frac{4kn}{mc_{12}} \left(\frac{kT}{\pi m}\right)^{1/2} \sum_{i_1,i_2} \frac{i_2 + 1}{Z_{12}}$$
$$\times \int \left(\frac{\Delta\varepsilon_{\mathrm{VT}_2}}{kT}\right)^2 g_0^3 \exp\left(-g_0^2 - \frac{i_1\varepsilon_{100}}{kT_{12}} - \frac{i_2\varepsilon_{010}}{kT_{12}}\right) \sigma_{\mathrm{VT}_2} d^2\Omega dg_0, \quad (14)$$

where c_{12} is the specific heat of the combined mode

$$c_{12} = \frac{\partial E_{12}}{\partial T_{12}},$$

we can write the linearized relaxation term in the form

$$R_{12}^{\mathrm{VT}_2,\,\mathrm{lin}} = \rho \frac{c_{12}(T - T_{12})}{\tau_{\mathrm{VT}_2}(T, T_{12})} \frac{T}{T_{12}}. \quad (15)$$

Assuming further that $E_{12}(T_{12}) \approx c_{12}T_{12}$ and $E_{12}(T) \approx c_{12}T$ (which holds only for the constant specific heat) we get

$$R_{12}^{\mathrm{VT}_2,\,\mathrm{lin}} = \rho\frac{E_{12}(T) - E_{12}(T_{12})}{\tau_{\mathrm{VT}_2}(T, T_{12})}\frac{T}{T_{12}}. \tag{16}$$

The last formula is similar to the Landau–Teller one but differs by the ratio T/T_{12}. Note that this result generalizes the result of [11] obtained for diatomic gases.

For $R_{12}^{\mathrm{VV}_{2-3}}$, the derivation is more complicated. Let us assume

$$\left|1 - \frac{T}{T_{12}}\right| \ll 1, \qquad \left|1 - \frac{T}{T_3}\right| \ll 1$$

which gives

$$\exp\left[\frac{1}{kT}\left(3\varepsilon_{010}\left(1 - \frac{T}{T_{12}}\right) - \varepsilon_{001}\left(1 - \frac{T}{T_3}\right)\right)\right] - 1 \approx \frac{3\varepsilon_{010}}{kT}\frac{T - T_{12}}{T_{12}} - \frac{\varepsilon_{001}}{kT}\frac{T - T_3}{T_3}$$

Introducing the relaxation time of VV_{2-3} exchange by

$$\frac{1}{\tau_{\mathrm{VV}_{2-3}}(T, T_{12}, T_3)} = \frac{4kn}{mc_{12}}\left(\frac{kT}{\pi m}\right)^{1/2}\sum_{i_1,i_2,i_3}\frac{i_2 + 1}{Z_{12}Z_3}$$
$$\times \int \left(\frac{\Delta\varepsilon_{\mathrm{VV}_{2-3}}}{kT}\right)^2 g_0^3 \exp\left(-g_0^2 - \frac{i_1\varepsilon_{100}}{kT_{12}} - \frac{i_2\varepsilon_{010}}{kT_{12}} - \frac{i_3\varepsilon_{001}}{kT_3}\right)\sigma_{\mathrm{VV}_{2-3}}d^2\Omega dg_0, \tag{17}$$

and assuming constant specific heat we obtain

$$R_{12}^{\mathrm{VV}_{2-3},\,\mathrm{lin}} = \rho\frac{E_{12}(T) - E_{12}(T_{12})}{\tau_{\mathrm{VV}_{2-3}}(T, T_{12})}\frac{3\varepsilon_{010}\frac{T - T_{12}}{T_{12}} - \varepsilon_{001}\frac{T - T_3}{T_3}}{(3\varepsilon_{010} - \varepsilon_{001})\frac{T - T_{12}}{T}}. \tag{18}$$

Finally, we can write the ratios of the linearized relaxation terms to those calculated using the Landau–Teller formula with multi-temperature relaxation times (14) and (17):

$$\frac{R_{12}^{\mathrm{VT}_2,\,\mathrm{lin}}}{R_{12}^{\mathrm{VT}_2}(LT)} = \frac{T}{T_{12}}, \tag{19}$$

$$\frac{R_{12}^{\mathrm{VV}_{2-3},\,\mathrm{lin}}}{R_{12}^{\mathrm{VV}_{2-3}}(LT)} = \frac{\left(3\varepsilon_{010}\left(1 - \frac{T}{T_{12}}\right) - \varepsilon_{001}\left(1 - \frac{T}{T_3}\right)\right)}{(3\varepsilon_{010} - \varepsilon_{001})\left(\frac{T_{12}}{T} - 1\right)}. \tag{20}$$

Moreover, using the multi-temperature relaxation times, we can calculate the ratios of $R_{12}^{\mathrm{VT}_2}$ and $R_{12}^{\mathrm{VV}_{2-3}}$ given by formulas (8) and (11) to the corresponding Landau–Teller terms

$$\frac{R_{12}^{\mathrm{VT}_2}}{R_{12}^{\mathrm{VT}_2}(LT)} = \frac{\left(\exp\left[\frac{1}{kT}\varepsilon_{010}\left(1 - \frac{T}{T_{12}}\right)\right] - 1\right)}{\frac{1}{kT}\varepsilon_{010}\left(1 - \frac{T_{12}}{T}\right)}, \tag{21}$$

$$\frac{R_{12}^{\mathrm{VV}_{2-3}}}{R_{12}^{\mathrm{VV}_{2-3}}(LT)} = \frac{\left(\exp\left(\frac{1}{kT}3\varepsilon_{010}\left(1 - \frac{T}{T_{12}}\right) - \frac{1}{kT}\varepsilon_{001}\left(1 - \frac{T}{T_3}\right)\right) - 1\right)}{\frac{1}{kT}(3\varepsilon_{010} - \varepsilon_{001})\left(\frac{T_{12}}{T} - 1\right)}. \tag{22}$$

It is worth noting that in the commonly used formulation, the Landau–Teller formula contains the relaxation times depending only on the gas temperature. Such times can be determined from experiments. In this case, the relaxation times can be calculated in the one-temperature approach using the method proposed in [9]:

$$\frac{1}{\tau_\gamma} = \frac{2kn}{mc_{\text{vibr}}} \left(\frac{kT}{\pi m}\right)^{1/2} \sum_{i_1,i_2,i_3} \frac{i_2+1}{Z_{\text{vibr}}(T)} \int \left(\frac{\Delta\varepsilon^\gamma}{kT}\right)^2 g_0^3 \exp\left(-g_0^2 - \frac{\varepsilon_{i_1 i_2 i_3}}{kT}\right) \sigma_\gamma d^2\Omega dg_0, \quad (23)$$

here $c_{\text{vibr}} = \frac{\partial E_{\text{vibr}}}{\partial T}$ is vibrational specific heat, $Z_{\text{vibr}}(T)$ is the equilibrium vibrational partition function, σ_γ and $\Delta\varepsilon^\gamma$ correspond to VT_2 and inter-mode VV exchanges. All transition probabilities are calculated using the SSH theory [15].

3. Results

In this section, we discuss the results of the relaxation terms evaluation using different models. In order to avoid an uncertainty introduced by using different experimental data for the relaxation times, we calculate both relaxation terms and relaxation times on the basis of the same SSH model for the transition probabilities. It is known that the accuracy of the SSH model is questionable under high-temperature conditions. Therefore, we do not compare the dimensional relaxation terms but rather their ratios to the relaxation terms obtained using the Landau–Teller model with either one-temperature or multi-temperature relaxation times. The cases studied below are summarized in Table 1.

Table 1. Cases studied.

Notation	Case description
Exact, LT (1T)	Ratios of relaxation terms calculated using formulas of rigorous kinetic theory (8), (11) to those calculated using the Landau–Teller formula with one-temperature relaxation time (23)
Exact, LT (MT)	Ratios of relaxation terms calculated using formulas of rigorous kinetic theory (8), (11) to those calculated using the Landau–Teller formula with multi-temperature relaxation times (14), (17)
Linear, LT (MT)	Ratios (19), (20) calculated with linearized formulas obtained using the Landau–Teller formula with multi-temperature relaxation times (14), (17)

Figures 1 and 2 present, respectively, the ratios of $R_{12}^{\text{VT}_2}$ and $R_{12}^{\text{VV}_{23}}$ obtained with the kinetic-theory algorithm (8), (11) to those obtained with the Landay–Teller model (12), (13) based on the one-temperature relaxation time (case Exact, LT (1T)). The ratios are given as functions of the vibrational temperature (furthermore we assume $T_{12} = T_3$) for various fixed gas temperature values. It is seen that for VT_2 relaxation, the error introduced by using the Landau–Teller approximation is not high, especially for low temperatures. Thus, for the gas temperature range 500–1000 K, the ratio does not exceed 2 and increases monotonically with T_{12}/T. For intermediate temperatures about 2000-3000 K, the ratio is non-monotonic with T_{12} and becomes less than unit for high vibrational temperatures. For high gas temperatures, the ratio is greater than unit and decreases with T_{12}; the discrepancy is not negligible for $T_{12}/T > 1$.

Situation is much worse for VV_{2-3} relaxation. The relaxation terms ratio varies within several orders of magnitude being much less than unit for $T = 500$ K and $T_{12}/T > 1$ and much greater than unit for $T > 1000$ K and low T_{12}. Under such conditions, using the Landau–Teller formula with one-temperature (experimental) relaxation times may cause a significant error in

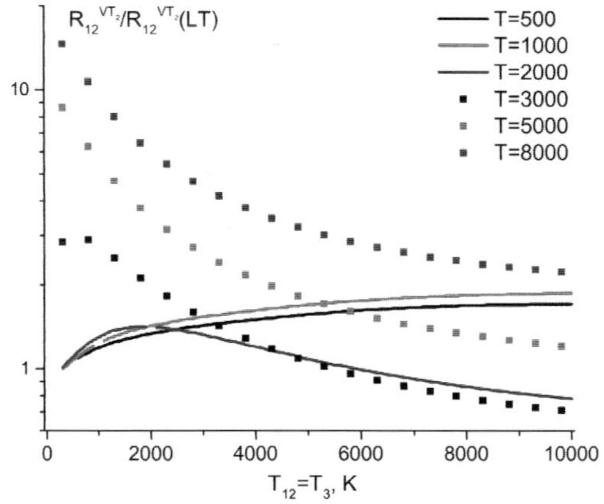

Figure 1. Ratio $R_{12}^{\mathrm{VT}_2}$ as a function of vibrational temperature for different gas temperatures. Case: Exact, LT (1T).

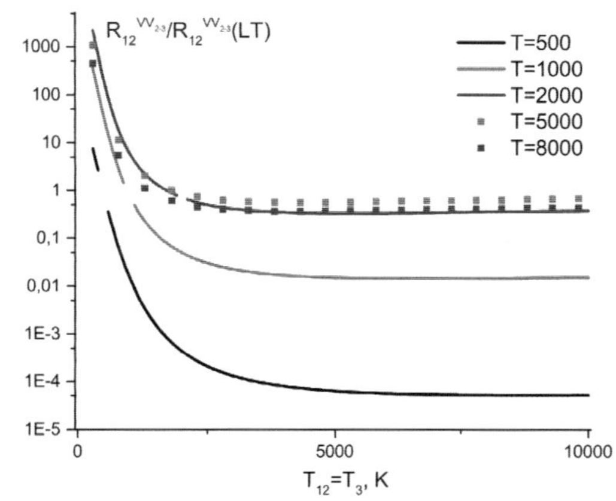

Figure 2. Ratio $R_{12}^{\mathrm{VV}_{2-3}}$ as a function of vibrational temperature for different gas temperatures. Case: Exact, LT (1T).

the predicted values of fluid-dynamic variables, and rigorous kinetic-theory methods have to be applied for the calculation of $R_{12}^{\mathrm{VV}_{2-3}}$.

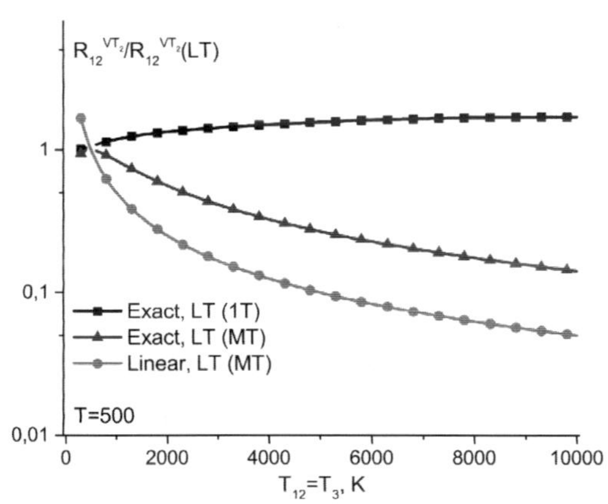

Figure 3. Ratio $R_{12}^{\mathrm{VT}_2}$ as a function of $T_{12} = T_3$ for different cases. $T = 500$ K.

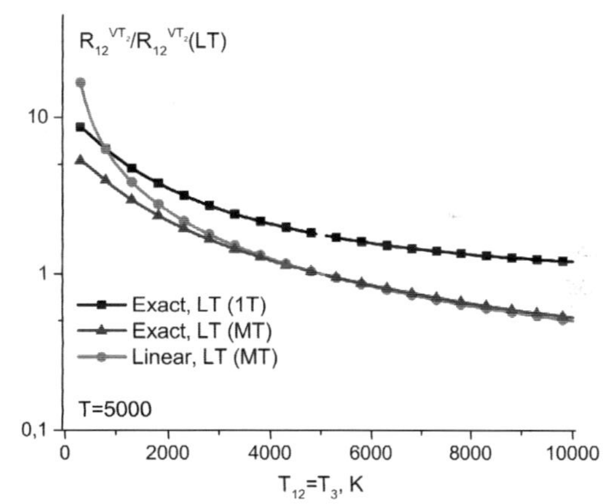

Figure 4. Ratio $R_{12}^{\mathrm{VT}_2}$ as a function of $T_{12} = T_3$ for different cases. $T = 5000$ K.

In Figures 3–4, the ratios of $R_{12}^{\mathrm{VT}_2}$ to those obtained using the Landau–Teller approximation are presented for various cases (see Table 1 for the case description). The ratios obtained using the multi-temperature definitions of the relaxation times in the Landau–Teller formula show similar behavior for different gas temperatures. The linearized formula (19) considerably under-predicts the relaxation terms at low gas temperature and high T_{12}/T and over-predicts them at high T and low T_{12}. It is interesting to note that for $T < 2000$ K, the ratio calculated using the one-temperature relaxation time (case Exact, LT(1T)) shows different behaviour compared to other cases and increases with T_{12}/T. This is explained by the fact that the one-temperature relaxation time in the Landau–Teller model does not depend on T_{12} whereas the multi-temperature relaxation times decrease with vibrational temperatures. Anyway, for

VT_2 exchanges, using the Landau–Teller approximation gives a satisfactory accuracy for all considered cases.

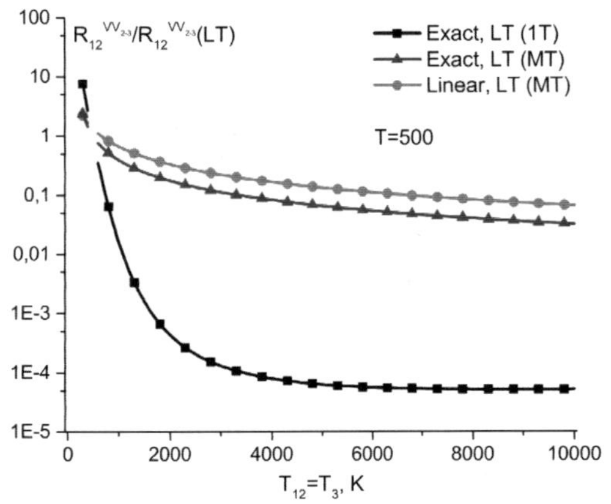

Figure 5. Ratio $R_{12}^{VV_{2-3}}$ as a function of $T_{12} = T_3$ for different cases. $T = 500$ K.

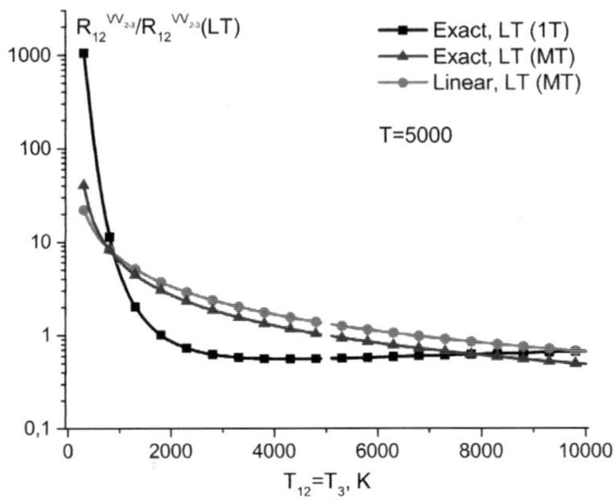

Figure 6. Ratio $R_{12}^{VV_{2-3}}$ as a function of $T_{12} = T_3$ for different cases. $T = 5000$ K.

Similar analysis can be done for VV_{2-3} relaxation starting from Figures 5–6. It is worth noting that for this case, the use of the one-temperature relaxation time yields very high errors in the relaxation terms. On the other hand, using the multi-temperature definition of the relaxation time provides much better accuracy, even when linearized formulas (20) are applied. Thus, for numerical simulations of $R_{12}^{VV_{2-3}}$ we can recommend using the Landau–Teller model with multi-temperature relaxation times taking into account the correction in the form (22) or even in the linearized form (20). This may considerably improve the accuracy and the computational efficiency of multi-temperature flow modeling.

4. Conclusions

A theoretical approach for the calculation of the multi-temperature vibrational energy production terms in non-equilibrium CO_2 flows is proposed. The method is based on the calculation of the integral collision operators for various types of vibrational transitions in the zero-order approximation of the generalized Chapman–Enskog method. The relaxation terms are expressed as functions of the temperature, vibrational temperatures of different modes and cross sections of corresponding collisions.

Based on the rigorous kinetic-theory expressions, the limits of applicability for the Landau–Teller equations are estimated. It is shown that, under certain assumptions, the Landau–Teller formula can be derived from the kinetic theory but contains the correction factor depending on the ratios of translational and vibrational temperatures of different modes. For VT relaxation, the Landau–Teller model provides a satisfactory accuracy; for VV_{2-3} relaxation, introducing relaxation times depending on both translational and vibrational temperatures is required.

Acknowledgments

This study is supported by the Russian Science Foundation, grant 19-11-00041.

References

[1] Giovangigli V 1999 *Multicomponent Flow Modeling* (Boston: Birkhauser)
[2] Nagnibeda E and Kustova E 2009 *Nonequilibrium Reacting Gas Flows. Kinetic Theory of Transport and Relaxation Processes* (Berlin, Heidelberg: Springer-Verlag)

[3] Armenise I and Kustova E 2014 *Chem. Phys.* **428** 90–104
[4] Armenise I and Kustova E 2018 *J. Phys. Chem. A* **122** 5107–120
[5] Taylor R and Bitterman S 1969 *Rev. Mod. Phys.* **41** 26
[6] Park C, Howe J, Howe R, Jaffe R and Candler G 1994 *J. Thermophys. Heat Transfer* **8** 9–23
[7] Kustova E, Nagnibeda E, Shevelev Y and Syzranova N 2011 *Shock Waves* **21** 273–287
[8] Kustova E and Nagnibeda E 2012 *Chem. Phys.* **398** 111–117
[9] Kustova E, Mekhonoshina M and Kosareva A 2019 *Phys. Fluids* **31** 046104
[10] Landau L and Teller E 1936 *Phys. Z. Sowjetunion* **10** 34–43
[11] Kustova E and Oblapenko G 2015 *Phys. Fluids* **27** 016102
[12] Kustova E and Oblapenko G 2016 *Phys. Rev. E* **93** 033127
[13] Kustova E and Nagnibeda E 2006 *Chem. Phys.* **321** 293–310
[14] Koura K 1992 *Phys. Fluids A* **4** 1782–1788
[15] Schwartz R, Slawsky Z and Herzfeld K 1952 *J. Chem. Phys.* **20** 1591

IOP Conf. Series: Materials Science and Engineering **927** (2020) 012048 doi:10.1088/1757-899X/927/1/012048

The influence of chemical reaction models on shock heated flow parameters in $CO_2/CO/O_2/O/C$ mixture

A Kosareva, E Nagnibeda and A Savelev

Saint Petersburg State University, 7/9 Universitetskaya nab., St. Petersburg, 199034, Russia

E-mail: kos-hellen@yandex.ru and aleksey.schumacher@gmail.com

Abstract. In the present paper non-equilibrium flows of $CO_2/CO/O_2/O/C$ mixture behind shock waves is studied taking into account vibrational excitation of CO_2, CO, O_2 molecules. The kinetic scheme includes vibrational energy transitions, dissociation, recombination and exchange chemical reactions. The problem is solved in the five-temperature kinetic theory approximation. The equations for vibrational temperatures of symmetric–bending and asymmetric CO_2 modes and for diatomic species are coupled to the gas dynamic equations and applied to study flow parameters in the relaxation zone behind shock waves. Different models for one-temperature and multi-temperature rates of chemical reactions are used in calculations and the influence of exchange reactions and reaction models on flow parameters in the relaxation zone behind the shock wave is studied in the paper.

1. Introduction

Investigations of the flows of chemically reacting mixtures containing carbon dioxide molecules are important for many practical problems such as for simulation of the flows near vehicles descending to the Mars and Venus atmospheres [1, 2], in laser physics [3] and chemical technologies. To select the optimal computational flow scheme, it is necessary to take into account the influence of different kinetic processes in CO_2 containing mixtures on the variation of flow parameters. This problem is studied by many authors considering various kinetic processes in CO_2 containing mixtures [4, 5]. In most of these papers the influence of vibrational energy distributions on CO_2 flow parameters is studied however the role of chemical reactions in relaxation of CO_2 containing mixtures is not so well investigated. In this paper, we consider the flows of shock heated mixture $CO_2/CO/O_2/O/C$ and study the influence of chemical reactions and different reaction models on changes in characteristic mixture parameters behind a shock wave. The problem is solved in the framework of the five-temperature approximation [6, 7].

2. Governing equations

In the paper, we consider the flows of the mixture $CO_2/CO/O_2/O/C$ in the relaxation zone behind shock waves taking into account vibrational energy transitions of mixture species and the following processes of CO_2, CO, O_2 dissociation (1), (2), (3) and exchange chemical reactions

(4), (5), (6):

$$\mathrm{CO_2}(i_1, i_2, i_3)\mathrm{+M} \leftrightarrow \mathrm{CO}(i)\mathrm{+O + M} \quad (1)$$

$$\mathrm{CO}(i)\mathrm{+M} \leftrightarrow \mathrm{C + O + M}, \quad (2)$$

$$\mathrm{O_2}(i)\mathrm{+M} \leftrightarrow \mathrm{O + O + M} \quad (3)$$

$$M = \mathrm{CO_2, CO, O_2, O, C},$$

$$\mathrm{CO_2+O} \leftrightarrow \mathrm{CO+O_2} \quad (4)$$

$$\mathrm{CO+O} \leftrightarrow \mathrm{C + O_2} \quad (5)$$

$$\mathrm{CO+CO} \leftrightarrow \mathrm{CO_2+C}, \quad (6)$$

where i_1, i_2, i_3 are vibrational level numbers of CO_2 symmetric, bending and asymmetric modes, index i corresponds to vibrational level numbers of CO and O_2 molecules.

Following the kinetic theory [7] and using the existing data concerning reaction rates in CO_2 containing mixtures [8], we divide the collision processes in the considered mixture into two groups – rapid and slow ones. The first group includes the VV_m (m =1,2,3) vibrational energy exchanges within three CO_2 modes and diatomic molecules (m =O_2, CO) as well as VV_{12} exchange between symmetric and bending CO_2 modes. The second group includes non-resonant VV and VT energy transitions at the collisions of mixture molecules and chemical reactions.

Taking into account the fast exchange of vibrational energies VV_{12} between the symmetric and bending CO_2 modes, as well as inside the asymmetric mode VV_3 and diatomic CO, O_2 molecules VV_{CO}, VV_{O_2} the vibrational temperatures T_{12}, T_3 of the combined and asymmetric CO_2 modes and T_{CO}, T_{O_2} of CO and O_2 molecules are introduced [7]. Under these conditions the expressions for CO_2 vibrational level populations are derived in the form [7]:

$$n_{i_1 i_2 i_3}(T_{12}, T_3) = \frac{n_{\mathrm{CO_2}}(i_2+1)}{Z^{\mathrm{CO_2}}_{\mathrm{vibr}}(T_{12},T_3)} \exp\left(-\frac{i_1\varepsilon_{100}+i_2\varepsilon_{010}}{kT_{12}} - \frac{i_3\varepsilon_{001}}{kT_3}\right), \quad (7)$$

here $Z^{CO_2}_{vibr}(T_{12}, T_3)$ is the vibrational partition function.

Level populations of diatomic species n^c_i ($c = \mathrm{CO, O_2}$) are described by the Boltzmann distributions with vibrational temperatures T^{CO}_v, $T^{O_2}_v$.

The set of macroscopic parameters giving a closed description of the flow of the considered mixture behind the shock front includes flow velocity $\mathbf{v}$, gas temperature T, number densities of mixture species n_c and vibrational temperatures T_{12}, T_3, T^{CO}_v, $T^{O_2}_v$. The system of governing equations for the one-dimensional non-viscous non-conductive flow of the mixture behind the shock wave includes equations for numerical densities of mixture species and vibrational temperatures coupled to impulse and total energy conservation equations [7]:

$$\frac{d(n_c\mathbf{v})}{dx} = R^{\mathrm{diss}}_c + R^{\mathrm{exch}}_c, \quad c = \mathrm{CO_2, CO, O_2, O, C}, \quad (8)$$

$$\frac{d(\rho_{\mathrm{CO_2}}E_{12}\mathbf{v})}{dx} = R_{12}, \quad (9)$$

$$\frac{d(\rho_{\mathrm{CO_2}}E_3\mathbf{v})}{dx} = R_3, \quad (10)$$

$$\frac{d(\rho_c E^c_{vibr}\mathbf{v})}{dx} = R^c_W, \quad c = \mathrm{CO, O_2}, \quad (11)$$

$$\rho\mathbf{v} = \rho_0\mathbf{v}_0, \quad (12)$$

$$\rho\mathbf{v}^2 + p = \rho_0\mathbf{v}_0^2 + p_0, \quad (13)$$

$$E + p/\rho + \mathbf{v}^2/2 = E_0 + p_0/\rho_0 + \mathbf{v}_0^2/2. \quad (14)$$

Here ρ_c, $(c = CO_2, CO, O_2)$, ρ are densities of mixture molecules and total mixture density, m_c are species masses, $p = nkT$, E is the total mixture energy per unit mass:

$$\rho E = \frac{3}{2}nkT + \rho_{mol}E_{rot} + \rho_{mol}E_{vibr} + \rho E_f, \quad (15)$$

where $n = \sum_{c=\mathrm{CO_2,CO,O_2,O,C}} n_c$, $\rho_{mol} E_{rot} = \sum_{c=\mathrm{CO_2,CO,O_2}} n_c kT$ is the rotational energy, E_f is the formation energy of species, parameters in the free stream are specified by the index "0". The vibrational energy of CO_2, CO, O_2 molecules are written as:

$$\rho_{CO_2} E_{vibr}^{CO_2} = \sum_{i_1 i_2, i_3} (i_1 \varepsilon_{100} + i_2 \varepsilon_{010} + i_3 \varepsilon_{001})\, n_{i_1 i_2 i_3}(T_{12}, T_3), \tag{16}$$

$$\rho_c E_{vibr}^{c} = \sum_i \varepsilon_i^c n_i^c\,(T_v^c), c = CO, O_2, \tag{17}$$

here ε_i^c is the energy of vibrational level i of species c.

For vibrational energies of combined and asymmetric modes one can write:

$$\rho_{CO_2} E_{12} = \sum_{i_1 i_2} (i_2 + 1)\,(i_1 \varepsilon_{100} + i_2 \varepsilon_{010}) \exp\left(-\frac{i_1 \varepsilon_{100} + i_2 \varepsilon_{010}}{kT_{12}}\right), \tag{18}$$

$$\rho_{CO_2} E_3 = \sum_{i_3} (i_3 \varepsilon_{001}) \exp\left(-\frac{i_3 \varepsilon_{001}}{kT_3}\right). \tag{19}$$

The right-hand sides of the kinetics equations R_c^{diss}, R_c^{exch}, R_{12}, R_3, R_W^c describe dissociation, recombination, exchange reactions and vibrational energy transitions. Changes in vibrational energy were expressed through the relaxation times measured experimentally [9]:

$$R_{12} - m_{CO_2}(R_{CO_2}^{diss} + R_{CO_2}^{exch}) = \rho_{CO_2} \frac{E_{12}^{eq}(T) - E_{12}(T_{12})}{\tau_{12}}, \tag{20}$$

$$R_3 - m_{CO_2}(R_{CO_2}^{diss} + R_{CO_2}^{exch}) = \rho_{CO_2} \frac{E_3^{eq}(T) - E_3(T_3)}{\tau_3}, \tag{21}$$

$$R_W^M - m_M(R_M^{diss} + R_M^{exch}) = \rho_M \frac{E_{vibr}^{eq,M}(T) - E_{vibr}^M(T_v^M)}{\tau_M}, M = CO, O_2. \tag{22}$$

Here τ_{12}, τ_3, τ_M (M =CO,O_2) are the characteristic times of processes in combined, asymmetric CO_2 modes and in diatomic species.

Relaxation terms in equations (8) describing changes of number densities due to exchange reactions are written in the form:

$$R_c^{exch} = \sum_{dc'd'} \left(k_{c' \to c}^{d' \to d}(T) n_{c'} n_{d'} - k_{c \to c'}^{d \to d'}(T) n_c n_d \right), \tag{23}$$

where $k_{c' \to c}^{d' \to d}(T)$ and $k_{c \to c'}^{d \to d'}(T)$ are rate coefficients of forward and backward reactions.

Dissociation- recombination reactions are described as follows:

$$R_{CO_2}^{diss} = \sum_M n_M \left(k_{rec,CO_2}^M (T)\, n_{CO} n_O - k_{diss,CO_2}^M (T, T_{12}, T_3)\, n_{CO_2} \right), \tag{24}$$

$$R_{CO}^{diss} = \sum_M n_M \left(k_{rec,CO}^M (T)\, n_C n_O - k_{diss,CO}^M (T)\, n_{CO} \right), \tag{25}$$

$$R_{O_2}^{diss} = \sum_M n_M \left(k_{rec,O_2}^M (T)\, n_O^2 - k_{diss,O_2}^M (T)\, n_{O_2} \right), \tag{26}$$

with corresponding rate coefficients of mixture species dissociation and recombination at the collisions with the partner M.

Non-equilibrium dissociation rate coefficients are found using the Treanor-Marron model [10] for CO and O_2 molecules and extension of this model for multi-temperature CO_2 dissociation [11]. The exchange reaction rate coefficients are calculated using the single-temperature equilibrium reaction models of Park [1], McKenzie and Arnold [12], Ibragimova with coauthors [13] and new nonequilibrium multi-temperature exchange reaction model [6, 14]. Below we present the comparison of equilibrium reaction rate coefficients found using expressions given in [1], [12] and [13].

In figure 1 comparison of rate coefficients for exchange reaction (4) and dissociation (1) is shown for the temperature range from 250 K to 25000 K with three different sets of the Arrhenius law parameters. It can be seen that for $T < 3000$ K the exchange reactions dominate but after

5000 K the situation changes, and the contribution of dissociation becomes more noticeable. But for all models the difference is within one and a half of the order of magnitude for the high temperature case. It is important to note that at low temperatures the use of McKenzie parameters leads to a significant overestimation of the rate coefficients for both reactions (up to tens of orders of magnitude), while the use of parameters from the works [1] and [13] gives close values. At temperatures above 5000 K the use of McKenzie parameters leads to the lowest reaction rate coefficient values in comparison with other considered models.

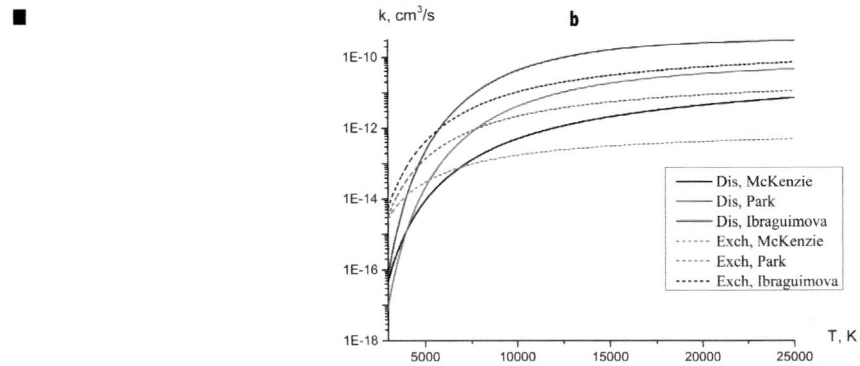

Figure 1. Reaction rate coefficients in dependence on T for temperatures less than 3000 K (a) and above 3000 K (b).

Recently we studied the influence of chemical reaction models on the time evolution of parameters in space-homogeneous CO_2 containing mixtures taking into account not only thermal equilibrium reactions but also non-equilibrium multi-temperature reaction rates [6]. In the next section we show the influence of the same reaction models on shock heated flow parameters.

3. The influence of chemical reaction models on the flow parameters behind the shock wave

First, we show the influence of considered thermal equilibrium reaction models on the evolution of flow parameters in the relaxation zone behind shock waves.

The equations (8-14) were solved using the chemical reaction models considered in the previous section. The flow in the free stream is supposed to be equilibrium under the conditions corresponding to the flows in the Mars atmosphere at the altitude 60 km: $T = 271$ K, density $\rho = 3.141 \cdot 10^5 kg/m^3$ and mixture composition $n_{CO_2} = 0.5n, n_{CO} = 0.05n, n_{O_2} = 0.025n, n_O = 0.4n, n_C = 0.025n$. Calculations were done for the Mach number in the free stream M =15,5.

The gas temperature and number densities of CO_2, O_2, CO molecules behind the shock wave obtained in the five-temperature approach are presented in figures 2-4. Comparison of temperature values shows that Park and Ibragimova models [1, 13] lead to close results for the gas temperature (the difference in T values is less than 11%) (see figure 2). The McKenzie model [12] caused a little shift in the temperature profile and later start of chemical relaxations, the difference in T values obtained using McKenzie and Ibragimova models reaches 23%.

In the most of considered cases, two phases of chemical relaxation in the mixture can be distinguished: the first one is characterized mainly by the rapid CO_2 dissociation with CO and O formation and in the second phase the dominant exchange reactions (4) (5) are observed (figures 3, 4).

It is worth noting that the first zone with the active CO_2 dissociation and then also O_2 dissociation plays the main role in the mixture relaxation and leads to the decrease of the gas temperature T, which accounts for up to 98% changes in numerical densities (relative to the

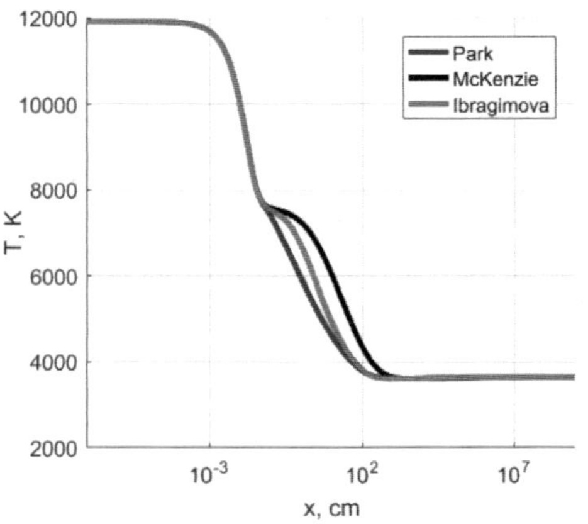

Figure 2. Gas temperature in dependence on x behind shock wave.

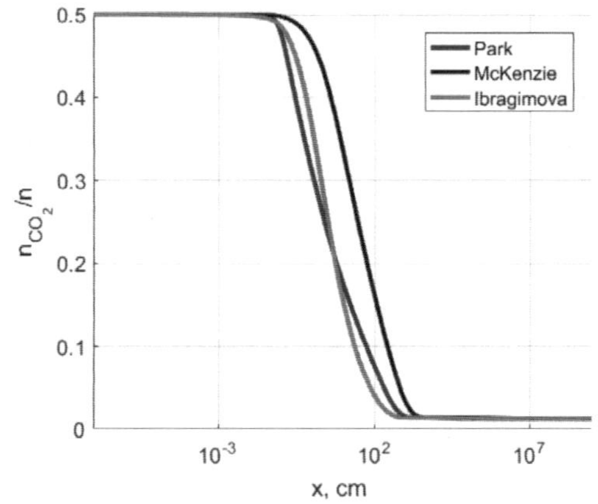

Figure 3. Number density of CO_2 molecules in dependence on x.

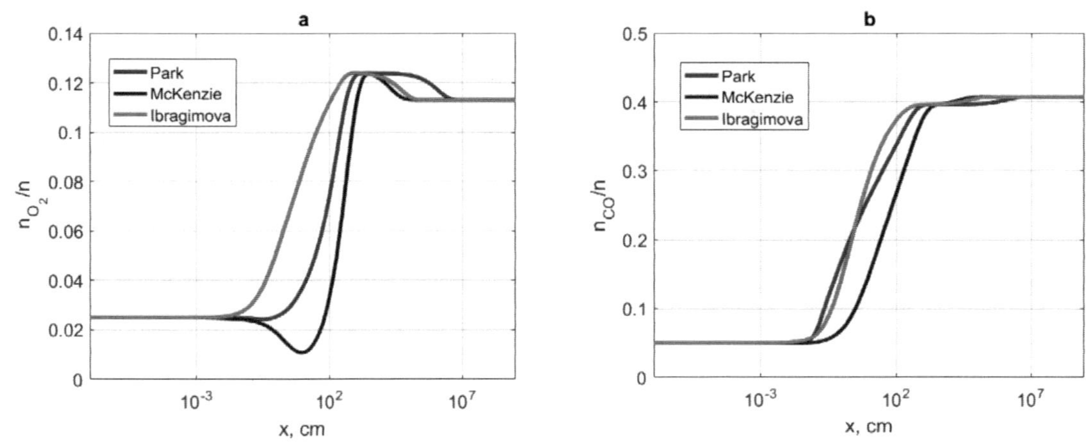

Figure 4. Number densities of O_2 (a) and CO (b) molecules in dependence on x.

initial values). Whereas in the second phase chemical relaxation due to exchange reactions leads to a change in species number densities by 2% of the initial values. Based on the presented results, it can be concluded that the choice of a model of equilibrium reaction rate coefficients has a slight effect on the gas temperature and numerical densities of main mixture species (CO_2 and CO), but can lead to some qualitative differences in the relaxation of individual components of the mixture.

To assess the effect of multi-temperature reaction models on flow parameters and relaxation rates, the following cases were considered: 1) multi-temperature (m-t) dissociation and equilibrium exchange reactions , 2) (m-t) dissociation and nonequilibrium exchange reactions, 3) (m-t) dissociation without exchange reactions. The case 3 with only dissociation-recombination was considered to show the contribution of exchange reactions to the relaxation processes in the mixture. In figure 5 relative number densities of mixture species are presented as functions of the distance from the shock front for three different models.

One can see that the equilibrium values n_c in the end of the relaxation zone obtained using different models are almost the same (the difference does not exceed 1%). Therefore, we can

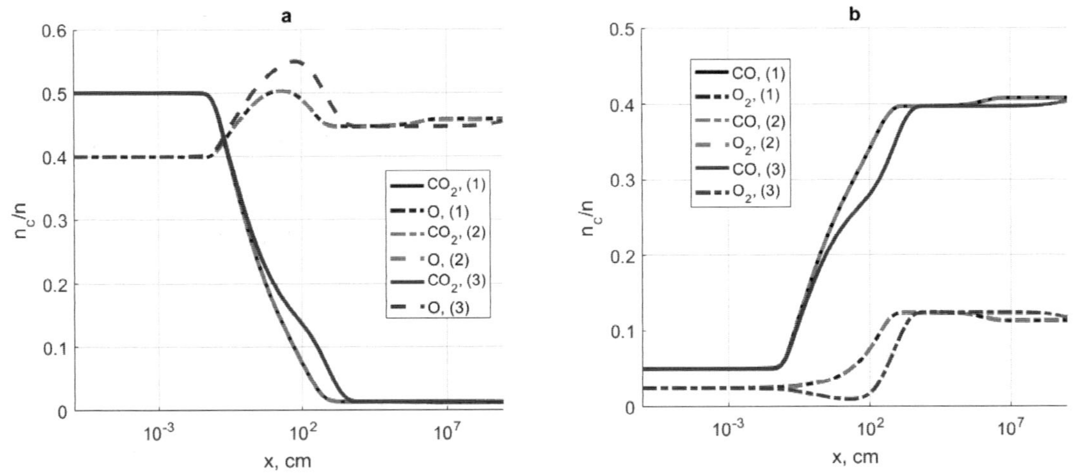

Figure 5. Number densities of CO_2, O (a) and CO, O_2 (b) species in dependence on x.

conclude that taking into account exchange reactions does not lead to significant changes in the equilibrium mixture composition behind the shock wave under the considered free stream conditions. However, in the zone of active chemical relaxation $10^{-2} \le x \le 10^2$, the absence of exchange reactions (the 3rd case) leads to visible differences in species number densities. Thus, a more active oxygen dissociation (figure 5 (b)) leads to an increase of atomic O concentration (see figure 5 (a)). When considering the complete scheme of chemical reactions, the number density of O_2 molecules does not decrease due to their dissociation because of the more active exchange reaction (4), which is accompanied by an increase in relative number densities of O_2, CO and a decrease in CO_2 content in the mixture. The minimum differences in number densities of mixture species when comparing the complete reaction scheme with the case without exchange reactions are found 12% for n_O/n, while O_2 relative number densities differ in 4 times.

4. Conclusions

The vibrational and chemical relaxation in multi-temperature shock heated flows of $CO_2/CO/O_2/O/C$ mixture is studied taking into account dissociation of mixture molecules and exchange chemical reactions. The results are obtained using different one-temperature models for equilibrium exchange reactions as well as with the use of multi-temperature reaction model recently applied in [6] for the space-homogeneous mixture. The comparison of the results obtained with the use of different models for exchange reactions is presented and the role of non-equilibrium chemical reactions and reaction models in the relaxation zone behind shock waves in the considered mixture is discussed in the paper.

Acknowledgments

This study is supported by the Russian Fund of Basic Research Grant 18-01-00493.

References

[1] Park C, Howe J and Jaffe R 1994 *J. Thermophys. Heat Transfer* **8** 9–23
[2] Kustova E, Nagnibeda E, Shevelev Y and Syzranova N 2011 *Shock Waves* **21** 273–287
[3] Taylor R and Bitterman S 1969 *Rev. Mod. Phys.* **41** 26
[4] Cenian A 1989 *Chem. Phys.* **132** 41–48
[5] Likalter A 1976 *Prikl. Mekh. Tekn. Fiz.* **4** 3 (in Russian)
[6] Kosareva A, Nagnibeda E and Savelev A 2020 *Chem. Phys.* **533** 110718

[7] Kustova E and Nagnibeda E 2012 *Chem. Phys.* **398(1)** 111–117
[8] Gordiets B, Osipov A and Shelepin L 1980 *Kinetic Processes in Gases and Molecular Lasers* (Nauka. Moscow)
[9] Losev S, Kozlov P, Kuznezova L, Makarov V, Romanenko Y, Surzhikov S and Zalogin G 1998 *Proceeding of the Third European Symposium on Aerothermodynamics for Space Vehicles, ESTEC, Noordwiik.* vol 426 (ESA) pp 437–444
[10] Marrone P and Treanor C 1963 *Phys. Fluids* **6** 1215–1221
[11] Kustova E and Nagnibeda E 2006 *Chem. Phys.* **321** 293–310
[12] Kenzie R L M and Arnold J 1967 *AIAA Thermophysics Specialist Conference* AIAA paper pp 67–322
[13] Ibragimova L, Smekhov G, Shatalov O, Eremin A and Shumova V 2000 *High Temperature* **38** 33–36
[14] Aliat A 2008 *Physica A* **387** 4163–4182

IOP Conf. Series: Materials Science and Engineering **927** (2020) 012049 doi:10.1088/1757-899X/927/1/012049

Effect of aspect ratio and end plates on rotational oscillations of thick plates in air flow

R V Shmigirilov and A N Ryabinin
Department of Hydroaeromechanics, St. Petersburg State University, St. Petersburg, 198504, Russia

E-mail: a.ryabinin@spbu.ru, robin_fin@mail.ru

Abstract. The steady oscillations of elastically fixed thick plates in the air flow are considered. The oscillation amplitudes are measured using the strain gauge method. In experiments, the aspect ratio varies. Some of the experiments were performed with end plates. Steady-state oscillation amplitude increases with an increase in plate aspect ratio. In the presence of end plates, the oscillation amplitude linearly depends on the Strouhal number. Increasing the aspect ratio of the thick plate from 5 to 7 does not lead to a significant change in the amplitude of oscillations.

1. Introduction

The aim of this work is to study the rotational oscillations of bluff bodies in air stream. The study of rotational oscillations is important for predicting of oscillations of the bridge span caused by wind load [1]. The end plates are installed to exclude air flow through the ends. The flow around short bodies with end plates is like the flow around bodies of big length. Long body cannot be directly tested because of the limited size of the test section of the wind tunnel. In an aerodynamic experiment, small elongation bodies are provided with end plates. The surface of the end plate is parallel to the flow velocity vector. End plates are used both in experiments with wings [2] and in experiments with bluff bodies. Using this method, the bridge segments are tested [3, 4]. Examples of studying bodies of various shapes with end plates can be found in [5, 6, 7]. The size of the end plates is larger than the size of the model under study. The presence of end plates changes the aerodynamic characteristics of the models. Recently, it was discovered [8] that end plates significantly change the dependence of the normal force coefficient of a square prism on the angle of attack. This phenomenon leads to a decrease in the critical speed at which the elastic fixed prism oscillates. In this paper, we investigate the effect of end plates on the aerodynamic characteristics of thick plates.

2. Experimental method

Rotational oscillations of thick plates with the thickness d = 20 mm are studied. The width of the plate w is approximately equal to 100 mm. The length of the plate l varies 200, 350, 500 and 700 mm. Thus, the aspect ratio of plate $\lambda = l/w$ is equal to 2, 3.5, 5 and 7. Experiments are performed both with and without end plates. The end plate is a disk with a diameter d = $2w$ = 200 mm.

Experiments are carried out in the wind tunnel AT-12 located on the territory of St. Petersburg State University. The wind tunnel has an open test section. The diameter of the outlet part of the round cross section nozzle is 1.5 m. The length of the test section is 2.25 m. The model is affixed with a wire suspension. The model can be rotated around a horizontal axis that is perpendicular to the velocity vector of the air movement. A steel tail holder is attached to the rear end of the model. Two steel springs are attached to the holder (figure 1). A semiconductor strain gauge measures the tension of the lower spring.

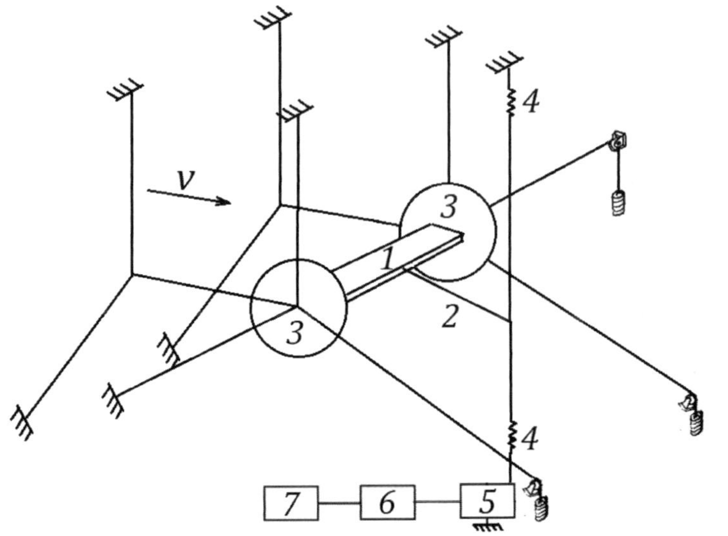

Figure 1. Experimental setup diagram. *1* – thick plate, *2* – tail holder, *3* – end plates, *4* – springs, *5* – strain gauge, *6* – PC-oscilloscope, *7* – computer.

The Velleman-PCS500 PC oscilloscope converts the analog signal of the strain gauge into a digital readout and transmits it to the computer. The measurement frequency is 100 Hz or 1250 Hz. The recording duration is 17 seconds at 100 Hz or 3.3 seconds at 1250 Hz. We obtain that the frequency of oscillations does not depend on the flow rate. If there is no flow in the test section, the oscillations of the plates on the elastic suspension are damped. The frequency of oscillations remains the same as for oscillations due to air flow.

We assume that the spring tension at the extremes of the signal dependence on time is equal to the spring tension under the action of a constant force that causes a deviation equal to the amplitude of the oscillations. This assumption allows us to associate the oscillation amplitude with the maximum or minimum tension of the lower spring. Calibration experiments were performed. During the recording of the strain gauge readings, the known load is suspended at the point of attachment of the tail holder to the wire. According to the results of measurements, the change in the device readings caused by a known force is determined. At the same time, the displacement of the load is measured. The ratio of the amplitude of vibrations of the lower spring tension to the amplitude of rotational vibrations of the model is obtained. The flow rate in the test section was determined by the pressure drop on the wind tunnel nozzle.

3. Data processing

The signal processing technique is described in [9, 10].

Let inclination angle β_i measured at the time t_i be the sum of harmonic function with amplitude A, constant E and experimental error ξ_i:

$$\beta_i = A\cos(\omega t_i - \varphi) + E + \xi_i = A\cos\varphi\cos\omega t_i + A\sin\varphi\sin\omega t_i + E + \xi_i. \tag{1}$$

Experimental error is random variable with zero expectation. Let us consider a sample consisting of n readouts ($i = 1, 2, 3, \ldots, n$). We assume that

$$\sum_{i=1}^{n} \xi_i \cos \omega t_i = \sum_{i=1}^{n} \xi_i \sin \omega t_i = \sum_{i=1}^{n} \xi_i = 0. \tag{2}$$

Multiplying the formula (1) sequentially by $\cos \omega t_i$, $\sin \omega t_i$ and 1, summing the resulting expressions from $i=1$ to n, taking into account the expression (2), we get the system of equations (3).

$$\begin{pmatrix} \sum_{i=1}^{n} \cos^2 \omega t_i & \sum_{i=1}^{n} \cos \omega t_i \sin \omega t_i & \sum_{i=1}^{n} \cos \omega t_i \\ \sum_{i=1}^{n} \cos \omega t_i \sin \omega t_i & \sum_{i=1}^{n} \sin^2 \omega t_i & \sum_{i=1}^{n} \sin \omega t_i \\ \sum_{i=1}^{n} \cos \omega t_i & \sum_{i=1}^{n} \sin \omega t_i & n \end{pmatrix} \begin{pmatrix} A \cos \varphi \\ A \sin \varphi \\ E \end{pmatrix} = \begin{pmatrix} \sum_{i=1}^{n} \beta_i \cos \omega t_i \\ \sum_{i=1}^{n} \beta_i \sin \omega t_i \\ \sum_{i=1}^{n} \beta_i \end{pmatrix}. \tag{3}$$

If n is the number of readouts in one oscillation period $T = 2\pi/\omega$ and this number is large, then the system of equations (3) can be simplified, considering that

$$\sum_{i=1}^{n} \cos^2 \omega t_i \approx \sum_{i=1}^{n} \sin^2 \omega t_i \approx {}^{n\omega}/_{2\pi} \int_0^{2\pi/\omega} \cos^2 \omega t \, dt = n/2.$$

$$\sum_{i=1}^{n} \cos \omega t_i \sin \omega t_i \approx \sum_{i=1}^{n} \sin \omega t_i \approx \sum_{i=1}^{n} \cos \omega t_i \approx 0.$$

Thus, instead of a system of equations (3), one can solve a simple system (4).

$$\begin{pmatrix} n/2 & 0 & 0 \\ 0 & n/2 & 0 \\ 0 & 0 & n \end{pmatrix} \begin{pmatrix} A \cos \varphi \\ A \sin \varphi \\ E \end{pmatrix} = \begin{pmatrix} \sum_{i=1}^{n} \beta_i \cos \omega t_i \\ \sum_{i=1}^{n} \beta_i \sin \omega t_i \\ \sum_{i=1}^{n} \beta_i \end{pmatrix}. \tag{4}$$

We calculate the amplitude of rotational oscillation using expression:

$$A = (A^2 \cos^2 \varphi + A^2 \sin^2 \varphi)^{0.5}.$$

4. Results

It is obtained that the readout frequency does not affect the result of determining the amplitude of oscillations. In almost all cases, oscillations with a constant amplitude are observed. The exception is the oscillations of a short thick plate with the aspect ratio $\lambda = 2$ at low air flow speeds v. Figures 2 and 3 show the dependences of the steady-state oscillation amplitude on the Strouhal number $\mathrm{Sh} = w/vT$, where T is the oscillation period.

The amplitude of oscillations grows with the decrease in the Strouhal number. Figure 2 shows that for a thick plate without end plates, the greater the aspect ratio the greater the oscillation amplitude. Oscillation of the plate $\lambda = 2$ with a constant amplitude takes place only at high air flow velocity. Figure 3 shows that if end plates are present then the range of Strouhal numbers in which there are oscillations with a constant amplitude for the cylinder $\lambda = 2$ increases significantly.

The dependencies of the oscillation amplitude for thick plates with end plates on the Strouhal number are almost linear. The presence of end plates makes the dependencies smoother. This phenomenon can be explained by the fact that the 3D vortex system around the body without end plates is more complex and changes with increase in the flow rate. In figure 3, the graphs of the amplitude dependence on the Strouhal number for plates with an aspect ratio λ of 5 and 7 are close to each other.

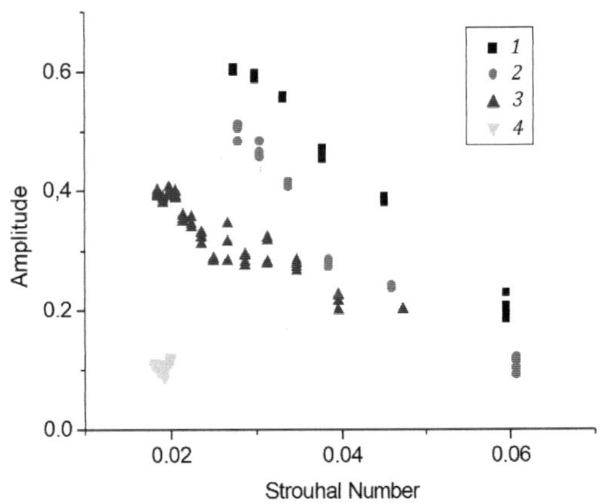

Figure 2. Dependencies of amplitude of oscillations on Strouhal number without end plates. *1* – $\boldsymbol{\lambda} = \mathbf{7}$, *2* – $\boldsymbol{\lambda} = \mathbf{5}$, *3* – $\boldsymbol{\lambda} = \mathbf{3.5}$, *4* – $\boldsymbol{\lambda} = \mathbf{2}$.

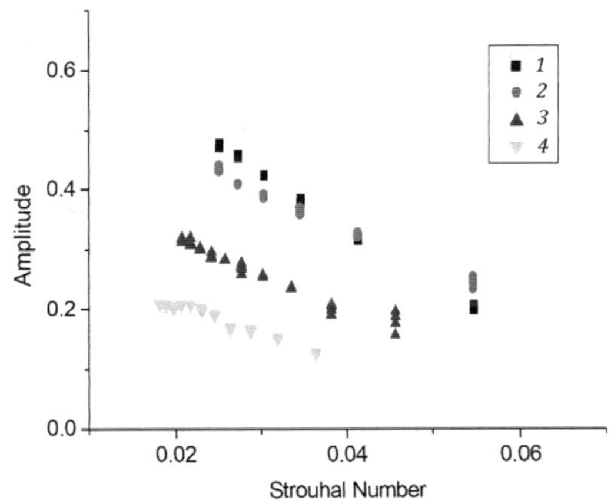

Figure 3. Dependencies of amplitude of oscillations on Strouhal number with end plates. *1* – $\lambda = 7$, *2* – $\lambda = 5$, *3* – $\lambda = 3.5$, *4* – $\lambda = 2$.

5. Conclusions

Experiments were performed with thick rectangular plates attached to an elastic suspension with one rotational degree of freedom in the air flow. It is found that the amplitude of steady-state oscillations increases with increasing of aspect ratio of the plates. Constant oscillation amplitude of the plate with aspect ratio $\lambda = 2$ is realized only at a sufficiently high flow rate. The presence of end plates changes the dependence of the oscillation amplitude on Strouhal number. For a plate with an aspect ratio $\lambda = 2$, the range of flow velocities at which oscillations with a constant amplitude are observed is expanded. In this case, the amplitude is much greater than the oscillation amplitude of a thick plate without end plates. The dependencies of the oscillation amplitude for thick plates with end plates on the Strouhal number are close to linear. Increasing the aspect ratio of the thick plate from 5 to 7 does not lead to a significant change in the amplitude of oscillations.

Acknowledgement

The financial support for the present project was provided by the Russian Foundation for Basic Research under the Grant No.19-38-90045.

References

[1] Simiu E and Scanlan R 1966 *Wind effects on structures: Fundamentals and applications to design* (New Jork, John Wiley) p 704
[2] Pelletier A and Mueller T J 2001 Effect of endplates on two-dimensional airfoil testing at low Reynolds Number *J Aircraft* **38** 1056–1059
[3] Zheng Y, Liu Q, Ma W and Liu X 2017 Effects of end plates on aerodynamic force of rectangular prisms in wind tunnel test *J Experiments in Fluid Mech.* **31** 38–45
[4] Wang Zh and Dragomirescu E 2016 Flutter Derivatives Identification and Aerodynamic Performance of an Optimized Multibox Bridge Deck *Advances in Civil Engineering* **2016**

8530154
[5] Kubo Yo and Kato K 1986 The role of end plates in two dimensional wind tunnel tests *Structural Eng./Earthquake Eng. Japan Society of Civil Engineers* **3** 167–174
[6] Li Sh, Wan R, Wang D and Guo P 2019 Effect of end plates on transiting test for measuring the aerodynamic coefficient of structures using wind generated by a moving vehicle *J Wind Engin. Industrial Aerodynamics* **190** 273–286
[7] Shmigirilov R and Ryabinin A 2018 Influence of end plates on aerodynamic characteristics of bluff bodies *AIP Conference Proceedings* **1959** 050030
[8] Ryabinin A N and Lyusin V D 2015 Galloping of small aspect ratio square cylinder *ARPN J Engin. Applied Sciences* **10** 134–138
[9] Ryabinin A and Bogomolov R 2018 Rotational oscillation study of the cylinders with end plates in airflow *AIP Conference Proceedings* **1959** 050027
[10] Ryabinin A and Kiselev N 2017 Rotational oscillation of a cylinder in air flow. *ARPN J Engin. Applied Sciences* **12** 6803–6808

IOP Conf. Series: Materials Science and Engineering **927** (2020) 012050 doi:10.1088/1757-899X/927/1/012050

On the formation of a vortex couple in fluid

V A Gushchin and I A Smirnova
Institute for Computer Aided Design RAS, 19/18, 2-nd Brestskaya str., Moscow, Russia

E-mail: gushchin47@mail.ru

Abstract. The task of forming a vortex couple in a homogeneous non-viscous fluid is investigated numerically. One of the latest variants of the Splitting Method for Incompressible Fluid (SMIF) developed by the authors is used to solve the problem. The finite difference scheme of the method has the following properties: the second order of approximation on spatial variables, minimal scheme viscosity and dispersion, monotony. It is shown that vortex couple is formed only at relatively short distances between the centers of vortexes. As this distance increases, the couple is formed longer in time. In the future, it is planned to take into account the impact of viscosity on this formation.

1.Introduction

Studying the dynamics of the vortexes in the environment is important both from a theoretical point of view and from a practical point of view. Similar challenges arise in meteorology, when the equations take into account the forces of Coriolis. The aviation security tasks explore the dynamics of vortexes that break from the aircraft surface. It also explores the aero acoustic noises generated by such vortexes [1]. In studying the pollution that arises on the free surface of large reservoirs, the tasks of interaction of vortex structures with such boundaries are also considered [2]. The tasks of modeling non acoustic signs of movement of underwater objects are also based on the dynamics of vortexes and their interaction with a free surface or a pycnocline.

The interaction between two vortexes depends on whether they are homostrophic or heterostrophic (of the same sign or of the opposite sign). Following literature we chose to call two vortices of the same sign a pair, and two vortices of opposite sign a couple. The task of moving a couple in a homogeneous non-viscous fluid is considered in many classic monographs [3]. In this work, we would like to start by forming a couple in a homogeneous non-viscous fluid.

2. Statement of the problem

The two vortexes of the radius r_0 and the same intensity of Γ, but with the opposite directions of rotation, are arranged symmetrically relative to the vertical axis (figure 1). Distance between the centers of vortexes is b. Due to the symmetry of the task it is possible to consider the flow in the right half-plain only. For certainty, let the right vortex rotate clockwise.

It is known [3] that such a vortex couple will rise up at speed $U = \Gamma/2\pi b$, where Γ is intensity of the vortex $\Gamma(\mathbf{v}) = \oint \mathbf{v} dl = \oint \omega r_0^2 d\varphi = 2\pi\omega r_0^2$, ω – the magnitude of the vorticity.

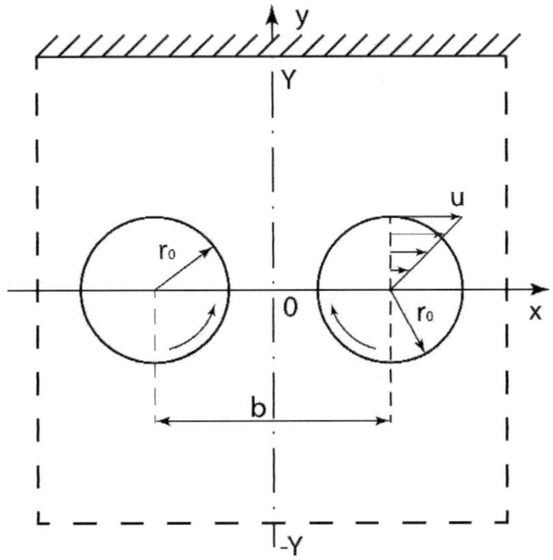

Figure 1. Statement of the vortex couple dynamics problem.

The task is described by the equations:

$$\frac{\partial \mathbf{v}}{\partial t} + (\mathbf{v} \cdot \nabla)\mathbf{v} = -\nabla p, \tag{1}$$

$$\nabla \cdot \mathbf{v} = 0 \tag{2}$$

where **v** is velocity vector with constituents u, v, accordingly, along the axes x and y at the Cartesian coordinate system, chosen as indicated in figure 1, p is pressure.

We intentionally use equations in primitive variables ($\vec{\mathbf{v}}$, p) to summarize the viscous and stratified fluid approach in the future.

The initial conditions. Let us assume that at the initial moment of time $t = 0$ inside the area of the initial vortex A particles of fluid move at speed $\mathbf{v} = (u, v)$, described by equations (3), and in the rest of the area, the fluid is at rest $\mathbf{v} = 0$ (no motion).

$$\begin{aligned} &\sqrt{(x-b/2)^2 + y^2} \le 1, \quad (x, y) \in A, \\ &u(x, y) = \sqrt{(x-b/2)^2 + y^2} \cdot sign(y), \\ &v(x, y) = -\sqrt{(x-b/2)^2 + y^2} \cdot sign(x-b/2). \end{aligned} \tag{3}$$

The boundary conditions. On the left boundary ($x = 0$) the conditions of symmetry are used. At the upper boundary is a condition taken in oceanology as an approximation of a solid cover (a hard impenetrable wall with slippage). Conditions of rest are set on the lower and right boundaries, considering them quite remote from the vortex.

3. Solution method

One of the latest variants of the Splitting Method for Incompressible Fluid (SMIF) [4] developed by the authors is used to solve the problem. The finite difference scheme of the method has the following properties: the second order of approximation on spatial variables, minimal scheme viscosity and dispersion, monotony.

IOP Conf. Series: Materials Science and Engineering **927** (2020) 012050 doi:10.1088/1757-899X/927/1/012050

3.1. Splitting scheme

Suppose that at a time $t_n = n\tau$, where τ is the time step and n is the number of steps, the fields of velocity $\mathbf{v}$, pressure p are known. Then, the scheme for finding the unknown functions at the time $t_{n+1} = (n+1)\cdot\tau$ can be written as

$$\frac{\tilde{\mathbf{v}} - \mathbf{v}^n}{\tau} = -(\mathbf{v}^n \cdot \nabla)\mathbf{v}^n, \tag{4}$$

$$\tau\Delta p = \nabla \cdot \tilde{\mathbf{v}}, \tag{5}$$

$$\frac{\mathbf{v}^{n+1} - \tilde{\mathbf{v}}}{\tau} = -\nabla p. \tag{6}$$

At stage I (4), it is assumed that the transfer of momentum occurs only due to convection. At stage II (5), given the intermediate velocity field $\tilde{\mathbf{v}}$ found earlier and taking into account the solenoidity of the velocity vector $\mathbf{v}^{n+1}$, the pressure field is found by solving the Poisson equation. At stage III (6), it is assumed that the transfer of momentum occurs only due to the pressure gradient (there are no convection).

3.2. Finite difference scheme

The flow region under examination is covered by grid of cells (the grid is uniform with respect to x and y)

$$\Omega = \begin{cases} x_{i+1/2} = ih_x, & h_x > 0, \quad i = 0, 1, \ldots, L; \quad Lh_x = X, \\ y_{j+1/2} = jh_y, & h_y > 0, \quad j = 0, 1, \ldots, M; \quad Mh_y = 2Y, \end{cases}$$

where h_x and h_y are the grid sizes, and L and M are the numbers of grid cells in the directions x and y, respectively (the point with the coordinates (i, j) coincides with the cell center). Here, as in the original splitting method, the staggered grid is used, i.e., the coordinates of the grid functions are spaced apart as shown in figure 2. This allows us to visually interpret each cell as an element of the medium volume; such an element is characterized by the pressure $p_{i,j}$, (and possibly by the temperature, energy, etc.), and the divergence $D_{i,j}$ (D determines the presence of a source or sink in the volume, depending on its sign) computed at the cell center. The knowledge of the normal velocity component on a cell side makes it possible to calculate directly the momentum flux through this side. The finite difference scheme for the problem of spot collapse in stratified fluid for 2D case was published in details in [5].

4. Calculation results

4.1. Testing the program

To verify the results produced by the program, we performed the computations in the absence of the vortex on various grids up to $t = 100$. The characteristic vortex points (the upper, lower, and right ones) remained at their places, the pressure was constant, and the velocity divergence was at the level of 10^{-3}. Thus, the results of these computations confirmed that the conservation laws hold with the desired accuracy.

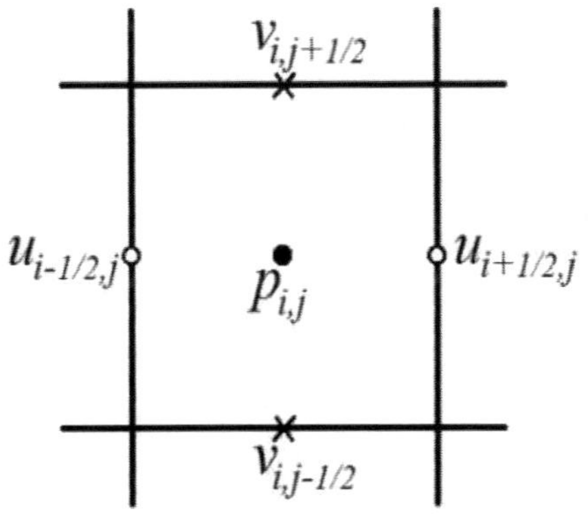

Figure 2. Grid stencil.

4.2. Study of the effect of distance between vortex centers

It is known that a vortex couple is not formed at any b. In order to study the effect of the distance b between the vortex centers on the formation of the couple the calculations were performed for the following set of parameters. In all test calculations, the vortex radius, the horizontal size of the area, the steps of the spatial grid and the step in time were as follows: $x = 0\text{-}10$, $h_x = h_y = 0.1$, $\tau = 0.01$, $r_0 = 1$, and the vertical size of the area and b varied according to the data in the table 1.

Table 1. Calculated variants.

№	*Y*	*b*
1	-5..5	4.0
2	-10..10	3.0
3	-10..10	2.5
4	-5..5	2.5
5	-5..5	2.0
6	-10..10	2.0

There is no pairing for option 1 ($b = 4$). In all other cases, a vortex couple is formed.

4.3. The formation of a vortex couple

The process of forming of a vortex couple and the beginning of its dynamics for option 5 is presented in figure 3, which shows the isoclines of vorticity in the moments of time $t = 1, 2, 3, 4, 5, 10, 15, 30$.

It is seen that single vortexes set at the initial moment of time by the velocity (3) form a couple only to $t = 5$. It should be noted that the distance between the centers of vortexes has decreased from 2 to 1.5. By the time of $t = 15$, this distance has been reduced to 1 and is further maintained up to a top boundary. Then the vortexes run under a top boundary in opposite directions.

Note also that the induced (lower) vortex creates a couple in the lower half-plane with its antipode and, keeping the original distance between the centers, moves to the lower boundary. A similar scenario is observed for option 6. With the increase in b a couple is formed "in agony" and for b = 2.5 (options 3 and 4) the couple is formed only to t = 25. For b = 3 (option 2), the couple is formed only to t = 60. Apparently, taking into account viscosity will only aggravate the process of pairing. But this is the subject of future research.

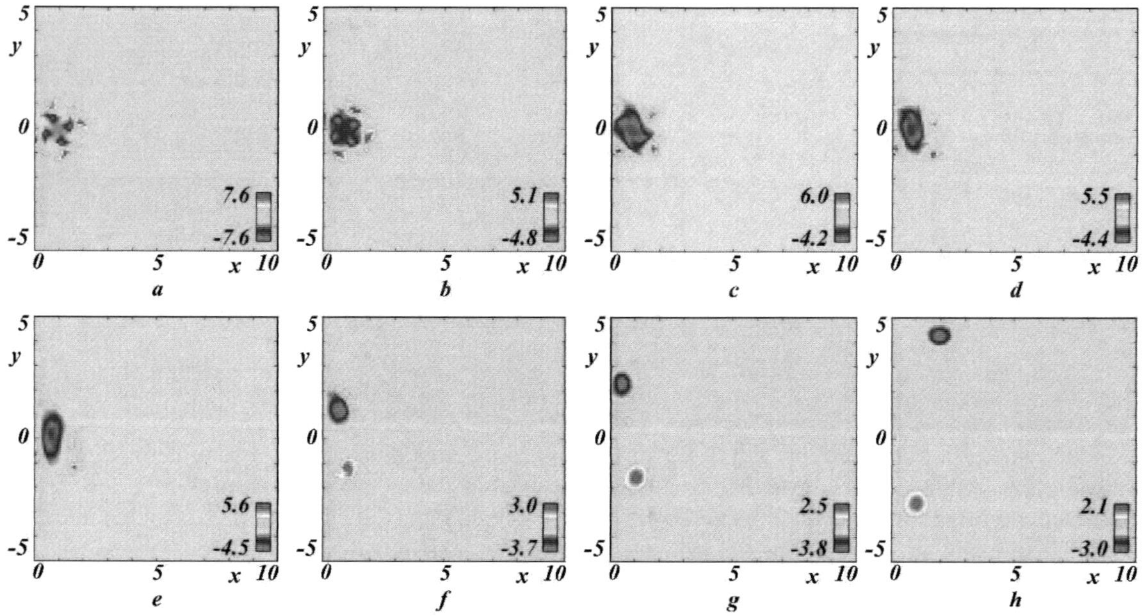

Figure 3. The isoclines of vorticity. Domain X = (0..10), Y = (-5..5), 100 × 100, $h_x = 0.1$, $h_y = 0.1$, and $\tau = 0.01$; (a) t = 1, (b) t = 2, (c) t = 3, (d) t = 4, (e) t = 5, (f) t = 10, (g) t = 15, (h) t = 30.

5. Conclusion

The task of forming a vortex couple, generated by vortex threads, in a homogeneous non-viscous fluid has been studied numerically. One of the latest variants of the Splitting Method for Incompressible Fluid (SMIF) [4] developed by the authors is used to solve the problem. The finite difference scheme of the method has the following properties: the second order of approximation on spatial variables, minimal scheme viscosity and dispersion, monotony. It was shown that a couple is formed only at relatively short distances between the centers of vortexes. As this distance increases, the couple is formed longer in time. In the future, we plan to take into account the impact of viscosity on this formation.

References

[1] Nakashima Y 2008 Sound generation by head-on and oblique collisions of two vortex rings *Physics of Fluids* **20** (5) 056102
[2] Wu-Ting T, Dick K P Y 1995 Effects of soluble and insoluble surfactant on laminar interactions of vortical flows with a free surface *J. Fluid Mech.* **289** 315-349
[3] Batchelor G K 1967 An introduction to fluid dynamics (Camb. Univ. Press)
[4] Gushchin V A 2016 On a one family of quasi monotone finite-difference schemes of the second order of approximation *Math. Models Comput. Simulation* **8** 487–496
[5] Gushchin V A, Smirnova I A 2020 Mathematical Modeling of Spot Dynamics in a Stratified Medium *Comput. Math. Math. Phys.* **60** (5) 879–894

Issues of controlling the motion of a space object by the impact of the ion beam

V M Kulkov [2], N N Markin [1] and Yu G Egorov [2]
[1] Moscow Aviation Institute, Volokolamskoye shosse 4, Moscow, A-80, GSP-3, 125993, Russia
[2] Research Institute of Applied Mechanics and Electrodynamics of the Moscow Aviation Institute (RIAME MAI), Leningradskoye shosse 5, 125080, POB 43, Moscow, Russia

E-mail: vmk_1@mail.ru

Abstract. The problem of controlling the trajectory motion of space objects (SO) using particle beams interacting with the surface of the SO is considered. The source of high-energy particles can be an ion thruster mounted on board a spacecraft (SC). To evaluate the control forces, a model is presented that reflects the mechanism of interaction of the ion beam with the surface of the SO. The motion control for the center of mass of the "SC - SO" system is provided due to the thrust vectoring capability of the service spacecraft electric propulsion system (EPS). Mutual motion control of the SC and SO is ensured by controlling the position of the service spacecraft relative to the SO in the orbital coordinate system. In this paper, expressions for an asymptotically stable control process ensuring the transportation of SO with the specified characteristics are derived.

1. Introduction

The study of the possibility of using spacecraft equipped with electric propulsion systems (EPS) to remove space debris (SD) from orbit and the development of proposals for the removal of space objects (SO) from orbit using a service spacecraft (SC), with EPS including, involve the use of methods for controlling both the mutual movement of the SC and SO, and the motion of "SC-SO" system as a whole.

An ion beam can be used to remove SD or SC fragments from orbits. For the SD removal from the near-Earth space, a method of non-contact impact on SO with a narrow beam of high-velocity ions (up to 30 km/s and more) is proposed. The ion beam source (IBS) is an electric propulsion thruster (EPT) of a service spacecraft located in the close vicinity to the SO (at a distance of about 10–20 meters) [1-5].

2. The mathematical model of the ion beam interaction with SO surface

A beam of accelerated quasineutral plasma, reaching the SO surface, affects it with a certain force, the value of which substantially depends on the angle of divergence of the ion beam, the interaction of ions with the surface of the SO, etc. The plasma flow is considered to be composed of ions and electrons, as well as to be completely ionized, quasi-neutral, and collisionless (the mean free path of the beam particles is significantly longer than the distance between the SC and the SO), and the inhomogeneity of the plasma beam is substantially smaller than the distance between the SC and the SO.

The mathematical model used to calculate the force impact of the ion beam on the SO is based on the hypothesis of local interaction of the particle flow with the SO surface [6]. The impact on the elementary area ΔAi of the surface of the space object depends only on the parameters of the incoming flow and the angle between the velocity vector of the incoming flow $\vec{V}_\infty$ and the local normal to the surface $\vec{n}$ (figure 1).

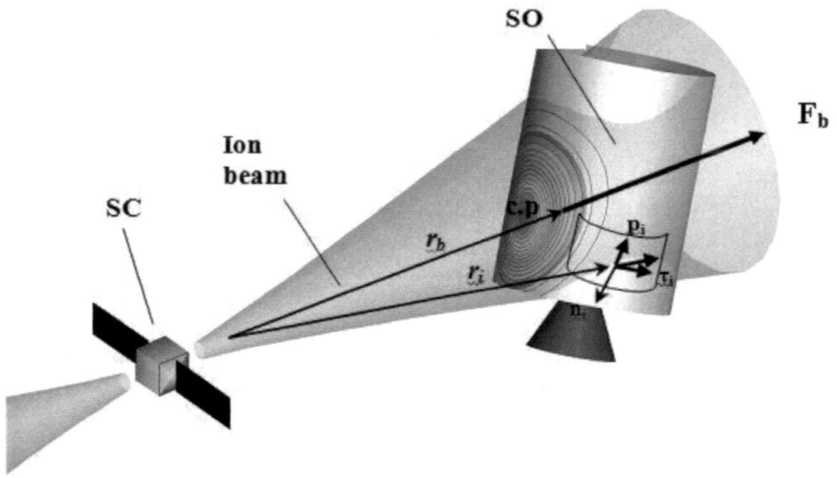

Figure 1. Scheme of the ion beam impact on the transported SO

The local force acting on the ΔA_i area is determined through the flows of the normal and tangential momentum P_n and P_τ, the unit vectors of the normal $\vec{n}$ and the tangent $\vec{\tau} = \frac{(\vec{n}\times\vec{V}_\infty)\times\vec{n}}{(\vec{n}\times\vec{V}_\infty)\times\vec{n}}$ to the area ΔA_j : $\Delta\vec{F}_i = (P_n\vec{n} + P_\tau\vec{\tau})\Delta A_i$.

To determine P_n and P_τ, the accommodation coefficients of the normal momentum and tangent momentum are used:

$$a_n = \frac{P_{in} - P_{rn}}{P_{in} - P_{wn}}, \qquad a_\tau = \frac{P_{i\tau} - P_{r\tau}}{P_{i\tau} - P_{w\tau}},$$

where P_{in}, $P_{i\tau}$ are the momenta delivered to the surface unit, P_{rn}, $P_{r\tau}$ are the momenta carried away from the surface unit, P_{wn} is the normal momentum of the reflected particles if they were reflected with a Maxwell distribution corresponding to the wall temperature T_w.

The values of local stresses from P_n and P_τ for each elementary area ΔA_i of the object are defined by the formulas [6]:

$$P_n = \frac{\rho_\infty V_\infty^2}{2}\left\{\left[\frac{\cos\theta}{S\sqrt{\pi}}\chi(S\cos\theta + \frac{1}{2} + \frac{1}{2}erf(S\cos\theta)\right](2-a_n) + \frac{a_n}{2S^2}\sqrt{\frac{T_r}{T_\infty}}\chi(S\cos\theta)\right\},$$

$$P_\tau = \frac{\rho_\infty V_\infty^2}{2} a_\tau \frac{\sin\theta}{\sqrt{\pi}S}\chi(S\cos\theta),$$

where ρ_∞ is the incident flow density; V_∞ is the incident flow velocity modulus; T_∞ is the incident flow absolute temperature; T_r is the absolute temperature of the reflected particles; $R = 8.314$ J / (K • mol) is the universal gas constant; μ is the molecular weight of the flow particles; θ is the angle between the vectors $\vec{V}_\infty$ and $\vec{n}$; $S = \frac{|V_\infty|}{(2RT_\infty/\mu)^{1/2}}$, $\chi(x) = e^{-x^2} + \sqrt{\pi}x(1+erf(x))$; $erf(x) = \frac{2}{\sqrt{\pi}}\int_0^x e^{-t^2}dt$ is the probability integral.

The actual momentum fluxes will depend on the parameters of the ion beam (particle velocity, density, temperature, etc.), the interaction of the beam ions with the surface of the SO (reflection schemes, accommodation coefficients, etc.), surface characteristics (material composition, temperature, roughness, etc.). To determine the real momentum fluxes, the laboratory experiments are necessary.

As S → ∞, $\chi(x) = 2\sqrt{\pi}S\cos\theta$, then the normal and tangential specific impulses for each elementary area ΔA_i are defined by the formulas:

$$p_{ni} = \rho_i V_i^2 (2-\alpha_n)\cos^2\theta_i;\; p_{\tau i} = \rho_i V_i^2 \alpha_\tau \sin\theta_i \cos\theta_i,$$

where ρ_i is the beam density, V_i is the velocity of the ion beam particles.

The particle beam is characterized by the magnitude of the momentum flux $dP_i = \rho_i V_i^2 f_n(\beta,\varphi)d\Omega_i$, $f_n(\beta,\varphi)$ is the direction distribution function of the particle flow in the spherical coordinate system, $d\Omega_i = \cos\theta_i dA_i / r_i^2$ is the solid angle corresponding to the elementary area dA_i (here r_i is the modulus of the radius vector from the source to the elementary area dA_i, θ_i is the angle between the radius vector $\vec{r}_i$ and the normal to the area dA_i $\vec{n}$).

The function $f(\beta,\varphi)$ can be approximated by a function of the form $f(\beta) = k_1\cos^m\beta\cdot\exp(-k_2\beta^n)$, where k_1, k_2, m, n are empirical coefficients [5].

The total force impact of the beam on the SO are found by summing up the local forces acting on the elementary areas ΔA_i ($i = 1 \ldots N$) over the entire surface of the SO:

$$\vec{F} = \sum_i^N \Delta\vec{F}_i .$$

The force acting in the direction of the axis of the ion beam is:

$$F = r_0^2\rho_0 V_0^2 \iint_{\beta,\varphi} \bar{V}^2(\beta) f(\beta)[(2-a_n)\cos^2\theta(\beta,\varphi) + a_\tau \sin^2\theta(\beta,\varphi)]\sin\beta\cos\beta d\beta d\varphi ,$$

where ρ_0, V_0 are the density and velocity of the beam at a distance r_0 from the IBS, respectively; $\bar{V} = V/V_0$.

Under $a_n = a_\tau = 1$, the expression in square brackets is 1 and the SO geometry is defined by its projection in the coordinates β and φ, bounded by the curve β = β (φ):

$$F = r_0^2\rho_0 V_0^2 \int_0^{\beta_k} \bar{V}^2(\beta) f(\beta)\sin\beta\cos\beta d\beta \sum_j \int_{\varphi_{0j}(\beta)}^{\varphi_{kj}(\beta)} d\varphi ,$$

where the summation is performed over j arcs corresponding to the intersection of the SO projection boundary curve β = β (φ) with the circle β = const.

The effectiveness of the beam during transportation of SO is defined by the expression:

$$\eta = F/F_0 = F/(\dot{m}_b V_{eff}),$$

where $\dot{m}_b$ and V_{eff} are the mass flow rate and effective velocity of the ion beam, respectively.

For a cylindrical SO (figure 1) with a diameter d_0 = 2 m and a height h_0 = 2 m, with a distance between SC and SO r_b = 10 m and beam divergence β_b = 10° (at 95% of the propellant mass flow rate), the efficiencies with longitudinal and transverse orientations of the SO axis relative to the ion beam axis are: $\eta_{\uparrow} = 0.74$, $\eta_{\rightarrow} = 0.82$.

3. Motion control for the SC-SO system

The motion control for the center of mass of the "SC-SO" system is secured by controlling the vector $\vec{R}_{EPS} - \vec{R}_{IBS}$ (here $\vec{R}_{EPS}$ is the thrust of the compensating EPS, $\vec{R}_{IBS}$ is the thrust of the IBS). To ensure equality of accelerations of the SC and SO, the thrust generated by the compensating electric propulsion system must satisfy the relation [3]:

$$\vec{R}_{EPS} = \left(1 + \eta \frac{m_{SC}}{m_{SO}}\right) \vec{R}_{IBS} ,$$

where $\eta = F / R_{IBS}$ is the ion beam utilization coefficient.

Calculations were performed for a SO with the mass of 1000 kg and midsection area of 5 m^2, removed from a near-geostationary orbit with the semi-major axis a = 42164 km and eccentricity e = 0.001 to a higher orbit, using an ion beam with the force of 50 mN. Time dependence of the orbit radius is presented in figure 2.

Figure 3 shows the change in the orbit radius of SO with the mass of 100 kg and midsection area of 1 m^2 that is removed from the orbit 450 km high with an eccentricity e = 0.001 using the 50 mN ion beam into the inner orbit before entering the atmosphere.

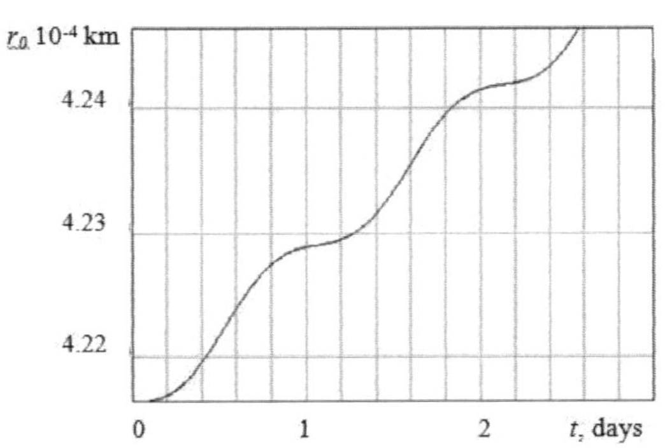

Figure 2. Time dependence of the orbit radius

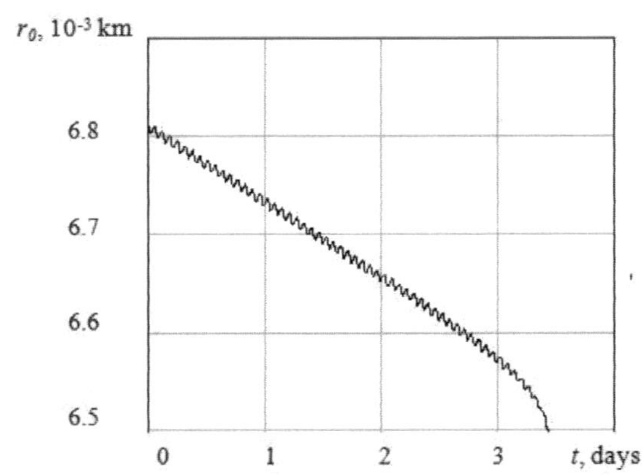

Figure 3. Time dependence of the orbit radius

The mutual motion of the SC and SO is controlled by active maneuvering of SC in the orbital coordinate system by controlling the module and the direction of the vector $\vec{R}_{EPS} - \vec{R}_{IPS}$ [8]. The origin of the concomitant orbital coordinate system is in the center of mass of the SO. The 0x axis is directed transversally against the velocity vector of the orbital motion, the 0y axis is directed along the radius vector, the 0z axis complements the coordinate system to the right.

The equations for relative motion of objects are given in [7]. For the perturbed relative motion of two objects, the motion in the orbit plane and along the normal to it are independent and can be considered separately [7]. In polar coordinates, the equations of SC motion in the plane of a circular orbit are written as follows [9]:

$\ddot{l} = l(\dot{\alpha}^2 + 2\omega_0\dot{\alpha} + 3\omega_0^2 \sin^2 \alpha) - w_n;$

$\ddot{\alpha} = -2\frac{\dot{l}}{l}(\omega_0 + \dot{\alpha}) + 3\omega_0^2 \sin\alpha\cos\alpha - \dot{\omega}_0 - w_\tau,$

where w_n, w_τ are the longitudinal and transverse acceleration, l is the distance between the objects; ω_0 is the angular velocity of the orbital motion.

The linearized system of differential equations for an arbitrary programmable trajectory of the movement of objects is represented in the form:

$\Delta\ddot{l} = [\dot{\alpha}_0^2 + 2\omega_0\dot{\alpha}_0 + 3\omega_0^2 \sin^2 \alpha_0]\Delta l + 3\omega_0^2 \sin 2\alpha_0 \cdot l_0\Delta\alpha + 2l_0[\dot{\alpha}_0 + \omega_0]\Delta\dot{\alpha} - \Delta w_n;$

$\Delta\ddot{\alpha} = -\frac{2}{l_0}\left[\Delta\dot{l} - \frac{\dot{l}_0}{l_0}\Delta l\right](\omega_0 + \dot{\alpha}_0) - 2\frac{\dot{l}_0}{l_0}\Delta\dot{\alpha} + 3\omega_0^2 cos2\alpha_0 \cdot \Delta\alpha - \Delta w_\tau,$

where l_0, α_0, $\dot{l}_0, \dot{\alpha}_0$ are the coordinates of the programmed trajectory; $\Delta l = l - l_0$, $\Delta\alpha = \alpha - \alpha_0$, $\Delta\dot{l} = \dot{l} - \dot{l}_0$, $\Delta\dot{\alpha} = \dot{\alpha} - \dot{\alpha}_0$, $\Delta w_n = w_n - w_{n.0}, \Delta w_\tau = w_\tau - w_{\tau.0}$.

During the orbital movement of the system "SC-SO", the specified distance between the SC and SO ($l_0 = l_{SC\text{-}SO}$, $\dot{l}_0 = 0$) and the transverse direction of the line of sight "SC-SO" ($\alpha_0 = 0$, $\dot{\alpha}_0 = 0$) are maintained.

The equations of relative motion in deviations from the programmed trajectory take the form:

$\Delta\ddot{l} = 2l_0\omega_0\Delta\dot{\alpha} - \Delta w_n;$

$\Delta\ddot{\alpha} = -\frac{2\omega_0}{l_0}\Delta\dot{l} + 3\omega_0^2\Delta\alpha - \Delta w_\tau,$

The program for controlling mutual motion of SC and SO is formed by varying the longitudinal and transverse components of the service SC EPS thrust relative to the programmed value and is represented as:

$$\Delta w_n = \omega_0^2 a\Delta l + \omega_0 b\Delta\dot{l} + \omega_0 l_0 c\Delta\dot{\alpha}\ , \quad \Delta w_\tau = \omega_0^2 d\Delta\alpha + \omega_0 e\Delta\dot{\alpha} + (\omega_0 / l_0) f\Delta\dot{l}$$

where a, b, c and d, e, f are the gain factors. The coefficients a, b, c and d, e, f are selected from the condition of stability of the transition process.

To analyze the asymptotic stability of a linear system, a characteristic equation is compiled:

$$\lambda^4 + a_3\lambda^3 + a_2\lambda^2 + a_1\lambda + a_0 = 0\ ,$$

where $a_3 = \omega_0(b+e)$; $a_2 = \omega_0^2(a+d+be-2c+2f-cf+1)$; $a_1 = \omega_0^3(-3b+bd+ae)$; $a_0 = \omega_0^4 a(d-3)$.

A necessary and sufficient condition for the asymptotic stability of the system according to the Hurwitz criterion is the positivity of all coefficients of the characteristic equation and an additional condition: $a_1(a_3a_2 - a_4a_1) - a_0a_3^2 \geq 0$.

From these conditions, we find the ranges of gain coefficients in the law of controlling the mutual motion of the SC and SO: $a \geq 0$; $b \geq 0$; $c \geq 2$; $d \geq 3$; $e \geq 0$; $f \geq -2$.

The on-board implementation of the control law assumes the availability of information on the relative motion of objects to be obtained from the results of measuring the range, radial velocity, the angle and angular velocity of the line of sight. Control acceleration is formed as a linear or relay function of the measured parameters: $\Delta w = k \cdot u$, $\Delta w = k \cdot \text{sign}(u)$.

With the constant control acceleration, solution of the system has the form [7]: $\mathbf{x} = \mathbf{M}\cdot\mathbf{x}_0 + \mathbf{W}\cdot\mathbf{w}$,

where
$$
\boldsymbol{x}=\begin{vmatrix}\Delta l\\ l_0\Delta\alpha\\ \Delta\dot{l}\\ l_0\Delta\dot{\alpha}\end{vmatrix};\quad
\boldsymbol{x_0}=\begin{vmatrix}\Delta l_0\\ l_0\Delta\alpha_0\\ \Delta\dot{l}_0\\ l_0\Delta\dot{\alpha}_0\end{vmatrix};\quad
\boldsymbol{M}=\begin{vmatrix}1 & -6(\sin\omega_0 t-\omega_0 t) & (4/\omega_0\sin\omega_0 t-3t) & 2/\omega_0(1-\cos\omega_0 t)\\ 0 & (4-3\cos\omega_0 t) & -2/\omega_0(1-\cos\omega_0 t) & 1/\omega_0\sin\omega_0 t\\ 0 & 6\omega_0(1-\cos\omega_0 t) & 4\cos\omega_0 t-3 & 2\sin\omega_0 t\\ 0 & 3\omega_0\sin\omega_0 t & -2\sin\omega_0 t & \cos\omega_0 t\end{vmatrix}
$$

$$
\boldsymbol{W}=\begin{vmatrix}1/\omega_0^2[4(1-\cos\omega_0 t)-3/2\omega_0^2 t^2] & 2/\omega_0^2(\sin\omega_0 t-\omega_0 t)\\ -2/\omega_0^2(\sin\omega_0 t-\omega_0 t) & 1/\omega_0^2(1-\cos\omega_0 t)\\ 1/\omega_0[4\sin\omega_0 t-3\omega_0 t] & 2/\omega_0(\cos\omega_0 t-1)\\ 2/\omega_0(1-\cos\omega_0 t) & 1/\omega_0\sin\omega_0 t\end{vmatrix};\ \boldsymbol{w}=\begin{vmatrix}w_n\\ l_0 w_\tau\end{vmatrix}.
$$

4. Conclusions

The problem of controlling the motion of space objects while removing them from orbit using an ion beam is investigated. A model representing the mechanism of interaction of the ion beam injected from the service spacecraft with the SO surface is presented. The modes for the SC trajectory motion control during the SO transportation are developed. The laws of spacecraft motion control are defined that provide asymptotic stability of the system.

The calculations were performed for the SO removed by the ion beam from the geostationary orbit to the outer orbit, and from LEO down to the earth's atmosphere re-entry. The calculation results show that the proposed control method allows for the SO transportation by an ion beam.

This work was financially supported by a grant from the Government of the Russian Federation for state support of scientific research conducted under the supervision of leading scientists at Russian universities and research organizations (project No. 075-15-2019-1894 of December 3, 2019).

References

[1] Nadiradze A B, Obukhov V A and Svotina V V 2019 Effects of an ion beam on large-sized objects of space debris *Proceedings of the XXI International Conference on Computational Mechanics and Modern Applied Software Systems* (CMMAPS'2019), Alushta (Moscow: Publishing House of the Moscow Aviation Institute) pp 734–737

[2] Nadiradze A B, Obukhov V A , Pokryshkin A I, Popov G A and Svotina V V 2016 Modeling the power and erosion effects of an ion beam on a large object of space debris of technogenic nature *News of the Academy of Sciences (Energetics* No. 2) pp 146–157

[3] Obukhov V A, Petukhov V G, Pokryshkin A I and Popov G A The method of contactless transportation of space objects. *Patent for invention* RU No. 2568960 C1.

[4] Bombardelli Claudio and Pelaez Jesus 2011 Ion Beam Shepherd for Contactless Space Debris Removal *Journal of Guidance Control and Dynamics* pp 1-9

[5] Bombardelli Claudio, Merino Mario, Ahedo Eduardo, Pelaez Jesus, Urrutxua Hodei, Herrera-Montojo Javier and Itturi-Torrea Ander 2011 Ion Beam Shepherd for Contactless Space Debris Removal. *Final Report Ariadna Call for Ideas: Active Removal of Space Debris (ESA)* pp 1-90

[6] Bondarev E N, Dubasov V T, Ryzhov Yu A et al. 1993 *Aerohydromechanics: Textbook for Students of Higher Technical Educational Institutions* (Moscow: Mechanical Engineering) p 608

[7] Konstantinov M S, Kamenkov E F, Perelygin B P and Bezverby V K 1989 *The mechanics of space flight: a Textbook for Technical Colleges* Edited by Mishin V P (Moscow: Mechanical Engineering) p 408

[8] Alpatov A, Khoroshilov S and Bombardelli C 2018 Relative Control of ion Beam Shepherd Satellite Using the Impulse Compensation *Acta Astronautica* **151** 543-554

[9] Egorov Yu G, Kulkov V M, Terentyev V V, Tuzikov S A and Firsyuk S O 2015 Rational design of the orientation system of small spacecraft using mathematical models of motion in gravitational field *Applied Mathematical Sciences* **9** **137** 6835-48

IOP Conf. Series: Materials Science and Engineering **927** (2020) 012052 doi:10.1088/1757-899X/927/1/012052

Solving equations of 3D elasticity for orthotropic bodies

V P Revenko[1] and V N Bakulin[2]
[1] Pidstryhach Institute for Applied Problems of Mechanics and Mathematics National Academy of Sciences of Ukraine, 3-b Naukova str., Lviv, 790060 Ukraine
[2] Laboratory of Non-Classical Models of Mechanics of Composite Materials and Structures, Institute of the Applied Mechanics Russian Academy of Science, St. Leningradsky 7, Moscow, 125040 Russia

E-mail: victor.rev.12@gmail.com, vbak@yandex.ru

Abstract. A mathematical model of the three-dimensional orthotropic body was used. A technique has been developed for integrating equilibrium equations in displacements that describe the stress state of orthotropic materials. We find the solution of equilibrium equations in displacements for orthotropic bodies. The expression of displacements, strains and stresses is obtained through the introduced displacement function, which satisfies the sixth-order equation for partial derivatives. We develop the algorithm of the analytical and numerical solutions of boundary value problem for orthotropic rectangular prism. Solution of boundary value problems is reduced to minimize a generalized quadratic form.

1. Introduction

Orthotropic materials are widely used in aerospace and other engineering, therefore, the solution the equations the theory elasticity and the study of the stress-strain state (SSS) for such elastic bodies is an important scientific problem [1–3]. Using the semi-inverse Saint-Venant method [1, 2], simple stress states of bending and torsion of orthotropic prisms have been studied without using the general representation of solutions of equations of an orthotropic body. In [4, 5], a finite-body method was developed for calculating the stress state of a cylindrical orthotropic shell with a free and reinforced rectangular hole. Currently, solutions of the equations of the linear theory of elasticity are found only for transversely isotropic bodies [6–8]. A complete general solution is constructed only for an isotropic body: in [3, 9] it is expressed in terms of four harmonic functions and in [10] in terms of three harmonic functions in an orthogonal curvilinear coordinate system. Therefore, the construction of a general representation of three-dimensional stresses in orthotropic bodies is of current interest problem, the solution of which allows us to simplify the calculation of their strength and mechanical behavior.

The purpose of research is to construct stress state components of the three-dimensional theory elasticity through the introduced displacement function for the widest class of orthotropic materials in locally loaded orthotropic prisms.

2. A general solution of equations of equilibrium in linear elasticity orthotropic bodies

When constructing a system of equations that describe the stress state of an elastic body, we will proceed from the physical relations for an orthotropic material in a Cartesian coordinate system $x_1 = x,\ x_2 = y,\ x_3 = z$:

$$\sigma_j = \sum_{k=1}^{3} B_{jk}\varepsilon_k, \quad \tau_{kj} = G_{kj}\gamma_{kj}, \; k \neq j, \tag{1}$$

where orthotropic coefficients B_{jk} are determined as follows:

$$B_{jj} = \frac{(1-\nu_{km}\nu_{mk})E_j}{D}, \qquad B_{jk} = B_{kj}, \; B_{jk} = \frac{(\nu_{kj} + \nu_{mj}\nu_{km})E_k}{D}, \quad m \neq k \neq j,$$

$$m,\, k,\, j = \overline{1,3}, \;\; D = 1 - \nu_{23}\nu_{32} - \nu_{12}(\nu_{21} + \nu_{31}\nu_{23}) - \nu_{13}(\nu_{31} + \nu_{32}\nu_{21}) > 0,$$

E_k are Young's moduli; ν_{kj} are Poisson's ratios; G_{kj} are shear moduli of orthotropic material [1, 2]; $\nu_{jk}E_j = \nu_{kj}E_k$, $G_{kj} = G_{jk}$. Components of deformations γ_{kj}, $\varepsilon_j \equiv \gamma_{jj}$ had been expressed [3]

$$\gamma_{jj} \equiv \varepsilon_j = \frac{\partial u_j}{\partial x_j}, \quad \gamma_{kj} = \frac{\partial u_k}{\partial x_j} + \frac{\partial u_j}{\partial x_k}, \; k,\, j = \overline{1,3}, \; k \neq j. \tag{2}$$

In the absence of volumetric forces, the solution of the problems of the linear theory of elasticity of orthotropic bodies, after using the known equilibrium equations [2, 3] and relations (1), (2), have been reduced to integration of such equations:

$$L_j u_j + \frac{\partial}{\partial x_j} \sum_{k=1, k\neq j}^{3} D_{jk} \frac{\partial}{\partial x_k} u_k = 0, \quad j = \overline{1,3}, \tag{3}$$

where

$$L_k = \sum_{j=1}^{3} T_{kj} \frac{\partial^2}{\partial x_j^2}, \; T_{kk} = B_{kk}, \; T_{kj} = G_{kj}, \; D_{kj} = B_{kj} + G_{kj}, \; k \neq j, \; k, j = \overline{1,3}.$$

After eliminating from the relations (3) alternately the displacements u_3, u_2, u_1 a system of three differential equations in partial derivatives have been obtained

$$\frac{\partial}{\partial x_j} L_k^1 u_k = \frac{\partial}{\partial x_k} L_j^1 u_j = 0, \quad k \neq j, \; k, j = \overline{1,3}, \tag{4}$$

where differential operators L_k^1 had been expressed

$$L_j^1 = D_{km} L_j - D_{jk} D_{jm} \frac{\partial^2}{\partial x_j^2}, \quad k \neq j \neq m, \; j, k, m = \overline{1,3}. \tag{5}$$

In equations (1), (3)-(5), for the independent orthotropic coefficients, we take: G_{kj}, B_{kj}, B_{kk}, $k \neq j$, $k, j = \overline{1,3}$, only nine of independent elastic constants. If all the operators L_j^1, $j = \overline{1,3}$ given by relations (5) are not equivalent to each other: $L_j^1 \neq c L_m^1$, $j \neq m$, where $c \in R$ is unknown coefficient, then the solution (4) can be represented

$$u_j = \frac{\partial}{\partial x_j} \prod_{k \neq j} L_k^1 \Phi, \tag{6}$$

where Φ is a displacement function. We had substituted the displacements (6) into the system of equations (3) and had found the defining equation for the displacement function Φ

$$L\Phi = \{L_1 L_2^1 L_3^1 + D_{12}\frac{\partial^2}{\partial x_2^2} L_1^1 L_3^1 + D_{13}\frac{\partial^2}{\partial x_3^2} L_2^1 L_1^1\}\Phi = 0, \tag{7}$$

where the differential operator L has the form

$$L = d_1\frac{\partial^2}{\partial x_1^6} + d_2\frac{\partial^6}{\partial x_1^4 \partial x_2^2} + d_3\frac{\partial^6}{\partial x_1^2 \partial x_2^4} + d_4\frac{\partial^6}{\partial x_1^4 \partial x_3^2} + d_5\frac{\partial^6}{\partial x_1^2 \partial x_3^4}$$
$$+ d_6\frac{\partial^6}{\partial x_2^6} + d_7\frac{\partial^6}{\partial x_2^4 \partial x_3^2} + d_8\frac{\partial^6}{\partial x_2^2 \partial x_3^4} + d_9\frac{\partial^6}{\partial x_1^2 \partial x_2^2 \partial x_3^2} + d_{10}\frac{\partial^6}{\partial x_3^6}, \tag{8}$$

and its coefficients take the following values:

$$d_1 = T_{11}T_{12}T_{13}, \quad d_2 = [T_{11}(T_{12}T_{23} + T_{13}T_{22}) + T_{12}^2 T_{13} - D_{12}^2 T_{13}],$$

$$d_3 = [T_{22}(T_{11}T_{23} + T_{12}T_{13}) + T_{12}^2 T_{23} - D_{12}^2 T_{23}],$$

$$d_4 = [T_{11}(T_{12}T_{33} + T_{23}T_{13}) + T_{12}T_{13}^2 - D_{13}^2 T_{12}],$$

$$d_5 = [(T_{11}T_{23} + T_{13}T_{12})T_{33} + T_{23}T_{13}^2 - D_{13}^2 T_{23}], \quad d_6 = T_{12}T_{22}T_{23},$$

$$d_7 = [(T_{13}T_{23} + T_{12}T_{33})T_{22} + T_{12}T_{23}^2 - D_{23}^2 T_{12}],$$

$$d_8 = [(T_{12}T_{23} + T_{13}T_{22})T_{33} + T_{13}T_{23}^2 - D_{23}^2 T_{13}],$$

$$d_9 = \{[2D_{12}D_{13}D_{23} - D_{23}^2 T_{11} - D_{12}^2 T_{33} - D_{13}^2 T_{22}] + (T_{11}T_{22} + T_{12}^2)T_{33} +$$

$$+ T_{11}T_{23}^2 + T_{22}T_{13}^2 + 2T_{13}T_{12}T_{23}\}, \quad d_{10} = T_{13}T_{23}T_{33}.$$

We have reduced the construction of the general solution equations of elasticity theory to integration of equations (7), (8) in partial derivatives of the sixth order with respect to three variables. The function Φ must be at least six times continuously differentiable on all variables.

Note that (unlike equations that depend only on two variables), the operator (8) depends on three variables x_k, $k = \overline{1,3}$, and, in the general case, cannot be decomposed into three or two factors.

We used the displacement representation (6) and have determined the deformations

$$\varepsilon_m = \prod_{i \neq m} L_i^1 \frac{\partial^2}{\partial x_m^2}\Phi, \quad m = \overline{1,3}, \quad \gamma_{ni} = \frac{\partial^2}{\partial x_n \partial x_i} L_m^1 (L_n^1 + L_i^1)\Phi, \quad m \neq n, i, \; n \neq i. \tag{9}$$

Thus, the obtained strain expressions (9) give us the possibility to find stress components behind formulas (1) and determine the SSS of orthotropic bodies through only one displacement function Φ.

We show how to find a solution to equation (7) and construction of the stress-strain state for canonical bodies made of orthotropic materials.

3. Three-dimensional stress state of an orthotropic rectangular prism

Let us consider the SSS of a rectangular prism $\Pi = \{(x, y, z) \in ([-a, a] \times [-b, b] \times [0, \infty])\}$. The prism had been loaded only on a face $z = 0$ symmetrical with respect to the axes Ox, Oy. Here are no tangent loads. We write the boundary conditions on the face

$$\sigma_3(x, y, 0) = f(x, y), \quad \tau_{31}(x, y, 0) = \tau_{32}(x, y, 0) = 0, \tag{10}$$

where $f(x, y) = \sigma_M \frac{(a_1^2 - x^2)(b_1^2 - y^2)}{a_1^2 b_1^2}$, if $x \in [-a_1, a_1]$, $y \in [-b_1, b_1]$; and $f(x, y) = 0$, $f(x, y)$ is continuous local function, σ_M is its maximum value, a rectangle $\Pi_1 = \{(x, y) \in ([-a_1, a_1] \times [-b_1, b_1])\}$ defines the scope of local load. To maintain equilibrium on the opposite side of the prism, constant stresses must be set at infinity σ_3^∞, $\sigma_3^\infty = \sigma_0$: $\sigma_0 = \frac{P_3}{S}$, where $P_3 = \iint f_1(x, y) dS = \frac{16}{9} a_1 b_1 \sigma_M$ is the force on the face $z = 0$, σ_0 is the average stresses of the basic SSS and $S = 4ab$ is the face area.

Thus, we separate of the stress state of the prism by the sum of the basic SSS ($\sigma_0 = \sigma_3^\infty$) with one component of the forces, and the perturbed self-balanced state with zero basic vectors of forces and moments [10, 11].

Suppose that $a_1 << a$, $b_1 << b$ then on the lateral faces of the prism, the stresses of the perturbed state will be insignificant, and we will search for it's in the form of such a series

$$\Phi = \sum_{k=1}^{\infty} \sum_{j=1}^{3} a_k^j \exp(-k\lambda_j z) \cos(k\omega_1 x) \cos(k\omega_2 y), \tag{11}$$

where Φ is the displacements function, decreasing as you move away from the face $z = 0$, $\omega_1 = k_1 / a$, $\omega_2 = k_1 / b$, $2.8 < k_1 \le \pi$; $\{\cos(k\omega_1 x)\cos(k\omega_2 y)\}$ are the complete system of pair functions in the rectangle $\Pi_2 = \{(x, y) \in ([-a, a] \times [-b, b])\}$. The eigenvalues $\mu_j > 0$, $j = \overline{1,3}$ are determined after the substitution of function (11) in equations (7), (8), and are determined as the roots of such a characteristic equation:

$$d_{10}\mu^3 + g_1\mu^2 + g_2\mu + g_3 = 0, \tag{12}$$

where $g_1 = -(d_5\omega_1^2 + d_8\omega_2^2)$, $g_2 = d_4\omega_1^4 + d_7\omega_2^4 + d_9\omega_1^2\omega_2^2$, $g_3 = -(\omega_1^6 d_1 + \omega_1^4\omega_2^2 d_2 + \omega_1^2\omega_2^4 d_3 + \omega_2^6 d_6)$, $\mu_j = \lambda_j^2$. Depending on the values of the orthotropy coefficients of the material, equation (12) can have three real or one real and two complex roots. Consider various real roots $\lambda_j = \sqrt{\mu_j} > 0$, $j = \overline{1,3}$. We write the displacement function (11)

$$\Phi = \sum_{k=1}^{\infty} \sum_{j=1}^{3} d_k^j \varphi_k^j, \tag{13}$$

where $\varphi_k^j = \exp(-k\lambda_j z)\cos(k\omega_1 x)\cos(k\omega_2 y)$, $\lambda_j = \sqrt{\mu_j}$. If we have two identical roots $\mu_1 = \mu_2$, then instead of function $\exp(-k\lambda_2 z)$ it is necessary to use an additional solution $z\exp(-k\lambda_1 z)$. If there are three identical roots $\mu_1 = \mu_2 = \mu_3$, then such solutions $(a_0 + a_1 z + a_2 z^2)\exp(-k\lambda_1 z)$ should be used.

We substitute function (13) into relation (5) and find the value of the operators

$$\mathrm{L}^1_{\mathrm{m}}\varphi^j_k = k^2\chi^j_m\varphi^j_k\,,\ m=\overline{1,3}, \tag{14}$$

where

$$\chi^j_m = D_{3-m,m}(\mathrm{G}_{m3}\mu_j - \mathrm{G}_{12}\omega^2_{3-m}) - [D_{3-m,3}\mathrm{T}_{mm} - D_{j3}D_{12}]\omega^2_m\,,\ m=\overline{1,2},$$

$$\chi^j_3 = [D_{12}\mathrm{T}_{33} - D_{13}D_{23}]\mu_j - D_{12}(\mathrm{G}_{13}\omega^2_1 - \mathrm{G}_{23}\omega^2_2)\,.$$

We have used the relations (6), (14) and have expressed the displacements

$$\mathrm{u_m} = \sum_{k=1}^{\infty} k^4 \sum_{j=1}^{3} d^j_k \prod_{n\neq m}^{3} \chi^j_n \frac{\partial}{\partial x_m}\varphi^j_k\,. \tag{15}$$

The strains are determined from relations (9), (15) and we have obtained

$$\varepsilon_m = \sum_{k=1}^{\infty} k^4 \sum_{j=1}^{3} d^j_k \prod_{n\neq m}^{3} \chi^j_n R^2_{1,\mathrm{j}}\varphi^j_k\,,\ m=\overline{1,3},$$

$$\gamma_{mi} = \sum_{k=1}^{\infty} k^4 \sum_{j=1}^{3} S^j_{m,i} R_{m,\mathrm{j}} R_{i,\mathrm{j}} d^j_k \frac{\partial^2\varphi^j_k}{\partial x_m \partial x_i}\,,\ m\neq n,i\,,\ n\neq i\,, \tag{16}$$

and from relations (1), (16) we have found the stress components

$$\sigma_{\mathrm{i}} = \sum_{j=1}^{3} \mathrm{T}^j_i \sum_{k=1}^{\infty} k^6 d^j_k \varphi^j_k\,,\ \mathrm{i}=\overline{1,3},\quad \tau_{mi} = G_{mi}\gamma_{mi}\,,\ m\neq i\,, \tag{17}$$

where

$$R_{1,\mathrm{j}} = \omega_1,\ R_{2,\mathrm{j}} = \omega_2,\ R_{3,\mathrm{j}} = \lambda_{\mathrm{j}},\ \mathrm{T}^j_i = \sum_{m=1}^{3} B_{\mathrm{im}} t_m \prod_{n\neq m}^{3} \chi^j_n R^2_{m,\mathrm{j}},$$

$$S^j_{m,i} = \chi^j_n(\chi^j_m + \chi^j_i)\,,\ t_1 = t_2 = -1,\ t_3 = 1\,.$$

In numerical calculations, we restricted relations (13), (15)–(17) to the first N members of the series relative to the index k and had obtained $M = 3N$ unknown coefficients that describe the stress-strain state of prism.

To find the perturbed stress state, it suffices to satisfy the boundary conditions (10), which, after substituting the stresses (17) in them, have been reduced to a compact form

$$\sum_{k=1}^{M} c_k A_{m,k}(x,y) = P_m(x,y)\,,\ x\in[\text{-a, a}],\ y\in[\text{-b, b}]\,,\ m=\overline{1,3}, \tag{18}$$

where c_k, $k=\overline{1,M}$ are unknown coefficients, $A_{1,k}(x,y) = a_{1,k}\cos(k\omega_1 x)\cos(k\omega_2 y)$, $A_{2,k}(x,y) = a_{2,k}\sin(k\omega_1 x)\cos(k\omega_2 y)$, $A_{3,k}(x,y) = a_{3,k}\cos(k\omega_1 x)\sin(k\omega_2 y)$ are known functions, $c_{\mathrm{k+jN-N}} = d^j_k$, $j=\overline{1,3}$, $k=\overline{1,N}$, $P_1(x,y) = f(x,y) - \sigma_0$, $P_2(x,y) = P_3(x,y) = 0$.

To solve equations (18), we had used the analytical-numerical method developed in [4, 5, 10–12] to satisfy the boundary conditions. The given method allows us have been reduced numerical solution of equations (18) to determination of the generalized quadratic form of a minimum (function F(N)):

$$\sum_{m=1}^{3}\left\|\sum_{k=1}^{M} c_k A_{m,k}(x,y) - P_m(x,y)\right\|^2 = \sum_{k,j=1}^{M} c_k c_j W_{kj} - 2\sum_{k=1}^{M} c_k V_k + P^2, \tag{19}$$

where

$$\|f(x,y)\| = \sqrt{\int_{-a}^{a}\int_{-b}^{b} f^2(x,y)dxdy} \text{ is norm, } (f,g) = \int_{-a}^{a}\int_{-b}^{b} f\,g\,dxdy \text{ is scalar product,}$$

$$W_{kj} = \sum_{m=1}^{3}(A_{m,k}, A_{m,j}),\ V_k = \sum_{m=1}^{3}(A_{m,k}, P_m),\ k,j=\overline{1,M},\ \mathrm{P}^2 = \sum_{\mathrm{m}=1}^{3}\|\mathrm{P_m}\|^2.$$

In [4, 5, 10], the existence and uniqueness of a solution, the convergence of the method, and its accuracy, which is determined by a unique minimum value $F(N)$, were proved. In [10–12], numerical calculations and an applied analysis of the accuracy and convergence of the method for isotropic rectangular plates and cylindrical bodies were made. The performed calculations confirmed the mathematical conclusion about the uniqueness of the solution.

Note that the proposed approach allows solving the corresponding boundary value problems for orthotropic bodies and determining their three-dimensional stress-strain state.

4. Conclusions

Mathematical and physical rigor has been maintained in the construction of calculation set of formulas of orthotropic elasticity theory. It has been established that the general solution of the equations of the elasticity theory of the orthotropic body is expressed through one function that satisfies the equation in sixth-order partial derivatives. On the basis of the general solution of the equilibrium equations, an expression of deformations and stresses in the Cartesian coordinate system has been built. We develop the algorithm of the analytical and numerical solutions of boundary value problem for orthotropic rectangular prism. Universal way of reduction of all boundary conditions to minimize the generalized quadratic forms is proposed.

References

[1] Lekhnitskii S G 1981 *Theory of elasticity of an anisotropic body* (Moscow: Mir Publishers)
[2] Rand O and Rovenski V 2005 *Analytical Methods in Anisotropic Elasticity* (Boston: Birkhäuser)
[3] Sadd M.H. 2009 *Elasticity. Theory, applications, and numerics* (Burlington: Academic)
[4] Bakulin V N and Revenko V P . *2016 Russian Mathematics* **60**(6) 1-11 DOI: 10.3103/S1066369X16060013
[5] Revenko V P and Bakulin V N 2019 *Journal of Physics: Conference Series* **1392** 012021 IOP Publishing doi:10.1088/1742-6596/1392/1/012021
[6] W. T. Chen *J. Appl. Mech.* 1966, **33**(2) 347–355 https://doi.org/10.1115/1.3625048
[7] Hu H C *Data Sci. Sinica.* 1953 **2** 145–151
[8] W T Chen *J. Appl. Mech.* 1966 **33** 347–355 doi: https://doi.org/10.1115/1.3625048
[9] Timoshenko S P and Goodier J N 1970 *Theory of Elasticity* (New York: McGraw-Hill)
[10] Revenko V P 2009 *Int. Appl. Mech.* **45**(7) 730-741 https://doi.org/10.1007/s10778-009-0225-4
[11] Revenko V P 2008 *Materials Sci.* **44** 482–488 https://doi.org/10.1007/s11003-009-9118-2
[12] Bakulin V N and Revenko V P 2017 *Materials physics and mechanics* **31**(1-2) 93–96

Study of the low-power krypton-operated stationary plasma thruster plume

V P Kim, D V Merkurev, E A Shilov, V S Zakharchenko and V G Kalyazin
Research Institute of Applied Mechanics and Electrodynamics of the Moscow Aviation Institute (RIAME MAI), Leningradskoye shosse 5, 125080, POB 43, Moscow, Russia

E-mail: riame4@sokol.ru, merckuriev.denis@yandex.ru

Abstract. The paper presents the results of experimental studies of the plume of low power stationary plasma thruster SPT-40 model, operating with krypton at discharge power of about 250 W, and with xenon at discharge power of about 200 W. A comparative analysis of the SPT -40 model plume characteristics was carried out that revealed significant differences in the distributions of current density of accelerated ions in the angle of deviation of the measurement directions from the thruster axis and in the distributions of average ion energies in the plume when operating with the above propellants. The flow parameters were measured by a multigrid probe – the retarding potential analyzer (RPA). This devise moved along the semicircular trajectory in a plane that included the thruster axis. The distance from the thruster exit plane to the RPA was equal to 6-7 outer diameters of the thruster-accelerating channel; at such distance, we can consider the thruster as a point source. The measurements were carried out at different operating modes in the propellant consumption and discharge voltage, and it is shown that during the SPT-40 model operation with xenon and krypton the half-angle of the plume divergence was about 70 degrees in the discharge voltage range 150-200 V and it decreased with the discharge voltage growth.

1. Introduction

Stationary plasma thrusters (SPT) have been successfully used in space technology for many years and during this time, they have proven to be efficient and reliable actuators for various maneuvers of spacecraft (SC). Therefore, they are regularly used in the orbit correction systems for geostationary spacecraft, and recently they have also been used for the final insertion of spacecraft to target orbits. In light of the fact that one of the trends in the modern astronautics is the miniaturization of spacecrafts with the aim of reducing their cost, as well as of using them as parts of multi-satellite constellations the manufacturers of small spacecraft are showing an increased interest in electric propulsion systems based on the low-power SPT. The fact is that the aforementioned electric propulsion systems should provide a sufficiently long period of active spacecraft existence, have high reliability and relatively low cost. The extensive experience in the development and application of electric propulsion systems based on SPT shows that they can meet the above set of requirements. Therefore, the development of low power SPT for small SC motion control systems is relevant. At the same time, for constellations of small SC, it is of interest to develop low-power SPT operating with propellants alternative to xenon, which was the main propellant for SPT and ion thrusters. The fact is that xenon is produced in small quantities and is a very expensive propellant. Therefore, it is problematic to load xenon to EPS of all SC of such a constellation as Starlink with a design SC number of more than 10,000. Probably with

IOP Conf. Series: Materials Science and Engineering **927** (2020) 012053 doi:10.1088/1757-899X/927/1/012053

this in mind, krypton, which is more than 10 times cheaper than xenon and is produced in significantly larger quantities, is already used in the electric propulsion systems for this constellation [1].

In consideration of the foregoing, the Research Institute of Applied Mechanics and Electrodynamics of the Moscow Aviation Institute (RIAME MAI) has been developing competitive low-power SPT operating with xenon and krypton for several years already. This paper presents the results of study of beam characteristics of one of the SPT-40 laboratory models in order to identify the plume features of low-power SPT operation with krypton.

2. Method for determining the SPT-40 plume characteristics

The integral characteristics of the SPT-40 model and the characteristics of its plume were investigated at the test facility comprising vacuum chamber with an internal diameter of 2 m and a useful internal volume of 10.5 m^3. The vacuum chamber is equipped with cryogenic high-vacuum pumps and an additionally installed turbo-molecular pump to provide "oil-free" deep vacuum. Such pumping means allow us to provide the residual pressure in the vacuum chamber at a level of $2\cdot10^{-6}$ Torr, while with the thruster operating at the mass flow rate of 2 mg/s - not worse than $2\cdot10^{-5}$ Torr that being acceptable to study the beam of low-power electric propulsion.

During the experiments, the discharge current, power, and thrust of the model were recorded at different operating modes of the thruster. In addition, the distribution of the current density of accelerated ions in the thruster beam in the angle β was determined at a distance R from it along the spherical reference surface centered at the intersection of the thruster axis with the exit plane of its accelerating channel (figure 1). The distance R from the thruster to the reference surface should be equal to 5-10 outer diameters of the thruster accelerating channel, i.e. for a thruster of the SPT-40 scale, this distance should be R = 0.2 ... 0.5 m (in practice, the probe was located at a distance of 0.5 m from the exit plane). Another important characteristic of the beam defined in that study was the energy distribution of ions moving in the beam in directions with different angles of deviation from the thruster axis at the same distance from it. The results of determining the integral parameters and the mentioned characteristics of the SPT-40 model beam are presented in the next section of the paper. The next beam characteristics were recorded in the paper: the total current I_i of accelerated ions in the beam, the half-angle $\beta_{0.95}$ of the beam divergence, and the distributions of the average energy of ions moving in the beam in different directions.

The results of determining the total current of ions leaving the thruster allow us to evaluate the efficiency of the propellant utilization, i.e. the efficiency of transformation of the initial neutral atoms of propellant entering the SPT accelerating channel to ions. The importance of the beam divergence half-angle is also high, since the divergence of the accelerated ion flow affects the thrust efficiency. This parameter is also crucial for determining the thruster interaction with the SC elements, which can be exposed to the peripheral part of the beam.

The distribution of the average ion energy in the angle of deviation of their motion directions from the axis is also an important characteristic of the beam, since it defines the flows of momentum and of ion energy in the beam in different directions. The mentioned parameter determines the contribution of different parts of the beam to the thrust and possible mechanical, thermal and erosive effects of the beam on the structural elements of the SC exposed to it, which could be correlated with modeling [2].

IOP Conf. Series: Materials Science and Engineering **927** (2020) 012053 doi:10.1088/1757-899X/927/1/012053

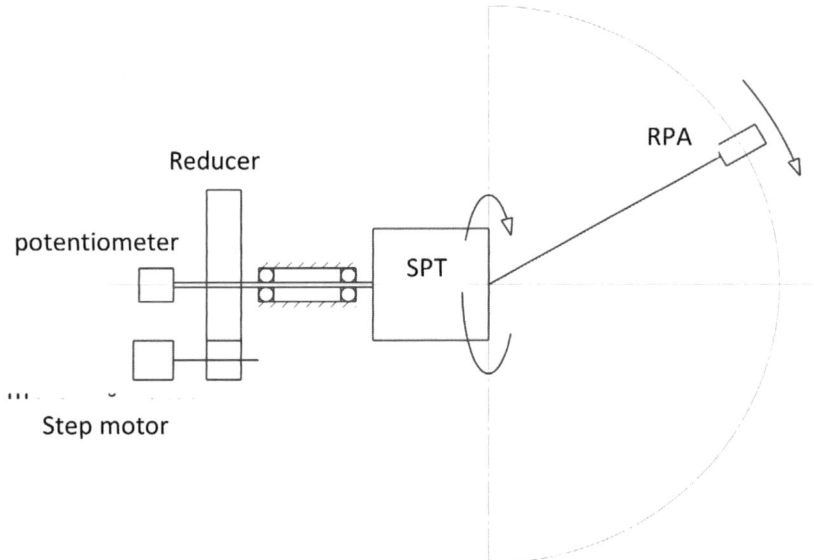

Figure 1. Scheme of the RPA displacement

A RPA was used as the main diagnostic tool for SPT plumes, for the calibration of which a Langmuir flat probe was used [3].

The distribution of the current density of accelerated ions in the directions of their flow out of the thruster axes and the energy distribution of ions moving in different directions are the standard characteristics of the SPT plume [3].

In view of the above, in this study we determined the current density distribution of accelerated ions in the angles of their motion direction deviation from the axis and the energy distributions of ions moving in different directions, at different modes of thruster operation with krypton and xenon.

3. Integral parameters and plume characteristics of the SPT-40 laboratory model

The integral parameters of the SPT-40 laboratory model during its operation with krypton and xenon are presented in tables 1 and 2.

Table 1. Parameters of the krypton-operated SPT-40 laboratory model in different modes

	m=1.5 mg/s, U_d = 200 V	m=1 mg/s, U_d = 200 V	m=1 mg/s, U_d = 300 V	m=2 mg/s, U_d = 200 V	m=2 mg/s, U_d = 175 V	m=2 mg/s, U_d = 150 V
Power, W	288	144	246	460	411	300
Current, A	1.4	0.7	0.8	2.3	2.4	2.0
Thrust, mN	12.05	4.21	7.45	15.68	13.72	10.78
Efficiency	0.17	0.06	0.11	0.13	0.11	0.1

Table 2. Parameters of the xenon-operated SPT-40 laboratory model in different modes

	m=1.5 mg/s, U_d = 200 V	m=1 mg/s, U_d = 200 V	m=1 mg/s, U_d = 300 V	m=1.5 mg/s, U_d = 250 V	m=1.5 mg/s, U_d = 175 V	m=1.5 mg/s, U_d = 150 V
Power, W	290	162	261	337	245	225
Current, A	1.5	0.8	0.9	1.4	1.4	1.5
Thrust, mN	15.48	8.43	11.86	17.15	12.74	10.74
Efficiency	0.28	0.22	0.27	0.29	0.22	0.17

As can be seen from the data presented, the efficiency of the krypton-operated SPT-40 is significantly lower than that of the same thruster model operating with xenon. This corresponds to similar results obtained previously for the IPPLM and TUD-H3-P models [4, 5]. At the same time, in the latter paper it was concluded that the main reason for the low efficiency of krypton-operated SPT is in the low degree of ionization of the krypton flux entering the accelerating channel.

To clarify the peculiarities of low-power SPT operation with krypton, during the execution of the work presented the distributions of accelerated ion current density were determined in the angle of the measurement directions deviation from the thruster axis, as well as the energy distribution of ions moving in different directions, for different modes of thruster operation with krypton and xenon. Typical distribution of the accelerated ion current density in the angle of the measurement directions deviation from the thruster axis is shown in Figure 2 (the other distributions were similar to that shown), where the thruster axis corresponds to 0 degrees. Processing of these distributions made it possible to obtain estimates for the accelerated ion current and half-angle of the accelerated ion flow divergence, shown in Figures 3-6.

The data obtained demonstrates:

- the ion current I_i increased with the propellant consumption and discharge voltage, but for krypton this growth with an increase in discharge voltage was sharper (Figure 3);

- the ratio of the ion current I_i to the "flow rate" current $I_m = \dot{m}_a e / M$, where $\dot{m}_a$ is the mass flow rate of the propellant through the accelerating channel, e is the electron charge and M is the mass of the propellant atom, was much lower for krypton than for xenon (Figure 4);

- the ratio of the ion current I_i to the discharge current I_d with the krypton use in the investigated operating modes was lower than during the thruster operation with xenon (Figure 5);

- the flow divergence half-angle was wider during the thruster operation with krypton than with xenon (Figure 6).

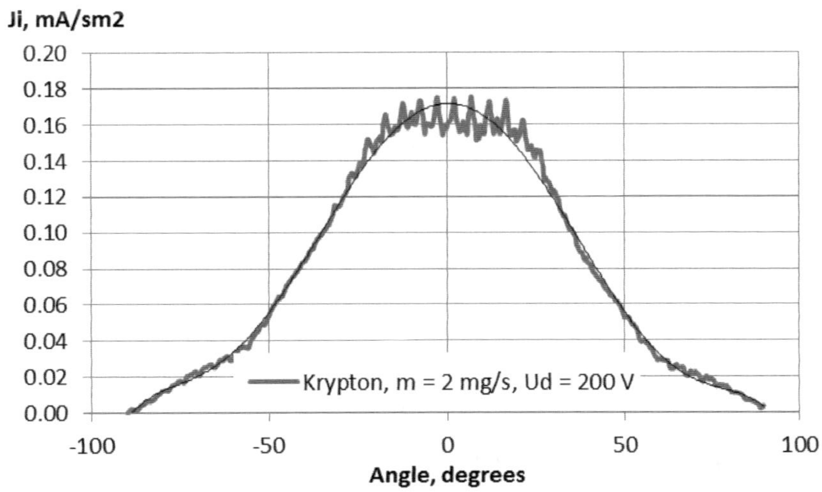

Figure 2. Distribution of the current density of accelerated ions in the beam of the SPT-40 model during its operation with krypton at the flow rate of 2 mg/s and discharge voltage of 200 V (measured and smoothed values)

The energy distributions of ions and the corresponding distributions of average ion energies depending on the angle of the measurement direction deviation from the thruster axis (Figures 7, 8) indicate that under comparable conditions of operation with krypton and xenon the differences in the obtained energy characteristics of ions in the SPT-40 plumes are negligible.

IOP Conf. Series: Materials Science and Engineering **927** (2020) 012053 doi:10.1088/1757-899X/927/1/012053

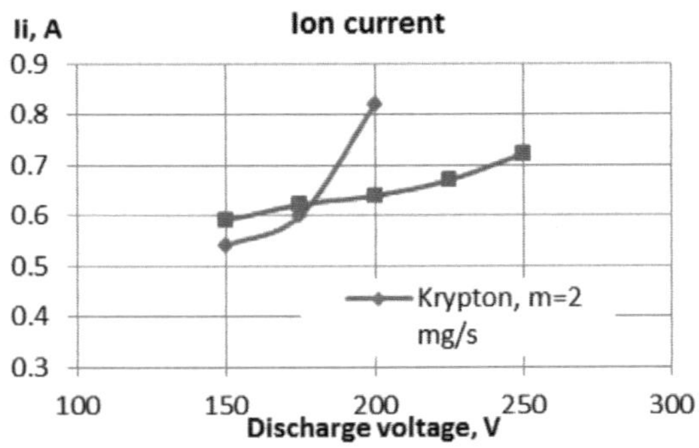

Figure 3. Ion current dependence on the discharge voltage for the SPT-40 model operating with krypton and xenon

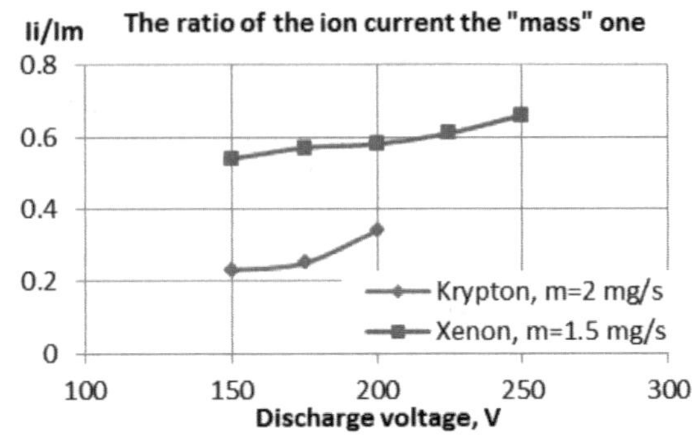

Figure 4. Dependence of the ion current to the "flow rate" current ratios on the discharge voltage for the SPT-40 model operating with krypton and xenon

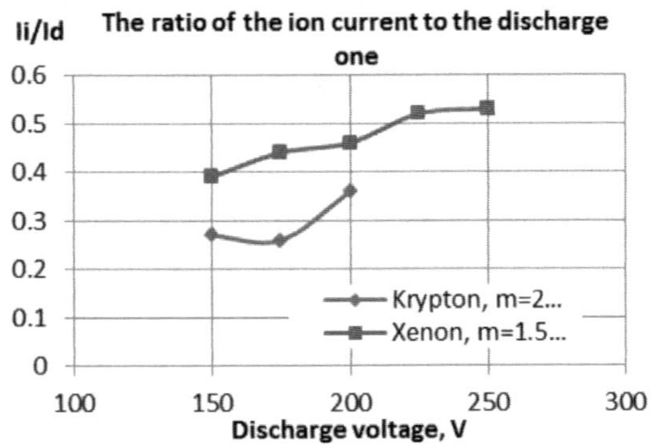

Figure 5. Dependence of the ion current to the discharge current ratio on the discharge voltage for the SPT-40 model operating with krypton and xenon

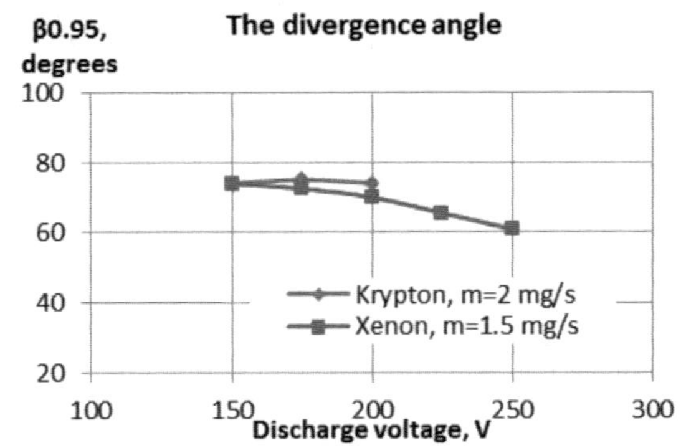

Figure 6. Dependence of the accelerated ion flow divergence half-angle on the discharge voltage for the low-power SPT model operating with krypton and xenon

4. Discussion of the results

In general, the data obtained allowed concluding that the main reason for the decrease in the thrust efficiency of low-power SPT operating with krypton compared to its operation with xenon is the decrease in the propellant utilization efficiency, since the greatest difference was obtained for the ratio I_i/I_m as in [4,5] for thrusters of larger sizes. This is explained by the fact that in the low-power krypton-operated models it is more difficult to organize optimal operating modes. Indeed, the krypton-operated thrusters, due to the higher ionization potential, reach modes with enhanced characteristics at higher discharge voltages, and in view of the power limitations, it is not possible to ensure their operation at higher discharge voltages. In addition, even at comparable ion currents, ion densities and discharge voltages (ion energies), the velocities of krypton atoms and ions are higher than the corresponding velocities of xenon atoms and ions due to their lower mass, while the density of krypton ions is lower. Under the above conditions, while using krypton as propellant, the electron density n_e equal to the ion density should also be lower, as well as the factor of the krypton atoms ionization rate $\langle \sigma_i V_e \rangle$ [6]. The action of all these factors leads to the fact that the mean free path of krypton atoms $\lambda_i = V_a / \langle \sigma_i V_e \rangle n_e$ when operating with krypton is significantly longer under the specified conditions. Accordingly, the ionization probability of krypton atoms and their utilization factor in the low-power SPT discharge appear to be much lower.

Thus, the obtained results confirm that the main reason for the low thrust efficiency of the low-power SPT operating with krypton is the low krypton utilization efficiency.

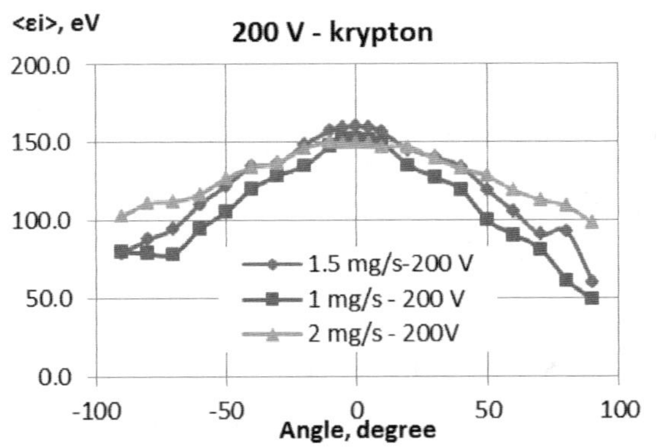

Figure 7. Distribution of the average ion energy relative to the axis of the SPT-40 model operating with krypton in various modes

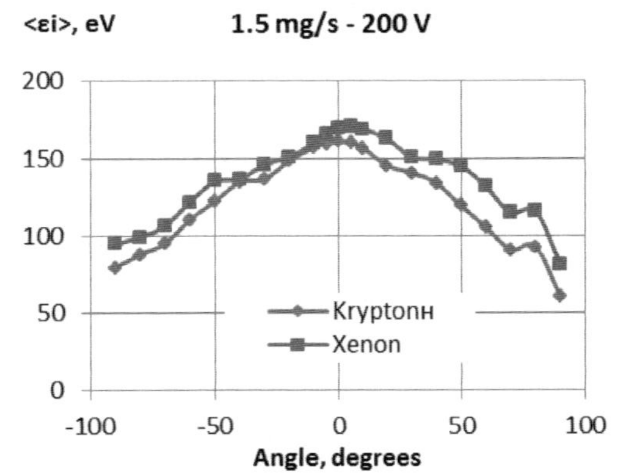

Figure 8. Distribution of average ion energy relative to the axis of the SPT-40 model operating with xenon and krypton in equivalent modes

5. Conclusion

The study confirmed that the low-power SPT models operating both with xenon and krypton have a low level of thrust efficiency due to the complexity in organizing their operation in optimal modes caused by power limitations. At the same time, such limitations affect thrust efficiency of the krypton-operated SPT more significantly due to the difficulty of ensuring high propellant utilization efficiency, primarily due to the higher velocities of krypton atoms and ions and lower plasma density insufficient to ensure a high probability of krypton atom ionization in the accelerating channel of the thruster.

Acknowledgments

This work was supported by the Russian Science Foundation (project No. 16-19-10429).

References

[1] Caleb H 2019 SpaceX launches 60 Starlink satellites, begins constellation buildout *Spacenews* https://spacenews.com/spacex-launches-60-starlink-satellites-begins-constellation-buildout

[2] Kotelnikov V. A., Kotelnikov M. V. 2020 Computational modeling of rarefied plasma and neutral gas effusion into vacuum *Smart innovation, systems and technologies,* **173**, 119-132

[3] King L B 1998 *Transport – property and mass spectral measurements in the plasma exhaust plume of a Hall-effect space propulsion system.* (PhD Thesis).

[4] Zhasan V S, Kim V P, Merkurev D V, Murashko V M, Nesterenko A N, Popov G A, Potapenko M Yu, Smirnov P G and Shilov E A 2016 *Proc. of the Russian Academy of Sciences. Energ.* **2**, 66-79.

[5] Kurzuna J, Jakubczak M and Szelecka A 2018 Performance tests of IPPLM's krypton Hall thruster. *Laser and Particle Beam s.* **36** 1 105-114

[6] Morozov A I and Melikov I V 1977 *Phys. of Plasma* **44** 3 544-49.

IOP Conf. Series: Materials Science and Engineering **927** (2020) 012054 doi:10.1088/1757-899X/927/1/012054

Natural convection in horizontal layers with lateral heating at various Rayleigh and Prandtl numbers

A I Fedyushkin
Ishlinsky Institute for Problems in Mechanics of Russian Academy of Sciences,
Prospekt Vernadskogo, 101-1, Moscow, 119526, Russia

E-mail: fai@ipmnet.ru

Abstract. The paper presents the results of numerical solution of the problem of thermal convection of an incompressible liquid in a horizontally elongated layer with lateral heat supply. In this paper the influence of Rayleigh and Prandtl numbers of boundary conditions on the flow structure and on the existence of counter-flows are considered. It is also shown that for laminar convection modes, flows with Prandtl numbers less than 0.01 differ qualitatively from the flow of a liquid with a zero Prandtl number, so the approximation of the zero value of the Prandtl number may not always be correct.

1. Introduction

Convective circulation flows in horizontal layers can be observed in the ocean, in the asthenosphere, in the atmosphere and they are already known to researchers more than 300 years ago (Halley-1686r, Hadley-1735r.). This problem is one of the fundamental problems of the theory of thermal convection (as well as the well-known Rayleigh-Benard problem about the occurrence of convection when the layer is heated from below). This problem is essentially different from the Rayleigh-Benard problem in that the convective flow occurs (loses stability) at an arbitrarily small value of the Rayleigh number [1, 2].

The problem of convection in horizontal layers has been studied in many papers: for example, analytically in [3-5], experimentally in [6-9] and numerically in [2, 9-12]. The study of this problem has not lost interest to this day, because it has a wide range of applications in geophysical (flows in the atmosphere, in the ocean and in the earth's asthenosphere [8]), thermophysical (heat collectors and coolers), environmental (the spread of pollution in reservoirs and the atmosphere) and technological processes (in the production of materials and horizontal Bridgman crystal growth).

This problem is characterized by boundary conditions and three dimensionless numbers: Rayleigh (Ra), Prandtl (Pr) and the ratio of the sides of the layer (L/H>>1). In case of medium and large Rayleigh numbers the convective flows in horizontal layers may have features that are not described by the analytical solutions. In these cases, numerical modeling is used. In this paper, numerical results are obtained for various values of Rayleigh, Prandtl numbers and types of boundary conditions. The appearance of counter-flows and layered structures caused by vertical stratification of density created only by thermal convection is shown. Features of convective flow at values of the Prandtl number close to zero are presented.

2. Problem statement and mathematical model

The problem of thermal convection of an incompressible liquid in a horizontally elongated layer ($L>>H$, were L - is length and H - is height), laterally heated in the field of gravity (with acceleration of free fall g), is considered. At lateral heating, constant values of temperature T_1 and T_2 ($T_1 < T_2$) on the side walls are set; for velocities, non-slip conditions are set. The following boundary conditions are considered for horizontal walls: for velocity – the non-slip condition (at rigid wall) or the slip condition (at free surface), for temperature – either the condition of thermal insulation, or an isothermal condition with a linear temperature profile is set. Further in the paper, if the temperature boundary conditions on the horizontal walls are not specified, it means that the horizontal walls were adiabatic (by default) $\partial T / \partial y |_{y=0, y=H} = 0$, and in the case of isothermal boundaries $T|_{y=0, y=H} = T_1 + x(T_2 - T_1)H / L$, this will be said specifically. In this paper, the results will be presented for layers with elongation $L/H = 12.71$.

The scheme of the calculated geometry, boundary conditions and simulation results in the view of isolines of the stream function (a) and isotherms (b) of a convective flow with a single-vortex structure at $Ra = g\beta(T_2 - T_1)H^3 / \nu a = 2 \cdot 10^5$, $Pr = a/\nu =$ 5.8 are shown in figure 1.

The mathematical model is based on the numerical solution of a system of non-stationary planar 2D Navier-Stokes equations for natural convection of an incompressible liquid in the Boussinesq approximation (1-4) [13]:

$$\frac{\partial u}{\partial t} + u\frac{\partial u}{\partial x} + v\frac{\partial u}{\partial y} = -\frac{1}{\rho}\frac{\partial P}{\partial x} + \nu\left(\frac{\partial^2 u}{\partial x^2} + \frac{\partial^2 u}{\partial y^2}\right) + F_1, \tag{1}$$

$$\frac{\partial v}{\partial t} + u\frac{\partial v}{\partial x} + v\frac{\partial v}{\partial y} = -\frac{1}{\rho}\frac{\partial P}{\partial y} + \nu\left(\frac{\partial^2 v}{\partial x^2} + \frac{\partial^2 v}{\partial y^2}\right) + F_2, \tag{2}$$

$$\frac{\partial u}{\partial x} + \frac{\partial v}{\partial y} = 0, \tag{3}$$

$$\frac{\partial T}{\partial t} + u\frac{\partial T}{\partial x} + v\frac{\partial T}{\partial y} = a\left(\frac{\partial^2 T}{\partial x^2} + \frac{\partial^2 T}{\partial^2 y}\right), \tag{4}$$

where x, y - horizontal and vertical Cartesian coordinates; u, v - components of the velocity vector; t - time; T – temperature; P- pressure; ρ - density; $F_1 = 0$ and $F_2 = -g\beta(T - T_1)$ - components of the vector of external forces; g –gravitational acceleration of the earth's free fall; β, ν, a - coefficients of temperature expansion of the liquid, kinematic viscosity and thermal conductivity, respectively; in the future, we will use the notations v_x, v_y of dimensionless components of the velocity and t' of dimensionless time (which was made dimensionless through the viscosity ν and height of the layer *H*).

The results presented in this paper were obtained using different numerical methods: the finite-difference scalar method [2, 14], the fully implicit matrix finite-difference method [15], and the conservative control volume method [16]. The good accuracy of numerical results was confirmed by comparison with experimental data and comparison of numerical results obtained by different numerical models [2, 9, 12, 14, 15].

3. The results of numerical simulation

At low Rayleigh numbers, a single-vortex up-and-down convective stationary flow occurs in a closed horizontal layer filled with water with the lateral heating (figure 1), which can be described analytically under certain assumptions [1, 3, 4]. At large Rayleigh numbers ($Ra \geq 10^6$), some nonlinear features of convective flows begin to appear as a result of the nonlinear nature of the dependence of the flow characteristics on the determining its parameters [12]. Features of stationary laminar flows can consist, for example, in the formation of various stationary horizontal layered flow structures with counter-flows inside the layer (circulation flows opposite directed to the main one), as shown in figures 2-6, or in the formation of a large-scale multicellular structure with a certain wave number [1], for example, such as shown in figure 9. The effect of the Rayleigh number on the flow structure for the same parameters can be seen by comparing the simulation results for water at $Ra = 2\cdot10^5$ (figure 1) and at $Ra = 10^9$ (figure 2). At $Ra = 2\cdot10^5$ the flow has a single-vortex structure, which begins to change as increase the Rayleigh number $Ra \geq 10^6$ with the formation of counter-flows inside the layer (which are shown in figures 2-4).

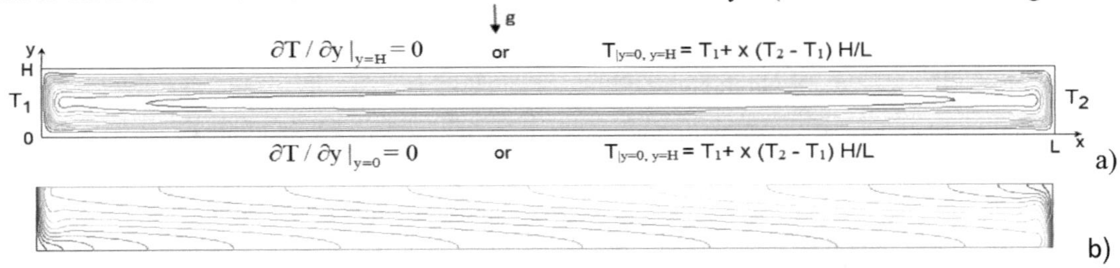

Figure 1. Scheme of the calculated geometry, boundary conditions and isolines of the stream function (a) and isotherms (b) at $Ra = 2\cdot10^5$, $Pr = 5.8$.

The existence of convective counter-flows in closed horizontal layers heating from the side was shown experimentally and numerically in paper [9]. The numerical results were compared with experimental local velocity and temperature profiles in [9]. This comparison showed good accuracy of the numerical model. The intensity and number of secondary flows (counter-flow layers) in the center of the layer increases with increasing Rayleigh number, as shown in figure 2, and this was also noted in [9, 12].

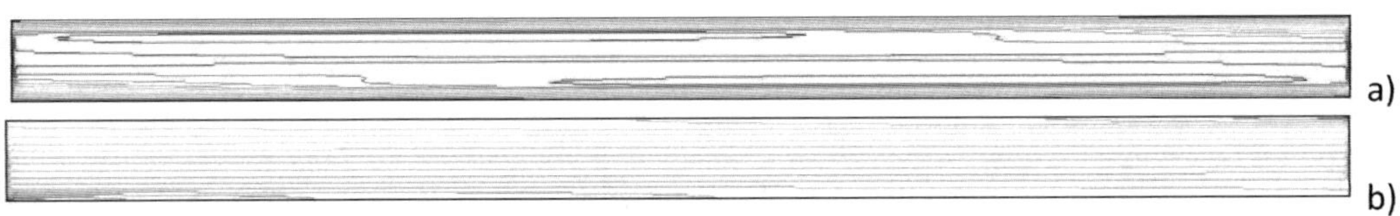

Figure 2. The isolines of the stream function (a) and the isotherm (b) at $Ra = 10^9$, $Pr = 5.8$.

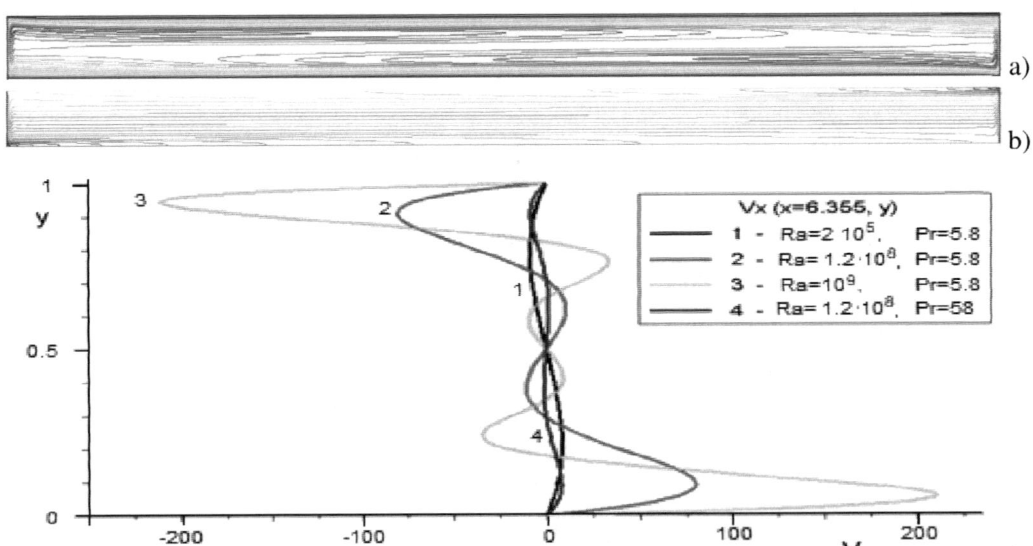

Figure 3. The isolines of the stream function (a), the isotherms (b) at $Ra = 10^9$, $Pr = 58$ and the profiles of the horizontal velocity in the middle vertical cross section (x=6.355) (c) for $1 - Ra = 2\cdot10^5$, $Pr = 5.8$, $2 - Ra = 1.2\cdot10^8$, $Pr = 5.8$, $3 - Ra = 10^9$, $Pr = 5.8$, $4 - Ra = 1.2\cdot10^8$, $Pr = 58$.

The profiles of the horizontal velocity in the middle vertical cross section (x=6.355) for follow parameters: $1-Ra=2\cdot10^5$, $Pr=5.8$, $2-Ra=1.2\cdot10^8$, $Pr=5.8$, $3-Ra=10^9$, $Pr=5.8$, $4-Ra=1.2\cdot10^8$, $Pr=58$ are shown in figure 3c. The comparisons of profiles show the influence of Rayleigh and Prandtl numbers on the existence and characteristics of counter-flows. The counter-flow exists for all the specified parameters, except case $1-Ra=2\cdot10^5$, $Pr=5.8$, and for case $3-Ra=10^9$, $Pr=5.8$ inside the counter-flow there is a second one. The number of inner counter-flows increases as increase Rayleigh number. The height of zone with counter-flow becomes bigger as the Prandtl number increases (Pr>5.8). This show in figures 3 for *Pr*=58.

The results of the modeling convective fluid flow with a Prandtl number less than unit are shown in figures 4-9. The simulations of convection in a horizontal layer for a liquid with values of Prandtl numbers less than unit have shown that counter-flows are exist. The presence of a layered structure with counter-flow is more clearly visible at low Prandtl than at large Prandtl numbers (figures 4).

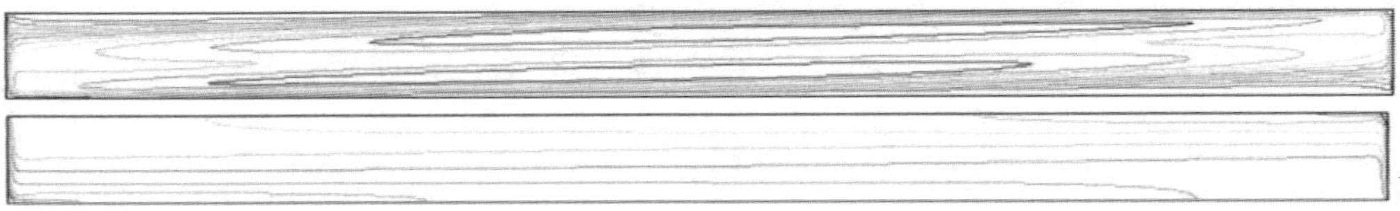
a)
b)

Figure 4. The isolines of the stream function (a) and the isotherm (b) in a layer with a solid top boundary for $Ra=1.7\cdot10^7$, $Pr=0.1$.

a)
b)

Figure 5. The isolines of the stream function (a) and the isotherm (b) in layer with free top boundary for $Ra=1.7\cdot10^7$, $Pr=0.1$.

If the horizontal boundary is free, this does not influence on the existence of secondary flows, but it affects only on the symmetry of the flow in the layer (figure 5). In figure 6a the profiles of dimensionless velocities in the middle vertical section for the cases of a solid top boundary (red line 1) and for the case of a free top boundary (black line 2) and in figure 6b horizontal temperatures profiles are shown for $Ra=1.7\cdot10^7$, $Pr=0.1$.

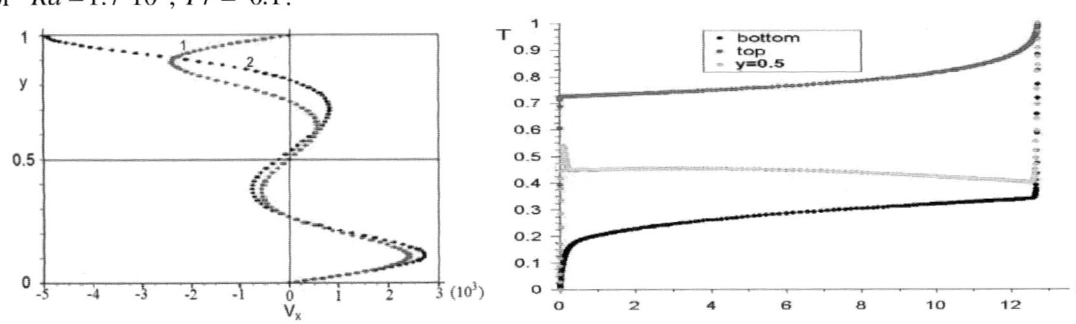

a) b)

Figure 6. Velocity profiles in the middle vertical cross section (x=6.355) with a solid (line 1) and free (line 2) upper boundary of the layer (a); and horizontal temperature profiles (y=0; 0.5; 1) for the case of free top boundary (b); for $Ra=1.7\cdot10^7$, $Pr=0.1$.

The simulation results showed that convective counter-flows exist both in the case of a free boundary and in the case of small Prandtl numbers (figures 4-6). If the top horizontal boundary of the layer is free, this influence on the symmetry of the flow in the layer (figures 5-6), but it does not effect on the appearance of counter-flows. If the Prandtl number is less than unit, the intensities of flow and counter-flow becomes greater than when the Prandtl number is greater than unit.

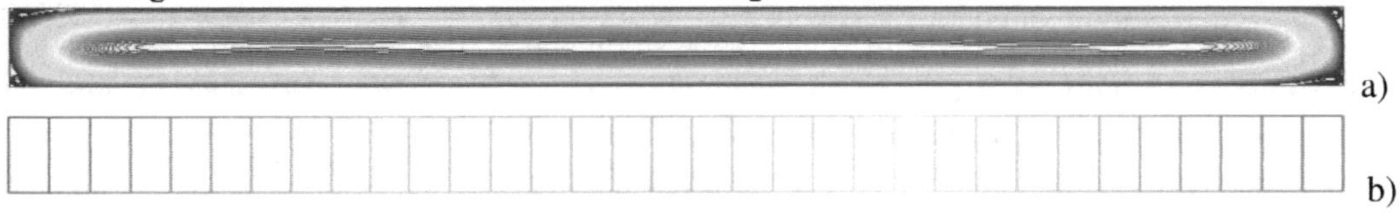

Figure 7. The isolines of the stream function (a) and the isotherm (b) for $Ra = 10^9$, $Pr = 0$.

Consider the difference between the flow of a liquid with a zero and low Prandtl number. At the Prandtl number equal to zero the temperature field keeps unchanged (for our model have linearly distributed to horizontally as in figure 7b). The structure of the convective flow up to the values of the Rayleigh numbers equal to $Ra = 10^9$ stay single-vortex (with small secondary vortices in the corners of the layer) and practically stationary (figure 7a). If the Prandtl value is equal to $Pr = 10^{-2}$, in this case even with the Rayleigh number equal to $Ra = 2 \cdot 10^5$, the flow velocity and temperature field become unsteady and eventually acquire an oscillatory character, see figure 8. The type of boundary conditions at horizontal walls plays an important role in the flow regime for small Prandtl numbers. The isolines of the stream function (a) and the isotherms (b) of stationary flow for $Ra = 2 \cdot 10^5$, $Pr = 0.01$ at isothermal boundary conditions: $T|_{y=0, y=H} = T_1 + x(T_2 - T_1)H / L$ are shown in figure 9.

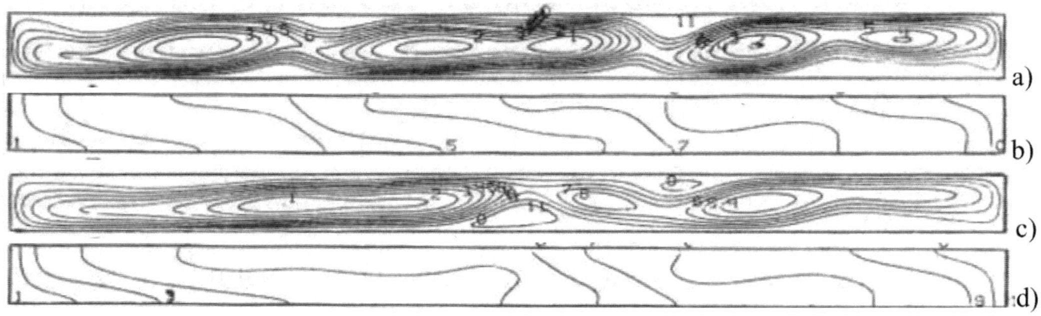

Figure 8. The isolines of the stream function (a, c) and the isotherms (b, d) for two moments of dimensionless time: $t' = 0.82$ (a, b) and $t' = 0.84$ (c, d); for $Ra = 2 \cdot 10^5$, $Pr = 0.01$.

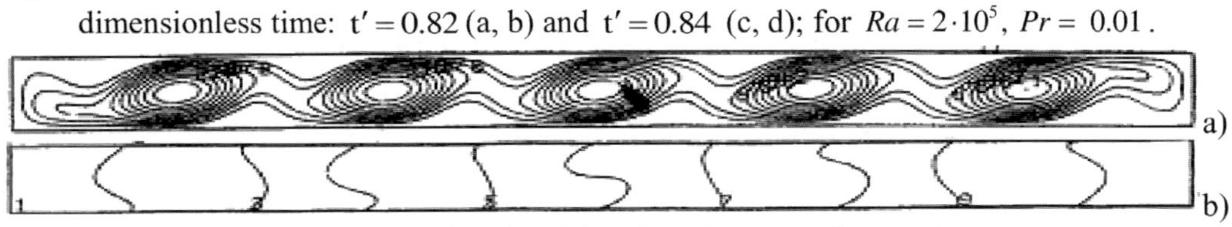

Figure 9. The isolines of the stream function (a) and the isotherm (b) at $T|_{y=0, y=H} = T_1 + x(T_2 - T_1)H / L$ for $Ra = 2 \cdot 10^5$, $Pr = 0.01$.

4. Summary of results

The reasons for convective flows with counter-flows can be following: firstly, large Rayleigh numbers and the nonlinear nature of dependence of the flow velocity on temperature; secondly, the formation of a steady vertical stratification of temperature (density) along the layer (except for zone near the heated side ends), and thirdly, the influence of end walls, as well as the properties of the fluid (Prandtl number) and boundary conditions. Near vertical walls convective flows have a complex character in the zones of thin boundary layers with the presence of stationary secondary vortices, which also participate in the formation of horizontal counter-flows. The number of counter-flows in a horizontal layer filled with water rise as increase the Rayleigh number ($Ra > 5 \cdot 10^8$) (inside the counter-flow a secondary counter-flow appears with the opposite direction).

The simulation results showed that for laminar convection modes (with Rayleigh numbers equal or greater than $Ra = 2 \cdot 10^5$), flows with Prandtl numbers less than $Pr \leq 10^{-2}$ differ qualitatively from the flow of a liquid with a zero Prandtl number, so the assumptions of approximation of the zero value of the Prandtl number may not always be correct. In the case of isothermal boundary conditions at the horizontal walls of the layer, the flow structure can become stationary and multicellular (figure 9). In this case, the multi-vortex flow structure significantly influences on the temperature distribution and it can lead to difficulty mass transfer along the layer, what must be taken into account in applied problems.

5. Conclusions

At large Rayleigh numbers of stationary convection flows may have non-linear features. These features of the flow structure can be, for example: in the form of an ordered large-scale multicellular structure of flow with a certain wave number, or in the form of stationary horizontal layered stream structures with counter-flows.

The steady vertical temperature stratification induced by convection (in the entire layer with the exception of zone near the heated side ends) increases with increasing Rayleigh numbers, which contributes to the delaying of the laminar stationary regime of convective water flows up to the values of the Rayleigh number equal $Ra = 10^9$.

As the Prandtl number decreases (at $Pr<1$), the intensity of counter-flows increases, and flow direction tends turn closer to the diagonal direction. The flows of fluid with Prandtl numbers equal or less than 0.01 differ qualitatively from the flow of fluid with a zero Prandtl number, and so the approximation of the zero value of the Prandtl number may not always be correct for an accurate description of the fluid flow and distribution of temperature and impurity.

Acknowledgements

The study was supported by the Government program (contract # AAAA-A20-120011690131-7).

References

[1] Gershuni G Z and Zhukhovsitskii E M 1972 *Convective Stability of an Incompressible Liquid* M: GRFML 392
[2] Polezhaev V I, Bello M S, Verezub N. A, et al 1991 *Convective Processes in Weightlessness* M: «Nauka» 240
[3] Gershuni G Z, Zhukhovitsky E M 1970 *On slow flow of viscous fluids in a closed region* In: «Hydrodynamics» issue **2** Perm 207-217
[4] Birikh R V 1966 *J Appl. Mech. Tech. Phys.* **7** 3 43-47
[5] Cormack D E, Leal L G and Imberger J 1974 *J. Fluid Mech* **65** 209–229
[6] Zimin V D, Lyakhov Yu N, Shaidurov G F 1971 *Experimental study of the temperature field under natural convection of a liquid in a closed rectangular cavity* In: «Hydrodynamics» No. 248 **3** Perm 126-138
[7] Bejan A, Al-Homoud A A and Imberger J 1981 *J. Fluid Mech.* **109** 283–299
[8] Kirdyashkin A G 1989 *Thermal gravitational flows and heat transfer in the asthenosphere.* Novosibirsk: "Nauka" Siberian branch 81
[9] Kurdyashkin A G, Polezhaev V I and Fedyushkin A I 1983 *J. of Appl. Mech. Tech. Phys.* **24** 876–882
[10] Drummond J E and Korpella S A 1987 *J. Fluid Mech.* **182** 543-564.
[11] Cormack D E, Leal L G, Seinfield J H 1974 *J. Fluid Mech.* **65** 231–246
[12] Fedyushkin A I, Puntus A A 2018 *Trudy MAI* **102** 20
[13] Landau L D, Lifshitz E M 1987 *Course of Theoretical Physics. Fluid Mechanics* **6**. Oxford: Pergamon Press 2nd ed. 556
[14] Polezhaev V I, Bune AV, Verezub N A, et al. 1987 *Mathematical modeling of convective heat and mass transfer based on Navier-Stokes equations* M.: «Nauka» 272

[15] Fedyushkin A I 1990 *Research of a matrix method for solving convection equations. Complex of programs "MARENA".* Preprint No 471 M.: IPM of the USSR Academy of Sciences 32
[16] Patankar S V 1980 *Numerical Heat Transfer and Fluid Flow* Hemisphere 214

IOP Conf. Series: Materials Science and Engineering **927** (2020) 012055 doi:10.1088/1757-899X/927/1/012055

A coalescence of the droplets

A I Fedyushkin and A N Rozhkov
Ishlinsky Institute for Problems in Mechanics of Russian Academy of Sciences, Prospekt Vernadskogo, 101-1, Moscow, 119526, Russia

E-mail: fai@ipmnet.ru

Abstract. The paper considers the dynamics of the coalescence of two drops of Newtonian fluid. Using numerical simulations for a two-phase liquid-air system, the change in droplet shapes over time is shown for different properties of liquids. Numerical simulation data are compared with experimental data.

1. Introduction

The study of droplet coalescence is of great fundamental scientific importance for studying the laws of multiphase hydrodynamics in various physical processes and phenomena, since droplet coalescence processes have a wide range of applications.

The pioneering work on the study of surface tension forces in the process of droplet coalescence was the work [1]. In [1], an analysis of droplet coalescence times was carried out, the stages of sintering of metal droplets were indicated, which laid the scientific basis for powder metallurgy.

In recent years, a large number of numerical and experimental studies on the coalescence of droplets under various conditions have appeared. From these works, some works can be indicated, for example, [2–9]. The dynamics of droplet coalescence substantially depends on the forces of surface tension and the rheological properties of the liquid. In [2, 3], the results of studying the coalescence of drops of non-Newtonian fluids in comparison with Newtonian fluids are presented.

An important condition is in what environment the drops coalescence. Many studies have been carried out on the coalescence of droplets in vacuum and in air. The work [4] is devoted to the study of the coalescence of gas bubbles and liquid droplets in oil and in brine (in a denser medium than air). It was shown in [4] when the viscosity of the surrounding fluid is important in the coalescence of droplets in a liquid. The work [5] devoted to the study of meniscus formation at the initial stage of droplet coalescence, gives correlation dependences for cases of viscous and inviscid liquids. In [6], the coalescence of droplets of an ideal fluid without the influence of the environment was studied. The results of a study of meniscus formation using the method of integral boundary equations are presented; the possibility of the formation of a toroidal bridge is shown. In [7], a theoretical and experimental work on coalescence was presented, scale factors and approximation expressions for the time of droplet coalescence were analyzed, and two different mathematical models were compared with the experiment. In [7–10, 12, 13], the existence of oscillating regimes of change in the shape of the resulting droplet upon the coalescence of two droplets is indicated. The purpose of this work is to study the dynamics of the coalescence of two drops of one liquid in air, changes in the shape of the drops and the bridge between them over time.

2. Problem statement and mathematical model

At the initial moment, two drops of liquid are in contact in the air and the dynamics of changes in the shape of the drops and the bridge between the drops is considered. Figures 1 and 2 show a photograph of the initial arrangement of the droplets in the experiment and a bridge scheme at the initial instant of the arrangement of the droplets in the experiment (the figures were taken from [2]). Figure 3 shows the enlarged zone of contact of the droplets at the initial instant in the computational domain of the mathematical model. In the mathematical model, it is assumed that at the initial moment of time an instantly minimal bridge was formed between the drops, for example, the radius of the initial bridge for drops with a diameter of d=1 mm was $r_m(t=0)=5\ 10^{-4}$ mm (figure 3).

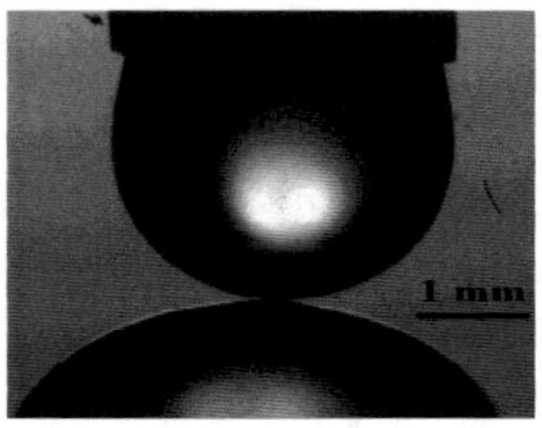

Figure 1. The initial arrangement of the droplets in the experiment [2, 3]

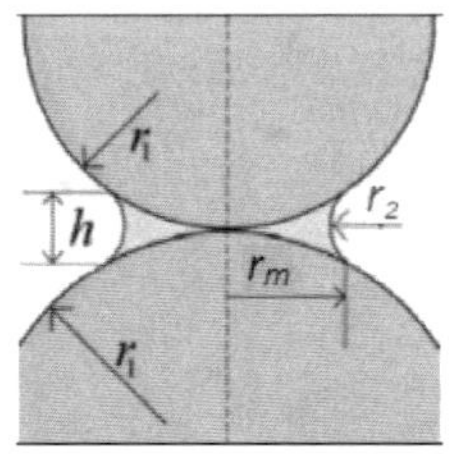

Figure 2. The scheme of the bridge in the experiment [2, 3]

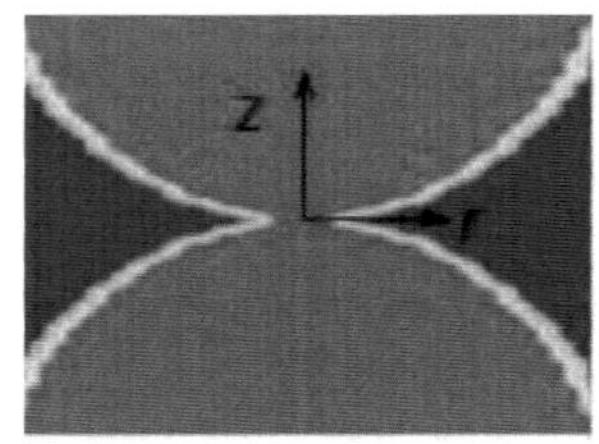

Figure 3. Scheme of the initial arrangement of drops in a mathematical model

The mathematical model is based on solving the system of two-dimensional Navier-Stokes equations in the axisymmetric approximation. The coalescence of two drops of the same liquid in air is considered without taking into account gravity (g=0). In a cylindrical coordinate system, without rotation, for a two-phase gas-liquid system in the approximation of the “mixture” model [1, 10] can be written in the form:

$$\frac{\partial u_1}{\partial x_1}+\frac{u_1}{x_1}+\frac{\partial u_2}{\partial x_2}=0 \quad (1)$$

$$\frac{d(\rho u_1)}{dt}=-\frac{\partial p}{\partial x_1}+\frac{1}{x_1}\frac{\partial}{\partial x_1}\left(x_1\mu\frac{\partial u_1}{\partial x_1}\right)+\frac{\partial}{\partial x_2}\left(\mu\frac{\partial u_2}{\partial x_2}\right)-\mu\frac{u_1}{x_1^2}+F_1 \quad (2)$$

$$\frac{d(\rho u_2)}{dt}=-\frac{\partial p}{\partial x_2}+\frac{1}{x_1}\frac{\partial}{\partial x_1}\left(x_1\mu\frac{\partial u_2}{\partial x_1}\right)+\frac{\partial}{\partial x_2}\left(\mu\frac{\partial u_2}{\partial x_2}\right)+F_2 \quad (3)$$

where $df/dt=\partial f/\partial t+v_r\partial f/\partial r+v_z\partial f/\partial z$ is the substantial derivative, $\mathrm{x}_1\equiv\mathbf{r}$, $\mathrm{x}_2\equiv\mathrm{z}$ are radial and axial coordinates, $\mathrm{u}_1\equiv\mathrm{v}_r$, $\mathrm{u}_2\equiv\mathrm{v}_z$ are velocity vector components $\mathbf{u}(u_1,u_2)$, t is the time, p is the pressure, ρ is the density, μ is the dynamic viscosity coefficient, F_1, F_2 are radial and axial components of external force $\mathbf{F}(F_1,F_2)$, operating in a narrow zone along the gas-liquid interface.

To describe the two-phase gas-liquid system, we used the system of equations (1-3) with one equation for momentum transfer under the assumption of a “mixture” model [10] with averaged velocities $\mathbf{u}=\varepsilon\mathbf{u}_{gas}+(1-\varepsilon)\mathbf{u}_{liquid}$, density $\rho=\varepsilon\rho_{gas}+(1-\varepsilon)\rho_{liquid}$, and viscosity $\mu=\varepsilon\mu_{gas}+(1-\varepsilon)\mu_{liquid}$, where the values with the index *gas* refer to gas, and with the index *liquid* refer to liquid. The volume fraction of liquid ε $(0<\varepsilon<1)$ was determined from the solution of the transfer equation: $\partial\varepsilon/\partial t+u_1\partial\varepsilon/\partial\partial x_1+u_2\partial\varepsilon/\partial x_2=0$.

The boundary conditions at the gas–liquid interface were determined from the equilibrium condition of surface forces and pressure [10]:

$$(p_1-p_2+\sigma\kappa)n_i=(\tau_{1ij}-\tau_{2ij})n_j \quad (4)$$

where σ is the surface tension coefficient is assumed to be constant, p_1, p_2 are the fluid and gas pressures, $\kappa=1/R_1+1/R_2$ is the surface curvature where R_1, R_2 are the radii of curvature for liquid and gas, **n** is the unit normal vector directed into the second fluid, $\tau_{\alpha ij}$ is the viscous stress tensor index α denotes: α=1 – liquid, α=2 – air. Condition (4) is written for a constant coefficient σ.

The conditions for the absence of friction are set at the external boundaries of the computational domain, the condition of symmetry on the axis, and condition (4) at the interface of the two-phase liquid-air system. The modeling of the change in the shape of the gas-liquid interface was performed using the model of liquid volumes (VOF - Volume of Fluid method). The interface was determined by the VOF method with increased resolution and taking into account surface forces by the CSF method (Continuum Surface Force) [11]. The CSF method allows one to remove the singularity in the case of turning the radius of curvature of the interface surface to zero and increase the accuracy of the calculations [11]. When solving the problem, condition (4) at the interphase boundary was taken into account through an additional local bulk force **F** on the right side of the momentum transfer equation (2)-(3). The force **F** acts only in a very narrow area, enclosed along an interface line of width Δh. At every point l_s interface lines l when striving Δh to zero force **F** can be written as $\mathbf{F}(l_s)=\sigma\kappa(l_s)\mathbf{n}(l_s)$, where $\kappa(l_s)$ is the interface curvature at a point l_s, $\mathbf{n}(l_s)$ is the normal to the site at the interface point l_s [11].

In the process of numerical solution of the system of equations (1-4), the conservative method of control volumes with the approximation of spatial derivatives of the second order and first order in time was used.

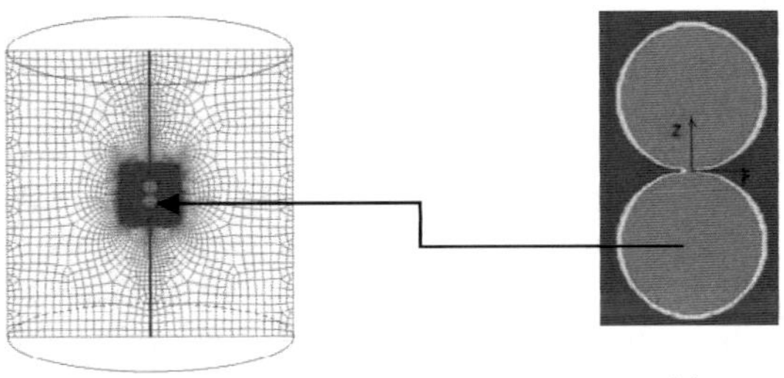

a) b)

Figure 4. Scheme of the computational domain with drops at the initial moment of time,
a) - design grid
b) - the increased part of the calculation area

Figure 4a presents a scheme of an axisymmetric computational domain (r, z) and a computational grid with the arrangement of two identical drops. Figure 4b shows the enlarged central part of the computational domain with drops at the initial time. Axial symmetry was assumed in the model; therefore, the computational domain was equal to half of the domain shown in figure 4a and was limited by the axis of symmetry 0z. To exclude the influence of external boundaries, the dimensions of the computational domain were chosen 10–20 times the diameter of the droplets involved in the coalescence. Figure 4a shows droplets with a diameter of 1 mm at the initial time with the dimensions of the computational domain equal to: $\{0 \le r \le 10$ mm, $-10 \le z \le 10$ mm$\}$. To increase accuracy, the computational grid in the central part of the region was rectangular and strongly detailed (figure 4a).

3. The results of numerical simulation

Based on the numerical solution of the Navier-Stokes equations for the two-phase air-liquid system (1) - (5), the coalescence of liquid droplets was simulated for different properties. The coalescence processes of two identical drops of the following liquids were considered: water, glycerin, mercury, silicone oil and silicone liquid (the properties are shown in table 1). The time and nature of the coalescence of the droplets is characterized by the properties of the liquid, more precisely by the ratio of viscous and surface forces. In the present work, the droplet diameter is chosen as the length scale. The coalescence process can be characterized by two time scales: for domination of viscosity forces by the time scale $T_{visc}=d\mu/\sigma$, and for domination of surface forces by the time scale: $T_{inert}=(d\rho^3/\sigma)^{1/2}$ [7]. The dimensionless numbers of Ohnesorge $Oh=\mu/(\rho\sigma d)^{0.5}$ which presents the ratio of viscous forces to inertial forces, can be (or by other words, this is the ratio of the time scale of viscous forces T_{visc} to the time scale of inertial forces T_{inert}: $Oh=T_{visc}/T_{inert}$).

The coalescence of identical drops of various liquids was simulated. Table 1 shows: the properties of Newtonian fluids used in the simulation in this paper and the time scales T_{visc} and T_{inert}, values of Ohnesorge $Oh = T_{visc} / T_{inert}$ and Reynolds $Re = (d\sigma\rho)^{0.5} / \mu$ numbers for drops with a diameter d=1 mm.

Table 1. Properties of liquids.

Liquid	Density ρ [kg/m^3]	Dynamic viscosity μ [kg/(m s)]	Surface tension σ [10^{-3} N/m]	Time scale $T_{visc}=d\mu/\sigma$ [s]	Time scale $T_{inert}=(\delta\rho^3/\sigma)^{0.5}$ [s]	Ohnesorge number $Oh=T_{visc}/T_{inert}$	Reynolds number Re
Mercury	13529	$1.5\cdot10^{-3}$	486.5	10^{-6}	$4\cdot10^{-3}$	$4\cdot10^{-3}$	268
Water	998	10^{-3}	72.6	10^{-5}	$5\cdot10^{-3}$	3	0,34
Silicone fluid	2000	$1.7\cdot10^{-5}$	59	10^{-5}	$6\cdot10^{-3}$	$5\cdot10^{-5}$	$2\cdot10^{4}$
Glycerol	1260	0.8	59.4	10^{-2}	$6\cdot10^{-3}$	77	0,01
Silicone oil	760	10	22	10^{-1}	$5\cdot10^{-3}$	$6\cdot10^{-4}$	1680

The mathematical model assumes that at the initial moment of time the droplets were already in contact in a small area, like solid spheres (at the initial moment, the meniscus and radius of curvature were determined only by the size of the mesh cell, and the meniscus formed during counting during the first microsecond). At the initial time, all the droplets were d = 1 mm in diameter and the initial radius of the common part of the contact of the droplets (bridge) was the same, approximately equal to r_m (t = 0) = 0.05 mm. All the velocities at the initial moment were equal to zero, and the pressure was constant, equal to atmospheric. The step of the computational grid near the bridge was no more than 10-3 mm. To increase the accuracy of calculating the air-liquid interface, the calculations were carried out with a small time step limited by the Courant number equal to 2. The results of numerical modeling showed that the formation of the natural configuration of the liquid bridge with the meniscus (according to the condition of equal pressure forces and surface forces (4)) occurs very quickly (in the first steps in time up to 10-7 seconds). Subsequently, a numerical solution showed that the width of the bridge varies monotonously, as shown in figure 6. Thus, at each time step, changes were obtained in the width of the bridge between the drops and a change in the shape of the drops.

For liquids with the properties presented in Table 1, mathematical modeling of the coalescence of droplets with a diameter of 1 mm was carried out and the time dependences of the change in the size of the liquid bridge and the dynamics of the change in the shape of the surfaces of the droplets were obtained. In figure 5 shows the experimental dependences of the radius of the liquid bridge r_m (mm) on time t (ms). In figure 5 presents data for three fluids: water, silicone oil, and a 4% solution of potylene oxide (PEO) with a viscosity close to that of silicone oil. The experimental data show a very rapid bridge formation at the initial stage of droplet coalescence, after which the movement of the bulk of the droplets begins. This property of droplet coalescence is confirmed by numerical calculations of figures 6-7. begins. The width of the bridge varies with time and is approximated by a power-law dependence with an exponent ½. In figure 5 shows the effect of surfactant additives (4% PEO) on droplet coalescence. Polymer additives slow down the coalescence of droplets sometime after the onset of coalescence [2, 3].

Figure 6 shows graphs of the change in the radius of the bridge r_m from time at the initial stage of the coalescence of water droplets to a time t <2.5 μs. In figure 6, line 1 is the numerical results, the maximum radius of the bridge in the section is z = 0, line 2 is the experimental data [2]. The results of mathematical modeling showed that the transition from the initial to the final stationary form of droplets (liquids with low viscosity) is of periodic decaying nature, as shown in figure 7.

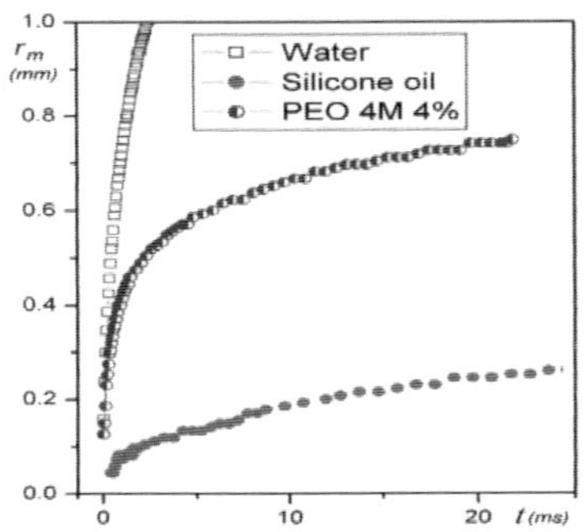

Figure 5. Experimental dependences of the radius of the liquid bridge r_m (mm) on time t (ms).

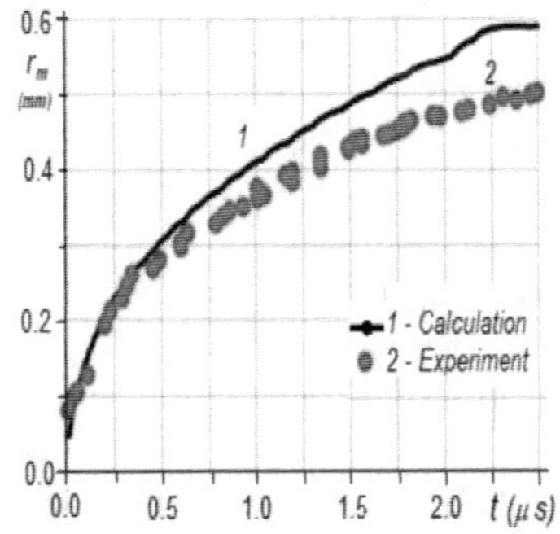

Figure 6. Change in the radius of the bridge r_m (mm) of coalescing water droplets over time t (μs) (1 - calculation, 2 - experiment)

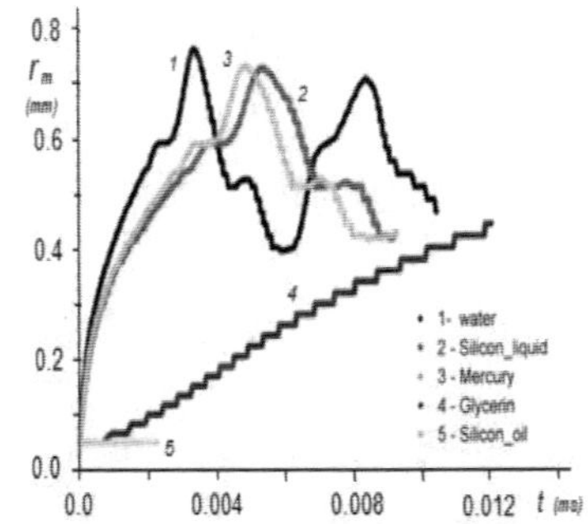

Figure 7. Change in time t (ms) of the radius of the bridge r_m (mm) during the merging of droplets of different liquids

Table 2. Changing the shape of coalescing two identical drops for various liquids over time.

Time (ms)	Mercury	Water	Silicone (liquid)	Glycerin	Silicone (oil)
t=0	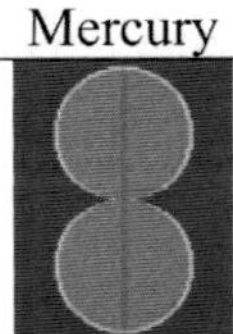	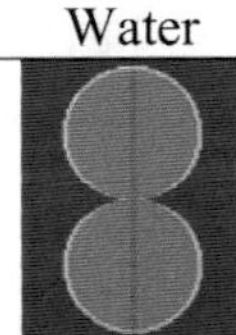	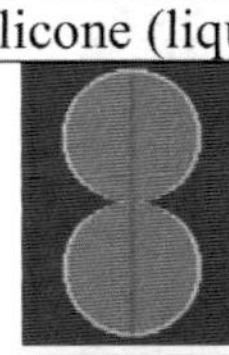	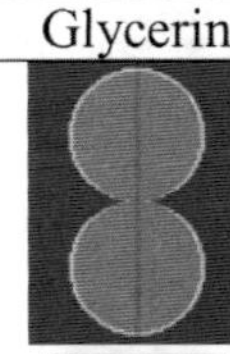	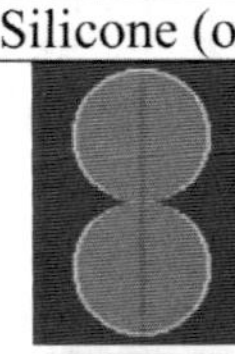
1	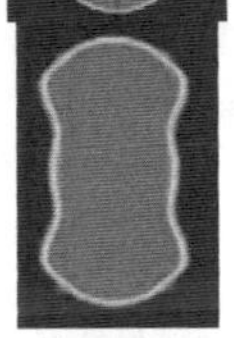	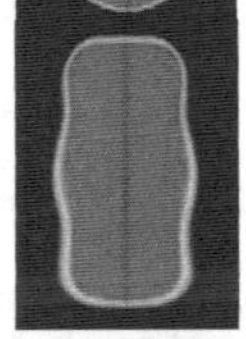	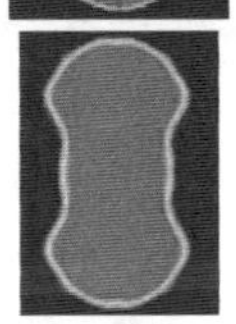		
3	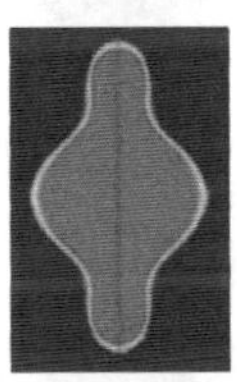	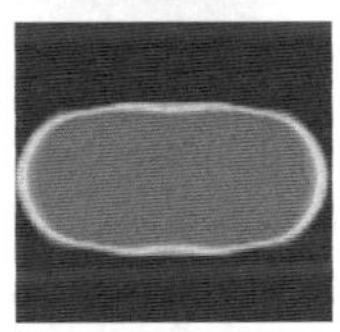	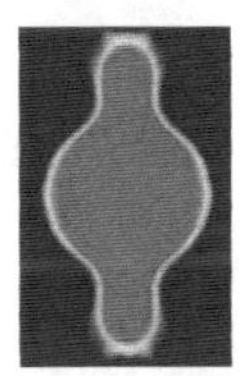	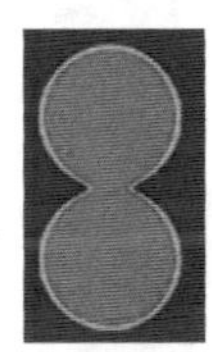	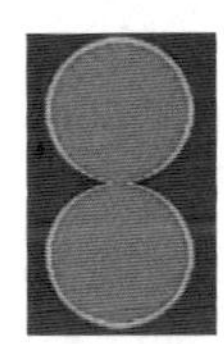
5	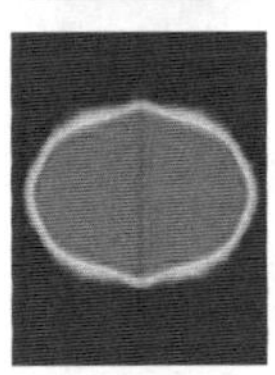	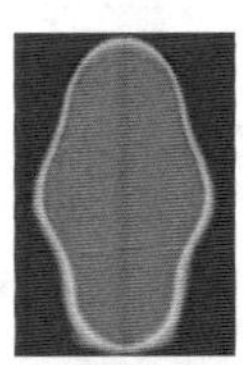	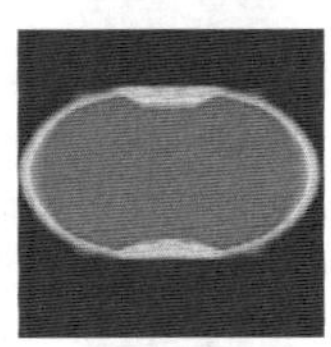	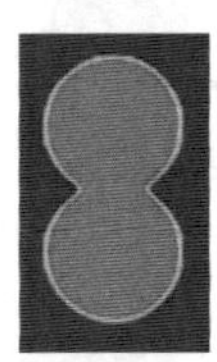	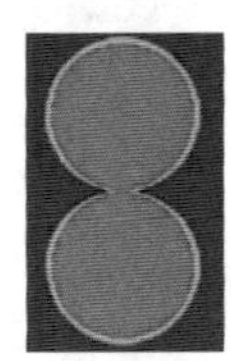
7	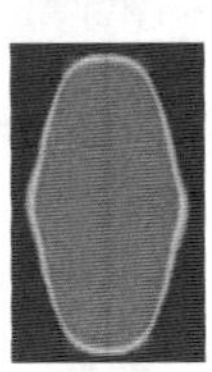	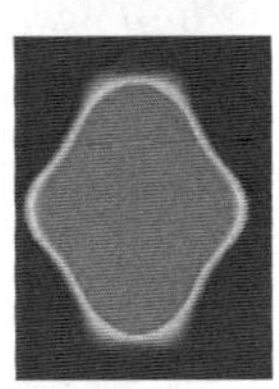	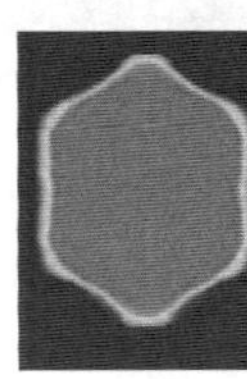	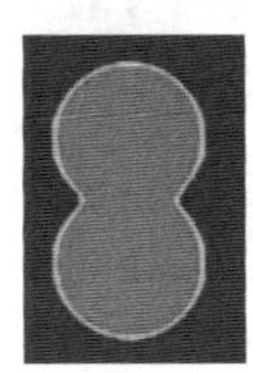	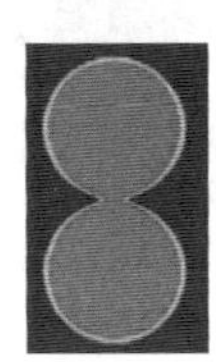
t>10	t=11 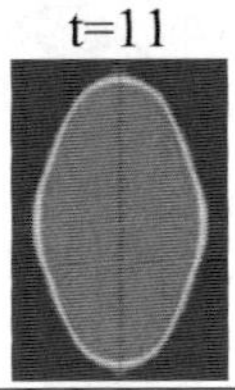	t=12 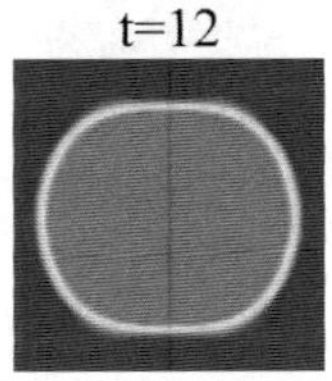	t=13 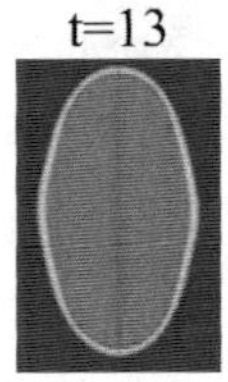	t=20 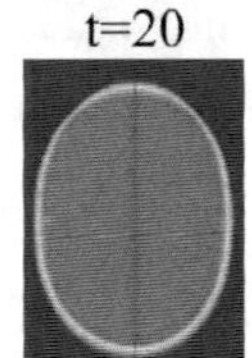	t=50 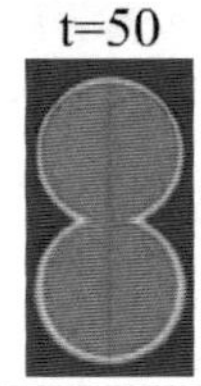

IOP Conf. Series: Materials Science and Engineering **927** (2020) 012055 doi:10.1088/1757-899X/927/1/012055

The change in droplet shapes over time is shown in the figures in Table 2 (the time t (ms) is indicated in the first column of Table 2. The shapes in the droplets for various liquids and at different times in the interval from t = 0 to 50 ms are shown in the figures in Table 2. A more detailed picture of the change in the shape of the droplets is presented in [12].

From the results presented in the figures of Table 2, it can be seen that oscillating waves are formed on the surfaces of the merging liquid droplets with a small viscosity, which lead to the oscillatory nature of the dynamics of the formation of the surfaces of the resulting drops. This is also illustrated by the dependences in figure 7 and is consistent with experimental results [2, 3, 13]. In [13], photographs were presented showing periodic changes in droplet shapes during the coalescence of two drops of dibutyl phthalate levitating in sodium chloride solution. The results are shown in the table 2, also agree with the results obtained by another numerical method in [14] on the oscillating change of the shape of two identical droplets during their coalescence.

4. Conclusions

The dependences of the time variation of the radii of bridges between coalescing drops for various liquids are presented. The effect of small polymer additives on the coalescence of droplets is strongly shown. For the initial moments of the coalescence of water droplets, a comparison was made of the numerical and experimental data [2] of the change in the width of the bridge, which showed good agreement between the results.

The periodic nature of the change in the shape of droplets over time is shown for liquids with low viscosity. For identical instants of time, patterns of droplet shape changes for different liquids during their coalescence are presented, showing the non-stationary nature and scale of the coalescence of droplets.

Acknowledgements

The study was supported by the Government program (contract # AAAA-A20-120011690131-7).

References

[1] Frenkel J 1945 Journal of Physics **9** 385-391

[2] Rozhkov A, Mitkin V and Theofanous T 2008 *The dynamics of visco-elastic bridges in drop coalescence XXII Int. Cong. of Theor. and Appl. Mech. Adelaide.* Australia 24–29 Aug. 2p.

[3] Rozhkov A N, Mitkin V and Theofanous T 2008 *Coalescence of elastic-viscous drops* 24 Symp. on rheology Abstracts of reports. Karacharovo. June 3-7 5

[4] Paulsen J D, Carmigniani R, Kannan A, Burton J C and Nagel S R 2014 *Coalescence of bubbles and drops in an outer fluid.* Nature communications **5**:3182 DOI: 10.1038/ncomms4182 1-7

[5] Eggers J, Lister J R and Stone H A 1999 *J. Fluid Mech.* **401** 293–310

[6] Duchemin L, Eggers J and Josserand C 2003 *J. Fluid Mech.* **487** 167–178

[7] Sprittles J E and Shikhmurzaev Y D 2012 *Phys. of fluids* **24** 122105 27

[8] Becker E, Hiller W J and Kowalewski T A 1991 *J. Fluid Mech.* **231** 189-210

[9] Watanabe T 2010 *Int. Journal of geology* Issue1 **4** 5-13

[10] Landau L D and Lifshitz E M 1987 Fluid Mechanics **6** Oxford: Pergamon Press 2nd ed. 556p

[11] Brackbill J U, Kothe D B and Zemach C 1992 *J. Comp. Phys.* **100** 335–354

[12] Fedyushkin A I and Rozhkov A N 2014 *Coalescence of Newtonian liquid droplets* Preprint 1087 M.: IPMech RAS 27p.

[13] Babalyan G A 1974 *Physical and chemical processes in oil production* M.: "Nedra" 200p

[14] Sprittles J E and Shikhmurzaev Y D 2012 *Coalescence of liquid drops: Different models versus experiment* Physics of Fluids **24**, 122105; https://doi.org/10.1063/1.4773067

IOP Conf. Series: Materials Science and Engineering **927** (2020) 012056 doi:10.1088/1757-899X/927/1/012056

Biharmonic spline interpolation of spatial dose rate distribution in the minimum dose trajectory optimization problem

A E Starchenko
Research Institute of Applied Mechanics and Electrodynamics of the Moscow Aviation Institute (RIAME MAI)
Leningradskoye shosse 5, 125080, POB 43, Moscow, Russian Federation

E-mail: aleksandr.starchenko@phystech.edu

Abstract. In this paper issues relating to geostationary orbit (GEO) insertion trajectory optimization is considered. The article discusses the GEO transfer trajectory optimization for a spacecraft with an electric propulsion (EP) system with the purpose of reduction of ionizing radiation dose absorbed by onboard systems from the Earth's radiation belts. The considering trajectory optimization technique is based on the Pontryagin's maximum principle and the method of numerical continuation. Much attention is paid to smooth approximation of the dose rate dependence on spatial coordinates using biharmonic splines interpolation. The proposed method makes it possible to abandon the methodological simplification of averaging the radiation dose rate over circular orbits to effectively solve this class of problems and directly consider the dependence of the dose rate on time and geographical longitude. Using as an example the minimum time problem for a test transfer, the errors in calculating the dose at the end of the transfer both by the proposed method and by the method of averaging over circular orbits are estimated. The proposed method turned out to be more than 13 times more accurate than the analogue. It can be applied in transfer trajectory optimization problems of an EP spacecraft in order to minimize the total accumulated ionizing radiation dose.

1. Introduction

When considering a transfer of a spacecraft into high working orbits using an electric propulsion system, there arises a problem of large ionizing space radiation doses accumulation by onboard systems. In addition to the structural methods for solving this problem, an alternative approach uses the fact that, for example, during the geostationary orbit (GEO) insertion the Earth's radiation belts make the most significant contribution to the dose of accumulated radiation. Since the Earth's radiation belts have an inhomogeneous spatial distribution of charged particles fluxes, this approach consists in finding a transfer trajectory path through the radiation belts from the initial orbit to the GEO with the minimal possible accumulated dose. However, in this case the insertion time and the propellant consumption increase relative to the minimum time trajectory. There are several works [1-3] devoted to the study of this approach. This approach allows reducing the radiation dose by up to 40% of the dose on the minimum time trajectory. At the same time, transfer duration increases by 10-15% of the minimal duration.

To solve the above mentioned optimization problem different approaches could be utilized [4,5]. In this paper the Pontryagin's maximum principle is used [1,2]. The boundary value problems that occur in this case are solved by utilizing the method of numerical continuation and the embedded integration of differential equations of optimal motion [6,7]. In order for the continuation method to work effectively, it is necessary to use high-order methods of numerical integration of ordinary differential equations, which, in its turn, imposes restrictions on the degree of smoothness of the radiation dose rate dependence on spatial coordinates. Existing models of spatial distribution of radiation belts particles are not intended for use in the right-hand sides of the equations of motion when integrated by high-order methods, since they have discontinuities of both partial derivatives and values. To construct a smooth dependence of the radiation dose rate on spatial coordinates, high-order and smoothing splines are used in [1,2], and the method for averaging the dose rate over circular orbits is proposed. The latter method allows to reduce the number of input parameters of the interpolating spline from four (time and radius vector components) to two (radius and inclination of a circular orbit). Despite the simplification of the interpolating spline construction, this method introduces some systematic error in the calculation of radiation-safe transfer trajectories.

The goal of this study is to construct an interpolating radiation dose rate function that depends on all input variables (time and radius vector components) and has a sufficient degree of smoothness for its use in the right-hand sides of the equations of motion for effective numerical integration by high-order methods. To reach the goal of the study the radial basis functions interpolation is used, in particular biharmonic splines [8,9].

2. Dose rate approximation

The reference data for constructing an approximation dependence of the dose rate on spatial coordinates were calculated using the IRENE software package (International Radiation Environment Near Earth, version 1.50.001, https://www.vdl.afrl.af.mil/programs/ae9ap9/downloads.php). The AE8/AP8 MAX models of the Earth's radiation belts protons and electrons fluxes distribution and the Shieldose2 dose calculation program were used in calculations. Since the AE8/AP8 MAX proton and electron, flux models can be used only with Earth's magnetic field models, which only consider internal sources, the distribution of particle fluxes will be constant in any coordinate system rotating with the Earth. It should be noted that in reality at high altitudes the Earth's magnetic field is less dependent on internal sources and is more determined by external currents and, therefore, radiation belt particles will no longer have a constant flux distribution at high altitudes in rotating coordinate systems. But this phenomenon is neglected in this paper.

To construct an approximation dependence, the dose rate value was calculated by the above method on a set of fixed reference points in the Earth-centered, Earth-fixed (ECEF) coordinate system. The set of points is obtained by mapping a regular three-dimensional grid from spherical coordinates to Cartesian coordinates. The distribution of grid nodes is uniform in angular coordinates and uneven in radius. The grid has dimensions $28\times15\times16$ and contains $n=6720$ reference points.

2.1. Biharmonic spline interpolation

One of the methods for approximating scattered data is approximation by radial basis functions [8,9]. In this case, the approximated dependence of dose rate on spatial coordinates could be written as

$$N_D(\vec{r})=\sum_{i=1}^{n}c_i\phi\left(\|\vec{r}-\vec{r}_i\|_2\right), \qquad (1)$$

where $\phi(s)$ is a real-valued on $[0;+\infty)$ radial basis function, $\|\cdot\|_2$ is the L_2-norm, c_i are the approximation coefficients, $\vec{r}_i=(\rho_i,\theta_i,\varphi_i)^T$ are the spherical coordinates of the centers, $\vec{r}=(\rho,\theta,\varphi)^T$ are the spherical coordinates of a point of interest. In this paper, biharmonic spline is used as a radial basis function, i.e. $\phi(s)=s$. It is assumed that the centers coincide with the reference points, but in the general it might not be the case. The problems of reconstructing the coefficients from equation (1) are often solved by the least squares method (LSM). However, if there are no restrictions on the LSM coefficients, it can give a solution such that at some points the resulting function will be negative, which is impossible for dose rate from a physical point of view. Therefore, to reconstruct from the reference data $(\rho_i, \theta_i, \varphi_i, N_{Di}), i = 1 \ldots n$, unknown coefficients c_i, $i = 1 \ldots n$ the modified LSM with an additional penalty term is used. The penalty term is selected so that when negative values occur at the calculated points; this term rapidly grows, thereby correcting the values of the coefficients in the solution of the LSM. The proposed LSM error minimization problem is written as follows

$$\Phi = \left\| A\vec{c} - \vec{b} \right\|_2 + \gamma \sum_{i=1}^{n} \exp\left(-\sum_{j=1}^{n} A_{ij} c_j \right) \to \min, \quad (2)$$

where $A_{ij} = \left\| \vec{r}_i - \vec{r}_j \right\|_2$, $b_i = N_{Di}$, $\gamma > 0$. Setting gradient of function (2) to zero and following the study [10], an iterative process of the following form is constructed

$$\vec{c}^{(k+1)} = \left(A^T A + \gamma A^T R^{(k)} \right)^{-1} A^T \vec{b}, \quad k = 0,1,2\ldots, \quad (3)$$

$$R^{(0)} = 0_{n\times n}, \quad (4)$$

$$R^{(k)} = diag\left(\frac{1}{c_1^{(k)}} \exp\left(-\sum_{j=1}^{n} A_{1j} c_j^{(k)} \right), \ldots, \frac{1}{c_n^{(k)}} \exp\left(-\sum_{j=1}^{n} A_{nj} c_j^{(k)} \right) \right), \quad k = 1,2,3\ldots, \quad (5)$$

where $diag(a_1, \ldots, a_p)$ is a $p \times p$ diagonal matrix with $a_1, \ldots, a_p$ coefficients on the main diagonal. The zero step corresponds to the pure LSM, and the subsequent steps take into account the penalty for the dose rate negativity at the centers. It turned out that the proposed iterative process with $\gamma = 0.001$ allows one to obtain positive values of the resulting function at all centers already at the first step, after which the process is stopped.

A slice of the obtained dose rate function in the ECEF coordinates $Ox_g y_g z_g$ at $x_g = 0$ is shown in figure 1 in shades of gray. It can be seen that the obtained function reflects the effect of the tilt of the Earth's magnetic axis relative to the Earth's rotation axis, and, consequently, the tilt of the Earth's radiation belts. It should be noted that the method of averaging over circular orbits from [1] cannot reproduce this effect.

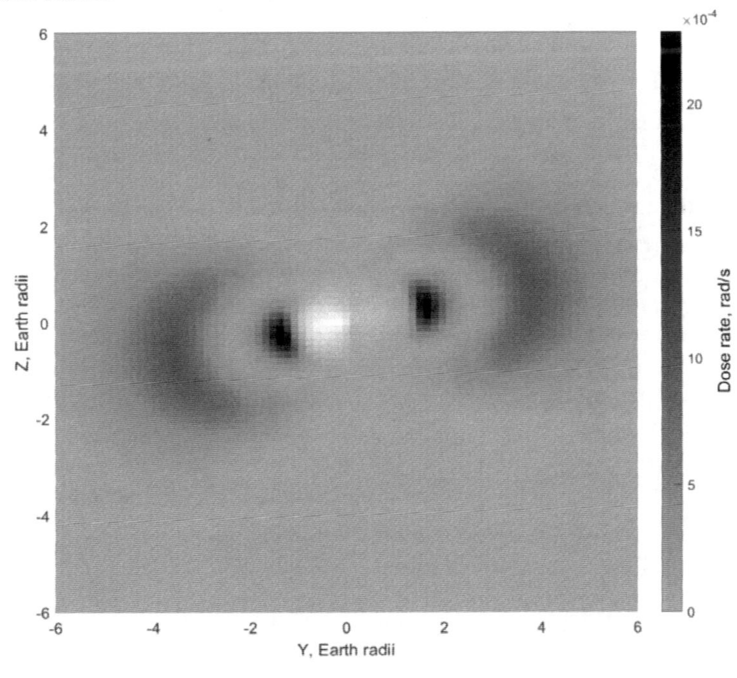

Figure 1. The dose rate in rad/s (shown in shades of gray) calculated using (1)

and the obtained coefficients, depending on ECEF y_g and z_g coordinates at $x_g = 0$.

3. Numerical results

Consider a test transfer of a reusable orbital transfer vehicle from [1], equipped with a megawatt class nuclear electric propulsion system from an 800 km circular orbit into the GEO. Initial inclination is 11°, the longitude of the ascending node is zero. The total thrust of the electric propulsion system is 27 N and the specific impulse is 4762 s. For this transfer the minimum time problem is considered. In order to solve this problem, the optimization technique from [1], based on the Pontryagin's maximum principle and the method of numerical continuation, is used. However, instead of using the method of averaging over circular orbits, the following equation is used as an additional motion equation for the dose:

$$\frac{dD}{dt} = N_D(\rho(t), \theta(t), \varphi(t)), \tag{6}$$

where t is time and

$$\rho(t) = \sqrt{x_g^2(t) + y_g^2(t) + z_g^2(t)}, \quad \theta(t) = \frac{\pi}{2} - atan2\left(z_g(t), \sqrt{x_g^2(t) + y_g^2(t)}\right), \tag{7}$$

$$\varphi(t) = \begin{cases} atan2(y_g(t), x_g(t)), & if\ atan2(y_g(t), x_g(t)) \geq 0 \\ atan2(y_g(t), x_g(t)) + 2\pi, & if\ atan2(y_g(t), x_g(t)) < 0 \end{cases}, \tag{8}$$

and $atan2$ is the arctangent function with the quadrant determination. The coordinates $x_g(t)$, $y_g(t)$, $z_g(t)$ of the spacecraft in the ECEF coordinate system (the Earth pole precession, nutation and other perturbations [11] are neglected) are expressed through the spacecraft coordinates in the Earth-centered inertial system x, y, z as follows:

$$\begin{pmatrix} x_g(t) \\ y_g(t) \\ z_g(t) \end{pmatrix} = \begin{pmatrix} \cos(GMST_0 + \omega_e t) & \sin(GMST_0 + \omega_e t) & 0 \\ -\sin(GMST_0 + \omega_e t) & \cos(GMST_0 + \omega_e t) & 0 \\ 0 & 0 & 1 \end{pmatrix} \begin{pmatrix} x \\ y \\ z \end{pmatrix}, \tag{9}$$

where $GMST_0$ is the Greenwich mean sidereal time corresponding to 00:00:00 on January 1, 2020, and ω_e is the Earth's mean angular velocity. The parameters $GMST_0$ and ω_e were selected to reproduce the implemented in the IRENE conversion between $x_g(t)$, $y_g(t)$, $z_g(t)$ and x, y, z as accurately as possible. Coordinates x, y, z were calculated from equinox elements [12] used in the considered equations of motion from [1].

For the above mentioned test transfer, the maximum principle boundary value problem was posed and solved using averaging of the differential equations of optimal motion. Then, the obtained initial values of the conjugate variables and the transfer duration were used to integrate the non-averaged equations of optimal motion. For the obtained non-averaged trajectory, the absorbed ionizing dose was calculated in three ways: using the IRENE software package, using equations (6)–(9), and using the method of averaging over circular orbits from [1]. The error in calculating the dose in the latter two ways relative to the dose calculated using IRENE, depending on the transfer time, is shown in figure 2.

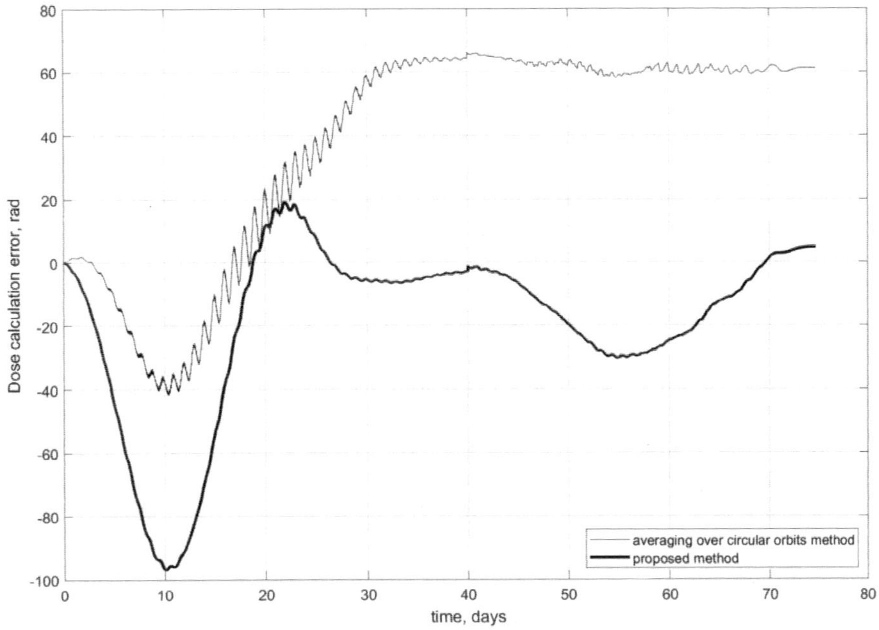

Figure 2. The error in calculating the dose by the proposed method and the method of averaging over circular orbits from [1] depending on the transfer time. As the true value, the dose calculated using the IRENE software package is used.

It can be seen that the error in calculating the dose at the end of the transfer by the proposed method is at least 13 times less (4.67 rad vs. 61.1 rad) than that of the method from [1]. The calculated errors make up 0.12% and 1.5% of the dose calculated using IRENE (3987 rad, 4 mm Al spherical shielding), respectively.

4. Conclusion

The paper describes a significantly more accurate method for calculating the total ionizing dose from Earth's radiation belts in the problems of orbit transfer trajectory optimization in electric orbit raising operations. The approximation error in the accumulated dose of the proposed method is 13 times less than for the averaging over circular orbits method [1]. The advantages of the proposed method also include the infinite differentiability of the approximating function, in contrast to the finite order splines used in [1]. The proposed method for ionizing dose calculation can be used in spacecraft accumulated dose minimization problems where the transfer trajectory variation is used to reduce space radiation exposure of onboard systems during electric orbit raising to high orbits.

Acknowledgments

This study was supported by the grant of the Government of the Russian Federation allocated from the federal budget for state support of scientific research conducted under the guidance of leading scientists in Russian educational institutions of higher education, research institutions and state research centers of the Russian Federation (Contest 7, Resolution No.220 of the Government of the Russian Federation of April 9, 2010). Agreement No. 075-15-2019-1894 of December 3, 2019.

References

[1] Starchenko A E 2019 Trajectory optimization of a low-thrust geostationary orbit insertion for total ionizing dose decrease *Cosmic Res* **57** 289–300

[2] Starchenko A E 2019 Minimizing the degradation of triple junction solar array of a spacecraft during geostationary orbit insertion *Cosmic Res* **57** 364–77
[3] Dutta A, Kasdin N J and Choueiri E 2016 Minimizing proton displacement damage dose during electric orbit raising of satellites *J. Guid. Control Dyn* **39** 960–7
[4] Panteleev A and Stanovskaya Ya 2019 The dragonfly algorithm for parametric optimization of open loop nonlinear dynamic systems control *AIP Conf. Proc.* **2181** 020027-1–8
[5] Karane M and Panteleev A 2019 Hybrid multi-agent optimization method of interpolation search *AIP Conf. Proc.* **2181** 020028-1–14
[6] Petukhov V G 2004 Optimization of multi-orbit transfers between noncoplanar elliptic orbits *Cosmic Res* **42** 250–68
[7] Petukhov V G 2012 Method of continuation for optimization of interplanetary low-thrust trajectories *Cosmic Res* **50** 249–61
[8] Buhmann M D 2004 *Radial Basis Functions: Theory and Implementations* (Cambridge, UK: Cambridge University Press)
[9] Carr J C, Beatson R K, Cherrie J B, Mitchell T J, Fright W R, Mccallum B C and Evans T R 2001 Reconstruction and representation of 3D objects with radial basis functions *SIGGRAPH'01: Proc. 28th Conf. on Computer Graphics and Interactive Techniques* pp 67–76
[10] Farquharson C G and Oldenburg D W 1998 Non-linear inversion using general measures of data misfit and model structure *Geophys. J. Int.* **134** 213–27
[11] Krylov S S, Perepelkin V V and Filippova A S 2020 Long-period lunar perturbations in Earth pole oscillatory process: theory and observations *Advances in Theory and Practice of Computational Mechanics (Smart Innovation, Systems and Technologies* vol 173) eds Jain L C, Favorskaya M N, Nikitin I S and Reviznikov D L (Singapore: Springer) pp 315–31
[12] Battin R H 1999 *An Introduction to the Mathematics and Methods of Astrodynamics* (Reston, VA: AIAA).

The use of CFD calculations for the solution of the control problem of applying protective coatings of varying thickness

S A Borisov, Ju I Glukhovskaya, S V Dobrovolskiy and I V Podporin
Moscow Aviation Institute (National Research University), Volokolamskoe shosse 4, 125993, Moscow, Russia

E-mail: dobrovolskiy_s@mail.ru

Abstract. The paper analyzes the results of simulation of a supersonic air flow about simple 3D forms in order to use them for applying protective coatings to the surface with varying thickness by means of cold gas-dynamic spaying. To calculate the results, an aerodynamic analysis is performed in SolidWorks Flow Simulation software environment. Air flow zones on a streamlined surface with various characteristics (pressure, temperature, geometric shape, etc.) are analyzed, and possible methods of determination the boundaries of these zones are considered. For the solution of the control problem of applying protective coatings of varying thickness, the displacement behaviour of the movable nozzle with a small cross section is considered.

1. Introduction

When solving problems related with the rapid movement of objects, the estimation of forces acting on the object is often required. Any moving vehicle (air, water or land transport) is affected by aerodynamic forces, and determination of magnitudes of these forces allows to evaluate their impact on the technical characteristics of the vehicle [1]. The analysis of zones of various stresses on a streamlined surface is performed on examples of model problems, such as a supersonic air flow around a cylinder and flat disk along the normal to a surface and with a deviation from the normal [2].

We will show how CFD calculations allow one to determine areas on a streamlined surface with various stresses created by the air flow in order to develop a method for applying protective coatings to the surface with varying thickness to ensure the uniform erosion of the surface under the impact of a heterogeneous flow [3].

2. Simulation of a supersonic flow about the cylinder

We simulate a supersonic flow about the cylinder in order to determine zones on a surface with various pressures (Figure 1). There are the following basic initial conditions:

- Thermal condition on the wall: adiabatic wall;
- Thermodynamic parameters: pressure = 100 kPa, temperature = 20° C;
- Velocity parameters: velocity along the X axis = 360 m/s;
- Turbulence intensity: 0.1%.

The numerical results are exported to Excel, and they consist of several tables (one for each flow line) and graphics. Each row in the table contains the data for one vector: coordinates of the vector and the pressure values for the vector.

The following pressure zones are considered: zone "A" (pressure values more than 180 kPa), zone "B" (pressure values from 150 kPa to 180 kPa), and zone "C" (pressure values from 100 kPa to 150 kPa). Zone boundaries are the pressure contour lines that can be extracted by means of SolidWorks API or another software, or we can select some resulting points manually (Figure 2) and then connect them by chords to create a closed boundary curve. In this task, the values of the X coordinates are practically unchanged and can be considered as constant ones. Each chord is described by the equation of the line $F(z, y)$ [4].

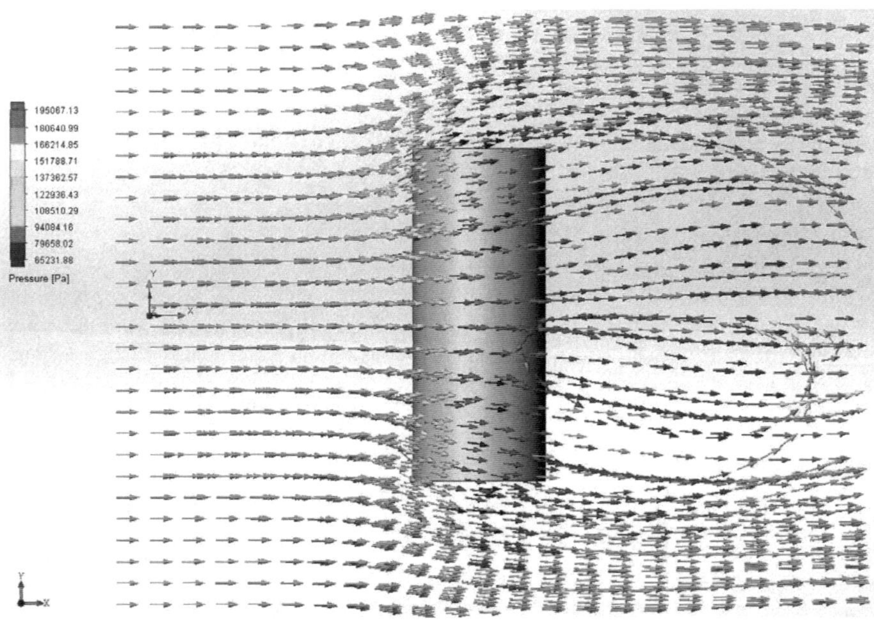

Figure 1. Flow about the cylinder.

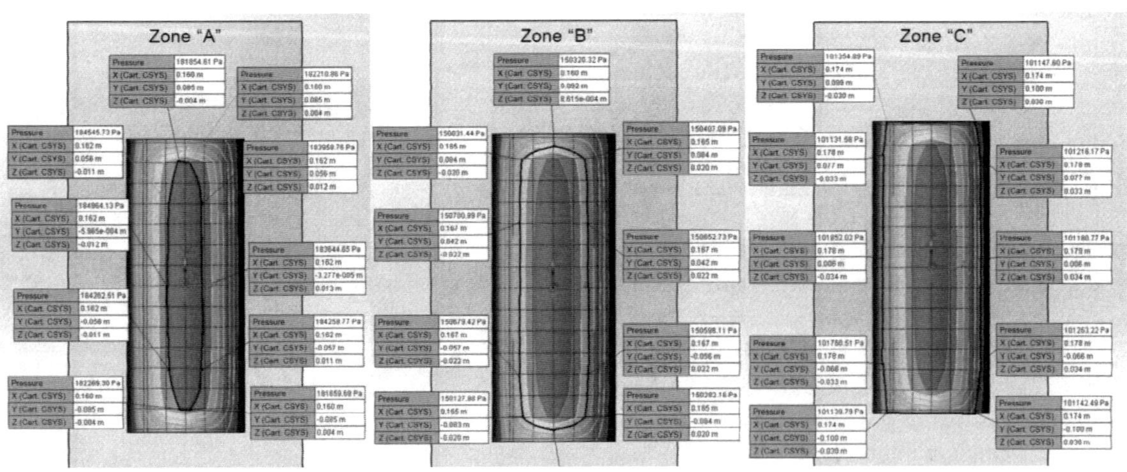

Figure 2. Boundaries of the pressure zones on the streamlined cylinder surface.

3. Simulation of a supersonic flow jet from a small nozzle along the normal to the cylinder and flat disk

We simulate a flow jet blowing on a cylinder from a nozzle with a cross section area of 5 mm^2 to determine the diameter of the jet after exiting the nozzle, and the shape of a jet spot that occurs when the flow collides with a surface (Figure 3). To determine these characteristics, it is necessary to determine the dependence of coordinates of vectors on the pressure created along the way of the flow: before a contact with an obstacle, at the contact, on streaming about an obstacle, and after passing through it. Plots of the coordinates *X, Y, Z* of the flow depending on pressure were created. Based on the analysis of the results for the outmost flow lines in the jet, the flow jet radius on the exit of the nozzle is calculated, as well as the shape of the flow jet spot on the surface. After exiting the nozzle, *Y* and *Z* coordinates of the flow line vectors change until a certain point, but then remain almost constant until the contact with a surface. In the interval where the coordinates do not change, we can determine the radius of the jet, and under these conditions it is ≈ 3.2 mm [5].

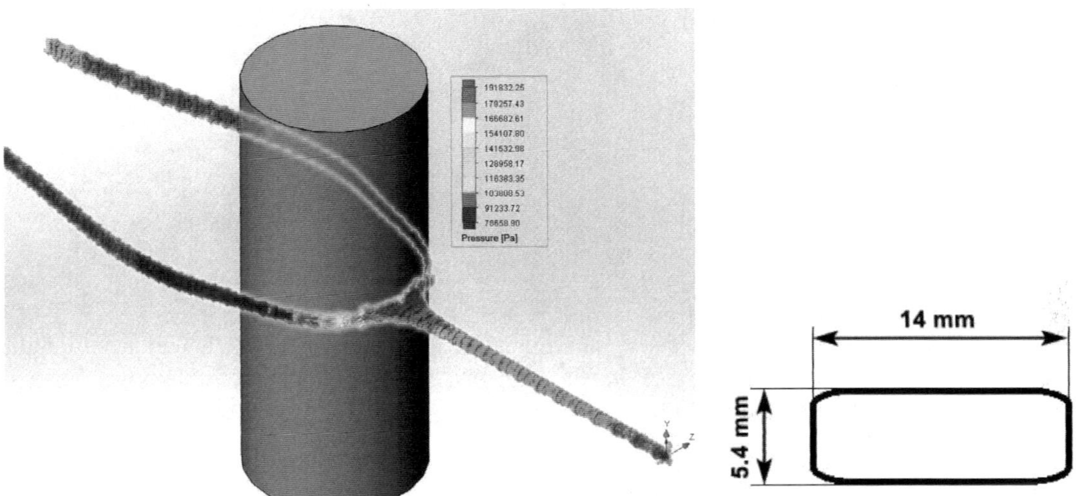

Figure 3. Flow jet stream about the cylinder and the spot shape on the surface.

To determine the shape of spot on the cylinder surface, the outmost flow lines along *Y* and *Z* axes were used, and the plots established at what point the pressure is maximum. At this point, the vectors of the outmost flow lines collide with the cylinder surface. Under these flow conditions, for outmost positive flow lines, the pressure has a maximum at $Z \approx 0.007$ and $Y \approx 0.0027$, and for outmost negative ones, the pressure has a maximum at $Z \approx - 0.007$ and $Y \approx - 0.0027$. The spot boundary on the cylinder surface in a projection is a rectangle of height ≈ 5.4 mm and width ≈ 14 mm with rounded corners (Figure 3).

Figure 4 shows the collision of a jet with a flat disk and the shape of a spot on the surface under the same flow conditions as in previous case. Using the same technique, it was found that for outmost positive flow lines, the pressure has a maximum at $Z = Y \approx 0.0076$ m ± 0.0005 m, and for outmost negative flow lines, the pressure has a maximum at $Z = Y \approx - 0.0076$ m ± 0.0005 m. The spot shape on a flat surface of the disk is a circle with a diameter of ≈ 15.2 mm.

Figure 4. Flow jet stream about the flat disk and the spot shape on the surface.

4. Simulation of a supersonic flow jet from a small nozzle at the angle to the cylinder

We simulate a flow jet with a deviation of 15° from the normal to the cylinder surface under the same flow conditions, and analyze the behavior of the jet. Output numerical results are in the same form as in the previous cases (tables and plots) (Figure 5).

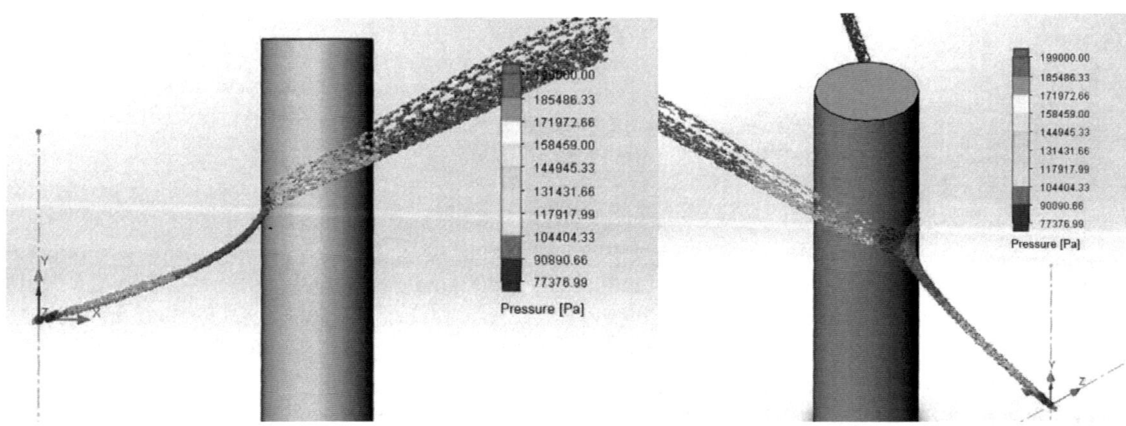

Figure 5. Flow jet stream about the cylinder at the angle of 15° from the normal to the cylinder.

Figure 5 shows that the flow jet, without reaching the surface, begins to shift along the *Y* axis towards the deviation of the nozzle from the normal to the cylinder. This is due to the occurrence of reflective (wall) pressure from the surface. Thus, when applying a protective coating, it is necessary to strictly maintain the normal to the surface. Otherwise, when deviating from the normal, it is difficult to determine the shape and area of the jet spot on the surface to take them into account when applying a protective coating [6].

5. Control the movement of the nozzle over zones on the surface when applying protective coatings of varying thickness

The thickness of the applied a protective coating depends on the consumption of the sprayed material and the speed of movement of the nozzle. Now we consider a method to control the speed of movement of the nozzle in zones "A" and "B" in a projection of a top view (Figure 6), which provides a protective coating of varying thickness.

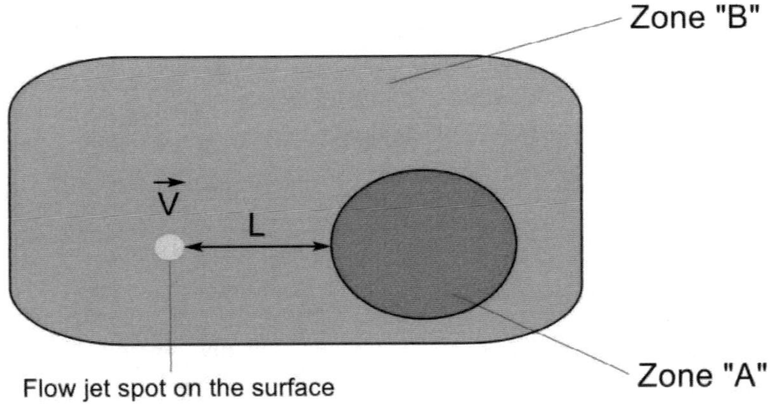

Figure 6. Scheme of the jet spot movement with the speed V to the boundary of zone "A".

In one coordinate system we have a surface model and a movable nozzle, the flow from which forms a spot on the surface. The nozzle is initially located in zone "B". The speed of movement of the nozzle is $V = k_B * v$, where k_B is the factor of the speed change in zone "B". When the nozzle moves, its coordinates *(x, y)* change, and it is necessary to control the distance from the boundary of the spot to the boundary of zone "A" at each moment. When the spot moves from zone "B" to zone "A", other coefficient should be used instead of k_B to change the speed of movement, and, consequently, to change the thickness of protective coating in zone "A". So, let R be the radius of the jet spot on the surface and L be the distance from the boundary of the spot to the boundary of the next zone. If $L>R$, then the speed is calculated as follows: $V = k_B * v$, otherwise $V = k_A * v$ (where k_A is the factor of the movement speed change for the nozzle in zone "A") [7].

6. Conclusion

To calculate a trajectory of the nozzle movement within the boundaries of each zone of pressure on the surface, it is proposed to use adapted NC programs for machining tool trajectories generated by the CAM system. To determine the required thickness of the protective coating, we can use the results of the assessment of surface erosion wear using the installation, the concept of which is considered in [3]. As a result of simulations of a supersonic air flow blowing some simple models, the tools and methods was shown to describe zones of various stresses on the surface. The data from simulations of a supersonic air flow jet streaming from a nozzle with a small cross section allowed to evaluate the flow jet behaviour in a collision with the surface of the model. This made it possible to describe the flow jet spot on the surface and to show that it is necessary to keep the flow direction along the normal to the surface on which a protective coating is applied. The idea of the method for applying protective coatings of varying thickness on the surface is proposed.

Acknowledgements

This work was financially supported by the Ministry of Education and Science of the Russian Federation; the agreement number: 05.607.21.0299 of December 3, 2019; and the unique project identification number: RFMEFI60719X0299.

IOP Conf. Series: Materials Science and Engineering **927** (2020) 012057 doi:10.1088/1757-899X/927/1/012057

References

[1] Nikitin P V, Payko V V and Frolov Yu P 1996 *Method and device for applying protective coatings by cold gas-dynamic method* (Moscow: Institute of Scientific and Technical Information)

[2] Alyamovsky A A 2019 *SolidWorks Simulation and FloEFD. Practice, Methodology, Ideology* (Moscow: DMK Press)

[3] Borisov S A, Dobrovolsky S V, Glukhovskaya Yu I and Nikitin P V 2019 The study of erosion wear of coatings using a moving jet of a heterogeneous flow of a limited area *Surf. X-ray, synchr. and neutr. Stud.* **3** 106–13

[4] Alkhimov A P, Kosarev V F, Klinkov S V, Sova A A, Trubacheev G V and Zaykovsky V N 2012 The formation of conical separation zones when a supersonic jet flows onto an obstacle under conditions of cold azodynamic spraying *Therm. and aer.* **19(2)** 257-64

[5] Profizi P, Combescure A and Ogawa K 2016 Numerical analysis of single particle impact in the context of Cold Spray: a new adhesion model *IOP Conf. Ser.: Earth and Env. Scien.* **32**

[6] Ryabinin A N, Irissu E, Legou J G and Moreau K 2009 Heat transfer of a supersonic jet with an obstacle during cold gas-dynamic spraying *Bull. of St. Pet. State Un.* **1(1)** 123-8

[7] Kiselev S P and Kiselev V P 2015 Numerical simulation by molecular dynamics of the collision of a nanoparticle with an obstacle taking into account the effect of an oxide film *Phys. Mes.* **18(6)** 50-6

IOP Conf. Series: Materials Science and Engineering **927** (2020) 012058 doi:10.1088/1757-899X/927/1/012058

On solving the warehouse procession optimization problem using a tuple of heuristics

D N Gainanov[1], **I G Korablev**[1] **and V A Rasskazova**[2]
[1]Ural Federal University, Yekaterinburg, Russia
[2]Moscow Aviation Institute (National Research University), Moscow, Russia

E-mail: damir.gainanov@gmail.com, varvara.rasskazova@mail.ru

Abstract. The paper is a review of the present-day problem about the logistic optimization of both two-dimensional and three-dimensional warehouses with m cranes. A tuple of heuristic algorithms is proposed to solve the NP-hard problem with minimizing the total execution time of a set of jobs using m cranes. The described optimization approaches are applicable to both two- and three-dimensional warehouses. As part of the numerical results, results regarding the solution of applied problems are shown. The ability to train the proposed model allows one to select the best heuristic settings for a specific warehouse based on real world data sets. Another important feature of the model is the ability to distribute the power and parallel execution of three independent processes: the ant colony optimization algorithm to construct a successful sequence of jobs, the algorithm to search the optimal set of heuristic weights of k_i, and the algorithm to solve the planning problem.

1. Introduction

The general features of problems about warehouse management are described in [1-3, 5, 6]. Warehouse logistics are described in detail in [4]. One of the most important indicators of efficient warehouse management is the average time needed to complete technological processes as well as the overall time needed to complete a daily job for a set of warehouse processes. At the same time, there are many different ways to execute warehouse processes, which use different types of equipment and navigation within the warehouse. One of the important classes of warehouse systems are the warehouses where processes are executed using cranes. To improve the efficiency of this type of warehouses, a set of cranes is often used with an identical crane span. For such a class of warehouses, there is an important problem about cranes management and control to allow m cranes to execute the job flow without conflicts. In the framework of this paper, we will review this problem as a mathematical programming problem with a criterion in the form of minimizing the overall time needed for unloaded cranes movement.

The optimization theory and mathematical programming have a great number of applications in various technical fields. There are both exact and approximate algorithms allowing to solve different applied problems, but usually exact algorithms are very time-consuming and that is why different approximation approaches are used so efficiently. A geometric method for solving optimization problems was proposed in [7-9], different guaranteed algorithms could be found in [10-12], and finally, heuristic and metaheuristic methods were developed in [13,14]. To solve the optimization problem in question we use a heuristic approach which includes the well-known genetic heuristic and a number of original approaches as components.

Suppose we are given a set Z of jobs for moving I materials in a warehouse. The warehouse has one base x coordinate for moving cranes. The length of the warehouse is equal to l. The set of jobs Z is shown as ordered coordinates of the starting and ending points of cargo movement:

$$Z = \left\{ z_i = \left\{ x_i^{(1)}, x_i^{(2)} \mid x_i^{(1)}, x_i^{(2)} \in R, i = [1, n], 1 < x_i^{(1)} \leq l, 1 \leq x_i^{(2)} \leq l \right\} \right\}.$$

The cranes are job executors and can move along the x axis at the speed of v_x. For all cranes, the v_x speed is constant when moving with or without a load. The set of cranes is denoted by $K = \{K_1, K_2, \dots, K_m\}$. The current coordinate of a crane at any time is $X = \left\{ x_u^{(t)} : u = [1, m] \right\}$. At the initial time $t = 0$, the cranes' starting points of are $A = \left\{ x_u^{(0)} : u = [1, m] \right\}$. The cranes compete for the use of the same path on the x axis, at any time t for any crane, the inequality $x_{u=1}^{(t)} < x_{u=2}^{(t)} < \cdots < x_{u=m}^{(t)}$ will be satisfied. This constraint means that the cranes' movement cannot be independent.

Each crane can move with load as well as without one, or be static (paused). A load can only be carried in one lift, and any stops of a crane with a load are prohibited. Moving a crane without a load or stopping it is called *idling*. When developing a plan, the number of idle moves in the crane's movement is unlimited.

Figure 1 below shows an example of movement through a warehouse, for one or three cranes.

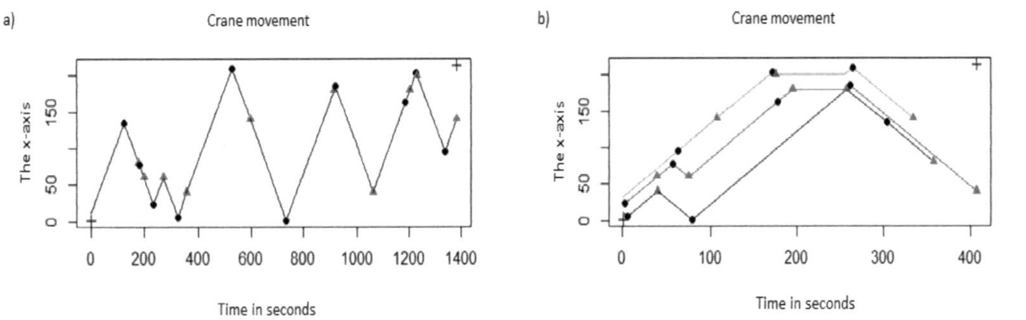

Figure 1. Examples of traffic trends for a) one crane and b) three cranes

In figure 1, the black circles indicate storage areas for materials, the triangles indicate the places where materials are delivered, and the connecting lines indicate routes for cranes.

The optimization purpose of warehouse logistics with m cranes is to build such a movement plan for m cranes that all the jobs from Z are completed and, moreover, the maximum time for movement, among all cranes, is minimal.

2. A formalized statement of the problem

The job on the movement of crane u from starting point $x_u^{(0)}$ to point $x_i^{(1)}$ for the first job z_i will be denoted by z_0. To execute job z_0 one needs the time as follows:

$$t_u(z_0, z_i) = \frac{\left| x_i^{(1)} - x_u^{(0)} \right|}{v_x}.$$

To execute job i and move the load from point $x_i^{(1)}$ to point $x_i^{(2)}$, crane u needs the time as follows:

$$t_u(z_i) = \frac{\left| x_i^{(2)} - x_i^{(1)} \right|}{v_x}.$$

To change job z_i to job z_j, one needs to move from point $x_i^{(2)}$ to point $x_j^{(1)}$, and this operation will need the time as follows:

$$t_u(z_i, z_j) = \frac{\left| x_i^{(2)} - x_j^{(1)} \right|}{v_x}.$$

Such movement of crane between jobs i and j is idle movement.

If there is more than one crane, the cranes are regularly forced to give way x to other cranes. Let's call such movement *forced idling* and denote it by $t_u^{empty}(z_i, z_j)$. Forced time $t_u^{empty}(z_i, z_j)$ is not regulated and can have values from zero to infinity, because the crane could make multiple infinite movements or stand idle without a load.

With accounting to the forced movement when moving from job z_i to the job z_j, we have a fully idle movement of the crane that's equal to $\left(t_u^{empty}(z_i, z_j) + t_u(z_i, z_j)\right)$.

Let us suppose that as a result of the planning, set Z of jobs is distributed by a sequence generator between several cranes and $(z_{i_1}, z_{i_2}, \dots, z_{i_n})$ is one of such permutations. The splitting of sequence Z by m sequences into $n_1, n_2, \dots, n_m$ elements will have the form $(z_1, \dots, z_{n_1})$, $(z_{n_1+1}, \dots, z_{n_1+n_2})$, $(z_{n_1+n_2+\dots+n_m+1}, \dots, z_n)$, where $n_1 + n_2 + \dots + n_m = n$. For simplicity, we denote this permutation by $(z_1, z_2, \dots, z_m)$. So, to execute a given set of jobs by unique crane u one needs to spend the overall time as follows:

$$t_u = t_u(z_0, z_i) + \sum_{i=n_1+1}^{n_1+n_2} \left(t_u(z_i) + t_u^{emty}(z_i, z_j) + t_u(z_i, z_j)\right).$$

Problem 1. The problem about the assignment of m cranes needs to be reduced to the problem of searching for such a set of combinations of set $(z_1, z_2, \dots, z_m)$ and such planning of idle and forced idle movements where $\max\{t_u : u = [1, m]\} \to \min$.

Further, it will be shown that to solve problem 1 a number of heuristics could be used. The first of these heuristics is to find an optimal route for the unique crane.

Problem 2. The problem about the optimization of unique crane logistics is to sequence the execution of jobs from set Z starting from beginning point $x_1^{(0)}$ and using the minimum possible time.

When there is only one crane, there are no problems regarding routes and movements for other cranes, and that is why the time for idle and forced idle movements will be equal to zero: $\sum_{i=1}^{n} t_1^{empty}(z_i, z_j) = 0$.

The summed time for execution of n jobs of the transportation of materials from point $x_i^{(1)}$ to point $x_i^{(2)}$ could not be changed for any generated sequence of jobs $\sum_{i=n_1+1}^{n_1+n_2} t_1(z_i) = const$. Thus, the optimization problem with only one crane could be reduced to the minimization of the problem of idle crane movement, and searching for such a sequence of jobs execution where $t_1(z_0, z_i) + \sum_{i=1}^{n} t_1(z_i, z_j) \to \min$.

The problem mentioned above could be reformulated in terms of graph theory.

Let the set Z be a set of vertices of directed graph $G = (Z, E)$, and let us suppose each pair of $z_i, z_j, i \neq j, i = [1, n], j = [1, n]$ holds $(z_i, z_j) \in E$. For each arc $(z_i, z_j) \in E$ there is a value $t_1(z_i, z_j)$ of weight function given, which is equal to overall time of transportation of the corresponding crane from point $x_i^{(2)}$ to point $x_j^{(1)}$, i.e. from the finishing point of job z_i to the starting one.

Let there be given a generator of sequences of elements of set Z. And let $(z_{i_1}, z_{i_2}, \dots, z_{i_n})$ be one of such sequences. In this case the movement of the crane from the starting point $x_i^{(0)}$ to the point $x_i^{(1)}$ could be considered job z_0, and that is why we will add such a job to set Z.

So, problem 2 corresponds to the following one.

Problem 3. We need to find a Hamiltonian path in the graph $G = (Z \cup \{z_0\}, E)$ which starts from point z_0 and has a minimum total weight of the included arcs among all possible ones, i.e.

$$\left\{t(z_0, z_{i_1}) + \sum_{k=1}^{n-1} t(z_{i_k}, z_{j_k})\right\} \to \min.$$

If it is required that a crane is returned to the starting state, then we have a classic travelling salesman problem (TSP).

There are many exact and approximate methods to solve problem 3 that provide effective and reliable results. In the framework of this paper, there was used the Ant Colony Optimization algorithm (ACO)

IOP Conf. Series: Materials Science and Engineering **927** (2020) 012058 doi:10.1088/1757-899X/927/1/012058

to solve problem 3. Further, it will be shown how the results of solving problem 3 could be used to solve problem 1.

The ACO formula for probability of transition from vertex i to vertex k is the following: $P_{ik} = \frac{\eta_{ik}^{\beta} * \tau_{ik}^{\alpha}}{\Sigma \eta_{ij}^{\beta} * \tau_{ij}^{\alpha}}$, where $\eta = \frac{1}{L_{ij}}$ is a greedy heuristic and $L_{ij} = t(z_i, z_j)$ is the weight function for transition from vertex (i.e., job) i to vertex (i.e., job) j. As a result of the ACO running, there formed a pheromone matrix τ_{ij} that shows possible transitions and the recommended probability to transit from any vertex i to vertex j. If we got to a random vertex i and need to get an effective result with high probability at the next step, we need to select a corresponding chain with high weight coefficients τ_{ij}. It also needs to be mentioned that there could exist several successful transitions from each vertex i. Thus, matrix τ_{ij} forms all possible and optimal chains to transit from any job to another.

Let us implement the following heuristic to solve problem 1: when we got to vertex i and need crane u to move further, we recommend to use successful slots of matrix τ_{ij} that was formed by ACO with only one crane. Thus, we could expect an increase in probability for crane u, which will move using an ideal route formed by ACO with only one crane. In case there are no possible routes for crane u to execute any next job of matrix τ_{ij}, there could exist alternative routes from vertex i to another potential job j for execution.

At the moment when crane u needs to vacate the route for another crane, a heuristic will try to return crane u to the current or alternative successful sequence of jobs execution. Let us denote such a successful route by P_{ij} and a current coordinate of the crane's location by x_u^t. So, to calculate the weight value of transportation from the previous job to the next one at the time moment t, we will use the following formula:

$$P_{ij} = \frac{\tau_{ij}}{\Delta x}$$

where $\Delta x = \left| x_u^t - x_i^{(1)} \right|$ is the distance from current state of crane to the next job j, i.e. it is the distance between crane and the nearest successful route.

We will normalize the described successful route heuristic with $P_{ij} = \frac{P_{ij}}{max(\tau_{ij})}$.

When there is more than one crane m>1, the following idea arises - to exclude the crane conflicts we need to require the crane remove to its own management zone after each job completion. Such a heuristic is extremely relevant for a high number of cranes. It is obvious that in case of normal distribution of material among the warehouse the first crane needs to be used closer to $x_{min} = 0$, and the last crane m needs to be used closer to the $x_{max} = l$ coordinate. Similarly, to pheromone matrix τ_{ij} we create a matrix τ_{ui} of successful zones for using each crane.

Each element of the successful zone matrix will be calculated with the following distribution density formula:

$$\tau_{ui} = \frac{1}{\sigma\sqrt{2\pi}} e^{\frac{-\left(x_i^{(2)} - \mu_u\right)^2}{2\sigma^2}},$$

where $\mu_u = l$ is the median for crane u and $\sigma = \frac{l}{2m}$ is the mean square deviation for all cranes.

Let us denote such a heuristic by $P_{ui} = \tau_{ui}$. An example of distribution density P_{ui} graphics is shown at Figure 2 with the length of warehouse equal to $l = 100$.

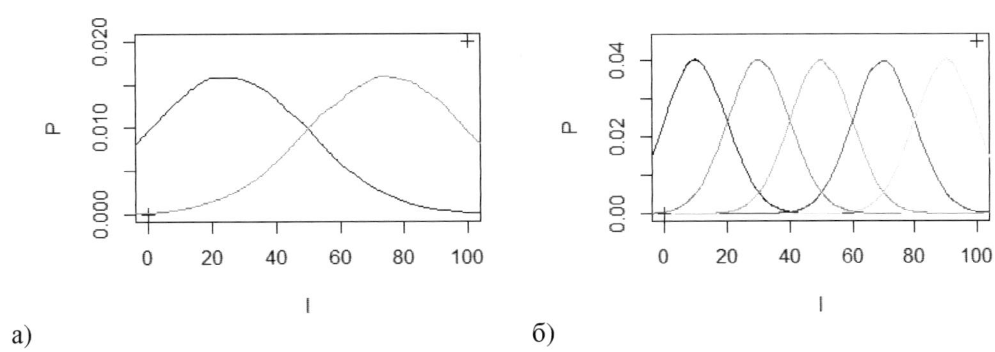

Figure 2. Graphics of distribution density P_{ui} for a) one and b) five cranes

It is experimentally established that the P_{ui} and P_{ij}heuristics allow one to develop an effective plan of movements, but there are cases when these heuristics could be improved with additional logic. One can use the greedy algorithm and take the material closest to the crane and, if possible, first of all, take the most time-consuming jobs, with a large ratio of the loaded movement of the crane to the idle one. The logic of the greedy algorithm does not require explanations. The logic of priority execution of the most time-consuming jobs is based on the fact that, firstly, big jobs should be planned, and small ones - on the residual principle, last, because they are easier to include in the plan.

We will denote a weight of the greedy heuristic by P_η, and state $P_\eta = \frac{1}{x_u^t - x_i^{(1)}}$.

The weight of the time-consuming jobs selection heuristic will be equal to $P_{volume} = \frac{|x_i^{(2)} - x_i^{(1)}|}{|x_u^t - x_i^{(1)}|}$, and we also normalize such a heuristic, i.e. $P_{volume} = \frac{P_{volume}}{max(P_{volume})}$.

Thus, four heuristics have been considered to select the next job for execution at each step of planning: $P_{ij}, P_{ui}, P_\eta, P_{volume}$. Let us introduce a corresponding weight coefficient k_1, k_2, k_3, k_4 for each mentioned heuristic. The total weight of heuristics for selecting the next job j will be defined by their sum with accounting to the weight coefficient for each heuristic:

$$P_j = k_1 P_\eta + k_2 P_{ij}, +k_3 P_{ui} + k_4 P_{volume}.$$

At each step of planning we select a vacant job j with the maximum value of P_j and assign such a job for execution by crane u with constraint $x_{u=1}^t < x_{u=2}^t < \cdots < x_m^t$. In contrast with the ACO algorithm, the stochastic component of selecting the next job is excluded.

Change of weight coefficients k_1, k_2, k_3, k_4 provide a variety of optimizers running for different tuples of Z and K.

3. Numerical results

The main purpose of experimental investigation is to learn how proposed heuristics and planning efficient are connected. The great number of possible plan combinations with several cranes lead to difficulties in getting low bounds for results, but it will be shown further that the implementation of the proposed heuristics provides better results in comparison with the ACO algorithm for only one crane.

Table 1 below shows how the changes of heuristic weight coefficients k_1, k_2, k_3, k_4 influence the final result. The search for extrema was carried out for data set Z, K, X by the gradient descent method with step +/-0.05 for each variable k_1, k_2, k_3, k_4.

The value of the total idle movement for the ACO algorithm with a single crane on the test data set is equal to 970 meters, which could be used as a lower bound for the model and is shown in the last column of the table.

IOP Conf. Series: Materials Science and Engineering **927** (2020) 012058 doi:10.1088/1757-899X/927/1/012058

Table 1. Connection of heuristics and optimization results

Cranes	k1	k2	k3	k4	Time	Loaded movement	Idle movement	Lower bound in comparison with ACO
1	0	0	0	0	12912	6449	6463	6.66
1	0.15	0.1	0	0.05	7364	6449	915	0.94
2	0	0	0	0	5747	6449	5045	5.20
2	0.3	0.1	0	0.05	3737	6449	1025	1.06
3	0	0	0	0	3997	6449	5542	5.71
3	0.4	0.15	0	0.1	2654	6449	1513	1.56
4	0	0	0	0	3146	6449	6135	6.32
4	0.35	0.2	0.05	0.2	1974	6449	1447	1.50
5	0	0	0	0	2526	6449	6181	6.37
5	0.35	0.1	0.05	0.05	1631	6449	1706	1.76

Heuristics are used when $k_1 = 0, k_2 = 0, k_3 = 0, k_4 = 0$, and available jobs are selected in order, with a queue from Z.

Using a different input data sets Z and K optimizer provides better results when the values of k_1, k_2, k_3, k_4 are different.

Given that in practice materials are usually stored in the same storage areas and have the same type of delivery points, it makes sense to train the model by selecting k_1, k_2, k_3, k_4 with typical sets Z and K of the warehouse where the optimizer will be used.

The average time to construct a single version of the plan-graph for one and five cranes is 10.5 and 25 seconds, respectively, of which 10 seconds are spent on forming pheromone matrix τ_{ij} by the ACO method.

4. Conclusion

The paper provides a method for solving the NP-hard problem using several heuristics. The proposed approaches are applicable to both two- (x, y) and three-dimensional (x, y, z) warehouses. As part of numerical results, the method has shown high efficiency. The ability to train the model allows one to select the best heuristic settings for a specific warehouse based on real-world data sets. Another advantage of the model is the ability of power distribution and parallel execution of three independent processes: the ACO algorithm for constructing a successful sequence of jobs, the algorithm for finding an optimal set of heuristic weights k_i, and, in fact, the planning.

Due to the great number of combinations of input data Z, K,and A, a neural network can be used to train the model with specific data.

References

[1] Gadzhinskiy A M 2009 The modern warehouse. Organization, technology, control and logistics (M.: Prospect)

[2] Dybskay V V 2009 Warehouse control for supply chains (M.: Alpha press) p 720

[3] Kanke A A and Koshevaya I P 2008 Logistics (M.: Forum) p 284

[4] Kanke A A 2014 Logistics of a warehouse *Marketing* N1 134

[5] Lukinsky V S 2003 Models and methods of logistic theory: study book (Piter, St. Petersburg)

[6] Nikolaychuk V E 2008 Transport and warehouse logistics (M.: Dashkov & Co) p 452

IOP Conf. Series: Materials Science and Engineering **927** (2020) 012058 doi:10.1088/1757-899X/927/1/012058

[7] Ibragimov D N 2019 About the optimal speed problem for the class of linear autonomous infinite-dimensional discrete-time system with bound control and a degenerated operator. *Autom. Remote Control.* **80** 3 393-412
[8] Pardalos P M, Li Y and Hager W W 1995 Linear Programming Approaches to the Convex Hull Problem in R^m *Computers Math. Applic.* **29 7** 23-29
[9] Gainanov D N et. al. 2019 Convex hulls and convex separable sets in the multiclass pattern recognition *Trudy MAI.* **109**
[10] Mingozzi A, Maniezzo V, Ricciardelli S and Bianco L 1996 An Exact Algorithm for the Resource Constrained Project Scheduling Problem Based on a New Mathematical Formulation. *Management Science.* **44** 10.1287/mnsc.44.5.714
[11] Khachai M Y and Neznakhina E D 2015 A polynomial-time approximation scheme for the Euclidean problem on a cycle cover of a graph. *Proc. Steklov Inst. Math.* **289** 111–125
[12] Todosijevic R, Mladenovic N et. al. 2016 Nested general variable neighborhood search for the periodic maintenance problem. *Eur. J. Oper. Res.* **252** 385-396
[13] Pei J, Liu X, Fan W, Pardalos P M and Lu S 2017 A hybrid BA-VNS algorithm for coordinated serial-batching scheduling with deteriorating jobs, financial budget, and resource constraint in multiple manufacturers. Omega, 2017a, DOI 10.1016/j.omega.2017.12.003.
[14] Han Z, Zhang Q, Shi H and Zhang J 2019 An Improved Compact Genetic Algorithm for Scheduling Problems in a Flexible Flow Shop with a Multi-Queue Buffer. *Processes* **7** 302

IOP Conf. Series: Materials Science and Engineering **927** (2020) 012059 doi:10.1088/1757-899X/927/1/012059

H-infinity approach to design optimal output control for the aggregated interpolation model of helicopter motion

I Kudryavtseva, A Efremov and A Panteleev
Moscow Aviation Institute, 4 Volokolamskoe shosse, 125993, Moscow, Russia

E-mail: kudryavtseva.irina.a@gmail.com

Abstract. This paper presents the problem of designing a feedback control of the aggregated model describing helicopter flight dynamics under the influence of exogenous factors based on the H-infinity approach. The original nonlinear equations of dynamics are assumed to be linearized regarding to several flight modes that characterize with a set of precisely specified flight conditions such as an attitude of an aircraft, its forward speed and turn rate. Matrices of the linearized systems are dependent on the speed of the aircraft. Due to that the aggregated model comprises eight submodels corresponding to flight modes when the speed varies from hover to 140 kts in a certain regular interval. The proposed principle of designing includes two stages: on the first stage the feedback control input is selected for each submodel separately so that the H-infinity norm of the value functional with respect to the H-infinity norm of disturbances is minimum; on the second one the interpolation technique lying in the expansion of matrices in the basis of finite functions is applied. The solution method is experienced on three helicopter models: ZD559-Lynx, the twin engine, utility helicopter in the 4.5 -ton category; S123-Bo105, the twin engine helicopter in the 2.5 ton class; SA330-Puma, the twin engine, medium-support helicopter in the 6-ton category.

1. Introduction

The complexity of helicopter flight dynamics makes the problem of a regulator design to stabilize a helicopter system under the effect of disturbances tough to solve [1,8]. Due to importance of tasks performed by helicopters this problem gains a considerable attention from many researches. To provide the good robustness and handling quality of a system especially if uncertainties of different types are needed to take into account the H-infinity strategy is useful [2,4-6]. The H-infinity design problem can be stated as the problem of searching a maximum value of an uncertainty such that the desired regulator provides the stability of a closed-loop system with bounded disturbances in accordance with the H-infinity norm. There exist several implementation of the H-infinity approach among which algorithms where more than one associated Ricatti equations are required to solve or the equivalent LMI strategy can be distinguished [6-7]. In the paper, this approach is implemented as the recursive procedure of solving one Riccati equation without a preliminary initialization of the system gain that produces the desired regulator [2].

2. System model

Consider the dynamical model of helicopter motion containing the state equation that describes the evolution of a state vector $x(t)$ for a given control input $u(t)$ provided that the motion is undergone by exogenous disturbances $d(t)$ of a known intensity like strong wind, wind gusts, low level wind shear and the output equation for an output vector $y(t)$:

$$\begin{aligned}\dot{x}(t) &= A_i x(t) + B_i u(t) + D_i d(t), \;\; i = 1,...,m \\ y(t) &= C_i x(t)\end{aligned} \tag{1}$$

where $x(t) \in R^n$, $u(t) \in R^k$, $d(t) \in R^s$, $A_i = A(V_i) \in R^{n\times n}$, $B_i = (V_i) \in R^{n\times k}$, $D_i \in R^{n\times s}$, $C_i \in R^q$. The subscript i indicates a mode of the motion. In the paper one considers 8 modes $(m=8)$ which correspond to the flight on sea level with zero sideslip and turn rate provided that the helicopter forward speed V_i varies from hover to 140 kts: $V_i = (i-1)\Delta V_i$, $\Delta V_i = 20\text{kts}$, $i = 1,..,8$.

The equations in the system (1) are linearized equations of helicopter flight dynamics in a neighborhood of the aforementioned trim modes [1]. With this regard letting the state vector $x = (u, w, q, \theta, v, p, \varphi, r)^T$ and the control input $u = (\theta_0, \theta_{1s}, \theta_{1c}, \theta_{0T})^T$ the matrices A_i and B_i are represented as

$$A_i = \begin{bmatrix} X_u & X_w & X_q - W_e & -g\cos\Theta_e & X_v + R_v & X_p & 0 & X_r + V_e \\ Z_u + Q_e & Z_w & Z_q - U_e & -g\cos\Phi_e\sin\Theta_e & Z_v - P_e & Z_p - V_e & -g\sin\Phi_e\sin\Theta_e & Z_r \\ M_u & M_w & M_q & 0 & M_v & \begin{matrix} M_p - 2P_e I_{xz}/I_{yy} \\ -R_e(I_{xx} - I_{zz})/I_{yy} \end{matrix} & 0 & \begin{matrix} M_r + 2P_e I_{xz}/I_{yy} - \\ P_e(I_{xx} - I_{zz})/I_{yy} \end{matrix} \\ 0 & 0 & \cos\Phi_e & 0 & 0 & 0 & -\Omega_a\cos\Theta_e & -\sin\Phi_e \\ Y_u - R_e & Y_w + P_e & Y_q & -g\sin\Phi_e\sin\Theta_e & Y_v & Y_p + W_e & g\cos\Phi_e\cos\Theta_e & Y_r - U_e \\ L'_u & L'_w & L'_q + k_1 P_e - k_2 R_e & 0 & L'_v & L'_p + k_1 Q_e & 0 & L'_r - k_2 Q_e \\ 0 & 0 & \sin\Phi_e\tan\Theta_e & \Omega_a\sec\Theta_e & 0 & 1 & 0 & \cos\Phi_e\tan\Theta_e \\ N'_u & N'_w & N'_q - k_1 R_e - k_3 P_e & 0 & N'_v & N'_p & 0 & N'_r - k_1 Q_e \end{bmatrix}$$

$$B_i = \begin{bmatrix} X_{\theta_0} & X_{\theta_{1s}} & X_{\theta_{1c}} & X_{\theta_{0T}} \\ Z_{\theta_0} & Z_{\theta_{1s}} & Z_{\theta_{1c}} & Z_{\theta_{0T}} \\ M_{\theta_0} & M_{\theta_{1s}} & M_{\theta_{1c}} & M_{\theta_{0T}} \\ 0 & 0 & 0 & 0 \\ Y_{\theta_0} & Y_{\theta_{1s}} & Y_{\theta_{1c}} & Y_{\theta_{0T}} \\ L'_{\theta_0} & L'_{\theta_{1s}} & L'_{\theta_{1c}} & L'_{\theta_{0T}} \\ 0 & 0 & 0 & 0 \\ N'_{\theta_0} & N'_{\theta_{1s}} & N'_{\theta_{1c}} & N'_{\theta_{0T}} \end{bmatrix},$$

where u, v, w are components of the transitional velocity; p, q, r are components of the rotational velocity; φ, θ are Euler angles (roll, pitch angles) defining a position of the aircraft with respect to earth; θ_0 and θ_{0T} are the main and tail rotor collective pitch angles; θ_{1s} and θ_{1c} are longitudinal and lateral cyclic pitch; X, Y, Z and L, R, N are components of applied external forces and moments respectively, g is a gravitational acceleration, I_{xx}, I_{yy}, I_{zz} are moments of inertia about the corresponding axes; I_{xz} is a product of inertia about the x and z axes; $L_v, M_q, ..$ are moment derivatives; $X_u, X_p, ...$ are X - force derivatives; $Y_v, Y_r, ...$ are Y - force derivatives; $Z_w, Z_q, ...$ are Z - force derivatives; Θ_e, Φ_e are trim Euler angles; Ω_a is the aircraft angular velocity in trim flight; P_e, Q_e, R_e are trim angular velocities in fuselage axes system; U_e, V_e, W_e are trim velocities in fuselage axes system; k_1, k_2, k_3 are inertia coupling parameters; $X_{\theta_0}, X_{\theta_{1s}}, X_{\theta_{1c}}, Z_{\theta_0}, Z_{\theta_{1s}}, Z_{\theta_{1c}}, M_{\theta_0}, M_{\theta_{1s}}, M_{\theta_{1c}}$ are control derivatives regarding to main rotor longitudinal; $Y_{\theta_0}, Y_{\theta_{1s}}, Y_{\theta_{1c}}, L'_{\theta_0}, L'_{\theta_{1s}}, L'_{\theta_{1c}}, N'_{\theta_0}, N'_{\theta_{1s}}, N'_{\theta_{1c}}$ are control derivatives regarding to main rotor literal; $X_{\theta_{0T}}, Z_{\theta_{0T}}, M_{\theta_{0T}}, Y_{\theta_{0T}}, L'_{\theta_{0T}}, N'_{\theta_{0T}}$ are control derivatives regarding to tail rotor. The yaw angle is omitted because of the considered flight conditions.

Form the value functional (quality criterion) for each subsystem given above:

$$\|z(t)\|^2 = x^T(t)\, S_i\, x(t) + u^T(t)\; G_i\; u(t), \tag{2}$$

where $S_i \in R^{n\times n}$ is a nonnegative definite matrix, $G_i \in R^{k\times k}$ is a positive definite matrix.

Define L_2 gain of a subsystem as

$$\frac{\int_0^\infty \|z(t)\|^2 dt}{\int_0^\infty \|d(t)\|^2 dt} = \frac{\int_0^\infty [x^T(t) S_i\, x(t) + u^T(t) G_i\, u(t)]dt}{\int_0^\infty d^T(t) d(t)\, dt} \tag{3}$$

The purpose of the paper is to find the output feedback control $u(t)$ for the aggregated model (1) presented as the combination of subsystems for flight conditions with different values of the forward speed. The desired $u(t)$ should be designed so that the system is stable and L_2 gain is attenuated by some prescribed value γ^2.

3. Solution algorithm

The wished $u(t)$ is formed basing on the aggregation principle with use of output feedback control $u_i(t)$ for subsystems. It is proposed that $u_i(t)$ should be searched in the form

$$u_i(t) = -K_i^* C_i x(t) = F_i\, x(t),$$

where the gain K_i^* such that the i th subsystem is stable and L_2 gain is bounded by γ^2.

Then according to the result in [2] for the disturbance $d^*(t) = \frac{1}{\gamma^2} D_i^T P_i\, x(t)$ there exists K_i^* such that

$$K_i^* C_i = G_i^{-1}(B_i^T P_i + L_i),$$
$$L_i = G_i K_i^* C_i - B_i^T P_i,$$
$$A_i^T P_i + P_i A_i + \frac{1}{\gamma^2} P_i D_i D_i^T P_i - P_i B_i G_i^{-1} B_i^T P_i + S_i + L_i G_i^{-1} L_i = 0.$$

If it is required to solve the problem for a flight mode when the forward speed takes values differing from the mentioned above the interpolation concept based on using the basis of finite functions are applied. The utilized finite functions are [3,9]

$$S_p^*(\tau) = \begin{cases} 2^{p-1}(1+\tau)^p, \tau \in \left[-1, -\frac{1}{2}\right], \\ 1 - 2^{p-1}|\tau|^p, \tau \in \left[-\frac{1}{2}, \frac{1}{2}\right], \\ 2^{p-1}(1-\tau)^p, \tau \in \left[\frac{1}{2}, 1\right], \\ 0, \qquad \tau \notin [-1,1] \end{cases}$$

where $p = 2,3,4$ is an interpolation parameter. The finite functions take zero values outside the interval $[-1,1]$.

Then the original aggregated system can be expressed in the form

$$\dot{x}(t)=(A+BF)x(t),$$

if the expansion of the system matrices in the basis of $S_p^*(\tau)$ are used: $A=\sum_i A_i S_p^*\left(\frac{V}{\Delta V_i}-(i-1)\right)$, $B=\sum_i B_i S_p^*\left(\frac{V}{\Delta V_i}-(i-1)\right)$, $F=\sum_i F_i S_p^*\left(\frac{V}{\Delta V_i}-(i-1)\right)$. Thus the output feedback $u(t)=-F\,x(t)$.

The detailed algorithm [2] is given below. It should be noted that this algorithms doesn't contain the step of setting initial gain K_i^0.

Algorithm

1. Set $k=0,\ L_i^0=0,\ i=1,..,m.$ Define small number ε.
2. Solve the Riccati equation

$$A_i^T P_i^k + P_i^k A_i + \frac{1}{\gamma^2} P_i^k D_i D_i^T P_i^k - P_i^{\,k} B_i G_i^{-1} B_i^T P_i^k + S_i + L_i^k G_i^{-1} L_i^k = 0$$

and find P_i^k.

3. Find $K_i^{k+1}, L_i^{k+1}, i=1,..,m$:

$$K_i^{k+1}=Q_i^{-1}(B_i^T P_i^k + L_i^k)C_i^T(C_i C_i^T)^{-1},\ L_i^{k+1}=G_i K_i^{k+1} C_i - B_i^T P_i^k,$$

4. Stop and set $K_i^*=K_i^{k+1}$ if $\left\|K_i^{k+1}-K_i^k\right\|\le\varepsilon$. Otherwise set $k=k+1$ and go to step 2.
5. Find $F_i=G_i^{-1}B_i^T P_i + L_i,\ i=1,..,m$.
6. Find interpolated matrices A,B,F applying the expansion in the basis of the finite functions for a given mode.
7. Solve the equations of aircraft dynamics

$$\dot{x}(t)=(A+BF)x(t)$$

used any of numerical methods of appropriate order, for example the Runge-Kutta methods.

4. Analysis of helicopters' motion

The proposed technique is demonstrated on the example of modeling dynamics for three helicopter models: ZD559-Lynx, the twin engine, utility helicopter in the 4.5 -ton category; S123-Bo105, the twin engine helicopter in the 2.5 ton class; SA330-Puma, the twin engine, medium-support helicopter in the 6-ton category. Fig.1 visualizes poles location for three considered helicopters models before the H-infinity design is performed. It is clearly seen the aforementioned systems are originally unstable because some of poles are positioned on the right half plane and as consequence the synthesis of a regulator is required.

Figure 2 demonstrates the evolution of trajectories produced by the obtained control for three systems. The presented behavior of components of the vector state points out on damping transition processes and stabilizing systems. It should be noted that for the computation of the output feedback gain K_i^* the initialization is not needed. The parameter ε should be chosen in conformity with values of entries of system matrices. The γ^2-value is selected experimentally. The results are obtained provided that $0.2\le\gamma^2\le 0.4$ in dependence on a model of helicopter. It should be underlined that the very small γ^2-value is not acceptable because of impossibility of solving the Riccati equation. The closed-loop eigenvalues (λ_i) given in tables 1-3 for each of helicopter models verify the fact that the systems became stable. The solution algorithm is enough quickly implemented in MATLAB.

IOP Conf. Series: Materials Science and Engineering **927** (2020) 012059 doi:10.1088/1757-899X/927/1/012059

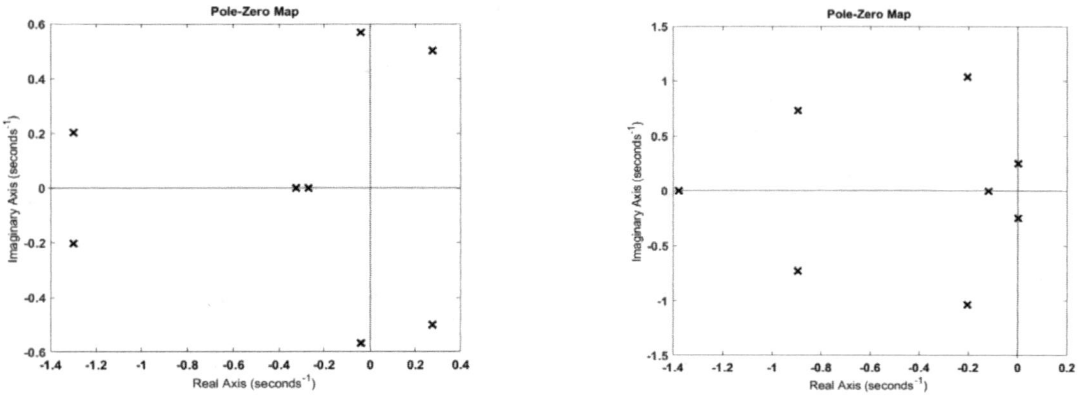

a)

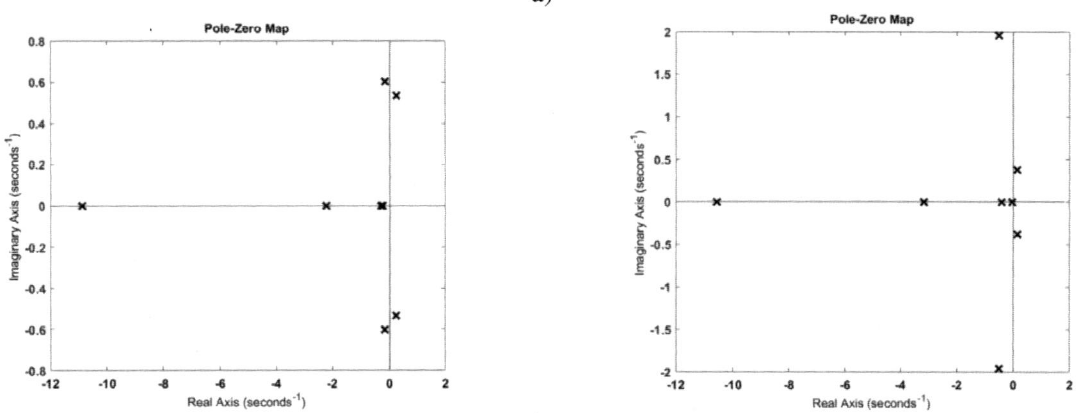

b)

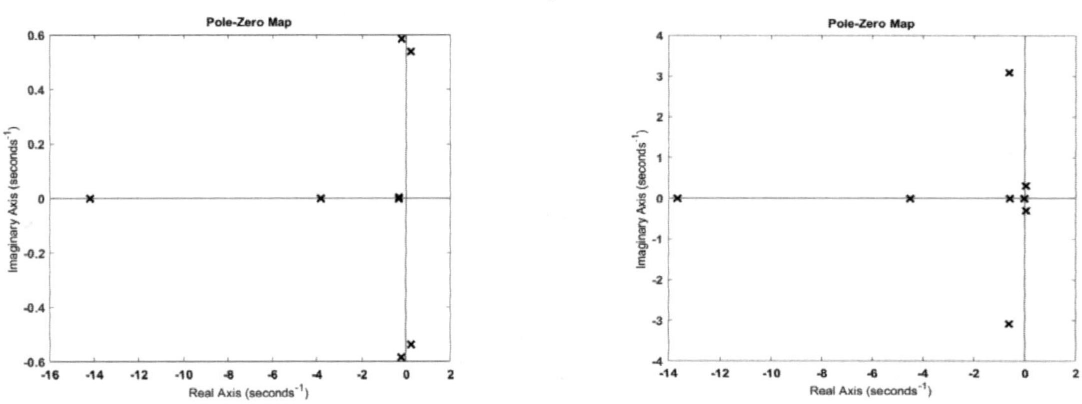

c)

Figure 1. Poles location for the helicopter systems: a) SA330-Puma b) ZD559-Lynx c) S123-Bo105 without the regulator if $V = 0$ (on the left) and $V = 40$ (on the right)

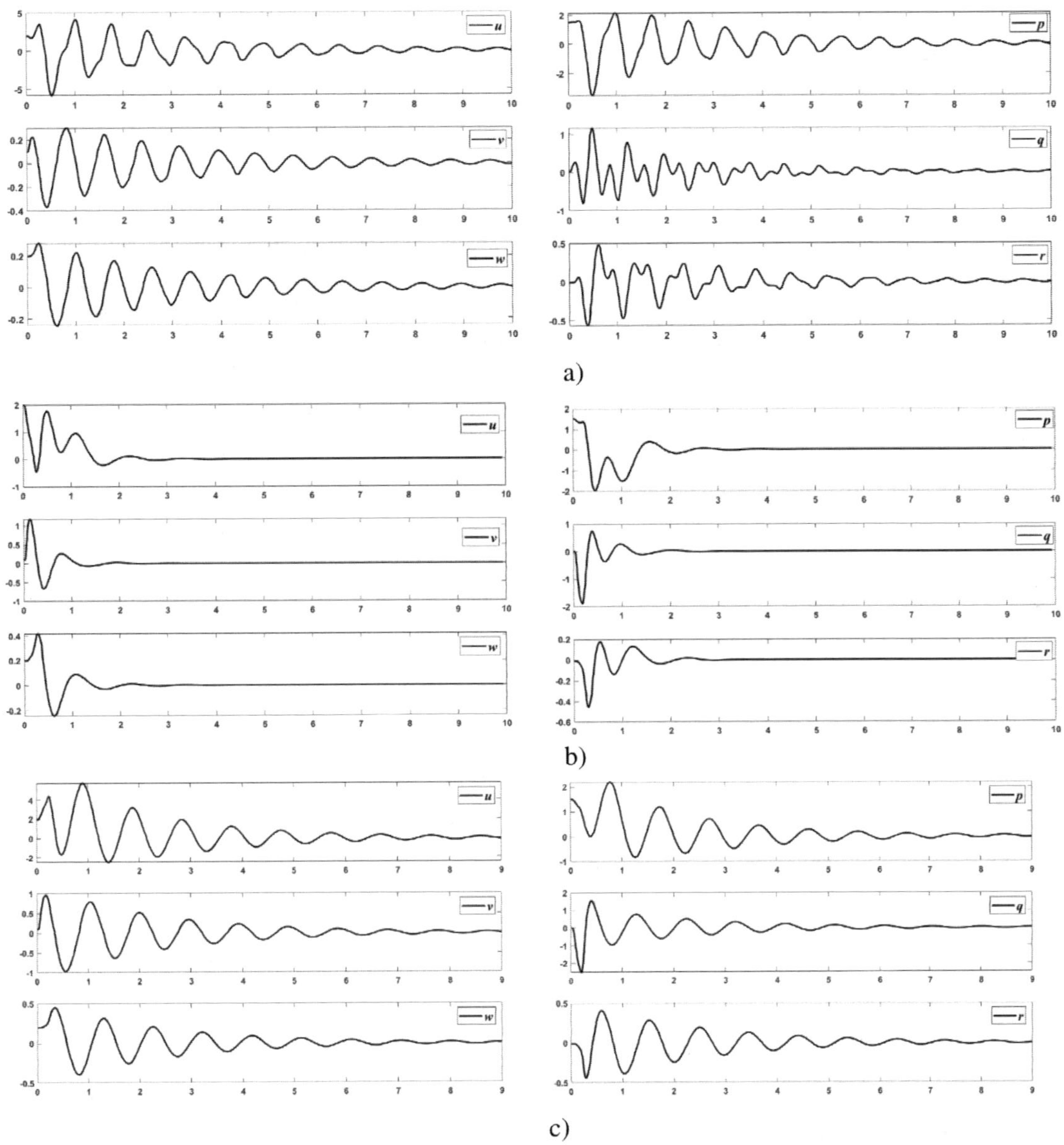

Figure 2. Graphs of evolution of phase coordinates for a) SA330-Puma b) ZD559-Lynx c) S123-Bo105 if $V = 0$

Table 1. The closed-loop eigenvalues for SA330-Puma

V_i	0 (kts)	20 (kts)	40 (kts)	60 (kts)	80 (kts)	100 (kts)	120 (kts)	140 (kts)
	-1,20787	-1,15176	-1,13578	-1,06783	-1,0432	-1,03161	-1,02533	-1,02133
	-1,20787	-1,15176	-1,13745	-1,15431	-1,22637	-1,42084	-1,86443	-1,34891
	-2,12702	-1,47852	-1,13745	-1,15431	-1,22637	-1,42084	-1,86443	-4,32008
λ_i	-2,12702	-5,59646	-10,7964	-11,2438	-10,481	-9,82581	-9,07914	-7,731
	-11,1628	-11,9279	-10,7964	-13,9798	-17,7763	-20,6381	-22,764	-24,4443
	-11,9261	-11,9279	-14,1489	-17,4252	-20,6256	-24,0752	-27,9953	-33,0807
	-31,0569	-34,5725	-35,0073	-35,3097	-35,6274	-35,9804	-36,2652	-35,6286
	-120,933	-116,769	-125,582	-142,776	-159,393	-175,524	-191,877	-208,899

Table 2. The closed-loop eigenvalues for ZD559-Lynx

V_i	0 (kts)	20 (kts)	40 (kts)	60 (kts)	80 (kts)	100 (kts)	120 (kts)	140 (kts)
	-0.11889	-0.33728	-0.27424	-0.1652	-0,0721	-0,04907	-0,03786	-0,02224
	-0.65317	-2.34263	-1.92651	-1.78388	-1,92051	-2,03603	-2,27176	-2,77745
	-8.09829	-8.88193	-11.2115	-14.2157	-17,1709	-19,9274	-22,5653	-25,1583
	-11.0215	-8.88193	-11.2115	-14.2157	-17,1709	-19,9274	-22,5653	-25,1583
λ_i	-30.268	-24.544	-26.1545	-31.3221	-45,1835	-56,9563	-68,8041	-84,5709
	-30.268	-24.544	-26.1545	-31.3221	-45,1835	-56,9563	-68,8041	-84,5709
	-292.341	-460.351	-471.048	-563.535	-775,599	-888,523	-1007,12	-1170,21
	-1848.68	-832.02	-862.584	-931.541	-1185,1	-1272,59	-1348,46	-1464,44

Table 3. The closed-loop eigenvalues for S123-Bo105

V_i	0 (kts)	20 (kts)	40 (kts)	60 (kts)	80 (kts)	100 (kts)	120 (kts)	140 (kts)
	-0,65317	-2,34263	-1,92651	-1,78388	-1,92051	-2,03603	-2,27176	-2,77745
	-8,09829	-8,88193	-11,2115	-14,2157	-17,1709	-19,9274	-22,5653	-25,1583
	-11,0215	-8,88193	-11,2115	-14,2157	-17,1709	-19,9274	-22,5653	-25,1583
	-30,268	-24,544	-26,1545	-31,3221	-45,1835	-56,9563	-68,8041	-84,5709
λ_i	-30,268	-24,544	-26,1545	-31,3221	-45,1835	-56,9563	-68,8041	-84,5709
	-292,341	-460,351	-471,048	-563,535	-775,599	-888,523	-1007,12	-1170,21
	-1848,68	-832,02	-862,584	-931,541	-1185,1	-1272,59	-1348,46	-1464,44

In Figure 3 the graphs of evolution of phase coordinates for the case when the velocity changes linearly are given. To perform interpolation with use of the finite functions the value of the parameter p is 2. As is seen the results illustrate the effectiveness of the proposed solution algorithm. The system becomes stable. However it's worthwhile to say that the time which the system takes for damping processes increases (see Figure 2 b).

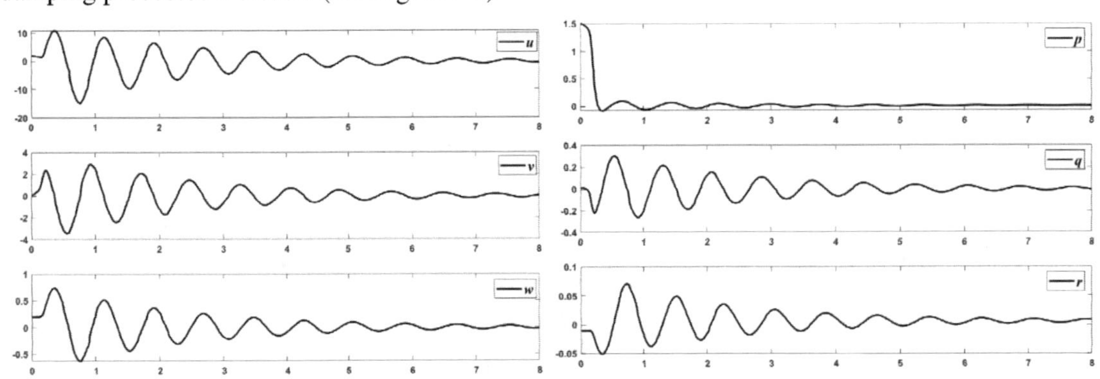

Figure 3. Graphs of evolution of phase coordinates for ZD559-Lynx if V changes like a linear function

5. Conclusion

In this paper, the H-infinity approach with combination of the interpolation principle is proposed to design the state feedback control of helicopter motion described by the aggregated model with uncertainties. The interpolation technique that consists in the use of the expansion of system matrices in the basis of finite functions permits to obtain the solution for modes of helicopters motion where the forward speed changes according to different types of functional correspondences. The described strategy is examined on the example of three helicopter models: ZD559-Lynx, S123-Bo105, SA330-

IOP Conf. Series: Materials Science and Engineering **927** (2020) 012059 doi:10.1088/1757-899X/927/1/012059

Puma. The presented numerical results verify satisfaction of requirements for system transition processes.

References

[1] Padfield D 2007 Helicopter Flight Dynamic. The Theory and Application of Flying Qualities and Simulation Modelling (Blackwell Publishing)

[2] Gadewadikar J, Lewis F L and Abu-Khalaf M 2006 Necessary and Sufficient Conditions for H-infinity Static Output-Feedback Control **29(4)** 915-920

[3] Kudryavtseva I, Efremov A and Panteleev A 2019 Optimization of helicopter motion control based on the aggregated interpolation model *AIP Conference Proceedings* **2181**

[4] Luo C C, Liu R F, Yang C D and Chang Y H 2003 Helicopter H∞ control design with robust flying quality. *Aerospace Science and Technology* **7(2)** 159–169

[5] Yue A, Postlethwaite I 1990 Improvement of helicopter handling qualities using H∞-optimisation. *IEE Proceedings D Control Theory and Applications*, 137(3), 115–129.

[6] Veremey E I, Knyazkin Y V 2017 H∞-Optimal synthesis problem with nonunique solution. *CEUR Workshop Proceedings*, **2064** 270-276

[7] Boyd S, Ghaoui L, Feron E, Balakrishnan V 1994 Linear Matrix Inequalities in System and Control Theory (SIAM, Philadelphia)

[8] Prouty R W 2002 Helicopter performance, stability and control (Krieger pub. co. 2002), 45, 46, 657, 65.

[9] Panteleev A V and Rybakov K A 2018 *Automation and Remote Control* **79**(1) 103–116

IOP Conf. Series: Materials Science and Engineering **927** (2020) 012060 doi:10.1088/1757-899X/927/1/012060

Use of fuzzy areas of preference in tasks of multi – criteria evaluation of freight air transport

V Sudakov[1] and T Sivakova[2]
[1] *Moscow Aviation Institute*, 4, Volokolamskoe shosse, 125993, Moscow, Russia
[2] *Plekhanov Russian University of Economics*, 36 Stremyanny lane, 117997, Moscow, Russia

E-mail: sudakov@ws-dss.com

Abstract. This paper shows how, based on the theory of fuzzy sets, the problem of multi-criteria selection of cargo planes is solved. It is proposed to determine fuzzy preference intervals on the scales of aircraft efficiency criteria. At the same time it is allowed to intersect such intervals taking into account the limitations on the maximum and minimum degree of belonging of criterion values. For combinations of fuzzy intervals, Decision Maker makes its judgments on a fuzzy preference scale. The resulting aircraft efficiency model is checked for coverage of all points of the criterion space with a level of preference not lower than the specified one. This model can be used to estimate any number of cargo aircraft in automatic mode. Both clear and fuzzy criterion values can be submitted to the model input, and then the method allows to build the functions of aircraft belonging to all areas of preference. The software implementation of this model can be used to make decisions about the purchase, manufacture, repair of certain types of aircraft and to determine which types of aircraft should be decommissioned.

1. Introduction

There is a steady upward trend in the number of transported cargo on the air transport market. The value of the statistical indicator of the number of tons carried per thousand kilometers by Russian airlines over the past two years has steadily increased by 8% [1]. However, according to official data, the fleet of the main domestic cargo aircraft is decreasing. Over the last three years, the number of civil transport aircraft AN-124 has not changed, and the number of IL-76Ts has decreased by 26% from 2017 to 2019 [2]. Therefore, an urgent task is to develop tools for evaluating various types of aircraft to support decision-making in their production, modernization, repair and operation.

Integral assessment of the prospects of using different types of aircraft is a complex multi-criterion task. To solve the problem, criteria corresponding to technical characteristics, efficiency and import dependence were used. A part of the criteria has an objective numerical evaluation (for example, technical characteristics), but there is also a significant share of qualitative indicators, evaluation of which is difficult due to the weak structure of the task [3]. To solve this problem the original method of fuzzy areas of preferences has been used [4].

2. Literature review

There is a developed theory of decision making with numerous methods and techniques allowing for multi-criteria analysis of aircraft [5]. There are quite a few works related to the methods of decision theory, including those by Saati and other authors [6-10].

Attempts to create a fully universal method for multi-criteria analysis of alternatives have proved ineffective. And the developed and implemented Decision Support Systems (DSS) are not based on the whole range of available methods and techniques. Nor are they productive to create specialized methods for each emerging problem, so to speak "from scratch".

An actual scientific problem is the creation of a new combined method of fuzzy multi-criteria analysis of alternatives for DSS, which allows to combine methods of qualitative and fuzzy multi-criteria assessment of alternatives in a single criteria space. There is a need to effectively (quickly and with less resources) customize these methods to address a variety of specific high-dimensional multi-criteria aviation industry tasks.

Among widely used methods, only the Saati Hierarchy Analysis Method is suitable for working with a high-dimensional criterion [11]. However, V.V. Podinovsky's works show that in some problems it can give an incorrect result [12].

Other methods in theory do not limit the dimensions of the criterion space, but in practice they are usually not applicable to more than 7±2 criteria. It is caused by features of a human brain on analysis of large information arrays [13] which is shown in works of G. Simon. In practice in qualitative methods, for example in Pareto-optimality method, there are quite a lot of non-dominant alternatives with a large number of criteria. Developers of DSS should not insist on choosing any single method of them. The main thing is that the software implementation of DSS contains tools for fast integration of methods that are necessary for Decision Maker (DM). In [4,14] an analysis of existing methods and approaches to fuzzy estimation of alternatives is made, but without linking to the problems of integration of fuzzy estimations in a single DSS.

Among the DSS created on the principle of shells, we can distinguish the following systems, which are presented on the commercial market and are successfully operated:

- Expert Choice - a system created with the participation of the famous American scientist T. Saati, the author of the method of analysis of hierarchies, and is widely used. Expert Choice works by the method of hierarchy analysis. Charts in the form of circles and columns are widely used for paired comparison of criteria. In Expert Choice it is possible to connect an ODBC source to access a relational database. •
- Super Decisions - the system is based on the method of analytical networks (MAS) - generalization of the method of hierarchy analysis.
- Criterium Decision Plus - American DSS. It allows you to work both by the method of analysis of hierarchies, and using multi-dimensional utility theory. Program interface is a lot like the Expert Choice interface.
- Decision Lens - the system is based on the method of hierarchy analysis.
- IAS EVALUATION & SELECTION - the development of domestic scientists D.A. Abdrakhimov, and A.I. Iofin, applies the method of weighted sum and the ideal point, allows you to build a hierarchy of indicators and visualize the evaluated solutions.
- Fast Prototype" - works only in MS Excel environment and allows taking into account the opinion of several experts. It conducts alternative ranking using the hierarchy analysis method. The openness of "Fast Prototype" DSS is achieved due to the availability of all source code written in Visual Basic for Application.
- DASS is a domestic development that uses the method of qualitative consideration of the importance of individual components of V.V. Podinovsky's Vector Criterion [12].
- FEASIBLE GOALS - software for solving linear multi-criteria optimization problems uses A.V. Lotov's method to visualize the Pareto boundary.
- PC MOVI - the software complex used for setting and solving tasks of multi-criteria optimization in interactive mode uses R.B. Statnikov's method.

- Web-HIPRE - a system designed to solve multi-criteria problems by methods of hierarchy analysis and utility theory is developed in Finland and oriented to work on the Internet.

Software complexes developed in Riga Polytechnic Institute under the guidance of Prof. A.N. Borisov [15] can be considered one of the first domestic DSS based on fuzzy estimations, namely:

- DELTA software system, designed to select multi-criteria alternatives with fuzzy restrictions using utility theory methods;
- SDMIS (Simulation and Decision Making Interactive System) dialogue system, designed for multi-criteria evaluation and selection of alternatives with unclear background information on characteristics of alternatives and preferences of DM;
- Dialogue system "Alternative", designed to solve problems of support and decision making in conditions of risk or uncertainty in the presence of several criteria.

As the analysis has shown, the automated support of decisions in multi-criteria problems is being developed actively enough. However, it is possible to point out a number of common disadvantages typical for many systems. The majority of DSS implements any single method of convolution of vector criterion from a set of already known and, as a rule, it is a method of weighted summation. Most DSS cannot work with lexical indicators. They require pre-digitalization, i.e. conversion to a quantitative scale. However, it is difficult for DM to make quantitative judgments about lexical variables. Most DSS do not allow to consider the dependence of indicators by preference. Most DSS are not open source (their sources are not published in open sources and cannot be used or modified by third parties), do not have a specific database.

The preferences that the DM tells us are actually inaccurate, "vague" in nature. In addition, the assessments of alternatives by criteria themselves are often given with a certain margin of error that must be taken into account. In order to take into account vague judgments, it is advisable to use the apparatus of "soft" calculations, the theory of fuzzy sets.

In tasks of decision-making support, the fuzzy sets apparatus has been used for quite a long time. There is a lot of research on this subject [16-18]. However, authors usually try to use classical approaches of decision making theory by "mechanically" replacing clear numbers with fuzzy ones. For example, instead of clear values of criterion weights, indistinct weights are used. This leads to the following problems:

- it is not clear how the functions of the weights should be defined;
- the result of fuzzy weights is difficult to explain and to adjust the weights in case of inadequate estimates of alternatives;
- the degree of "blurring" of assessments of some criteria may depend on what values the other criteria have adopted.

3. Preference modeling

Let U_i be the scale of the i-th criterion consisting of fuzzy g_{ij} grading and the number of such grading q_i, and $\mu_{ij}(x)$ is the function that determines for the i-th criterion the fuzzy belonging of the value of x j-th grading. The combinations of the grading values of all criteria divide the criterion space into areas. A set of all such areas is a Cartesian product of U_i sets:

$$Y=\{g_{11},g_{12},\ldots,g_{1q_1}\}\times\{g_{21},g_{22},\ldots,g_{2q_2}\}\times\ldots\times\{g_{n1},g_{n2},\ldots,g_{nq_n}\}.$$

DM sets preferences in some subset of set Y: $k = 1..K$ is the number of preference area for which DM set preferences, p_k - fuzzy preference level of k-th area, $\rho_k(y)$ - function of belonging of rank y for preference level p_k. $M_k = (j_1, j_2, \ldots, j_n)$ is a vector of numbers of gradation, for k-th fuzzy preference area.

Multi-criteria analysis was performed for a number of cargo types of aircraft: IL-76, AN-124, AN-12, Tu-204-100C. The task of determining the level of preference for cargo aircraft was performed using given values of criteria (x_1, x_2, ... x_n). The values of the criteria were brought to a fuzzy form by assigning the function of belonging to a specific criterion value.

Fuzzy areas of preference are recorded in the form of fuzzy type implication: **IF** fuzzy criterion values belong to the area M_k, **THEN** aircraft preference level of a given fuzzy preference level p_k. This implication is recorded in a form:

$$p_k(y, x_1, x_2, \dots, x_n) = \min\left(\min_i \left[\sup(\min\left\langle \lambda_{ix_i}(x), \mu_{ipr_iM_k}(x) \right\rangle \right], \rho_k(y) \right).$$

The combination of fuzzy rules for all areas of fuzzy preference determines the integral fuzzy aircraft evaluation:

$$p(y, x_1, x_2, \dots, x_n) = \max_k p_k(y, x_1, x_2, \dots, x_n).$$

After building the final accessory function for some aircraft it may appear that $p(y, x_1, x_2, \dots, x_n) = 0$ or close to 0. This indicates that there are points of the criterion space not covered by preference areas. In this case you should change the scales and/or add new areas (increase k).

Next, you can define a fuzzy ratio of domination between planes, which is laborious and can lead to ambiguous rankings.

In the case of a large number of comparable aircraft, it is advisable to defuzzify preferences using the center-of-gravity method:

$$\tilde{p}(x_1, x_2, \dots, x_n) = \frac{\int_{y_1}^{y_2} y p(y, x_1, x_2, \dots, x_n) dy}{\int_{y_1}^{y_2} p(y, x_1, x_2, \dots, x_n) dy}.$$

If all the original accessory functions were set as a polyline, the resulting accessory function of the preference level $p(y, x_1, x_2, \dots, x_n)$ is also a polyline. In this case, the integration is carried out using a formula:

$$\tilde{p}(x_1, x_2, \dots, x_n) = \frac{\sum_{l=1}^{s-1} \left(a_l \frac{y_{l+1}^3 - y_l^3}{3} + b_l \frac{y_{l+1}^2 - y_l^2}{2} \right)}{\frac{1}{2} \sum_{l=1}^{s-1} \left((y_{l+1} - y_l)(p(y_{l+1}, x_1, x_2, \dots, x_n) + p(y_l, x_1, x_2, \dots, x_n)) \right)},$$

s – is the number of tops polylines,
y_l – is the coordinate of the top polyline,
$p(y, x_1, x_2, \dots, x_n)$ – value of the membership function at the top of a polyline,
a_l and b_l – are the coefficients of the polyline segment equation.

Further the ranking of aircraft types is performed according to the obtained ranks $\tilde{p}(x_1, x_2, \dots, x_n)$.

When operating the software implementation of the system based on this method, DM is responsible for the input of criteria, sets the scales, adjusts the scales, and introduces preferences. After the criteria and scales are set an expert in the field of air transport enters assessments of aircraft on primary criteria. The decision support system software then implements the required mathematical methods, including operations with fuzzy sets and numbers. If the result is unsatisfactory, DM returns to the input and editing of the scales and preferences. The stage of setting preferences and input of fuzzy grades of alternatives can pass in parallel.

If for some type of aircraft there is not at least one point with a sufficient level of belonging more than 0.5, it is necessary to enter a new preference area on the gradations that provide maximum belonging of the given type of aircraft. If there is no such combination of grading that provides at least 0.5 accessory for the AF type the new grading must be introduced.

The software implementation of this ranking method is publicly available on the WS-DSS portal [19]. WS-DSS allows to allocate software implementations of decision support and optimization methods in global and corporate web environments. It is written in Python, C++, Ruby, R. The portal provides the ability to connect third-party web services and organize a sequence of calls to mathematical models with the transfer of input / output parameters between them.

IOP Conf. Series: Materials Science and Engineering **927** (2020) 012060 doi:10.1088/1757-899X/927/1/012060

4. Conclusion

Realization of the given technique of estimation of cargo aircraft allows to raise integral efficiency of use of domestic aviation equipment. The method of fuzzy preference areas contributes to the solution of a whole set of tasks to improve the efficiency of innovation management aimed at increasing the competitiveness of Russian civil transport aircraft in the world market.

Acknowledgments
The reported study was funded by RFBR according to the research project No 19-01-00520.

References

[1] Unified interdepartmental information and statistical system 2020 Done tonnage Retrieved from https://www.fedstat.ru/indicator/37018 [in Russian]
[2] Unified interdepartmental information and statistical system 2020 Number of aircraft Retrieved from https://www.fedstat.ru/indicator/37012 [in Russian]
[3] Nigam N, Ayyalasomayajula S, Tang Y, Urbanczyk P and Alonso J 2017 Design Optimization of Advanced Exhaust Systems *AIAA AVIATION Forum: Proc. 18th AIAA/ISSMO Multidisciplinary Analysis and Optimization Conference (Denver, Colorado, 5-9 June 2017)*
[4] Dutov A, Nesterov V, Sudakov V and Sypalo K 2018 Fuzzy Preference Domains and Their Use for Selecting an Electronic Flight Bag for Flight Crews *Journal of Computer and Systems Sciences International* **57(2)** 230–238
[5] Sun X 2014 Incorporating Multicriteria Decision Analysis Techniques in Aircraft Conceptual Design Process *J. Of Aircraft* **51** 861-869
[6] Saaty T, Rokou E 2017 How to prioritize inventions *World Patent Information* **48** 78-99
[7] Siebert J and Keeney R 2015 Creating More and Better Alternatives for Decisions Using Objectives *Operations Research* **63** 1144-1158
[8] Saaty T, Vargas L 2017 Origin of Neural Firing and Synthesis in Making Comparisons *European Journal of Pure and Applied Mathematics* **10(4)** 602-613
[9] Saaty T 2017 Neurons the decision makers, Part I: The firing function of a single neuron, *Neural Networks* **86** 102-114.
[10] Saaty T 2017 Part 2—The firings of many neurons and their density; the neural network its connections and field of firings *Neural Networks* **86** 115-122
[11] Saaty T, Pierfrancesco P 2017 Rethinking Design and Urban Planning for the Cities of the Future, *Buildings* **7(76)** 1-22
[12] Podinovski V and Podinovskaya O 2014 An approach of the criteria importance theory to decision making problems with hierarchical criterial structures *Autom. Doc. Math. Linguist* **48** 1–5
[13] Simon H 1974 How Big Is a Chunk? By combining data from several experiments, a basic human memory unit can be identified and measured *Science* **183** 482-488
[14] Sudakov V and Kurennykh A 2019 Decision Making in Case of Inconsistent Pair Comparison Matrixes *Proc. 2019 International Russian Automation Conference RusAutoCon* (Sochi, Russia, 6-16 September 2019)
[15] Valishevsky A and Borisov A 2003 Using adaptive networks to process fuzzy-granular information *Proc. of Seventh International Conference on Cognitive and Neural Systems (Boston, USA, 29-31 May 2003)*
[16] Noghin V 2017 Pareto set reduction based on an axiomatic approach with application of some metrics *Computational Mathematics and Mathematical Physics* **57**(4) 645-652
[17] Noighin V 2018 Ultimate Possibilities of Pareto Set Reduction Based on Quanta of Fuzzy Information *Scientific and Technical Information Processing* **45(6)** 452-457.
[18] Noghin V 2019 Multicriteria Choice on a Fuzzy Set as a Problem of Searching for Compromise *Scientific and Technical Information Processing* **46(6)** 397–403.
[19] Sudakov V 2020 Web Services for Decision Support Systems Retrieved from https://ws-dss.com/?locale=en

Influence of the nozzle shape on sound generation in a jet-driven Helmholtz oscillator

A A Abdrashitov[1], E A Marfin[1,2] and E A Plakhova[2]
[1] Institute of Power Engineering and Advanced Technologies, FRC Kazan Scientific Center, Russian Academy of Sciences, 2/31 Lobachevsky Street, Kazan, Russia
[2] Kazan Federal University, 18 Kremlyovskaya Street, Kazan, Russia

E-mail: abdary@mail.ru

Abstract. The excitation of pressure fluctuations by an air jet in the model of a Helmholtz oscillator, consisting of a cylindrical chamber with a nozzle in the front cover and an outlet opening in the rear one was studied. The influence of the shape of the nozzle in combination with the length of the chamber and the diameter of the outlet opening was considered. The amplitude-frequency spectra when changing the configuration of the oscillator and the Reynolds number in the allowable range was analyzed. The formation of the reverse flow area in a cylindrical nozzle and its influence on the intensity of the interaction of the jet with the flooded chamber volume are considered. The advantage of a nozzle with a profiled inlet providing a continuous flow in the nozzle and a top-hat profile of velocity at the nozzle exit are determined. The jet-driven Helmholtz oscillator can be used as a downhole emitter for intensifying oil production.

1. Introduction
To increase oil recovery in productive formations, the method of wave action by pressure fluctuations on the bottomhole formation zone is used. When pumping technical fluid into the formation, flowing hydrodynamic emitters are used, which are installed in injection wells. All liquid flows through these emitters, and part of the energy of the pressure is converted into the energy of elastic vibrations during deformation of the flow in the profiled channel of the emitter.

The highest efficiency of energy conversion is provided by the configuration of the channel, made according to the scheme of the jet-driven Helmholtz oscillator (JDHO). Structurally, the JDHO is a cylindrical chamber with two flat covers. There are through openings in the covers. The opening in the front one – the nozzle – serves to supply fluid to the chamber, and the hole in the back one serves to remove the spent fluid from the chamber. A stream of liquid inside the chamber develops along the axis, in the interval between the openings in the covers.

From a physical point of view, JDHO is a combination of two relatively independent devices: a jet generator (JG) and an acoustic resonator (AR) [1]. The jet generator consists of a nozzle, a jet of liquid and a sharp annular edge of the outlet opening. It is known that such a JG "nozzle - jet - sharp edge" is capable of producing small pressure fluctuations at discrete frequencies f_n in local space when the jet interacts with a sharp edge [2]. A jet generator is placed inside the chamber of the acoustic resonator. The acoustic resonator is a cylindrical chamber with two openings. When one of the discrete frequencies f_n coincides with the natural frequency of oscillations (NF) f_0 of the acoustic resonator, resonance occurs ($f_R \approx f_0$), where f_R is the resonance frequency. At the same time, harmonic-type acoustic waves are emitted from the outlet opening.

Historically, JDHO evolved from a flow-obstacle-generating system capable of producing pressure fluctuations of small amplitude in the surrounding space [3] when a stream flows at a certain speed W onto an obstacle in the form of a rod [4], which leads to the formation of the Karman vortex street behind the obstacle (see the corresponding section in the Van Dyke album). The frequency of vortex formation f in the wake behind a rod of diameter D smoothly changes in accordance with the Strouhal number ($Sh_D = f \cdot D/W$) over a wide range of parameters D and W.

The development of jet sound sources is the "nozzle-jet-wedge" generating system with an active element in the form of a flat jet, the speed of which is conveniently controlled by adjusting the pressure drop ΔP on the nozzle ($W^2 = 2\Delta P/\rho$), where ρ is the density of the liquid. It turned out that the wedge value should be proportional to the jet thickness, and the jet length L ($Sh_L = f \cdot L/W$) [5] is the size that determines the generation frequency f. The generation frequency f is intermittent and is related to the feedback one that determines the tone frequency f_T.

In the axisymmetric JG "nozzle - jet - hole" with a round nozzle, the outlet opening must be proportional to the nozzle, and the jet length L ($Sh_L = f \cdot L/W$) is also the size that determines the generation frequency f. The theory of sound generation in JG "nozzle - jet - opening", proposed by Kruger and Schmidtke [6], explains the discrete frequency of generation by the formation of a chain of vortex disturbances on the jet shell (Kelvin – Helmholtz vortices).

Kelvin-Helmholtz vortices are formed with stable periodicity in the mixing layer of a free stream flowing out of the hole into the flooded space. The vortex intensity is determined by the velocity gradient dW/dy in the mixing layer [7], where dy is the transverse coordinate, in other words, the velocity distribution in the cross section of the jet at the outlet of the hole.

The formation of a velocity diagram with gently sloping edges provides a low velocity gradient in a thick mixing layer and a low vortex formation intensity. A rectangular velocity diagram at the exit of the hole with a sharp change in velocity at the jet boundary provides a high velocity gradient in a thin mixing layer and a high vortex formation intensity. The authors of [7] reported on the influence of the shape of the nozzle on the formation of Kelvin – Helmholtz vortices. In the case of a profiled nozzle with a rectangular velocity diagram at the outlet opening, a chain of vortices begins to form directly in the plane of the nozzle exite. Lengthening the profiled nozzle with a cylindrical section leads to smoothing (rounding) of the velocity diagram at the edges and a thickening of the mixing layer. As a result, the velocity gradient becomes less sharp, and the onset of vortex formation shifts further from the nozzle exit.

In [8], the influence of the shape of the velocity diagram on the flat nozzle exit in the jet system "nozzle – jet – wedge" on the generation spectrum was studied. A significant difference is noted in the amplitude-frequency spectra of generation in the case of using two nozzles: forming jets with a rectangular and parabolic exit velocity profile.

The combination of JG with an acoustic resonator allows pressure fluctuations of significant amplitude in the surrounding space. A resonator with a single neck can be excited by a sliding fluid stream flowing along the plane of the throat and touching the sharp edge of the throat [9], or by a sliding jet also touching the sharp edge of the throat [10]. An ordinary referee whistle works in a similar way [11].

In [12], a detailed study of both the influence of the configuration of individual elements of the device and their mutual arrangement on the generation amplitude in the air flow is presented. Unfortunately, in his many experiments, the author used a cylindrical nozzle of constant length equal to its diameter (one gauge). The author compared a cylindrical nozzle with a profiled one, but did not notice a noticeable advantage in a profiled nozzle.

In our work, we present the results of experiments with the JDHO model, in which the cylindrical nozzle had a different length, or the rounding of the input edge.

2. Model, methodology and experiment

The model JDHO was a cylindrical chamber 1 (figure 1), limited on both sides by covers. A through round opening was made in the center of the front cover 2 — a nozzle 3, and a round outlet opening 5 was made in the center of the rear one 4. A measuring microphone 6 was mounted in the front cover and there was an opening 7 for measuring the static pressure in the chamber. The camera was made from pieces of plastic pipe. The nozzle is designed to accelerate the flow of air before feeding it into the chamber. The outlet opening is designed to remove exhaust air from the chamber.

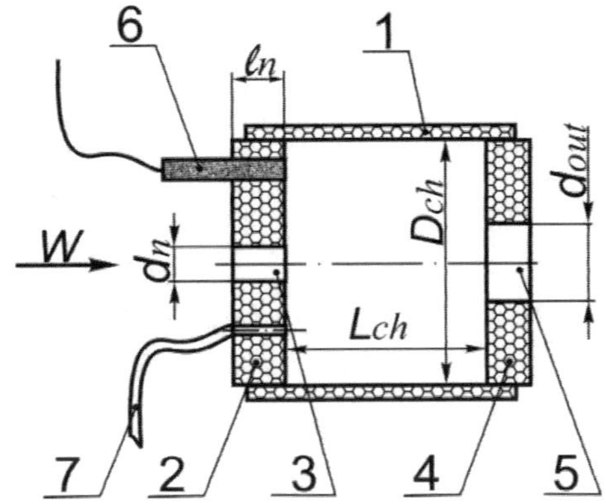

Figure 1. The scheme of the model of the jet-driven Helmholtz oscillator.

The experimental test stand was a vacuum chamber connected to a vacuum pump. The model JDHO was mounted on the lid of the camera, with an outlet inside the camera. Air was sucked into the nozzle from the laboratory. Installation of the model, measuring equipment and the algorithm for processing the measurement results are described in detail in [13-14]. The jet length ℓ_{JET} corresponded to the chamber length L_{CH} and was determined by the interval between the lids (L_{CH}/d_N=0.5-3.5), where d_N is the nozzle diameter. The outlet opening was made with a diameter of d_{OUT} from 10 to 30 mm (d_{OUT}/d_N =1-2.5) and a length of 10 mm.

Microphone RFT MV 201 Robotron measured the pressure fluctuation in the chamber with a frequency above 20 Hz. The static pressure in the chamber was measured by the strain gauge pressure transducers PD 150-DV250 and PD 150-DV2500 OVEN in the corresponding pressure variation intervals. The signals from the microphone and pressure transducer were fed into the E14-140 14 module of an analog-to-digital converter (ADC) with a sampling frequency of 10 kHz and then to a personal computer (PC). For registration and signal processing, the program Power Graph 3.3.8 (PG) was used. The total error of the acoustic measuring system when recording the signal frequency was about 0.4%. The error of the PD is declared by the manufacturer within 1.5%.

The Strouhal numbers ($Sh_L=f{\cdot}L_{CH}/W$) and Reynolds numbers ($Re_L= L_{CH}{\cdot}W/\nu$), where ν is the kinematic viscosity of air, in this work were calculated from the ideal jet velocity $W^2 = 2\Delta P/\rho$. Air was considered to be an incompressible fluid at a speed of 70 m/s [15]. The generation amplitude is considered to be the one of the pressure fluctuations inside the cavity chamber, measured with a microphone.

3. Results and analysis

The results of experiments with the JDHO model are presented in figure 2, where it is seen that the generation amplitude curves have a pronounced upper extremum, in the region $L_{CH}/d_N \approx 2$, and a decrease or increase in the length of the chamber leads to a decrease in the generation amplitude. The optimal length of the cylindrical nozzle is $\ell_N/d_N \approx 2$.

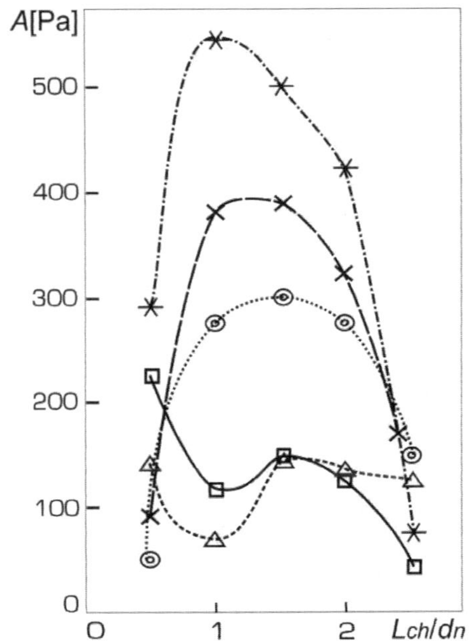

Figure 2. The influence of the nozzle length ℓ_N and the rounding of the input R_N on the generation amplitude A at different chamber lengths L_{CH}/d_N and d_{OUT}/d_N=1.33 in JDHO.
Δ – $\ell_N = d_N$; □ – ℓ_N = 1.37 d_N; x – ℓ_N = 2 d_N; ʘ – ℓ_N = 2.9 d_N; * – $R_N = d_N/2$ (ℓ_N = 1.37 d_N)

From consideration of figure 2, it becomes obvious that in the manufacture of the JDHO model, the length of the cylindrical nozzle is important to ensure a high generation amplitude. As one can see, the nozzle length should be approximately two diameters ($\ell_N \sim 2\ d_N$).

The velocity profile at the nozzle exit varies with its length, since the jet inside the nozzle accelerates and contracts to the minimum cross section, approximately one gauge distant from the entrance plane (ℓ/d = 1). Further, the jet gradually expands and carries adjacent air layers into its motion, forming a region of reduced pressure inside the nozzle. Inside the nozzle, a reverse flow area (RFA) is formed, communicating with the chamber volume behind the nozzle exit. The shape of the RFA determines the velocity profile at the nozzle exit and the intensity of the interaction of the jet with the surrounding air.

As experiments show, the presence of a RFA in the nozzle leads to a decrease in the generation amplitude in the device, since the thickness of the mixing layer increases and the velocity gradient at the jet boundary decreases. A nozzle with a profiled inlet provides an uninterrupted flow along the walls and forms a stream with a top-hat velocity distribution over the cross section at the outlet.

Acknowledgments
Part of the work, namely the review of the current state of the problem was carried out as a part of the state task FRC Kazan Scientific Center, Russian Academy of Sciences, part of the work on experimental study was carried out with the financial support of Russian Foundation for Basic Research and the Government of the Republic of Tatarstan as a part of the scientific project 18-48-160001.

References
[1] Nyborg W L 1954 *J. Acoust. Soc. Am.* **26** 174
[2] Richardson E G 1931 *Proc. Phys. Soc.* **43** 394
[3] Sondhauss C 1852 *Ann. der Phys. und Ch.* **85** 58
[4] Strouhal V 1878 *Wied. Ann.* **5** 216
[5] Curle N 1953 *Proc. Roy. Soc. Lond.* **216A** 412
[6] Krüger F and Schmidtke E 1919 *Ann. der Ph. und Ch.* **60** 701
[7] Kozlov G V, Grek G R, Sorokin A M and Litvinenko Yu A 2008 *Thermophys. Aeromech.* **15** 59
[8] Vaik I, Varga R and Paal G 2014 *Polytech. Mech. Eng.* **58** 55

IOP Conf. Series: Materials Science and Engineering **927** (2020) 012061 doi:10.1088/1757-899X/927/1/012061

[9] Chanadi F, Arjomandi M, Cazzolato B and Zander A 2014 *Exp. Therm. Fluid Sci.* **58** 80
[10] Meissner M 2005 *Arch. Acoust.* **30** 57
[11] Chanaud R C *Sci. Am.* **222** 40
[12] Morel Th 1979 *J. Fluids Eng.* **101** 383
[13] Abdrashitov A A, Marfin E A and Chachkov D V 2018 *Acoust. Phys.* **64** 237
[14] Abdrashitov A A, Marfin E A, Chachkov D V and Chefanov V M 2018 *Acoust. Phys.* **64** 492
[15] Idelchik I E and Fried E 1986 *Handbook of hydraulic resistance: Second edition* (United States: N. p., Web)

IOP Conf. Series: Materials Science and Engineering **927** (2020) 012062 doi:10.1088/1757-899X/927/1/012062

Stabilization of a diffusion flame by a non-stationary electric field

A V Tupikin[1] and V V Zamashchikov[2,3]

[1]Khristianovich Institute of Theoretical and Applied Mechanics, Russian Academy of Sciences, Siberian Branch, 630090 Novosibirsk, Russia
[2]Voevodsky Institute of Chemical Kinetics and Combustion, Russian Academy of Sciences, Siberian Branch, 630090 Novosibirsk, Russia
[3]Novosibirsk State University, 630090 Novosibirsk, Russia

E-mails: tupikin@itam.nsc.ru, albor@kinetics.nsc.ru

Abstract. The work is devoted to the study of stabilization of a diffusion torch by an electric field with a time-varying field line configuration. The stability of combustion was studied using this method of stabilization of the flame, namely the position of the ignition points in the presence and absence of an electric field. The experiments proved that the stabilizing effect of an electric field with a time-varying field line configuration is observed when the torch is stabilized near and above the plane of the electrodes. When the direction of electric forces coincides with the direction of the fuel jet, the effect of the field will strengthen the fluctuation of flame.

1. Introduction

Improving gas burner devices is one of the tasks of the power industry, the solution of which is aimed at improving the environmental and economic conditions of energy production. The possibility of creating electrical systems for stabilizing and controlling the process of burning hydrocarbon fuels is associated with the presence of free charges in their flames. The main carriers of the negative charge are electrons. Among the positive ions H3O+ should be distinguished, its concentration in the flame front is comparable to the concentration of electrons [1]. Therefore, the electric field is most effective for controlling the combustion of hydrocarbon fuels. In [2], a method for stabilizing a diffusion flame by an electric field with a time-varying configuration is presented, and the effectiveness of this approach is shown.

2. Results and discussion

Below there are the results of studying the proposed method of combustion control under the same conditions as in [2], in particular, the stability of the arson points in the presence and absence of an electric field. The experiments were carried out on a setup that included: a diffusion burner, inner diameter $d0$ = 2 mm, outer diameter $d1$ = 6 mm, voltage (U = 1200 V) was applied in pairs ("+", "-") to 8 thin electrodes located uniformly in a circle (diameter D = 14 mm) at a height of h = 18 mm from the burner. Figure 1 shows a schematic diagram of the experiment and the stabilization region in terms that describe a raised diffusion flame when it flows into atmosphere [3], a dimensionless torch lift height $H/d = f(\mathrm{Ho}, \rho_j/\rho_e)$, where ρ_j is the density of the combustible gas and ρ_e is the density of the medium into which the jet flows. $\mathrm{Ho} = \frac{u \cdot \tau_{\Gamma\min}}{d}$ is the criterion of homochronism, u is the average rate of expiration of the fuel jet, $\tau_{\Gamma\min}$ is the minimum characteristic burning time (time spent in the laminar flame front).

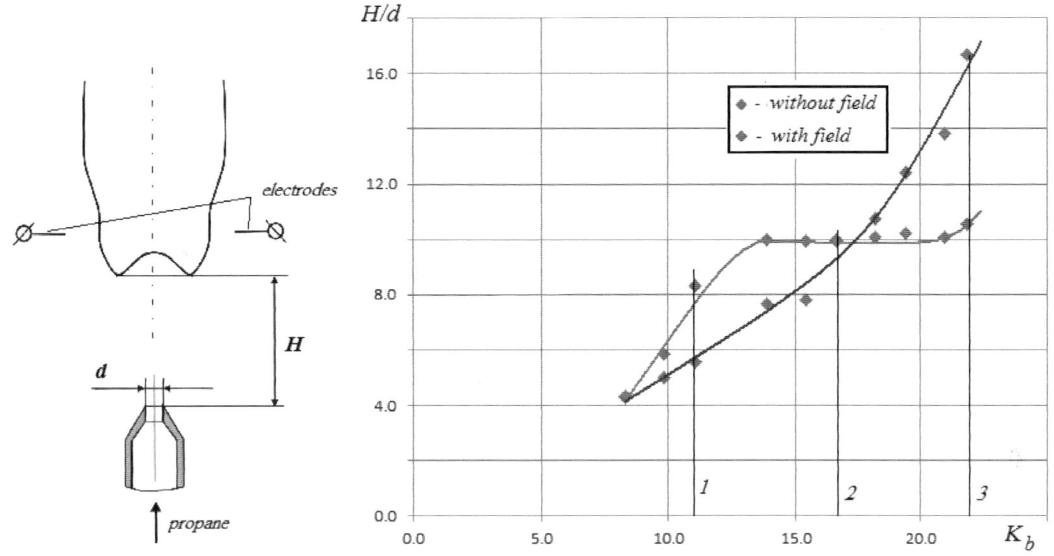

(a) (b)

Figure 1. Experimental scheme (a) and lifting height (b): 1 – K_b=11; 2 – K_b=16.5 и 3 – K_b=22.

Since the electron mobility is many times higher than the ion mobility, the electric forces acting on the charges in the area of the flame stabilization are, on average, always directed to the plane of the electrodes. In Figure 1b, the lifting height is represented by the dependence on the dimensionless criterion $K_b = \mathrm{Ho}\sqrt{\rho_j / \rho_e} = \frac{u \cdot \tau_{\Gamma\min}}{d}\sqrt{\rho_j / \rho_e}$. The upper and lower positions of the ignition points were determined from multispectral images at the wavelength of the CH* radical in the presence and absence of an electric field with a time-varying configuration. Figure 2 shows the results for three areas of flame stabilization: below, near and above the plane of the electrodes. As you can see, when the flame is below the plane of the electrodes, the activation of the field increases the amplitude of oscillation of the ignition points. In this case, on average, electric forces are co-directed with gas-dynamic ones (flow).

If the stabilization regions in the presence and absence of a field coincide, then the effect of the field on the flame stability is almost not shown. When stabilizing above the plane of the electrodes, the electric field and the flow velocity have opposite directions, and the field reduces the amplitude of fluctuations of the ignition points. The graphs show the average relative height of the flame rise. The rms deviation depends on the position of flame stabilization and the presence of the field. Without a field, the dispersion increases with an increase in the flow rate of the fuel jet from 4.5% to 8% within the considered speed range. In an electric field, when stabilizing below the plane of the electrodes, the rms deviation of the position of the ignition point can reach 20%; in other stabilization areas, the dispersion rate is about 5%.

IOP Conf. Series: Materials Science and Engineering **927** (2020) 012062 doi:10.1088/1757-899X/927/1/012062

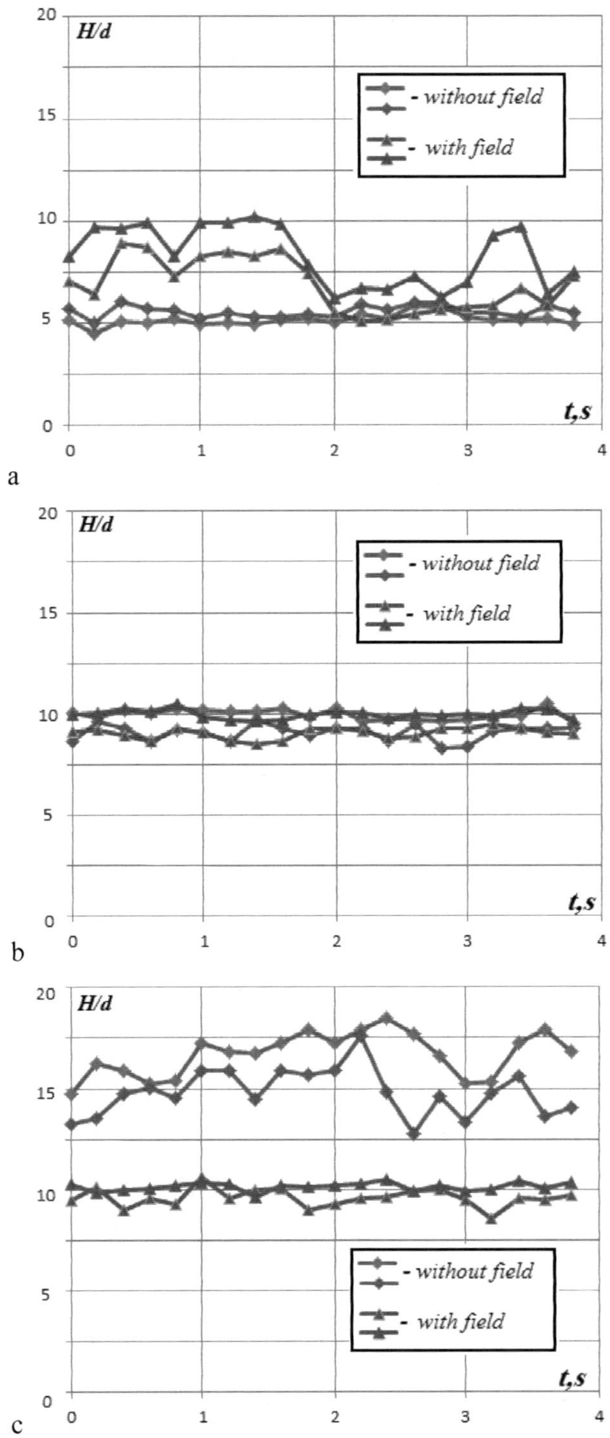

Figure 2. The upper and lower position of the ignition points in the presence and absence of an electric field: (a) below; (b) - near and (c) - above the plane of the electrodes (see Figure 1b).

3. Conclusions

Thus, the experiments proved that the stabilizing effect of an electric field with a time-varying configuration is observed when the torch is stabilized near and above the plane of the electrodes. When the direction of electric forces coincides with the direction of the fuel jet, the effect of the field will strengthen the fluctuation of flame.

Acknowledgments. The research was (partly) carried out within the framework of the Program of Fundamental Scientific Research of the state academies of sciences in 2013-2020 (projects No. AAAA-A17-117030610135-6 and AAA-A17-117112 00018-4).

References

[1] Colcote H F 1979 *Combustion & Flame* **1**
[2] Tupikin A V, Tretyakov P K, Denisova N V, Zamashchikov V V and Kozulin V S 2016 *Combustion, Explosion, and Shock Waves* **52**(2)
[3] Baev V K and Yasakov V A 1969 *Izvestiya. SO AN SSSR* **3**(1) [in Russian]

Numerical study of gas-dynamic and condensation processes in an ejector device used to purify gas-air emissions

V S Nazarov[1], I E Ivanov[1,2] and I A Kryukov[1,3]

[1]Moscow Aviation Institute (National Research University), Russian Federation
[2]Federal State Budget Educational Institution of Higher Education M.V.Lomonosov Moscow State University (Lomonosov MSU or MSU), Russian Federation
[3]Ishlinsky Institute for Problems in Mechanics of the Russian Academy of Sciences, Russian Federation

E-mail: naz.vladislav@yandex.ru

Abstract. The article presents a numerical study of vapor condensation appearing in an ejector. The ejector is a part of a device developed by the authors and used to purify air emissions. The article considers condensation of pure water vapor and condensation of vapor in a mixture of vapor and carrier gas. The system of axisymmetric Navier-Stokes equations and the system of moment equations are considered as a mathematical model for describing two-dimensional non-stationary gas-dynamic processes and processes of phase transitions in the device. The general system of equations is numerically solved using the finite volume method.

1. Introduction

The treatment facilities as scrubbers [1] are designed to purify gas with a fine solid fraction (smoke emissions, dusty air). When in devices, the gas or air medium is accelerated aerodynamically in the Venturi pipe, thus, gas collides with sprayed water or other liquid. The high speed and turbulence of the flow that occurs in the narrow part of the Venturi tube contribute to the interaction of water with solid impurities. However, particles smaller than 10 microns do not coagulate well and cannot be separated effectively.

Ejectors are mainly used in various pumps [2] and cooling systems [3-7]. In the ejector, the gas flow exits the nozzle and creates a low pressure medium in the tapering section, thus, causing the aspiration of other medium. Condensate in a jet can significantly affect jet parameters [8, 9].

In the proposed device, gas is purified by increasing the size of solid particles used as condensation nuclei. In our system, we want to use this phenomenon to purify exhaust gases using water vapor.

The article focuses on the potential possibility of increasing foreign particles to the size required for their separation in the drop catcher. Thus, one of the most important elements of the device for purifying gas-air emissions is a steam-air ejector, which scheme is shown in Figure 3. The actuating medium (water vapor) enters the inlet of the Laval nozzle (nozzle). No. 1). Vapor flowing in this nozzle expands, accelerates, and becomes cool. Water condensation occurs inside the supersonic part of the nozzle (nucleation and cluster growth), while condensation heat generation occurs and gas dynamic flow parameters change. A two-phase jet consisting of water vapor and water droplets flowing out of the nozzle ejects the contaminated gas mixture, which is injected into the holding chamber of the ejector and enters the diffuser or mixing chamber with it. The latter consists of conical and cylindrical parts. In the diffuser, gases flowing from the first and second nozzles are mixed, then the supersonic flow slows down at the shock wave, and the speeds gradually align with the flow.

2. Mathematical model

We consider a viscous, turbulent, thermodynamically perfect multicomponent gas where only one of the gaseous phases may condense, as a two-phase medium. For the liquid fraction, it is assumed that droplets do not fill the volume, do not create additional pressure, while the distance between them prevents their coagulation with each other.

2.1. Equation system

Two-dimensional unsteady-state Navier-Stokes equations in an axisymmetric formulation, which include moment equations [10-12], are used to simulate the gas flow. A combination of algebraic models is considered for turbulence modeling. In addition, it is assumed that the mass fraction of the condensing substance is variable and changes as a result of condensation.

2.2. Moment method

The result is an infinite chain of moment equations, which is the Hill chain [13]:

$$\frac{\partial}{\partial t}(\rho Q_k)+\frac{\partial}{\partial x_i}(\rho U_i Q_k)=(r_*)^k \rho J+k\rho Q_{k-1}\dot{r}, \qquad k=1,\infty \tag{1}$$

$$Q_n=\int_{x_*}^{\infty} r^n f(x,t,r)dr \tag{2}$$

Q_n - moments of the n-th order. The right-hand sides of moment equations in the system (1) are determined using parameters of the modified classical nucleation theory, such as J – nucleation rate (the number of nuclei formed per unit volume per unit time) [12,14], $\dot{r}$ – average size of the cluster (drop) growth rate, r_* – critical radius of a drop (the lower bound of drop radii in the distribution function f)

$$J=\frac{q_c}{(1+\eta)}\left(\frac{2\sigma}{\pi m^3}\right)^{\frac{1}{2}}\frac{\rho_v^2}{\rho_l}\exp\left(-g\frac{4\pi}{3}\frac{r_*^2\sigma}{R_V mT}\right) \tag{3}$$

$$\eta=2\frac{\gamma_f-1}{\gamma_f+1}\frac{L}{R_vT}\left(\frac{L}{R_vT}-\frac{1}{2}\right) \tag{4}$$

$1/1+\eta$ – correction accounting factor for the process non-stationarity [16],

q_c – condensation coefficient ($q_c \approx 1$),

$\sigma=k_\sigma\sigma_\infty$, σ_∞ – surface tension of a flat film,

k_f – correction factor taking into account the drop flexion,

g – multiplier that corrects the beginning of nucleation,

$S=\frac{p_V}{p_S}$ – saturation parameter.

To determine the value of the drop growth rate $\dot{r}$, the Hertz-Knudsen model is used for the free-molecular and continuous flow mode of the cluster-drop with a stream of condensed matter:

$$\frac{dr}{dt}=\frac{\beta}{\rho_l}\frac{p_V-p_{S,r}}{\sqrt{2\pi R_V T}} \tag{5}$$

$$p_{S,r}=p_S\exp\frac{2\sigma}{\rho_l R_V T r_{Hill}} \tag{6}$$

$$r_{Yill}=\begin{cases}\sqrt{\dfrac{Q_2}{Q_0}}, \alpha>10^{-6}\\ 0, \alpha\le 10^{-6}\end{cases} \tag{7}$$

$$r_*=\begin{cases}\dfrac{2\sigma}{\rho_l R_V T\ln S}, S>1\\ \infty, S\le 1\end{cases} \tag{8}$$

$p_{S,r}$ – saturation pressure on the surface of a medium-sized radius drop
β – multiplier that corrects the drop growth rate

2.3. Constitutive relationship
Thermodynamic properties of a mixture are determined by the composition of components and the presence of a liquid therein. The mixture is considered to be in thermodynamic and calorific equilibrium [15]. The viscosity is divided into turbulent μ_t and laminar μ_l.

2.4. Turbulence model
In this work, the "zone" turbulence model is used which is a combination of the Baldwin-Lomax algebraic turbulence model and the Smagorinsky sub-grid model [16,17]. At the same time, each of these models is used in its own field, namely, the Baldwin-Lomax model - in the near-wall region, where it shows good results, while the Smagorinsky model - in the flow core and separation zones. The combined model can be used for detached flows with high Reynolds numbers ($Re=10^5-10^7$).

3. Validation
To validate the model, we used the results of a series of experiments described in [13]. The pressure at the inlet to the nozzle is 60 kPa, while the temperature is 287.6 K, and the partial pressure of water vapor at the inlet to the nozzle is 1 kPa. The origin of the coordinate system is located in the critical section of the nozzle. A schematic image is shown in Figure 1, while the comparison of results of the numerical calculation and experiment is shown in Figure 2. To prevent the formation of shock waves, the nozzle was rounded in the transonic region.

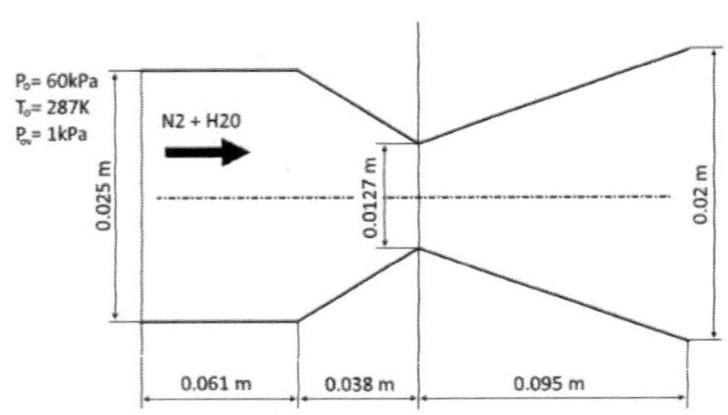

Figure 1. Geometry of nozzle.

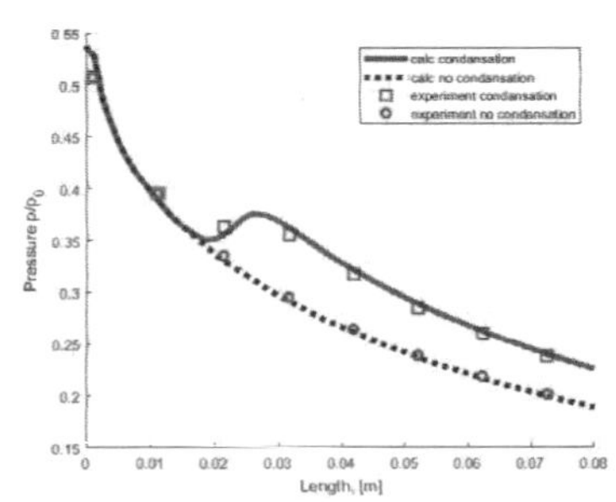

Figure 2. Calculation result of pressure in numerical simulation and in experiments.

4. Device under consideration

The authors consider the case when the input of the device receives products of incomplete combustion of fuel (soot) with air (nozzle No. 2). The work simulates only the part of the device where gas is ejected by water vapor flowing from nozzle No. 1.

4.1. Geometry

The geometry of the calculated device consists of two coaxially arranged conical nozzles shown in Figure 3. The corresponding dimensions are given in Table 1. Water vapor comes from the region designated in_1 in the first nozzle and ejects air from region in_2 of the second nozzle.

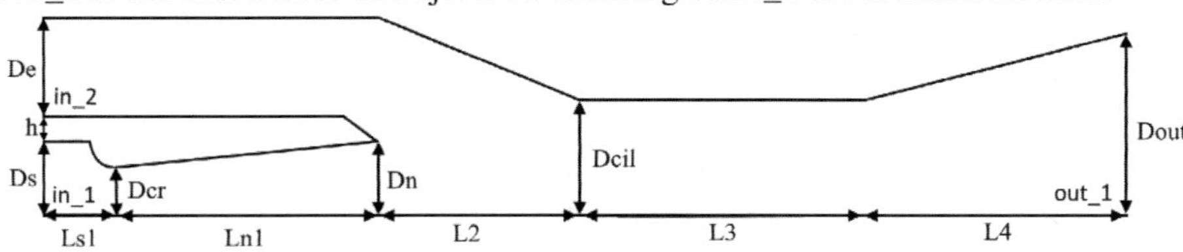

Figure 3. Geometry of the ejection filtration system.

The dimensions figure are shown in table 1.

Table 1. Ejection System sizes

term	value, mm
D_e	4.0
D_s	4.3
D_{cr}	3.0
D_n	3.5
D_{cil}	5.5
D_{out}	7.0
h	0.5
L_{s1}	20
L_{n1}	60
L_2	80
L_3	350
L_4	80

4.2. Computational grid

The numerical simulation uses the block-structured computational grid shown in Figure 4. For better "operation" of boundary conditions, an external region is added to nozzles. The grid consists exclusively of quadrilateral cells and has about 40,000 computational cells.

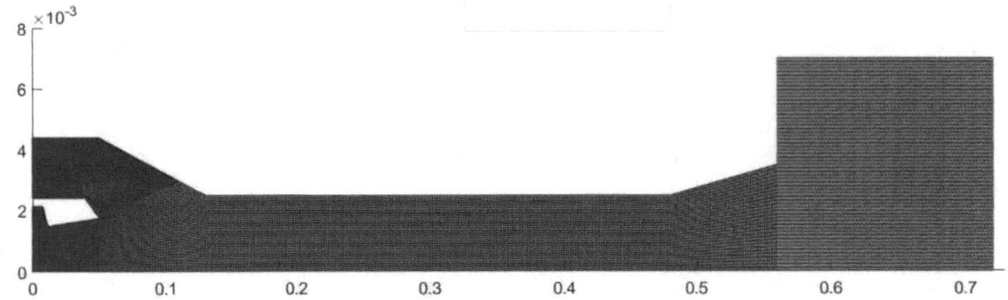

Figure 4. Computational grid of the ejection system.

4.3. Boundary conditions

Pressure and temperature values are set at the nozzle inlet, in the in_1 and in_2 regions. A constant pressure is set in the outer region behind the out_1 nozzle. Temperature and pressure values are shown in Table 2.

Table 2. Boundary conditions

Area	Pressure, Pa	Temperature, K
in_1	8.0×10^5	450
in_2	1.5×10^5	500
out_1	1.0×10^5	-

4.4. Results

To ensure an effective device operation, it is necessary to effectively mix the components and form as much condensate as possible in the mixture and mix as well as possible. To ensure an effective mixing, a very long cylindrical part of the second nozzle is selected in this system. The first nozzle has a small degree of expansion to prevent the formation of a large amount of condensate inside the nozzle.

Figure 5 and Figure 6 show that there are 3 regions of active condensate formation. The first is located inside the nozzle, the second - behind the first nozzle, and the third - in the confusor of the second nozzle. More than 11% of water condensate is formed in the confusor of the second nozzle. It is expected that in the future, gas mixed with vapors and condensate will enter the sludge collector, where its speed will fall, thus, leading to the formation of even more condensate on harmful impurities in gas.

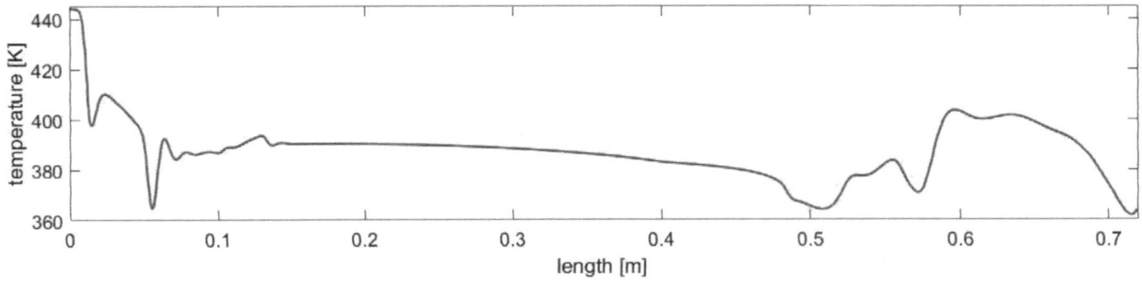

Figure 5. Distribution of temperature along the axis.

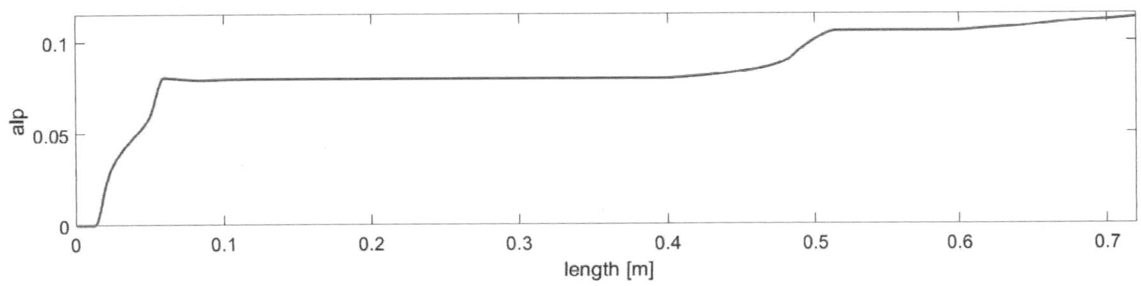

Figure 6. Distribution of mass fraction of the liquid phase along the axis.

Additionally, vapor condensation is strongly affected by the mass of the condensed substance in the mixture (Figure 9). The accounting for its distribution is introduced in equation 9 of the system (1).

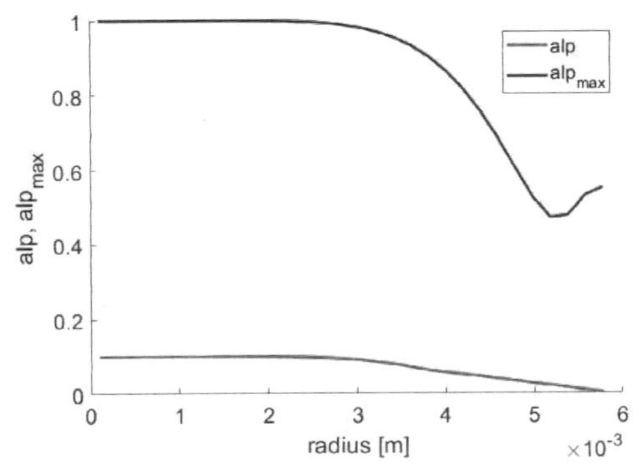

Figure 7. Distribution α_{max} at a distance of 0.5 m from the inlet section of the first nozzle. The axis is located at (0,0).

In numerical experiments, the ratio of injected gas to water vapor was approximately 25%, with a water consumption of approximately 120 liters per hour. Presumably, this system can be used for purifying from harmful impurities, if the percentage of harmful impurities does not exceed 5%.

5. Conclusion

The change in the concentration phase fraction as a result of mixing water vapor with injected gas has a significant impact on condensation. To ensure purification it is necessary to provide heat removal from the walls of the second nozzle. In this case, the mass fraction of condensate can reach 11%. The initial temperature of water vapor in the first nozzle also has a significant effect. The purification system requires that the condensation shock in the first nozzle is as small as possible. For these purposes, the nozzle has a low degree of expansion. This helps to achieve the second condensation shock behind the nozzle as a result of expanding water vapor. In addition, gas flowing out of the second nozzle expands even more, thus, leading to the formation of additional condensate, but already in the mixture in the confusor of the second nozzle. Due to this and heterogeneous condensation, solid particles of injected gas become heavier and, as a result, they can be removed from the flow in the separator.

Acknowledgments
The reported study was funded by RFBR, project number 19-31-90130.

References

[1] Kousalya devi S, Mr Venkatesh S, Mr Chandrasekaran 2015 Performance improvement of venturiwet scrubber *IJRDO-Journal of Mechanical and Civil Engineering* **1(4)** 127-135
[2] Deberne N, Leone J F, Duque A, Lallemand A 1999 A model for calculation of steam injector

performance. *International Journal of Multiphase Flow* **25** 841-855

[3] Scott D, Aidoun Z, Bellache O and Ouzzane M 2008 CFD Simulations of a Supersonic Ejector for Use in Refrigeration Applications. *International Refrigeration and Air Conditioning Conference at Purdue, July 14-17, 2008* **2159** 2-9

[4] Scott D A, Aidoun Z 2015 CFD analysis of an ejector for cooling applications. ID: 347 pp 1-8

[5] Saito Yu, Ito T, Matsuo A and Sato H 2014 Ejector Configuration for Designing a Simple and High Performance Solar Cooling System. *Energy Procedia* **57** 2564-71

[6] Cizungu K, Mani A and Groll M 2001 Performance comparison of vapour jet refrigeration system with environment friendly working fluids. *Applied Thermal Engineering* **21(5)** 585-598

[7] Sun D W and Eames I W 1996 Performance characteristics of HCFC-123 ejector refrigeration cycles. *International Journal of Energy Research* **20(10)** 871-885

[8] Ao Li , Anthony Chun Yin Yuen, Timothy Bo Yuan Chen, Cheng Wang, Hengrui Liu, Ruifeng Cao, Wei Yang, Guan Heng Yeoh and Timchenko V 2019 Computational Study of Wet Steam Flow to Optimize Steam Ejector Eciency for Potential Fire Suppression Application. *Appl. Sci.* 9 1486

[9] Giacomellia F, Biferia G, Mazzellia F and Milazzoa A 2016 CFD modeling of the supersonic condensation inside a steam ejector. Energy Procedia 101 (2016) 1224 – 1231

[10] Ivanov I E, Nazarov V S, Gidaspov V Yu and Kryukov I A 2020 Numerical Simulation of the Process of Phase Transitions in Gas-Dynamic Flows in Nozzles and Jets. *Smart Innovation, Systems and Technologies* **173** 133-150

[11] Luo X, Cao Y, Xie H and Qin F 2017 Moment method for unsteady flows with heterogeneous condensation *Computers and Fluids* **146** 51–58

[12] Wyslouzil B E, Heath C H, Cheung J L and Wilemski G 2000 Binary condensation in a supersonic nozzle *J. Chem. Phys.* **113 17** 7317-29

[13] L. Sova, G. Jun, M. Stastny Modifications of steam Condensation model implemented in commercial solver AIP Conference Proceedings 1889, 020039 (2017)

[14] A. Kantorowitz "Nucleation in Very Rapid Vapor Expansions", The Journal of Chemical Physic 1951, 19(9), 1097-1100

[15] N.Y. Bykov, Yu.E. Gorbachev Cluster formation in copper vapor jet expanding into vacuum: the direct simulation Monte Carlo. Vacuum, Vol. 163, 2019, P. 119-127

[16] Smagorinsky J. General circulation experiments with the primitive equations. I. The basic experiments. Mon. Weather Rev. 1963, vol. 91, pp.99-164.

[17] Camelli F., Lohner R. Combining the Baldwin-Lomax and Smagorinsky turbulence models to calculate flows with separation regions. AIAA-Paper №2002-0426, 2002.

IOP Conf. Series: Materials Science and Engineering **927** (2020) 012064 doi:10.1088/1757-899X/927/1/012064

Analyzing models of air equilibrium composition calculation

V N Emelyanov[1], **A G Karpenko**[2], **S S Tolstoguzov**[1, 2] **and K N Volkov**[1, 2]
[1] Baltic State Technical University, St Petersburg, ulitsa Pervaya Krasnoarmeiskaya 1, 190005, Russia
[2] St Petersburg State University, St Petersburg, Universitetskii prospekt 28, 198504, Russia

E-mail: semen.tolstoguzov96@mail.ru

Abstract. The atmospheric composition behind a normal shock ahead of a hypersonic vehicle differs from the atmosphere ahead of the shock. Various models of air equilibrium composition are compared, analysis of areas of these models applicability is performed. Numerical simulations of flow in a channel with a step are performed using models of perfect gas and high-temperature air. The results computed with both models are compared.

1. Introduction

The perfect gas model quite adequately describes the state of the medium at low temperatures and pressures. Real gas properties are taken into account at high temperatures and pressures.

To calculate the equilibrium composition of the reacting gas mixture, it is necessary to solve a system consisting of the equations of mass conservation, the equations of acting masses, and the Dalton equation. The number of equations of mass conservation is equal to the number of chemical elements in the mixture N_e. The number of equations of the acting masses N_l is equal to the difference between the considered components of the mixture N_s and the number of chemical elements considered N_e (N_l=N_s–N_e). This system equations is as follows

$$\begin{cases} \sum_k^{N_s} \phi_i^k e^{x_k} - \chi_i^0 x_{N_{s+1}} = 0, \quad i = 1 \dots N_e, \\ \sum_k^{N_s} \eta_k^l x_k + \beta^l \ln p_\Sigma \quad - \ln K_p^l = 0, \ \beta^l = \sum_k^{N_s} \eta_k^l, \ l = 1 \dots N_l, \\ \sum_k^{N_s} e^{x_k} - 1 = 0, \end{cases} \qquad (1)$$

where $x_k = \ln\chi_k$, $x_{N_{s+1}} = \frac{\nu_\Sigma^0}{\nu_\Sigma}$, χ_k is the molar fraction of component of the mixture, ϕ_i^k is the number of particles of chemical element i in the component of the mixture k, η_k^l is the stoichiometric coefficient of reaction l, which is negative for initial components, ν_Σ is the number of moles of chemical elements, ν_Σ^0 is the number of moles of chemical elements in the initial state, $\ln K_p$ is the natural logarithm of constant of chemical equilibrium on pressure, which is taken from [1, 2]. Superscript k is related to component of the mixture.

The purpose is to study the effect of accounting for different components on the thermodynamic parameters of air, as well as to consider the effect of accounting for chemical reactions on the flow of gas.

The system of algebraic equations (1) is non-linear and it is solved with Newton-Raphson iterative methods [1].

2. Models

Air is a mixture of gases whose properties differ at different temperatures and pressures. The equilibrium composition of air changes with temperature and pressure. The following air models are considered.

1. The model of four components taking into account dissociation reactions of oxygen and nitrogen (O_2, N_2, N, O).
2. The model of five components taking into account dissociation reactions of oxygen and nitrogen and formation reaction of nitrogen oxide (O_2, N_2, N, O, NO).
3. The model of eight components taking into account the possibility of formation of electrons as a result of ionisation of oxygen and nitrogen (O_2, N_2, N, O, NO, e^-, N^+, O^+).
4. The model of nine components taking into account the possibility of formation of electrons as a result of ionisation of oxygen, nitrogen and nitrogen oxide (O_2, N_2, N, O, NO, e^-, N^+, O^+, NO^+).
5. The model of nine components taking into account argon and its ionisation (O_2, N_2, N, O, NO, e^-, N^+, O^+, NO^+, Ar, Ar^+).

The results computed with these models are compared with the results from [2], which are computed with the model taking into account the components O_2, N_2, N, O, NO, e^-, N^+, O^+, NO^+, , Ar, Ar^+, O_2^+, N_2^+, NO^+, N^{++}, O^{++}, Ar^{++}.

Figure 1 shows a graph of the relative errors of the molar mass at a pressure p=100 atm, obtained using 4-component (without NO) and 5-component (with NO) models. The graph shows that with the neglecting formation of nitrogen oxide, the error starts to increase significantly at 3000 K (at a pressure of p=100 atm) and at 6000 K reaches its maximum of 3%. After that, the error decreases and at 12,000 K it is compared with the results of 5-component model. Figure 2 shows the intensive process of dissociation of nitrogen oxide into oxygen and nitrogen atoms starts at temperature of 6,000 K.

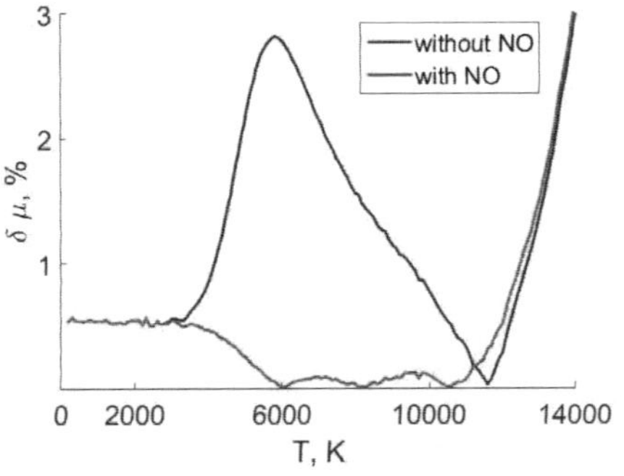

Figure 1. Relative error of molar mass of air components computed with 4-component (blue line) and 5-component (red line) models.

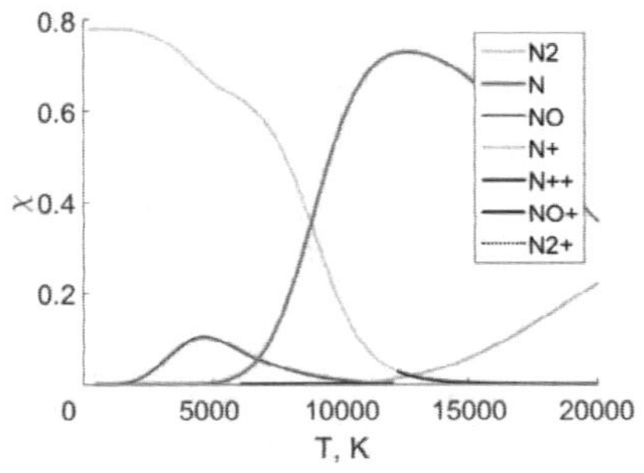

Figure 2. Molar concentration of different components for pressure p=100 atm.

With decreasing pressure, the error decreases (Figure 3) due to the fact that the formation of the maximum concentration of nitrogen oxide decreases. A sharp increase in the error is observed at temperatures from 7,000 K to 12,000 K depending on pressure. It is associated with not taking into account the processes of a single ionization of the components. It can be concluded that not taking into account the formation of nitrogen oxide at high pressures gives a significant error in the calculations. The 5-component model is applicable only up to 7,000 K. At high temperatures, account of the ionization of the components must be taken otherwise the error increases up to 150% at a temperature of 20,000 K (at p=0.001 atm).

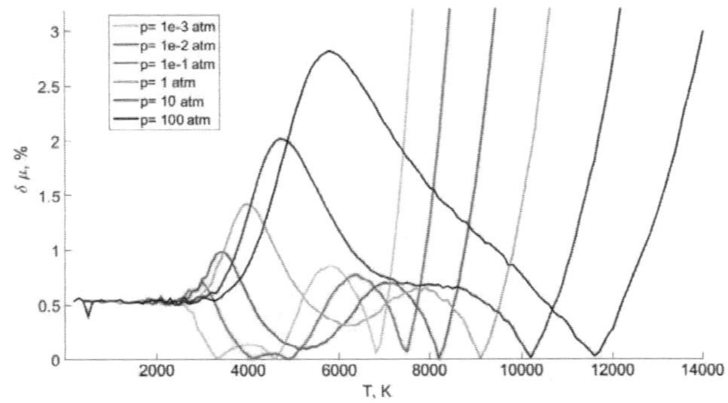

Figure 3. Relative errors of molar mass computed with 4-component model at various pressures.

Figures 4 and 5 show the results computed with 8-, 9- and 11-component models that take into account the single ionization of various air components. Taking into account the formation of NO^+ weakly affects the changes in the relative error of the molar mass, which means that it can be ignored in the calculations. It is also seen that taking into account argon ionization reduces the error, but not significantly, since the 8-component model up to 16,000 K does not give an error above 1.3%.

Figure 6 shows the graphs of the relative errors of the molar mass at different pressures obtained using the 11-component air model. A sharp increase in the error is observed at low pressures. It is associated with an increase in double ionization of air components. It is worth noting that this process makes a significant contribution to changes in molar mass only at low pressures lying in the range from 0.001 to 0.01 atm. At high pressure values, the relative error will not exceed 3%.

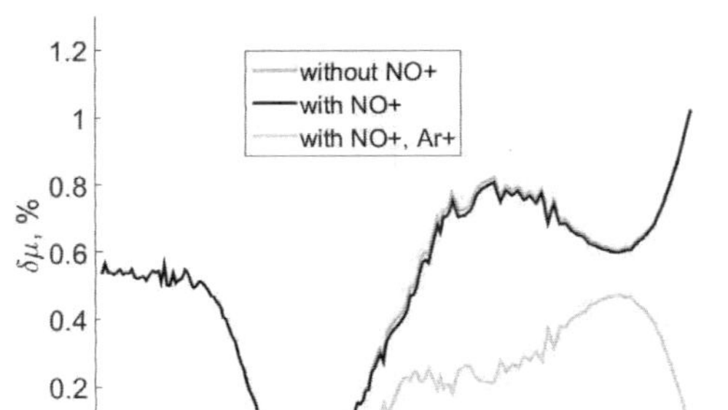

Figure 4. Relative error of molar mass for pressure p=100 atm.

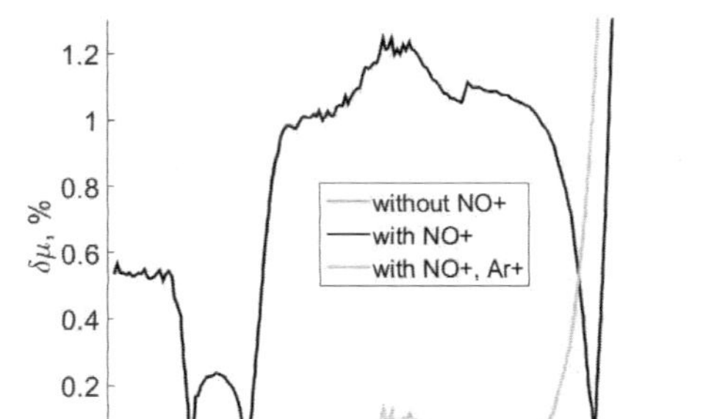

Figure 5. Relative error of molar mass for pressure p=0.001 atm.

Although the method considered gives a good solution in large ranges of temperatures and pressures when taking into account a large number of components, a significant resources are used for the solution, since the system of equations is non-linear, and its solution is performed numerically. This means that the time spent on a numerical solution increases significantly. Quite interesting and successful are the models for air as for real gas proposed by Kraiko [3]. The great advantage of these models is their analytical presentation, which is an absolute plus for including models in large-scale calculations. Figures 7 and 8 compare Kraiko's models based on 5 and 13 components of air.

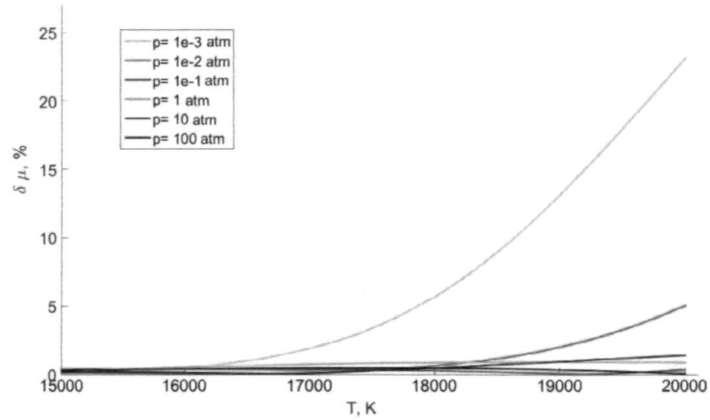

Figure 6. Relative error of molar mass for different pressures based on the model of single ionisation of components.

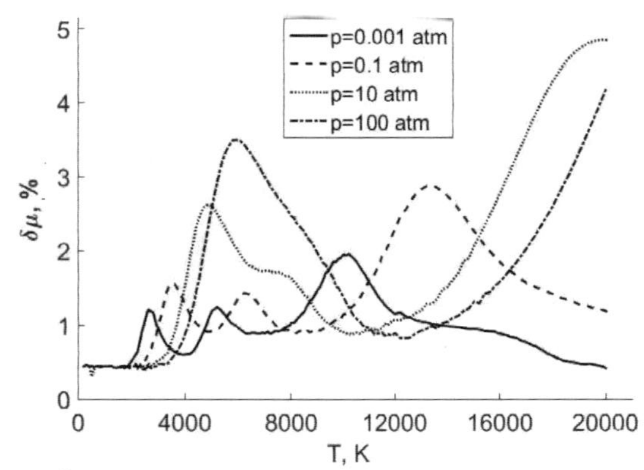

Figure 7. Relative error of molar mass computed with 5-component Kraiko model.

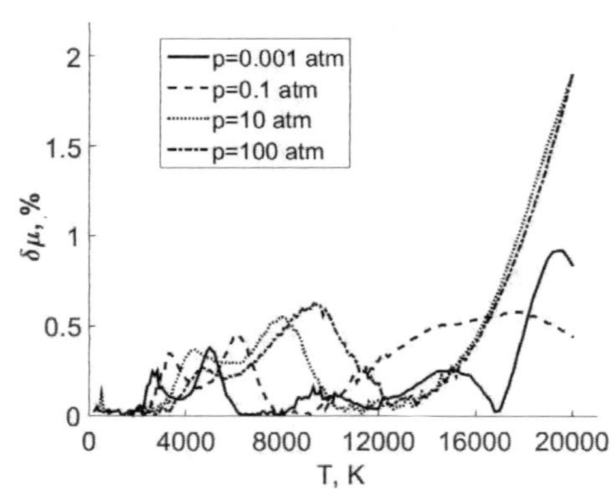

Figure 8. Relative error of molar mass computed with 11-component Kraiko model.

3. Numerical simulation

3.1. Mathematical model

Unsteady flow of inviscid compressible gas is described with the system of equations

$$\frac{\partial}{\partial t}\oint_V \mathbf{U}\,\mathbf{dV} + \oint_{\partial V} \mathbf{F}\cdot\mathbf{dS} = \mathbf{0}, \qquad (2)$$

where **t** is the time, *U* is the vector of conservative variables at point *x* and time *t*, *F* is the flux tensor, *V* is the closed gas volume with the boundary **∂V**, $dS=\boldsymbol{n}\cdot d\boldsymbol{S}$ is the vector of element of area $d\boldsymbol{S}$ on the boundary **∂V** with external normal ***n***. The vector of conservative variables and the flux tensor are

$$\mathrm{U} = \begin{pmatrix} \rho \\ \rho u \\ \rho e \end{pmatrix},\ \mathrm{F} = \begin{pmatrix} \rho u \\ \rho uu + pI \\ (\rho e + p)u \end{pmatrix},$$

where ρ is the density, *p* is the pressure, u is the velocity vector, *I* is the unit tensor, $e = \varepsilon + 0.5\,u^2$ is the total energy per units mass, ε is the internal energy. The equation of state is

$$p = \rho\frac{R_0}{M_\Sigma\ (p,T)}T,$$

where R_0 is the universal gas constant, M_Σ is the molar mass which is a function of pressure and temperature. The molar mass and specific heat capacity are constant for perfect gas model.

3.2. Numerical scheme

The equation (2) is solved with the finite volume method [4] implemented on a structured grid. In this approach, updating the solution to the limit of a fixed control volume includes the contributions of all fluxes through the cell boundaries in one step. To calculate the fluxes on the edges of the control volume, the Roe scheme is applied [5]. For approximate accounting of complex physical and chemical processes in the Riemann solver for real gases, a methodology for the effective adiabatic index has been developed, which allows calculations for these gases [6].

3.3. Flow in a channel with a step

A two-dimensional flow is considered, which is formed in a channel with a step when a supersonic uniform flow with a Mach number M=15 flows into it. The computational domain is a section of a channel of unit height and length three times the height. Inside the channel there is a step starting at a distance of 0.6 m from the left inlet boundary of the region and having a height equal to 20% of the channel height. As initial conditions, it is assumed that the entire channel is filled with an air stream having the following parameters: $\boldsymbol{\rho = 1.16\ kg/m^3}$, $\mathbf{p = 10^5\ Pa}$, $\mathbf{T = 300\ K}$, $\mathbf{u_x = 5208\ m/s}$, $\mathbf{u_y = 0\ m/s}$. The unsteady flow generated from such an initial distribution (somewhat artificial for a two-dimensional problem) tends in the process of numerical solution to its stationary limit, which is reached under stationary boundary conditions. The condition of supersonic inflow through the left boundary with the above-mentioned parameter values and the conditions of the reflective type on impermeable sections of the channel walls and steps are used. The conditions in the outlet section are not significant, since the flow is evacuated there at a supersonic speed.

Figures 9 and 10 show that the difference in the flow patterns is different from each other. The temperature in an equilibrium gas is 2 times less than that of a perfect gas. If the maximum temperature of an equilibrium gas was about 9,000 K, then for a perfect gas it was about 17,000 K.

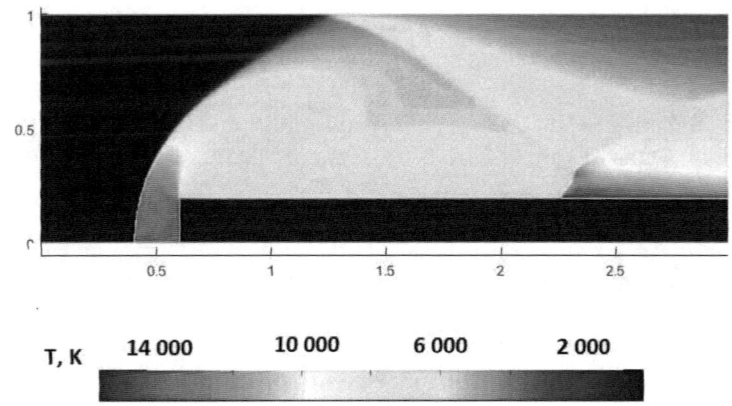

Figure 9. Temperature contours computed with model of perfect gas.

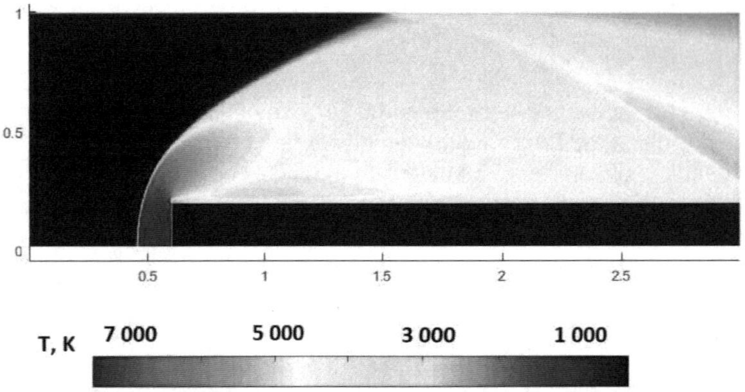

Figure 10. Temperature contours computed with model of real gas.

4. Conclusion

With strong shock waves, not taking into account changes in the composition of air can lead to incorrect results. The processes of dissociation and ionization strongly affect the flow pattern. Failure to take into account the formation of NO strongly affects the error in the calculation of thermodynamic quantities at high pressures. Up to 7,000 K, ionization processes can be ignored, since these processes almost do not occur. The processes of double ionization must be taken into account from 15,000 K. However, at high pressures, the double ionization of the components can be ignored, since the error does not exceed 1% at a temperature of 20,000 K and a pressure of 1 atm.

Acknowledgments

The study was financially supported by the Russian Science Foundation (project No. 19-71-10019).

References

[1] Anderson J D 2006 Hypersonic and high temperature gas dynamics Second Edition *AIAA Education Series*
[2] Chase M W 1985 NIST-JANAF Thermochemical tables *J. Phys. Chem. Ref. Data* **14**
[3] Kraiko A N and Makarov V E 1996 Explicit analytic formulas defining the equilibrium composition and thermodynamic functions of air for temperatures from 200 to 20000 K *High Temp.* **34** 2 202–13
[4] Toro E F 1999 Riemann solvers and numerical methods for fluid dynamics: a practical introduction *Berlin: Springer*
[5] Roe P L 1981 Approximate Riemann solvers, parameter vectors, and difference schemes. *J. Comput. Phys.* **43** 357-72
[6] Brykov N A, Emelyanov V N, Karpenko A G, Volkov K N 2019 Flows of real gas in nozzles with unsteady local energy supply *Comp. Math. Appl.* doi: 10.1016/j.camwa.2019.12.015.

IOP Conf. Series: Materials Science and Engineering **927** (2020) 012065 doi:10.1088/1757-899X/927/1/012065

Studying the applicability of a pulsed magnetic field to biological tissues

N Podolskaya[1], K Kovalev[1] and V Popov[2]
[1]Moscow Aviation Institute (National Research University), Moscow Volokolamsk highway, 4, Russia
[2]Aeroelectromash, Moscow, Bolshaya Novodmitrovskaya street, 12, Russia

E-mail: natalidmi18@ya.ru

Abstract. In biomedicine, there are still a number of issues to be solved and studied. One of them is the treatment of cancer and the problem of gene modification by a non-invasive method. The authors propose a method of using a magnetic field (MF) to solve these issues, in particular, the use of magnetic field strength and eddy currents when exposed to biological tissue. The article describes the mathematical calculations and an experiment model necessary to consider the reactions of healthy and cancer cells to MF. The article highlights prospects for the development of this area.

1. Introduction

In many countries, cancer treatment research is a priority in healthcare. The relevance of this topic allows countries to invest in the research and production of drugs, the development and testing of technical systems for high-quality diagnosis, treatment and maintenance of life in patients with cancer.

Moreover, the existing methods of treating oncological diseases have a narrowly targeted range of effects on the disease, many side effects and an exceedingly small part of patients experience complete remission. Such common methods as chemotherapy, surgery, radio and phototherapy are included in this list. In this article, the subject of research is one of the methods of cancer treatment - magnetic therapy, namely, the use of magnetic field strength when exposed to biological tissue.

The use of magnetic fields (MF) has many advantages, including non-invasiveness, safety, presumably small harm to healthy tissues due to the absence of ionization and the selection of field parameters in such a way as to elicit a response only in the tumor tissue. Also, this method has a greater penetration depth compared to ionizing radiation and the possibility of combining it with other treatment methods. MF has already been used successfully «in vitro» and «in vivo» in the treatment of various types of cancer, including lymphomas, breast, prostate, liver, lung, ovary, pancreas, bladder, thyroid, and colon cancer [1 - 4]. The disadvantages include insufficient knowledge of the mechanisms of MF effect on biological tissue, the relationship between the observed effects and field parameters, lack of research of MF in oncology, especially concerning the human body.

2. Materials and Methods

The research goal is to determine the effect of MF parameters (amplitude, frequency and pulse duration) on cancerous and healthy cells and the regenerative potential of cells after exposure to a pulsed magnetic field. In this treatment method with the creation of a pulsed magnetic field (PMP), the cellular material is not considered as a dielectric, but as a conductive material with real electrical conductivity, which acts on weak motor forces.

The interaction of microcurrents I of a cancer cell of size L with the magnetic field of the concentrator is the excitation of pulses of the magnetic field strength H (r), which changes in space. This process is possible to describe using a mathematical model and calculate the necessary parameters.

The origin of the occurrence of ponderomotive pressure in biological objects is carried out in two ways. The first is due to the creation of a pressure of the electric field strength (E [V / m]) defined by equation (1):

$$P = ½\cdot\varepsilon\cdot\varepsilon_0\cdot E^2 \quad (1)$$

where, ε is the relative dielectric constant, for water $\varepsilon = 81$; $\varepsilon_0 = 8.8542\cdot10^{-12}$ [F/m] – is the electric constant.

The second way is associated with the pressure of the magnetic field strength (H [A/m]) calculated by the formula (2):

$$P = 0.5\cdot H^2\cdot(1 - e^{-2h/\Delta})\cdot\mu_0 \quad (2)$$

where, E – is the intensity of the electric field pulses, [V/m]; h – is the thickness of the malignant and adjacent healthy tissue, m; Δ – is the penetration depth of the magnetic field, m; $\mu_0 = 4\cdot\pi\cdot10^{-7}$, [H·m] – is the magnetic permeability.

The ponderomotive suppression of the functions of malignant tissues [2] was carried out by pulses of electric field strength: $E = 40$ [$V\cdot10^3/m^{-2}$], which exerts pressure on biological tissues regardless of their electrical conductivity (3):

$$P = ½\cdot\varepsilon\cdot\varepsilon_0\cdot E^2 = 6000\ H/m^2 \quad (3)$$

The resulting heat excitement and ponderomotive forces in cancer cells are greater than in healthy cells. This determines the selectivity of the magnetic pulse therapy method in terms of force and thermal effects on cancer cells (Figure 1).

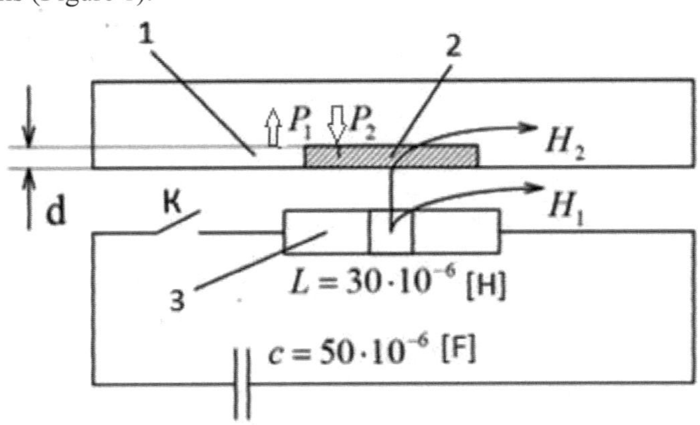

Figure 1. Scheme of the effect on the biological tissue of the magnetic field strength H_1, H_2, excited by a discharge L-C circuit; 1 - healthy biological tissue, 2 - oncological biological tissue, 3 – Inductor.

The calculation of the pressure of the magnetic field on the biological tissue is represented by the formula (4):

$$P = \frac{(H_1^2 - H_2^2)}{2}\cdot\mu_0 = \frac{H_1^2\cdot(1-K^2)}{2}\cdot\mu_0 \quad (4)$$

IOP Conf. Series: Materials Science and Engineering **927** (2020) 012065 doi:10.1088/1757-899X/927/1/012065

where, H_1 – is the magnetic field strength under the biological tissue; H_2 – the magnetic field strength in the biological tissue, d - is the thickness of the biological tissue.

The factor of penetration of a magnetic field into biological tissue is calculated by the formula (5):

$$K = \frac{H_2^2}{H_1^2} = e^{-\frac{2d}{\Delta}} \tag{5}$$

The equipment for the experiment consists of a medical and technical part.

The technical part of the equipment (Fig. 1) is represented by an interconnected system consisting of such units as an inductor with a cooling fan, the purpose of which is to excite pulses of magnetic field strength; high voltage current pulse generator, designed to power the inductor.

Besides, the system contains a high voltage capacitor for storing electrical energy. Management and programming console, with functions: set pulse repetition periods; setting the number of pulses; pulse amplitude regulation; inductor temperature measurement; inclusion of installation; installation of locks.

Medical equipment (required for further stages of the experiment) includes an LSM confocal laser scanning microscope which gives fluorescent images of cancer cell organelles. The microscope is used to determine changes in the activity of cancer cell organelles after pulsed electromagnetic exposure. The medical part also includes the technology "Tumor in a Petri dish", which will be located on the concentrator of the electromagnetic field.

3. Results

The magnetic pressure on the malignant tissue is 2.73 times greater than another one on healthy tissue when the capacitor is discharged to the inductor (Figure 2).

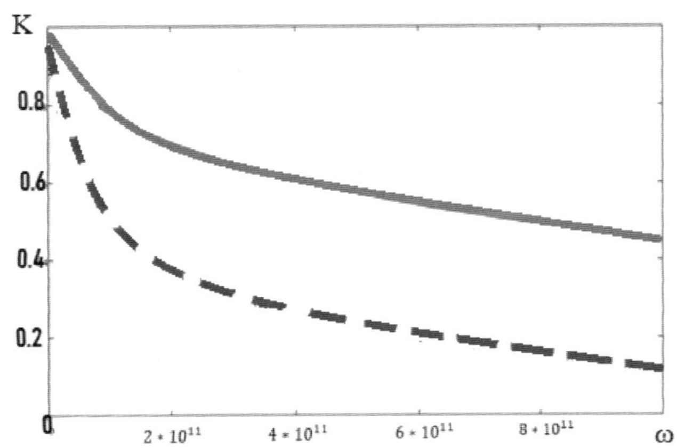

Figure 2. $K_1(\omega)$ – a graph of the coefficient of penetration of the magnetic field in healthy tissue, (red line, γ=1); $K_2(\omega)$ – a graph of the coefficient of penetration of the magnetic field into the malignant tissue (blue - dotted line, γ=7.5).

There is a difference in electrical conductivity in healthy and cancer tissue.

Aluminum and copper plates, which have different electrical conductivity like bio cells, were used as a model for healthy and oncological cells. This made it possible to prove that the effect of a pulsed electromagnetic field on the compared plates is different, which means that biological tissue will show a different response to MF.

The specific fundamental tasks to be solved are as follows:

1. The detection of emerging effects in a biological cell (biological tissue) when exposed to a pulsed magnetic field. This will allow structuring the existing information about the parameters of cells

(permeability, thermal conductivity, etc.) when exposed to physiotherapeutic methods (in particular, a magnetic field).

2. Determination of the dependence of the degree of exposure to a pulsed magnetic field and the appearance of the regenerative potential of healthy and cancer cells.

The solution of the assigned tasks will provide the next answers to the following questions:

• whether the method has side effects;

• how much heating of oncological cells is greater than heating of normal cells, how does the effect of general hyperthermia decrease;

• selectivity of ponderomotive effect on cancer cells;

• improving the ergonomics of the installation for a non-contact method of supplying magnetic field strength to the target.

4. Conclusion

Based on the data obtained, it was suggested that in the treatment of MF oncology there are the following advantages: heating of cancer cells will be more than heating of normal cells; non-contact method of supplying magnetic field strength to the object of influence; stimulation of the regenerative potential of cells. All these advantages are associated with the choice of device parameters: frequency, pulse shape, pulse duration, magnetic field induction.

The development of research on this topic will be in the following areas:

- creation of mathematical models to describe the behavior of healthy and malignant cells when exposed to a pulsed magnetic field;
- performing additional calculations and adjusting the selected parameters, analyzing the results of the experiment and further stages of the research;
- replacement of copper and aluminum plates with the technology "Tumor in a Petri dish". Obtaining and analysis of the results of this experiment.

These models, together with further experiments, will confirm the advantages of this method and will allow you to move on to the next stage - conducting in vivo experiments.

Besides, it is proposed to introduce a method and device for inactivating vaccines by eddy currents that excite the intensity of an alternating magnetic field in the smallest pathogens of viruses. It is supposed to use equipment designed for ponderomotive suppression of the functions of malignant tissues. With the hyperthermic action of eddy currents excited inside viruses by the smallest infectious agents, the temperatures reach values many times higher than the temperature of the envelopes of viruses.

The proposed method of ponderomotive effects of MF on microorganisms can be used to solve the problems of genetic technologies without the introduction of foreign genetic materials.

References

[1] Vamsi K 2018 A review on the use of magnetic fields and ultrasound for non-invasive cancer treatment, *Journal of Advanced Research* **14** 97-111

[2] Crocetti S, Beyer C, Schade G, Egli M, Frohlich J and Franco-Obregon A 2013 Low Intensity and Frequency Pulsed Electromagnetic Fields Selectively Impair Breast Cancer Cell Viability *PLOS ONE* **8** 44

[3] Ulashchik S 2015 Some results of a study of the antitumor effect of magnetic fields in an experiment, *Questions of balneology, physiotherapy and physiotherapy* (Moscow) **4** 49-52

[4] Goudra B, Shah D, Balu G, Gouda G, Balu A, Borle A and Singh P M 2017 Repetitive Transcranial Magnetic Stimulation in Chronic Pain: A Meta-analysis *Anesth Essays Res.* **11** 751–757

Investigation of the Stress-Strain State of the Composite Cylindrical Shell with Rectangular Cutouts

V N Bakulin
Laboratory of Non-Classical Models of Mechanics of Composite Materials and Structures, Institute of Applied Mechanics, Russian Academy of Sciences, 7 Leningradsky Avenue, Moscow 125040, Russia

E-mail: vbak@yandex.ru

Abstract. An algorithm for constructing a model of effective finite elements of natural curvature for the stress-strain analysis of thin composite cylindrical bending shells, including those weakened by rectangular cutouts, is given. The reliability, accuracy, and model efficiency of the considered finite element and the approximations, which it is based on, are shown by comparison with the most commonly used finite elements. Experimental data for a composite shell with rectangular cutouts are obtained and shown. The computation results obtained using the considered finite element are compared with the experimental data for a composite shell with rectangular cutouts.

1. Introduction

In modern aeronautical engineering, shell structural elements, especially cylindrical shells, are widely used. The introduction of composite materials with high specific characteristics (strength, stiffness) and other properties necessary for critical products allows you to create highly efficient structures with specified parameters.

At the same time, having high specific strength and stiffness, composite materials have a pronounced anisotropy of properties and are sensitive to the onset of internal damage, which reduces the strength and life of the product. Internal defects occur, as a rule, in areas of high stress. Such an overstressed areas occurs near various cutouts, hatches, which have a structural, technological or other purpose. Insufficient knowledge of the stress-strain state in the areas of increased stress may lead to the structural failure. Therefore, the development of adequate models to consider the design features of products made of composite materials and to determine the stress-strain state with the necessary accuracy, including the stress-strain state in areas of high stress, is an urgent scientific problem, which has an important applied value.

A great attention is paid to the study of the stress-strain state of shells with cutouts. Brief reviews of such papers are given in introductions of [1,2]. The analysis of these and other works shows that the application of analytical methods faces mathematical difficulties, especially for the case of shells with rectangular cutouts. It is noted in the monograph [3] that it is possible to obtain exact solutions to the problems of stress concentration only for the case of the circular hole in the spherical shell, i.e. for the problems with separable variables. It is impossible to obtain exact solutions for noncircular holes within the framework of the theory of transversal isotropic shells [3]. For that reason, numerical, numerical analytical [2,4,5], and other methods have been used for the stress-strain analysis of shell structures with cutouts.

The publications on the analysis of shells with cutouts mainly consider isotropic homogeneous shells [6]. Much smaller number of publications are devoted to the study of the stress-strain state of composite shells with holes [7,8] and very few of them to experimental works [9].

Therefore, the creation of new models and the improvement of existing ones that allow for the stress-strain analysis of composite shells with rectangular cutouts (which is the most common shape of cutouts) with high accuracy and level of detail is an important scientific and technical task [10].

The most suitable method for solving these problems is the finite element method (FEM), which allows for the analysis of structures, including composite shells with rectangular cutouts, and the study of overstressed regions in the vicinity of cutouts.

2. Formulation of problem

In this paper, the author considers an effective finite element model for a refined stress-strain analysis, in the general case of irregular cylindrical shells, including those weakened by rectangular cutouts.

The considered model for the stress-strain analysis of irregular thin bending cylindrical shells is based on the construction of two-dimensional shell finite elements of natural curvature. Finite elements of natural curvature simulate the geometric shape of structures, which avoids error in the geometry approximation that arise when, for example, using plane or other shapes of finite elements. The considered finite elements have a rectangular shape in plain view, which facilitates the modeling of cylindrical shells weakened by rectangular cutouts. Cutout sizes can vary widely.

The shape functions obtained by approximation of the displacement fields by first and second order polynomials are frequently used for the building of finite elements [11, etc.], which often leads to cumbersome finite element models. Thus, the number of finite element model degrees of freedom grows considerably, and, consequently, computation error, computing resource requirements, and computation time grow. This makes it difficult to analyze such structures with proper accuracy, and therefore the problem of reducing the dimensionality of finite element models of shells with high-stress areas arises. In this regard, it is important to build effective models with shape functions that make it possible to increase the rate of convergence of the results, which leads to a decrease in the order of equation systems in comparison with the known approaches.

A few quite widely used rectangular finite elements of cylindrical shells are known [12, etc.]. However, they do not show high rate of convergence of the obtained results.

The effective shape functions may be obtained by approximation of the generalized strains [13]. With this approach, it is necessary that the approximating functions satisfy the strain compatibility equations.

The considered effective model for the stress-strain analysis of cylindrical shells with rectangular cutouts is based on this approach. Below is a brief summary of such an approach for obtaining the shape functions for the considered finite element.

3. Research methods

Since the structural shells of modern aeronautical engineering structures are sufficiently thin and rigid, to model the stress-strain state in them we will use two-dimensional finite elements of natural curvature with twenty degrees of freedom. These finite elements are developed on the basis of the bending theory of shells based on the Kirchhoff - Love hypotheses.

An important requirement for the approximating displacement functions of the shell finite elements is to account for the rigid-body displacements [12,13], which makes it possible to increase the rate of convergence of the numerical procedures and, therefore, to reduce the number of finite elements necessary for the analysis.

Integration of the Cauchy equations, which relate strains and displacements, at zero strains yields the rigid-body displacement functions. Then, by integrating the Cauchy equations with the obtained expressions for the generalized strain approximating functions, the functions of the displacements caused by the deformation of the finite elements are calculated.

The nodal displacements of the finite elements are the displacements of the mid-surface points $\delta = \{u, v, w\}^T$ and the rotation angles of the mid-surface normal related to the axial and circumferential coordinates *x*, $y(\varphi)$ $\vartheta=\{\vartheta_x,\vartheta_y\}^T$. The approximating functions of the generalized strains and displacements are written with the use of indefinite coefficients α, whose number is equal to the number of the finite element degrees of freedom. Six out of the twenty coefficients α are used to describe the rigid-body displacement. The remaining fourteen indefinite coefficients are distributed between the generalized strains. In this case, the terms obtained from the strain continuity equation are added to the expressions for the generalized strains. As a result, the expression for the vector of generalized strains can be written as follows:

$$\varepsilon=\Omega\,\alpha \qquad (1)$$

where $\alpha=\{\alpha_1\ldots\alpha_{20}\}^T$ is the vector of indefinite coefficients, Ω is the matrix of the generalized strain approximating functions.

Table 1. Matrix Ω for the considered finite elements

						1	φ												
								1	x		$-x^2/2R$	$-x^3/6R$	$-x^2\varphi/2R$	$-x^3\varphi/6R$					
										1									
											1	x	φ	$x\varphi$					
															1	x	φ	$x\varphi$	
							$-x/R^2$						x/R	$x^2/2R$		$R\varphi$		$R\varphi^2/2$	1

In obtaining the approximating function of generalized strains, the incorrectness allowed in [13] and caused by the replacement of three strain continuity equations by four equations was eliminated.

Summing up the obtained approximations of the rigid-body displacements and the displacements caused by deformation, we obtain approximating functions of finite element displacements

$$\delta=T\alpha,$$

where *T* is the matrix of the approximating displacement functions for the shell finite elements.

Knowing the physical law, taking into account (1), we can write expressions for stresses (forces and moments) using the vector of indefinite coefficients α. The stiffness matrixes of finite elements are determined from the condition of the minimum total potential energy (Lagrange variation principle).

Figure 1 shows the comparison of the convergence of the numerical results obtained with the help of the considered finite element approximations and the constructed finite element of the cylindrical shell (line 2), the plate finite element from MSC NASTRAN (line 1), and the known finite elements in the context of S.P. Timoshenko's problem for free circular cylindrical shell loaded with two diametrically opposite radial point forces [14]. Figure 1 shows the dependences of the dimensionless shell deflection $-wDl/PR^3$ on the order *N* of the system of equations being solved for the point of application of force *P* directed to the shell axis in the middle of its length. The following notation is used: *w* is the deflection; *D, l, R, h* are the flexural rigidity, length, radius, and thickness of the shell respectively. The figures show the solutions obtained using the common finite elements with different approximations [12].

The analysis of dependencies given in figure 1 confirms the reliability and accuracy of the results obtained using the considered finite element, and the effectiveness of this model. Efficiency is evidenced by the high accuracy of the results, which is achieved with a small number of finite elements.

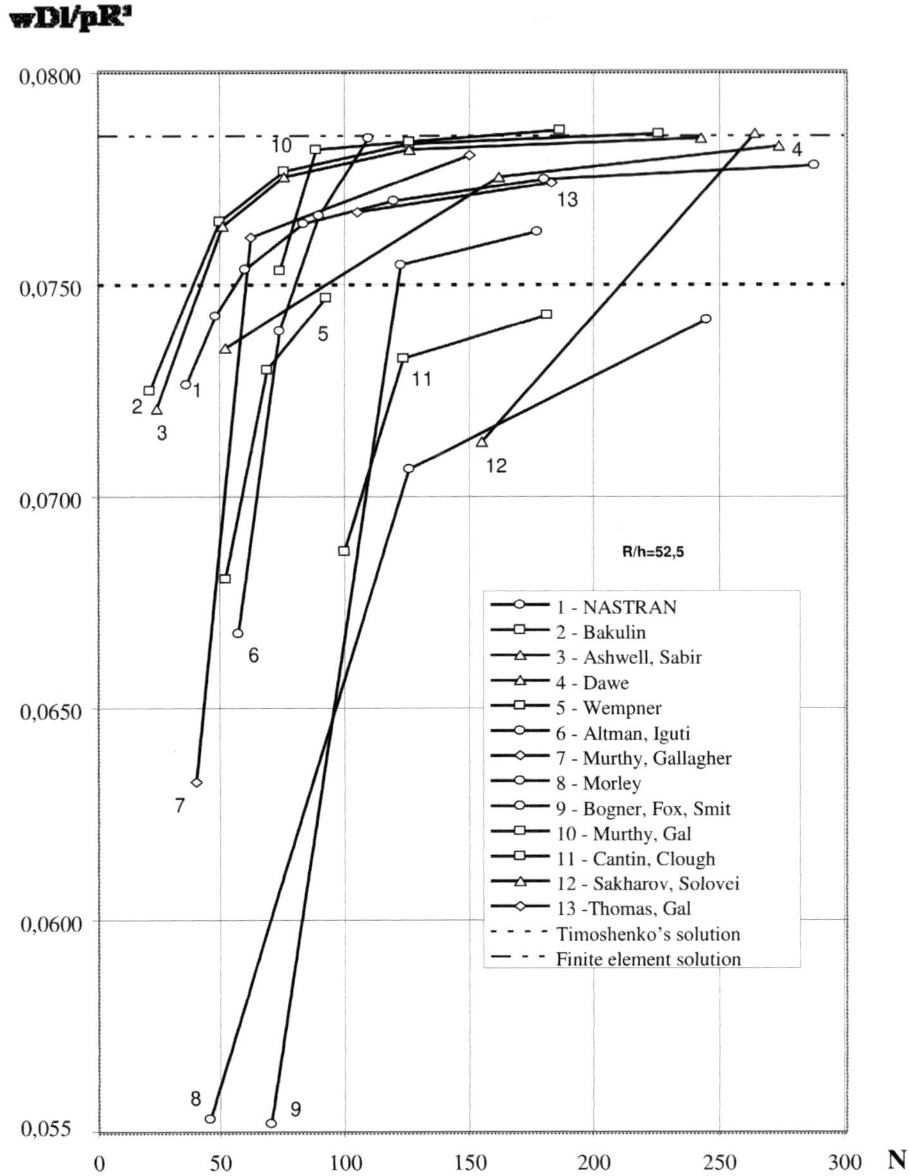

Figure 1. Graphs of convergence of numerical solutions obtained with the most common finite elements with different approximations

4. Results and discussion

The study of the stress-strain state of the hinged fiberglass shell (Figure 2) with the following parameters is carried out:

$R=0.18$ m, $l=2H=0.9$ m, $h=0.003$ m, $E_1=1.8\cdot10^4$ MPa, $E_2=3.25\cdot10^4$ MPa, $G=0.4\cdot10^4$ MPa, $\mu_1=0.12$, where E_1, E_2, G, μ_1 - moduli of elasticity and Poisson's ratio.

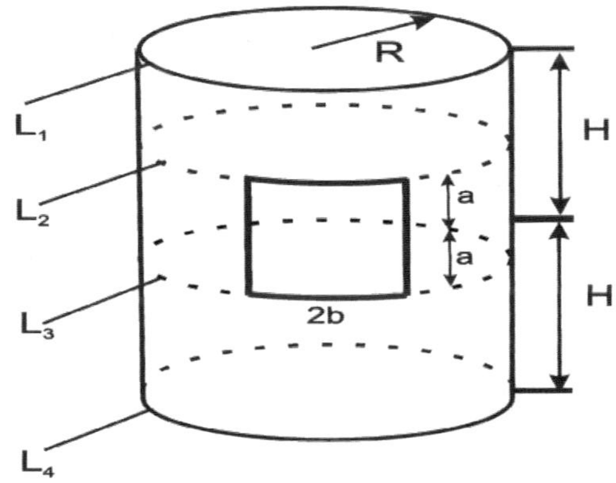

Figure 2. Cylindrical shell with rectangular cutout

Rectangular cutouts have dimensions of 2b = 0.08m, 2a = 0.1m. The longer side of the cutout is parallel to the shell axis.

The obtained solutions for distribution of the stress concentration coefficient $K\sigma = \sigma/\sigma_0$,($\sigma_0$ is the stress in the area away from the cutout (diagrams 2,3 figure 3) is compared with experimental results (diagrams 1) (along straight (a) and curved (b) edges of the cutouts) along the contour of rectangular cutouts).

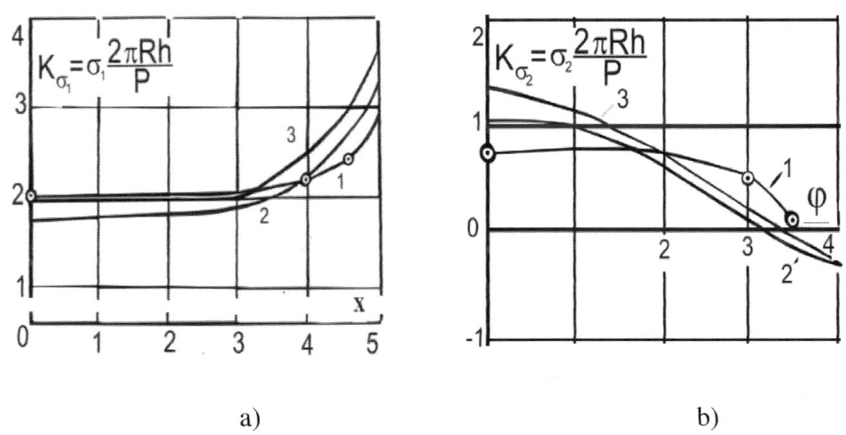

Figure 3. Distribution of stress concentration coefficient $K\sigma$ along the contour of rectangular cutouts (1 - experiment; 2,3 - calculation) for hinged fiberglass shells under axial compressive force loading

Two cases of axial force loading were considered: the load is evenly distributed at the ends of the shell (line 2), and the load is applied through the rigid plate which models of the equality of the axial displacements at the shell ends (line 3). There is a satisfactory agreement between the computational and experimental results, especially for the second loading case, corresponding to the compression of the shell by the base plates in the experiment.

The discrepancy between the calculation results and experimental data, in our opinion, is caused by the inaccuracy of determining the physical and mechanical characteristics of the composite material of the shell and the insufficient number of sensors (points with experimental data).

Acknowledgments
The work has been performed under the state order. Official registration number: AAAA - A19-119012290177-0.

References

[1] Bakulin V N 2019 Layer-by-Layer Analysis of the Stress-Strain State of Three-Layer Shells with Cutouts *Mechanics of Solids* **54** **3** 448–460 DOI 10.3103/S 0025654419050054
[2] Revenko V P 2016 Analytical and Numerical Method of Finite Bodies for Calculation of Cylindrical Orthotropic Shell with Rectangular Hole *Russian Mathematics (Iz. VUZ)* **60** 6 1-11 DOI: 10.3103/S1066369X16060013
[3] Pelekh B L, Syaskiy A A 1975 *Stress distribution near the holes in anisotropic shear resistant casings*. (Kiev: Naukova Dumka), p 198
[4] Bakulin V N 2019 Block Finite-Element Model of Layer-by-Layer Analysis of the Stress-Strain State of Three-Layer Generally Irregular Shells of Double-Curvature Revolution *Doklady Physics* **64** 1 9-13 DOI: 10.1134/S1028335819010026
[5] Revenko V P 2019 Method of Finite Bodies for Mathematical Modeling of the Stress-strain State of Cylindrical Orthotropic Shell with the Reinforced Rectangular Hole *Journal of Physics: Conference Series* **1392**
[6] Lee S E, Sahin S, Rigo P, Park M, & Paik J K 2017 Ultimate strength of cylindrical shells with cutouts *Ships and Offshore Structures*, **12** (sup1) 153-173
[7] Bakulin V N 2020 Investigation of the Influence of the Cutout Dimensions on the Stress-strain State of Three-layer Shells with Load-bearing Layers of Composite Materials *IOP Conf. Ser.: Mater. Sci. Eng.* **714** 012002
[8] Sun F, Wang P, Li W, Fan H and Fang D 2017 Effects of circular cutouts on mechanical behaviors of carbon fiber reinforced lattice-core sandwich cylinder. Composites *Part A: Applied Science and Manufacturing* **100** 313-23
[9] Hatami H and Shariati M 2019 Numerical and Experimental Investigation of SS304L Cylindrical Shell with Cutout Under Uniaxial Cyclic Loading. *Iran J Sci Technol Trans Mech Eng* **43** 139–153
[10] Bakulin V N 2019 A Model for Refined Calculation of the Stress-Strain State of Sandwich Conical Irregular Shells *Mechanics of Solids*, **54** 5 786-796
[11] Kaledin V O, Aulchenko S M, Mitkevich A B, et. al. 2014 *Modeling Statics and Dynamics of Shell Structures Made of Composite Materials* (Fizmatlit, Moscow) p 196
[12] Golovanov A I, Tyuleneva O N, and Shigabutdinov A F 2006 *The Finite Element Method in Statics and Dynamics of Thin-Walled Structures* (Fizmatlit, Moscow) p 392
[13] Ashwell D G, Sabir A B. 1972 A new cylindrical shell finite element, based on simple independent strain functions *International Journal of Mechanical Sciences*. **14** 3 171-183
[14] Timoshenko S and Woinovsky-Krieger S 1969 *Theory of Plates and Shells* Second Edition (New York: McGraw-Hill Book Company INC) p 636

Analytical Review of Account Methods and Experimental Approaches to Stress-Strain State Investigation of Structurally-Anisotropic Aircraft Panels Made from Composite Materials

L M Gavva[1] and V V Firsanov[2]
[1]«Aircraft Design» Department, «Machine components» Department at Moscow Aviation Institute (NRU), Moscow, Russia
[2]«Machine components» Department at Moscow Aviation Institute (NRU), Moscow, Russia

E-mail: rva101@mail.ru, nio1asp@mail.ru

Abstract The different approaches were analyzed to investigate the stress-strain state problems of structurally-anisotropic panels made from composite materials. The classification of mathematical models, analytical methods of calculations, numerical methods of calculations, testing results are presented in this review. New mathematical model relations for the stress-strain state investigation of structurally-anisotropic panels made of composite materials are presented. The primary scientific novelty is the further development of the theory of thin-walled elastic ribs related to the contact problem for the skin and the rib with an improved rib model. A promising area of research is the design of composite aircraft structures in the field of production technologies. We consider the residual thermal stresses and the preliminary tension of the reinforcing fibers with respect to the panel production technology. The stress-strain state problem results in the boundary value problem when solving for the eighth order partial derivative equation in the rectangular field. A computer program package is developed in MATLAB. The computer program package has been utilized for multi-criteria optimization of the design of structurally-anisotropic aircraft composite panels. The results of testing series are presented as the verification of the proposed mathematical model.

1. Introduction

A new design problem – the design to cost – is possible to be solved combining the high-precision models and modern computer technologies and decreasing the test amount. A promising area of research is the design of composite aircraft structures in the field of production technologies.

The different approaches were analyzed to investigate the stress-strain state (SSS) of structurally-anisotropic panels made from composite materials. The authors considered the studies of scientific schools from 2000 year to present time, mainly. The classification of mathematical models, analytical methods of calculations, numerical methods of calculations, testing results are presented in this review.

The static problems of a flat rectangular multilayer panel made from polymer fiber composite materials with the eccentric longitudinal and lateral stiffening set are considered. The SSS of a flat rectangular composite panel being anisotropic due to non-symmetric package structure over the thickness is also discussed. The panels are subjected to the normal loading per panel square in the stationary temperature field. The boundary conditions at the contour are assumed to be of general type.

One should take into consideration the technological factors occurring in the fabrication of composites, namely, residual thermal stresses arising during cooling after hardening and pre-stressed tension of reinforcing fibers that is performed in order to increase the bearing strength of the structure [1] Gavva L M, Endogur A I, 2018.

The refined statement of the static problems has been formulated for the flat rectangular multiplied panels made of polymer fiber composite materials, the casing of which is eccentrically supported by the longitudinal-lateral stiffening set.

2. Survey manuscripts

The stress-strain state (SSS) problem investigation of flat rectangular structurally-anisotropic composite panels is relevant for the design of the bearing surfaces of FA.

The following papers (Table 1) are presented as survey manuscripts:

Table 1. Classification of survey manuscripts: (a) - (k)

a) [2] Obraztsov I F, Sirotkin O S, Litvinov V B, 2000

Panel	Mathematical model	Solution methods
-Integral CM structures -CM multilayered panels -Stiffened CM panels	Contribution of MAI (NRU) science school to the creation of science basis for the design mechanics and production technology of composite panels	Analytical methods Numerical methods (FEM) Experimental investigations

b) [3] Dudchenko A A., 2013
[4] Ugrimov S V, 2002
[5] Zenkour A M, Fares M E, 2001

Panel	Mathematical model	Solution methods
-CM multilayered panels -Stiffened CM panels	Approach analysis to the SSS account	Analytical methods

c) [6] Akhmedov A B, 2011

Panel	Mathematical model	Solution methods
-CM multilayered plates	Analysis of the theory current state Classical theory Transverse shear first order theory Transverse shear third order theory – Reddy theory	Comparison of the results

d) [7] Fomin V P, 2004

Panel	Mathematical model	Solution methods
-CM multilayered plates	Classical theory	Comparison of the results

	Various transverse shear third order theories (Reddy, Merfy, Levinson, Ambartsumyan) Theory difference estimation	

e) [8] Nemirovskiy Yu V, Gorynin G L, 2011
[9] Golushko S K, Nemirovskiy Yu V, 2008
[10] Nemirovskiy Yu V, Yankovskiy A P, 2004
[11] Andreev A N, Nemirovskiy Yu V, 2001

Panel	Mathematical model	Solution methods
-CM multilayered panels	3-D SSS approach Non-classical mathematical model Transverse shear deformations Normal line compression	Comparison of the results

f) [12] Carrera E, 2001
[13] Carrera E, Demasi L, 2003
[14] Carrera E, 2003

Panel	Mathematical model	Solution methods
-CM multilayered panels	2-D SSS approach 2-D high order theories in view of transverse shear deformations 3-D solutions	Analytical methods Comparison of the results

g) [15] Motygin O V, Nazarov S A, 2000
[10] Nemirovskiy Yu V, Yankovskiy A P, 2004

Panel	Mathematical model	Solution methods
-CM multilayered plates	2-D models The justification for the use of the Kirchhoff hypothesis Inaccuracy estimation	Comparison of the results

h) [16] Sokolkin Yu V, Makarova E Yu, 2002

Panel	Mathematical model	Solution methods
-Stiffened CM panels	Current tendency analysis of averaging methods for mechanics of composites	Comparison of the results

k) [17] Grigolyuk A I, Kulikov G M, 2005

Panel	Mathematical model	Solution methods

-CM multilayered panels	Deformation models with the continuity conditions for the displacements and transverse stresses at the layer interface	Comparison of the results

3. Main directions of stress-strain state theory development

Most authors of the theoretical studies during the latest twenty years have been paying attention to the features of the composite panel deformation behavior. The classification and survey of the main directions of the stress-strain state (SSS) theory development of the structurally-anisotropic composite panels, the proposed mathematical models and equations are of the most interest.

The SSS Static Problem Statement, SSS Problem Statement with Thermal and Force Loading, SSS Problem Statement subject to Production Technology, Analytical Methods to solve SSS Problems, Numerical Methods to solve SSS Problems and Test Investigations are considered.

The exact 3-D solutions of the elasticity theory equations are presented as the mathematical modeling. The refined SSS mathematical models with the various degrees of approximation in view of the transverse shear deformations are proposed. The main equations for the multiplied orthotropic plates and stiffened composite plates are formulated basing the high order transverse shear deformation Reddy theory. The asymptotic analysis of the refined "layer models" for the plates with non-symmetric package structure are performed. The normal line compression is additionally taken into consideration for the applied theory based on the classical Kirchhoff hypotheses theory. The classical Kirchhoff theory is used; the accuracy estimation of the Kirchhoff hypotheses for the multilayer plate theory is realized with the help of asymptotic method. The boundary value problem statement for thermo-elasticity theory resolving equations is considered.

The thermo-elasticity theory for the multiplied anisotropic composite plates is designed using the general 3-D thermo-elasticity theory equations with the asymptotic small parameter decomposition and proposed as the mathematical model, while the thermal and mechanical loading. There are no assumptions relative to the displacement and stress distribution per panel thickness here. The estimation of the high order theories for the multilayer composites upon the thermal field gradient is performed. The approach to solve the thermal bending problems for the multiplied composite thermo-elastic plates based on the "layer modeling" is given.

The mathematical modeling in view of the production technology process includes the methodology development for the evaluation of the out-of plane displacements and residual thermal stresses connected with the plane composite panel curvature in cooling after hardening. The residual displacements and thermal stresses essentially change the stress distribution after mechanical static loading. The classic mathematical model to investigate the residual thermal stresses of the composite package in cooling after hardening exists, but the boundary value problem solution is not obtained.

The analytical methods to solve SSS static problems involve the generalized multilayer plate theory and solutions based on the expansion of each layer displacement vector in the power series. The total solution of the mix problem state for the anisotropic layered plates within the coupled bending-extension problem is designed as the analytical complex value functions. The double trigonometric series analytical solution is presented in closed form; the single trigonometric series solution is developed for the bending problem. The Lurie-Vorovich method is proposed for the SSS investigation of the 3-D composite plates. The effective version of the uniform solution method is developed, e.g. the exact boundary value problem solutions for the stiffened panels are designed as the uniform solution Fadl-Papkovich functions.

The numerical methods to solve SSS static problems are the numerical procedure combining the exact 3-D equations of the continuum mechanics and semi-analytical finite element method (FEM).

The effective method for the numerical solution of the non-classical theory boundary value problems is proposed and realized using the invariant immersion ideas. The FE model with the uninterrupted transverse shear stresses is developed. The high order shear element is designed, the coupled bending-tension element is proposed, the FEM version with the general element combining the tension-compression, bending and first order transverse shear deformations is presented. The analogue FE is developed for the thermo-elasticity problems. The different FE model variants are compared and discussed. The modified FE with the stepwise-changing displacements and free from the Kirchhoff hypotheses is used. The FE modeling for the polymer composite material panels is performed. The FEM is used to the SSS coupled problems of the composite stiffened panel under the thermal loading.

The test investigations analyze the composite box structure deformations under the bending and torsion loads. The tests of the wing box made from polymer composite materials are the part of the experimental works to ensure the strength of the MS-21 aircraft bearing surface structures. The experimental investigation of the anisotropic composite plate thermal behavior is carried out. The test estimation of the stiffened composite panel displacements under the mechanical and thermal loads is performed.

The aim of this study is the stress-strain state problem statement and the approach to solve this problem in view of the production technology. The SSS problem results in the boundary value problem when solving for the eighth order partial derivative equation in the rectangular field**.** The SSS problem statement and the proposed approach to solve this problem are of the interest from the point of view of designing and manufacturing an aircraft specimen made from modern composite materials.

4. Refined problem statement of Stress-Strain State of structurally-anisotropic composite panels

New mathematical model relations for the stress-strain state investigation of structurally-anisotropic panels comprising composite materials are presented. The mathematical model of a stiffening rib being torsioned under one-sided contact with the skin is refined. The scientific novelty of this research reflects the further development of the theory of thin-walled elastic ribs related to the contact problem for the skin and rib with an improved rib torsion model. The SSS problem is integral, it cannot be divided into a plane part and a plate bending**.** The schematization of the panel as structurally-anisotropic has been proposed as a design model when the stress-strain state is determined.

The eighth order differential equation is resolved for the SSS [18] Boytsov B V, Gavva L M, Endogur A I and Firsanov V V, 2018, [19] Gavva L M, 2017

$$\frac{K_{80}}{a^8}\frac{\partial^8\Phi}{\partial x^8}+\frac{K_{71}}{a^7 b}\frac{\partial^8\Phi}{\partial x^7\partial y}+\frac{K_{62}}{a^6 b^2}\frac{\partial^8\Phi}{\partial x^6\partial y^2}+\frac{K_{53}}{a^5 b^3}\frac{\partial^8\Phi}{\partial x^5\partial y^3}+\frac{K_{44}}{a^4 b^4}\frac{\partial^8\Phi}{\partial x^4\partial y^4}+$$
$$+\frac{K_{35}}{a^3 b^5}\frac{\partial^8\Phi}{\partial x^3\partial y^5}+\frac{K_{26}}{a^2 b^6}\frac{\partial^8\Phi}{\partial x^2\partial y^6}+\frac{K_{17}}{a b^7}\frac{\partial^8\Phi}{\partial x\partial y^7}+\frac{K_{08}}{b^8}\frac{\partial^8\Phi}{\partial y^8}=q \qquad (1)$$

The deflection function $w(x, y)$ is coupled with the potential function $\Phi(x, y)$ as

$$w=\left(R_{40}\frac{\partial^4}{\partial x^4}+R_{31}\frac{\partial^4}{\partial x^3\partial y}+R_{22}\frac{\partial^4}{\partial x^2\partial y^2}+R_{13}\frac{\partial^4}{\partial x\partial y^3}+R_{04}\frac{\partial^4}{\partial y^4}\right)\Phi \qquad (2)$$

The coefficients R_{ij} , $_{i=4,3,...,0}$, $_{j=0,1,...,4}$ in the relation formulas (2) and the coefficients K_{ij} , $_{i=8,7,...,0}$, $_{j=0,1,...,8}$ in the resolving equation (1) are the constant values, which depend on the elastic characteristics of the material and geometrical structure parameters.

For the composite panel of orthotropic structure the left-hand part of the eighth order differential equation is limited by the even-numbered partial derivatives of $\Phi(x, y)$

$$L^{(8)}\Phi = q. \tag{3}$$

$$L^{(8)} = \frac{K_{80}}{a^8}\frac{\partial^8}{\partial x^8} + \frac{K_{62}}{a^6 b^2}\frac{\partial^8}{\partial x^6 \partial y^2} + \frac{K_{44}}{a^4 b^4}\frac{\partial^8}{\partial x^4 \partial y^4} + \frac{K_{26}}{a^2 b^6}\frac{\partial^8}{\partial x^2 \partial y^6} + \frac{K_{08}}{b^8}\frac{\partial^8}{\partial y^8}.$$

$x = x/a$, $y = y/b$ are the dimensionless coordinates per the panel length a, its width b.

All components of the stress-strain state including the inner force factors are related with the potential function $\Phi(x, y)$ as

$$\left.\begin{aligned} N_x &= L_{Nx}\Phi - N_x^T - N_x^H, \quad N_y = L_{Ny}\Phi - N_y^T - N_y^H \\ N_{xy} &= L_{Nxy}\Phi - N_{xy}^T - N_{xy}^H, \quad N_{yx} = L_{Nyx}\Phi - N_{yx}^T - N_{yx}^H \end{aligned}\right\} \tag{4}$$

The linear differential operator for the orthotropic structure, for example

$$L_{Nx} = P_{60}^x \frac{\partial^6}{\partial x^6} + P_{42}^x \frac{\partial^6}{\partial x^4 \partial y^2} + P_{24}^x \frac{\partial^6}{\partial x^2 \partial y^4} + P_{06}^x \frac{\partial^6}{\partial y^6},$$

The thermal forces and moments are N_x^T, N_y^T, N_{xy}^T, N_{yx}^T, the tension forces and moments of the composite fibers are N_x^H, N_y^H, N_{xy}^H, N_{yx}^H. The coefficients P_{ij}^X, $i = 6,4,2,0$, $j = 0,2,4,6$ as the coefficients in the relation formulas (2), depend on the elastic characteristics of the material and geometrical structure parameters. The linear differential operators of the relations between the SSS kinematic and static components and the resolving potential function $\Phi(x, y)$ are known.

The general boundary value problem for the resolving equation (3) is proposed as elastic fixity constrains as

$$x = \text{const} \quad (\gamma_1 u_0 + \delta_1 N_x) = (\gamma_2 v_0 + \delta_2 N_{xy}) = (\gamma_3 w + \delta_3 \overline{Q}_x) = (\gamma_4 w'_x + \delta_4 M_x) = 0,$$

$$y = \text{const} \quad (\alpha_1 u_0 + \beta_1 N_{yx}) = (\alpha_2 v_0 + \beta_2 N_y) = (\alpha_3 w + \beta_3 \overline{Q}_y) = (\alpha_4 w'_y + \beta_4 M_y) = 0 \tag{5}$$

The coefficients γ_i, δ_i, α_i, β_i, $i = 1, 2, 3, 4$ are from 0 to1, $u_0(x, y)$ and $v_0(x, y)$ are the displacements u and v while $z = 0$.

The integral of the equation (3) that satisfies the edge conditions (5) is proposed by the single trigonometric series and uniform solution method.

One can estimate the influence of the production technology factors on the bearing strength of structurally-anisotropic composite panels if the boundary value problem is considered and boundary conditions are non-conformable as (5), the solution is designed by a single trigonometric series or uniform solution method.

5. Stress-strain state of structurally-anisotropic composite panels. Numerical results

A computer software package was developed in the MATLAB operating environment according to the algorithm presented. Computer programs are intended for the stress-strain state analysis and computer multi-criteria optimization of the design for structurally-anisotropic aircraft composite panels.

As an example, Figure 1 presents the epure of the normal stresses of a flat rectangular panel made from carbon plastic eccentrically stiffened in the longitudinal direction. The panel is under the action of uniform normal loading per unit square.

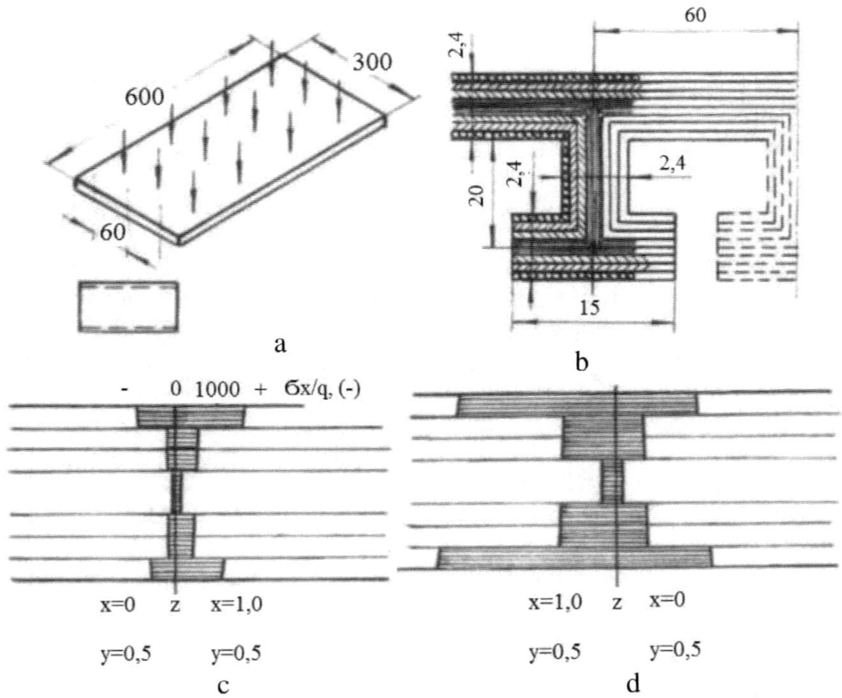

Figure 1. Stringer panel under the normal load: (a) panel sizes; (b) layout of the package and stringer geometry; (c) the normal stresses in the skin layers; (d) the normal stresses in the stringer flange layers

6. Verification of the mathematical model. Experimental investigations of structurally-anisotropic composite panel buckling

A computer program package is developed using the MATLAB operating environment. The computer program package has been utilized for the determination of the critical forces subject to the uniform and non-uniform pre-buckling stress state and for multi-criteria optimization of the design of structurally-anisotropic aircraft composite panels.

The testing series of uniform compressed stiffened composite panels for carrying the objects to the moment of stability loss have been made for the proposed mathematical model verification [1] Gavva L M, Endogur A I, 2018, [20] Gavva L M, Firsanov V V and Korochkov A N, 2020.

The refined theoretical results and the experimental data are in agreement qualitatively with respect to buckling modes and quantitatively with respect to critical stresses within 12 - 13% if pre-buckling stress state is considered uniform. And they reveal a higher precision of 8 - 10% in view of non-uniform main pre-buckling stress state. Thus, it confirms the authenticity of the presented mathematical model.

7. Conclusion

The presented classification and survey of the main directions of the stress-strain state theory development of the structurally-anisotropic composite panels demonstrate the need and scope for the refined theory paying the attention to the features of the composite panel deformation behavior. The results of the stress analysis calculations in the refined statement, as well as the results of the buckling analysis calculations offer opportunities for reducing and optimizing the weight characteristics of aircraft elements. Since the solution is obtained by analytical methods, the calculation time is minimal. This is of interest from the point of view of practical design using parametric analysis.

Acknowledgements
The study was performed in the framework of the Ministry of Science and Higher Education (Russia), *the project № FSFF-2020-0013.*

References
[1] Gavva L M and Endogur A I 2018 Static and buckling problems of aircraft structurally-anisotropic composite panels with the influence of production technology *IOP Conference Series: Materials Science and Engineering* **312(1)** 012009
[2] Obraztsov I F, Sirotkin O S and Litvinov V B 2000 Integral structures from composite materials and perspectives of application *J. Structures from Composite Materials* **2** 78-84 (in Russian)
[3] Dudchenko A A 2013 Stress-strain state account of aircraft and engineering structure elements from composite materials *Moscow Fed. Inf. System* p 232 (in Russian)
[4] Ugrimov S V 2002 Generalized theory of multilayer plates *Int. J. Solids and Struct.* **4 (39)** 819-839
[5] Zenkour A M and Fares M E 2001 Bending, buckling and free vibrations of non-homogeneous composite laminated cylindrical shells using a refined first-order theory Compos. A. **3 (32)** 237-247
[6] Akhmedov A B 2011 Quality algorithm of theories for plates from composite materials *J. Probl. Mech.* 81-84 (in Russian)
[7] Fomin V P 2004 Transverse bending of multilayer plates *J. CAGI Proceed.* **2664** 122-135 (in Russian)
[8] Nemirovskiy Yu V and Gorynin G L 2011 Method of stiffness functions for account problems of multilayer ribs and plates *J. Vestn. of Nigegorod Univ. by N I Lobachevskiy* 1654-1656 (in Russian)
[9] Golushko S K and Nemirovskiy Yu V 2008 Direct and inverse problems of elastic composite plate and shell mechanics *Moscow Fizmatlit* p 432 (in Russian)
[10] Nemirovskiy Yu V and Yankovskiy A P 2004 About limits of applicability of some account theories for bended reinforced plates *J. Sci. vestn. NGTU* **3** 91-113 (in Russian)
[11] Andreev A N and Nemirovskiy Yu V 2001 Multilayered anisotropic shells and plates *Novosibirsk Science* p 288 (in Russian)
[12] Carrera E 2001 Developments, ideas and evaluations based upon Reissner mixed variation theorems in the modeling of multilayered plates and shells *J. Appl. Mech. Rev.* **4 (54)** 301-329
[13] Carrera E and Demasi L 2003 Two benchmarks to assess two-dimensional theories of sandwich composite plates *AIAA J.* **7 (41)** 1356-1362
[14] Carrera E 2003 Historical review of Zig-Zag theories for multilayered plates and shells *J. Appl. Math. Rev.* **3 (56)** 287-308
[15] Motygin O V and Nazapov S A 2000 Justification of the Kirchhoff hypotheses and error estimation for two-dimensional models of anisotropic and inhomogeneous plates, including laminated plates *IMA J. Appl.* **1 (65)** 1-28
[16] Sokolkin Yu V and Makarova E Yu 2002 Stochastic models of composite mechanics for design problems of aircraft structure elements *J. Compos. Mater. Mech.* **5 (38)** 591-606 (in Russian)
[17] Grigolyuk A I and Kulikov G M 2005 Ways of development of multilayered plate and shell theory *J. Vestn. TGTU* **2 (11)** 439-448 (in Russian)
[18] Boytsov B V, Gavva L M, Endogur A I and Firsanov V V 2018 Stress-strain state and buckling problems of structurally-anisotropic aircraft panels made of composite materials in view of production technology *J. Russian Aeronautics* **61 (4)** 524-532

[19] Gavva L M 2017 Parametric analysis of stress-strain state of structurally-anisotropic aircraft composite panels in view of production technology using MATLAB operating environment *J. MAI Proceed.* **93** 27 (in Russian)
[20] Gavva L M, Firsanov V V and Korochkov A N 2020 Buckling problem statement and approaches to buckling problem investigation of structurally-anisotropic aircraft panels made from composite materials. *IOP Conference Series: Materials Science and Engineering* **714(1)** 012007

The strength tests of thin-walled composite shells under non-stationary loads of different physical nature

A V Ostrik[1,*], **V N Bakulin**[2] **and D N Nikolaev**[1]
[1]Institute of Problems of Chemical Physics of the Russian Academy of Sciences, Academician Semenov Avenue 1, Chernogolovka, Moscow Region 142432, Russia
[2]Department of Aircraft Design and Certification, Moscow Aviation Institute (National Research University), 4 Volokolamskoe shosse, Moscow 125993, Russia

*ostrik@ficp.ac.ru

Abstract. Mechanical action of energy fluxes of different physical nature is proposed to be simulated by one-side non-stationary loading. A set of gas-dynamic devices for generation of loads of the required space-time distribution is described. Two new explosive devices for formation of low-impulse loads of microsecond duration are proposed. The methods to measure the parameters of thin-walled composite constructions under dynamic and pulse loads are considered. Time deformation profiles of fiberglass cylindrical shell under low-impulse load of microsecond duration are given. The use of wire or foil sensors for the measurement of deformations gives close results when investigating the shell stage of deformation of fiberglass thin-walled constructions.

1. Introduction

A wide application of composite materials in load-carrying thin-walled shell structures of rocket and space equipment (RSE) leads to considerable interest in studying of regularities of its deformation and destruction [1, 2], including the case of non-stationary loads [3, 4]. Awareness of the need to keep the operability of these constructions under intensive pulse and dynamic loads of different physical nature [5, 6] required the development of gas-dynamic devices [3, 4, 7], which allow to perform a tests of full-scale elements of RSE constructions with reproduction of flight conditions [4, 8, 9].

The purpose of this work is to present the results of studying the impact of one-side non-stationary loading on composite shells of RSE. These results are a part of the data obtained within the framework of the common methodology [4, 9] for testing the strength of composite shells to the mechanical action of radiation (MAR). The information presented in the article can serve as an illustration of the variety and complexity of problems that occur during testing the strength of the elements of RSE constructions to non-traditional types of actions having high energy density.

2. Types of non-stationary loads and reasons of constructions destruction

Thin-walled composite RSE constructions may be affected by the action of flight loads [10, 11] and external loads of different physical nature [6]. Flight loads are generally axisymmetric and quasi-stationary. Methods of strength calculations for loads of this type are well developed and implemented in the practice of design of RSE constructions. However, external loads are one-sided and non-stationary in the most cases [3-5]. The results of the strength calculations in this case are not so reliable and strength tests are required.

Non-stationary actions are conventionally divided into pulse and dynamic loads depending on the time duration of their actions and the reasons of constructions damages. Dynamic loading conditions are realized in case of action durations comparable to period of free oscillations of thin-walled construction and at least in 10... 15 times longer time of disturbance propagation through the shell thickness (this condition ensures the prevalence of deformation shell stage). Destruction of composite constructions in case of dynamic deformation occurs due to development of unacceptable deflections and formation of delaminations and cracks.

In case of pulse loading, when duration of action not only does not exceed a quarter of period of oscillations but is comparable or less than time of disturbance propagation through thickness of thin-walled construction element, the main cause of destruction is development of wave processes accompanied by formation of delamination (it is typical for composite material) and spalls (it is typical for metals). An important factor facilitating the investigation of the effects of pulse loading is the possibility to select the fragment of construction with results of testing close to that of testing the whole structure. An obvious necessary condition for the correctness of such selection is the low propagation time of the compression wave through the thickness of the fragment compared to the arrival times of the unloading waves from the edges of the fragment. This condition is not difficult to realize for full-scale thin-walled construction.

However, the deformation process is completed by the shell stage also in the case of pulse loading of thin-walled structures when the stress waves are damped. The attenuation results in an increase in the spatial size of the wave impulse. The shell stage occurs when the size of the wave becomes comparable with the thickness of the construction. Moreover, the use of porous sheetings and heterogeneous protective packs significantly reduces the role of wave processes in the destruction of the material and practically does not protect against the formation of cracks and the development of unacceptable displacements. Therefore, the final strength tests of the composite construction as a whole to the dynamic loads causing the shell stage of deformation are also necessary when only pulse action is expected during RSE flight.

3. Gas-dynamic devices for reproducing the mechanical action of radiation

Experimental studies of the effects of MAR on the elements of the RSE constructions by direct irradiation are not possible due to the absence of powerful laboratory radiation sources capable of generating the necessary energy densities on surfaces with dimensions of several meters [2, 4, 10, 11]. In most cases, it is not possible to obtain sufficiently reliable results by methods of similarity [12], because the requirement for matching of criteria parameters of full-scale construction and its model is reduced to the identity in dimensions and material properties. For example, if the geometrically similar model of solid propellant rocket engine have the fuel charge with characteristic dimensions less than the critical detonation diameter of propellant, than it is not possible to reproduce the detonation caused by the mechanical impulse of radiation, although it may occur for a real engine. Similar difficulties arise when trying to simulate the non-stationary destructions of thin-walled composite constructions made by the method of filament winding. The condition of equality of relative thickness of h/R (h and R are thickness and characteristic radius of shell curvature) for a full-scale construction and its model keeping invariable the thickness of reinforcing filament leads to reduction of the number of reinforcing layers in model that distorts the conditions and sequence of destruction of these layers. Therefore, the main method of studying the MAR effects is a test of the full-scale RSE constructions at non-stationary loading by gas-dynamic devices for reproduction of mechanical action of pulsing radiation [4, 7, 8]. The capabilities [4] of these devices are shown in table 1 (τ_p is pressure impulse duration and I_p is pressure impulse). As can be seen from this table, the presented set of modeling devices do not provide generation of low-impulse loads of microsecond duration with high simultaneous action on the RSE surface.

IOP Conf. Series: Materials Science and Engineering **927** (2020) 012068 doi:10.1088/1757-899X/927/1/012068

Table 1. Gas-dynamic devices for reproduction of MAR.

Type of explosive device	Loading parameters		Purposes (the modeled actions)
	τ_p, μsI_p,	kPa×s	
the charge located in contact with the surface by separate sectors	1...10	0.8…5	the monopulse action of UV or X-ray radiations
the charge initiated by light and placed in contact with the surface	0.2...1	0.05...2	the monopulse action of X-ray radiation
the charge equidistantly placed above the surface	30...200	0.3...3	the monopulse action of UV radiation or visible light in air
the charge placed with the specified volume distribution above the surface	100...500	0.1...2	the monopulse action of IR radiation or visible light in air
the cumulative charge placed with the specified volume distribution above the surface	100...500	0.1...2	the monopulse cumulative action of IR radiation or visible light in air
the multi-channel blast tube	50…200	0.5...2	the pulse-frequency resonant action of IR radiation or visible light in air
the multi-channel blast tube with nozzles	50…200	0.5...2	the pulse-frequency resonant action of IR radiation or visible light in air having a complex space-time profile
the system of the rotating charges	100…200	0.5...2	the multiple action of radiation on high-temperature flows for initiation of engine surge

However, such non-stationary loads are required for the strength tests of thin-walled constructions. Therefore, the tape charge [13] and the controlled initiation charge [14] have been developed for generating short-time loadings having $\tau_p < 10$ μs and $I_p < 0.5$ kPa×s.

Figure 1. Tape explosive charge.

Figure 2. Controlled initiation explosive charge.

IOP Conf. Series: Materials Science and Engineering **927** (2020) 012068 doi:10.1088/1757-899X/927/1/012068

The tape charge is made of explosive tapes which are placed on circular tubes (figure 1). Tubes are equidistant from surface of loaded object. This device generates loadings having τ_p =5...10 μs and I_p = 0.1…1 kPa×s. Multipoint initiation system of explosive tape ends is used to ensure simultaneous loading of surface.

The charge with the controlled initiation is produced in the form of thin (≤3mm) shell from fibrous material. The shell has milled channels filled with plastic explosive (figure 2). Placement of HE in channels offers a number of advantages over other methods of localization of explosive (e.g. by solid layer or sectors on a porous substrate). Controlled initiation is implemented by the method taken from the technology for generation of megabar level pressure [15, 16]. This device generates loads having τ_p = 3...5 μs, I_p = 0.05…1 kPa×s.

Note that the set of devices [4, 7], supplemented by these new developments, offers more capabilities. The use of this set is planned for testing of thin-walled RSE constructions for strength to thermal and mechanical actions of radiation and particles fluxes of different physical nature and impact of compact solids.

4. Measurements of non-stationary parameters of construction response

4.1. The wave stage of deformation.

In the study of wave processes in materials, kinematic (particle velocity and shock wave (SW) velocity) and force (pressure between layers of the tested package of materials) characteristics are usually chosen as measured parameters.

The electromagnetic method is used to measure the particle velocity of the material inside the sample [17]. The essence of the method is the generation of electromotive force in a conductor (sensor) moving in a magnetic field. The sensor is located in the sample perpendicular to its axis (loading direction). The sample is placed in the magnetic field during the tests. When the sample is loaded, the sensor is driven to a speed equal to the particle velocity of the surrounding substance. As a result, the working plane of the sensor crosses the magnetic field lines and the difference of potentials is generated that is proportional to the particle velocity of the material. The use of a pair of sensors makes it also possible to measure the SW velocity.

The profiles of particle velocity of the back surface of the sample are measured by capacitive sensors (in the case of metal samples) [18] and laser Doppler velocimeter [19]. The analysis of profiles of the particle velocity of the sample back surface allows to obtain its shock Hugoniot and spall strength.

One of the essential response parameters determining the strength of the multilayer sample at the wave stage is the pressure at the boundary of the different layers. Closed type resistive sensors are used for pressure measurements. For example, a N2T-TR-A06-00370 pressure sensor is suitable (manufacturer is VPG MicroMeasur). The initial resistance of the sensor is 37 Ohm (this value varies from 36 to 38 ohms). The material of the sensor is pure nickel. The relative change of resistance vs pressure is 5% per 1 GPa for this material. Estimations show that the error of pressure measurement by this gauge does not exceed 25%.

4.2. The shell stage of deformation.

The displacement, deformation, and accelerations (*g*-load) are typically measured during the study of deformation and destruction of a thin-walled constructions at shell stage of deformation.

The shell stage of deformation of thin-walled constructions under mechanical action of radiation and particles is characterized by small (up to hundreds of m/s) speeds and relatively large (up to several centimeters) displacements. The requirements for time resolution of the displacement measurement methodology are moderate. A disadvantage of traditional laser Doppler interferometry is the small depth of field of the optical system. The contrast of the interference picture is degraded at displacements greater than 1 cm and registration becomes impossible. Therefore, the method of measuring the non-stationary displacements is based on registration of the change in illumination of the moving surface is developed and used [20]. Total error of this method for displacement measurement does not exceed 15%.

Deformations are measured by electrotensometric technique with the balanced single bridge scheme. The wire strain sensor KB-10-200 (manufactured by CNIIMASH, Korolev, Russia) and the foil strain sensors BF120-3AA, BF350-3AA (manufactured by ZIMEC, China) are used. Sensors allow measuring of the relative non-stationary deformations up to 4% with error not exceeding 15% [4].

High-frequency vibration-resistant accelerometers with high conversion coefficient and maximum values of measured accelerations up to $10^4 g$ (for example, it is ADP-10-1 accelerometer having measurement error of not more than 20% [4]) are used to measure g-loading.

5. Fiberglass construction test results for shell stage

Strength tests of the fiberglass cylindrical shell ($h/R = 0.06$, $L/R = 3$) were done to compare the performance of various (production of CNIIMASH and production of ZIMEC) resistive strain sensors. The choice of fiberglass as a composite shell material is due to the fact that strain measurements on fiberglass give more reliable results than for many other composites. For example, organoplastic composite produced by means of filament winding [1, 2] are characterized by substantial non-uniformity of structure. In this case, the correct measurement of deformations becomes very problematic. One-sided and non-stationary load was generated by a controlled initiation explosive device. Angular dimension of the loaded cylindrical surface was 115$^{\rm O}$ and pressure impulse was $I_{\rm p} = 0.4$ kPa×s.

Shell was firmly connected with two steel bottoms by means of bolts and steel straps. The whole construction was mounted in a ballistic pendulum to measure the pressure impulse. The shell at the pendulum with the explosive device is shown in figure 3.

Figure 3. Shell with explosive device in pendulum

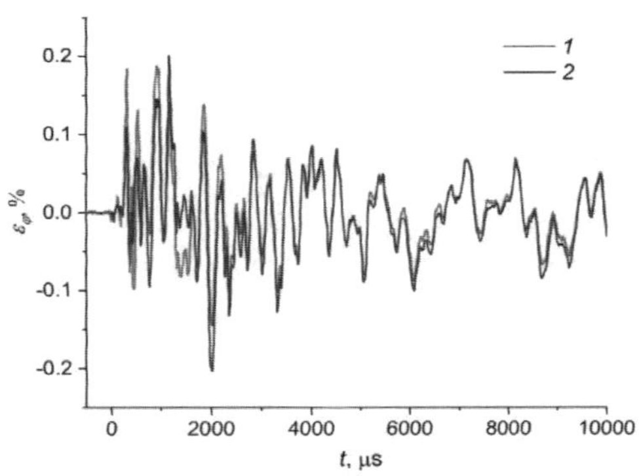

a)

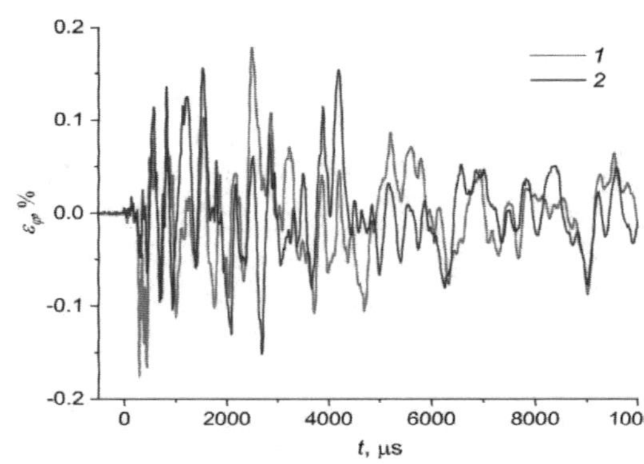

b)

Figure 4. Circular deformations ε_φ vs time t: a) comparison of circular deformations at $\varphi = 0^{\rm O}$ measured by various sensors (1st is by the KB-10-200 strain sensor, 2nd is by the BF350-3AA strain sensor); b) 1st is $\varphi = 90^{\rm O}$, 2nd is $\varphi = 180^{\rm O}$

Strain sensors were glued at an inner surface at points (φ is a circular coordinate from the center of a loaded surface): $\varphi = 0^{\rm O}$, $\varphi = 90^{\rm O}$ and $\varphi = 180^{\rm O}$. In a point $\varphi = 0$ the wire sensor KB-10-200 was paired by the foil sensor BF350-3AA. The results of comparison of circular deformations measured by different sensors are shown in figure 4a. It can be seen that wire and foil sensors in one experiment register deformations that differ by 5... 15 % relative to each other. Time dependences are also consistent with good accuracy, but the signal of sensor KB-10-200 is noisier.

Data of strain sensors in points φ=90, and φ=180 (wire sensors KB-10-200 were used at these points) are shown in figure 4b.

Note that the load parameters were insufficient to significantly destroy of the test shell (cracks, delaminations and spalls were not observed). Relatively small (up to 0.2 %) deformations did not result in delamination and separation of strain sensors.

6. Conclusion

This article considers the problem of strength testing of RSE under mechanical action of intense energy fluxes of different physical nature. It is proposed to simulate this action by means of a set of explosive devices. Time deformation profiles of fiberglass cylindrical shell under pulse loading are given. This data can be useful for testing by numerical methods to calculate the response of composite shells. When investigating the shell stage of deformation of fiberglass thin-walled constructions, the use of wire or foil strain gauges to measure deformations leads to close results. As a result, the authors developed a set of explosive gas-dynamic devices and methods for measuring response parameters to study the strength of thin-walled composite RSE constructions in a wide range of changes in space-time characteristics of loads.

Acknowledgments
Work is made with financial support from the Russian Foundation of Basic Researches (grant No. 18-08-00094-a, 19-08-00606-a).

References

[1] Protasov V D (Editor) 1992 *Mechanics of constructions made of composite materials. Collection of scientific papers* (Moscow: Mechanical engineering)
[2] Vasiliev V V 2009 *Composite pressure vessels — Analysis, design and manufacturing.* (Blacksburg: Bull Ridge Publ.)
[3] Karmishin A, Skurlatov E, Startsev V and Felshtein V 1982 *Non-stationary aeroelasticity of thin-walled constructions* (Moscow: Mechanical engineering)
[4] Bakulin V, Gribanov V, Ostrik A, Romadinova E and Cheprunov A 2008 *Mechanical action of X-ray radiation on thin-walled composite constructions* (Moscow: FML)
[5] *Physics of Nuclear Explosion, Vol. 2: The Action of an explosion* ed V M Loborev B A Shilobreev, V E Fortov and et al. (Moscow: FML, 2010)
[6] Bakulin V and Ostrik A 2015 *Complex action of radiation and particles on thin-walled constructions with heterogeneous sheetings*(Moscow: FML)
[7] *Physics of nuclear explosion,* Vol. 3:*Reproduction of factors of explosion* ed V M Loborev, B A Shilobreev, V E Fortov and et al. (Moscow: FML, 2013)
[8] Ostrik A, Bakulin V and Cheprunov A 2000 *Collection of Problems of Strength and Plasticity* (NizhniiNovgorod: Nizhegorod. Univ.) 117–121
[9] Ostrik A 2002 *J. Composite materials constructions* **1** 41–55
[10] Balabukh L, Alfutov N and Usyukin V 1984 *Structural mechanics of rockets* (Moscow: Higher. School)
[11] Fakhrutdinov I and Kotelnikov A 1987 *The construction and design of solid fuel rocket engines* (Moscow: Mechanical engineering)
[12] Sedov L 1977*Methods of similarity and dimension in mechanics*, (Moscow: Science)
[13] Girin Yu, Martynov A, Cheprunov A and et al. 2015 *Patent for invention RUS 2557298*
[14] Ostrik A, Nikolaev D, Bugay I2019 *Book of Abstracts of XXXIV Inter.Conf. on Interaction of I ntense Energy Fluxes with Matter* (Moscow: JIHT RAS) 87
[15] Nikolaev D, Ternovoi V, Kim V and Shutov A 2014 *J. Phys.: Conf. Ser* **500** 142026

[16] Nikolaev D, Ostrik A, Ternovoi V and Shutov A 2017 *Patent* for invention RUS 2680506
[17] Dremin A, Pershin S and Pogorelov V 1965 *Combust. Explos. Shock Waves* **1** 1–4
[18] Kanel G, Petrova E 1981 *Proc. of Workshop on detonation* ed Dremin A N et al. (Chernogolovka, Russia) 136–42
[19] Kanel G, Razorenov S, Utkin A and V Fortov V 1996 *Shock-wave phenomena in condensed matter* (Moscow: Yanus–K)
[20] Ostrik A, Nikolaev D and Cheprunov A 2019 *Composite materials constructions* **3** 55-63

How to reduce the probability of erroneous decisions in the systems of collective intelligence

V Protasov[1], Z Potapova[1] and G Akhobadze[2]
[1] Moscow Aviation Institute, Volokolamskoe shosse, 4, Moscow, Russia, 125993
[2] V. A. Trapeznikov Institute of Control Sciences of the Russian Academy of Sciences, Profsoyuznaya st., 65, Moscow, Russia 117997

E-mail: protvlad@gmail.com

Abstract. This work discusses and solves the methodological problem of reducing the probability of erroneous decisions when solving local problems by groups of intelligent agents. Intellectual agents can be carriers of both natural and artificial intelligence. The basic method of the new information technology is based on the theory of evolutionary decision reconciliation developed by the authors. The method is built on the original use of genetic algorithms in which intelligent agents perform the fitness function based on their knowledge and skills. The technology is based on the knowledge model proposed during the work, the Rasch model, and the mathematical apparatus of the Condorcet theorem. A theorem that allows finding the conditions under which the probability of an erroneous solution tends to zero, is formulated and proved. The results of using the theory for objects recognition from images by a group of neural networks are presented. A significant decrease in the probability of errors of the first kind is obtained in comparison with the classical methods. It is proposed to use the technology of evolutionary decision reconciliation to solve local problems in medical diagnostics, banking, and other fields.

1. Introduction

The tasks that arise in various areas of human activity are becoming more complicated, and the available resources are not always sufficient to maintain the required pace of development [1 - 4]. This has urged the process of finding and theoretical justification of new technologies, which should surely be able to solve the arising tasks by the available specialists, computer and software systems (hereinafter referred to as, *actors*) [5 - 7]. This new resource is a properly organized work of specialists groups or committees of neural networks of deep learning, supported by modern information technologies [8 - 11].

In decision-making practice, it was observed that with a well-chosen system of collective work, a group of actors always gives a smaller percentage of erroneous decisions than a single actor [9]. Nevertheless, even in this case, the main drawback is the unpredictability of the results of the group's work when solving new, previously unmet problems, and when solving such problems there is no guarantee that they are solved correctly.

Recently, the authors have been conducting research on the creation of new systems of collective intelligence based on the methodology of evolutionary coordination of decisions. In the proposed system, there is an increase in the probability of correct solutions to problems of medium difficulty and a significant decrease in the probability of errors in difficult cases [12].

IOP Conf. Series: Materials Science and Engineering **927** (2020) 012069 doi:10.1088/1757-899X/927/1/012069

2. Method of the evolutionary decision matching. Terms and Definitions

Since the new field of interdisciplinary studies of collective intelligence has developed relatively recently, it is necessary to introduce the terms and definitions specific to the collective intelligence systems, as well as metric characteristics and procedures that allow to measure parameters of both collective intelligence systems and their individual components. The introduction of such definitions and terms will make it possible to distinguish clearly the subject of research of the new scientific field from the ones developed earlier, for example, in the Decision theory.

In the present work, the following terms and definitions are used:

Actor (Lat. *doer*): an individual, community group, institute, intellectual agent, neural network, program or another entity that performs specific actions within the framework of its role. In collective intelligence systems, an actor can be the carrier of natural or artificial intelligence and can play two roles: generate a text that is the solution to a project or part of it, or evaluate the correctness of other actor's decisions.

There are three types of actors:

Single actor: a separate individual, hereinafter referred to as an actor of the zero rank or simply an *actor*.

Group actor of the *i-th* rank: a group that unites several actors of the i *-1* rank. The group actor of the 1st rank is hereinafter referred to simply as a *group actor*.

Workshop: the aggregate of all actors of different backgrounds who have knowledge in a particular area.

Workshop's knowledge is the totality of all established knowledge presented in the repositories of knowledge and partially among the carriers of intelligence

Collective intelligence: the ability of a group of actors to find solutions that are more effective than the best individual solutions in this group. This ability depends on the ability of individual actors in the group, as well as on the rules or procedures of actors' interaction in the process of working on a project.

Project: a task that has a local character, a clearly defined goal, understood by the actors, and a small number of factors affecting its resolution.

Slot: a data cell that actors fill out with some information. The slot is an independent component of the project and partially fulfils the goals of the project. During the work on a project, the slot may be filled as correct or incorrect, or may remain empty (the answer is "I don't know").

The system of collective intelligence: a set of tools and procedures that allow, under certain conditions, to organize the collective work of actors to fill the project slots with the right decisions.

EDR method (method of the evolutionary decision reconciliation): the procedure of filling out the project slots by the group actor in accordance with the interaction rules taken from the genetic algorithms [13]. The process consists of two stages: generation of decisions and evaluation of decisions. The generation stage always consists of only one iteration, while the coordination stage can consist of one or several iterations. At the decision generation stage, the actor fills the slots with his decisions either in full, or partially, or does not make decisions at all (if he does not know). At the stage of decisions evaluation, each actor receives L decisions from the other actors and, in accordance with his ability to evaluate decisions, selects the ones that are correct in his opinion and fills them into the slots, or leaves the slots empty again. At the next iterations, the process of filling the slots continues until more than half of the actors for each slot receive the same decisions, or the number of iterations reaches a predetermined number T. If more than half of the same decisions are not collected for any slots, then it is considered that the problem is either solved incorrectly, or the group actor gave the answer "I don't know."

Examples of local tasks solved using the EDR method [13]:

1. Tasks, the solution of which can be described by a linear chromosome, for example, compiling a facial composite by a group of eyewitnesses; measuring an IQ; translation of texts, tasks of strategic planning, assignment problem, etc.

2. Tasks whose solutions are represented by the decision tree, such as a chess problem, compiling a computer program for a training sample, a traveling salesman problem, and a financial portfolio compilation task.

3. Creative tasks, the solution of which can be any text or a hypertext that meets the objectives of the project.

The most important in application are the tasks presented in the third bullet point, since many tasks facing groups of specialists have solutions in the form of some texts that satisfy the goal. This can be, for example, texts of technical specifications, strategic plans for the development of companies, a description of the totality of measures related to emergency response, scientific articles, etc.

3. Condorcet Jury Theorem

The probability of a correct solution to a problem using EDR method at the stage of generating solutions can be determined using a mathematical model related to the Condorcet jury theorem [14], published in 1785.

One of the statements of the theorem, widely used in a number of sources [15, 16], is as follows:

Assume that one of the two decisions proposed by the jury group is correct, and the jury, on average, will vote more often for the right decision. It is argued that the likelihood Q_0 of a correct decision by a majority of votes is growing with the increase in the number of jury M and approaches 1.

$$Q_0 = \sum_{i=0}^{\frac{M-1}{2}} C_M^i G_R^{M-i} (1 - G_R)^i \qquad (1)$$

where G_R is the likelihood of filling out the slot with the correct answer.

For G_R >0.5, we observe the Condorcet effect: the tendency of Q_o to 1 with an increase in the number of actors. For G_R<0.5, we observe the opposite effect — a decrease in Q_o to 0. As will be seen further, the group actor can overcome the "Condorcet barrier" only at the stages of EDR and get the right solution with a probability approaching 1, even if each actor fills in the decision generation stage correct answers less than half of the project slots.

4. Georg Rasch Model

We assume that an actor is characterized by the four parameters when working on a project with slots of the same degree of difficulty:

G_R - the probability of filling the slot with the correct answer,

G_N - the probability of filling the slot with the incorrect answer,

E_R – probability of correct slot evaluation,

E_N - probability of incorrect slot evaluation.

These parameters of the actor can be connected to the following four auxiliary parameters that will be used in the future:

$G_S = G_R + G_N$ - the probability of filling the slot at the decision generation stage,

$G_V = 1 - G_S$ - the probability of getting the "I don't know" answer at this stage,

$E_S = E_R + E_N$ - the probability that the slot will be filled during the evaluation,

$E_V = 1 - E_S$ - the probability that the slot will not be filled during the evaluation.

Next, we introduce the auxiliary concept of an *ideal actor*, which is used to formulate the statements and conclusions of the theory. The ideal actor is a hypothetical actor for whom $\mathrm{G_S} = \mathrm{G_R}$, $\mathrm{G_N} = 0$ and $\mathrm{E_R} = 1$.

Let us find the probability Q of the correct slot filling by the group of ideal independent actors M, provided that for all actors thus probability is the same.

The probability Q is determined as follows. The probability that the ideal actor will not solve the problem is equal to $1 - G_R$. The probability that none of the M actors will solve the problem, according to the theorem on the probability of the simultaneous occurrence of independent events, is equal to the product of the probabilities of these events $(1 - G_R)^M$. Since $E_R = 1$, then

$$Q = 1 - (1 - G_R)^M \qquad (2)$$

As will be shown below, subject to the principle of equal pay for equal work, expression (2) can be used to introduce units of measurement of the relative labor costs of single actors and their groups.

For the precision of expression, hereafter we will consider *project* as a set of test questions of varying difficulty (slots) with the only known correct answers in advance and that when ideal actors work, the Rasch model is fulfilled [17].

The relationship between the level of difficulty of test questions and the degree of preparedness of the actors in determining the probability of a correct answer was established in the most general theory of test design, based on measurement theory - Item Response Theory (IRT). The one-parameter Rasch model is suitable for our purposes, as the simplest model connecting the probability of obtaining the correct answer G_R of the test subject with the level of his preparedness θ_{GR} and the difficulty of the task β. Actor's preparedness and task difficulty are latent variables. It is believed that they cannot be directly measured, but can only be obtained by using mathematical models that operate on observable variables called indicator variables.

Indicators in the Rasch model are the results of testing of a group of actors using test tasks of various difficulty levels.

The probability G_R of the correct filling of the slot of difficulty β according to this model is determined by the formula:

$$G_R = \frac{1}{1+e^{\beta-\theta_{GR}}}. \quad (3)$$

According to the Rasch model, the probability G_S that the actor will fill out the slot is:

$$G_S = \frac{1}{1+e^{\beta-\theta_{GS}}}, \quad (4)$$

where θ_{GS} is the preparedness of the actor to fill the slots with both correct and incorrect answers.

Then the probability of making an incorrect decision by the actor will be

$$G_N = \frac{1}{1+e^{\beta-\theta_{GS}}} - \frac{1}{1+e^{\beta-\theta_{GR}}}. \quad (5)$$

Similar expressions are obtained for the probabilities of a correct and erroneous examination:

$$E_R = \frac{1}{1+e^{\beta-\theta_{ER}}}, \quad E_S = \frac{1}{1+e^{\beta-\theta_{ES}}}, \quad (6)$$

$$E_N = \frac{1}{1+e^{\beta-\theta_{ES}}} - \frac{1}{1+e^{\beta-\theta_{ER}}}, \quad (7)$$

The dependency graphs for (4 - 7) are shown in Figure 1. The graph also shows the dependence of the probability of an incorrect solution $G_{N2} = 1 - G_R$ in the case when binary logic is used in evaluating the correctness of decisions, and the actor is trying to give an answer, even if he does not know it and tries to guess. It is seen that with increasing difficulty of the problem this probability significantly exceeds the probability G_N. The decrease in the probability of making the wrong decision G_N with increasing difficulty β is due to the fact that the actor gives the answer "I don't know" in the cases that are difficult for him.

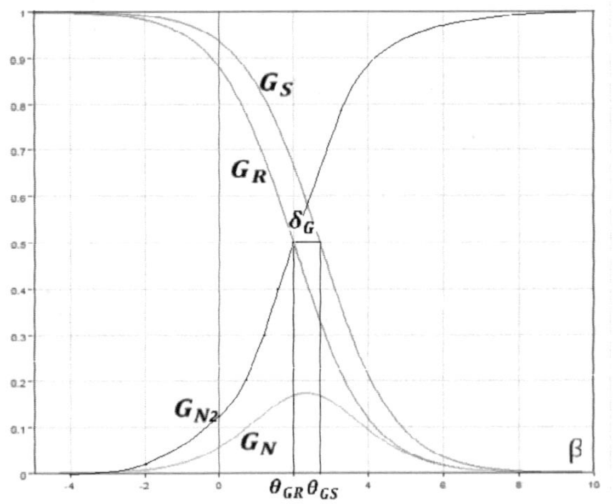

Figure 1. Graphs of the dependencies of $G_R(\beta)$, $G_S(\beta)$ and $G_N(\beta)$ on the difficulty of the task

Let us determine δ_G and δ_E which characterize the "inclinations" of the actors to the corresponding mistakes:

$\delta_G = \theta_{GS} - \theta_{GR}$,

$\delta_E = \theta_{ES} - \theta_{ER}$.

For the ideal actor, these values are considered to equal zero.

5. The model of the workshop's and actors' knowledge

Let us break up the knowledge in the workshop into two subsets: a subset of "knowledge quanta" $\mathbb{W}$– of some facts, results, definitions, etc., and a subset of $\mathbb{P}$ - procedures with algorithms for deriving new knowledge. Each subset, in turn, is divided into $\mathcal{K} \in \mathbb{N}$ ($\mathbb{N} = \{1,2,...\} \cup \{0\}$) subsets corresponding to knowledge levels k=0,1,2,..., $\mathcal{K}$. Level 0 corresponds to the initial knowledge quanta $W_{0,i} \in \mathbb{W}$ ($k = 0$, i=1,2,...n_0 , n_0 – the number of quanta knowledge of the 0 level), corresponding to the primary concepts, definitions, etc., and initial procedures for deriving new knowledge $P_{0,j} \in \mathbb{P}$ ($k = 0$, j=1,2,...m_0, m_0 – the number of procedures, corresponding to the 0 level).

From knowledge, quanta of the zero level $W_{0,i}$ using procedures of the zero level $P_{0,j}$ new quanta of knowledge of this level are derived. With the help of zero-level procedures, zero-level problems are also solved - the initial data of the problem and the necessary knowledge quanta are substituted into the procedure and, following the algorithm of the procedure, the problem is found.

The procedures of the first level $P_{1,j}$ are constructed based on the procedures and quanta of knowledge of the zero level, and can be used to find the quanta of knowledge of the first level and to solve problems of this level. If there is no suitable procedure for solving the first level problem, then first you need to derive it from the procedures and knowledge quanta of the zero level. The same goes for the rest of the levels.

The second level procedures are built from knowledge elements of the zero and first levels and so on to the achieved level of knowledge of the $\mathcal{K}$ workshop - the horizon of knowledge.

Thus, over time, the number of elements of the subsets $\mathbb{W}$ and $\mathbb{P}$ continuously increases in the workshop, leading to an increase in n_k and $m_k$and an increase in the level of the knowledge horizon in workshop $\mathcal{K}$.

It is clear that the individual actor of the workshop possesses only part of all the knowledge of the workshop and its level of knowledge is often below the knowledge horizon of the workshop. Moreover, some of the quanta of knowledge and procedures were learned incorrectly and, using them in the process of deriving new knowledge or solving new problems, he will receive erroneous knowledge and solutions without knowing it. The introduction discussed the occurrence of four types of errors received by actors or their groups in solving new problems. In the light of this model of knowledge, some of them get a fairly simple explanation.

Next, we introduce two subsets of the attributes of mastering some workshop knowledge quanta $\mathbb{H}_W$ and procedures $\mathbb{H}_P$. Elements $(H_W)_{j,i} \in \mathbb{H}_W$ and $(H_P)_{j,i} \in \mathbb{H}_P$ (i=1,2,..., n_k , $j = 1,2,\ldots,m_k$, k=0,1,2,..., $\mathcal{K}$) get the value of:

$$\mathcal{V} = \begin{cases} 1 \\ 0 \\ -1 \end{cases},$$

that means with $\mathcal{V} = 1$ the correct learning, $\mathcal{V} = 0$ a gap in knowledge, and $\mathcal{V} = -1$ – erroneous learning.

Let $\mathcal{M}$ be the number of actors in the workshop that form the set of $\mathbb{M}$. For each m-actor (m=1,2,...,$\mathcal{M}$), we will define the set of knowledge that consist of the two subsets $\mathbb{w}_m$ and $\mathbb{p}_m$, that are defined in the following way: into each of the elements $(w_m)_{j,i} \in \mathbb{w}_m$ and $(p_m)_{j,i} \in \mathbb{p}_m$ when $\mathcal{V} = 1$ a correct entry from $\mathbb{W}$ or $\mathbb{P}$ is introduced, an empty value when $\mathcal{V} = 0$ or an erroneous entry of the knowledge quanta or procedure when $\mathcal{V} = -1$.

The actor's knowledge model presented above correlates with the Rusch model (3). Let us show it. Since argument (3) is the excess of the difficulty of the problem over the preparedness of the $\beta - \theta$

actor, then for the analysis it is necessary to introduce a reference level. To do this, we introduce the notion of the average difficulty of the workshop task and set $\beta = 0$ for it. We also introduce the notion of the averagely prepared workshop's actor and set $\theta = 0$ for him. According to (3), the averagely prepared actor in the workshop solves a task of an average difficulty with a probability of 0.5. The simple workshop tasks, close to zero, where the average actor has no gaps or false knowledge, are solved by him with the probability of a correct solution G_R close to 1.

As the tasks become more complicated, the actor gaps in knowledge and false knowledge in the corresponding knowledge area increase, and the likelihood of getting the correct solution decreases. He solves difficult tasks with a probability close to zero. This is to be expected, since an intermediate-level actor does not have knowledge of the procedures of high levels for him. As the actor's preparedness increases, his knowledge horizon approaches the workshop's knowledge horizon. Thus, the logistic curve line (that characterizes the probability of a correct solution) shifts to the right when the preparedness of the actor increases, and to the left, when the preparedness becomes lower than average.

There is also a correlation between the number of false quanta of knowledge and / or procedures and the values of δ_G and δ_E - the actor's tendency to generate erroneous and, accordingly, to incorrect examination of decisions.

6. Possible structural diagrams of EDR systems

All the main characteristics of the actor have been presented and formalized above, allowing us to understand the features and capabilities of his work in solving problems. It was found that a single actor, with task difficulties commensurating with or greater than his preparedness, is susceptible to making erroneous decisions and even training him on tasks of varying degrees of difficulty in order to avoid mistakes when making decisions by transferring them to the "I don't know" zone does not save him from mistakes. This is clearly shown in Figure 1 (curve lines G_N and G_{N2}).

In this regard, the question arises: is it possible in principle to reduce the likelihood of making erroneous decisions for a group actor in solving difficult problems while maintaining the ability of a group actor to give better decisions than a single actor?

To answer this question, it is necessary to conduct research for different structural schemes for organizing the work of groups of actors, taking into account the characteristics of actors.

Here are the main structural features of the group actor of the first and second rank. A description of the work of a group actor of the first rank is given above in the definition of the EDR. The group actor of the first rank is formed by the set of M actors, taken from the actors set $\mathbb{M}$ and having the knowledge $(w_m)_{j,i} \in \mathbb{W}_m$ and $(p_m)_{j,i} \in \mathbb{P}_m$. Here, m equals the values of the indexes from $\mathbb{M}$, where $i=1,2,...,n_k$, $j = 1,2,\dots,m_k$, $k=0,1,2,...,\mathcal{K}$, $\mathcal{K}$ – the number of the levels of knowledge in the workshop.

Group actor of the second rank consists of M actors divided into M_1 groups, each of which consists of M_2 actors, so that $M = M_1 M_2$. Thus, in the general case, we can say that the group actor of the first rank is structurally characterized by the two numbers: M_1 and M_2. When $M_1 = 1$, we are getting the group actor of the first rank, consisting of M actors.

Actor of the first rank is researched in this work.

7. Finding conditions under which the probability of an erroneous decision by an actor of the first rank tends to zero

Theorem

Assume that the group actor consists of M single actors with preparedness θ_{GR}, θ_{GS}, θ_{ER} and θ_{ES}, numerical values of which lie in the interval $[\theta - \delta/2, \theta + \delta/2]$, $\theta \in R$ and $\delta \in R$.

It is argued that if $M \to \infty$ and $\delta < ln\, 2$, then the probability of the incorrect decision for the task of an unspecified difficulty β by the group actor approaches zero.

Prof

From all the set of the preparedness values θ_{GR}, θ_{GW}, θ_{ER} and θ_{EW} we choose those for which the probability of getting an incorrect decision by the group actor will be the highest, that is

$$\theta_{GR} = \theta - \delta, \theta_{GW} = \theta + \delta, \theta_{ER} = \theta - \delta, \theta_{EW} = \theta + \delta. \quad (8)$$

According to the Rasch model, for the corresponding probabilities it is possible to write the following expressions:

$$G_R = \frac{1}{1+e^{\beta-\theta+\delta}}, \quad G_W = \frac{1}{1+e^{\beta-\theta-\delta}}, \quad G_V = 1 - G_W = \frac{1}{1+e^{-\beta+\theta+\delta}},$$

$$G_N = G_W - G_R = \frac{1}{1+e^{\beta-\theta-\delta}} - \frac{1}{1+e^{\beta-\theta+\delta}}. \quad (9)$$

The same expressions we can write accordingly for E_R, E_S and E_N, that is

$$E_R = G_R, E_W = G_W \; and \; E_N = G_N \quad (10)$$

It is easy to show that in the end of the EDR iteration cycle when $M \to \infty$ the probability of getting an incorrect decision of one actor within the group will be

$$P_N = G_N + \frac{E_N}{E_W} G_V \quad . \quad (11)$$

Inserting (9) and (10) into (11), we get after the calculations, the dependency $P_N(\beta)$:

$$P_N = \frac{1}{1+e^{\beta-\theta-\delta}} + \frac{1+e^{\beta-\theta-\delta}}{2+e^{-\beta+\theta+\delta}+e^{\beta-\theta-\delta}}. \quad (12)$$

Differentiating ((12) with respect to β and equating the expression with the zero, we will get that with the only value $\beta = \theta$ for any δ the function (12) has an extremum. Because the second derivative in this point is negative, we conclude that (12) gets the maximum in this point.

It is clear that if the probability of getting an incorrect decision of a single actor in the point of maximum is less than 0.5, than for the whole interval of change β from $-\infty$ till $+\infty$ this probability is also less than 0.5. Thus, at the whole interval of change of β when $M \to \infty$, according to the Condorcet theory the probability of getting the incorrect decision by the group actor Q_N approaches zero. The maximum permitted value $\delta = ln2$ is derived from the solution of the formula (12) when $P_N = 0.5$, which was needed to be proven.

In the same fashion, it is possible to prove that if preparednesses θ_{ER} and θ_{ES} are located in the interval $[\theta - \delta + 1, \theta + \delta + 1]$ with the same values of θ_{GR} and θ_{GS}, then the maximum value of $\delta_1 = ln\sqrt{5}$.

Based on this theorem, the following conclusion can be made. If for a certain class of tasks we have a set of certified experts and we know their levels of preparedness, we can unite them into a group actor which will either solve similar tasks correctly, or will give the answer "I cannot solve this" with the afore given randomly small probability of an incorrect decision. In the second case, the unsolved task can be given to another group of actors with a higher level of preparedness.

In [12], there are the results of object recognition by a group actor consisting of the neuron networks with a small probability of error (table 1).

Table 1. Results of testing of the neuron networks

Degree of difficulty	Singular neuron network			Group neuron actor		
	G_R	G_N	G_V	Q_R	Q_N	Q_V
Low	0.95	0.03	0.02	0.98	0.00	0.02
Medium	0.64	0.27	0.09	0.87	0.01	0.12
High	0.22	0.47	0.31	0.46	0.02	0.52

Figure 2 contains the summary results of calculating the probabilities of the unknown quantities for group actors of the first rank, consisting of 15 and 51 actors and the probabilities for a single actor with predetermined values $\delta = 0.99 * ln2$, θ_{GR}=11, $\theta_{GS} = \theta_{GR} + \delta$, $\theta_{ER} = 11$, $\theta_{ES} = \theta_{ER} + \delta$ for different actors. In all experiments, it was assumed that the number of variants of erroneous answers does not exceed four.

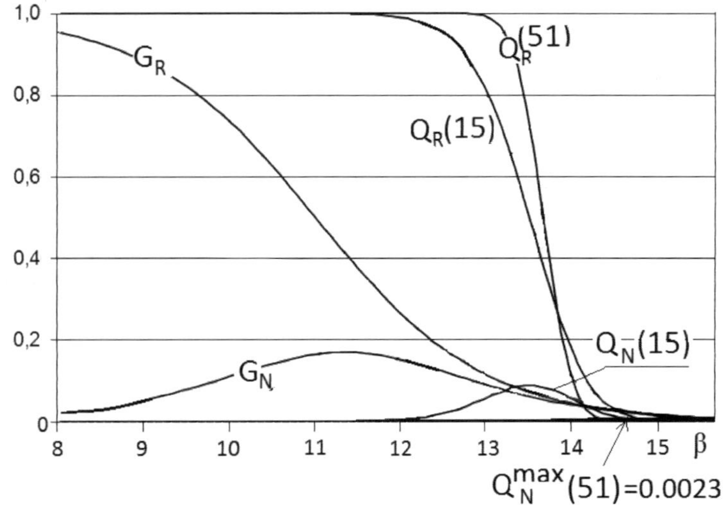

Figure 2. Results of computer experiments of determining the probabilities of the correct (index R), incorrect (index N) decisions.

Two groups of experiments were conducted. In the first grout, the curve lines for how probabilities are dependent of the difficulty of the task for the group actor of the first rank when M=51. The probability of the incorrect decision in maximum was in this case 0.0023, which correlates well with the theorem given above.

In the second group of experiments with M=15, the maximum of mistake was about 0.1, which in some practical cases is not acceptable.

8. Results discussion

The results obtained in this work are of fundamental importance for the development and application of the theory of collective intelligence systems based on the use of the evolutionary decision matching method, the Condorcet theorem, and the Rush model. For the first time, it becomes possible to solve local problems of arbitrary difficulty with a low probability of an erroneous solution. With the right scheme of organization of work of a group of certified actors, it becomes possible to guarantee the correct solution of problems of known difficulty.

Application of the methodology to reduce drastically the likelihood of making erroneous decisions using specially constructed neural network committees with deep learning will solve the problem of distrust in the results obtained with their help. It is known that, despite the impressive successes of deep training of neural networks in recognizing objects that in some cases exceed human capabilities, they are widely introduced into a number of promising areas, such as, for example, medical diagnostics, banking, or searching for specific individuals in streaming video, received. This is due to the fact that the mechanisms for obtaining a solution are not transparent and the developers of this kind of intelligence themselves acknowledge that it is impossible to achieve the required practice of reducing the probability of false recognition to almost zero in difficult cases and other approaches need to be sought.

The methodology presented in the work, although it does not directly solve the problem of understanding how deep learning neural networks solve problems, at least it allows solving local problems of arbitrary difficulty with a predetermined sufficiently low probability of error, giving the answer “I don’t know” in difficult cases.

IOP Conf. Series: Materials Science and Engineering **927** (2020) 012069 doi:10.1088/1757-899X/927/1/012069

References

[1] Trahtenherz E 1998 *Computer Support in Decision Making* (Moscow, SINTEG)
[2] Trahtenherz E 1994 Distributed computer-aided systems for the design of complex engineering systems: peculiarities of the system software *Automation Remote Control* **55** (11) pp 1681-93
[3] Trahtenherz E 1994 Sequential-parallel computer-aided design of complex engineering systems *Automation Remote Control* **55** (2) pp 290–302
[4] Chikersal P, Tomprou M, Young Ji Kim, Williams Woolley A and Dabbish L 2017 *Deep Structures of Collaboration: Physiological Correlates of Collective Intelligence and Group Satisfaction. Proceedings of the Conference on Computer Supported Cooperative Work and Social Computing* pp 873-88
[5] Wiener N 1967 *I am a mathematician* Brief transl. from English by Yu. Roman (Moscow: Nauka)
[6] Wiener N 2002 *Cybernatics and Society* (Moscow: Taydeks Co.)
[7] Chmait N 2016 *Factors of collective intelligence: How smart are agent collectives?* Proceedings of the Twenty-second European Conference on Artificial Intelligence pp 542-550
[8] Grudin J 1994 Computer-supported cooperative work: its history and participation. *Computer 27* **4** 19-26
[9] Herrera F, Herrera-Viedma E and Verdegay J 1996 Direct approach process in group decision making using linguistic OWA operators. *Fuzzy sets and systems* **79** 175-90
[10] McGrath J 1997 Small group research, that once and future field: An interpretation of the past with an eye to the future. *Group Dynamics: Theory, Research, and Practice* **1** (1) 7-27
[11] Chmait N 2017 Understanding and measuring collective intelligence across different cognitive systems: an information-theoretic approach. Proceedings of the Twenty-Sixth International Joint Conference on Artificial Intelligence (*IJCAI-17*) 5171-2
[12] Protasov V, Mirakhmedov R, Potapova Z, Sharnin M and Sharonov A 2018 Reducing of Errors of the First-Rank Diring Recognition of the Shapes of Airplanes Using Collective Intelligence of the Unmanned Aerial Vehicle (Herald of the Kabartino-Nalkarsky Scientific Center of the Russian Academy of Sciences) **6** (86) 70-82
[13] Protasov V 2009 The construction of metasystematic transitions (Protvino, Herald of the Moscow Institute of Physics and Technology) 186
[14] Condorcet, Marquis Marie-Jean-Antoine-Nicolas de Caritat 1785 Essai sur l'application de l'analyse à la probabilité des décisions rendues à la pluralité des voix (Imprimerie Royale, Paris)
[15] Baharad E, Goldberger J, Koppel Moshe and Nitzan Shmuel 2011 Beyond Condorcet: optimal aggregation rules using voting records (The Open Access Publication Server of the ZBW – Leibniz Information Centre for Economics). Cesifo Working Paper No. 3323 Category 2: Public Choice January 2011
[16] Koriyama Y, Szentes B 2009 A resurrection of the Condorcet jury theorem *Theoretical Economics* **4** 227–52
[17] Rasch G 1980 *Probabilistic Models for Some Intelligence and Attainment Tests* (Chicago: University of Chicago Press)

IOP Conf. Series: Materials Science and Engineering **927** (2020) 012070 doi:10.1088/1757-899X/927/1/012070

Distributed computing based on Container-component model

A Semenov
Department of Computational Mathematics and Programming,
Moscow Aviation Institute (National Research University), MAI,
Moscow, 125993 Russia

E-mail: Semenov_Alex@yahoo.com

Abstract. Cloud computing focuses on scalability, elasticity, and cost-efficiency of the service. In this context, shifting towards the concept of "algorithm as a service" is a contemporary issue. The three main drivers of it includes, machine learning, artificial intelligence, and Digital Earth. The following assumption was made throughout the article, that algorithm as a service is endowed with a key property of fractal, which provides the corresponding scalability and elasticity. The article focuses on the design of elastic algorithms based on Container-component simulation model. The model has fractal property. The application of the model to design gradient ascent algorithm as a service is presented.

1. Introduction

Cloud computing focuses on scalability, elasticity, and cost-efficiency of the service. The cloud computing evolves from distributed and grid computing.

Physically a distributed system consists of a number of nodes (autonomous computers) interconnected by a network. The nodes may be computers, physical servers, agents, virtual machines, containers, or any other entity that can connect to the network, have local memory, and communicate by passing messages [1, 2]. The basic models of a distributed system are messages passing and shared memory models. The message is the unit of communication of a distributed system. In the synchronous and asynchronous message passing models, nodes of the distributed system communicate by messages only [3,4].

Typically, distributed system lacks global control. This requires means to model and to verify distributed algorithms. Distributed algorithms are designed to run on different computers of the network. Parts of an algorithm run concurrently and independently, each with a limited amount of information [5]. There are two problems: how to adequately describe a distributed algorithm and how to prove its properties. Designing appropriate distributed [6] and formation control [7] is generally a problem, especially for multi-agent systems with complex dynamics. A distributed algorithm is adequately described if it takes into account property of the computing environment. A graph [8] and an adjusted version of Petri nets [9] used to model a distributed system, and distributed algorithms that make use of some property of these means to solve problems.

Designing efficient distributed algorithms is very important for analyzing big data. Three properties of big data, velocity, volume and variety must be considered. Technically, distributed algorithms to process big data are designed for a specific application and mainly based on the MapReduce framework [10]. The framework implements two functions Map and Reduce. Map function divides the input data into data partitions that constitute key-value pairs. Each partition is assigned to a unique compute node. Nodes outputs are one or more intermediate key-value pairs. The framework collects all the intermediate key-value pairs, sorts and groups them by key. The Reduce function aggregates the values associated to the key according to a predefined program and stores all the output key-value pairs in a file. Hadoop is an open source project written in Java that implements MapReduce framework [11]. It is possible to write some MapReduce jobs in Python and then run them in Hadoop Streaming that operates like the pipes in Linux. Hadoop jobs are running on Amazon Web Services that provides on-demand cloud computing platforms. This is fundamental support to Big Data [12] that are unprecedented content for Digital Earth [13].

In cloud computing systems an on-demand network model is used to provide access to shared pool of configurable IT resources. One of the biggest features of cloud computing is their scalability and elasticity [14]. The elasticity of an application is a measure of its transformation that depends on fluctuating demands [15]. The concept of "everything as a service" has been associated and used with many core components of cloud computing including communication, infrastructure, data, platforms, and software. In this context, shifting towards the concept of "algorithm as a service" is a contemporary issue. The three main drivers of it includes, machine learning, artificial intelligence, and Digital Earth. The following assumption was made throughout the article, that algorithm as a service is endowed with a key property of fractal, which provides the corresponding scalability and elasticity. The article focuses on the design of elastic algorithms based on Container-component simulation model [16]. The model has fractal property.

The approach presented in the article is an integral part of the analysis, design and fractal programming of elastic systems [17]. Fractal programming is a programming paradigm based on the concept of "elastic objects", which can transformed (unfolded and folded) dynamically at runtime by using strategy planning model, production and restricting rules. These rules are keeping the object structure self-similar thus it has fractal property: parts similar to the whole. The paradigm aims to optimize searching suitable workflow structure of the object at run time.

The article is organized as follows. At the beginning, in the context of fractal programming, a description of the Container-component model and its properties are given.
In the next section the application of the Container-component model to design gradient ascent algorithm [18] as a service is presented.
The contribution of the article and future works are discussed in conclusion.

2. Fractal programming and Container-component model

In this section Fractal programming of elastic objects and Container-component model are discussed. The elastic property of Container-component model provides the basis for determining the fractal complexity of distributed computing process. Fractal complexity is characterized by computational accuracy, computational time, and required computing resources.

2.1. Fractal programming

The most important fractal properties of an elastic object are self-similarity, scaling, and fractal dimension. The term fractal is used to highlight properties of elastic objects: abstraction, ability for unfolding and folding, pattern ability, similarity, symmetry, and scalability. In many works on fractals, self-similarity is used as defining property. In this article, the ordered bag of self-similar prototyping objects is used. The ordered bag makes possible to define position of object in it.

The bag (for example, implemented by container) could include repeatable objects a number of times. The assumption was made that self-similarity describes object in which the same objects are repeated over and over again on different levels of the scalability, for example, binary tree: the part looks like the whole.

2.2. Container-component model

The fundamental concept of a model is an elastic container object. A container object has the following properties: it contains other containers, contains IT resources (data, functions and methods), manages the storage space of its elements and provides access to them, moves from one computing environment to another. Containers can be used very effectively to identify non-serving, non-interactive programs that simply accept input and return a result. Let each container run on its own server. Server do not share global memory and communicate exclusively through messaging. The model algorithm is constructed as follows.

Let upper indexes R (Receive), S (Send), respectively, be the inputs and outputs for the container e_0, then the container program

$$E = {}^{R}\{e_0\}^{S} \quad (1)$$

A distributed program E consists of a set of n asynchronous container programs ${}^{R}\{e_0\}^{S}, {}^{R}\{e_1\}^{S}, {}^{R}\{e_2\}^{S}, \dots, {}^{R}\{e_n\}^{S} \in E$, which interact through containers transmitted over the network, and the number of containers n varies over time in accordance with the requirements.

Components fulfill the role of communication programs between container programs. The component program is designated as ${}^{R}\{c\}^{S}$ and cannot be divided.

Depending on each specific case, the operation to receive (R) and to send (S) can be redefined which is indicated by the appropriate upper index for the curly brace. If there is no index, then the operation is not applied.

2.3. Definition

Container-component model $E = f^n({}^{R}\{e_0\}^{S}, {}^{R}\{c\}^{S}, \Delta)$, where

f^n: $E \rightarrow E$ – is a recursive mapping of a *set E* in itself by Δ operations, n = 0,1,...,k – is a step of mapping;

${}^{R}\{e_0\}^{S}$ – is an initial container program;

${}^{R}\{c\}^{S}$ – is a component program;

$\Delta = \{\equiv (1/r, N), \downarrow, n{+}{+}, n{-}{-}\}$ – is an ordered set of uniquely invertible operations:

$\equiv (1/r, N)\ {}^{R}\{e\}^{S}$ – is an operation of prototyping container program ${}^{R}\{e\}^{S}$ with parameters:

r – is a scaling ratio. It means that container program is divided in r containers and component programs,

N – is a number of self-similar container programs ${}^{R}\{e\}^{S}$ after division, $1/r \neq N$;

$\downarrow : E' \rightarrow e$ – is an insert set E' in the container e.

$n{+}{+}$ – is an increment operation, n = n+1;

$n{-}{-}$ – is a decrement operation, n = n –1.

Components obtained by the prototyping operation are used to simulate message passing between neighboring container programs and the container program from which they are prototyped. Figure 1. shows intercommunication between initial program ${}^{R}\{e_0\}^{S}$ and subprograms ${}^{R}\{e^1_1\}^{S}$ and ${}^{R}\{e^1_3\}^{S}$ if r = 1/3 and n = 1. Graphically container is depicted by rectangle. Nested containers are depicted by nested rectangles. Component is depicted by grey rectangle.

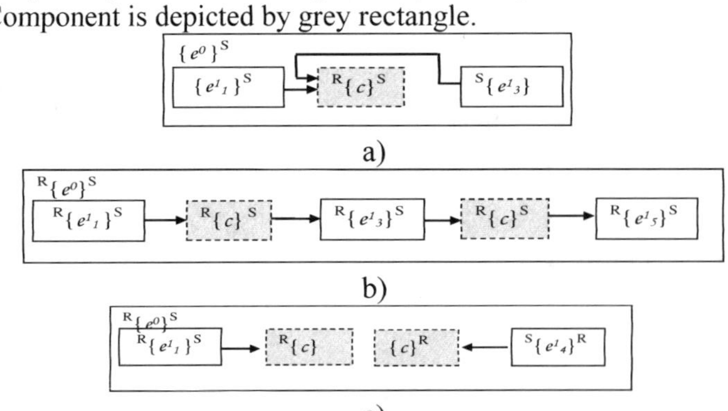

Figure 1. Container-component model pattern in the first iteration step, n = 1.
a) $1/r = 3, N = 2$ b) $1/r = 5, N = 3$ c) $1/r = 4, N = 2$

IOP Conf. Series: Materials Science and Engineering **927** (2020) 012070 doi:10.1088/1757-899X/927/1/012070

Elastic Container-component model for the Figure 1. a). is written as follows:

$$E = f^n (E^0 = \{e_0\}^S, {}^R\{c\}^S, [n{+}{+}, \forall \{e\}^S \in E^{n-1} \{e\}^S \downarrow (E^n = \equiv (3,2) \{e\}^S) \mid 0 < n =< k], [n{-}{-}, \forall \{e\}^S \in E^{n-1} \{e\}^S \downarrow^{-1} (E^n = \equiv^{-1} (3,2)\{e\}^S) \mid 0 < n >= k]) \quad (2)$$

Container-component model is based on the concept of "elastic object". Such an elastic object is a container. Elastic Container-component model makes possible to design gradient ascent algorithm as a service.

3. Elastic Container-component algorithm of gradient ascent

In this section elastic Container-component algorithm for gradient ascent is introduced. At the beginning the gradient ascent algorithm [8] is explained. The distributed version based on elastic Container-component model is introduced.

3.1. Gradient ascent algorithm for big data ordered in time

Let big data be ordered at time T with equal intervals τ. Figure 2 roughly illustrates this. The idea is to identify the slope and move up it. This method doesn't require to compute or even know f(x), but it compute the slope, that is, $v = v_{\tau 2} - v_{\tau 1}$, where $v_{\tau 2} - v_{\tau 1}$ is a slop. If the slope is positive, max(v) will increase. If the slope is negative, max(v) will decrease.

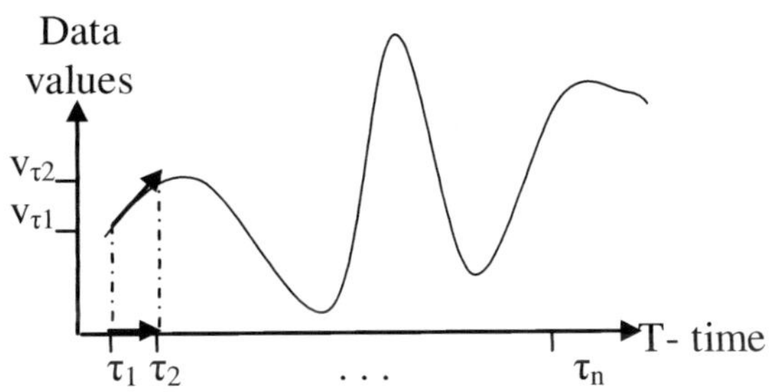

Figure 2. Gradient ascent algorithm for Big data ordered in time.

The idea is to identify the slope between the neighboring points. Here is the algorithm:

Algorithm 1. Gradient Ascent for big data ordered in time:

1. V – is a random initial vector ordered in time T
   $\max(v_\tau) = 0$ – is an initial value of random vector max/min $v_\tau \in V$
2. foreach τ in T
3. if $\max(v_\tau) < v_\tau$
4. $\max(v_\tau) = v_\tau$
5. end foreach
6. return $\max(v_\tau)$

The algorithm runs until found $\max(v_\tau)$ or $\min(v_\tau)$.

3.2. Elastic Container-component gradient ascent algorithm for Big data ordered in time

Our objective is to design elastic Container-component gradient ascent algorithm that captures the relationship between the computing time (Θ) of Algorithm 1, and the capacity requirement (c). For solution of this issue the following assumptions are made:

- Let vector have about V = 2 000 M (Millions) values ordered in time.
- Let Θ be the computation time, and c - be the capacity of the system that handles by Θ.
- Let a server be encapsulated in container e_0. The capacity of the server c = 2 MWIPS (Millions of Whetstone Instruction Per Second [19, 20]) is a constant and each server is encapsulated in container has the same capacity.

- The algorithm 1 consists from three operations in one loop (w = 3): foreach τ in T, condition less then <, and assignment =.

Thus, the computation time Θ of the algorithm 1 by C containers will be written as follows:

$$\Theta = V / (c * C / w) \tag{3}$$

Figure 3 shows Elastic Container-component tree. Containers are marked by capacity.

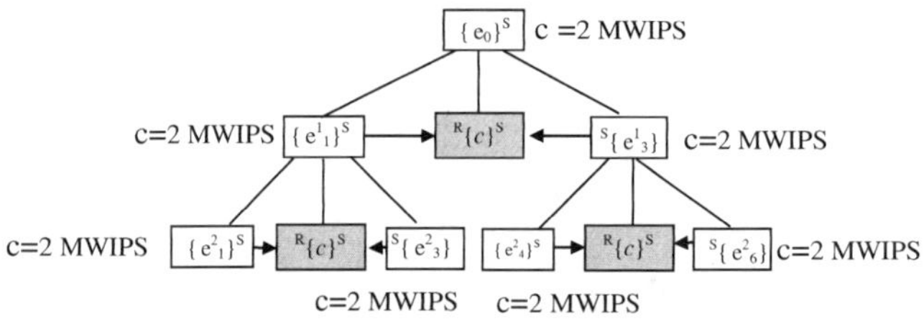

Figure 3. Elastic Component-container tree, $1/r = 3$, $N = 2$, $n = 2$.

The number of containers C for this elastic Container-component model can be calculated by the formula:

$$C = \frac{N^{n+1} - 1}{N - 1} \tag{4}$$

With the computation time Θ required for vector V processing by Algorithm 1 (A1) will be needed $C = V \cdot w / (\Theta \cdot c) = 2\ 000\ M \cdot 3 / (2\ MWIPS \cdot \Theta)$ containers. If Θ is equal to $\approx$ 429 sec, C = 7. Then the number of steps k of the elastic model is calculated by the formula:

$$k = \lceil \log_N((N-1) \cdot C + 1) \rceil \tag{5}$$

There is no loss of generality in assuming that each container E runs A1, with next fractal rules for each container program**:**

- r_l – *foreach loop* of the program $\{e_0\}^S = A1$ is transformed to *for* $\tau_k, \dots, \tau_l$ *loop*, where a range $\tau_k, \dots, \tau_l \in T$ in accordance with operation of prototyping $\equiv (1/r, N)\ \{e\}^S$. This transformation of the program E is written as $r_l\ (\{e\ \}^S)$**.**

The elastic Container-component model for gradient ascent algorithm for big data ordered in time is presented as follows:

$$E = f^n(E^0 = \{e_0\}^S = A1\ ,\ ^R\{c\}^S,$$
$$[n{+}{+}, \forall \{e\}^S \in E^{n-1}\ \{e\}^S\!\downarrow\ (E^n = \equiv (3,2)\ r_l\ (\{e\ \}^S))\quad |\ 0 < n =< k],$$
$$[n{-}{-}, \forall \{e\}^S \in E^{n-1}\ \{e\}^S\!\downarrow^{-1} (E^n = \equiv^{-1} (3,2)\ r_l\ (\{e\ \}^S))\ |\ 0 < n >= k\ ]) \tag{6}$$

When Elastic Component-container tree folded the result will be in the initial container component. One of the potential problem is that elasticity takes time.

4. Conclusion

The main contribution of the article is the application of the elastic Container-component model to design gradient ascent algorithm as a service.

As a result of this current study, several conclusions can be drawn:

- one of the tools for designing algorithms as services is the Container-component model;
- the patterns of organizing the Big data processing are given;
- the computation time Θ of algorithm can be controlled by containers;
- essential role of the fractal programming and Container-component model are shown.

Future work concerns problem of designing different elastic algorithms, the patterns of organizing the big data processing and creating methodology based on fractal programming.

References

[1] Tel G 2000 *Introduction to Distributed Algorithms* (Cambridge University Press) p 596

[2] Santoro N 2007 *Design and Analysis of Distributed Algorithms* (New Jersey: John Wiley & Sons, Inc) p 589
[3] Raynal M 2013 *Distributed Algorithms for Message-Passing Systems* (Springer-Verlag Berlin Heidelberg) p 470
[4] Raynal M 2018 Fault-*Tolerant Message-Passing Distributed Systems an Algorithmic Approach* (Springrer) p 459
[5] Lynch N A 1996 Distributed *Algorithms* (San Francisco, California: Morgan Kaufmann Publishers, Inc.) p 872
[6] Zhongkui L, Zhisheng D 2015 *Cooperative Control of Multi-Agent Systems a Consensus Region Approach* (Taylor & Francis Group) p 250
[7] Rastgoftar H 2016 *Continuum Deformation of Multi-Agent Systems* (Springer International Publishing AG) p 217
[8] Erciyes K. 2013 *Distributed Graph Algorithms for Computer Networks* (London: Springer-Verlag) p 325
[9] Reisig W 2013 *Elements of Distributed Algorithms: Modeling and Analysis with Petri Nets,* (Springer Science & Business Media) p 302
[10] Oussous A, Benjelloun F, Lahcen A, Belfkih S 2018 Big Data technologies: A survey Journal of King Saud University – Computer and Information Sciences pp 431 - 448
[11] Harrington P 2013 *Machine Learning in Action* (Manning Publications) p 354
[12] Yanga C, Huangb Q, Lic Z, Liua K and Hua F 2017 Big Data and cloud computing: innovation opportunities and challenges *International Journal of Digital Earth* **10** 1 13–53
[13] Guo H, Goodchild M, Annoni A (Editors) 2016 *Manual of Digital Earth* (Singapore Springer) p 852
[14] Herbst N R, Kounev S, and Reussner 2013 Elasticity in Cloud Computing: What It Is, and What It Is Not Proceedings of the Autonomic Computing 10th International Conference, San Jose, CA, pp 23-27
[15] Becker S, Brataas G, and Lehrig S (Eds.) 2017 *Engineering Scalable, Elastic, and Cost-Efficient Cloud Computing Applications. The CloudScale Method.* (Springer International Publishing AG) p 187
[16] Semenov A S 2019 Prototype based Programming with Fractal Algebra *AIP Conference Proceedings* **2181**, 020009 https://doi.org/10.1063/1.5135669
[17] Semenov A S 2020 Essentials of Fractal programming L. C. Jain et al. (eds.), Advances in Theory and Practice of Computational Mechanics, *Smart Innovation, Systems and Technologies* **173** 373-386
[18] Luke S 2015 *Essentials of Metaheuristics* Second Edition Online Version 2.2
[19] Rob W 2014 Core i7-5960X Extreme Edition Review: Intel's Overdue Desktop 8-Core Is Here. (Techgage)
[20] Chiappetta M 2020 AMD Threadripper 3990X Review: A 64-Core Multithreaded Beast Unleashed. HotHardware.

IOP Conf. Series: Materials Science and Engineering **927** (2020) 012071 doi:10.1088/1757-899X/927/1/012071

Influence of the angular momentum on definition of pressure

E Prozorova
Mathematical-Mechanical Department St. Petersburg State University Av. 28, Peterhof, 198504, Russia

E-mail: e.prozorova@spbu.ru

Abstract. The paper studies the impact of angular momentum in aerodynamic on the problem of flowing liquid near the surface. In a modified theory, the laws of conservation in integral form remain the same, but the divergent and rotational components are not separate. In this theory, the order of different equations is increased. The correspondence between the potential flow with velocity circulation, Newton law and Euler equations with different values of pressure on coordinate axis are discussed for potential flow. As an example of the influence of the angular momentum, the problem of fluid motion between two closely spaced parallel plates is considered.

1. Introduction

The work is devoted to the development of a new mathematical model of continuum mechanics, the elimination of inaccuracies in the description, and bringing the model correspondence with the principles of classical theoretical mechanics. The number of analytical methods used in physics to study mathematical models is limited: dimensional qualitative analysis of the problem, asymptotic methods for large (small) distances or times, for large (small) values of a certain dimensionless parameter of the problem, and others. These methods are applicable for a limited class of tasks. Since we do not have many laws, so we do not have many equations. They express conservation laws in each particular science. Currently, there are no adequate mathematical models for very important physical processes. For example, the theory of turbulence, virial equations of state, etc. Many of the principles of classical theory contradict theoretical mechanics and mathematics. A possible reason is poor mathematical models and calculation algorithms, despite a lot of work in this area of research. Such provisions include the decomposition of the projections of velocities into divergent and rotational components with respect to the axis passing through an arbitrary point. According to the rules on theoretical mechanics, the axis must pass through the center of inertia of the elementary volume. Therefore, we do not take into account the possibility of rotation of an elementary volume, since the moment of momentum is discarded, which leads to a loss of continuity of the medium. For example, the paper [1] discusses the "discontinuity of matter" in the mechanics of a continuous medium. The role of the moment for a material without structure was discussed in [2].

Some problems of mechanics cannot be solved at all in the framework of models with a symmetric stress tensor [3]. If we recall the derivation of the equations of mathematical physics [4] or the Boltzmann equation, then associating the derivation with the theory of difference schemes, we can see the correspondence of the derivation of conservation laws with the first order of approximation of numerical calculation. Consequently, the question arises of the required approximation order for difference schemes. N. N. Yanenko posed the question at a conference on problem of numerical methods. Currently, a field point of view is widespread; elementary volumes are considered as points. As a result, we have a symmetric stress tensor, a symmetric velocity tensor, etc. In classical theory to construct conservation laws using the Ostrogradsky-Gauss theorem and replace the surface integral with a volume integral by volume, we neglect the integral term, i.e. neglect the circulation on the sides of the elementary volume (in the two-dimensional case, this is clearly visible). Circulation means the presence of rotation, which in turn means the presence of a moment of force (angular momentum). Thus, the hypothesis of the absence for rotation of an elementary volume lays in all of continuum mechanics.

The speeds of various processes at the time of writing the equations were relatively small compared to modern ones. Subsequently, the scope of the theory developed for potential flows into flows with significant gradients of physical parameters was expanded. It was based on the laws of equilibrium of forces, the law of conservation of moment was considered as a consequence of the fulfillment of the law of equilibrium of forces. Regarding the difference methods, a number of remarks can be made: the experiment deals with material objects, i.e. there is an integral formulation of conservation laws.

In going over to differential equations, smooth functions are required. Therefore, one will work with other functions. Piecewise continuous functions were initially required, now smooth. The author proposes a model that takes into account the joint fulfillment of laws and the corresponding equation of state, including the contribution of the rotational velocity component. It is shown that after integration by parts, the transition from an integral formulation to a differential one is difficult. However, one can take advantage of physical and geometric considerations and obtain equations based on them, since the gradient of the moment of force is the force itself. The stress tensor obtained in theory turned out to be asymmetric. The lack of symmetry leads to an increase in the orders of the Navier-Stokes equations. Accordingly, the role of boundary conditions increases. Their specific form is determined by the task in question. Since the initial formulation obtained in the experiments is integral, numerical methods using conservation laws in integral form are preferred. These methods include the finite volume method. The implementation of the method is achieved in various ways. Within the volume, processes are not taken into account if there are no additional sources. There are many particles inside the volume, so that the statistical properties are satisfied. It follows that one has no right to reduce the volume to infinitely small. The Runge-Kutta method is widely used with the stated approximation order up to the fourth. The paper provides an estimate of the formal approximation order of various versions of the finite volume method with time integration by the Runge-Kutta method. Newton's experimentally confirmed law on the relation between the stress tensor and the strain tensor concerned the total velocity without distinguishing the symmetric term. In the kinetic theory, the coefficients of viscosity and thermal conductivity, even in the nonequilibrium version, use only the equilibrium part of the Chapman-Ensky function; the remaining terms in the function do not contribute. Consequently, the full account of the viscous contribution is ensured by the total speed, in fact, by its corresponding first derivatives. However, the macro parameters included in the equilibrium distribution function should not be calculated from the zeroth approximation from the Euler equations, but from the equations of the first approximation (Navier-Stokes equations). Similar equations are obtained from modified kinetic theory. The most important characteristic of an elementary volume is its position of the center of inertia and the correct design of the boundaries of an arbitrary volume on the coordinate axes. Due to the arbitrariness of the directions of motion, the viscous velocity components must be taken into account even in one-dimensional motion. With any restructuring of the flow, the position of the center of inertia changes, a moment of force is created, the gradient of which is the force itself. Additional force is added to the stresses, changing the direction of motion and the speed of the molecules. The result is a change in the equation of state even under equilibrium conditions and the formation of fluctuation phenomena. The equation of state (Newton's law) in classical mechanics provides the symmetry of the stress tensor. However, the equation does not include all the speeds that should be considered. The existing classical theory is constructed so that the main role is played by pressure forces. This is probably why, at low speeds, results coincidences with experiment are obtained. However, calculations even for a potential flow lead to the formation of a vortex sheet. For Euler's equations, this means that Pascal's law in the nonequilibrium case does not work and it is necessary to consider separately not p, but p_x, p_y, p_z.-

pressure on the normal to the surface of coordinate axis. An example of the importance of separating the individual components is the study of the processes of wave propagation in a rotating fluid [5] with the introduction of an item related to rotation in pressure gradient. The used values of the functions in the nodes of the difference schemes, no matter how one reduces the step, give asymmetric pressure values. The result of the study is a new closed model based on classical theoretical mechanics and mathematics.

2. Equations

The paper shows the necessity of considering the influence of the distributed moment in the well-known Taylor experiment [6] and, in general, in atmospheric physics, where gravity affects it. Stratification and gravity lead to a density gradient and the creation of a distributed moment, creating the formation of vortices relative to the horizontal coordinate axes. In [7] was obtained the equations with accounting influence of angular momentum

$$\frac{\partial \rho}{\partial t}+\frac{\partial \rho u_i}{\partial x_i}+\frac{\partial}{\partial x_i}\left(x_i\frac{\partial \rho u_i}{\partial x_i}\right)=0.$$

$$\frac{\partial \rho u_i}{\partial t}+\frac{\partial}{\partial x_i}\left(\rho u_i u_j+P_{ij}+x_i\frac{\partial P_{ij}}{\partial x_i}\right)-\frac{X_i}{m}\rho=0.$$

$$\frac{\partial}{\partial t}\rho\left(\frac{3}{2}RT+\frac{1}{2}u^2\right)+\frac{\partial}{\partial x_j}\left[\rho u_j\left(\frac{3}{2}RT+\frac{1}{2}u^2\right)+u_k P_{kj}+q_j\right]+$$

$$+\frac{\partial}{\partial x_i}x_i\frac{\partial}{\partial x_j}\left[\rho u_j\left(\frac{3}{2}RT+\frac{1}{2}u^2\right)+u_k P_{kj}+q_j\right]=0$$

$$y\,(\frac{\partial P_{xz}}{\partial x}+\frac{\partial \sigma_{yz}}{\partial y}+\frac{\partial \sigma_{zz}}{\partial z})-z(\frac{\partial \sigma_{xy}}{\partial x}+\frac{\partial \sigma_{yy}}{\partial y}+\frac{\partial \sigma_{zy}}{\partial z})+\sigma_{zy}-\sigma_{zy}=0$$

$$x\,(\frac{\partial \sigma_{xy}}{\partial x}+\frac{\partial \sigma_{yy}}{\partial y}+\frac{\partial \sigma_{zy}}{\partial z})-y(\frac{\partial \sigma_{xx}}{\partial x}+\frac{\partial \sigma_{yx}}{\partial y}+\frac{\partial \sigma_{zx}}{\partial z})+\sigma_{yx}-\sigma_{xy}=0$$

$$x\,(\frac{\partial \sigma_{xz}}{\partial x}+\frac{\partial \sigma_{yz}}{\partial y}+\frac{\partial \sigma_{zz}}{\partial z})-z(\frac{\partial \sigma_{xx}}{\partial x}+\frac{\partial \sigma_{yx}}{\partial y}+\frac{\partial \sigma_{zx}}{\partial z})+\sigma_{zx}-\sigma_{xz}=0$$

The viscous component should be determined by full speed.

$$V'_{kt}=V+\frac{1}{2}\,rot\,V$$

t- time, x_i- coordinates, V, u_i-speeds, μ - viscosity, ρ-density, T-temperature, q –thermal-flow, P_{ij}-tensor of viscous pressure, X-force. The system is complemented by the law of conservation.

Let us try to consider the potential flow in the plane and compare the Euler equations in the record of potential functions. We have equations for the potential flow (φ is the velocity potential, Ψ is the current function)

$$\frac{\partial^2\varphi}{\partial x^2}+\frac{\partial^2\varphi}{\partial y^2}=0,$$

$$u=\frac{\partial\varphi}{\partial x},\quad v=\frac{\partial\varphi}{\partial y}\quad \text{и}\ \frac{\partial^2\psi}{\partial x^2}+\frac{\partial^2\psi}{\partial y^2}=0,$$

$$\frac{\partial\varphi}{\partial x}=\frac{\partial\psi}{\partial y},\ \frac{\partial\varphi}{\partial y}=-\frac{\partial\psi}{\partial x}.$$

Euler equations in new variables

$$u\frac{\partial u}{\partial x}+v\frac{\partial u}{\partial y}=-\frac{\partial p}{\partial x}\Rightarrow\frac{\partial\varphi}{\partial x}\frac{\partial^2\varphi}{\partial x^2}+\frac{\partial\varphi}{\partial y}\frac{\partial^2\varphi}{\partial y\partial x}=-\frac{\partial\frac{\partial y}{\partial x}p}{\partial x}$$

$$u\frac{\partial v}{\partial x}+v\frac{\partial v}{\partial y}=-\frac{\partial p}{\partial y}\Rightarrow\frac{\partial\varphi}{\partial x}\frac{\partial^2\varphi}{\partial y\partial x}+\frac{\partial\varphi}{\partial y}\frac{\partial^2\varphi}{\partial y^2}=-\frac{\partial p}{\partial y}.$$

Let us express in terms of the function ψ

$$\frac{\partial\psi}{\partial y}\frac{\partial^2\psi}{\partial x^2} - \frac{\partial\psi}{\partial x}\frac{\partial^2\Psi}{\partial y\partial x} = -\frac{\partial p}{\partial y}$$

We have a different equation and other boundary conditions. If we take the average pressure function, the pressure is obtained depending on the method of its determination. Hence, the impossibility of the same values of pressure p and pressure is a vector as the force. In this case, the Lamb equation has the form

$$\frac{\partial V}{\partial t} + grad\left(\frac{V^2}{2}\right) + rotV \times V = F - \frac{1}{\rho}\, grad\,\vec{P}$$

and there is a dependence of density as an average value over an elementary volume, then the Bernoulli equation has the form. For the numerical solution of the Euler and Navier-Stokes equations, the finite volume method is often used. However, its formulation is based on differential equations with a symmetric stress tensor. In the mechanics of liquid and gas, it is not customary to average the calculation results over the calculated cells, in solid mechanics this is practiced, with a large number of steps this gives significant errors.

3. The influence of the moment in the problem of fluid motion between two close parallel plates

Considering the flow of a very viscous fluid between two parallel plates, the distance h will be considered very small. Reynolds number is small, external forces are absent. Initial statement of the problem is

$$\frac{\partial p}{\partial x} = \mu\left(\frac{\partial^2 v_x}{\partial x^2} + \frac{\partial^2 v_x}{\partial y^2} + \frac{\partial^2 v_x}{\partial z^2}\right),$$

$$\frac{\partial p}{\partial y} = \mu\left(\frac{\partial^2 v_e}{\partial x^2} + \frac{\partial^2 v_y}{\partial y^2} + \frac{\partial^2 v_y}{\partial z^2}\right),$$

$$\frac{\partial p}{\partial z} = \mu\left(\frac{\partial^2 z}{\partial x^2} + \frac{\partial^2 v_z}{\partial y^2} + \frac{\partial^2 u_z}{\partial z^2}\right).$$

The axes Ox, Oy lie in one of the boundary planes, the axis Oz is directed perpendicular to these planes and the equations of the boundary planes z = 0, z = h. It is assumed that the velocity is directed parallel to the boundary planes, so that

$$v_z = 0.$$

Due to the small distance between the plates, the order of the derivative $\frac{\partial v_z}{\partial x}$ is large compared to the derivatives $\frac{\partial v_x}{\partial x}$ and $\frac{\partial v_x}{\partial y}$. The order of the derivatives $\frac{\partial^2 v_x}{\partial z^2}$ is large compared to the orders of the derivatives $\frac{\partial^2 v_x}{\partial x^2}$ and $\frac{\partial^2 v_x}{\partial y^2}$. Then the equations take the form

$$\frac{\partial p}{\partial x} = \mu\frac{\partial^2 v_x}{\partial z^2},\quad \frac{\partial p}{\partial y} = \mu\frac{\partial^2 v_y}{\partial z^2},\quad \frac{\partial p}{\partial z} = 0,\quad \frac{\partial v_x}{\partial x} + \frac{\partial v_н}{\partial н} = 0.$$

The author shows the influence of the moment in this problem.

The author takes into account that the Euler equations provide the main order. In classical theory the pressure is selected in the center of the cell, is the same for all coordinate axes, and is equal to p. Suppose that the angular momentum acts and creates a small additional force.

$\tilde{p}_1 = p + \epsilon f_1$, $\tilde{p}_2 = p + \epsilon f_2$, The result of the additional force will be new velocity

$\tilde{v}_x = v_x + \epsilon V_1$, $\tilde{v}_y = v_y + \epsilon V_2$.

$$\frac{\partial\tilde{p}_1}{\partial x} = \mu\frac{\partial^2\tilde{v}_x}{\partial z^2},\quad \frac{\partial\tilde{p}_2}{\partial y} = \mu\frac{\partial^2\tilde{v}_x}{\partial z^2},\quad \frac{\partial\tilde{p}_1}{\partial z} = 0,\quad \frac{\partial\tilde{p}_2}{\partial z} = 0,\quad \frac{\partial\tilde{v}_x}{\partial x} + \frac{\partial\tilde{v}_y}{\partial x} = 0.$$

$$f_1 = -y\frac{\partial\tilde{p}_1}{\partial x},\quad f_2 = x\frac{\partial\tilde{p}_2}{\partial y}.$$

$$\tilde{v}_x = \frac{z^2}{2}\frac{\partial \tilde{p}_1}{\partial x} + z\,A_1\,(x,y) + B_1\,(x,y).$$

$$\tilde{v}_y = \frac{z^2}{2}\frac{\partial \tilde{p}_1}{\partial x} + z\,A_2\,(x,y) + B_2\,(x,y).$$

Under the boundary conditions, we have

$$V_1 = -\frac{y}{2\mu}\frac{\partial^2 \tilde{p}_1}{\partial x^2} z(h-z)\,,\; V_2 = -\frac{y}{2\mu}\frac{\partial^2 \tilde{p}_1}{\partial y^2} z(h-z)\,,$$

Under the assumption of a small influence of the moment

$$V_1 = -\frac{y}{2\mu}\frac{\partial^2 p}{\partial x^2} z(h-z),\; V_2 = -\frac{y}{2\mu}\frac{\partial^2 p}{\partial y^2} z(h-z)\,.$$

We get two equations that pressure must satisfy

$$\frac{\partial^2 p}{\partial x^2} + \frac{\partial^2 p}{\partial y^2} = 0, \text{ and } y\frac{\partial}{\partial x}\left(-\frac{\partial^2 p}{\partial x^2}\right) + x\,\frac{\partial}{\partial y}\frac{\partial^2 p}{\partial y^2} = 0.$$

Therefore, the following equation must be satisfied

$$y\frac{\partial}{\partial x}\left(\frac{\partial^2 p}{\partial x^2}\right) - x\,\frac{\partial}{\partial y}\frac{\partial^2 p}{\partial x^2} = 0.$$

4. Conclusion

The author investigated the influence of angular momentum in aerodynamic on the problem of flowing. The influence of angular momentum is investigated for the problem of liquid motion between two closely spaced parallel plates. In a modified theory, the laws of conservation in integral form remain the same, but the divergent and rotational components are not separate. In suggested theory, the order of differential equations increases. The correspondence between a potential flow with velocity circulation, Newton law and Euler's equations with different values of pressure along the coordinate axes is discussed. The results allow one to hope that the pressure distribution along the corresponding axes is responsible for large eddies. Dispersion creates small waves. Viscosity is responsible for dissipation.

References

[1] Ilyushin A A 1996 *Asymmetry of strain and stress tensors in the mechanics of a continuous medium.* (WEST. Moscow. University. Ser. 1. Mathematics-mechanics)

[2] Prozorova E V 2013 The Effect of Angular Momentum and Ostrogradsky-Gauss Theorem in the Equations of Mechanics. *WSEAS Transaction on fluid Mechanics* DOI: 10.37394/232013.2020.15.2

[3] Bulanov E A 2012 *Moment stresses in the mechanics of a solid, bulk and liquid body.* M .; University book.

[5] Loytsyansky L G 1970 *Fluid Mechanics* (M: Nauka)

[6] Kochin N E, Kibel I A, Rose N V 1963 Theoretical hydromechanics. Part 2. M: State publisher. Physical and mathematical literature.

[7] Prozorova. E 2018 *Influence of the dispersion in model of continuous mechanics.* (Lambert)

IOP Conf. Series: Materials Science and Engineering **927** (2020) 012072 doi:10.1088/1757-899X/927/1/012072

Research of a two-phase steam system during its flow in the pipe and a device for measuring dryness steam

B Y Kaplan[1], A A Motorin[2] and E L Stupitsky[2]
[1] Federal State Budget Educational Institution of Higher Education «MIREA - Russian Technological University», 18, Prospect Vernadskogo, Moscow, 119454, Russian Federation
[2] Institute for Computer Aided Design, Russian Academy of Sciences, 19/18, 2nd Brestskaya str., Moscow, 123056, Russian Federation

E-mail: vansp91@gmail.com, stup@bk.ru

Abstract. Water steam is widely applied in various fields of industry and heat power engineering. The article shows a wide application of water steam in heat-and-power engineering. In this regard, the issue of preserving and controlling the dryness of steam during its transportation via steam lines remains relevant. Quite detailed theoretical studies of steam parameters during its flow in the pipe have been performed. A comparative analysis of steam velocity and temperature plots depending on certain flow parameters and accepted approximations is performed. The analysis presented that gas-dynamic and thermodynamic processes can be considered approximately. This does not play a fundamental role in the general patterns of dropping liquid. As a result of numerical studies of the steam condensation kinetics, the specific volume and density of drop condensate along the pipe section are determined. The obtained results and the proposed methodological approach for determining the parameters of steam were used in the development and further improvement of the device for measuring the dryness of steam, the original construction of which was proposed earlier by one of the authors. The article presents the functional diagram and layout of the measuring unit of the gauge.

1. Introduction

Water steam of various temperatures and pressures is widely used in industry, the production of building materials (concrete products, polystyrene, etc.), power engineering (steam turbines for electric generators drive), food and pharmaceutical industries.

Steam plays a special role in oil production and petrochemistry. High-pressure steam is injected into wells to reduce its viscosity and increase the pressure in the reservoir; during oil refining, steam is injected into rectification columns to obtain light hydrocarbon fractions; during the transportation of hydrocarbons, steam is used to dilute products for washing pipeline, tanks etc.

Water steam has several positive properties, which explains its wide practical application in various fields:

- high heat capacity
- full explosion and fire safety
- chemical safety of components (steam turns to water during condensation);

IOP Conf. Series: Materials Science and Engineering **927** (2020) 012072 doi:10.1088/1757-899X/927/1/012072

- easy regulation of steam parameters, by changing the pressure or by throttling.

A significant part of the fuel consumed in our country is spent on steam generation. Therefore, from an economic point of view, it is important that during its transportation through steam pipelines to the consumer, losses in its heat content are minimal. Poor-quality and worn-out thermal insulation of steam pipelines leads to cooling of the steam and its partial transformation into droplet moisture as a result of condensation, which can significantly reduce the amount of heat transferred. Therefore, the important issue is the tooling measurement of steam dryness during its transportation, instead of the existing indirect methods of estimation. Steam dryness k is meant as the relation of the mass of steam m_s to the sum of the masses of steam and the droplet liquid $m_d : k = m_s / (m_s + m_d)$.

The current approaches to estimating the flow of a two-phase steam-water mixture in a channel are purely empirical and have been adapted mainly to specific large-scale installations and systems [1,2]. In [3], an original system was proposed for automatic measurement of steam dryness. The system has a simple design, but it has the accuracy needed for industrial steam pipelines. However, to ensure the assumptions accepted in [3], a more detailed theoretical analysis of a number of processes affecting the moisture content in water steam is required. This is the purpose of this work.

The physical content of the idea of measuring steam dryness is as follows [3]. A sample of wet steam is taken from the steam pipe and its temperature and pressure are measured. Then, a sample of constant mass of steam is subjected to isochoric heating, and the pressure and temperature of the sample are measured after heating. Measurement of temperature and pressure are performed sequentially before and after sampling and by original means. Fig. 1 shows a diagram of the main elements of the device for implementing the proposed method of measuring the dryness of steam and the main parameters of the steam-gas mixture.

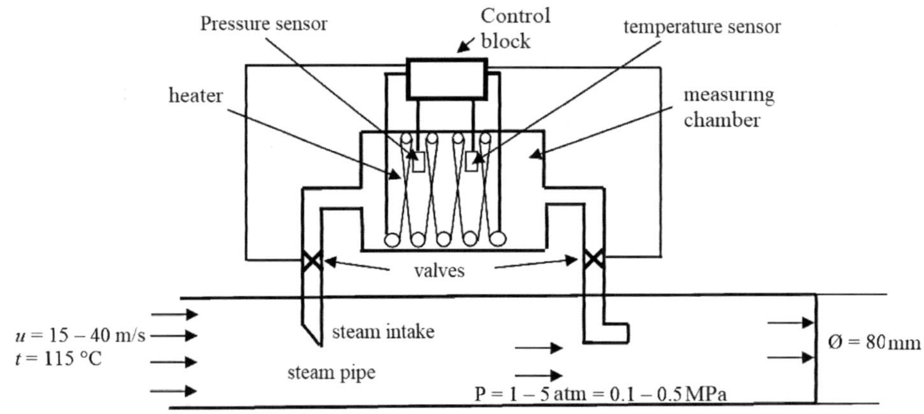

Figure 1. Scheme of the device design for measuring the dryness of steam and its main elements: $u - m/s$; $t - {}^\circ C$; $\% - mm$; P – atm.

As follows from the proposed idea of the device for measuring dryness steam and its scheme, in order to create adequate algorithms for the functioning of individual blocks, within the acceptable accuracy of determining steam parameters, detailed computational and theoretical researches of a number of issues are required:

- Estimate the distribution of the hydrodynamic characteristics of a two-phase flow along the radius of the pipe;
- To develop an approximate method for calculating the temperature distribution in the main channel and in the diagnostic chamber.
- To research the occurrence and evaporation of fine condensate, i.e. a two-phase steam-drop system when it moves in the pipe.

Special engineering analysis requires for questions related to transients in the diagnostic chamber itself, in particular, during steam intake, and others.

2. Comparative analysis of gas-dynamic flows in a pipe

To justify the assumptions made in [3], a more detailed analysis of a number of processes affecting the moisture content in water steam is required. At the same time, gas- and thermodynamic processes can be considered approximately, since they do not introduce fundamental distortions in the overall picture of changes in the fine-dispersed liquid phase during the flow of the steam-water media in the pipe.

If the internal friction force at the boundary with the pipe wall and at the boundaries between the layers of incompressible gas are balanced by the effect of pressure forces, and the movement is steady and laminar, then for the radial distribution of velocity, it is easy to obtain the well-known Poiseuille formula

$$u(r)=\frac{P_1-P_2}{4\eta L}R^2\left(1-\frac{r^2}{R^2}\right)=u_0\left(1-\frac{r^2}{R^2}\right), \tag{1}$$

where $\Delta P=P_1-P_2$ – pressure drop inside a pipe of length L; R – pipe radius; η – the coefficient of viscosity of the flow; $u_0=\Delta PR^2/4\eta L$ – maximum speed along the pipe axis. For the average speed we get:

$$\bar{u}=\int_0^R\frac{u(r)2\pi rdr}{\pi R^2}=\frac{1}{2}u_0=\frac{\Delta PR^2}{8\eta L}. \tag{2}$$

For the problem under consideration, the range $\bar{u}=15\div 40m/s$ is of interest when $R\approx 0.04m$; $\eta=1.12\times 10^{-5}kg/m\cdot s$; $\rho=1.1kg/m^3$.

It is known from the experiment that the flow pattern changes fundamentally and quite sharply if the dimensionless parameter is the Reynolds number

$$Re=\frac{2R\bar{u}\rho}{\eta} \tag{3}$$

becomes greater than a certain critical value, which in the case of a large cross-section pipe is approximately equal to $Re_{cr}\approx 2200$. At $Re>Re_{cr}$, the flow becomes turbulent, in which the transport processes are determined not by the molecular length of the free path, but by the spectrum of pulsations and their average scale [4]. However, near the wall, the velocity is small, so the friction against the wall is determined by the laws of laminar flow, which, however, gives a logarithmic divergence. Therefore, a power dependence for u(r) is often used, which gives similar results (Fig. 2)

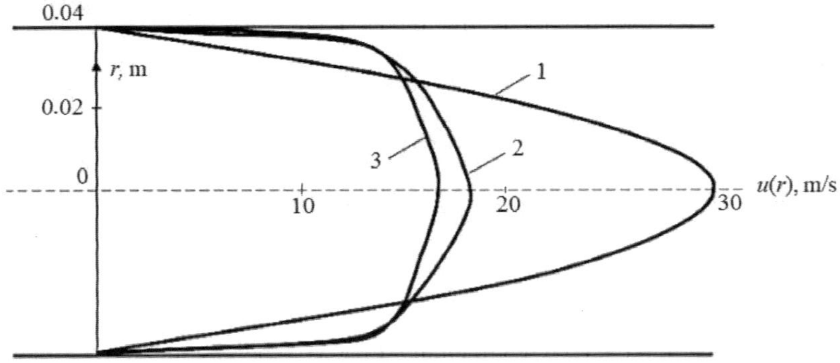

Figure 2. Velocity profiles: 1-laminar; 2 – power turbulent; 3 – logarithmic turbulent.

3. Comparative analysis of the temperature field during the motion of steam in the pipe.

Under the considered conditions $(t\le 150\text{ s})$, the equation for temperature has the form

IOP Conf. Series: Materials Science and Engineering **927** (2020) 012072 doi:10.1088/1757-899X/927/1/012072

$$\frac{3\rho k}{m_0}\left(\frac{\partial T}{\partial t}+\mathbf{u}\nabla T\right)+\frac{\rho kT}{m_0}\operatorname{div}\mathbf{u}=\operatorname{div}æ\cdot\operatorname{grad}T\ , \tag{4}$$

where $æ$ is the thermal conductivity of steam, $W\ /\ m\cdot K$.

The steam velocity has a component only along the pipe axis, i.e. at the x coordinate. After entering the pipe, the temperature profile along the r, as estimates show, is produced slightly longer than the velocity profile, but the initial transition process can be ignored and considered a stationary problem, i.e. $u=u(r); T(t,x)$, then

$$\operatorname{div}\mathbf{u}=0;\ (\mathbf{u}\nabla)T=u\frac{\partial T}{\partial x}.$$

Moreover, equation (4) can be written as

$$u(r)\frac{\partial T}{\partial x}=\frac{1}{r}\frac{\partial}{\partial r}\left(r\chi\frac{\partial T}{\partial r}\right)+\frac{\partial}{\partial x}\left(\chi\frac{\partial T}{\partial x}\right), \tag{5}$$

where $\chi = æm_0/3\rho k$ – is the thermal conductivity, m^2/s.

Using the method of splitting by coordinates $(r,\ x)$ with constant τ_0, the solution of the Bessel equation for the radial Temperature dependence in the form of a rapidly diverging series in $\frac{r}{(\chi\tau_0)^{1/2}}$, where χ is the temperature conductivity [5], we obtain an approximate solution of the temperature equation in the form

$$T(r,x)\approx T_0e^{-\lambda x}\left(1-\frac{r^2}{4\chi\tau_0}\right),$$

where $\lambda=\frac{u}{2\chi}\left[(1+\frac{4\chi}{\tau_0u^2})^{1/2}-1\right]$; and the value τ_0 is obtained from the condition of non-equilibrium heat flow through the pipe walls.

Then, setting $x = 0$ (i.e., not far from the beginning of the flow), we obtain

$$\tau_0=\frac{\dfrac{R^2}{4\chi_s}+\dfrac{R\rho_sC_p}{2\alpha}+\dfrac{R^2\rho_sC_p}{2æ_c}\ln\left(\dfrac{R+d}{R}\right)}{1-\dfrac{T_e}{T_0}}. \tag{6}$$

Taking C_p = 1.84×10^3 J/kg·K; ρ_s = 1,0 kg/m^3; α = 15 W/m^2K; $æ_c(F_e)$ = 74 W/$m\cdot K$; $R=0.04m$; d = 0.0003 m; T_e = 273 K; T_0 = 388 K; $\chi_s = 3.4\times10^{-5}\ m^2/s$, we obtain that $\tau_0 = 47.4\,s$.

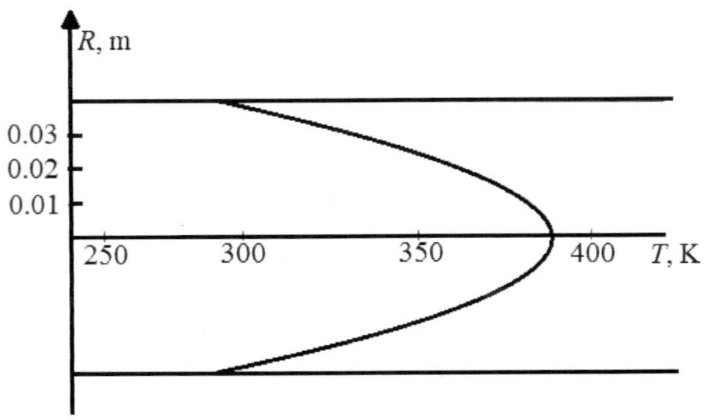

Figure 3. Temperature distribution over the entire section.

4. Kinetics of phase transitions and calculation of density and spatial distribution of moisture during steam motion in the pipe

The supersaturated substance can pass to the condensed phase in a homogeneous way, when the nuclei of the new phase are the centers of condensation, or heterogeneous, when condensation occurs at the existing centers (ions, solid particles), as well as in a mixed way. The condensation mechanism largely depends on the rate of change in the gas-dynamic characteristics of the media. These characteristics, as well as the characteristics of the introduced particles, are the initial data for the considered question of the dryness of steam during its motion in the pipe. The conditions of this problem are such that the process of homogeneous condensation is the most important.

The foundations of the theory of homogeneous concentration were developed by Zeldovich and Frenkel [7,8] and further development and application of this theory is contained in [2,6]. After the formation of condensation nuclei, the further description of the droplets growth can be strictly described by the function of the distribution of condensate particles along the radius. In this problem, the dryness of steam is determined by the total moisture content, so the most convenient model of the volume condensation process is given by the moment method [9.10], based on the distribution function. In [10], an algorithmic basis for this method is created, which is used in this work.

Figure 4 shows the change in the specific volume of condensate and its density along the pipe section. It should be noted a significant change of ρ_c depending on the distance to the pipe axis (6 times)

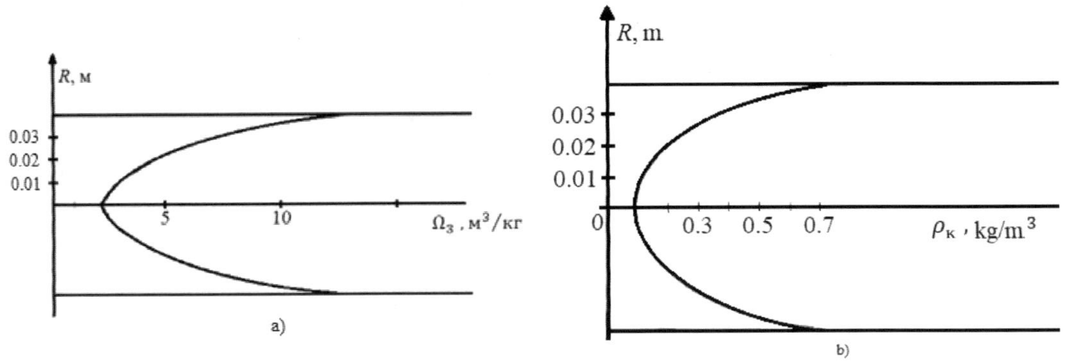

Figure 4. Specific volume of condensate (a). The density of drop condensate across the pipe section (b).

5. Conclusion

A fairly comprehensive physical and theoretical analysis of the processes that determine the behavior of the parameters of a steam-gas mixture during its motion in the pipe, carried out in this work, allowed us to start developing a device for measuring the dryness of steam. Currently, a prototype of such a device has been created (Figure 5).

Further improvement of individual nodes of this prototype, based on the research carried out in this work, and clarification of the quantitative values of the system parameters will allow us to begin its implementation in the industrial thermal power engineering.

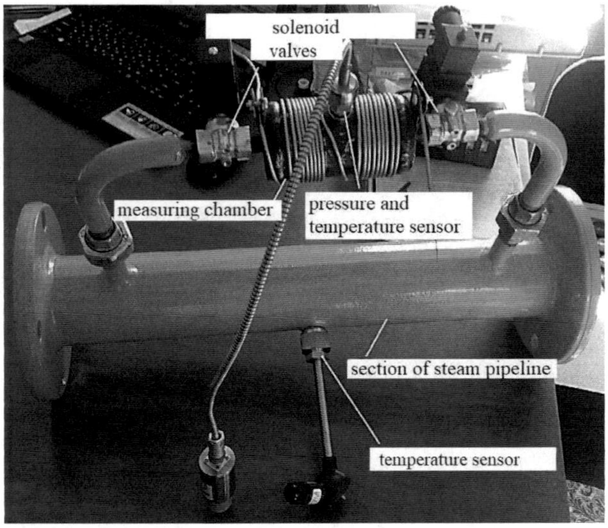

Figure 5. Prototype of dryness steam measurer

References

[1] G A Filipov and O A Povarov 1980 *Moisture Separation in NPP Turbines* (Moscow: Energiya)

[2] M E Deich and G A Filipov M E 1981 *Gas Dynamics of Two-Phase Media* (Moscow: Energoizdat)

[3] B Y Kaplan 2015 *Steam Dryness Meter* Russian patent for invention no. 2568050 Bulletin of inventions no. 31

[4] L D Landau and E M Lifshits 1953 *Mechanics of Continuous Media* (Moscow: Nauka)

[5] E Ianke, F Emde and F Lesh 1960 *Special Functions* (Moscow: Nauka)

[6] A A Lushnikov and A G Sutugin 1976 Present state of the theory of homogeneous nucleation *J. Usp. Khim.* **45 3** 385-415

[7] Ya I Frenkel 1939 *J. JETP* **9** 952

[8] Ya B Zeldovich 1942 Theory of new-phase formation. Cavitation *J. JETP* **12** 525

[9] L E Sternin 1974 *Fundamentals of the Gas Dynamics of Two-Phase Nozzle Flows* (Moscow: Mashinostroenie)

[10] N M Kortsenshtein, A A Kudryavtsev, V I Molotkov, A P Nefedov, A Yu Repin, E V Samuilov, E L Stupitskii 1998 The effect of finely divided particles and ionizing radiation on the characteristics of low-temperature plasma *J. High Temp.* **36** 854-9

IOP Conf. Series: Materials Science and Engineering **927** (2020) 012073 doi:10.1088/1757-899X/927/1/012073

Multiple antenna gyro-GNSS attitude determination system

M V Zharkov, K K Veremeenko, I M Kuznetsov and A N Pronkin
Automated Complexes of Attitude and Navigation Systems Department, Moscow Aviation Institute (MAI) (National Research University), Volokolamskoe shosse, 4, Moscow, Russian Federation

E-mail: mv_zharkov@mai.ru

Abstract. Attitude determination system is one of the most important system to control the moving object of any type. GNSS attitude determination systems have been actively developed in the last decade. Such systems have obvious drawbacks associated with the need for the visibility of a sufficient number of satellites. GNSS/Inertial integration is the best way to reduce satellite visibility dependence. The new concept of multiple antenna integrated inertial-satellite system design is proposed. System structure allows to exclude accelerometers from inertial measurement unit replacing it with gyro unit only. An important feature of the system is the availability of the initialization mode in motion and in conditions of limited satellite visibility. System performance was estimated in simulation and car test.

1. Introduction

Moving object attitude determination, especially of an aircraft, is an important task, without which it is almost impossible to control the object. The traditional source of measurement information in attitude determination systems are gyroscopic devices.

Strapdown attitude and heading reference system (AHRS) generally is one of the modern aircraft airborne attitude determination system variety. To create an integral correction, accelerometers are included in the AHRS structure along with gyroscopes. Depending on type of inertial sensors used, the AHRS accuracy and cost can vary significantly. Systems meeting the civil aviation accuracy requirements are one of the most expensive avionics systems. At the same time, attitude determination systems based on measuring global satellite navigation systems (GNSS) signals have been actively developed in the last decade [1, 2].

In order to determine attitude parameters using GNSS it is required to measure the difference in the ranges between navigation satellites and, in a minimal configuration, two antennas installed on object. Due to small baseline distance (the vector connecting two onboard antennas) it is reasonable to use interferometric measurement model [3]. In consequence, the range difference $D_{ANT1,2}^{SAT1}$ between antennas *ANT*1, *ANT*2 and satellite *SAT*1 (Figure 1) can be found as follows [4]:

$$D_{ANT1,2}^{SAT1} = -(\boldsymbol{l}_{ANT1,2}\ ,\ \boldsymbol{u}_{ANT2}^{SAT1}), \tag{1}$$

where $\boldsymbol{l}_{ANT1,2}$ – baseline vector between *ANT*1, *ANT*2; $\boldsymbol{u}_{ANT2}^{SAT1}$ – unit vector from *ANT*2 to the *SAT*1; $(\boldsymbol{l}_{ANT1,2}\ ,\ \boldsymbol{u}_{ANT2}^{SAT1})$ – vectors $\boldsymbol{l}_{ANT1,2}$ and $\boldsymbol{u}_{ANT2}^{SAT1}$ scalar product operator.

Hence attitude can be found after receiving the measurements of range and getting the projections of vector $\boldsymbol{u}_{ANT2}^{SAT1}$ to the Earth-fixed coordinate system (being calculated with the use of any antennas coarse coordinates and ephemeris) and vector $\boldsymbol{l}_{ANT1,2}$ in the body frame. For instance, the vector matching technique could be more suitable in this case.

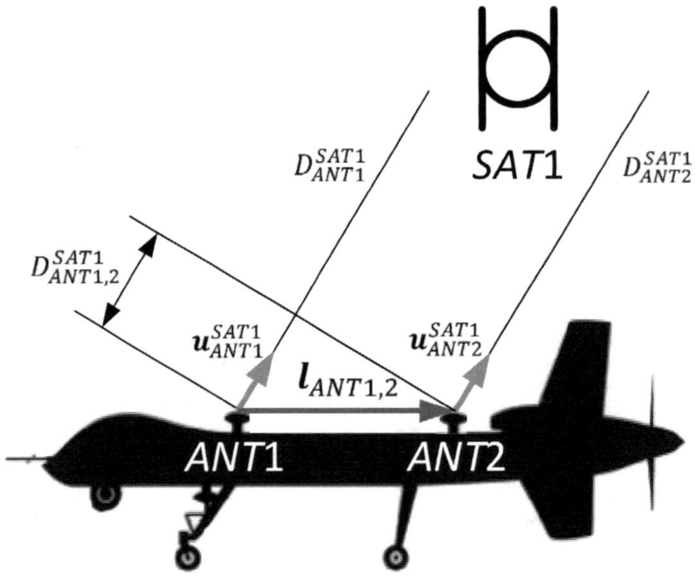

Figure 1. Interferometric measurement model.

To meet attitude accuracy requirements, it is necessary to use carrier phase measurements of range. Most phase measurements errors can be eliminated due to difference measurements (using double differences). The key task in developing an attitude determination satellite system is to eliminate the integer ambiguity of phase measurements. Moreover, the need of solving this problem arises both at the first epoch of measurement, and after the loss of tracking the phase of the so-called cycle slip.

2. System concept and structure

The system consists of two or three GNSS receivers and a gyro unit that measures the angular rates of an object in body frame. Note that when using the proposed system design concept, there is no accelerometer measurements necessity. If the conditions of GNSS signals measuring (satellite visibility, geometric factor, multipath, presence of cycle slip, etc.) allow resolving the integer ambiguity of phase measurements using so-called iterative methods [4] based only on information from GNSS receivers, the system initialization is performed (the first attitude angles determination). Moreover, initialization is also possible when the object is moving. Thereafter, at each measurement epoch, complex processing of measurements from GNSS receivers and the gyro unit is performed to calculate attitude parameters and control the integrity of the phase measurements. The gyro measurements are used to solve the modified Poisson matrix equation [5] in order to obtain a transformation matrix between the body frame and local level coordinate systems. In the short-term absence of GNSS measurements, the solution of the Poisson equation allows to support the inertial solution of attitude determination. In this case, the rate of error increase will obviously depend on the accuracy of the gyros.

An important feature of the system is the availability of the initialization mode in motion and in conditions of impossible use of iterative methods. Such conditions occur when the visibility of the satellites is limited, when their number in the constellation is less than five. To implement the initialization in these conditions, a new method is used for attitude parameters calculation using interferometric ambiguous GNSS phase measurements and absolute angular rate measurements [6].

3. Attitude determination in gyro-GNSS system [6]

In order to exclude receiver clock errors, equation (1) could be modified to:

$$D_{ANT1,2}^{SAT1,2} = D_{ANT1,2}^{SAT2} - D_{ANT1,2}^{SAT2} = -\left(\boldsymbol{l}_{ANT1,2}\ ,\ \boldsymbol{u}_{ANT2}^{SAT1,2}\right), \tag{2}$$

where $\boldsymbol{u}_{ANT1}^{SAT1,2} = \boldsymbol{u}_{ANT2}^{SAT2} - \boldsymbol{u}_{ANT2}^{SAT1}$.

The double difference phase measurement (2) at time *0* is:

$$F_{ANT1,2}^{SAT1,2}(0) = D_{ANT1,2}^{SAT1,2}(0) + I_{ANT1,2}^{SAT1,2} + E_{ANT1,2}^{SAT1,2}(0), \tag{3}$$

where $F_{ANT1,2}^{SAT1,2}(0)$ is double difference phase measurement; $D_{ANT1,2}^{SAT1,2}(0)$ is double difference real range; $I_{ANT1,2}^{SAT1,2}$ is the double difference integer ambiguity; $E_{ANT1,2}^{SAT1,2}(0)$ is the double difference phase measurement error.

Measurements (3) obtained for time *i* with no cycle slips can be found as:

$$F_{ANT1,2}^{SAT1,2}(i) = D_{ANT1,2}^{SAT1,2}(i) + I_{ANT1,2}^{SAT1,2} + E_{ANT1,2}^{SAT1,2}(i). \tag{4}$$

On the other hand, equation (2) could be written for time 0 and *i*:

$$D_{ANT1,2}^{SAT1,2}(0) = -\left(\boldsymbol{l}_{ANT1,2}(0)\ ,\ \boldsymbol{u}_{ANT2}^{SAT1,2}(0)\right), \tag{5}$$

$$D_{ANT1,2}^{SAT1,2}(i) = -\left(\boldsymbol{l}_{ANT1,2}(i)\ ,\ \boldsymbol{u}_{ANT2}^{SAT1,2}(i)\right). \tag{6}$$

Relationship between the projections of vector $\boldsymbol{u}_{ANT2}^{SAT1,2}(0)$ from the local level (*L*) coordinate system to the body frame (*B*) is direction cosine matrix:

$${}_B\boldsymbol{u}_{ANT2}^{SAT1,2}(0) = \left[M_B^L(0)\right] \cdot {}_L\boldsymbol{u}_{ANT2}^{SAT1,2}(0). \tag{7}$$

where $\left[M_B^L(0)\right]$ - direction cosine matrix; the left-side subscript denotes the coordinate system the vector is projected to.

Equation (2) for time *0* could be rewritten as:

$$D_{ANT1,2}^{SAT1,2}(0) = -\left({}_B\boldsymbol{l}_{ANT1,2}\ ,\ \left(\left[M_B^L(0)\right] \cdot {}_L\boldsymbol{u}_{ANT2}^{SAT1,2}(0)\right)\right). \tag{8}$$

Vector $\boldsymbol{u}_{ANT2}^{SAT1,2}(i)$ in (6) can be represented in the body frame:

$${}_{B(i)}\boldsymbol{u}_{ANT2}^{SAT1,2}(i) = \left[M_{B(i)}^{B(0)}\right] \cdot {}_{B(0)}\boldsymbol{u}_{ANT2}^{SAT1,2}(0), \tag{9}$$

where $\left[M_{B(i)}^{B(0)}\right]$ is the rotation matrix between the coordinate system coinciding with the body frame at time *0* and the coordinate system coinciding with the body frame at time *i*.

Matrix $\left[M_{B(i)}^{B(0)}\right]$ can be found using the differential equation [5] (Poisson matrix equation):

$$\left[\dot{M}_{B(i)}^{B(0)}\right] = -[\Omega] \cdot \left[M_{B(i)}^{B(0)}\right], \tag{10}$$

$$[\Omega] = \begin{pmatrix} 0 & -\Omega_{3_B} & \Omega_{2_B} \\ \Omega_{3_B} & 0 & -\Omega_{1_B} \\ -\Omega_{2_B} & \Omega_{1_B} & 0 \end{pmatrix}$$

where skew-symmetric matrix $[\Omega]$ elements Ω_{1_B}, Ω_{2_B}, Ω_{3_B} are the projections of the absolute angular rate (measured by the gyroscopes).

Thus, Equation (6) can be represented for phase measurements:

$$F_{ANT1,2}^{SAT1,2}(i) = -\left({}_{B}\boldsymbol{l}_{ANT1,2}\ ,\ \left(\left[M_{B(i)}^{B(0)}\right]\cdot\left[M_{B(0)}^{L(0)}\right]\cdot\left[M_{L(0)}^{L(i)}\right]\cdot {}_{L(i)}\boldsymbol{u}_{ANT2}^{SAT1,2}(i)\right)\right) - I_{ANT1,2}^{SAT1,2},\quad(11)$$

where $\left[M_{L(0)}^{L(i)}\right]$ is the direction cosine matrix, which can be taken as unit matrix for small periods of time.

Equation (11) can be extended in the case that more than two satellites and more than two antennas are used. The set of linear equations (11) can be solved by numerical methods. The unknowns are elements of matrix $\left[M_{B(0)}^{L(0)}\right]$ and the double difference of integer ambiguities $I_{ANT1,2}^{SAT1,2}$.

Thus, it is possible to calculate heading, pitch and roll for the first epoch *0* as well as floating values of integer ambiguities. The integer values of ambiguities can be verified by baseline length and contrast ratio tests [4].

4. Simulation results

The calculation accuracy of attitude parameters and integer ambiguities on the basis of proposed method depends on the condition number of coefficient matrix in set of linear equations (11). Other factors are the error in the GNSS phase measurements and the gyro drifts. The condition number of coefficient matrix depends mainly on the magnitude of the angular rate and the number of measurement epochs.

Numerous simulation scenarios were used to estimate multiple antenna gyro-GNSS attitude determination system characteristics. First of all, the simulation was done in terms of the effectiveness of the new method of attitude determination based on ambiguous GNSS phase measurements and absolute angular rate measurements. Simulation technique was the same as described in [6], but the range of simulated parameters values in this research is more extensive and specific.

The main parameters that were estimated during the simulation are: the value of the coefficient matrix condition number, the float values of integer ambiguities calculations error, heading, pitch, roll accuracy, number of measurement epochs, the error in the GNSS phase measurements, and gyro drifts.

The GNSS phase measurements error model is represented by specified intensity white noise and using GPS ephemeris. The gyro errors model are represented by the constant angular rate drift and specified intensity white noise.

Simulation results obtained in most stringent scenario are presented in Table 1. Scenario conditions: 100 km/h velocity driving; heading, pitch and roll rotations at a rate of 1 deg/s; two antennas with baseline length 1 m; phase measurement noise intensity – 5 mm; systematic gyro drift and the intensity of random gyro drift – 1 deg/h, GNSS output rate – 1 Hz, gyros output rate – 200 Hz.

Table 1. Simulation results

Time, s.			10			60			90		
Rotation[a]			H	H,R	H,R, P[c]	H	H,R	H,R, P	H	H,R	H,R, P
Sat[b]	PDOP										
4	1.85	μ[d]	10^6	10^5	10^4	9522	265	131	1578	127	46
		N[e], %	-	-	-	23	20	27	30	26	10
		ψ[f]	-	-	-	2.59	0.25	5.21	1.60	1.73	1.16
		υ	-	-	-	1.45	3.17	4.53	4.80	2.76	1.25
		γ	-	-	-	1.21	1.18	1.23	0.27	0.21	0.22
5	1.42	μ	10^7	10^5	10^4	10^4	179	145	1744	142	50
		N, %	-	-	-	17	34	17	10	28	7
		ψ	-	-	-	2.17	7.43	3.87	1.66	0.31	0.77
		υ	-	-	-	4.12	2.21	0.09	0.58	5.10	1.28
		γ	-	-	-	1.52	0.22	0.48	1.16	0.70	0.70
6	1.31	μ	10^7	10^5	10^4	8704	359	129	1447	126	44
		N, %	-	-	-	70	29	23	34	36	16

IOP Conf. Series: Materials Science and Engineering **927** (2020) 012073 doi:10.1088/1757-899X/927/1/012073

ψ	-	-	-	5.7	1.66	2.98	4.85	0.42	1.36
υ	-	-	-	11.9	1.03	3.44	3.70	3.82	2.15
γ	-	-	-	1.10	0.61	0.53	0.09	0.27	0.31

a Rotation corresponds to heading, roll and pitch.
b Number of satellites.
c H – heading, R – roll, P - pitch
d Coefficient matrix condition number.
e *N* is the RMS error in the calculation of floating ambiguities of phase measurements as a percentage of the wavelength.
f ψ, υ, γ is the RMS error of the heading, pitch and roll angles in degrees

5. Car test results

Car test was carried out with the use of small-sized integrated navigation system (SSINS) manufactured by MAI [7] (Figure 2) based on inertial module Silicon Sensing DMU02 and two GLONASS/GPS receivers NovAtel OEMV-1G. SSINS was mounted in the Moscow Aviation Institute Navigation System Automobile Testbed (Figure 3) inside a car roof rack [8]. Reference attitude parameters were obtained from Novatel SPAN GNSS Inertial Navigation System measurements and geodetic base GNSS receiver measurements postprocessed by NovAtel Inertial Explorer software. During the test the car moved through the streets of Moscow with different motion parameters: speed – from 0 to 110 km/h, pitch – from -5 to 2 deg., roll – from -5 to 5 deg., angular rate around longitudinal axis – from -30 to 30 deg/s., angular rate around lateral axis – from -12 to 12 deg/s., angular rate around normal (vertical) axis – from -25 to 15 deg/s. Test time is about 10 000 seconds.

Car test results (heading, pitch and roll errors) are shown in Figure 4. Roll error estimation results are the same as for the pitch.

Figure 2. MAI Small-sized integrated navigation system.

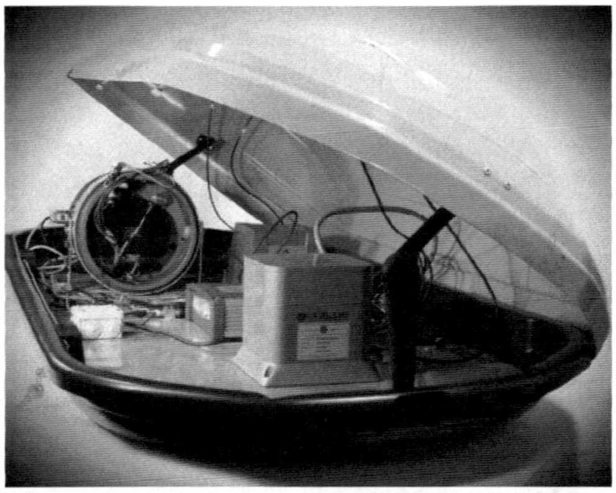

Figure 3. Moscow Aviation Institute Navigation System Automobile Testbed.

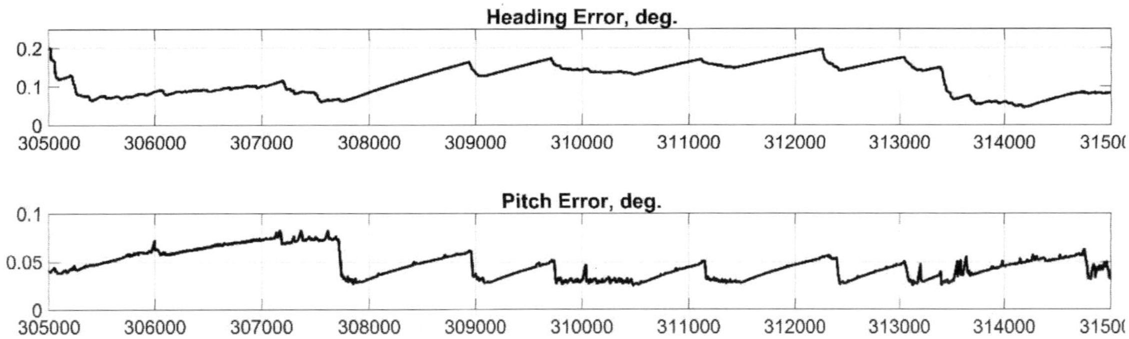

Figure 4. Attitude determination errors.

6. Conclusions

Simulation and car test results show gyro-GNSS attitude determination system performance. It was shown in [6] that integer ambiguity resolution can be resolved in case of rotation at rate of 5 deg/s. New simulation results obtained in most stringent scenario (rotation at rate of 1 deg/s) potentially expand system applicability. Pitch and roll maximum errors in the car test are 0.1 deg. and heading error is 0.2 deg. when using low-cost MEMS gyros and only two OEM L1 GPS receivers. Integer ambiguities was successfully resolved using iterative method (LAMBDA) or new method proposed in [6] during all test time (maximum GPS outage time was 60 seconds).

Acknowledgments

The reported study was funded by RFBR according to the research project № 19-08-01223a.

References

[1] Emel'yantsev G I, Stepanov A P, Blazhnov B A, Radchenko D A, Vinokurov I Y and Petrov P Y 2017 "Using satellite receivers with a common clock in a small-sized GNSS compass," *2017 24th Saint Petersburg International Conference on Integrated Navigation Systems (ICINS), St. Petersburg, 2017*, pp 1-2, doi: 10.23919/ICINS.2017.7995619

[2] Schnaufer B, McGraw G, Phan H, Joseph A 2016 "GNSS-Based Dual-Antenna Heading Augmentation for Attitude and Heading Reference Systems," *Proceedings of the 29th International Technical Meeting of the Satellite Division of The Institute of Navigation (ION GNSS+ 2016), Portland, Oregon, September 2016*, pp 3669-3691 https://doi.org/10.33012/2016.14603

[3] Emel'yantsev G I, Stepanov A P, Blazhnov B A and Semenov I V 2015 Glonass data processing in GPS compass with antenna base equal to carrier wavelength. *22nd Saint Petersburg International Conference on Integrated Navigation Systems, ICINS 2015 – Proceedings 2015*, pp 9-19

[4] Teunissen P J 2017 Carrier Phase Integer Ambiguity Resolution. In: Teunissen P J, Montenbruck O (eds) *Springer Handbook of Global Navigation Satellite Systems*. Springer Handbooks. Springer, Cham

[5] Savage Paul G 1998 Strapdown Inertial Navigation Integration Algorithm Design Part 1: Attitude Algorithms. *Journal of Guidance, Control, and Dynamics* **21(1)** 19-28

[6] Zharkov M V, Veremeenko K K, Antonov D A, and Kuznetsov I M 2018 Attitude Determination Using Ambiguous GNSS Phase Measurements and Absolute Angular Rate Measurements *Gyroscopy and Navigation* **9(4)** 275–283

[7] Veremeenko K K, Antonov D A, Zharkov M V, Zimin R Yu, Kuznetsov I M and Pron'kin A N 2012 Small-sized integrated navigation and landing complex of a variable structure. *19th Saint Petersburg International Conference on Integrated Navigation Systems, ICINS 2012 – Proceedings*. pp 279-282

[8] Antonov D A, Veremeenko K K, Zharkov M V, Zimin R Yu, Kuznetsov I M and Pron'kin A N 2016 Test Complex for the Onboard Navigation System of an Airport Ground Vehicle. *Journal of Computer and Systems Sciences International* **55(5)** 832–841

IOP Conf. Series: Materials Science and Engineering **927** (2020) 012074 doi:10.1088/1757-899X/927/1/012074

Dynamics of initiation of the pseudoshock combustion mode in a constant cross section channel

A V Tupikin[1], P K Tretyakov[1], A L Kuranov[2] and V M Abashev[3]

[1]Khristianovich Institute of Theoretical and Applied Mechanics, Russian Academy of Sciences, Siberian Branch, 630090 Novosibirsk, Russia
[2]Hypersonic Systems Research Institute of "Leninetz" Holding Company, 196066 St. Petersburg, Russia
[3]Moscow Aviation Institute, 125993 Moscow, Russia

E-mails: tupikin@itam.nsc.ru, mail@hypersonics.ru, abashevvm@mai.ru

Abstract. The results of a study on the dynamics of initiation of a pseudo-shock combustion mode in a channel of constant cross section under pulsed action on the flow are presented. Various conditions are considered (kerosene flow rate, change in channel length and working gas in control pulses). Pulsating combustion mode is implemented. It is shown that the reduction of the length of the constant section leads to a decrease in combustion stability. When applying control pulses, two scenarios of the development of disturbances can be implemented: with intensification of combustion and the formation of wave structures, or with attenuation without intensification of combustion. In case of combustion, the motion of the disturbance along the channel can conditionally be divided into two stages: $0 < t < 4$ ms and $t > 4$ ms. The duration of the first stage can be interpreted as a delay in the ignition of kerosene; in the second stage, the perturbation velocity relative to the channel walls is almost constant. The use of kerosene (synthesis gas) conversion products in pulses accelerates the transition from diffusion to pseudo-shock combustion, and the rate of disturbance propagation increases by about 2 times.

1. Introduction

Organization of an efficient working process in a ramjet engine with supersonic flow rate in the combustion chamber is the main fundamental task, solution of which will allow creating a new type of hypersonic aircraft. The efficiency of an engine cycle is directly related to the amount of the pressure loss along the engine path. At the same time, it is not always possible to minimize pressure losses while ensuring high combustion completeness. One possible solution is proposed in the patent [1]. The organization of combustion in wave structures, as in a pseudoshock, can provide high efficiency of the engine cycle in the combustion chamber. The passage of the fuel jet through the shock waves provides high-quality mixing of fuel with an oxidizer without additional devices. The optimal mode of operation of a high-speed ramjet will be when the average flow rate over the cross section of the path remains supersonic, and behind the pseudoshock the flow rate is close to the speed of sound across the combustion products. The most promising are the ramjet schemes operating on kerosene, since hydrocarbon fuel is preferable to hydrogen under the conditions of operation and storage, and can be used in the cooling system of an engine [2]. In the constant cross section area of the combustion chamber of variable geometry, it is possible to use kerosene conversion products to control the pseudoshock combustion mode. The effectiveness of such schemes requires experimental confirmation, and the processes taking place in the combustion chamber require detailed study. Information on the time characteristics of the phenomena is important.

The purpose of the work is to study the dynamics of initiation of pseudoshock combustion mode in a variable geometry combustion chamber. In the experiments, the parameters at the inlet of the model combustion chamber corresponded to the flight Mach number M≈6.0, and to intensify the combustion of kerosene, a pulse generator (*PG*) was used, operating on cold air or on the products of steam conversion of kerosene (synthesis gas).

2. Experimental setup

Figure 1 shows the experimental setup. The model combustion chamber consisted of two constant cross-section segments: the diameter of the first $d1$ = 50 mm, and the second $d2$ = 90 mm. combustion chamber with two lengths of the first section were tested: L = 423 (short) and 499 mm (long). Kerosene was supplied along the flow axis at the inlet of the combustion chamber. In both cases, the input location of the control pulses was at the end of the first section. At the combustion chamber inlet, the following flow parameters were set: total temperature $T^* \approx 1650$ K; total pressure $P^* \approx 0.7$ MPa; Mach number M = 2.2. The repetition frequency of the *PG* pulses was f = 45–52 Hz with a duration of $\tau \approx 12$–15 ms.

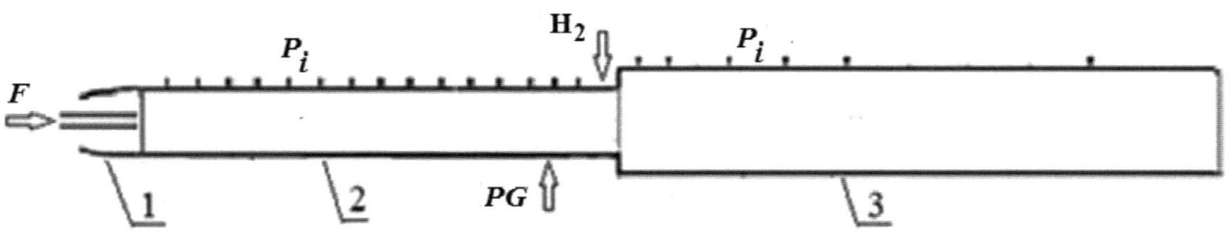

Figure 1. . Experimental setup: 1 – nozzle with fuel injector (*F*); 2 – constant cross section with gas-dynamic pulses input point (*PG*) and hydrogen supply (H_2); 3 – sudden expansion part; *Pi* – position of static pressure sensors.

3. Results and discussion

Figures 2 and 3 show the shape of gas-dynamic pulse (a) and change in the distribution of static pressure along the channel: b - at growth and c - at decrease of disturbance intensity. Fig. 2 demonstrates the case where the pulse does not cause the initiation and development of intense combustion (air to fuel equivalence ratio $\alpha = 1/\varphi = 0.6$), the pseudoshock wave structure does not form. The propagation of pressure and disturbance mainly occurs in the part of pulse growth (t_0–t_1), at the decline of the pulse there is an attenuation to the state of undisturbed current (t_1-t_2). Fig.3 is fundamentally different. Combustion intensification takes place, and a wave structure, such as pseudoshock, is formed. As a result, the disturbance front propagates during some period of pulse descent. In this case, the wave structure between pulses will not break and the current disturbance caused by pulses will not disappear.

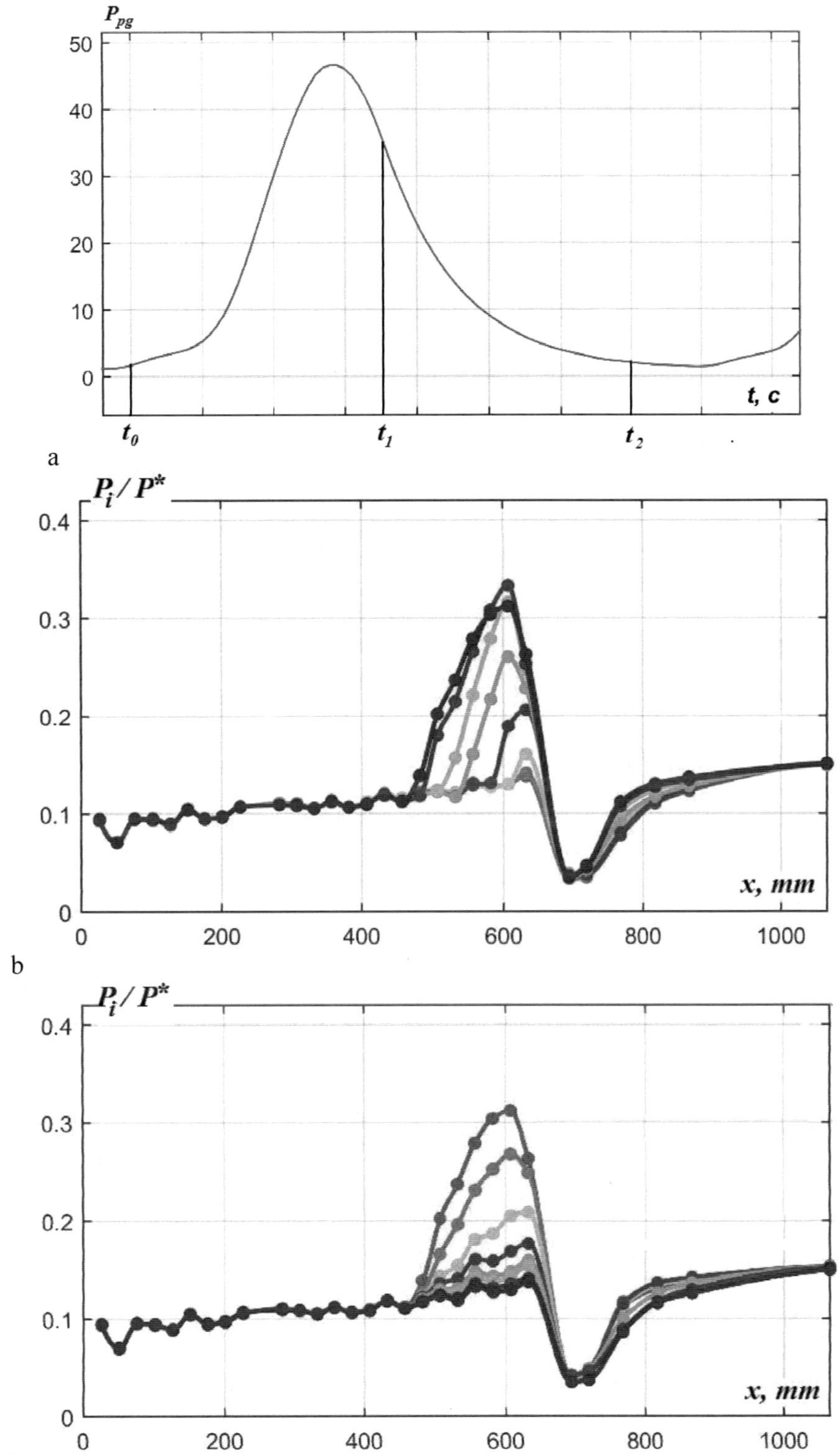

Figure 2. Pulse shape (a) and change in static pressure distribution along the channel (b – at growth and c – at decrease of disturbance intensity)

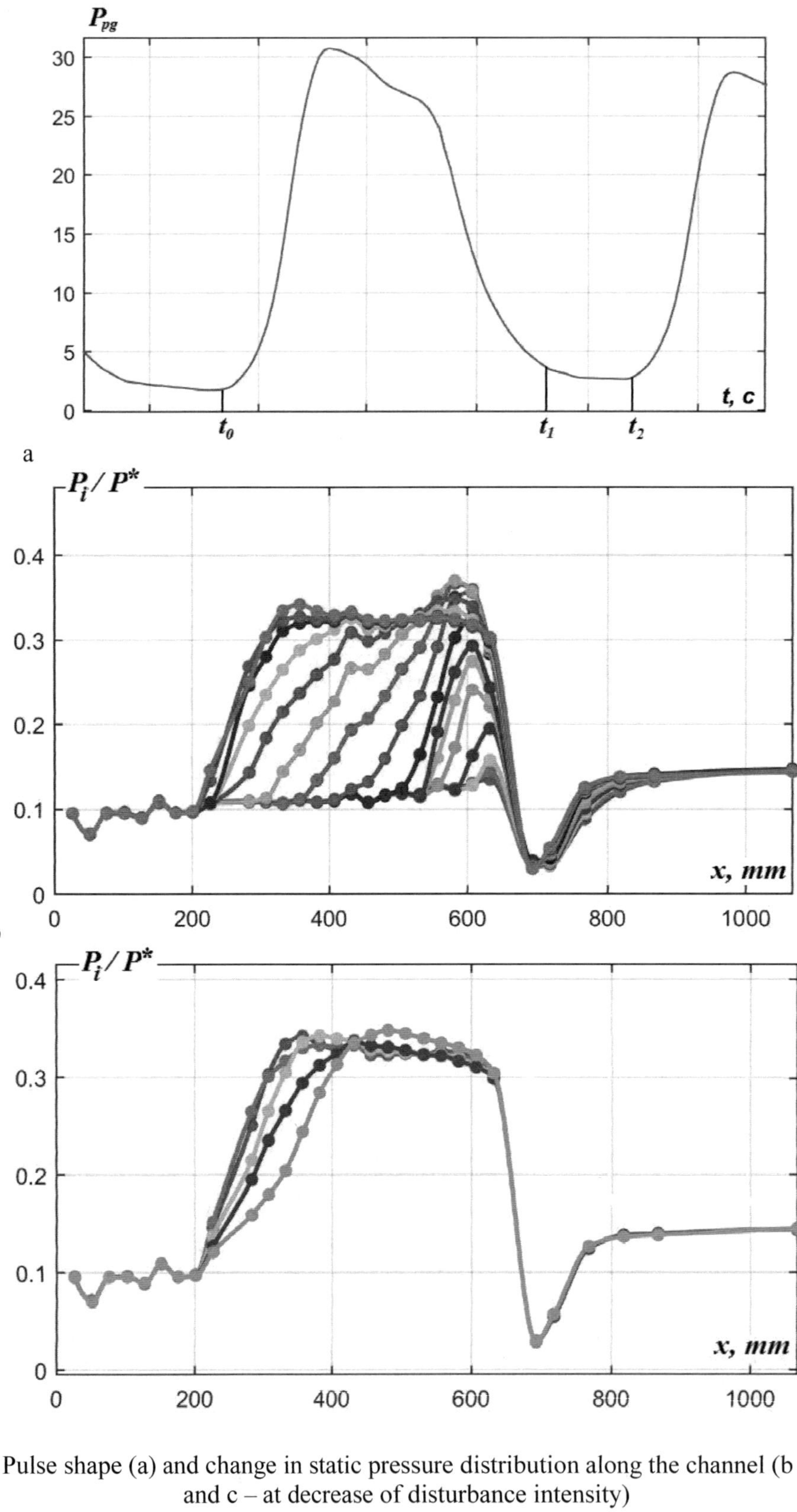

Figure 3. Pulse shape (a) and change in static pressure distribution along the channel (b – at growth and c – at decrease of disturbance intensity)

The graphs in Figure 4 represent the dynamics of initiation of a pseudoshock combustion mode in a channel of constant cross section. The air to fuel equivalence ratio varied from $\alpha = 0.6$ to $\alpha = 0.75$, while the amount of kerosene was not enough to realize thermal chocking of the flow in the channel. The distance Δx ($\delta x = \pm$ 1 mm) is calculated from the point of pulses supply. The time of arrival of the perturbation front was determined with an accuracy of $\delta t = \pm$ 0.2 ms. Under combustion conditions, the motion of the disturbance along the channel can roughly be divided into two stages: $0 < t < 4$ ms and $t > 4$ ms. The slow motion of the disturbance front in the first stage can be associated with the initiation of combustion which depends on the delay in the ignition of kerosene. At the second stage, the front velocity relative to the channel walls increases and remains almost constant ($u_{avg} \approx$ 60-75 m/s). In the absence of kerosene supply, the front movement is significantly lower (see Figure 4, $u_{avg0} \approx$ 9 m/s). The creation of gas-dynamic pulses by the generator from kerosene conversion products (synthesis gas) accelerates the transition from diffusion to pseudoshock combustion (see Figure 5), the rate of disturbance propagation increases by about 2 times.

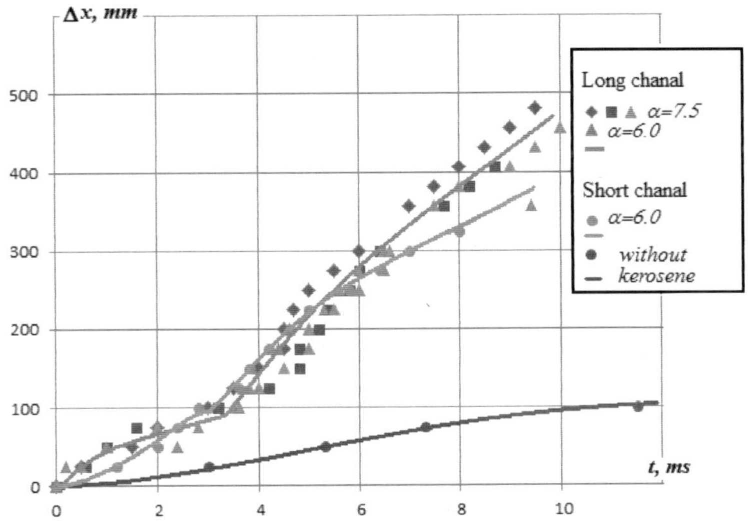

Figure 4. Propagation of disturbance front at different length of the 1st section.

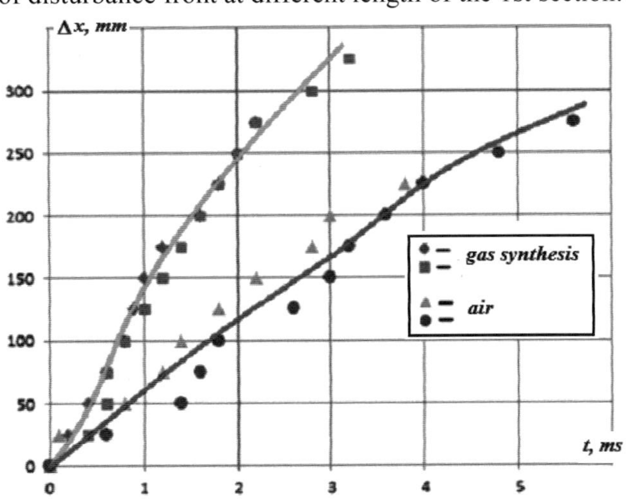

Figure 5. Propagation of disturbance front at different gas compositions in the PG.

4. Conclusions

Studies of the working process of a model combustion chamber are carried out under various conditions (kerosene flow rate, change in channel length and working gas in control pulses). Pulsating combustion mode is implemented. It is shown that the reduction of the length of the constant section leads to a decrease in combustion stability. When applying control pulses, two scenarios of the development of disturbances can be implemented: with intensification of combustion and the formation of wave structures, or with attenuation without intensification of combustion. In case of combustion, the motion of the disturbance along the channel can conditionally be divided into two stages: $0 < t < 4$ ms and $t > 4$ ms. The duration of the first stage can be interpreted as a delay in the ignition of kerosene; in the second stage, the perturbation velocity relative to the channel walls is almost constant. The use of kerosene (synthesis gas) conversion products in pulses accelerates the transition from diffusion to pseudo-shock combustion, and the rate of disturbance propagation increases by about 2 times.

Acknowledgments

The work was partially carried out within the framework of the Program of Fundamental Scientific Research of the State Academies of Sciences for 2013-2020 (project AAAA-A17-117030610135-6) and partly with financial support from the Russian Federal Property Fund (grant No. 20-08-00245).

References

[1] Tretyakov P K and Prokhorov A.N. 2018 *Patent RF,* **No. 2651016**
[2] Kuranov A L, Korabelnikov A V and Mikhailov A M 2016 *High Temperature* **54**(3)

IOP Conf. Series: Materials Science and Engineering **927** (2020) 012075 doi:10.1088/1757-899X/927/1/012075

Low-Power Ion Thrusters. State of Development at Research Institute of Applied Mechanics and Electrodynamics, MAI

R V Akhmetzhanov, A V Bogatyi, D A Kashirin, V A Obukhov, G A Popov, V V Svotina and S A Khartov
Research Institute of Applied Mechanics and Electrodynamics of the Moscow Aviation Institute (RIAME MAI)
Leningradskoye shosse 5, 125080, POB 43, Moscow, Russia

E-mail: ahmetzhanov1991@mail.ru

Abstract. This article describes the activities carried out by RIAME MAI in the field of low-power ion thrusters in 2010-2020. The results of study of a laboratory model of low power RIT using a different shape of GDC are presented. The results of work on creating an experimental model of such RIT with elements of the propulsion system are shown. The article also presents the research on the process of erosion of the accelerating electrode of the RIT – a critical element of the thruster design from the point of view of the lifetime.

1. Introduction

In recent years, the development of space technology has posed new challenges to the developers of electric propulsion thrusters (EPT). All countries actively involved in the development of space technologies have shown increased interest in the development of satellite systems based on small spacecraft (SSC). Based on the new requirements, various EPT options for SSC, including ion thrusters (IT), are currently under active consideration on a competitive basis.

There are different IT types, differing in the use of a particular method of transferring propellant to the plasma state. By now, ITs providing plasma formation based on the DC gas discharge (ITDC), RF gas discharge (RIT) and microwave discharge (MW IT) have been developed.

Technological problems of developing ITs of different types were already successfully solved in the early 70ies of the past century in the USA, Europe, Japan, and USSR and later in China. In the U.S., since 1997, the ITDC XIPS-13 has been used on the HS-601 platform [1], and the ITDC XIPS-25 [2] has been used as a standard system. In Europe, ITDCs T-5 and T-6 [3], and modifications of RIT-10, made according to the RIT scheme [4], have been tested in space. In Japan, along with ITDC, the microwave ITs are being successfully developed: μ-10 and μ-20. With the μ-10 thruster, the spacecraft successfully approached asteroids under the programs Hayabusa 1 and Hayabusa 2 [5]. In China, LIPS-200 ITDC with a rated power of about 1 kW is being developed and tested in space [6], and a more powerful ITDC LIPS-300 is tested in space also. In the USSR, ITDCs were successfully used in the 70ies in space experiments made by the Kurchatov Institute and TsAGI. Research and development of ITDC was conducted by MAI and EDB “Fakel”. In recent year, in Russia, ITDC have been successfully developed at the Keldysh Research Center [7], and RIT – at MAI [8].

In general, with reasonably chosen parameters for the specific impulse and power, all three IT schemes have proven their operability and have been successfully tested in various programs in Europe, USA and Japan.

2. Experimental

Over the past 10 years, the RIAME MAI has been designing, developing and studying radio-frequency ion thrusters of various power levels. Works in the field of RIT started in 2010, when the MAI won the tender and received a grant to establish a Laboratory of Radio-Frequency Ion Thrusters under the supervision of Professor Horst Wolfgang Loeb from the Giessen University to perform R&D: "Research and development of high-impulse radio-frequency electric ion thrusters for space applications" [8]. Under that grant, the laboratory models of ion sources with internal diameters of the gas discharge chamber (GDC) of 100 mm and 150 mm were designed, developed and studied. Work on the low-power RIT was continued in 2013, when RIAME MAI together with the JSC KBKhA started development of the radio-frequency ion thruster (RIT-8) with the power consumption of about 300 W. The laboratory model of RIT-8 was developed and tested at MAI (table 1) [9].

Table 1. Performance of RIT-8 laboratory model

No. of experiment	1	2
Specific impulse, s	2865	3775
Thrust, mN	8.1	8.1
Power consumption, W	273.1	281.4
Efficiency	0.417	0.533

In the course of research tests of RIT-8 laboratory model, three different types of GDC were used [9]:
GDC No. 1 – hemispherical GDC made of the mixture of alumina and silicon nitride;
GDC No. 2 – hemispherical GDC with the cylindrical part made of the same materials as the GDC No. 1;
GDC No. 3 – hemispherical GDC with the cylindrical part made of alumina ceramics VK-94.

Experiments were carried out, during which the dependence of the RFG power on the mass flow rate of xenon was obtained at fixed levels of the ion beam current (figure 1);

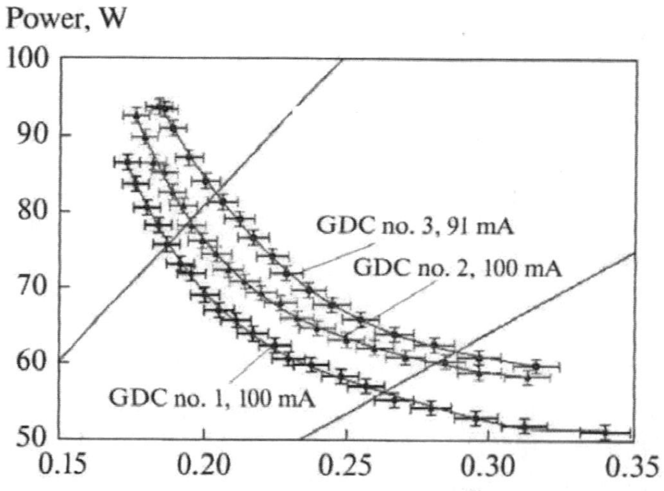

Figure 1. The dependence of RF power on the mass flow rate

Dependences of the power input into the discharge on the propellant mass flow rate for the close values of ion beam current for different GDCs (ref. to figure 1) allowed to draw a conclusion about the efficiency gain of the GDC No. 1 as of the part of RIT under study.

Based on the experience gained from the laboratory model study, the RIT-8 experimental model was developed and tested. In addition to the thruster itself, the RIT experimental model comprised a power processing unit (PPU) with radio-frequency generator (RFG) developed by the MAI Research and |Production Center "FRELA", as well as the gas distribution control unit (GDCU) and cathode-neutralizer (CN) developed by the Keldysh Research Center. The general arrangement of the RIT-8 experimental model is shown in figure 2 [9].

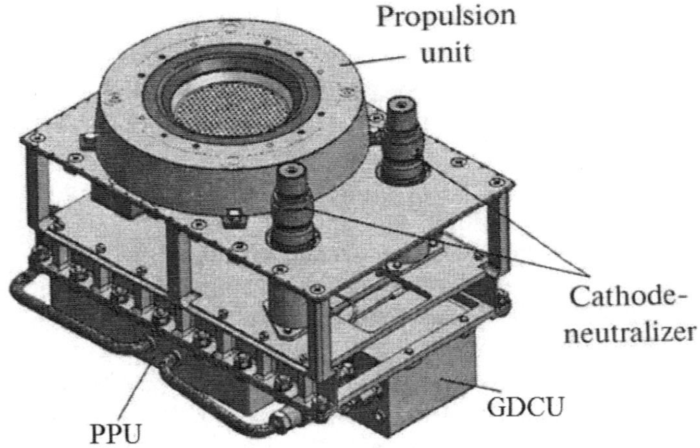

Figure 2. General view of the RIT-8 experimental model

The RIT-8 experimental model was tested by RIAME MAI at its test facility. The main performance of the RIT-8 thruster obtained in experimental studies at different mass flow rates are shown in table 2 [9]. For all operation modes, the accelerating voltage was constant and equal to 2 kV. Characteristics presented in the table are given without taking into account losses in GDCU, CN and PPU.

Table 2. Main performance data of RIT-8 at different mass flow rates

Mass flow rate, mg/s	0.174	0.193	0.213	0.227	0.242	0.262	0.282
Power consumption, W	316.9	321.8	312	321.5	319.3	313.8	298.3
Thrust, mN	8.44	8.81	8.81	9.17	9.17	9.17	8.81
Specific Impulse, s	4987	4692	4251	4155	3898	3600	3222
Efficiency, %	65.1	63.0	58.9	58.1	54.9	51.6	46.7

Currently, on the basis of RIT-8, a unit of ion propulsion system (IPS) is being developed as a part of the scientific equipment of IPS-S, which is planned to be installed on board the Russian segment (RS) of the International Space Station (ISS) [10]. The aims of space experiment are the IPS testing and obtaining experimental data on the peculiarities of its operation under space conditions, as well as the study of the propagation of the space-charge compensated ion beam in the ionospheric plasma taking into account the impact of the Earth's magnetic field [11]. Currently, the IPS prototype is being produced.

One of the prospective objectives of IT application on board spacecraft is to compensate for the aerodynamic drag of the upper atmosphere. [12]. The value of aerodynamic drag depends on solar and geomagnetic activity and can vary by more than an order of magnitude during the operation of the spacecraft. In order to compensate for aerodynamic drag, it is necessary to use an adjustable propulsion system that allows at least a 20-fold change in thrust. One of the RIT-8 studies was aimed at determining the range of thrust throttling. The results of that study are presented in table 3.

Table 3. RIT-8 performance at different power consumption levels

Power consumption, W	Thrust, mN	Specific impulse, s	Efficiency, %
433	11.4	4070	53
310	8.9	3790	53
257	7.3	3010	42
195	5.6	2590	36
116	3.4	1930	28
79	2.2	1420	19
45	1.1	1200	14
31	0.5	610	5

During the experiment, a 23-fold thruster throttling range was shown.

One of the main tasks faced by the EPT developers at present is to increase the thruster lifetime. ITs can have the lifetime of tens of thousands of hours. The critical element of the IT from the point of view of its lifetime is the accelerating electrode (AE), which is exposed to bombardment by charge-exchange ions, causing its erosion. The rate of erosion depends on the coefficient of ion sputtering of the material from which the accelerating electrode is made. The graphite-based materials have the lowest ion sputtering coefficient. However, such materials have disadvantages. The mass-produced carbon-carbon composite materials (CCCM) on a woven-laying, rod and thread basis do not meet the requirements for surface roughness due to the structural heterogeneity of carbon frames [13]. Such heterogeneity is inherited by the structural materials based on such frames. The carbon nonwoven fabrics of felt type have another structure. But the frames on the basis of carbon felts produced by the industry have a density of about 0.2 g/cm^3, which does not allow their use for achieving high structural strength in the composite.

Pyrolytic graphite is one of the most promising materials for use in ion-extraction systems (IES) of IT. Nevertheless, this material has much higher coefficient of linear thermal expansion compared to CCCM. In addition, the process of manufacturing pyrolytic graphite parts is time-consuming and costly.

The JSC "KOMPOZIT" jointly with the LLC "Niagara" developed a method of manufacturing a porous frame being the basis for the composite material [14], which allows with the help of needle-punching technology to obtain fine-mesh reinforcing carbon frame of Ipresscon® brand with the density of (0.5-0.6) g/cm3, and then – the CCCM based on it (CCCM-Ipresscon). The JSC "KOMPOZIT" produced AE of such material for RIT-8. Currently, RIT-8 with the AE made of CCCM-Ipresscon is being tested. At the first stage, it is planned to accumulate 1000 hours of operating time; at the time of article writing, the operating time is 690 hours. The estimated lifetime of the AE made of CCCM-Ipresscon, calculated using the IOS-3D software package developed by the Keldysh Research Center [15], is more than 25 thousand hours.

With such lifetime values, conducting full-scale lifetime tests requires significant amount of time. One of the tasks being solved by RIAME MAI is the development of computational and experimental procedure for determining the rate and nature of erosion of the RIT IES accelerating electrode on a shortened time base. AE erosion is caused by its bombardment by charge-exchange ions formed in the neutralization zone and in the volume of the primary ion beam. Bombardment by charge-exchange ions from the neutralization zone leads to the so-called "pits and grooves" erosion from the side facing the neutralization zone [16].

The proposed procedure is based on sputtering a multilayer coating pre-applied to the outer surface of the AE and consisting of optically contrasting layers of copper and chromium [17]. In this case, every 2 hours of the experiment, the surface of the electrode facing the neutralization zone is photographed by a digital camera, and the erosion dynamics is determined by counting the copper and chrome strips at each interval. The data obtained in the experiment are used as the initial data for numerical simulation of multilayer coating erosion with parameters corresponding to the experimental conditions.

At the Bauman Moscow State Technical University, a multilayer coating was applied to the AE of the RIT-8 thruster, and a 14-hour experiment was conducted at the RIAME MAI, during which a multilayer coating of the AE was sputtered. A photo of the multilayer coating erosion pattern is shown in the figures 3a (the photo was taken with a digital camera) and 3b (the photo was taken with an electron microscope).

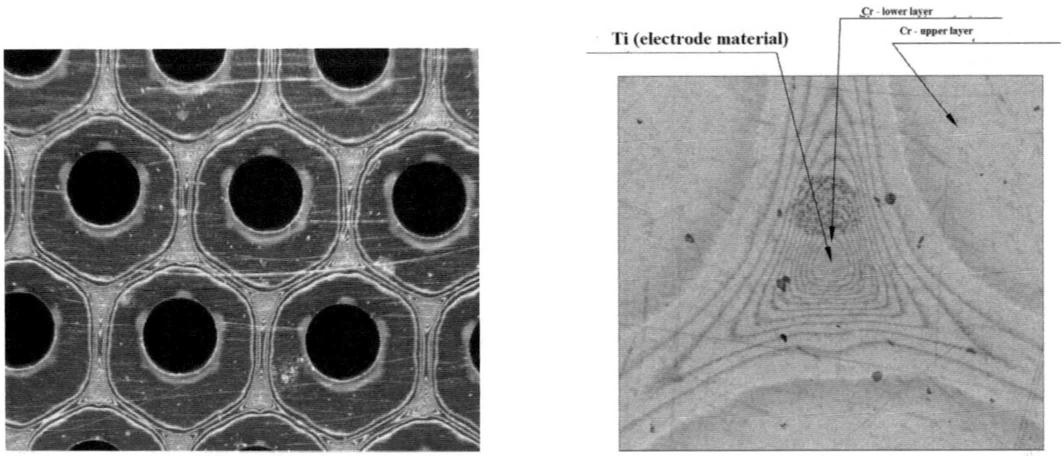

a b

Figure 3. Photos of the AE surface with multilayer coating after fourteen hours of operation, taken with digital camera (a) and electron microscope (b)

Based on the obtained pattern, the depth of erosion of the accelerating electrode is determined and then the maximum current density of the charge-exchange ions is calculated [17]. After that, by calculation in the software package IOS-3D environment, it is possible to estimate the service life of an electrode made of a particular material.

3. Conclusions

The radio-frequency ion thruster RIT-8 was developed and tested at the RIAME MAI. PPU for that thruster was developed at the MAI Research and |Production Center "FRELA", and GDCU and CN were developed at the Keldysh Research Center. Based on the RIT-8 development experience, the RIAME MAI is developing the ion propulsion system, which is planned to be tested on board the ISS.
A computational and experimental procedure for estimating the rate and nature of erosion of the IT accelerating electrode on the shortened time base was proposed and tested at the RIAME MAI.

References

[1] Beattie J R, Williams J D and Robson R R 1993 Flight Qualification of an 18-mN Xenon Ion Thruster IEPC-93-1015, *Proc. of the 23rd International Electric Propulsion Conference*, Seattle, Washington

[2] Goebel D M, Wirz R and Katz I 2006 Analytical Ion Thruster Discharge Performance Model *Proc. of the 42nd AIAA/ASME/SAE/ASEE Joint Propulsion Proceedings*, Sacramento

[3] Wallace N, Jameson P, Saunders C, Fehringer M, Edwards C, and Floberghagen R 2011 The GOCE Ion Propulsion Assembly – Lessons Learnt from the First 22 Months of Flight Operations IEPC-2011-327, *Proc. of the 32nd International Electric Propulsion Conference*, Wiesbaden, Germany

[4] Wirz R E, Sullivan R, Przybylowski J and Silva M 2008 Hollow Cathode and Low-Thrust Extraction Grid Analysis for a Miniature Ion Thruster *International Journal of Plasma Science and Engineering* pp 1-11

[5] Takanao S, Hirotaka S, Chisato O, Hajime Y, Yasuhiko T, Yasuhiro A and Makoto Y 2013 Small carry-on impactor of Hayabusa 2 mission *Acta Astronautica* **84** 227—236

[6] Kun-Long Wu, Hao Huang, Jun-Xue Ren, et al. 2019 Design and Experimental Study of a Miniature Ion Thruster IEPC-2019-175 *Proc. of the 36th International Electric Propulsion Conference* University of Vienna, Vienna, Austria

[7] Kravchenko D, Lovtsov A and Madeev S 2019 Development of 5-cm ion thruster and preliminary numerical simulation of plasma in its discharge chamber. *AIP Conference Proceedings*

[8] Loeb H W, Popov G A and Obukhov V A 2012 Radio-Frequency Ion Thruster Laboratory of Moscow Aviation Institute – a New Form of Russian-German Cooperation» *Trudy MAI" journal* **60**

[9] Antropov N N, Akhmetzhanov R V, Bogatyy A V, Grishin R A, Kozhevnikov V V, Plokhikh A P, Popov G A and Khartov S A 2016 Experimental Research of Radio-Frequency Ion Thruster *Thermal Engineering* **63** 13 957-963

[10] Akhmetzhanov R V, Bogatyi A V, Cherkasova M V, Konoshenko V P, Lapshinova O V, Obukhov V A, Plokhikh A P, Popov G A, Shemukhin A A and Zhilchenko K S 2019 Preparation of Space Experiment with Electric Propulsion System Based on Radio-Frequency Ion Thruster aboard the International Space Station *Proc. of the 36th International Electric Propulsion Conference* Vienna Austria

[11] Pushkin N M, Mednikov B A, Gomilka L A, Lapshinova O V, Mashkov A S and Antropov N N 1997 Dynamics of electric fields in the near-surface region of spacecraft during plasma ejection from the Mir orbital station *Cosmic Research* **35** 4 414-416

[12] Wallace N, Jameson P, Saunders C, Fehringer M, Edwards C, and Floberghagen R 2011 The GOCE Ion Propulsion Assembly – Lessons Learnt from the First 22 Months of Flight Operations IEPC-2011-327 *Proc. of the 32nd International Electric Propulsion Conference*, Wiesbaden, Germany

[13] Akhmetzhanov R V, Balashov V V, Bogachev Ye A, Yelakov A B, Kashirin D A, Svotina V V, Spivak O O and Cherkasova M V 2018 An Ion Thruster Accelerating Electrode Made of Carbon-Carbon Composite Material *Thermal Engineering* **65** 13 986-993

[14] Patent RF No. 2620810 Bogachev Ye A, Elakov A B, Beloglazov A P, Denisov Yu A and Timofeev A N Method of manufacturing a porous frame – the basis of a composite material. publ. 29.05.2017, Bul. No. 16

[15] Shagayda A, Nikitin V and Tomilin D 2016 Three-dimensional analysis of ion optics with misalignments of apertures *Vacuum* **123** 140-150

[16] Bundesmann C, Tartz M, Scholze F, Neumann H, Leiter H J, Scortecci F, Feili D, Frigot P-E and Gonzalez del Amo J 2009 In–Situ Temperature, Grid Curvature, Erosion, Beam and Plasma Characterization of a Gridded Ion Thruster RIT–22 *Proc. of the 31st International Electric Propulsion Conference* IEPC-2009-160 University of Michigan, USA

[17] Akhmetzhanov R V, Vorob'ev E V, Dukhopel'nikov D V, Ivakhnenko S G, Obukhov V A, Popov G A, Svotina V V and Cherkasova M V 2019 Procedure of Predicting the Lifetime of the Accelerating Electrode of an Ion-Extraction System of a Gas-Discharge Ion Source *Journal of Surface Investigation. X-ray, Synchrotron and Neutron Techniques* **13** 6 1217-1224

Effect of radiation on the contact thermal resistance in vacuum

A D Ezhov[1], L V Bykov[1] and S V Reznik[2]
[1]Moscow Aviation Institute (National Research University), Moscow, Russia
[2]Bauman Moscow State Technical University (National Research University of Technology), Moscow, Russia

E-mail: bykovlv@mai.ru

Abstract. The solution to most of the problems of ensuring the operating characteristics of units and assemblies of complex technical objects in a wide range of changes in temperature and mechanical loads is associated with ensuring the conditions of contacting surfaces, i.e. with the solution of the contact problem of determining the thermal and strength state of the connection of parts. This problem is also relevant to objects operated in space in a vacuum, where heat transfer between parts can occur only in contact spots (areas of direct contact of surfaces) due to thermal conductivity or due to radiation in places where there is no direct contact. Obviously, in this case, the total contact thermal resistance consists of the composition of these types of heat transfer, and largely depends on the profile and roughness of the contacting surfaces. An assessment of the effect of reemission on the value of contact thermal resistance when the temperature in the contact zone varies from 200 °C to 800 °C under vacuum conditions can allow a much more accurate prediction of the operational characteristics of the contact pair. In this paper, we compare the effect of reemission on the value of contact thermal resistance under various mechanical stresses.

1. Introduction.

A reliable assessment of the thermal state of the product during design is the key to long-term and trouble-free operation of equipment in space systems. Any equipment of modern space technology consists of various units and assemblies, contains the attachment points of various parts [1-5]. Due to the fact that the joints of the parts are not absolutely identical in geometry of the contacting surfaces, and can have in practice a completely different microrelief after processing, it must be taken into account that they can make significant changes in the nature of the heat flux distribution and, as consequence, the occurrence of thermal deformations.

To predict accurately the distribution of temperature fields, it is necessary to take into account thermal contact resistance (TCR) arising due to discreteness of contact, contraction and extension of heat flux lines to contact spots [6]. In outer space, under vacuum conditions, the problem under consideration is especially urgent, since heat transfer between contacting surfaces can occur only with the help of two mechanisms of heat transfer - heat conduction through actual contact zones and due to radiation in the intercontact region.

2. Method

Known analytical dependencies according to the determination of TCR are focused on specific contact conditions applicable in a limited range of intercontact pressures and temperatures, which could be realized during experiments [6-8]. As a consequence of this approach in determining the values of thermal contact resistances, the TCR calculated from these dependencies have a rather wide range of scatter. Some sources claim that the contribution of heat transfer by radiation at temperatures in the intercontact region up to 1000 K is not significant and is less than 1%. The studies conducted by the authors suggest the opposite.

The methodology used to calculate TCR taking radiation into account is based on the technique described in detail in [9-13], which boils down to a series of sequential actions, namely:

- determination of the roughness of the contacting surfaces;
- solid modeling of rough surfaces;
- three-dimensional strength analysis of the contact surface;
- three-dimensional thermal analysis of the contact of surfaces, taking into account the re-radiation in the contact zone.

The nonlinear dependence of the output power from the radiating surface temperature leads to the need for more detailed study of its impact on the overall heat transfer pattern in the contact region. In this case, the estimation of parameters of radiative heat transfer to be considered angles of incidence and reflection of radiant flux due microgeometry surfaces. This problem is solved with the help of an additional boundary condition between contact surfaces containing microgeometry shape parameters. Microgeometry forms are taken into account due to the use of the coefficient of the F_{ji} survey factor [14] (Figure 1).

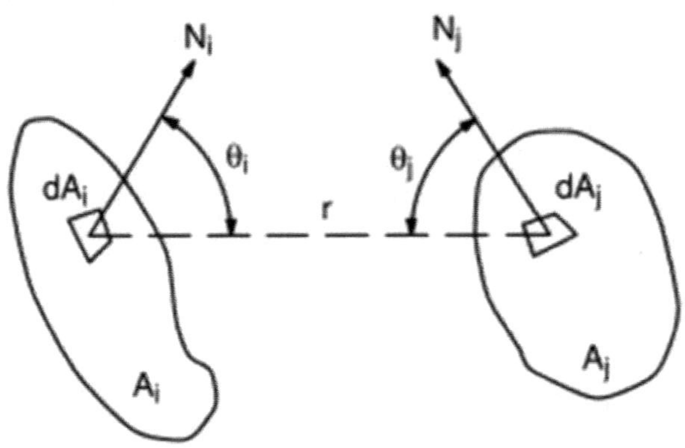

Figure 1. Graphic illustration of the interaction of two surfaces during radiant heat transfer.

The viewing factor is defined as part of the total radiant heat flux that originates from surface i and which flows directly to surface j. It is described by the following equation:

$$F_{ij} = \frac{1}{A_j}\int_{A_i}\int_{A_j}\frac{\cos\theta_i\cos\theta_j}{\pi r^2}d(A_j)d(A_i)$$

where:

A_i, A_j – area of surface I and surface j, r – distance between differential surfaces i and j, θ_i – angle between N_i and the radius line to surface $d(A_j)$, θ_j – angle between N_j and the radius line to surface $d(A_i)$, N_i, N_j –surface normal of $d(A_i)$ and $d(A_j)$.

To determine the heat flux due to re-radiation, use the Stefan-Boltzmann law for a gray body, taking into account the viewing factor:

IOP Conf. Series: Materials Science and Engineering **927** (2020) 012076 doi:10.1088/1757-899X/927/1/012076

$$Q_i = \sigma \, \epsilon_i F_{ji} A_i (T_i^4 - T_j^4)$$

where:
ϵ_1 – effective emissivity of surface i; F_{ji}- radiation view factors; A_i – area of surface i, σ – Stefan-Boltzmann constant; T_i, T_j – absolute temperature at surface i and j.

3. Results and Discussion

In this paper, as an example of such technique, we present the calculation results (Figure 2) of the contact thermal resistance of 12Kh18N10T materials with KMK-MS, roughnesses of 0.82 μm and 2.07 μm, respectively, with a compressive load of 0.14 MPa and 2 MPa. The temperature range is from 200 ° C to 800 ° C.

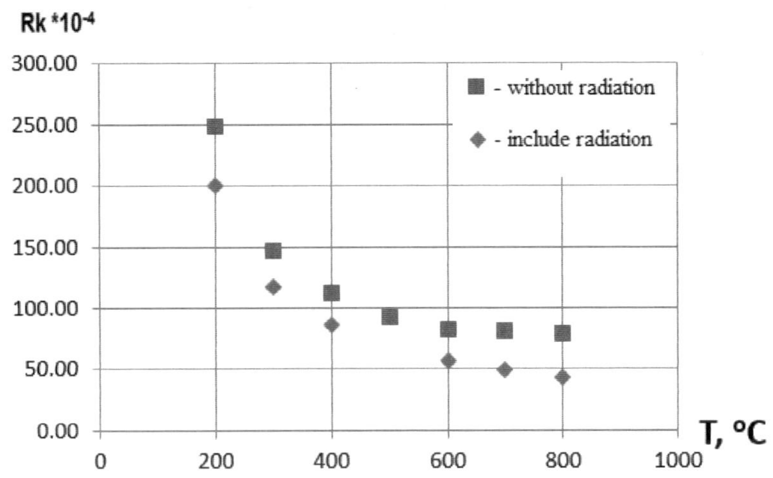

Figure 2. The dependence of the TCR on the temperature inside the contact at a contact pressure of 0.14 MPa.

Table 1. The results of the calculation of TCR with and without radiation at an intercontact pressure of 0.14 MPa

Temperature in the contact zone (° C)	200	300	400	500	600	700	800
TCR taking into account radiation (m K / W)	0.0200	0.0118	0.0086	0.0067	0.0056	0.0049	0.0043
TCR excluding radiation (m K / W)	0.0249	0.0146	0.0113	0.0092	0.0083	0.0081	0.0079
Difference in %	19.6	20.0	23.4	27.7	32.3	39.5	46.2

As can be seen from Figure 2 and Table 1, ignoring the contribution of heat transfer due to radiation leads to a significant error in the determination of TCR. This is especially pronounced at low values of compressive loads. This is because only a small part of the nominal contact area comes into contact and a significant amount of heat is transferred due to radiation, as shown in Figure 3.

IOP Conf. Series: Materials Science and Engineering **927** (2020) 012076 doi:10.1088/1757-899X/927/1/012076

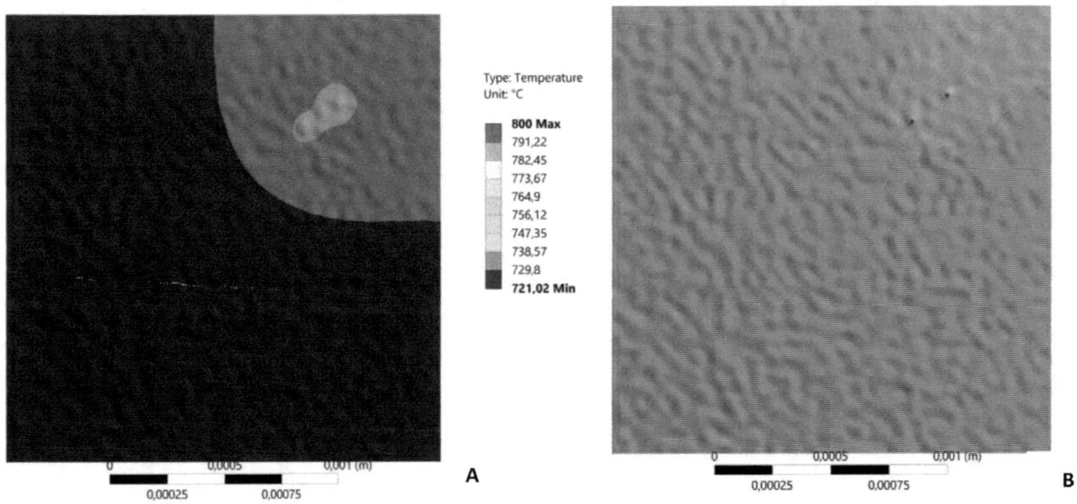

Figure 3. Temperature field distribution excluding (A) and taking into account (B) the radiation mechanism in the contact zone at a compressive load of 0.14 MPa.

A further increase in the compressive load leads to a sharp increase in the area of actual contact; new vertices of the contacting surfaces come into contact. This phenomenon is clearly illustrated in Figure 4.

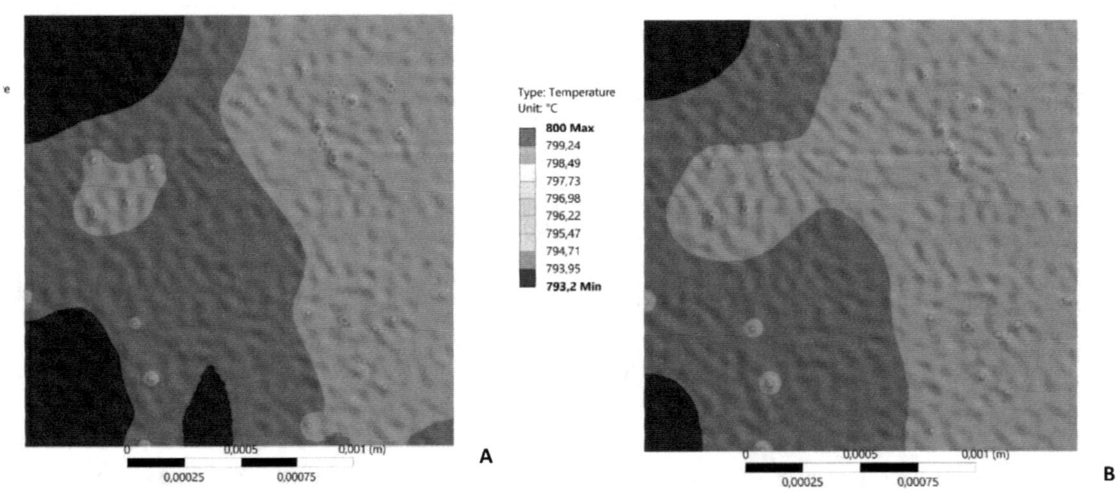

Figure 4. Temperature field distribution excluding (A) and taking into account (B) the radiation mechanism in the contact zone at a compressive load of 2.0 MPa.

The change in contact thermal resistance at an intercontact pressure of 2.0 MPa depending on the temperature in the contact zone is shown in Figure 5.

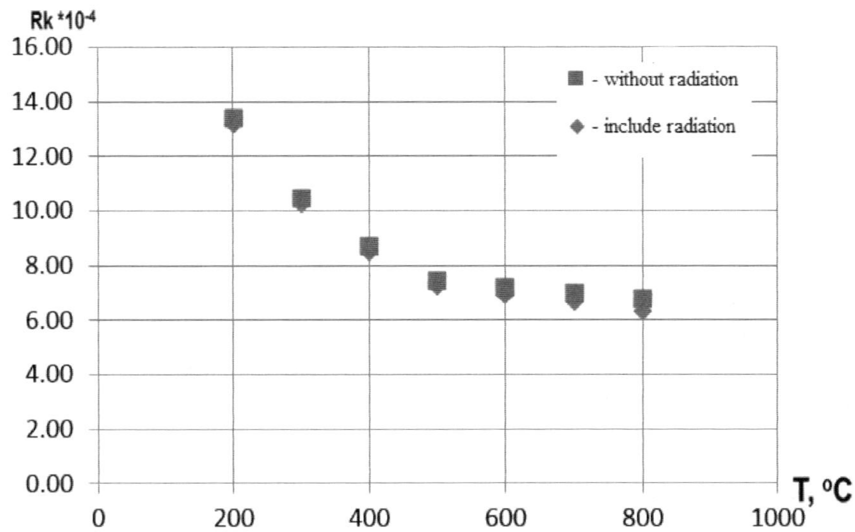

Figure 5. The dependence of the TCR on the temperature inside the contact at a contact pressure of 2.0 MPa.

Table 2. The results of the calculation of TCR with and without radiation at an intercontact pressure of 2.0 MPa

Temperature in the contact zone (° C)	200	300	400	500	600	700	800
TCR taking into account radiation (m K / W)	0.00131	0.00103	0.00085	0.00072	0.00069	0.00067	0.00064
TCR excluding radiation (m K / W)	0.00137	0.00104	0.00087	0.00074	0.00071	0.0007	0.00068
Difference in %	1.10	1.60	2.10	2.70	3.70	4.90	6.30

4. Conclusion

Studies have shown that in vacuum at high temperatures and low compressive pressures, heat transfer in the zone of the contact joint is largely determined by the amount of heat flux due to radiation. For structural elements that have limitations on acceptable assembly conditions, an error in determining the amount of heat transferred due to radiation can lead to a significant error in determining the contact thermal resistance, which may subsequently affect the calculation of the thermal and strength state of the entire spacecraft.

References

[1] Mesnyankin S Yu, Vikulov A G and Vikulov D G 2009 Solid-solid thermal contact problems: current understanding. *Phys.-Usp.* **52** 891
[2] Reznik S V ,Prosuntsov P V and Mikhailovskii K V 2018 Thermal regime of large space structure with transformable elements from hybrid composite *IOP Journal of Physics: Conf. Series* **1134** 012048
[3] Avduevskii V S 1992 *The basics of heat transfer in aviation and rocket and space technology* (Mashinostroenie, Moscow) [in Russian]
[4] Reznik S V, Prosuntsov P V, Mikhailovsky K V and Shafikova I R 2016 Material science problems of building space antennas with a transformable reflector 100 m in diameter *IOP Conference Series: Materials Science and Engineering* **153** 10

IOP Conf. Series: Materials Science and Engineering **927** (2020) 012076 doi:10.1088/1757-899X/927/1/012076

[5] Reznik S V 2018 Thermal regimes of space composite structures. Part I *MATEC Web of Conferences* **194** 01048
[6] Madhusudana, Chakravarti V 2014 *Thermal Contact Conductance* (Springer International Publishing)
[7] N B Demkin, E V Ryzhov 1981 *Surface quality and contact of machine components.* (Mashinostroenie, Moscow) [in Russian]
[8] Izmailov V V 2010 *Contact of solid bodies and their conductivity* (TGTU, Tver) [in Russian]
[9] Ezhov A, Bykov L and Mesnyankin S 2018 Numerical Method for Determining the Real Contact Area of Contacting Bodies *J. of Surface Investigation: X-ray, Synchrotron and Neutron Techniques* **12** 914
[10] Ezhov A, Bykov L and Artemchuk N 2019 Related problems of designing heat-loaded structures from ceramic matrix *IOP Conference Series: Materials Science and Engineering* **683** 012083
[11] Bykov L, Artemchuk N and Ezhov A 2020 Three-dimensional of modeling microgeometry of contact pairs in technical systems *IOP Conference Series: Materials Science and Engineering* **709** 022057
[12] Ezhov A D, Mesnyankin S Yu, Bykov L V 2017 V sbornike: Materialy KHKH Yubileynoy Mezhdunarodnoy konferentsii po vychislitel'noy mekhanike i sovremennym prikladnym sistemam (VMSPPS'2017) pp 236-238 [in Russian]
[13] Certificate of state registration of computer programs №2015618021 "Determination of coordinates of points on a rough flat surface" from 28.06.2015 [in Russian]
[14] Siegal R and Howell J R 2011 *Thermal Radiation Heat Transfer. Fifth Edition.* (CRC Press)

Spectral method for solving linear Caputo fractional stochastic differential equations

K Rybakov and A Yushchenko

Moscow Aviation Institute (National Research University),
Volokolamskoe highway 4, 125993 Moscow, Russian Federation

E-mail: rkoffice@mail.ru

Abstract. We concern on the problem of finding the solution to the linear Caputo fractional stochastic differential equation with additive and multiplicative noise. It is proposed to apply the spectral method based on the spectral form of mathematical description. This method provides both an explicit form of the solution as the orthogonal series with random coefficients and a continuous-time approximation of this solution as the partial sum. Earlier, the spectral method has been applied for solving linear (non-fractional) stochastic differential equations. The proposed method is demonstrated on the modeling fractional Ornstein – Uhlenbeck process described by a linear Caputo fractional stochastic differential equation with additive noise.

1. Introduction and problem formulation

The spectral method is proposed for solving the linear Caputo fractional stochastic differential equation (Caputo FSDE) [1,2]

$$ {}^{*}\mathcal{D}^{\mu}_{0+}x(t) = a(t)x(t) + b(t) + \big(c(t)x(t) + d(t)\big)v(t), \quad x(0) = x_0, \tag{1} $$

where $x(t)$ is the $\mathbb{R}$-valued random process, $t \in \mathbb{T} = [0,T]$, ${}^{*}\mathcal{D}^{\mu}_{0+}$ is the Caputo fractional differentiation operator of order $\mu \in (0.5, 1)$, $a(t)$, $b(t)$, $c(t)$, and $d(t)$ are given functions, $v(t)$ is the standard Gaussian white noise, x_0 is a random variable with a finite second-order moment.

The solution to the equation (1) is the random process $x(t)$ that satisfies the following integral equation (Itô – Volterra equation [3])

$$ x(t) = x_0 + \frac{1}{\Gamma(\mu)} \int_0^t (t-\tau)^{\mu-1}\big(a(\tau)x(\tau) + b(\tau)\big)d\tau $$
$$ + \frac{1}{\Gamma(\mu)} \int_0^t (t-\tau)^{\mu-1}\big(c(\tau)x(\tau) + d(\tau)\big) \circ dw(\tau), $$

in which the second integral is understood in the Stratonovich sense, $w(t)$ is the standard Wiener process, $\Gamma(\mu)$ is the gamma function. For $\mu = 1$ the equation (1) is the linear (non-fractional) stochastic differential equation written in the Langevin form. The existence of solutions for Caputo FSDEs is discussed in [4].

For the special case $a(t) = b(t) = c(t) = 0$ and $d(t) = 1$ the solution $x(t)$ is the Riemann – Liouville random process [5, 6]. Note that it differs from the fractional Brownian motion introduced in [7], however, it can also be used to describe a lot of real processes.

IOP Conf. Series: Materials Science and Engineering **927** (2020) 012077 doi:10.1088/1757-899X/927/1/012077

The main goal of this research is to develop the spectral method of finding the solution to the linear Caputo FSDE and to test this method by numerical experiments.

The proposed approach may be considered as an alternative to numerical methods for solving the linear Caputo FSDE. It allows to avoid a time discretization and to find the explicit form of the solution $x(t)$ as the orthogonal series with random coefficients. It also allows to obtain a continuous-time approximation of this solution as the partial sum. This spectral method is based on the spectral form of mathematical description for control systems [8,9], it is adapted for the representing random processes [10, 11]. The spectral collocation method has been used for the Itô – Volterra equations [12–15], but such approach indirectly includes a time discretization (collocation points), and it provides the approximate solution only in contrast to the proposed spectral method.

Obtained results may find the application in physics [16, 17], engineering [18, 19], economics and finance [20, 21], and other fields.

2. Definitions and notations

Let us introduce the necessary definitions and notations [10, 11], which will be used further.

An infinite column matrix $H = [H_i]$ is called the spectral characteristic of a square-integrable function $h(t)$ with respect to the basis $\{q_i(t)\}_{i=0}^{\infty}$ (similarly, the spectral characteristic can be defined for a mean square-integrable random process), if

$$H_i = \int_{\mathbb{T}} q_i(t)h(t)dt, \quad i = 0, 1, 2, \dots,$$

and

$$h(t) = \sum_{i=0}^{\infty} H_i q_i(t), \quad t \in \mathbb{T},$$

where $\{q_i(t)\}_{i=0}^{\infty}$ is a complete orthonormal system in $L_2(\mathbb{T})$. So, $H = \mathbb{S}[h(t)]$ and $h(t) = \mathbb{S}^{-1}[H]$, where $\mathbb{S}$ and $\mathbb{S}^{-1}$ denote the spectral transform of functions or random processes and the spectral inversion of spectral characteristics, respectively.

An infinite matrix $A = [A_{ij}]$ is called the spectral characteristic of a linear operator $\mathcal{A}\colon D_{\mathcal{A}} \subseteq L_2(\mathbb{T}) \to L_2(\mathbb{T})$ with respect to the basis $\{q_i(t)\}_{i=0}^{\infty}$, if

$$A_{ij} = \int_{\mathbb{T}} q_i(t)\mathcal{A}q_j(t)dt, \quad i, j = 0, 1, 2, \dots$$

The spectral transform of linear operators we will also denote by $\mathbb{S}$, i.e., $\mathbb{S}[\mathcal{A}] = A$.

We should also define the infinite three-dimensional matrix $V = [V_{ijk}]$ with entries

$$V_{ijk} = \int_{\mathbb{T}} q_i(t)q_j(t)q_k(t)dt, \quad i, j, k = 0, 1, 2, \dots, \tag{2}$$

and this matrix is called the spectral characteristic of the multiplier.

In [11] the main properties of the spectral transform for functions, random processes and linear operators are described in detail, they give a foundation for the spectral form of mathematical description. Its connection with the Karhunen – Loève expansion [22] and with the canonical representation of random processes [23] are discussed in [10]. Here we do not describe properties of the spectral transform to be short.

It is important to note that spectral characteristics can also be defined for functions, which are not square-integrable, and for random processes, which are not mean square-integrable. In fact, it is enough that expansion coefficients H_i can be calculated. For instance, we can define spectral characteristics for the Dirac delta function or for the Gaussian white noise [8, 10, 11].

3. Spectral method for solving linear Caputo FSDE

Rewrite the equation (1) as follows:

$$\mathcal{J}_{0+}^{1-\mu}\dot{x}(t) = a(t)x(t) + b(t) + \big(c(t)x(t) + d(t)\big)v(t), \quad x(0) = x_0, \tag{3}$$

where $\mathcal{J}_{0+}^{1-\mu}$ is the Riemann – Liouville fractional integration operator of order $1-\mu$. Then apply the spectral transform to the left-hand and right-hand sides of the equation (3) to obtain its solution using the linearity property of the spectral transform. So,

$$\begin{aligned}\mathbb{S}\big[\mathcal{J}_{0+}^{1-\mu}\dot{x}(t)|_{x(0)=x_0}\big] &= \mathbb{S}\big[a(t)x(t) + b(t) + \big(c(t)x(t) + d(t)\big)v(t)\big] \\ &= \mathbb{S}\big[a(t)x(t)\big] + \mathbb{S}\big[b(t)\big] + \mathbb{S}\big[\big(c(t)x(t) + d(t)\big)v(t)\big].\end{aligned} \tag{4}$$

Let X, B, and D be spectral characteristics of $x(t)$, $b(t)$, and $d(t)$, respectively, $\mathcal{V}$ be the spectral characteristic of a standard Gaussian white noise $v(t)$, i.e., $X = [X_i]$, $B = [B_i]$, and $D = [D_i]$, where

$$X_i = \int_{\mathbb{T}} q_i(t)x(t)dt, \quad B_i = \int_{\mathbb{T}} q_i(t)b(t)dt, \quad D_i = \int_{\mathbb{T}} q_i(t)d(t)dt, \quad i = 0, 1, 2, \dots,$$

as well as $\mathcal{V} = [\zeta_i]$, where ζ_i are independent Gaussian random variables with zero mean and unit variance.

Further, let A and C be spectral characteristics of multiplication operators with multipliers $a(t)$ and $c(t)$, respectively, P be the spectral characteristic of the differentiation operator taking into account the initial condition, i.e., $A = [A_{ij}]$, $C = [C_{ij}]$, and $P = [P_{ij}]$, where

$$A_{ij} = \int_{\mathbb{T}} a(t)q_i(t)q_j(t)dt, \quad C_{ij} = \int_{\mathbb{T}} C(t)q_i(t)q_j(t)dt, \quad P_{ij} = \int_{\mathbb{T}} q_i(t)\dot{q}_j(t)dt + q_i(0)q_j(0),$$
$$i, j = 0, 1, 2, \dots$$

In addition, let V be the spectral characteristic of the multiplier with entries (2) and let Δ be the spectral characteristic of the Dirac delta function, i.e., $\Delta = [\ q_0(0)\ \ q_1(0)\ \ q_2(0)\ \ \dots\ ]^{\mathrm{T}}$.

All above spectral characteristics have been used in [10, 11] to form the spectral method for solving linear (non-fractional) stochastic differential equations. Here we need to define the spectral characteristic of the Riemann – Liouville fractional integration operator of order $1-\mu$ [24]. It is denoted by $J^{1-\mu} = [J_{ij}^{1-\mu}]$, and using the definition of the spectral characteristic of a linear operator we have

$$J_{ij}^{1-\mu} = \frac{1}{\Gamma(1-\mu)} \int_{\mathbb{T}} q_i(t) \int_0^t \frac{q_j(\tau)}{(t-\tau)^{\mu}} d\tau dt, \quad i, j = 0, 1, 2, \dots$$

According to [10, 11],

$$\mathbb{S}\big[a(t)x(t)\big] = AX, \quad \mathbb{S}\big[b(t)\big] = B, \quad \mathbb{S}\big[\big(c(t)x(t) + d(t)\big)v(t)\big] = (V\mathcal{V})(CX + D),$$

consequently,

$$\mathbb{S}\big[a(t)x(t) + b(t) + \big(c(t)x(t) + d(t)\big)v(t)\big] = AX + B + (V\mathcal{V})(CX + D).$$

Next, we should apply the property of the spectral transform for the composition of linear operators [11]. Then

$$\mathbb{S}\big[\mathcal{J}_{0+}^{1-\mu}\dot{x}(t)|_{x(0)=x_0}\big] = \mathbb{S}\big[\mathcal{J}_{0+}^{1-\mu}\big]\,\mathbb{S}\big[\dot{x}(t)|_{x(0)=x_0}\big],$$

IOP Conf. Series: Materials Science and Engineering **927** (2020) 012077 doi:10.1088/1757-899X/927/1/012077

where

$$\mathbb{S}\big[\dot{x}(t)|_{x(0)=x_0}\big] = PX - x_0\Delta.$$

Hence, we can rewrite the equation (4) as

$$J^{1-\mu}(PX - x_0\Delta) = AX + B + (V\mathcal{V})(CX + D).$$

Further, represent the spectral characteristic Δ as $P\mathbf{1}$, $\mathbf{1}$ is the spectral characteristic of the unit step function, i.e., $\mathbf{1} = [\mathbf{1}_i]$, where

$$\mathbf{1}_i = \int_{\mathbb{T}} q_i(t)dt, \quad i = 0, 1, 2, \ldots,$$

consequently, using the property $J^{1-\mu}P = P^\mu$ we have

$$P^\mu X - AX - (V\mathcal{V})CX = x_0 P^\mu \mathbf{1} + B + (V\mathcal{V})D \tag{5}$$

and

$$X = \big(P^\mu - A - (V\mathcal{V})C\big)^{-1}\big(x_0 P^\mu \mathbf{1} + B + (V\mathcal{V})D\big). \tag{6}$$

In relations (5) and (6) P^μ is the spectral characteristic of the Riemann – Liouville fractional differentiation operator of order μ (spectral characteristics of the Riemann – Liouville and Caputo fractional differentiation operators are equal), i.e., $P^\mu = [P^\mu_{ij}]$, where

$$P^\mu_{ij} = \frac{1}{\Gamma(1-\mu)} \int_{\mathbb{T}} q_i(t) \frac{d}{dt} \int_0^t \frac{q_j(\tau)}{(t-\tau)^\mu} d\tau dt, \quad i, j = 0, 1, 2, \ldots$$

It is more preferable to redefine D as the spectral characteristic of the multiplication operator with a multiplier $d(t)$ when $c(t) = 0$. Then we have a new equation for the spectral characteristic X and its solution instead of relations (5) and (6)

$$P^\mu X - AX = x_0 P^\mu \mathbf{1} + B + D\mathcal{V} \tag{7}$$

and

$$X = (P^\mu - A)^{-1}(x_0 P^\mu \mathbf{1} + B + D\mathcal{V}). \tag{8}$$

For relations (7) and (8) the spectral characteristic V is not needed.

The solution $x(t)$ to the linear Caputo FSDE (1) can be expressed as the series with respect to the orthonormal basis $\{q_i(t)\}_{i=0}^\infty$. So,

$$x(t) = \mathbb{S}^{-1}[X] = \sum_{i=0}^\infty X_i q_i(t), \tag{9}$$

where expansion coefficients X_i depend on the initial state x_0 and entries of the spectral characteristic $\mathcal{V}$, i.e., countable set of independent random values having a standard normal distribution.

To obtain the approximate solution all spectral characteristics should be truncated on all dimensions to a certain order L [10, 11]. This means that all matrices in formulas (5)–(8) should be finite matrices with size L in each dimension, i.e., L for column matrices (spectral characteristics of functions and random processes, $L \times L$ for two-dimensional matrices (spectral characteristics of linear operators), and $L \times L \times L$ for the three-dimensional matrix (the spectral characteristic of the multiplier). In this case we can write the approximate solution to the linear Caputo FSDE (1) as the partial sum

$$x(t) \approx \sum_{i=0}^{L-1} X_i q_i(t), \tag{10}$$

where coefficients X_i depend on the initial state x_0 and entries of the truncated spectral characteristic $\mathcal{V}$, i.e., finite set of independent random values having a standard normal distribution.

The various systems of functions can be used as the basis $\{q_i(t)\}_{i=0}^{\infty}$, for example, the Legendre polynomials, trigonometric functions, the Walsh and Haar functions, etc. The uncomplete systems of functions, i.e., orthogonal finite functions, functions defined by wavelets or splines, can also be used, but these systems of functions imply the indirect use of a time discretization. All of them have been successfully applied to solve different problems by the spectral form of mathematical description for analysis and synthesis of linear and nonlinear deterministic and stochastic dynamical systems [9,24–26]. For the effective implementation of the spectral method the elementary and specialized algorithms to use the spectral form of mathematical description are developed, i.e., subroutines for calculating spectral characteristics of functions and linear operators including fractional integration and differentiation operators [24].

4. Numerical experiments

There are various Ornstein – Uhlenbeck (OU) process modifications [27–30]. Here we consider the fractional OU process described by the linear Caputo FSDE with additive noise

$${}^{*}\mathcal{D}^{\mu}_{0+}x(t) = \nu x(t) + \sigma v(t), \quad x(0) = x_0, \quad t \in \mathbb{T}, \tag{11}$$

where ν, σ and x_0 are given parameters, i.e., $a(t) = \nu$, $b(t) = 0$, $c(t) = 0$, $d(t) = \sigma$. Note that we have the classical (non-fractional) OU process when $\mu = 1$.

Using the relation (8) we can write the expression for the spectral characteristic X of the fractional OU process as follows:

$$X = (P^{\mu} - \nu E)^{-1}(x_0 P^{\mu}\mathbf{1} + \sigma\mathcal{V}), \tag{12}$$

i.e., $A = \nu E$ is the infinite diagonal matrix (the spectral characteristic of the multiplication operator with a multiplier $\alpha \in \mathbb{R}$ is αE), $B = \bar{O}$ is infinite zero column matrix, $C = O$ is infinite zero matrix, and $D = \sigma E$ is the infinite diagonal matrix similar to A.

In the relation (12) P^{μ} is the spectral characteristic of the Riemann – Liouville fractional differentiation operator of order μ and $\mathbf{1}$ is the spectral characteristic of the unit step function, $\mathcal{V}$ is the spectral characteristic of the Gaussian white noise $v(t)$. Then the fractional OU process can be expressed by the relation (9), and its approximation can be expressed by the relation (10) taking into account that all spectral characteristics are truncated on all dimensions to a certain order L.

Let $\{q_i(t)\}_{i=0}^{\infty}$ be the Legendre polynomials defined on $\mathbb{T}$. These polynomials are traditionally used primarily in the spectral method [8]. In addition, they have positively proven themselves in modeling repeated Itô and Stratonovich stochastic integrals [31,32], which is directly related to the solution of (non-fractional) SDEs. All needed spectral characteristics, i.e., the spectral characteristic of the Riemann – Liouville (Caputo) fractional differentiation operator and the spectral characteristic of the unit step function, are given in [24].

Sample paths of the fractional OU process are presented in the figure 1 for $T = 1$ (time interval $\mathbb{T} = [0,1]$), $\mu = 0.60$, $\nu = 1$, $\sigma = 0.5$, $x_0 = 0$ (parameters and the initial condition for the fractional OU process), $L = 100$ (truncation order for all spectral characteristics).

IOP Conf. Series: Materials Science and Engineering **927** (2020) 012077 doi:10.1088/1757-899X/927/1/012077

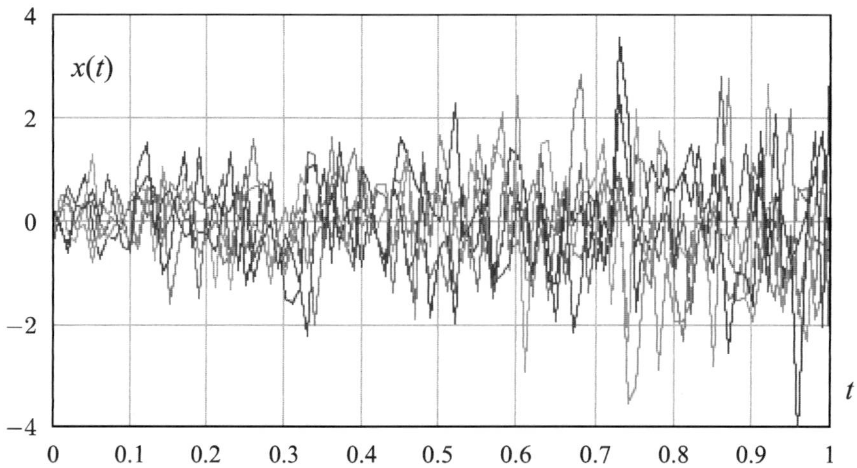

Figure 1. Sample paths of the fractional OU process, $\mu = 0.60$.

For $\mu = 0.75$ and $\mu = 0.90$ sample paths of the fractional OU process are shown in figures 2 and 3, respectively. Each figure contains six sample paths marked by various colors: red, blue, green, magenta, orange, and brown.

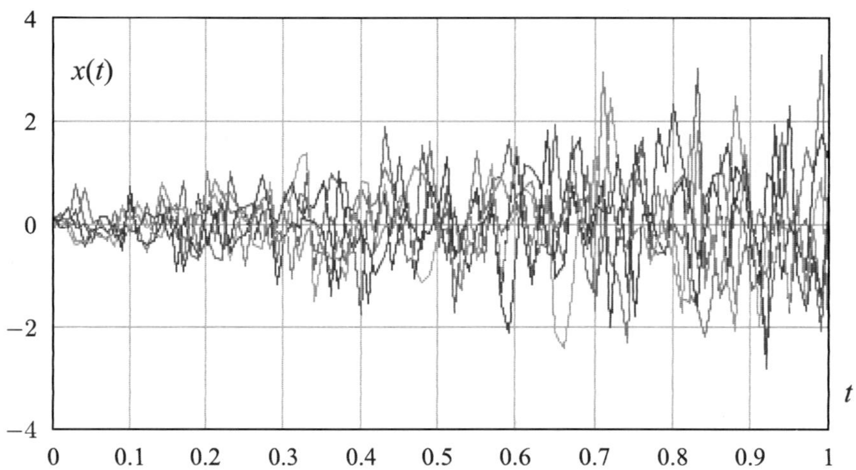

Figure 2. Sample paths of the fractional OU process, $\mu = 0.75$.

The above results show that with increasing parameter μ the variance and standard deviation decrease. The paths in figures 1–3 correspond to ones for centered random processes. A complete analysis of the obtained results requires the estimation of errors of the approximate solution. For instance, the paper [11] describes the technique of the exact error analysis for the mean and covariance as well as for the mean squared error of the classical OU process in detail. The similar technique can be applied for the considered fractional OU process.

5. Conclusion

In the paper the problem of finding the solution to the linear Caputo FSDE with additive and multiplicative noise has been considered. It is proposed to apply the spectral form of

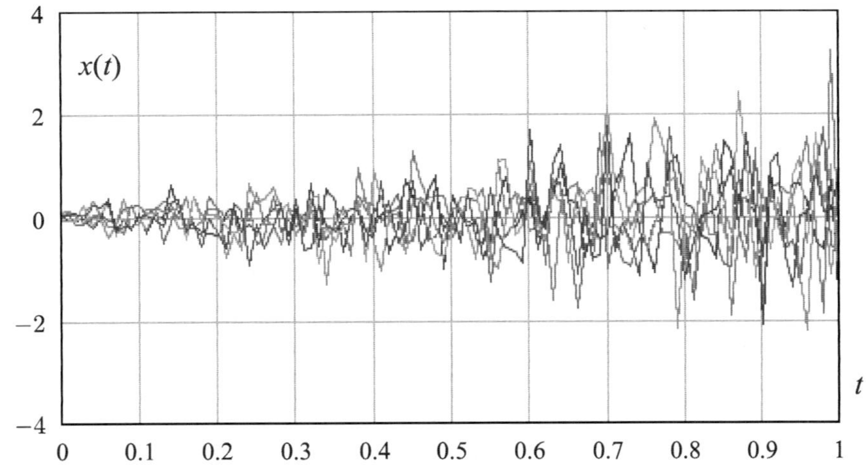

Figure 3. Sample paths of the fractional OU process, $\mu = 0.90$.

mathematical description and the spectral method for this problem. The main advantages of the proposed approach are both the explicit form of the solution as the orthogonal series with random coefficients and the convenient continuous-time approximation of this solution as the partial sum.

References

[1] Wang Y, Xu J and Kloeden P E 2016 Asymptotic behavior of stochastic lattice systems with a Caputo fractional time derivative *Nonlinear Anal. Theory Methods Appl.* **135** 205–222

[2] Doan T S, Huong P T, Kloeden P E and Tuan H T 2018 Asymptotic separation between solutions of Caputo fractional stochastic differential equations *Stoch. Anal. Appl.* **36**(4) 654–664

[3] Ito I 1979 On the existence and uniqueness of solutions of stochastic integral equations of the Volterra type *Kodai Math. J.* **2**(2) 158–170

[4] Sakthivel R, Revathi P and Ren Y 2013 Existence of solutions for nonlinear fractional stochastic differential equations *Nonlinear Anal. Theory Methods Appl.* **81** 70–86

[5] Jumarie G 2008 Modeling fractional stochastic systems as non-random fractional dynamics driven by Brownian motions *Appl. Math. Model.* **32**(5) 836–859

[6] Picard J 2011 Representation formulae for the fractional Brownian motion *Seminaire de Probabilites* vol XLIII *(Lecture Notes in Mathematics* vol 2006*)* ed C Donati-Martin et al. (Berlin: Springer-Verlag) pp 3–70

[7] Mandelbrot B B and Van Ness J W 1968 Fractional Brownian motions, fractional noises and applications *SIAM Review* **10**(4) 422–437

[8] Solodovnikov V V, Semenov V V, Peschel M and Nedo D 1979 *Design of Control Systems on Digital Computers: Spectral and Interpolational Methods* (Moscow: Mashinostroenie; Berlin: Verlag Technik) (in Russian; in German)

[9] Baghdasaryan G Y, Mikilyan M A, Panteleev A V and Rybakov K A 2020 Spectral method for analysis of diffusions and jump diffusions *Advances in Computational Mechanics and Numerical Simulation (Smart Innovation, Systems and Technologies* vol 173*)* ed L C Jain et al. (Singapore: Springer) chapter 21 pp 293–314

[10] Rybakov K A 2020 Modeling and analysis of output processes of linear continuous stochastic systems based on orthogonal expansions of random functions *J. Comput. Sys. Sc. Int.* **59**(3) 322–337

[11] Rybakov K A 2020 Spectral method of analysis and optimal estimation in linear stochastic systems *Int. J. Model. Simul. Sci. Comput.* **11**(3) 2050022

[12] Maleknejad K, Khodabin M and Rostami M 2012 A numerical method for solving m-dimensional stochastic Itô-Volterra integral equations by stochastic operational matrix *Comput. Math. Appl.* **63**(1) 133–143

[13] Maleknejad K, Khodabin M and Rostami M 2012 Numerical solution of stochastic Volterra integral equations by a stochastic operational matrix based on block pulse functions *Math. Comput. Model.* **55**(3–4) 791–800

[14] Mohammadi F 2015 A wavelet-based computational method for solving stochastic Itô-Volterra integral equations *J. Comput. Phys.* **298** 254–265
[15] Khan S U, Ali M and Ali I 2019 A spectral collocation method for stochastic Volterra integro-differential equations and its error analysis *Adv. Differ. Equ.* **2019** 161
[16] Metzler R and Klafter J 2000 The random walk's guide to anomalous diffusion: A fractional dynamics approach *Phys. Rep.* **339**(1) 1–77
[17] Borovik I N, Strokach E A and Severina N S 2019 Influence of the turbulent Prandtl number on numerical simulation reaction flow *AIP Conf. Proc.* **2181** 020029
[18] Panteleev A V, Letova T A and Pomazueva E A 2018 Parametric design of optimal in average fractional-order PID controller in flight control problem *Autom. Remote Control* **79**(1) 153–166
[19] Jain L C, Favorskaya M N, Nikitin I S and Reviznikov D L 2020 Advances in Computational Mechanics and Numerical Simulation *Advances in Computational Mechanics and Numerical Simulation (Smart Innovation, Systems and Technologies* vol 173*)* ed L C Jain et al. (Singapore: Springer) chapter 1 pp 1–8
[20] Önalan Ö 2010 Fractional Ornstein-Uhlenbeck processes driven by stable Levy motion in finance *Int. Research J. Finance Economics* **42** 129–138
[21] Chronopoulou A and Viens F 2012 Stochastic volatility and option pricing with long-memory in discrete and continuous time *Quant. Finance* **12**(4) 635–649
[22] Stark H and Woods J W 1994 *Probability, Random Processes, and Estimation Theory for Engineers* (Prentice-Hall)
[23] Pugachev V S and Sinitsyn I N 2002 *Stochastic Systems: Theory and Applications* (Singapore: World Scientific)
[24] Rybakov K A and Rybin V V 2016 *Modeling distributed integer-order and fractional-order processes and control systems by spectral method* (Moscow: MAI Publ.) (in Russian)
[25] Averina T A and Rybakov K A 2007 Comparison of a statistical simulation method and a spectral method for analysis of stochastic multistructure systems with distributed transitions *Rus. J. Numer. Anal. Math. Modelling* **22**(5) 431–447
[26] Panteleev A V and Rybakov K A 2011 Analyzing nonlinear stochastic control systems in the class of generalized characteristic functions *Autom. Remote Control* **72**(2) 393–404
[27] Maller R A, Müller G and Szimayer A 2009 Ornstein-Uhlenbeck processes and extensions *Handbook of Financial Time Series*, ed T Mikosch et al. (Berlin: Springer) chapter 18 pp 421–437
[28] Li L, Liu J and Lu J 2017 Fractional stochastic differential equations satisfying fluctuation-dissipation theorem *J. Stat. Phys.* **169** 316–339
[29] Ascione G, Mishura Y and Pirozzi E 2019 Fractional Ornstein-Uhlenbeck process with stochastic forcing, and its applications *Methodol. Comput. Appl. Probab.*
[30] Sandev T and Tomovski Z 2019 *Fractional Equations and Models: Theory and Applications* (Cham: Springer)
[31] Kuznetsov D F 2019 A comparative analysis of efficiency of using the Legendre polynomials and trigonometric functions for the numerical solution of Ito stochastic differential equations *Comput. Math. Math. Phys.* **59**(8) 1236–1250
[32] Kuznetsov D F 2019 Approximation of iterated Ito stochastic integrals of the second multiplicity based on the Wiener process expansion using Legendre polynomials and trigonometric functions *Differencialnie Uravnenia i Protsesy Upravlenia* (4) 32–52

IOP Conf. Series: Materials Science and Engineering **927** (2020) 012078 doi:10.1088/1757-899X/927/1/012078

Applying of the heat wave equation in the study of the movement of an aggressive environment front

Evgenii B Kuznetsov and Sergey S Leonov

Department of Dynamic Systems Modeling, Moscow Aviation Institute, Volokolamskoe Shosse, Moscow 125993, Russia

E-mail: kuznetsov@mai.ru

Abstract. The article investigates the problem of diffusion of an aggressive environment into a material of construction. This problem is especially relevant when the penetration of an aggressive environment into the structural elements of nuclear installations is investigated. For modeling the diffusion process, we use a hyperbolic wave equation describing the motion of a heat wave. Using the Fourier method, an analytical solution is obtained. An example illustrates the solution of the problem of hydrogen diffusion into a long, thin rectangular cross-section iron rod.

1. Introduction

A review of studies on modeling the processes of interaction of an aggressive environment with materials and structural elements shows a wide variety of research approaches and methods in this important area. Such interaction significantly affects the long-term strength of the structures as a whole and the safety of their operation throughout the entire service life. The safety of the operation of essential structures, including those used in nuclear energy, is not in doubt. Errors in reliability assessment can lead to industrial accidents and possible environmental disasters. We note several basic works. Of great interest is the article by Yu.N. Rabotnov [1] on the possible mechanisms of metal destruction in aggressive environment and a monograph by A.M. Lokoshchenko [2], which studies the long-term strength of metals in aggressive environments. Various aspects of the influence of an aggressive environment on the deformation and fracture characteristics of modern metals and alloys are considered in a large number of monographs. Among them, we note the monographs of A.M. Lokoshchenko [2, 3] and a monograph by I.V. Semenova et al. [4], devoted to corrosion and corrosion protection of metals. The monograph [3] considers the propagation of the front of aggression in a thin, long metal specimen of rectangular cross-section, which is placed in an aggressive environment. Lokoshchenko A.M. averages this equation by artificially introducing the movement of the front and obtains an approximate solution to the problem. Mathematical methods for describing diffusion processes are proposed in the monograph of the British scientist J. Crank [5], who considered various modifications of the classical parabolic diffusion equation. However, the parabolic diffusion equation modeling the process does not describe the movement of the front of aggression. In the present work, the movement of the front of aggression is described by the hyperbolic equation for the propagation of thermal waves [6, 7]:

$$\tau_r \cdot U_{TT} + U_T = D \cdot U_{XX}. \tag{1}$$

Here $U = U(X, T)$ is a concentration of the aggressive medium, τ_r is relaxation time, D is a diffusion coefficient of the environment in the source material, and T is time. The indices for the function U determine the partial derivatives. The diffusion process in a long rectangular cross-section rod is considered. The section width significantly exceeds its thickness $2H$, so that the influence of diffusion from the narrow sides of the rectangle can be neglected. We also assumed that the length of the rod is many times greater than its transverse dimensions, so the influence of the longitudinal coordinate of the rod on the diffusion process can also be ignored. The coordinate X is directed along the thickness of the bar so that the values $X = 0$ and $X = 2H$ correspond to the wide sides of the bar. From the symmetry condition, only half of the rod $0 \leq X \leq H$ is considered. Let U_0 be the concentration coefficient of the aggressive medium at the boundary with the metal. We rewrite equation (1) in the dimensionless form:

$$u_{tt} + 2b_1 \cdot u_t = u_{xx}, \tag{2}$$

where $u = \dfrac{U}{U_0}$, $x = \dfrac{X}{H}$, $t = \dfrac{T}{H}\sqrt{\dfrac{D}{\tau_r}}$, $b_1 = \dfrac{H}{2\sqrt{D \cdot \tau_r}}$.

We assume that at the initial moment, there is no aggressive medium inside the sample, and on the external surfaces, it remains constant U_0. Then the initial conditions will be homogeneous:

$$u(x, 0) = 0, u_t(x, 0) = 0, \tag{3}$$

and the boundary conditions take the form

$$u(0, t) = 1, u_x(1, t) = 0. \tag{4}$$

2. Methods

We will find the solution to the problem (2)-(4) by the separation of variables (the Fourier method). To solve the Sturm-Liouville problem, we make the boundary conditions (4) homogeneovs by introducing a new function $v(x, t) = u(x, t) - 1$. Then problem (2)-(4) takes the form

$$v_{tt} + 2b_1 v_t = v_{xx}. \tag{5}$$

$$v(x, 0) = -1, v_t(x, 0) = 0, \tag{6}$$

$$v(0, t) = 0, v_x(1, t) = 0. \tag{7}$$

According to the Fourier method, the solution to this problem is sought in the form

$$v(x, t) = X(x) \cdot T(t). \tag{8}$$

Solving the Sturm-Liouville problem, we obtain the eigenvalues of the problem (5)-(7) $\lambda_n = \dfrac{\pi}{2}(2n + 1)$ and the eigenfunctions $X_n(x) = \sin \dfrac{\pi(2n + 1)}{2} x$, where $n = 0, 1, 2, 3, \ldots$

The function $T_n(t)$ must satisfy the following equation:

$$\frac{d^2 T_n}{dt^2} + 2b_1 \frac{dT_n}{dt} + \lambda_n^2 T_n = 0. \tag{9}$$

Let $b_1 > \lambda_n$, which will be the case for $n \leq N$, where N is the integer part of the number $\dfrac{b_1}{\pi} - \dfrac{1}{2}$. Then the general solution of equation (9) will have the form

$$T_n(t) = A_n e^{k_1 t} + B_n e^{k_2 t},$$

IOP Conf. Series: Materials Science and Engineering **927** (2020) 012078 doi:10.1088/1757-899X/927/1/012078

where $k_1 = -b_1 + \sqrt{b_1^2 - \lambda_n^2}, k_2 = -b_1 - \sqrt{b_1^2 - \lambda_n^2}$ are the roots of the characteristic equation, and A_n, B_n are arbitrary constants.

If $b_1 < \lambda_n$, which will occur for $n > N$, then the general solution of equation (9) will take the form

$$T_n(t) = e^{-b_1 t}(A_n \cos\sqrt{\lambda_n^2 - b_1^2 t} + B_n \sin\sqrt{\lambda_n^2 - b_1^2 t}).$$

According to (8), the solution to the problem (5)-(7) will be determined by the series

$$v(x,t) = \sum_{n=0}^{\infty} T_n(t) X_n(x) = \sum_{n=0}^{N} (A_n e^{k_1 t} + B_n e^{k_2 t}) \sin\lambda_n x +$$
$$+ e^{-b_1 t} \sum_{n=N+1}^{\infty} (A_n \cos\sqrt{\lambda_n^2 - b_1^2 t} + B_n \sin\sqrt{\lambda_n^2 - b_1^2 t}) \sin\lambda_n x.$$

Here A_n, B_n are arbitrary constants that can be found from the initial conditions (3). The first condition leads to equality

$$v(x,0) = \sum_{n=0}^{N} (A_n + B_n) \sin\lambda_n x + \sum_{n=N+1}^{\infty} A_n \sin\lambda_n x =$$
$$= \sum_{n=0}^{\infty} A_n \sin\lambda_n x + \sum_{n=0}^{N} B_n \sin\lambda_n x = -1.$$

Due to the large value of N, this equality can be rewritten in the form

$$\sum_{n=0}^{\infty} (A_n + B_n) \sin\lambda_n x = -1,$$

therefore, the constants A_n, B_n, according to the representation of the function $\varphi(x) = -1$ by the Fourier series, must satisfy the equation

$$A_n + B_n = -\frac{4}{\pi(2n+1)}. \tag{10}$$

The second initial condition leads to the relation

$$\begin{aligned} U_t(x,0) &= \sum_{n=0}^{N} (A_n k_1 + B_n k_2) \sin\lambda_n x + \\ &+ \sum_{n=N+1}^{\infty} (-b_1 A_n + B_n \sqrt{\lambda_n^2 - b_1^2}) \sin\lambda_n x = 0. \end{aligned} \tag{11}$$

The sums in equality (11) relate to different sets of eigenfunctions, and therefore are linearly independent, therefore, expression (11) will be satisfied only when each amount is zero. For the first sum, taking into account (10), we obtain the system

$$\begin{cases} A_n + B_n = -\dfrac{4}{\pi(2n+1)}, \\ A_n k_1 + B_n k_2 = 0, \end{cases}$$

which we solve and receive

$$A_n = \frac{4k_2}{\pi(2n+1)(k_1 - k_2)}, B_n = -\frac{4k_1}{\pi(2n+1)(k_1 - k_2)}. \tag{12}$$

For the second amount, the system will take the form

$$\begin{cases} A_n + B_n = -\dfrac{4}{\pi(2n+1)}, \\ -A_n b_1 + B_n\sqrt{\lambda_n^2 - b_1^2} = 0. \end{cases}$$

Its solution can be represented in the form

$$A_n = -\frac{4\sqrt{\lambda_n^2 - b_1^2}}{\pi(2n+1)(b_1 + \sqrt{\lambda_n^2 - b_1^2})},\ B_n = -\frac{4b_1}{\pi(2n+1)(b_1 + \sqrt{\lambda_n^2 - b_1^2})}. \tag{13}$$

Then, taking into account (12), (13), the solution to problem (5)-(7) will be given by the function

$$U(x,t) = \frac{4}{\pi}\left(\sum_{n=0}^{N} \frac{k_2 e^{k_1 t} - k_1 e^{k_2 t}}{(2n+1)(k_1 - k_2)} \sin\lambda_n x - \right.$$
$$\left. -e^{-b_1 t} \sum_{n=N+1}^{\infty} \frac{\sqrt{\lambda_n^2 - b_1^2}\cos\sqrt{\lambda_n^2 - b_1^2}t + b_1 \sin\sqrt{\lambda_n^2 - b_1^2}t}{(2n+1)(b_1 + \sqrt{\lambda_n^2 - b_1^2})} \sin\lambda_n x\right)$$

and finally the solution of problem (2)-(4) takes the form

$$u(x,t) = 1 + \frac{4}{\pi}\left(\sum_{n=0}^{N} \frac{k_2 e^{k_1 t} - k_1 e^{k_2 t}}{(2n+1)(k_1 - k_2)} \sin\lambda_n x - \right.$$
$$\left. -e^{-b_1 t} \sum_{n=N+1}^{\infty} \frac{\sqrt{\lambda_n^2 - b_1^2}\cos\sqrt{\lambda_n^2 - b_1^2}t + b_1 \sin\sqrt{\lambda_n^2 - b_1^2}t}{(2n+1)(b_1 + \sqrt{\lambda_n^2 - b_1^2})} \sin\lambda_n x\right). \tag{14}$$

3. Results

As an example, we considered hydrogen diffusion into a fragment of iron rod thickness $H = 10^{-3}$ m [3]. In this case, the dimensionless coefficient b_1 in equation (2) is approximately equal to $\dfrac{5 \cdot 10^3}{3}$, that corresponds to the values of relaxation time $\tau_r = 10^{-10}$ s [7] and the diffusion coefficient D approximately equal to 10^{-3}. When calculating $u(x,t)$ according to formula (14) $N = 3 \cdot 10^4$ members of the series were taken. As the number N increased, the result did not change significantly.

The figure 1 shows graphs of the movement of the aggressive environment front for several dimensionless time moments. As the coefficient b_1 increases, the movement of the aggressive environment front decays rapidly, and when it decreases, the concentration coefficient of the aggressive medium $U(x,t)$ inside the sample quickly reaches the value U_0 equal to the concentration of the medium at the interface with the metal.

The solution to the problem (2)-(4) for several dimensionless time moments on a segment $x \in [0;1]$ is shown in figure 2. The graphs presenting in figure 2 show an increase in the concentration of the aggressive environment in the rod. After the front of the aggressive environment reaches the right border, its concentration monotonously increases until the concentration in the whole rod reaches the initial value given on the left border.

4. Discussion

An advantage of the proposed model for the propagation of an aggressive environment in a material is the fact that this model reflects the movements of the front of an aggressive environment. In other models (see, for example, [3]) this motion is specified artificially, and the diffusion equation of the parabolic type is averaged. This allows getting simple approximate technical formulas.

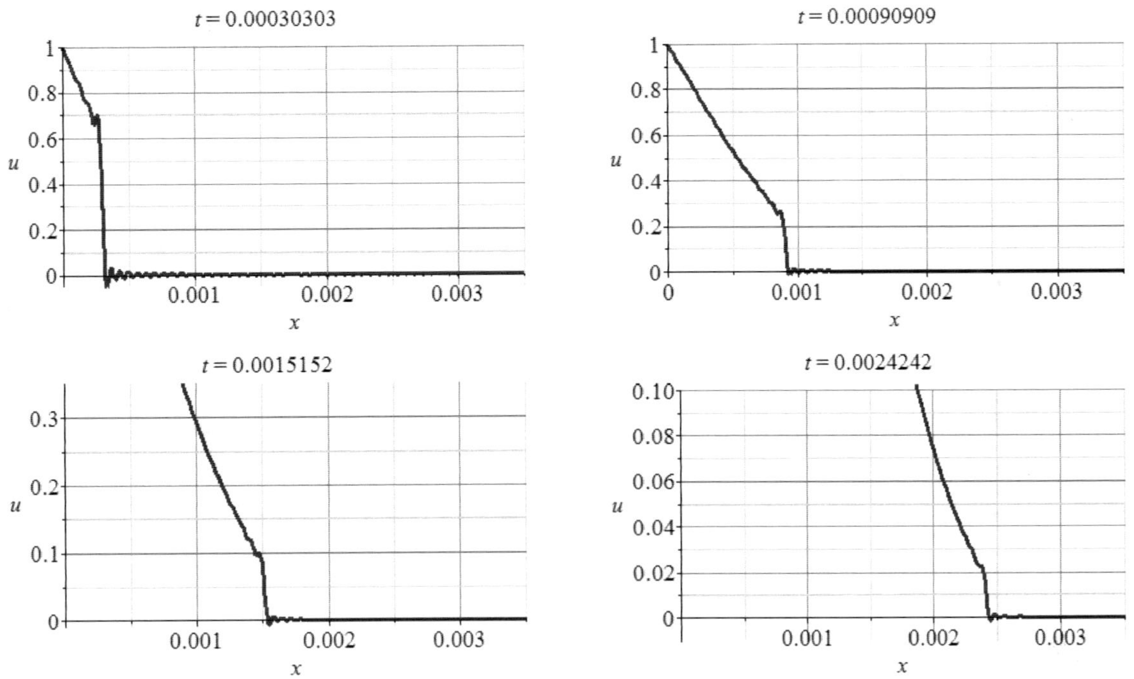

Figure 1. The movement of the front of aggression for several dimensionless time moments.

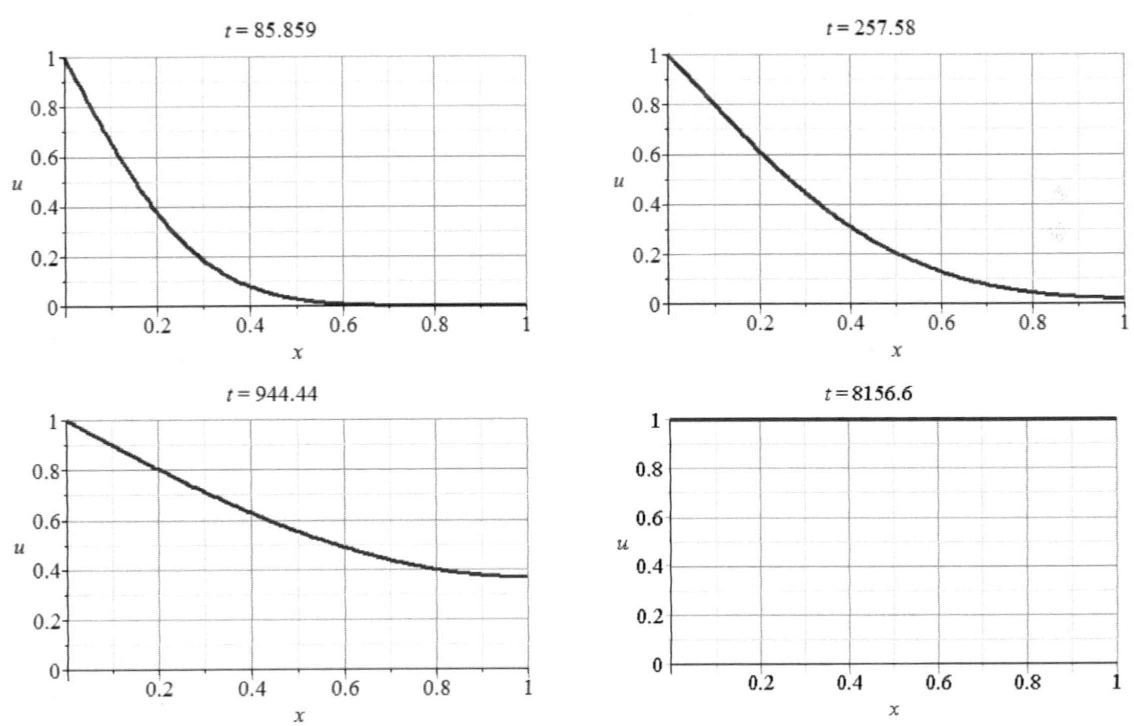

Figure 2. The solution of the problem (2)-(4) for several dimensionless time moments.

IOP Conf. Series: Materials Science and Engineering **927** (2020) 012078 doi:10.1088/1757-899X/927/1/012078

5. Conclusion

The authors construct the mathematical model that describes the propagation of an aggressive medium in a material using a hyperbolic type wave equation specifying the motion of a heat wave. The advantage of the model is that it defines the movement of the front of an aggressive environment, that is not possible for models based on the diffusion equation of a parabolic type.

Acknowledgments

This paper is based on research carried out with the financial support of the grant of the Russian Science Foundation (project 18-19-00474).

References

[1] Rabotnov Yu N 1954 *Izv. USSR Academy of Sciences. Sep. tech. sciences* 6 53
[2] Lokoshchenko A M 2000 *Creep and long-term strength of metals in aggressive environments* (Moscow: Moscow State University Press)
[3] Lokoshchenko A M 2018 *Creep and Creep Rupture of Metals* (Boca Raton, London, New York: CRC Press Taylor and Francis Group)
[4] Semenova I V, Florianovich G M and Khoroshilov A V 2002 *Corrosion and Corrosion Protection* (Moscow: Fizmatlit)
[5] Crank J 1975 *The Mathematics of Diffusion. Second ed.* (Oxford: Clarendon Press)
[6] Lykov A V 1967 *Theory of thermal conductivity* (Moscow: Higher school)
[7] Formalev V F 2015 *Heat transfer in anisotropic solids* (Moscow: Fizmatlit)

IOP Conf. Series: Materials Science and Engineering **927** (2020) 012079 doi:10.1088/1757-899X/927/1/012079

Numerical solving of initial value problems for ordinary differential equations with internal limiting singular points

A I Ivannikov, E B Kuznetsov and S S Leonov

Moscow Aviation Institute (National Research University), Volokolamskoe Shosse 4, 125993 Moscow, Russia

E-mail: powerandglory@yandex.ru

Abstract. In the present paper, we investigate the numerical methods for solving initial value problems for systems of nonlinear ordinary differential equations with several limiting singular points. Similar tasks have great practical importance in such areas as hydrodynamics, chemical kinetics, combustion theory, computational geometry, and nuclear reactors. However, the exact solution may be constructed to nonlinear initial value problems only in exceptional cases. Numerical methods also do not give the desired result. Explicit methods make it possible to obtain a solution only if there is one limiting singular point on the right boundary of the considered interval of the argument change. Implicit methods also have usage restrictions. They consist of the difficulty or impossibility of overcoming the internal limiting singular points. For solving the considered class of problems, the paper proposes to use the method of solution continuation, which allows passing any limiting singular points effectively. Features of the application of various numerical methods are shown by solving one test initial value problem. The reliability of the obtained numerical results is confirmed by comparison with the finding analytical solution of the considered test problem.

1. Introduction

In many applied fields, one has to deal with the solving of initial value problems for systems of ill-conditioned ordinary differential equations. Similar problems are characteristic of hydrodynamics, solid mechanics, chemical kinetics, the theory of catalytic reactions, problems of mechanical engineering, and computational geometry. All applied problems, as a rule, have a large dimension, and differential equations are nonlinear. Therefore, the exact solution to the problems under consideration can be obtained only in exceptional cases. However, approximate methods for solving ill-conditioned problems have many disadvantages associated with the presence of rapid growth or decrease sections on integral curves. Explicit methods [1] are hardly applicable for solving ill-conditioned initial problems. This fact led to the emergence of special methods for solving stiff problems [2], which include ill-conditioned problems. A large class of methods for solving stiff problems is based on implicit schemes. Asymptotic methods are another class of methods for solving ill-conditioned problems. These methods are used for solving problems with a small parameter at the senior derivative (with boundary layers [3] and contrast structures (interior layers) [4]). They consist of presenting the solution of the problem in the form of asymptotic series in powers of a small parameter.

IOP Conf. Series: Materials Science and Engineering **927** (2020) 012079 doi:10.1088/1757-899X/927/1/012079

This paper discusses one of the particular types of ill-conditioned problems associated with the presence on the integral curve limiting singular points at which the equations of the system are undefined. Most implicit and asymptotic methods experience significant difficulties when passing through limiting singular points. The authors showed [5] that many implicit schemes of various orders of accuracy require significant restrictions on the integration step. Moreover, implicit methods are hardly applicable when passing through internal limiting singular points, since iterative processes for solving nonlinear equation systems in a neighborhood of limiting singular points converge slowly or diverge. Similar problems of convergence of series in a neighborhood of limiting singular points arise in asymptotic methods. Therefore, despite the large and increasing number of new methods for solving ill-conditioned problems every year, there is an urgent need for an effective general approach.

For over 50 years in computational mathematics and mechanics, the method of solution continuation with respect to the best argument (the arc length method) has been developing. In the works of such scientists as I. I. Vorovich [6], E. Riks [7], E. I. Grigolyuk and V. I. Shalashilin [8], E. L. Allgower and K. Georg [9], M. A. Crisfield [10], V. I. Shalashilin and E. B. Kuznetsov [11], the method of solution continuation showed its effectiveness in solving ill-conditioned systems of algebraic, differential, differential-algebraic and functional-differential equations. The method of solution continuation with respect to the best parameter is the most effective method of constructing closed curves of the solution set and passing the limiting singular points for systems of nonlinear algebraic and transcendental equations. However, solving initial value problems for ordinary differential equations with several limit singular points has been insufficiently studied until now.

1.1. Aims of research

This paper aims to demonstrate the computational efficiency of the method of solution continuation with respect to the best argument as applied for solving initial value problems for nonlinear ordinary differential equations with several limiting singular points. Since the method of solution continuation shows its effectiveness in constructing the curve of the solution set of algebraic and transcendental equations with several limiting singular points, we can expect a similar situation for ordinary differential equations. We show the advantages and disadvantages of the solution continuation method compared to traditional explicit methods by the example of solving a test problem for which we obtain an exact solution. A comparison of the methods will be carried out by the computational time and the relative error of the numerical solution.

2. Statement of the problem

We consider the initial value problem for the first order ordinary differential equation of the form

$$\frac{dy}{dx} = \frac{F(x)}{(y-\alpha_1)(y-\alpha_2)\dots(y-\alpha_n)}, \quad y(0) = 0. \tag{1}$$

Here $y(x)$ is a desired solution, x is an argument taking values from a segment $[0; x_*]$, $\alpha_1, \dots, \alpha_n$ are given numbers, $\alpha_i \in [0; 1]$, $i = \overline{1, n}$, $F(x)$ is a predefined integrable function. It is supposed that $\alpha_i < \alpha_j$ at $i < j$.

The solution of the problem (1) using the Vieta's formulas can be written in the integral form

$$\int_0^x F(\xi)\, d\xi = \frac{y^{n+1}}{n+1} + a_1 \frac{y^n}{n} + \dots + a_{n-1}\frac{y^2}{2} + a_n y, \tag{2}$$

where the coefficients $a_1, \dots, a_n$ have the form

$a_1 = -(\alpha_1 + \dots + \alpha_n), \quad a_2 = \alpha_1\alpha_2 + \dots \alpha_1\alpha_n + \alpha_2\alpha_3 + \dots + \alpha_{n-1}\alpha_n,$

$a_3 = -(\alpha_1\alpha_2\alpha_3 + \dots + \alpha_1\alpha_2\alpha_n + \alpha_2\alpha_3\alpha_4 + \dots + \alpha_{n-2}\alpha_{n-1}\alpha_n),$

$\vdots$

$a_{n-1} = (-1)^{n-1}(\alpha_1\alpha_2\dots\alpha_{n-1} + \alpha_1\alpha_2\dots\alpha_{n-2}\alpha_n + \dots + \alpha_2\alpha_3\dots\alpha_n),$

$a_n = (-1)^n\alpha_1\alpha_2\dots\alpha_n.$

But, if the integral on the left-hand side of expression (2) cannot be calculated analytically, then the solution of problem (1) is possible only using numerical methods. Moreover, the points whose ordinates coincide with the zeros of the denominator of the right-hand side of the equation of the problem (1), depending on their location on the integral curve, can be limiting singular or essential singular points. Explicit schemes make it possible to construct a solution only up to the first limiting singular point; further progress along the integral curve is impossible. Implicit schemes for small-dimensional problems allow finding a solution to the initial problem, but are inferior to explicit schemes in terms of speed, which only intensifies for large-dimensional problems.

2.1. Particular case

We consider the features of the numerical solution of initial value problems of the form (1) using the following typical problem as an example

$$\frac{dy}{dx} = \frac{\cos x}{y(y-0.5)(y-1)}, \quad y(0) = 0. \tag{3}$$

For the problem (3), the use of explicit schemes is ineffective, since the equation of the problem (3) contains three limiting singular points on the integral curve at which the equations lose their meaning. Moreover, one of the limiting singular points coincides with the initial one. In this case, we use the reversion method, which consists in replacing the argument x with y. This transformation allows eliminating all limiting singular points of the problem (3). The reversed problem to the problem (3) has the form

$$\frac{dx}{dy} = \frac{y(y-0.5)(y-1)}{\cos x}, \quad x(0) = 0. \tag{4}$$

To solve the problem (3) we will also use a more efficient solution continuation method [11], which is based on the idea of replacing the original argument of the problem with a new one, so that the transformed problem does not have limiting singular points and can be solved by explicit methods.

For the problem (3) we will use the solution continuation with respect to the best argument, counted at each point along the tangent direction to the integral curve of the problem under consideration [11]. The differential of this argument satisfies the relation

$$(d\lambda)^2 = (dx)^2 + (dy)^2. \tag{5}$$

Assuming that the variables x and y are functions of λ, we represent relation (5) in the form

$$\left(\frac{d\lambda}{dx}\right)^2 = 1 + \left(\frac{dy}{dx}\right)^2.$$

Then we transform the equation of the problem (3) to the form

$$\frac{dy}{d\lambda} = \frac{\cos x}{y(y-0.5)(y-1)} \cdot \frac{dx}{d\lambda} = \pm\frac{\cos x}{\sqrt{y^2(y-0.5)^2(y-1)^2 + \cos^2 x}}. \tag{6}$$

Complementing it with the equation for the argument x

$$\frac{dx}{d\lambda} = \pm \frac{y(y-0.5)(y-1)}{\sqrt{y^2(y-0.5)^2(y-1)^2 + \cos^2 x}}, \tag{7}$$

we obtain a closed system of differential equations. The sign of the right-hand sides of equations (6) and (7) reflects the direction of motion along the argument λ. Hereinafter, we will assume that the sign of the right-hand sides is positive.

The initial conditions for the system (6)-(7) take the form

$$y(0) = 0, \quad x(0) = 0. \tag{8}$$

3. Numerical results

For solving the reversed problem (4) and transformed problem (6)-(8), we used the explicit Euler method. This method is the simplest explicit method, but all its disadvantages are also characteristic of other explicit schemes of a higher order of accuracy. To increase the accuracy of the numerical solution, we use variable step size varies by the Runge rule [1] with the accuracy parameter (permissible error) θ.

To assess the accuracy of the obtained numerical solutions, we will use the exact solution to the problem (3), obtained by formula (2) in the form

$$x = \arcsin\left(\frac{y^4}{4} - \frac{y^3}{2} + \frac{y^2}{4}\right). \tag{9}$$

Figures 1-2 show integral curves of the considered problems, where the solid black line is the exact solution, dashed green line is the numerical solution obtained by the explicit Euler method with a variable step size varies by the Runge rule with the permissible error $\theta = 10^{-4}$.

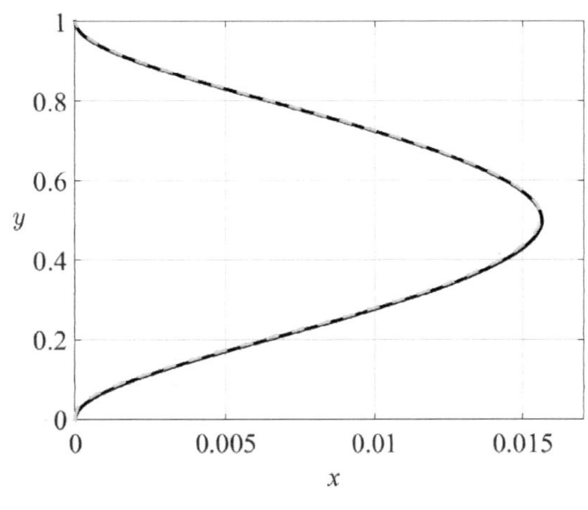

Figure 1. The numerical solution of the reversed problem (4).

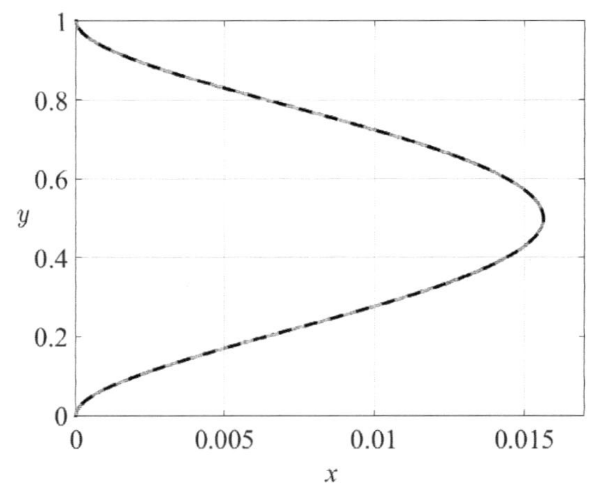

Figure 2. The numerical solution of the transformed problem (6)-(8).

Table 1 presents data on the relative error of the numerical solution and the computational time, where $\delta\varepsilon_{\text{max}}$ is a maximum relative error, $\delta\varepsilon_{\text{med}}$ is a median relative error (a middle element of all relative errors sorted by the value), $\delta\varepsilon_{\text{av}}$ is an average relative error [12], and t_c is a computational time. The initial step sizes for problems (4) and (6)-(8) we denote by h_0 and l_0 correspondingly. In the considered problems we accept $h_0 = 10^{-2}$ and $l_0 = 1$.

In this work, all calculations were carried out in the Matlab R2015a computing environment on a personal computer with processor Ryzen 5 (2500U, 2 GHz), RAM 8 GB, AMD Radeon graphics card (RX 560C, 4 GB), operating system Windows 10 Home.

Table 1. The relative error characteristics of the numerical solution and the computational time for the problems (4) and (6)-(8)

Problem statement	$\delta\varepsilon_{\text{max}}$, %	$\delta\varepsilon_{\text{med}}$, %	$\delta\varepsilon_{\text{av}}$, %	t_c, s
Permissible error $\theta = 5 \cdot 10^{-3}$				
Reversed problem (4)	746.3	29.05	96.42	0.00007
Transformed problem (6)-(8)	633.6	1.030	6.064	0.00074
Permissible error $\theta = 10^{-4}$				
Reversed problem (4)	688.6	1.669	5.055	0.00116
Transformed problem (6)-(8)	100.0	0.016	0.107	0.02523
Permissible error $\theta = 5 \cdot 10^{-6}$				
Reversed problem (4)	1397	0.243	1.097	0.00600
Transformed problem (6)-(8)	389.0	0.001	0.010	0.19086

4. Discussion

We will immediately note that explicit methods cannot solve the initial value problem (3) since the initial point is a limiting singularity. To eliminate this, we use the transformations of the original problem. The simplest and most obvious is the transition to the reversed problem (4). This eliminates all limiting singular points. However, from figure 1, it can be seen that in the neighborhood of the point $y = 0.5$, the error still grows. Lowering the permissible error in the Runge rule leads to a delayed increase in the accuracy of the numerical solution. Also, in a more general case, the numerator of the equation of the original problem (3) can vanish. This makes the reversion method ineffective, since the reversed problem (4), like the original one, will have one or more limiting singular points.

To increase the accuracy of the numerical solution and the generality of the reversion method, we use the method of solution continuation with respect to the best argument. A comparison of the graphs in figures 1 and 2 already shows that the numerical solution of the transformed problem (6)-(8) corresponds better to the exact solution. This is also confirmed by the calculated error values given in table 1. A comparison of the accuracy of the numerical solutions for a given value of the permissible error in the Runge rule shows that the relative error for the reversed problem (4) can be several times higher than the similar values for the transformed problem (6)-(8). So the maximum value, median, and average value of the relative error can be up to seven, 243, and 100 times greater, respectively. At the same time, the transformed problem (6)-(8) shows a better fit for the estimate of the local error according to the Runge rule, which allows us to obtain an approximate solution of the required accuracy. Also, in contrast to the reversed problem (4), there is a more rapid increase in the accuracy of the transformed problem solution while reducing the permissible error. Note that the computational time for a fixed value of the permissible error θ for the transformed problem can exceed up to 25 times the similar values for the reversed problem. At the same time, when comparing the computational time for both problems with the same error (as is observed in the case with the values $\theta = 5 \cdot 10^{-3}$ and $\theta = 10^{-4}$), the computational time for the transformed problem (6)-(8) becomes 35 percent less than for the reversed problem (4). All of the above indicates the higher effectiveness of the method of solution continuation with respect to the best argument for solving initial value problems with several limiting singular points, including internal ones.

The paper does not consciously suggest using implicit methods for solving the considered problems. The authors have already shown the disadvantages of implicit methods in solving this class of problems [5]. These disadvantages lie in need of numerically solving nonlinear algebraic and transcendental equations and their systems, which complicates the solving process and increases the computational time without a significant increase in accuracy.

5. Conclusion

The paper shows the features of various methods for passing the limiting singular points in numerical solving of initial value problems for ordinary differential equations. Using the example of solving one test initial value problem, we considered the explicit Euler method disadvantages, which are valid for all explicit schemes in applying to this class of problems. Explicit methods cannot solve the original problem since the first limiting singular point coincides with the initial. To eliminating the noted disadvantages, we proposed to use the reversion method and the method of solution continuation with respect to the best argument. The authors showed that reverse transformation does not allow achieving the effectiveness of the solution continuation method. This method allows to obtain higher accuracy when using a variable step size with control of the local error according to the Runge rule. Wherein, with the same accuracy of the numerical solution, the computational time for the transformed to the best argument problem is up to 35 percent less than for the reversed problem. Moreover, the accuracy of the numerical solution of the reversed problem has a delayed increase with decreasing the permissible error in the Runge rule. For the transformed problem, accuracy continues to rapidly increase.

Of course, for the method of solution continuation with respect to the best argument, there are many unexplored areas, such as estimating the local error of the solutions of transformed problems, numerically solving equations and systems with angular and bifurcation points. These problems will be investigated in the following papers. However, for today, the method of solution continuation remains the most comprehensive and effective tool for numerically solving ill-conditioned initial value problems with several limiting singular points.

Acknowledgments

The article was prepared on the basis of scientific research carried out with the financial support of the Russian Foundation for Basic Research grant (project No. 19-08-00718).

References

[1] Hairer E, Norsett S P and Wanner G 1987 *Solving Ordinary Differential Equations I: Nonstiff Problems* (Berlin: Springer)
[2] Hairer E and Wanner G 1996 *Solving Ordinary Differential Equations II: Stiff and Differential-Algebraic Problems* (Berlin: Springer)
[3] Lomov S A 1981 *Introduction to the General Theory of Singular Perturbations* (American Mathematical Society)
[4] Chang K W and Howes F A 1984 *Nonlinear Singular Perturbation Phenomena: Theory and Application* (New York: Springer)
[5] Kuznetsov E B and Leonov S S 2018 *Comput. Math. Math. Phys.* **58** 881
[6] Vorovich I I and Zipalova V F 1965 *J. Appl. Math. Mech.* **29** 1055
[7] Riks E 1972 *Trans. ASME* 4 1060
[8] Grigolyuk E I and Shalashilin V I 1991 *Problems of Nonlinear Deformation: The Continuation Method Applied to Nonlinear Problems in Solid Mechanics* (Dordrecht: Kluwer Academic Publishers)
[9] Allgower E L and Georg K 1990 *Introduction to numerical continuation methods* (Berlin / Heidelberg: Springer-Verlag)
[10] Crisfield M A 1981 *Computers & Structures* **13** 55
[11] Shalashilin V I and Kuznetsov E B 2003 *Parametric Continuation and Optimal Parametrization in Applied Mathematics and Mechanics* (Dordrecht: Kluwer Academic Publishers)
[12] Johnson N L and Leone F C 1977 *Statistics and Experimental Design: in Engineering and the Physical Sciences. Volume 1* (New York / London / Sydney / Toronto: John Wiley & Sons)

IOP Conf. Series: Materials Science and Engineering **927** (2020) 012080 doi:10.1088/1757-899X/927/1/012080

Application of Walsh series to represent iterated Stratonovich stochastic integrals

K Rybakov

Moscow Aviation Institute (National Research University),
Volokolamskoe highway 4, 125993 Moscow, Russian Federation

E-mail: rkoffice@mail.ru

Abstract. The goal of the paper is to represent iterated Stratonovich stochastic integrals using orthonormal expansions of functions by the Walsh series, to construct an algorithm for modeling iterated Stratonovich stochastic integrals, and to test this algorithm by numerical experiments. Obtained results may be used in the constructing high-order numerical methods based on the Taylor – Stratonovich expansion and the Taylor – Itô expansion. Therefore, they can be applied in a lot of problems of physics, engineering, economics and finance, when it is necessary to solve numerically stochastic differential equations.

1. Introduction

Iterated stochastic integrals play a fundamental role in the constructing high-order numerical methods for stochastic differential equations. These methods are based on the Taylor – Itô expansion and the Taylor – Stratonovich expansion for random processes. The first numerical method using iterated stochastic integrals of multiplicity $k = 2$ and orthogonal expansions into the trigonometric series was the Milstein's method [1, 2]. Iterated stochastic integrals of multiplicity $k = 2, 3$ and orthogonal expansions into the trigonometric series have also been used in [3]. In the Kuznetsov's method [4] an arbitrary complete orthonormal system may be applied for the representation of iterated stochastic integrals of multiplicity $k \geqslant 2$. Orthogonal expansions into the trigonometric series and the Haar series for the Milstein's method have been investigated in [5]. The paper [6] deals with orthogonal expansions for random processes by the spectral form of mathematical description.

In this paper the application of the Walsh series [7] for the representation of iterated Stratonovich stochastic integrals is considered. These functions form a multiplicative group and this property can be used, since the iterated Stratonovich stochastic integral of multiplicity $k \geqslant 2$ is a solution of the system of k bilinear stochastic differential equations including products of random processes that can be represented by the Walsh series. This representation is convenient due to it uses only one matrix of expansion coefficients of the unit step function for an arbitrary multiplicity $k \geqslant 2$.

Obtained results may be used in the constructing high-order numerical methods based on the Taylor – Stratonovich expansion for random processes [2, 3, 8–10], and also in numerical methods based on the Taylor – Itô expansion due to the known relationship between iterated Itô and Stratonovich stochastic integrals [9, 10]. High-order numerical methods may be applied for the modeling stochastic dynamical systems [3, 8, 10, 11], the solving optimal and suboptimal

IOP Conf. Series: Materials Science and Engineering **927** (2020) 012080 doi:10.1088/1757-899X/927/1/012080

filtering problems [12–14], and the optimizing dynamical systems of the joint estimation and control [15–17]. Iterated stochastic integrals can be used in the constructing high-order numerical methods for non-commutative semilinear stochastic partial differential equations [18].

2. Definitions and preliminary results

Iterated Stratonovich stochastic integrals of multiplicity $k \geqslant 2$ are defined as follows:

$$I_h^{*(j_1 j_2 \ldots j_k)} = \int_0^h \ldots \int_0^{\tau_3} \int_0^{\tau_2} dw_{j_1}(\tau_1) \circ dw_{j_2}(\tau_2) \circ \ldots \circ dw_{j_k}(\tau_k), \tag{1}$$

$$j_1, j_2, \ldots, j_k = 1, 2, \ldots, s,$$

where $h > 0$; $w_1(t)$, $w_2(t)$, $\ldots$, $w_s(t)$ are independent standard Wiener processes. The integral of multiplicity $k = 1$ is a centered Gaussian random variable

$$I_h^{*(j_1)} = \int_0^h dw_{j_1}(\tau) = w_{j_1}(h), \quad j_1 = 1, 2, \ldots, s,$$

with the second-order moment h. In this context h is an integration step in numerical methods for stochastic differential equations.

Note that for the case of pairwise distinct values $j_1, j_2, \ldots, j_k$ the iterated Stratonovich stochastic integral (1) coincides with the corresponding iterated Itô stochastic integral

$$I_h^{(j_1 j_2 \ldots j_k)} = \int_0^h \ldots \int_0^{\tau_3} \int_0^{\tau_2} dw_{j_1}(\tau_1) dw_{j_2}(\tau_2) \ldots dw_{j_k}(\tau_k).$$

In [9,10] it is proposed a general approach to the representation and modeling iterated Itô and Stratonovich stochastic integrals. This approach is based on finding the expansion coefficients of the function

$$K(t_1, t_2, \ldots, t_k) = \begin{cases} 1 & \text{for } t_1 < t_2 < \ldots < t_k, \\ 0 & \text{otherwise,} \end{cases}$$

with respect to the orthonormal basis $\{q(i_1, t_1) q(i_2, t_2) \ldots q(i_k, t_k)\}_{i_1, i_2, \ldots, i_k = 0}^{\infty}$ in $L_2([0, h]^k)$, where $\{q(i, t)\}_{i=0}^{\infty}$ is the orthonormal basis in $L_2([0, h])$. In [9, 10, 19] the Legendre polynomials and trigonometric functions are considered in detail. Moreover, the general case of the function $K(t_1, t_2, \ldots, t_k)$ and corresponding iterated stochastic integrals are discussed. Representations for iterated stochastic integrals of multiplicity $k = 1, 2$ have also been obtained for the Walsh and Haar functions [5, 10].

The spectral form of mathematical description has been applied in [6] for iterated stochastic integrals of multiplicity $k = 1, 2$ with respect to the Legendre polynomials and trigonometric functions as well as the Walsh and Haar functions. The case $k = 1$ is trivial, and for $k = 2$ we have $K(t_1, t_2) = \mathbf{1}(t_2 - t_1)$, where $\mathbf{1}(t)$ is the unit step function that is the impulse response function of the integrating element. Therefore, the iterated stochastic integral of multiplicity 2 may be represented using the spectral characteristic of the integration operator (two-dimensional nonstationary transfer function of the integrating element [20]). This spectral characteristic for the Walsh functions will be applied below for the case $k > 2$.

According to [9, 10, 19], the following representation of iterated Stratonovich stochastic integrals holds

$$I_h^{*(j_1 j_2 \ldots j_k)} = \sum_{i_1=0}^{\infty} \sum_{i_2=0}^{\infty} \ldots \sum_{i_k=0}^{\infty} C_{i_k \ldots i_2 i_1} \zeta_{i_1}^{(j_1)} \zeta_{i_2}^{(j_2)} \ldots \zeta_{i_k}^{(j_k)}, \tag{2}$$

where $C_{i_k\ldots i_2 i_1}$ are expansion coefficients of the function $K(t_1,t_2,\ldots,t_k)$, i.e.,

$$C_{i_k\ldots i_2 i_1}=\int_0^h\ldots\int_0^h\int_0^h q(i_1,\tau_1)q(i_2,\tau_2)\ldots q(i_k,\tau_k)K(\tau_1,\tau_2,\ldots,\tau_k)d\tau_1 d\tau_2\ldots d\tau_k$$
$$=\int_0^h q(i_k,\tau_k)\ldots\int_0^{\tau_3} q(i_2,\tau_2)\int_0^{\tau_2} q(i_1,\tau_1)d\tau_1 d\tau_2\ldots d\tau_k,\quad i_1,i_2,\ldots,i_k=0,1,2,\ldots, \tag{3}$$

and $\zeta_i^{(j)}$ are independent random variables having a standard normal distribution ($j=1,2,\ldots,s$; $i=0,1,2,\ldots$).

3. Application of Walsh series to represent iterated Stratonovich stochastic integrals

Let $\{q(i,t)\}_{i=0}^{\infty}$ be the system of the Walsh functions, which forms the orthonormal basis in the space $L_2([0,h])$. So,

$$q(i,t)=\frac{1}{\sqrt{h}}\begin{cases}1 & \text{for } i=0,\\ \prod\limits_{\{\kappa:\ \gamma_\kappa=1\}} r(\kappa,t) & \text{for } i=1,2,\ldots,\end{cases} \tag{4}$$

where γ_κ are coefficients in the binary representation of the number i, i.e., $i=\gamma_1 2^0+\gamma_2 2^1+\ldots+\gamma_\kappa 2^{\kappa-1}+\ldots+\gamma_{m+1}2^m$, m is the largest degree in this binary representation, $\gamma_\kappa\in\{0,1\}$, and $r(\kappa,t)$ are the Rademacher functions ($\kappa=1,2,\ldots$) defined by

$$r(\kappa,t)=\operatorname{sign}\bigl(\sin(2^\kappa\pi t)\bigr)=\begin{cases}1 & \text{for } \sin(2^\kappa\pi t)\geqslant 0,\\ -1 & \text{for } \sin(2^\kappa\pi t)<0.\end{cases}$$

These functions satisfy the following property [7] (the Walsh functions defined on $[0,1]$ form a multiplicative group)

$$q(i_1,t)q(i_2,t)=\frac{q(i_1\oplus i_2,t)}{\sqrt{h}}, \tag{5}$$

where $i_1\oplus i_2$ is the bitwise addition of numbers modulo 2 (XOR bitwise operation).

The function

$$\int_0^t q(i_1,\tau_1)d\tau_1$$

is represented as a series by the Walsh functions as follows [6, 20]:

$$\int_0^t q(i_1,\tau_1)d\tau_1=\sum_{j_1=0}^{\infty}P^{-1}_{j_1 i_1}q(j_1,t)$$

using the entries of the infinite matrix P^{-1} defined by

$$P^{-1}=h\cdot\begin{bmatrix}\frac{1}{2} & \frac{1}{4} & \frac{1}{8} & 0 & \ldots & M_{0m} & \ldots\\ -\frac{1}{4} & 0 & 0 & \frac{1}{8} & \ldots & M_{1m} & \ldots\\ -\frac{1}{8} & 0 & 0 & 0 & \ldots & M_{2m} & \ldots\\ 0 & -\frac{1}{8} & 0 & 0 & \ldots & M_{3m} & \ldots\\ \vdots & \vdots & \vdots & \vdots & \ddots & \vdots & \\ M_{m0} & M_{m1} & M_{m2} & M_{m3} & \ldots & M_{mm} & \ldots\\ \vdots & \vdots & \vdots & \vdots & & \vdots & \ddots\end{bmatrix},$$

where

$$M_{00} = \frac{1}{2}, \quad M_{\alpha,2^m+\alpha} = -M_{2^m+\alpha,\alpha} = \frac{1}{2^{m+2}}, \quad M_{ml} = 0 \quad \text{otherwise},$$
$$m, l = 0, 1, 2, \dots, \quad \alpha = 0, 1, \dots, 2^m - 1.$$

The infinite matrix P^{-1} is the spectral characteristic of the integration operator [6, 21] (two-dimensional nonstationary transfer function of the integrating element [20]).

According to the property (5), we have

$$q(i_2, t) \int_0^t q(i_1, \tau_1) d\tau_1 = q(i_2, t) \sum_{j_1=0}^{\infty} P^{-1}_{j_1 i_1} q(j_1, t) = \sum_{j_1=0}^{\infty} P^{-1}_{j_1 i_1} q(j_1, t) q(i_2, t)$$
$$= \frac{1}{\sqrt{h}} \sum_{j_1=0}^{\infty} P^{-1}_{j_1 i_1} q(j_1 \oplus i_2, t),$$

consequently,

$$\int_0^t q(i_2, \tau_2) \int_0^{\tau_2} q(i_1, \tau_1) d\tau_1 d\tau_2 = \frac{1}{\sqrt{h}} \sum_{j_1=0}^{\infty} P^{-1}_{j_1 i_1} \int_0^t q(j_1 \oplus i_2, \tau_2) d\tau_2$$
$$= \frac{1}{\sqrt{h}} \sum_{j_1=0}^{\infty} P^{-1}_{j_1 i_1} \sum_{j_2=0}^{\infty} P^{-1}_{j_2 j_1 \oplus i_2} q(j_2, t)$$

and

$$C_{i_2 i_1} = \int_0^h q(i_2, \tau_2) \int_0^{\tau_2} q(i_1, \tau_1) d\tau_1 d\tau_2 = \int_0^h \sum_{j_1=0}^{\infty} P^{-1}_{j_1 i_1} q(j_1, t) q(i_2, t) dt$$
$$= \sum_{j_1=0}^{\infty} P^{-1}_{j_1 i_1} \int_0^h q(j_1, t) q(i_2, t) dt = P^{-1}_{i_2 i_1}. \quad (6)$$

Further,

$$q(i_3, t) \int_0^t q(i_2, \tau_2) \int_0^{\tau_2} q(i_1, \tau_1) d\tau_1 d\tau_2 = \frac{1}{h} \sum_{j_1=0}^{\infty} P^{-1}_{j_1 i_1} \sum_{j_2=0}^{\infty} P^{-1}_{j_2 j_1 \oplus i_2} q(j_2 \oplus i_3, t),$$

therefore,

$$\int_0^t q(i_3, \tau_3) \int_0^{\tau_3} q(i_2, \tau_2) \int_0^{\tau_2} q(i_1, \tau_1) d\tau_1 d\tau_2 d\tau_3 = \frac{1}{h} \sum_{j_1=0}^{\infty} P^{-1}_{j_1 i_1} \sum_{j_2=0}^{\infty} P^{-1}_{j_2 j_1 \oplus i_2} \sum_{j_3=0}^{\infty} P^{-1}_{j_3 j_2 \oplus i_3} q(j_3, t)$$

and

$$C_{i_3 i_2 i_1} = \int_0^h q(i_3, \tau_3) \int_0^{\tau_3} q(i_2, \tau_2) \int_0^{\tau_2} q(i_1, \tau_1) d\tau_1 d\tau_2 d\tau_3 = \frac{1}{\sqrt{h}} \sum_{j_1=0}^{\infty} P^{-1}_{j_1 i_1} P^{-1}_{i_3 j_1 \oplus i_2}. \quad (7)$$

Implementing similar steps we find

$$q(i_k, t) \dots \int_0^t q(i_2, \tau_2) \int_0^{\tau_2} q(i_1, \tau_1) d\tau_1 d\tau_2 \dots d\tau_k$$
$$= \frac{1}{h^{\frac{k-1}{2}}} \sum_{j_1=0}^{\infty} P^{-1}_{j_1 i_1} \sum_{j_2=0}^{\infty} P^{-1}_{j_2 j_1 \oplus i_2} \dots \sum_{j_{k-1}=0}^{\infty} P^{-1}_{j_{k-1} j_{k-2} \oplus i_{k-1}} q(j_{k-1} \oplus i_k, t),$$

i.e.,

$$\int_0^t q(i_k,\tau_k)\ldots\int_0^{\tau_3} q(i_2,\tau_2)\int_0^{\tau_2} q(i_1,\tau_1)d\tau_1 d\tau_2\ldots d\tau_k$$
$$= \frac{1}{h^{\frac{k-1}{2}}}\sum_{j_1=0}^{\infty} P^{-1}_{j_1 i_1}\sum_{j_2=0}^{\infty} P^{-1}_{j_2 j_1\oplus i_2}\cdots\sum_{j_k=0}^{\infty} P^{-1}_{j_k j_{k-1}\oplus i_k} q(j_k,t)$$

and

$$C_{i_k\ldots i_2 i_1} = \int_0^h q(i_k,\tau_k)\ldots\int_0^{\tau_3} q(i_2,\tau_2)\int_0^{\tau_2} q(i_1,\tau_1)d\tau_1 d\tau_2\ldots d\tau_k$$
$$= \frac{1}{h^{\frac{k}{2}-1}}\sum_{j_1=0}^{\infty} P^{-1}_{j_1 i_1}\sum_{j_2=0}^{\infty} P^{-1}_{j_2 j_1\oplus i_2}\cdots\sum_{j_{k-2}=0}^{\infty} P^{-1}_{j_{k-2} j_{k-3}\oplus i_{k-2}} P^{-1}_{i_k j_{k-2}\oplus i_{k-1}}. \quad (8)$$

Nonzero entries in rows and columns of the infinite matrix P^{-1}, starting from some number, form the geometric progression with the common ratio 1/2. The maximum of absolute values of entries for P^{-1} is $P^{-1}_{00} = h/2$, and the sum of squares of all entries is $h^2/2$ (the square of the Hilbert – Schmidt norm). Moreover, any finite submatrix obtained from P^{-1} by first $L = 2^n$ rows and columns is the sparse matrix for $n > 1$. It has $1 + 2(2^n - 1) = 2L - 1$ nonzero entries. This guarantees the convergence of the series in the right-hand sides of relations (7) and (8).

In order to obtain relations for the expansion coefficients (3) in the matrix form and to represent the iterated Stratonovich stochastic integral (1) also in the matrix form, we introduce the following notation

$$V_{l_1 j_1 i_2} = \frac{1}{\sqrt{h}}\begin{cases} 1 & \text{for } j_1\oplus i_2 = l_1, \\ 0 & \text{otherwise.}\end{cases}$$

In fact, values $V_{l_1 j_1 i_2}$ are entries of the spectral characteristic of the multiplier [20, 21]. Note that the value $V_{l_1 j_1 i_2}$ does not depend on the order of indices l_1, j_1, i_2 because the condition $j_1 \oplus i_2 = l_1$ is equivalent to the condition $j_1\oplus i_2\oplus l_1 = 0$ ($l_1\oplus l_1 = 0$ for an arbitrary l_1), and the operation $\oplus$ has commutative and associative properties. Therefore,

$$P^{-1}_{i_3 j_1\oplus i_2} = \sum_{l_1=0}^{\infty} V_{l_1 j_1 i_2} P^{-1}_{i_3 l_1}$$

and the relation (7) can be written as follows:

$$\int_0^h q(i_3,\tau_3)\int_0^{\tau_3} q(i_2,\tau_2)\int_0^{\tau_2} q(i_1,\tau_1)d\tau_1 d\tau_2 d\tau_3 = \sum_{j_1=0}^{\infty}\sum_{l_1=0}^{\infty} P^{-1}_{j_1 i_1} V_{l_1 j_1 i_2} P^{-1}_{i_3 l_1}.$$

Representing similarly values $P^{-1}_{i_4 j_2\oplus i_3}, \ldots, P^{-1}_{i_k j_{k-2}\oplus i_{k-1}}$ we have another form for the relation (8):

$$\int_0^h q(i_k,\tau_k)\ldots\int_0^{\tau_3} q(i_2,\tau_2)\int_0^{\tau_2} q(i_1,\tau_1)d\tau_1 d\tau_2\ldots d\tau_k$$
$$= \frac{1}{h^{\frac{k}{2}-1}}\sum_{j_1=0}^{\infty}\sum_{l_1=0}^{\infty}\sum_{j_2=0}^{\infty}\cdots\sum_{l_{k-2}=0}^{\infty} P^{-1}_{j_1 i_1} V_{l_1 j_1 i_2} P^{-1}_{j_2 l_1}\cdots V_{l_{k-2} j_{k-2} i_{k-1}} P^{-1}_{i_k l_{k-2}}.$$

IOP Conf. Series: Materials Science and Engineering **927** (2020) 012080 doi:10.1088/1757-899X/927/1/012080

Denote the infinite identity matrix by E. Let E_i be the ith column of E ($i = 0, 1, 2, \dots$) and let $E_i^{\oplus}$ be the infinite matrix with entries $(E_i)_{m\oplus l}$ ($m, l = 0, 1, 2, \dots$). Hence, we can write relations (6)–(8) for expansion coefficients (3) in the matrix form

$$\begin{gathered} C_{i_2 i_1} = E_{i_2}^{\mathrm{T}} P^{-1} E_{i_1}, \quad C_{i_3 i_2 i_1} = E_{i_3}^{\mathrm{T}} P^{-1} E_{i_2}^{\oplus} P^{-1} E_{i_1}, \quad \dots, \\ C_{i_k \dots i_2 i_1} = E_{i_k}^{\mathrm{T}} P^{-1} E_{i_{k-1}}^{\oplus} \dots P^{-1} E_{i_2}^{\oplus} P^{-1} E_{i_1}. \end{gathered} \tag{9}$$

Next, let $\mathcal{V}_j$ be the infinite column matrix with entries $\zeta_i^{(j)}$ ($j = 1, 2, \dots, s$), and let $\mathcal{V}_j^{\oplus}$ be the infinite matrix with entries $\zeta_{m\oplus l}^{(j)}$ ($m, l = 0, 1, 2, \dots$). The iterated Stratonovich stochastic integral (2) can be obtained by summing up the products of expansion coefficients $C_{i_k \dots i_2 i_1}$ and random variables $\zeta_{i_1}^{(j_1)} \zeta_{i_2}^{(j_2)} \dots \zeta_{i_k}^{(j_k)}$ taking into account that

$$\mathcal{V}_j = \sum_{i=0}^{\infty} \zeta_i^{(j)} E_i.$$

Thus,

$$\begin{gathered} I_h^{*(j_1 j_2)} = \mathcal{V}_{j_2}^{\mathrm{T}} P^{-1} \mathcal{V}_{j_1}, \quad I_h^{*(j_1 j_2 j_3)} = \mathcal{V}_{j_3}^{\mathrm{T}} P^{-1} \mathcal{V}_{j_2}^{\oplus} P^{-1} \mathcal{V}_{j_1}, \quad \dots, \\ I_h^{*(j_1 j_2 \dots j_k)} = \mathcal{V}_{j_k}^{\mathrm{T}} P^{-1} \mathcal{V}_{j_{k-1}}^{\oplus} \dots P^{-1} \mathcal{V}_{j_2}^{\oplus} P^{-1} \mathcal{V}_{j_1}. \end{gathered} \tag{10}$$

4. Numerical experiments

The relations (9) and (10) give an exact representation for the expansion coefficients (3) and for the iterated Stratonovich stochastic integral (2). For an approximate representation and modeling we need to replace infinite matrices with corresponding truncated matrices in these relations by choosing the truncation order L. Then the column matrices E_i ($i = 0, 1, \dots, L-1$) and $\mathcal{V}_j$ ($j = 1, 2, \dots, s$) will be $L \times 1$, the matrix P^{-1} will be $L \times L$, therefore, the matrices $E_i^{\oplus}$ and $\mathcal{V}_j^{\oplus}$ will be $L \times L$. We denote by $\tilde{I}_h^{*(j_1 j_2 \dots j_k)}$ the random variable defined by the relation (10) taking into account the truncation of infinite matrices.

First, the case of pairwise distinct values $j_1, j_2, \dots, j_k$ is considered. According to the relation (10), for the approximate modeling iterated stochastic integrals of multiplicity k we need k random vectors $\mathcal{V}_{j_1}$, $\mathcal{V}_{j_2}$, …, $\mathcal{V}_{j_k}$ with entries having a standard normal distribution. So, we need kL realizations of the standard Gaussian random variable to draw one realization for the random variable $\tilde{I}_h^{*(j_1 j_2 \dots j_k)}$.

The exact values for the expectation and second-order moment for iterated stochastic integrals of multiplicity k are determined as follows [10]:

$$\mathbb{E} I_h^{*(j_1 j_2 \dots j_k)} = 0, \quad \mathbb{E}\big[I_h^{*(j_1 j_2 \dots j_k)}\big]^2 = \sum_{i_1=0}^{\infty} \sum_{i_2=0}^{\infty} \dots \sum_{i_k=0}^{\infty} C_{i_k \dots i_2 i_1}^2 = \frac{h^k}{k!},$$

where $\mathbb{E}$ means the expectation.

When truncated matrices are used, we have

$$\mathbb{E} \tilde{I}_h^{*(j_1 j_2 \dots j_k)} = 0, \quad \mathbb{E}\big[\tilde{I}_h^{*(j_1 j_2 \dots j_k)}\big]^2 = \sum_{i_1=0}^{L-1} \sum_{i_2=0}^{L-1} \dots \sum_{i_k=0}^{L-1} C_{i_k \dots i_2 i_1}^2.$$

Results of the numerical experiment are summarized in tables 1–4 for $k = 2, \dots, 5$ and $h = 1$. These results consist of the second-order moment for the random variable $\tilde{I}_h^{*(j_1 j_2 \dots j_k)}$

$$\tilde{M}_2(L) = \mathbb{E}\big[\tilde{I}_h^{*(j_1 j_2 \dots j_k)}\big]^2,$$

Table 1. Modeling results for $k = 2$.

L	$\tilde{M}_2(L)$	$\Delta_2(L)$	$\hat{M}_1(L)$	$\hat{M}_2(L)$
4	0.437500	12.500 %	0.000578	0.435486
8	0.468750	6.250 %	0.000079	0.466765
16	0.484375	3.125 %	−0.000136	0.483963
32	0.492188	1.563 %	−0.001043	0.493812
64	0.496094	0.781 %	−0.000965	0.497224

Table 2. Modeling results for $k = 3$.

L	$\tilde{M}_2(L)$	$\Delta_2(L)$	$\hat{M}_1(L)$	$\hat{M}_2(L)$
4	0.113281	32.031 %	0.000041	0.112692
8	0.137695	17.383 %	−0.000437	0.136612
16	0.151611	9.033 %	0.000229	0.151930
32	0.158997	4.602 %	−0.000040	0.159543
64	0.162796	2.322 %	−0.000344	0.162887

Table 3. Modeling results for $k = 4$.

L	$\tilde{M}_2(L)$	$\Delta_2(L)$	$\hat{M}_1(L)$	$\hat{M}_2(L)$
4	0.020264	51.367 %	−0.000147	0.020536
8	0.028656	31.226 %	−0.000117	0.029217
16	0.034527	17.136 %	−0.000146	0.034519
32	0.037931	8.966 %	0.000119	0.037860
64	0.039756	4.584 %	0.000195	0.039473

Table 4. Modeling results for $k = 5$.

L	$\tilde{M}_2(L)$	$\Delta_2(L)$	$\hat{M}_1(L)$	$\hat{M}_2(L)$
4	0.002823	66.125 %	−0.000006	0.002857
8	0.004540	45.515 %	0.000022	0.004544
16	0.006109	26.694 %	0.000026	0.006093
32	0.007130	14.446 %	−0.000030	0.007185
64	0.007707	7.513 %	−0.000005	0.007589

the relative error of the second-order moment for the iterated stochastic integral (1)

$$\Delta_2(L) = \left(\frac{h^k}{k!} - \tilde{M}_2(L)\right) \Big/ \frac{h^k}{k!} = 1 - \frac{k!}{h^k}\,\tilde{M}_2(L),$$

as well as the estimates of the expectation and second-order moment obtained by the statistical modeling the random variable $\tilde{I}_h^{*(j_1 j_2 \ldots j_k)}$ (the sample size is 10^6). They are denoted by $\hat{M}_1(L)$ and $\hat{M}_2(L)$, respectively.

These results means that the estimates of the expectation and second-order moment obtained by the statistical modeling are consistent with their theoretical values and the relative error of the second-order moment decreases with increasing the truncation order L. Moreover, $\Delta_2(L) \approx c/L$, where c is a positive constant depending on the multiplicity k only.

Obtained results for the case $h = 1$ can be exactly reduced to the case $h \neq 1$ by the multiplication with the factor $h^{k/2}$ for the expectation and with the factor h^k for the second-order moment.

Note that the relative error of the second-order moment increases with increasing the multiplicity k, but the absolute error for the numerical method for stochastic differential equations is small due to the factor h^k in the second-order moment of the iterated stochastic integral, since usually $h \ll 1$, where h is an integration step.

In the section 2 it is indicated that iterated Itô and Stratonovich stochastic integrals coincide for the case of pairwise distinct values $j_1, j_2, \ldots, j_k$. Moreover, the expectation for iterated Itô stochastic integrals is zero for arbitrary values $j_1, j_2, \ldots, j_k$. But it is not valid for iterated Stratonovich stochastic integrals in the general case.

Next, the case of equal values $j_1, j_2, \ldots, j_k$ is concerned. According to the relation (10), for the approximate modeling iterated stochastic integrals of multiplicity k we need one random vector $\mathcal{V}_j$ with entries having a standard normal distribution, $j = j_1 = j_2 = \ldots = j_k$. So, we need L realizations of the standard Gaussian random variable to draw one realization for the random variable $\tilde{I}_h^{*(j_1 j_2 \ldots j_k)} = \tilde{I}_h^{*(jj\ldots j)}$.

The exact values for the expectation for iterated stochastic integrals of multiplicity k are determined as follows [10]:

$$\mathbb{E} I_h^{*(j_1 j_2)} = C_{00} = \frac{h}{2}, \quad \mathbb{E} I_h^{*(j_1 j_2 j_3)} = 0,$$

$$\mathbb{E} I_h^{*(j_1 j_2 j_3 j_4)} = \sum_{i_1=0}^{\infty} \sum_{i_2=0}^{\infty} (C_{i_1 i_1 i_2 i_2} + C_{i_1 i_2 i_1 i_2} + C_{i_1 i_2 i_2 i_1}) = \frac{h^2}{8}, \quad \mathbb{E} I_h^{*(j_1 j_2 j_3 j_4 j_5)} = 0, \quad \ldots$$

When truncated matrices are used, we have

$$\mathbb{E} \tilde{I}_h^{*(j_1 j_2)} = C_{00} = \frac{h}{2}, \quad \mathbb{E} \tilde{I}_h^{*(j_1 j_2 j_3)} = 0,$$

$$\mathbb{E} \tilde{I}_h^{*(j_1 j_2 j_3 j_4)} = \sum_{i_1=0}^{L-1} \sum_{i_2=0}^{L-1} (C_{i_1 i_1 i_2 i_2} + C_{i_1 i_2 i_1 i_2} + C_{i_1 i_2 i_2 i_1}), \quad \mathbb{E} \tilde{I}_h^{*(j_1 j_2 j_3 j_4 j_5)} = 0, \quad \ldots$$

Results of the numerical experiment are summarized in tables 5–8 for $k = 2, \ldots, 5$ and $h = 1$. These results consist of the expectation for the random variable $\tilde{I}_h^{*(j_1 j_2 \ldots j_k)}$

$$\tilde{M}_1(L) = \mathbb{E} \tilde{I}_h^{*(j_1 j_2 \ldots j_k)},$$

the relative error of the expectation for the iterated stochastic integral (1) when $k = 4$ (this relative error equals zero for $k = 2$, and it is not defined for $k = 3$ and $k = 5$)

$$\Delta_1(L) = \left(\frac{h^2}{8} - \tilde{M}_1(L) \right) \Big/ \frac{h^2}{8} = 1 - \frac{8}{h^2} \tilde{M}_1(L).$$

The estimates of the expectation and second-order moment obtained by the statistical modeling the random variable $\tilde{I}_h^{*(j_1 j_2 \ldots j_k)}$ (the sample size is 10^6) are also given in tables 5–8. They are denoted by $\hat{M}_1(L)$ and $\hat{M}_2(L)$, respectively.

Here the relative error of the expectation decreases with increasing the truncation order L for the case $k = 4$. Moreover, $\Delta_1(L) = 2/L$. Estimates of the expectation obtained by the statistical modeling for $k = 2, \ldots, 5$ are consistent with their theoretical values.

Table 5. Modeling results for $k = 2$.

L	$\tilde{M}_1(L)$	$\Delta_1(L)$	$\hat{M}_1(L)$	$\hat{M}_2(L)$
4	0.500000	0.000 %	0.500666	0.750732
8	0.500000	0.000 %	0.498792	0.744745
16	0.500000	0.000 %	0.499033	0.747656
32	0.500000	0.000 %	0.498419	0.743239
64	0.500000	0.000 %	0.499987	0.749882

Table 6. Modeling results for $k = 3$.

L	$\tilde{M}_1(L)$	$\Delta_1(L)$	$\hat{M}_1(L)$	$\hat{M}_2(L)$
4	0.000000	———	−0.000306	0.497523
8	0.000000	———	−0.000432	0.454270
16	0.000000	———	0.000403	0.430723
32	0.000000	———	−0.000492	0.426879
64	0.000000	———	−0.000705	0.419813

Table 7. Modeling results for $k = 4$.

L	$\tilde{M}_1(L)$	$\Delta_1(L)$	$\hat{M}_1(L)$	$\hat{M}_2(L)$
4	0.187500	−50.000 %	0.189419	0.317023
8	0.156250	−25.000 %	0.155935	0.230116
16	0.140625	−12.500 %	0.140669	0.202712
32	0.132813	−6.250 %	0.132506	0.190662
64	0.128906	−3.125 %	0.128496	0.185300

Table 8. Modeling results for $k = 5$.

L	$\tilde{M}_1(L)$	$\Delta_1(L)$	$\hat{M}_1(L)$	$\hat{M}_2(L)$
4	0.000000	———	−0.000306	0.195595
8	0.000000	———	0.000057	0.108557
16	0.000000	———	−0.000055	0.081822
32	0.000000	———	0.000429	0.070984
64	0.000000	———	0.000309	0.068818

5. Conclusion

In the paper the representation problem for iterated Stratonovich stochastic integrals using orthonormal expansions of functions has been considered. For this purpose the Walsh series is proposed. This representation is convenient due to it uses only one matrix of expansion coefficients of the unit step function for an arbitrary multiplicity. The algorithm for modeling iterated Stratonovich stochastic integrals is tested by numerical experiments that confirm the theoretical results.

References

[1] Milstein G N 1974 Approximate integration of stochastic differential equations *Theor. Prob. Appl.* **19**(3) 583–588

[2] Milstein G N 1995 *Numerical Integration of Stochastic Differential Equations* (Dordrecht: Kluwer Academic Publ.)

[3] Kloeden P E and Platen E 1992 *Numerical solution of stochastic differential equations* (Berlin: Springer-Verlag)

[4] Kuznetsov D F 2002 Expansion of the Stratonovich multiple stochastic integrals based on the Fourier multiple series *J. Math. Sci.* **109**(6) 2148–2165

[5] Prigarin S M and Belov S M 1998 One application of series expansions of Wiener process *Preprint* 1107 (Novosibirsk: ICM & MG Publ.) (in Russian)

[6] Rybakov K A 2019 Applying spectral form of mathematical description for representation of iterated stochastic integrals *Differencialnie Uravnenia i Protsesy Upravlenia* (4) 1–31 (in Russian)

[7] Golubov B, Efimov A and Skvortsov V 1991 *Walsh series and transforms: Theory and applications* (Dordrecht: Springer)

[8] Artemiev S S and Averina T A 1997 *Numerical Analysis of Systems of Ordinary and Stochastic Differential Equations* (Utrecht: VSP)

[9] Kuznetsov D F 2017 Multiple Ito and Stratonovich stochastic integrals: Fourier-Legendre and trigonometric expansions, approximations, formulas *Differencialnie Uravnenia i Protsesy Upravlenia* (1) A.1–A.385

[10] Kuznetsov D F 2018 Stochastic differential equations: Theory and practic of numerical solution. With MATLAB programs *Differencialnie Uravnenia i Protsesy Upravlenia* (4) A.1–A.1073 (in Russian)

[11] Karachanskaya E and Tagirova T 2019 Construction of stochastic transport models with a constant function *IOP Conf. Ser.: Earth Environ. Sci.* **403** 012211

[12] Rybakov K A 2013 Solving approximately an optimal nonlinear filtering problem for stochastic differential systems by statistical modeling *Numer. Anal. Appl.* **6**(4) 324–336

[13] Rybakov K A 2017 Robust Duncan-Mortensen-Zakai equation for non-stationary stochastic systems *Proc. Int. Conf. on Engineering, Computer and Information Sciences (Novosibirsk)* (IEEE) pp 151–154

[14] Rudenko E 2019 Algorithms and programs of suboptimal nonlinear filtering for Markov processes *AIP Conf. Proc.* **2181** 020017

[15] Bortakovskii A S and Nemychenkov G I 2018 Suboptimal on average satellite attitude control in the presence of discrete inaccurate measurements *J. Comput. Sys. Sc. Int.* **57**(2) 197–207

[16] Khalina A S and Khrustalev M M 2019 Effect of displacement of optimal control in stabilization problems for quasi-linear diffusion-type stochastic systems *J. Comput. Sys. Sc. Int.* **58**(2) 159–166

[17] Davtyan L G and Panteleev A V 2019 Method of parametric optimization of nonlinear continuous systems of joint estimation and control *J. Comput. Sys. Sc. Int.* **58**(3) 360–373

[18] Kuznetsov D F 2019 Application of the method of approximation of iterated stochastic Ito integrals based on generalized multiple Fourier series to the high-order strong numerical methods for non-commutative semilinear stochastic partial differential equations *Differencialnie Uravnenia i Protsesy Upravlenia* (3) 18–62

[19] Kuznetsov D F 2019 Expansion of iterated Stratonovich stochastic integrals based on generalized multiple Fourier series *Ufa Math. J.* **11**(4) 49–77

[20] Solodovnikov V V, Semenov V V, Peschel M and Nedo D 1979 *Design of Control Systems on Digital Computers: Spectral and Interpolational Methods* (Moscow: Mashinostroenie; Berlin: Verlag Technik) (in Russian; in German)

[21] Rybakov K A 2020 Modeling and analysis of output processes of linear continuous stochastic systems based on orthogonal expansions of random functions *J. Comput. Sys. Sc. Int.* **59**(3) 322–337

The possibility of draining the boundary layer from the hypersonic nozzle

L B Ruleva and S I Solodovnikov

Ishlinsky Institute for Problems in Mechanics RAS, Vernadsky prospekt 101(1), Moscow, 119526, Russia

E-mail: ruleva@ipmnet.ru

Abstract. Hypersonic aerodynamic shock tube is used for comparison (validation) of the developed computer codes for the calculation of aerothermodynamics. Experiments in vacuum chamber of hypersonic aerodynamic shock tube showed the significant boundary layer in the nozzle. An increase in the boundary layer was observed with an increase in the length of the inner surface of the nozzle. Easy method to decrease the boundary layer of the nozzle was suggested. Areas of pressure field near hypersonic nozzle with a reduced pressure in the moment of quasi-stationary flow from the nozzle were determined experimentally.

1. Introduction

Aerothermodynamics of hypersonic aircraft is studied theoretically using computer modeling and developed calculation codes [1] when modeling the flight conditions of aircraft in the upper atmosphere and in outer space. This computer simulation involves validation with experimental data. Experiments require a high vacuum to be created in the area where the models are located, in front of the hypersonic nozzle. Hypersonic aerodynamic shock tube (HAST) is used for comparison (validation) of the developed computer codes for the calculation of aerothermodynamics. The pressure in the test chamber before the experiment is set as 10^{-7} atm. Models of simple geometric shapes in the gas flow at Mach numbers M=7, and Re=10^{-6} are studied experimentally. Short conical and elongated HAST nozzles are presented in computational and experimental studies [2]. An increase in the boundary layer was observed with an increase in the length of the inner surface of the nozzle. This fact distorts the experimental data for comparison. In this paper we suggest method decreasing the boundary layer of the nozzle.

The conical nozzle scheme and photos of real designs of short and elongated HAST nozzles are shown in figures 1 and 2, respectively.

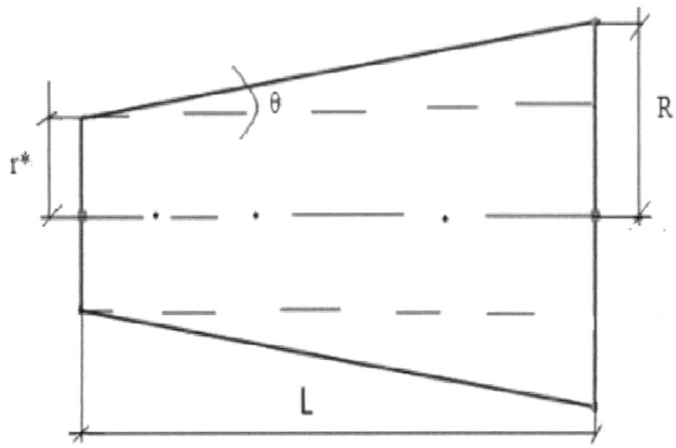

Figure 1. A HAST conical nozzle scheme.

a) b)

Figure 2. A HAST nozzle: short nozzle a), elongated nozzle b).

The elongated nozzle has higher Mach numbers. Elongated nozzle is orientated in the field of the viewing window. Due to this, the shock-wave structures of high-speed streams flowing from the nozzle section are recorded very well by a high-speed video camera. The thickness of the boundary layer in the elongated nozzle is greater than in a short nozzle, as can be seen from the calculated Mach lines of the short and elongated HAST nozzles [3] in figures 3 and 4.

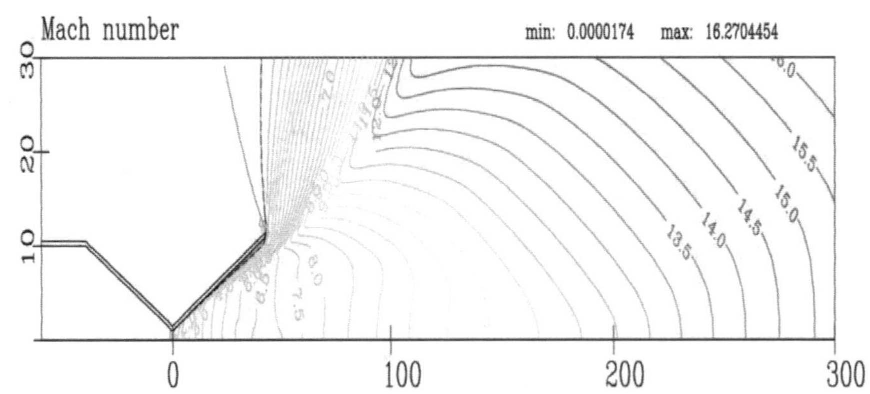

Figure 3. HAST short nozzle lines. L=100 mm, r=4mm.

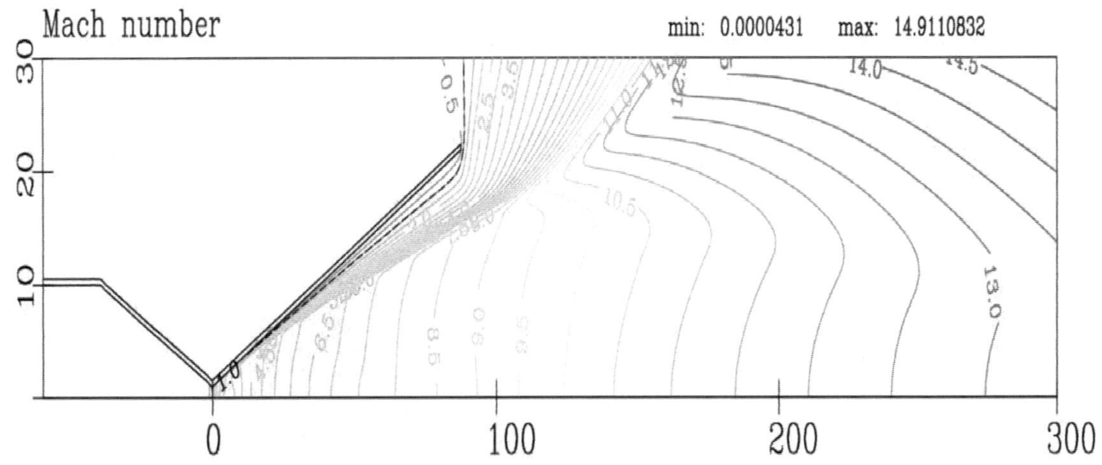

Figure 4. HAST elongated nozzle lines. L= 350 mm, r=4mm.

The presence of boundary layer in the nozzle complicates the calculation and distorts the results of the experiment. The boundary layer of the nozzle must be decreased. The boundary layer in a high-speed nozzle is experimentally investigated, for example, [4], by optical methods. The way to reduce the boundary layer in the nozzle in patent [5] is creating a vortex motion in the nozzle. The method [6] was connected with the HAST facility. This method was implemented, but it was technically difficult. This method has therefore been simplified. The simpler method to drain boundary layer from HAST elongated nozzle was discovered experimentally.

2. A method of reducing the boundary layer in the nozzle

In the HAST elongated nozzle, a groove of certain geometry is made along the inner cone, close to the section; the cavity is connected through a fitting and a flexible tube with a vacuum source, higher than in the vacuum chamber. The method for removing the boundary layer [6] includes an electromagnetic valve connected electrically to dynamic pressure sensors, located in the shock part of the HAST. When the hypersonic flow streams from the nozzle and the valve is activated, the boundary layer is removed. This action is short-term on a segment of the primary quasi-stationary streaming period, which is used in the experiment to compare the calculated and experimental flow structures.

It is assumed that in the vacuum chamber, where the models are located, it is the source of a higher vacuum. The device contains an electromagnetic valve that is triggered by a dynamic pressure sensor located in the shock part of the tube. The valve opens the source of a lower vacuum at a certain moment. It is necessary to calculate and organize experimentally the delay time between the operation of the dynamic pressure alarm sensor and the start time of the flow from the nozzle.

The beginning of the flow from the nozzle is characterized by a quasi-stationary process, on which validation is performed. This time is small (Figure 5)

IOP Conf. Series: Materials Science and Engineering **927** (2020) 012081 doi:10.1088/1757-899X/927/1/012081

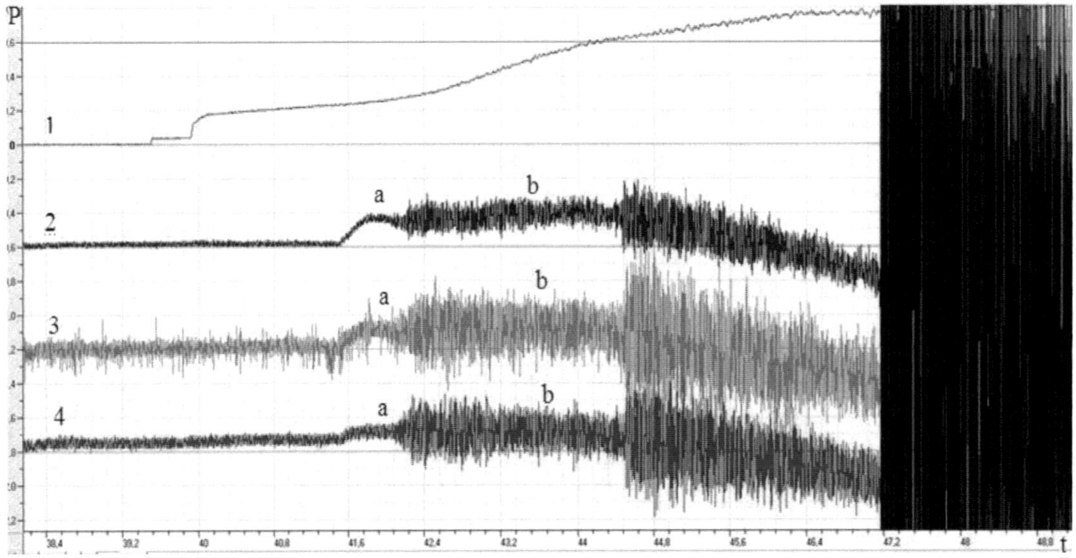

Figure 5. The signals of the pressure sensors. M=7. Re=10^6. Symbols: 1,2,3 - on the plate behind the nozzle for 20 mm, 40mm, 60mm, respectively. Symbols: 4, 5 - in the shock tube for 200 mm and 100 mm before entering into the nozzle.

Disadvantages are in the complexity of the device. It is necessary to accurately determine the time of flow from the nozzle. It is necessary to pump the balloon to the highest vacuum after each experiment. The high-speed valve must have a constant and short (units of ms) response time.

Observation of the pressure field near the nozzle made it possible to simplify this method. In a known device [9], we will exclude the valve and the cylinder with a higher vacuum. The annular groove, connected through a fitting with a branch pipe was left.

The annular groove is connected through a fitting with a branch pipe. The branch pipe open end of the nozzle is directed opposite to the flow direction on the outside of the nozzle.

3. Simplified method for reducing the boundary layer in the nozzle

Sensitive dynamic pressure sensors were located near the nozzle section (Figure 6 a) and at a distance of 80 mm (Figure 6 b). A flexible pipe connected the cavity on the nozzle and the sensor surface, which could be moved further to the base of the nozzle (Figure 6 c).

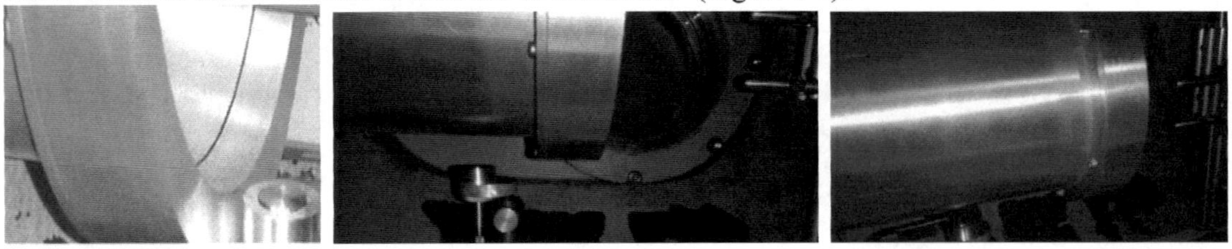

a) b) c)

Figure 6. Pressure sensors were located near the nozzle a), at a distance b), a flexible pipe connect the cavity on the nozzle and the sensor c).

A certified pressure sensor (type PCB 106B52) was used for the experiment. First, the time was determined between the shock wave front impact on the pressure sensor 100 mm before entering the nozzle and the response of the sensor located behind the nozzle section. This time was 0.6 ms. The corresponding graph is shown in figure 7.

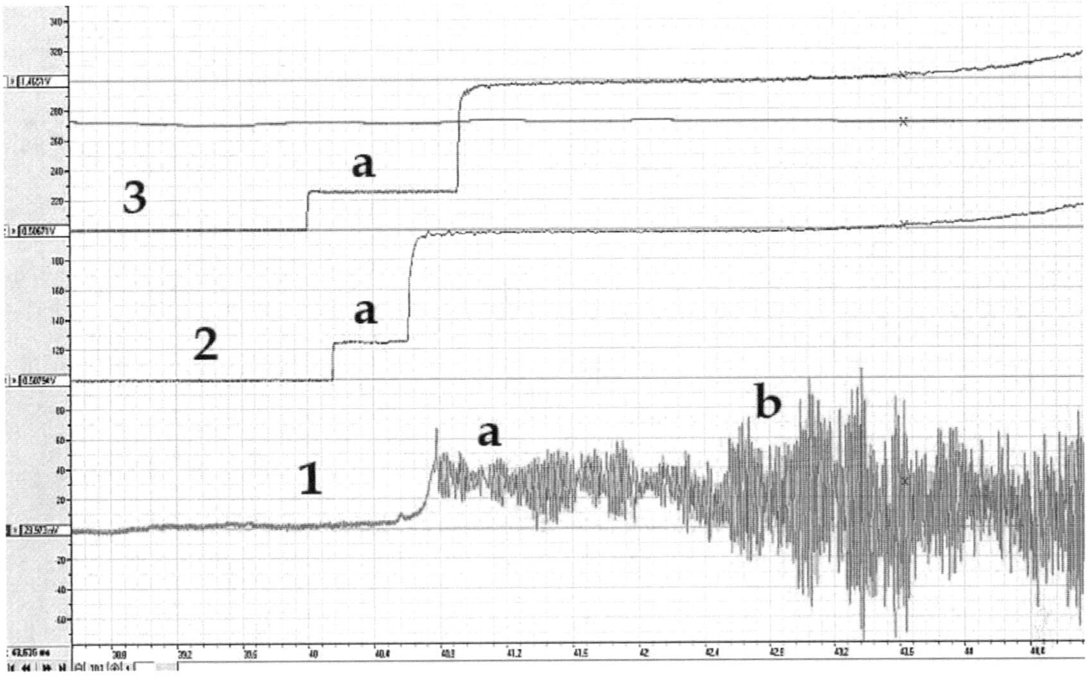

Figure 7. Pressure sensor graphs: 1- behind the nozzle section, 2, 3- for 100 mm and 200 mm before entering the nozzle. Areas: a- quasi-stationary, b- non-stationary.

The time of quasi-stationary area in figure 7 (1a) was 1,5 mc. Therefore, the registration of pressure on the sensor, when it is moving into deep of the nozzle outer surface showed at the same time delay of 0.6 ms and duration of 1.5 ms. The pressure in the 1a (figure 7) area is 44 Pa. If the sensor was moving deep on the outer line of the nozzle by 80 mm, the pressure will decrease to 4.38 Pa, and by140 mm will decrease to 3.18 Pa.

4. Conclusion

Experiments in vacuum camera of hypersonic aerodynamic shock tube are shown the significant boundary layer in the nozzle. Hypersonic aerodynamic shock tube (HAST) is used for comparison (validation) of the developed computer codes for the calculation of aerothermodynamics. An increase in the boundary layer was observed with an increase in the length of the inner surface of the nozzle. This fact distorts the experimental data for comparison. In this paper we suggest easy method decreases out the boundary layer of the nozzle.

The areas of the pressure field near a hypersonic nozzle with a reduced pressure (higher vacuum) at the moment of quasi-stationary flow from the nozzle were determined experimentally..With the flexible pipe connecting the cavity in the internal side of nozzle with the outer line of the nozzle you can reduce the pressure by an order of magnitude, without an additional balloon with the highest vacuum.

Acknowledgments
The presented study was supported by Government program of basic research of Russian academy of science (contract #AAAA-A17-117021310372-6).

References

[1] Surzhikov S T 2018 Analysis of the turbulent boundary layer on a flat plate at M=6 divided by 8.8 with the use of NERAT-2D code and algebraic turbulence models *J. of Physics: Conf. Series* **1009** 012016
[2] Ivanov I A, Kotov M A, Kryukov I A, Ruleva L B, Solodovnikov S I and Surzhikov S T 2019 Parameters of the air flow formed by the interchangeable nozzles with the transformed critical

part *J. of Physics: Conf. Series* **1250** 012016
[3] Ivanov I. A, Kotov M A, Kryukov I A, Ruleva L B and Solodovnikov S I 2018 The functioning of a removable elongated nozzle in the Hypersonic Aerodynamic Shock Tube *chemphys.edu.ru* **19** 3
[4] Brutyan M A Petrov A V Potapchik A B 2015 Method for studying the state of flow in the boundary layer *Patent* 2562276 (in Russian)
[5] Kiknadze G I, Talygin E A Gorodkov A Yu 2018 A method for sucking the boundary layer of a continuous medium from the surface of the body and a device for its implementation *Patent* 2691705 (in Russian)
[6] Kotov M A, Ruleva L B, Solodovnikov and S I Samokhvalov G V 2017 Shock hypersonic aerodynamic tube *Patent* 167762 (in Russian)

IOP Conf. Series: Materials Science and Engineering **927** (2020) 012082 doi:10.1088/1757-899X/927/1/012082

Increasing hypersonic aerodynamic shock tube working time duration

A V Panasenko, L B Ruleva and S I Solodovnikov

Ishlinsky Institute for Problems in Mechanics RAS, Vernadsky prospekt 101(1), Moscow, 119526, Russia

E-mail: ruleva@ipmnet.ru

Abstract. Longer working time of the hypersonic aerodynamic shock tube has been demonstrated by applying modelling and experiment. This facility is a short-term installation. It is used to validate the results of aerothermodynamic characteristics of aircraft modelling calculated using national computational codes. Numerical modeling and experimental data results are compared. Quasi-stationary state of the gas flow in operating time of the installation requires sufficient time. A review of publications about increasing the working time of shock tubes for initiation of high-speed gas flows was made. Some models of shock tubes with longer working time have been selected for discussion. The computational studies for hypersonic aerodynamic shock tube showed a positive trend for one of the operating modes. A simulation for a regular and lengthened shock tube facilities was carried out, taking the flow as in viscous medium and using Euler equations. The effectiveness of lengthening the low pressure channel section of the facility was shown by modelling. The low pressure chamber was elongated by 3m to the existing hypersonic aerodynamic shock tube.

1. Introduction

Hypersonic aerodynamic shock tube (HAST) [1] is a short-term installation. It is used to validate the results of aircraft aerothermodynamic characteristics modeling calculated using national computational codes [2] for given Mach and Reynolds numbers.

One of the important parameters for validation is the time of the quasi-stationary state of the flow in the shock part of the tube before and behind the nozzle cut. For installations of short-term action it is small and is estimated 1-2 ms. The quasi-stationary state of the flow behind the nozzle section directly depends on the time of existence of a constant pressure and density in the shock part of the HAST before entering the nozzle, the so-called "shelf". As the flow velocity increases, the shelf life decreases. For short-term installations, the characteristic parameter is the working time of the tube.

The possibility of increasing the working time of real installations of well-known research centers will be considered from publications. Here is an example of increasing the working time of the HAST due to the extension of the mounting sections of the low pressure channel.

2. A short review of the work on the study of the dependence of the working time of shock tubes on the length of sections

Studies of shock tubes working time, known from publications, are differing in parameters and designs from the HAST experimental installation, but give a general idea. Theoretical works on the calculation of the working time of shock tubes, for example, [3-8], using x-t diagrams, which involve parameters such as the length (x) of the tube sections and the shock wave propagation time (t), the rarefaction wave and the contact surface in these areas (Figure 1.) are known.

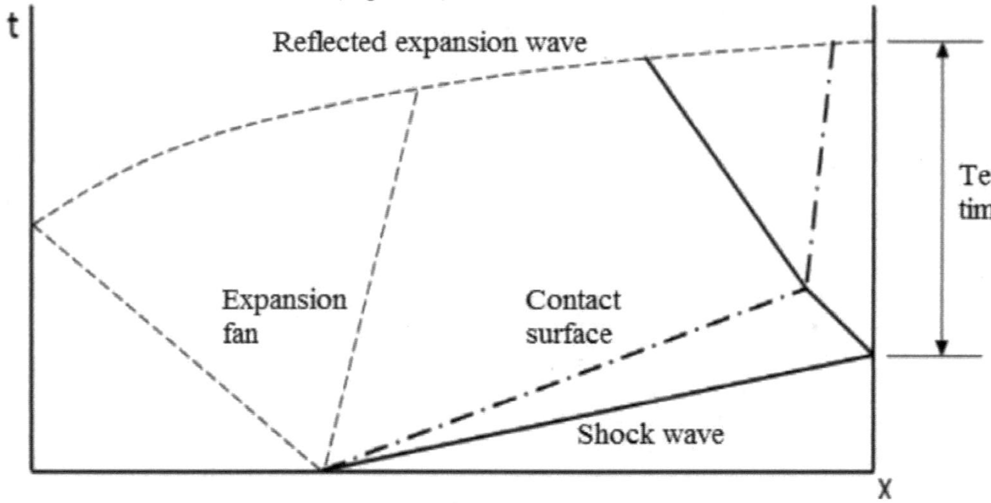

Figure 1. The x-t diagram.

The two chambers are filled to different pressures and initially separated by diaphragms. When the diaphragm between the driver and driven gas sections is broken, shock wave flows to the left, but rarefaction wave flows rightwards .Working time includes the time from the arrival of the shock wave to the end of the shock tube, the time of arrival of the contact surface and the time interval of the collision of the head of the rarefaction wave with the reflected shock wave and the contact surface.

Let's consider several papers, which suggest technical solutions how to increase the shock tube working time for existing experimental facilities.

There are works [3, 4] in which changes in the velocity of the shock wave, the contact surface, and the growth of the boundary layer were studied. Recommendations for the optimal length of this installation for a square-section shock tube are given in [4].

In [5] the analytical problem of shock tube working time in aspect of gas dynamics behind the starting shock wave was solved. It was noted that in the case of inviscid flow, the distance from the shock wave to the contact surface linearly increases with the distance from the diaphragm. This indicates a desirable increase in the low pressure channel.

In work [6] two driven cameras with double contact break of the diaphragms were used. Here, the graph on the time axis rises higher, and the difference between the arrival time of the shock wave and the contact surface is greater than at the first contact break (1 ms). The length of the low pressure channel is 16 m with an internal diameter of 100 mm, and the length of the high pressure channel is 3m.

Installation [7] has experiment duration of 5ms and large high pressure chamber of 6 m (6 sections of electric heaters of 2m each) at a pressure of 30 to 80 atm.

A practical approach to scientific center shock tube working time measuring is presented in [8]. A formula linking the measurement time, model size and free flow rate is given. The working time is 3 ms, while the energy spent for the test is 10 MJ. The x-t scheme in the study does not differ from well known. Implementations of this scheme are available at GASL, New York (HYPULSE), CUBRC in Buffalo, N.Y. (LENS-X). However, the opening time of the second valve leads to test time decrease on practice. The recommendations, given in the article for two diaphragm tubes, when applied to HAST: "the friction losses in the acceleration tube set a limit for length to diameter ratio. A reasonable maximum value is about 120", i.e. taking into account friction losses in the tube; the ratio of pipe length to diameter is about 120.

In relation the HAST low pressure chamber, with a diameter of 0.08 m when extended by 3 m to the length of 10.5 m, their ratio will be 125. The HAST design has single diaphragm and lower enthalpy compared to the considered shock tube designs.

IOP Conf. Series: Materials Science and Engineering **927** (2020) 012082 doi:10.1088/1757-899X/927/1/012082

Validation of calculated aerothermodynamic parameters with experimental data is performed not during the entire working time of the HAST, but during the quasi-stationary state of the gas flow. The quasi-stationary state is displayed in a straight line, parallel to the time axis, the so-called "shelf" on the pressure graph. The flow is replacing after the quasi-stationary state by a non-stationary and turbulent state, which are not used for HAST validation.

3. HAST installation working time simulation

Simulation for a regular and elongated HAST shock tube is carried out, taking the flow as in viscid and using Euler equations. Calculations predict an increase in "shelf" time by a third.

Calculations were made for the HAST shock tube. The initial pressure before opening the diaphragm in the high-pressure chamber is 36 atm, in the low-pressure chamber is 0.1 atm. The length of the high-pressure chamber is 2m. The calculation is made for two lengths of the low-pressure chamber: 7 m (regular) and 10 m (elongated).

The calculated comparison of flow parameters in a conventional shock tube (a) and an elongated shock part (b) of HAST low - pressure chamber is shown in figures 2, 3 and 4.

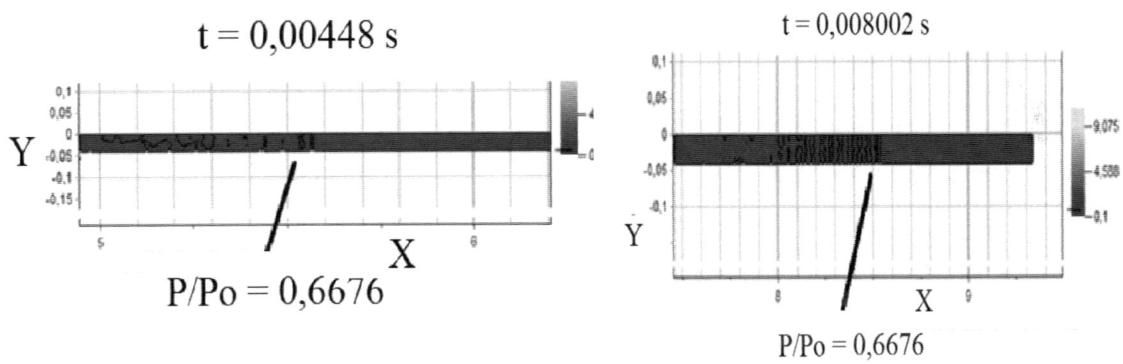

a) b)

Figure 2. Fragments calculation's

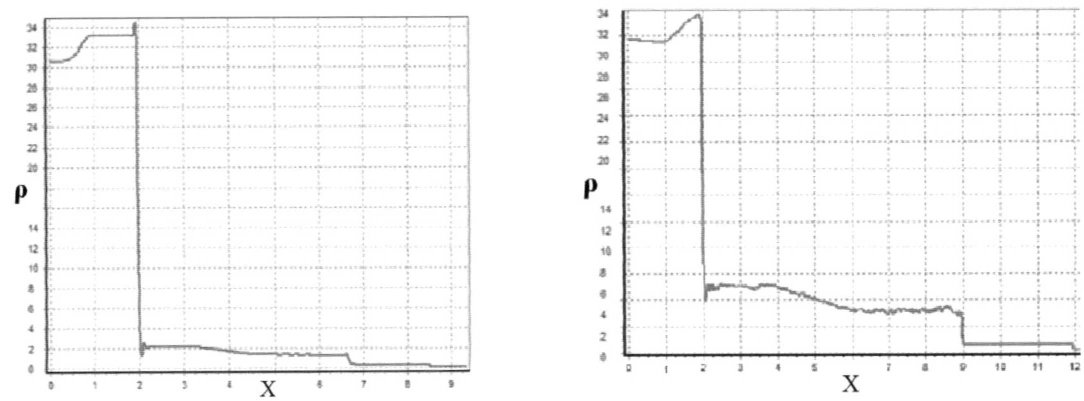

a) b)

Figure 3. The flow density calculated in regular (a) and elongated (b) of HAST low-pressure chamber

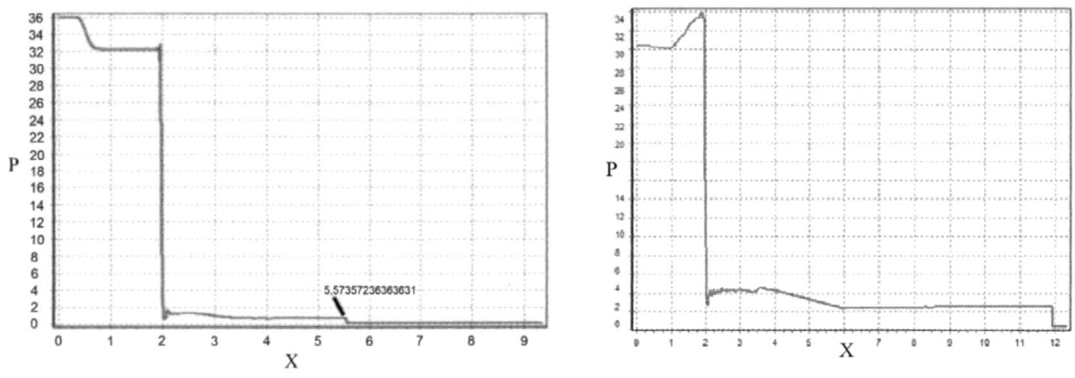

a) b)

Figure 4. The flow pressure calculated in regular (a) and elongated (b) of HAST low-pressure chamber

The x - axis are the length of the shock tube. The results of modeling the initial dimensions of the HAST sections and an increase of 3 m of the HAST low-pressure chamber are presented. The results of modeling the initial dimensions of the HAST sections and the low-pressure chamber enlarged by 3 m are presented. The graphs show that the time increased by a third.

4. Operating time for HAST facility

The scheme of HAST is shown in figure 5.

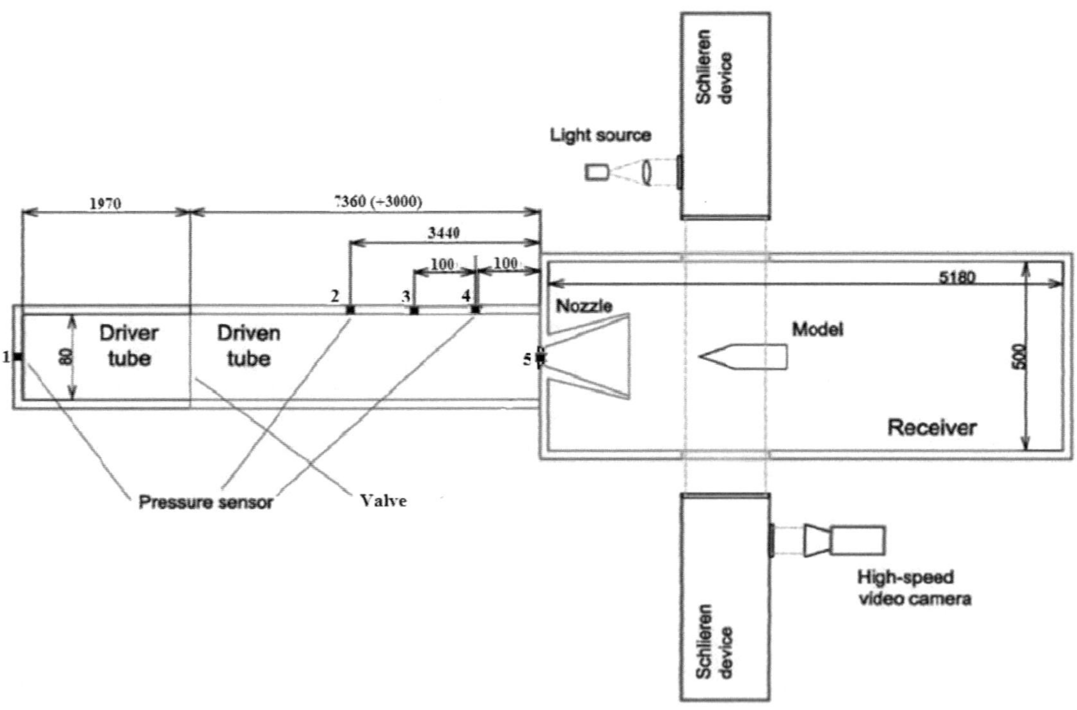

Figure 5. HAST scheme.

High-frequency certified dynamic pressure sensors (PCB, 500 kHz): 1, 2,3,4,5, are installed, respectively, at the end of the high pressure chamber, in the middle of the low pressure channel and inside, at 200 mm and 100 mm to the end of the and HAST low-pressure chamber. The sensors are connected via analog-to-digital converter units to the computer. The quick-acting valve: QV80-50 [9], instead of a ruptured membrane, designed for a maximum pressure of 50 atm, is installed between the high-pressure and low-pressure (0.01...0.5 atm) chambers. The aerodynamic block in experiments is pumped out to 10^{-6} atm. The initiation of the shock wave in the shock tube is achieved by filling the pushing gas of the high pressure chamber with a predetermined high pressure value, triggering a high-speed electromagnetic valve ($6*10^{-3}$s) and forming a traveling shock wave in the low pressure chamber. The shock wave front is formed after the decay of an arbitrary rupture at a distance of 40 d, i.e. 3.2 m. It is fixed through time t_C by a dynamic pressure sensor 2 (at a distance of 3.92 m from the rupture). Further, the incident shock wave is recorded by dynamic pressure sensors 3 for 0.2 m and 4 for 0.1 m. to the end (time t_{100} and t_{200}, respectively). In some experiments without an aerodynamic block, the sensor 5 (t_k) is installed at the rear end. The "shelf" with a steep front characteristic is in the charts.

The reflected shock wave with a steep front is recorded by the same sensors in reverse order. After the shock wave, the contact surface tends to the low pressure channel end (gas flow rate, the separation of the pushing and working gases). The contact surface collides with the high pressure area of the reflected shock wave.

Simultaneously with the shock wave propagating to the right, the rarefaction wave (the rarefaction head) runs to the left, after a time t_P reaching the end of the high pressure chamber, is registered by the sensor 1. Sensor 1 pressure begins to go down to a minimum (t_{PM}). The rarefaction head is reflected from the end of the high pressure chamber, rushes to the end of the low pressure channel. When approaching the end of the low pressure channel, the rarefaction head meets the reflected contact surface and the pressure on the sensors located at the end of the low pressure channel begins to fall. This moment is very important for the experiment, because it means the end of shock tube working time. Registration of stages described above on pressure graphs is given in figure 6.

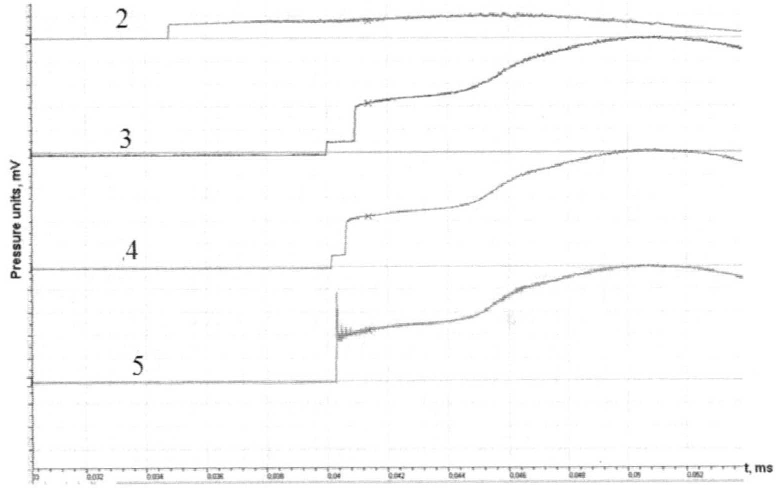

Figure 6. Pressure graphs of sensors 2,3,4,5 (from top to bottom).

After reflection from the right low pressure chamber flange, the shock wave comes to the left flange and re-reflections are repeated with the dissipation of shock wave energy due to friction against the tube wall and the increasing boundary layer. As a rule, secondary processes are not used in experiments in HAST, although shorter "shelves" and with less pressure will be repeated in the shock tube and, theoretically, it is possible to obtain short fragments of quasi-stationary flow at lower Mach numbers, than in the primary flows, but here the reflection of the flow behind the nozzle from the models and walls of the aerodynamic unit will affect.

5. The actual elongation of the tube and experimental results

The low pressure chamber of HAST was elongated by 3m, as it is shown in figure 7.

a) b)

Figure 7. Photos of the regular (a) and after modification the elongated (b) HAST low-pressure chamber

The results of experimental studies of regular (HAST-15) and enlarged (HAST18) tubes are shown in table 1.

Table 1. Parameters comparison for standard (HAST-15) and enlarged (HAST-18) tubes.

Pressure high / low pressure chamber, 36/0,1; atm	Time t_{2ref}- t_{in}, ms	Velocity V, m/s	Pressure $P_{3,4in}$, atm	Pressure $P_{3,4,5\ ref}$, atm	Work time, ms	"shelf"work $t_{5\ ref}$, ms
HAST -15	22	833	0.7	3.1	12.4	0.3
HAST-18	33	781	0.5	2.4	13.4	0.5
increase «+» decrease «-»	+11	-52	-0.2	-0.7	+1	+0.2

The HAST low-pressure chamber was increased by 3 m, the operating time of the unit was increased by 1 ms, and the "shelf" was increased by 0.2 ms at the Mach number M=2.4.

These experiments were carried out in accordance with one of HAST operational modes corresponding to the calculated. The tendency to increase the working time of the shock tube and the "shelf" with the extension of the shock tube section is observed. For a more accurate coincidence of the calculation with the experiment, it is necessary to take into account the viscosity of the gas and the time delay in the valve operation [9].

6. Conclusion

The questions of increasing the working time of the shock part of the hypersonic aerodynamic shock tube (HAST) at the initial set pressures and the expected length increase in its sections are considered. The results of modeling the initial dimensions of the HAST sections and their comparison with the modeling at an increase of 3 meters of the low pressure channel are presented. It is shown that an increase

IOP Conf. Series: Materials Science and Engineering **927** (2020) 012082 doi:10.1088/1757-899X/927/1/012082

in the length of the low pressure channel by 3 meters, increases the operating time of the facility by 1 ms and by 0,2 ms "shelf" at Mach number, M=2.4.

The simulation was carried out for ideal conditions in the absence of flow friction against the walls of the HAST sections. The importance of the problem is due to an increase in the experimental verification accuracy of aircraft aerodynamic characteristics modeling using computational codes developed in IPMeh RAS.

Acknowledgments
The presented study was supported by Government program of basic research of Russian academy of science (contract #AAAA-A17-117021310372-6).

References

[1] Kotov M A, Ruleva L B Solodovnikov S I and Surzhikov S T 2017 Computational and experimental studies of the structure of high-speed gas flow around of models of aircraft fragments *Bulletin of the Moscow state technical University. N. E. Bauman. Series Mechanical Engineering* **3** 18-30
[2] Surzhikov S T 2017 Validation of computational code UST3D by the example of experimental aerodynamic data *J. of Physics: Conf. Series* **815** 012023
[3] Bazhenova T V 1961 Changing the gas flow velocity after a jump in a shock tube (Moscow: Izdatelstvo USSR RAN) pp 31-5
[4] Bazhenova T V and Gvozdeva L G 1977 *Non-Stationary interaction the action of the shock waves* (Moscow: Nauka) p 274
[5] Chizhikov A S 2008 The working time determination in the shock tube *Experimental devices and techniques* **5** 118-22
[6] Maslennikov V G and Sakharov V A 1997 Dual-diaphragm shock tube of the Institute of Physics and technology *Fluid and Gas Mechanics* **11** pp 88-95 (in Russian)
[7] Verevkin A A and Tsirkunov Yu M 2008 The flow of a dispersed impurity in the Laval nozzle and the working section of a two-phase hypersonic shock tube *Applied mechanics and technical physics* **49** 102-13
[8] Hornung Hans 2010 Ground Testing for Hypervelocity flow *Capabilities and Limitations Graduate Aerospace Laboratories California Institute of Technology RTO-EN-AVT-186* 26 p
[9] Isakov S N and Yurkin S V 2007 Method of bringing to readiness an inflatable airbag of safety device for vehicle valve device *Patent US* 7.232.152 B2

IOP Conf. Series: Materials Science and Engineering **927** (2020) 012083 doi:10.1088/1757-899X/927/1/012083

Heat flow measurements on high-speed aircraft models

L B Ruleva and S I Solodovnikov

Ishlinsky Institute for Problems in Mechanics RAS, Vernadsky prospekt 101(1), Moscow, 119526, Russia

E-mail: ruleva@ipmnet.ru

Abstract. The functionality of "Hypersonic aerodynamic shock tube" facility was expanded to obtain more complete validation of developed computer codes for the prospective high-speed aircraft aerothermodynamics description. The method of luminescent temperature converters was applied in the laboratory installation. Measuring of temperature fields on the surface of models installed in the aerodynamic unit when they have ambient air flow at Mach numbers M=5-7, have been implemented. The calorimetric type heat flow sensor was selected among existing heat flow sensors as it based on microsecond inertia, miniaturization, and relative ease of manufacturing. The sensitive elements of the calorimetric sensors were manufactured for the laboratory installation. The ability of the manufactured sensors to work and the dependence of the response time from the thickness of the heat-receiving surface were revealed.

1. Introduction

The prospective high-speed aircraft aerothermodynamics is described by newly developed national computer codes [1, 2]. The reliability is verified by comparing calculated and experimental data. Data from full-scale flight tests, despite their high value, have not enough range of validated parameters and test modes. The most rapid validation of new computer codes is carried out on models in aerodynamic installations. A hypersonic aerodynamic shock tube is suitable for this purpose [3]. It can be used for operational validation of aerodynamic studies. The hypersonic aerodynamic shock tube (HAST) is shown in figure 1.

Figure 1. The laboratory hypersonic aerodynamic shock tube (HAST).

HAST is a multi-mode installation of short-term action, with a Mach numbers M=5...7 and length of 18 meters. Aircraft models are installed in front of the hypersonic nozzles in a vacuum (10^{-7} atm) block. A high-speed valve [4], of low inertia (0.006 s), serves to seal high and low pressure chambers. It ensures good repeatability of parameters from experiment to experiment. High-frequency certified dynamic pressure sensors (PCB) are installed throughout the HAST tract and in models. The model dimensions and its orientation relevant to specified flow during validation are the same for the calculation and experiment, and are independent of each other. For example, the model calculated with the author's codes [1] on structured multi-block grids with thickening in the problem areas of the air intake group is shown in figure 2a, the calculated temperature field is shown in figure 2b, respectively.

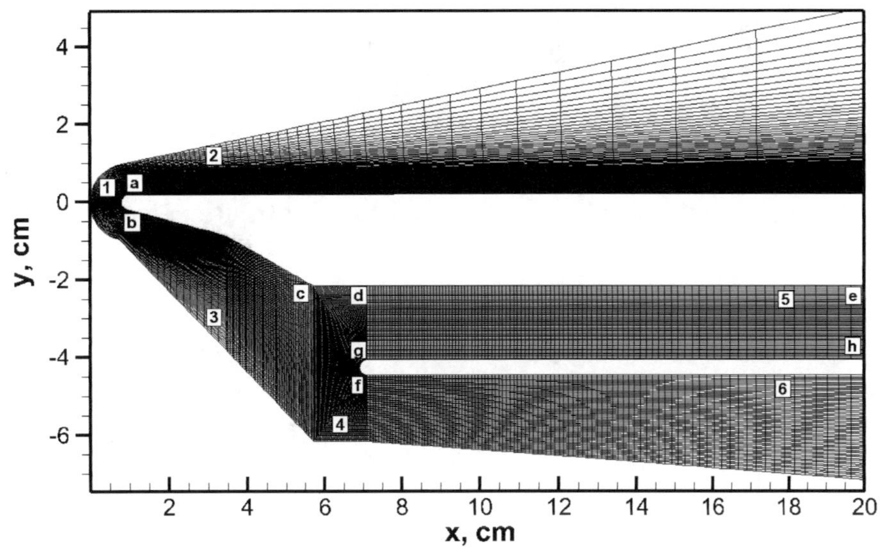

a)

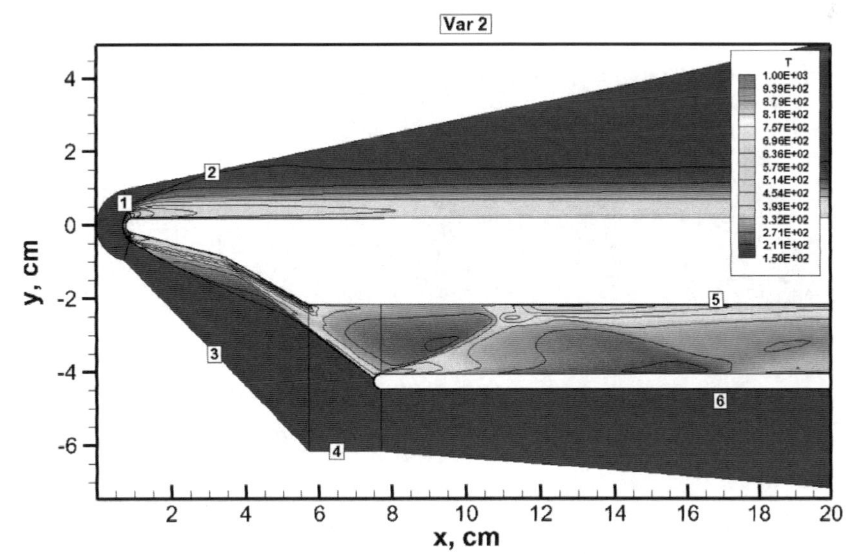

b)

Figure 2. The model calculated on structured grids a) temperature field; b) M-7, T=100K.

IOP Conf. Series: Materials Science and Engineering **927** (2020) 012083 doi:10.1088/1757-899X/927/1/012083

The calculation and experiment results are compared by visualization of flow density lines and shown in figure 3a. Model with pressure sensors located on the surface of the lower blunted edge is shown in figure 3b.

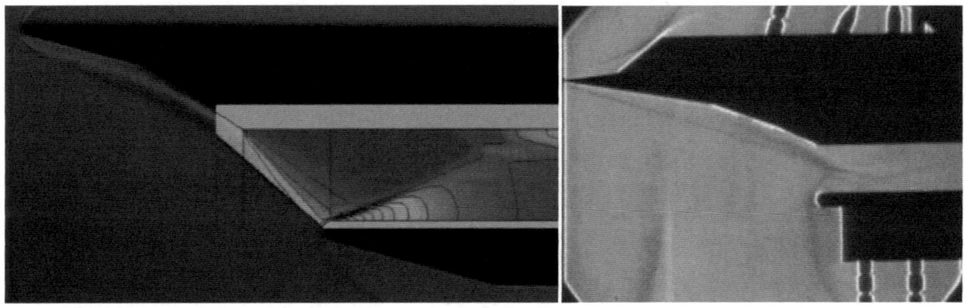

a) b)

Figure 3. Comparison of the experiment with the calculation by shlieren-foto of the lower edge of the model a) foto of pressure sensors on the lower edge; b) M=7. Re=10^6.

The area of quasi-constant pressure-the "shelf" (a), with a duration of t_{gf}=1ms is shown in figure 3c. It also indicates quasi-constant flow velocity and quasi-constant temperature. Under normal operating modes of the HAST, the time of the t_{gf} quasi-stationary flow from the nozzle is within 0.5÷ 3 ms. However, registering the temperature field on the surface of models with low inertia, which is much less than t_{gf} time, is an urgent and not idle task. The preferred measurement method for validation purposes is direct measurement without intermediate calculations.

2. Measurement of heat flow in HAST

The method of luminescent temperature converters (LPT) is successfully used in the world practice [5, 6]. Method of panoramic LPT registration based on the use of binary heat-sensitive paint was tested in HAST [7]. The acrylic model (figure 4a) was painted and installed in the HAST (figure 4b). The heat flux density was measured in experiments (figure 4c).

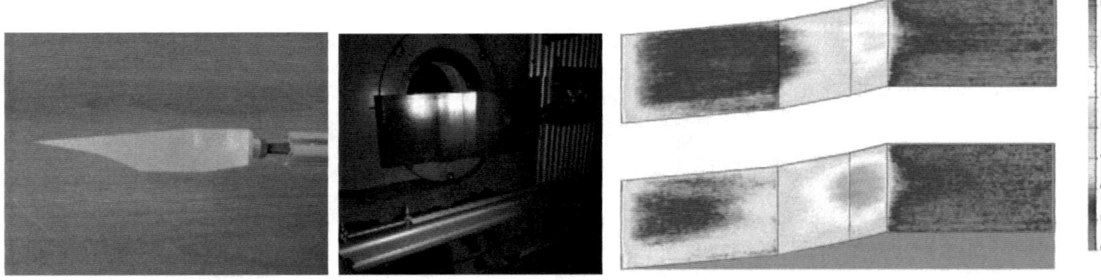

a) b) c)

Figure 4. The model a) is installed in the HAST, b) Heat flux densities in c) M=5.3 (top) and M=5.5 (bottom).

Regular application of this method in HAST is difficult, due to the complexity of conducting experiments. It is necessary to ensure the exact delay time of the sensor signal in the shock part of the HAST before the flash is triggered and the video frame is captured. The time delay for various modes of HAST operation must be selected. The accuracy of the method for validation purposes is limited by the presence of intermediate calculations. It requires high qualification in calculations and practical skills. The big advantage of this method is that it provides information over the entire surface of the model. However, for validation purposes, it is desirable to know the exact temperature at certain points on the surface of models with high frequency over a limited period of time. At the HAST installation, work is underway to equip models with sensors that measure the heat flow over the surface. Requirements for them are the same as for pressure sensors: low inertia at high flow rates, minimum size, direct measurements, without intermediate calculations, guaranteed accuracy. Sensors can also be located in the shock part of the HAST before entering the nozzle to measure the stagnation temperature. The operating time of the designed sensors is limited by the t_{gf} time.

3. Certified calorimetric sensor for HAST

Information about the use of certified temperature and heat flow sensors with a response time of milliseconds and microseconds, produced in Russian research centers, was not found . Sensors of our own production are widely used. Heat flow sensors: gradient [8], thin-film [9], thermocouple, calorimetric sensors [10,11] and others, more complex, which are not considered in this paper.

For complete information about the heat flow, many sensors on models surface are required of small sizes, in narrow channels, at the edges, as well as miniaturization of meters. An interesting design is a "thin wall", to which thermocouples are welded with a minimum step. By embedding a smooth wall flush in the model you can get a multi-point measurement of the temperature field. The disadvantage of the method is the flow of heat over the surface and the inability to insulate each thermocouple. The calorimetric type of sensor is the most attractive converter for HAST due to the simplicity of execution, low thermal inertia and dimensions.There are prerequisites to bring the design of such a sensor to the minimum thermal inertia, as in [12]. A certified temperature converter was purchased for the HAST installation, including the MST-48 temperature sensor with inertia of 3 mks (Mueller-Instruments, Germany) [13]. From the measured surface temperature signal the convective heat flux is deduced. An amplifier (k = 1000) and a low-pass filter must be used. MCTB 48 includes the coaxial bore hole for pressure transducer M60-1. This device may be used in wind tunnel experiments with high Mach numbers. The maximal measurement time for heat flux calculation is limited 40 to 100 ms. It is suitable in HAST. The characteristics of the MST-48 sensor are shown in table 1.

Table 1. The characteristics of the MST-48 sensor

Coaxial thermocouple	Type thermocouple	Temperature sensitivity 20^0C, mkV/^{0}K	Respond time, mks	External diameter, mm	Internal diameter for pressure sensor, mm
MCTB - 48	E	59.6	3	4.8	1.9

4. About calorimetric sensor for HAST

The thermal inertia of the heat flow Converter should not only not exceed the time t_{gf} of the quasi-stationary flow from the nozzle to the model, but also measure the local non-stationary heat flow density during this time. Non-stationary heat flow is caused by fluctuations in compaction. For example, on sharp and blunted edges of models, when shock-wave configurations interact on the surfaces of narrow channels of air intake groups in the areas of attachment of the torn boundary layer. The selected design of the tablet-type colorimetric sensor consists of a disk made from material with high thermal conductivity, such as copper. The diameter and thickness are small.

The scheme of heat flow calorimetric sensor sensitive element, manufactured for the HAST (a) models and its photo are shown in figure 5.

IOP Conf. Series: Materials Science and Engineering **927** (2020) 012083 doi:10.1088/1757-899X/927/1/012083

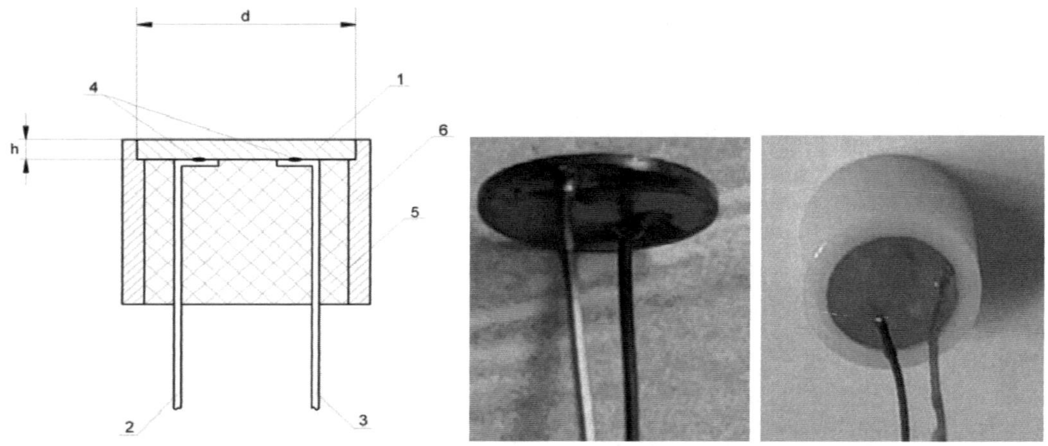

a) b) c)

Figure 5. Scheme of a calorimetric sensor a) manufactured sensor b) and c).

Here: 1-copper plate, 2-chromel electrode (-), 3-copel electrode (+), 4 - solder (silver) 5 - insulator sleeve, 6 - epoxy resin, d - diameter, h - thickness.

The working body of the sensor with diameter d=5 mm is made of copper sheet h= 0.15 mm thick on a lathe. At the laser spot welding plant, copper was welded to the copel and chromel wire elements (a diameter of 0.2 mm). The bushing body was made of fluoroplast, followed by pouring with epoxy resin. The "h" and "d" parameters (Figure 5a) should be reduced in the future to minimize the response time.

The heat Q brought to the calorimeter heat receiver of a known mass and geometric shape has the form:

$$Q = c \cdot m \frac{dT(x,\tau) \cdot \Delta\tau}{d\tau}, \tag{1}$$

where $\Delta\tau$ - is the time of measuring the heat flow, m is the mass of the heat-receiving tablet:

$$m = \rho \cdot F \cdot h, \tag{2}$$

where ρ, F, h -density, the size of the tablet and its thickness, respectively. In this case should be $\Delta\tau \le t_{gf}$.

With the miniature size of the sensor and the relative location of the electrodes in it, with a time limit of measurement, with some approximation, you can assume that there is no heat sink from the sensor to the housing. Then the specific heat flow is determined by expression:

$$q = \rho \cdot c \cdot h \frac{\Delta T}{\Delta\tau}, \tag{3}$$

where ΔT is the change in sensor temperature during the exposure time interval $\Delta\tau$.

Measurement errors are smaller, the smaller the size of the sensor, its heat capacity, and thermal resistance. Performance of the manufactured sensor according to the diagram (figure 5) was checked by exposure to a stationary laser. The laser power was 31 mW, which was determined by the average power and energy meter of laser radiation (IMO-2H). The sensor was connected to the oscilloscope via an MVA-10 amplifier with a low-pass filter from the MST-48 sensor. The rate of the pulse increases when exposed to the laser is 25 mV/s and 250 mV/s, with convective heating from the source (200^{0}C).

Preliminary experiments have showed the performance of the manufactured sensor. A signal amplifier and a fourth-order Butterworth filter were developed for the manufactured sensor. Calibration of this sensor and other prepared samples is assumed on quartz calibrated lasers, heat flow sensors, by comparing the readings with a certified MST-48 sensor on the table and on the surface of the model in HAST.

5. Conclusion

In order to validate the developed computational codes, the functionality of the laboratory installation "Hypersonic aerodynamic shock tube" is expanded. Test heat flow sensors are manufactured for located on the surface of aircraft models. From the analysis of existing sensors, the calorimetric type was selected based on microsecond inertia, miniaturization and relative ease of manufacturing. The efficiency of the sensors and the dependence of the response time on the thickness of the heat-receiving surface were revealed. The certified heat flow sensor was purchased for calibration.

Acknowledgments

The presented study was supported by Government program of basic research of Russian academy of science (contract #AAAA-A17-117021310372-6).

References

[1] Zheleznyakova A L and Surzhikov S T 2013 Application of the method of splitting by physical processes for the computation of a hypersonic flow over an aircraft model of complex configuration.*High Temperature (Teplofizika Vysokikh Temperatur)* **51** (6) 816–829

[2] Candler V G, Holder M S, Boyd L D, and Wang W L 2006 CFD validation for hypersonic flight: hypersonic double-cone flow simulations, *in 40th AIAA Aerospace Sciences Meeting & Exhibit* RTO-TR-AVT-007-V3 Reno NV USA

[3] Kotov M A, Ruleva L B Solodovnikov S I and Surzhikov S T 2017 Computational and experimental studies of the structure of high-speed gas flow around of models of aircraft fragments *Bulletin of the Moscow state technical University N E Bauman. Series Mechanical Engineering* **3** 18–30

[4] Isakov S N and Yurkin S V 2007 Method of bringing to readiness an inflatable airbag of safety device for vehicle valve device *Patent US* 7.232.152 B2

[5] Borovoy V Y, Mosharov V N, Noev A Y and Radchenko V N 2012 Temperature sensitive paint application for investigation of boundary layer transition in short-duration wind tunnels *Progress in Flight Physics* **3** 15-24

[6] Hubner J P, Carroll B F and Schanze K S 2002 Heat transfer measurements in hypersonic flow using luminescent coating techniques *J of Thermophysics and Heat Transfer, October–December* **16** (4) 516–522

[7] Mosharov V E, Radchenko V N, Senuev I V, Kotov M A, Ruleva L B, Solodovnikov S I and Surzhikov S T 2018 Preliminary experimental results of heat flux surface field registration at the hypersonic aerodynamic shock tube using temperature sensitive paint *J. of Physics: Conf. Series* **1009**(1) 012036

[8] Sapozhnikov S Z, Mityakov V Yu and Mityakov A V 2020 Gradient heat flux sensors gradient heat flux sensors *Heatmetr* pp 19–38 DOI: 10.1007/978-3-030-40854-1

[9] Irimpan K J, Mannil N, Arya H. and Menezes V 2015 Performance evaluation of coaxial thermocouple against platinum thin film gauge for heat flux measurement in shock tunnel *Measurement* **61** 291–298

[10] Sanderson S R and Sturtevant B 2002 Transient heat flux measurement using a surface junction thermocouple *Review of Scientific Instruments* **73** (7) 2781–87

[11] Kidd C T 1992 High heat-flux measurements and experimental calibrations/ characterizations, in: *Jag J, Singh R Antcliff (Eds) NASA CP-3161* NASA, Washington DC pp 31–50

[12] Zapryagaev V I et al 2019 Heat transfer in supersonic separated flow of the compression corner *J. Phys.: Conf. Ser.* **1382** 012049

[13] https://mueller-instruments.de/temperatursensoren/

IOP Conf. Series: Materials Science and Engineering **927** (2020) 012084 doi:10.1088/1757-899X/927/1/012084

Experimental investigation of the thermal loads upon electrodynamic modification of supersonic flow around axisymmetrical body

P A Popov[1], N A Monakhov[1], M A Kotov[2] and S A Poniaev[1]

[1] Ioffe Institute, Saint-Petersburg, Politechnicheskaya str. 26, 194021, Russia;
[2] Ishlinsky Institute for Problems in Mechanics of the Russian academy of sciences, Vernadsky prospekt 101(1), Moscow, 119526, Russia

E-mail: nikolay.monakhov@mail.ioffe.ru

Abstract. The results of experimental studies of the magnetohydrodynamic (MHD) effect on the flow around an axisymmetric model of a supersonic nitrogen stream are presented. The magnetic field was created by a solenoid located on the cylindrical part of the model under the influence of a pulsed electric current flowing through the solenoid. A pulsed gas discharge was ignited on the conical surface of the model, acquiring rotation in the solenoidal magnetic field around the model axis. The effect of the magnetic field on the gas-dynamic structure of the flow near the model and the heat flux to its surface were investigated. In the experiments, schlieren pictures,fast photographic scans of the discharge glow were recorded, and heat flux was measured. The effect of a magnetic field on the gasdynamic structure of the flow near the model and the heat load to its surface were found. Also, the dependence of MHD effects on the polarity of external voltage source connecting was found.

1. Introduction

During high-speed motion in the atmospheres of the Earth and other planets, the surface of the aircraft is subjected to intense mechanical and thermal stresses. The methods for the electrodynamic modification of the flow structure and thermal loads to the body surface are actively being developed [1-8]. Since the gas flow near the surface of the aircraft is ionized, magnetic hydrodynamics (MHD) methods can be considered as a way of influencing the heat load. The complexity of the phenomenon under consideration necessitates experimental studies. In experimental practice, shock tubes with a supersonic nozzle are common tool for such studies, which allow one to create high-speed flows with a characteristic process time of ~ 1 ms [9].

In this work, we experimentally investigated the effect of MHD effects on thermal loads during high-speed flow around an axisymmetric model. A feature of this work is the heating of gas in the shock layer using a high-current pulsed electric discharge. The discharge is ignited in a solenoidal magnetic field, which imparts rotation around the model axis and forms a high-temperature gas region near its surface.

2. Experimental setup

The studies were conducted on the big shock tube of the Physical Gasdynamics Lab of Ioffe Institute [10] with a nozzle. The shock tube consisted of a high pressure chamber (1) 1.5 m long with an external electric heater and a low pressure chamber (2). The inner diameter of the channel was of 110 mm, and the total length of the two pipe sections was of 16 m. Before the experiment, the high-pressure chamber was separated from the low-pressure chamber by a thin metal diaphragm. After the diaphragm destruction, a shock wave began to propagate through the working gas, which launches diagnostic tools. At the end of the low-pressure chamber there is a working section (3) with an internal rectangular section of 75x210 mm and a length of 600 mm with windows (4) on the side walls to visualize the flow. The other end of the working section is connected to a vacuum tank (5) with a volume of 6 m^3. The working section (3) was separated from the shock tube by a thin plastic diaphragm blocking the entrance to the nozzle. Supersonic nozzle was located inside the working section (3) and had dimensions of 75x95 mm. At the moment of the shock wave reflection from the end of the working section, the diaphragm bursts and a supersonic nozzle is launched. The model was located at the outlet of a supersonic nozzle. The nozzle was launched at the moment of the shock wave reflection from the end face. As a result, a flow was formed with a Mach number M = 4 with a duration of ~ 1.5 ms.

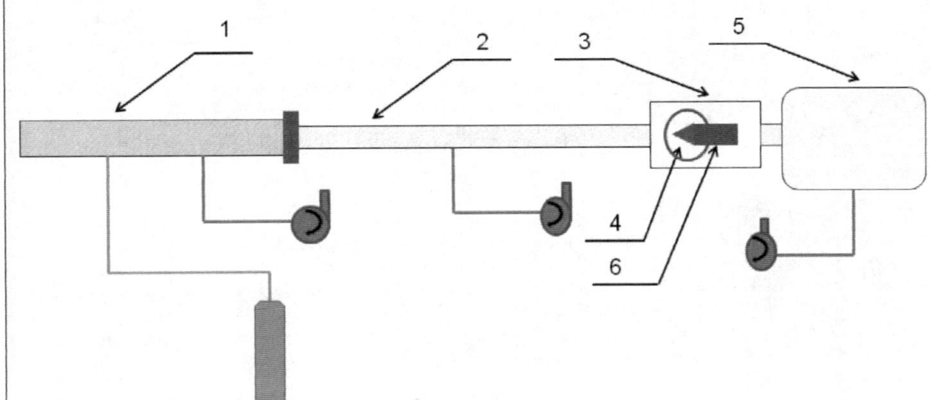

Figure 1. The experimental setup: 1 - high-pressure chamber, 2 - low-pressure chamber, 3 - working section, 4 - observation windows, 5 - vacuum tank, 6 - model.

The experiments were carried out using two models. Both models had the same shape of a cone conjugated with a cylinder. The shell of the models was made of caprolon, the cone opening angle was 60^0, the cylinder diameter was 34 mm, and its length was 38 mm. Inside the first model (Fig. 2-a), a magnetic steel core is coaxially aligned with it. The second model (without core) (Fig. 2-b) had a separate power supply circuit, which made it possible to independently change the current in the discharge circuit and the coil. A copper coil was wound on the cylindrical part of these models. One end of the coil was connected to a ring electrode, the other end of the coil was connected to one pole of an external pulsed voltage source. The second pole of the source is connected to a sharpened central electrode. The conical part of the central electrode and the surface of the ring electrode form an air gap that locks the external source circuit after the gas discharge initiation.

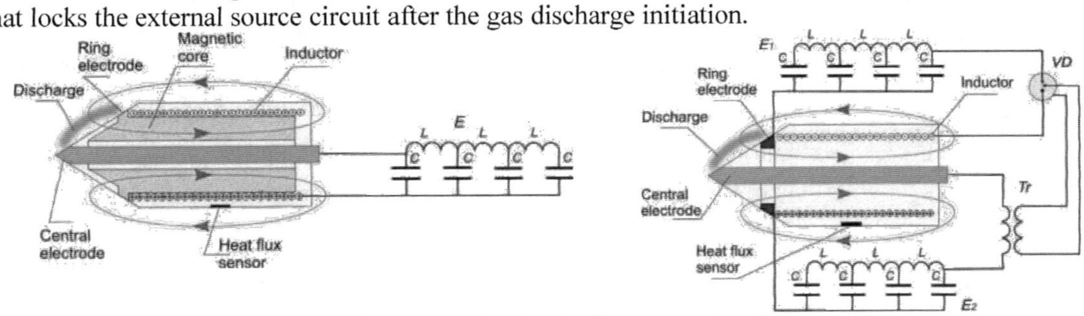

Figure 2. a - model with magnetic core and a single discharge circuit, b - model without a core and separate discharge and coil circuits.

An LC line of 14 cells, each of which has a capacitance C = 300μF and an inductance L = 3.5μH, was used as a source of pulse voltage. The discharge is initiated by electric breakdown in the discharge gap using a high-voltage transformer of small inductance, which creates a short (~ 1 μs) high voltage pulse (~ 30 kV). The arising pulse discharge with a duration of ~ 2 ms and a maximum current value of ~ 10^3 A (Fig. 3) is affected by the magnetic field of the inductor, which causes the discharge rotation around the model. In experiments on both models, the shape of the current pulse, its maximum value, and duration were approximately the same. With the same current strength in the coil circuit, the distribution of the magnetic field induction near the model surfaces significantly differed due to the presence of magnetic core inside one of them.

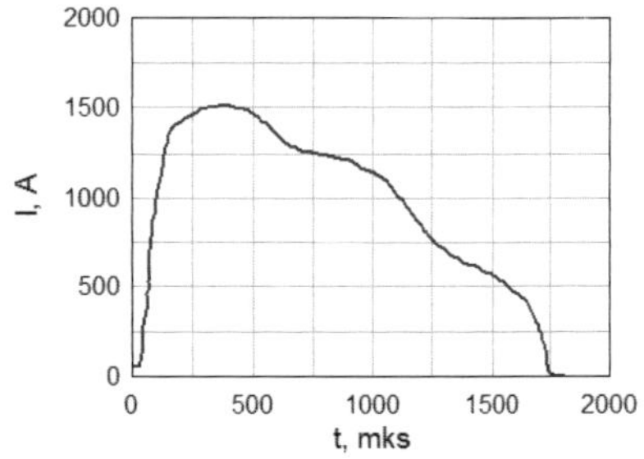

Figure 3. Gas discharge current in a magnetic core model.

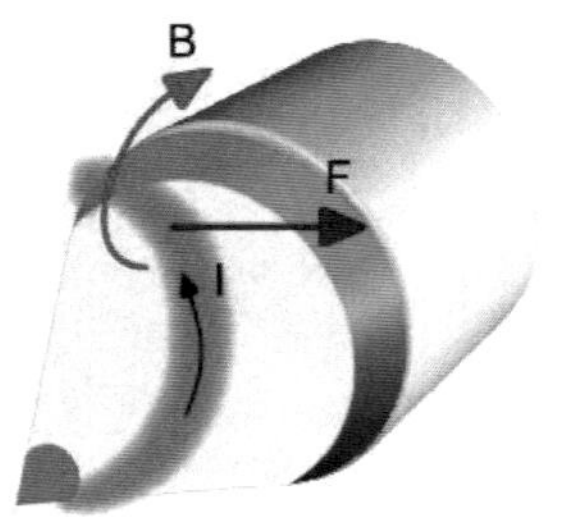

Figure 4. Formation of a ponderomotive force causing the discharge rotation in the bow of the model. B is the magnetic induction vector, I is the discharge current flowing through the discharge, F is the ponderomotive force vector.

The region of the most effective MHD interaction is located near the conical surface of the model between the central and ring electrodes. The maximum current density in the discharge is achieved at the electrodes and, to a greater extent, at the negative electrode. In the discharge region, the magnetic field created by the coil (Fig. 4) reaches its maximum value near the ring electrode. The current flowing through the discharge gap is affected by the coil magnetic field, causing the discharge channel rotation around the model axis of the model under the action of a ponderomotive force. The impact to the incoming gas flow is due to two factors: the ponderomotive force that rotates the discharge, and the Joule heat. Both factors cause a change in local gasdynamic parameters, which leads to the modification of the flow around the model and, in particular, the variation of heat flux to the model surface can take place.

Schlieren pictures of the flow around the model with an exposure of ~ 50 ns, photo-scan of the discharge luminescence, and current pulses in the gas discharge were recorded in the experiments. Heat sensor located on the cylindrical surface of the model was used for heat fluxes registration. The thermal sensor [11-12] is a battery of thermoelements connected in series from an anisotropic bismuth crystal cut at a certain angle to one of the crystallographic axes and having the shape of a parallelepiped with dimensions of 0.2 mm × 0.2 mm × 2 mm. Thermocouples are located on a 4x6 mm platform in such a way that the current direction in adjacent elements is opposite. This provides high noise immunity of the thermal sensor and allows thermal measurements in the presence of a magnetic field by induction of the order of 1 T.

When a thickness of thermocouples is equal to 0.2 mm, the time of reaching the stationary mode is ~ 1 s [13]. At shorter characteristic times of the thermal process, the use of stationary calibration leads to incorrect results; therefore, a different approach to the processing of measurement results is necessary. Therefore processing the results of thermal measurements was carried out using the technique described in [14].

3. Results

The use of high-speed photography allowed us to determine the features of the structure and dynamics of the discharge. It was established that in the model without the magnetic core the discharge glow has a diffuse structure and is distributed over a significant part of the conical surface. Due to the low brightness of the discharge, it was not possible to accurately determine the rotation frequency and its shape, because the rotation period was comparable with the duration of the experiment (~1 ms). On a model with a magnetic core, the current channel has a spiral shape, and its rotation was periodic.

Photographic registration of discharge rotation was carried out using the ZhFR-2 waiting photorecorder. The discharge image focused on the film was moved along it at a given speed using a rotating mirror prism, which made it possible to determine the discharge rotation frequency. The image size along the scan direction was limited by a slit diaphragm. A 2 mm wide recorder slit was oriented along the generatrix of the cone and covered its top and base (fig. 5-a).

Figure 5-b shows fragments of a photo-scan of the discharge luminescence on a model with a core corresponding to different polarities of the ring electrode. The horizontal direction corresponds to the time axis, and the direction from the bottom up corresponds to the movement from the cone top to its base. The dotted line (1) indicates the position of the central electrode, and the line (2) indicates the position of the ring electrode. A comparison of the presented photographic scans shows that in the case of a positive polarity of the ring electrode (top photo), each bright region has more blurred boundaries and is mainly distributed near the central electrode. In the bottom photo, corresponding to the negative polarity, each area is more compact and has sharper boundaries. It is seen that the shape of the bright areas on the upper photo scan is a mirror image relative to the vertical axis corresponding to the region on the lower photo scan. Inclined lines show the "front" of the discharge. Since in both cases the film was exposed in the same direction, it can be concluded that the direction of discharge rotation changes with the polarity changing. When the ring electrode is positive, the discharge rotation occurs in the direction of the ponderomotive force, and in the case of a negative polarity, the discharge rotates against the action of the force. Previous studies of this effect in a stationary gas showed that in this case the discharge rotation is determined by the peculiarities of the cathode spots motion in a magnetic field (so-called "anomalous motion of the cathode spot"), and not by the action of the ponderomotive force [15].

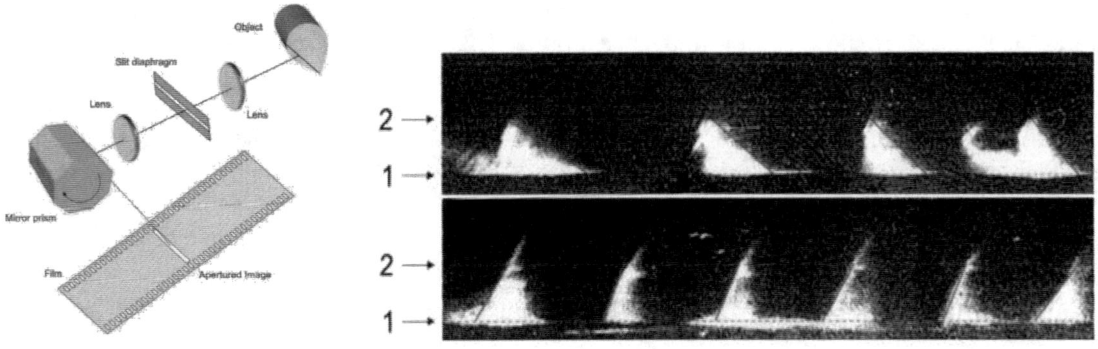

a b

Figure 5. a – optical scheme of discharge dynamics fast photoscanning; b - photographic scans of the discharge luminescence when the ring electrode is positive (upper photo) and negative (lower photo). Digits 1 and 2 on the left indicate the position of the central and ring electrodes, respectively, on the cone generatrix.

The processing of photographic scans obtained in the experiment at different polarity of the ring electrode was performed for the following experimental conditions: pressure at the nozzle exit p = $5\cdot10^3$ Pa, maximum value of the discharge current I = 1.5 kA, and magnetic induction B = 0.7 T. The following values of the discharge rotation frequency were obtained: for negative polarity, the discharge rotation frequency was f = 30 kHz, for positive polarity f = 15 kHz. In this way, the discharge rotation frequency substantially depends on the polarity of the ring electrode. For a model without a magnetic core, the rotation frequency was significantly lower compared to a model with a core - it turned out that the discharge makes one revolution around the model axis for a time comparable to the duration of the discharge t ~ 1 ms.

Figure 6 shows three schlieren pictures of supersonic flow around models. The bright region in Fig. 6b and 6c correspond to the region of the discharge rotating around the model axis. Schlieren pictures show that in the presence of MHD effects (Figs. 6b and 6c), the bow shock wave is located farther from the cone surface and its front is curved compared to the case of unperturbed flow (Fig. 6a). The glow in the discharge region near the model with the magnetic core is brighter, despite the fact that in both cases the electric power invested in the discharge was the same. Differences in the gasdynamic structure of the flow behind the shock wave are noticeable — in the model without magnetic core (Fig. 6c) turbulent disturbances are less pronounced. The rotation frequency of the discharge around a model with a magnetic core is higher compared to a model without a core.

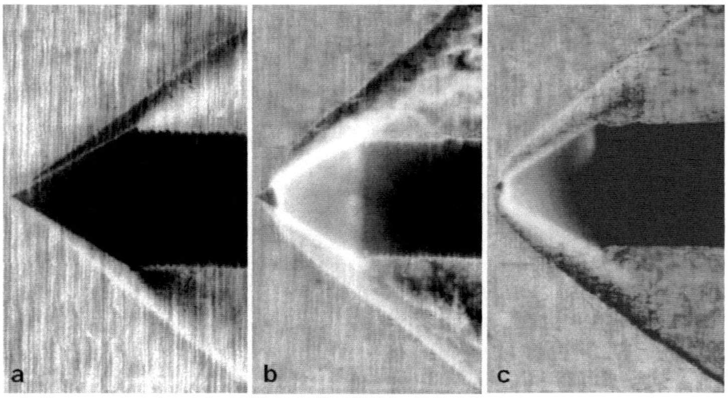

Figure 6. Schlieren pictures of the supersonic flow around the models: a – without MHD action; b, c - with MHD action on the model (b) with and (c) without the magnetic core

Figure 7 shows the signals of a thermal sensor when the magnitude of the discharge current variates. The characteristic pulsation frequencies on the signals correspond to the discharge rotation frequencies measured from photographic scans. Variation of the discharge current leads to the change in the heat flux. This change depends on the polarity of the ring electrode: when a ring electrode is anode, an increase in the discharge current leads to an increase of heat flux to the cylindrical surface of the model, when the connection is opposite, the signal of the heat sensor decreases. These results show the possibility of MHD effects on thermal loads to the body surface of a body in the high velocity flows of weakly ionized gas.

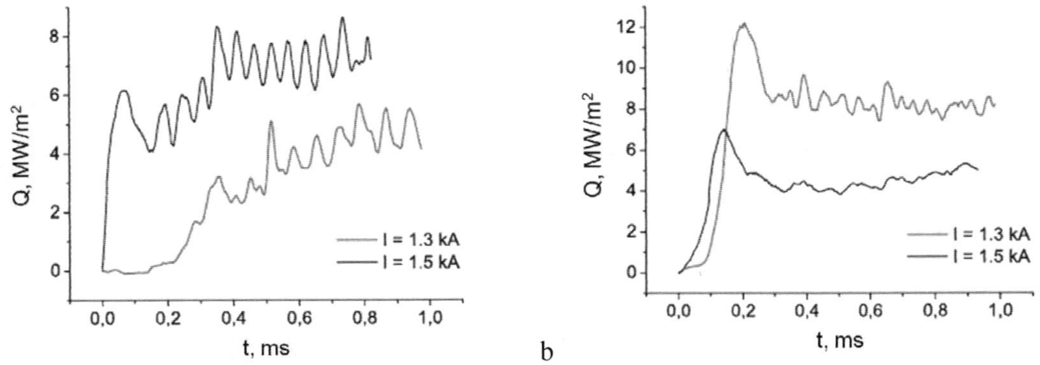

a b

Figure 7. Heat flux to the cylindrical surface of the model with a magnetic core at different discharge current magnitudes: a - ring electrode is anode; b – ring electrode is cathode

4. Conclusion

In this investigation, the MHD effect on a supersonic gas flow is implemented. The efficiency of MHD effect does not depend on the initial electrical conductivity of the medium. Two experimental models have been created, inside of which all the elements necessary for the realization of MHD effects are installed. Optical diagnostics has established that supersonic flow around models changes markedly with electrodynamic modification. The bow shock wave changes its position, and its front is noticeably curved in comparison with the unperturbed flow. The structure and features of the discharge rotation near the conical surface of the models are studied. It is established that in a model with a magnetic core, the discharge is localized and has clear boundaries, and its rotation is periodic. The rotation frequency is ~ 15-30 kHz, depending on the polarity of the ring electrode. In the model without magnetic core, the discharge is distributed over a significant part of the conical surface, and the rotation period is comparable with the experiment time. Heat flux measurements on the model cylindrical surface were carried out under MHD interaction. On the model with the magnetic core, a strong dependence of the heat flux on the polarity of the ring electrode was found. It has been established that in the case of a negative polarity of the ring electrode, an increase in the current in the discharge circuit by 15% is accompanied by an almost twofold decrease in the heat flux to the cylindrical surface of the model.

Acknowledgement
The work was partially supported by the Grant of President of Russian Federation (#MK-144.2020.1).

References

[1] Bletzinger P, Ganguly B N, Van Wie D and Garscadden A 2005 *J. Phys. D: Appl. Phys.* **38** R33.
[2] Adamovich IV 2010 Encyclopedia of Aerospace Engineering (John Wiley & Sons Ltd)
[3] Starikovskiy A and Aleksandrov N 2011 Nonequilibrium Plasma Aerodynamics Aeronautics and Astronautics, Ed. By M. Mulder (InTech).
[4] Bobashev S V, Mende N P, Popov P A and Sakharov V A 2010 *Tech. Phys.* **55** 1760–5
[5] Lapushkina T A, Erofeev A V and Poniaev S A 2011 *Tech. Phys.* **56** 616–22
[6] Fomichev V P, Yadrenkin M A 2017 *Tech. Phys. Lett.* **43** 1063–66
[7] Shang J S and Surzhikov S T 2018 Plasma Dynamics for Aerospace Engineering. Cambridge University Press
[8] Kotov M A *et al* 2018 *J. Phys.: Conf. Ser.* **1009** 012038
[9] O. Igra, F. Seiler, eds., Experimental Methods of Shock Wave Research, Springer International Publishing, Cham, 2016

[10] Maslennikov V G and Sakharov V A 1997 *Tech. Phys.* **42** 1322–28
[11] Sapozhnikov S Z, Mityakov V Y and Mityakov A V 2004 *High Temperature* **42** 629–638
[12] Mityakov A V, Sapozhnikov S Z, Mityakov V Yu, Snarskii A S, Zhenirovsky M I and Pyrhonen J J 2012 *Sens. and Act. A: Phys.* **176** 1–9
[13] Popov P A, Bobashev S V, Reznikov B I. et al. 2017 *Tech. Phys. Lett.* **43** 334–337
[14] Popov P A, Bobashev S V, Reznikov B I and Sakharov V A 2018 *Tech. Phys. Lett.* **44** 316–19
[15] Bobashev S V, Mende N P, Monakhov N A and Sakharov V A 2012 *Tech. Phys.* **57** 1637–40

AUTHOR INDEX